Micro and Nanoengineering of the Cell Microenvironment

Technologies and Applications

Artech House Series
Engineering in Medicine & Biology

Series Editors

Martin L. Yarmush, Harvard Medical School
Christopher J. James, University of Southampton

Advanced Methods and Tools for ECG Data Analysis,
Gari D. Clifford, Francisco Azuaje, and Patrick E. McSharry, editors

Advances in Photodynamic Therapy: Basic, Translational, and Clinical,
Michael Hamblin and Pawel Mroz, editors

Biomedical Surfaces, Jeremy Ramsden

Intelligent Systems Modeling and Decision Support in Bioengineering,
Mahdi Mahfouf

Translational Approaches in Tissue Engineering and Regenerative Medicine,
Jeremy Mao, Gordana Vunjak-Novakovic, Antonios G. Mikos, and Anthony Atala, editors

Micro and Nanoengineering of the Cell Microenvironment

Technologies and Applications

Ali Khademhosseini
Jeffrey Borenstein
Mehmet Toner
Shuichi Takayama

Editors

ARTECH HOUSE
BOSTON | LONDON
artechhouse.com

Library of Congress Cataloging-in-Publication Data
A catalog record for this book is available from the U.S. Library of Congress.

British Library Cataloguing in Publication Data
A catalogue record for this book is available from the British Library.

ISBN-13: 978-1-59693-148-0
ISBN-10: 1-59693-148-5

Cover design by Igor Valdman

10 9 8 7 6 5 4 3 2 1

Contents

Foreword

The development of new technologies has been a constant driving force for innovations in medicine and biology. For example, the invention of transmission electron microscopy, scanning probe microscopy, and atomic force microscopy has enabled the study of the realm of nanoscale phenomenon with unparalleled resolution. The applications of these technologies to biological systems have been used to examine and manipulate biomolecules and nanoscale structures. Similarly, the application of existing technologies to biological and biomedical problems has resulted in numerous advances that were previously not possible. For example, the application of microfabrication technologies, which were initially developed for microelectronics and computer industries, to biological and biomedical applications has resulted in a novel method of analyzing fundamental biology and fabricating diagnostics, screening, and therapeutic devices. This book will provide a comprehensive overview of both of these approaches in addressing one of the major challenges in biology: How do we manipulate cells and tissues with fine precision to direct the resulting behavior of individual cells or multicellular biological systems?

It is now well appreciated that the cells in the body are in the presence of a complex milieu that regulates the function of the cells. Although cell culture has been performed for over a century, much of the present tissue culture systems do not mimic the native microenvironment. This discrepancy between cells in the body and cells in tissue culture is the result of dramatic differences in cell behavior in vitro versus in vivo. For example, the liver has an amazing regenerative capability in the body by being able to regrow and restore its function even after losing nearly two-thirds of its mass. Despite this amazing regenerative capability, in tissue culture hepatocytes quickly lose their function. Many years of research in maintaining hepatocytes in culture have still not solved this dilemma. Thus, the ability to understand and re-create the native microenvironment of cells is of great interest in areas such as drug discovery, tissue engineering, and fundamental biological study.

In complex organisms nanoscale-length scales are relevant to cells, since many extracellular fibers and matrix components are nanoscale structures. Cells have been shown to be able to not only sense nanoscale features but also respond by changing their morphology, proliferation, and differentiation outcomes. Furthermore, cell-cell interactions and tissue architecture are controlled at the microscale. These interactions about contact between different cell types and the presence of a microvasculature are fundamental concerns in maintaining normal tissue function. In addition, spatial gradients such as morphogens or chemotactic signals that operate in the length scales of tens to hundreds of micrometers are also essential features of the native microenvironment. The use of micro- and nanoscale technologies is therefore of great promise in tissue engineering and biology.

My view of the science in this book comes from my more than 30 years of research in generating tissue engineering and drug delivery systems. Over this time, our understanding of biology has increased greatly, which has enabled us to better understand the constituents of the cellular microenvironment and tissue complexity. Furthermore, a plethora of approaches have been developed to control the various aspects of the cellular microenvironment such as cell-cell, cell-soluble factor, and cell-matrix interactions. In addition, many of these technologies are now being actively investigated in applications in aiming to re-create the complexity of various organs. With the explosion of this research over the past few years, a thorough amalgamation of this research is greatly needed. Therefore, this book provides a timely contribution to the emerging technologies and applications of micro- and nanoscale systems for controlling the cellular microenvironment. This book brings together an unparalleled number of experts in various aspects of the two-dimensional and three-dimensional engineering of the cellular milieu. The various technologies are introduced and supplemented by examples that provide an up-to-date context and reference to the seminal works in various fields. Furthermore, the section on applications provides examples that give updates on the state-of-the-art approaches used to engineer the microenvironment of various tissues. Thus, this book promises to be of great benefit for experts and novices alike.

Robert Langer
Institute Professor
Massachussetts Institute of Technology
Cambridge, Massachusetts
February 2008

CHAPTER 1

Micro- and Nanoengineering the Cellular Microenvironment

Ali Khademhosseini

1.1 Introduction

The ability to understand and manipulate biological systems has many benefits in diverse areas, ranging from renewable energy to biomedicine. Most biological systems are highly complex systems that have evolved over time to respond to various changes in a cell's environment. In general, cellular behavior is derived from intrinsic and extrinsic factors. The intrinsic factors are the internal genetic makeup of a cell defined by the transcription factors and gene regulatory networks. The gene expression profiles initiate the expression of messenger RNA and protein synthesis. Once a protein is transcribed, it undergoes a number of post-transcriptional modifications in which it obtains the proper glycosylation patterns. The regulation of the gene regulatory networks is made from stochastic and deterministic intercellular events. Although these events regulate the cell behavior at times in an autonomous manner, much of the signals that regulate cellular behavior are extrinsic and derived from the surrounding microenvironment.

The cellular microenvironment, or the cellular niche, highly regulates the resulting behavior of cells. For example, during the early stages of mammalian development, a blastocyst is formed, which is comprised of an inner cell mass (ICM) and a supporting layer of trophoblast cells. The ICM cells, from which embryonic stem cells (ESCs) are derived, give rise to the organism. Although during the early stages of development all the ICM cells have the same potential, they quickly differentiate into the three primary germ layers. The subsequent movement of these tissues induces morphogenesis, in which a series of highly ordered and directed differentiation and tissue movements shape the embryo [1, 2]. The process of embryonic development is highly conserved every time an egg is fertilized and is comprised of intrinsic and extrinsic factors. As an example, soluble factors generate soluble concentration gradients that signal cells to differentiate in a concentration-dependent manner. Furthermore, cells generate mechanical forces, which results in the movement of tissues within the developing embryo and morphogenesis.

Similarly, in adult tissues, a cell's microenvironment highly regulates the resulting cellular behavior. For example, in the liver, the hepatocytes are quiescent cells that perform much of body's metabolic functions. However, during times of stress, such as during partial liver failure, the hepatocytes convert from quiescent cells into highly proliferative cells that have the ability to regenerate the lost mass and function of the liver. The proliferative ability of native hepatocytes is so strong that if 2/3 of the mass of the liver is removed, the remaining cells can proliferate to regain the lost mass [3]. This amazing regenerative capability, which is highly dependent on the changes in the hepatocyte microenvironment, was known even in ancient Greek mythology, in which Prometheus was tortured each day by having his liver eaten by an eagle, after which his liver would regenerate by nightfall.

Another fascinating example of the significant regenerative capability of adult stem and progenitor cells to extrinsic signals is the regulation of hematopoietic stem cells (HSCs). HSCs reside in adult bone marrow and contribute to the normal turnover in the hematopietic system. HSCs are often slowly cycling and quiescent cells that self-renew and differentiate to all the blood lineages. New insight into the HSC niche has indicated that the bone marrow microstructure and the HSC niche is highly organized and complex [2, 4]. Specifically, HSCs interact with the surrounding environment through the exposure to a complex regulatory network of cytokines, chemokine and growth factors. Furthermore, other cells such as osteoblasts and endothelial cells have been implicated in maintaining the HSC niche. Interestingly, upon injury, HSCs obtain signals from their surroundings to proliferate and differentiate extensively to such a degree that an individual transplanted cell can be induced to repopulate the entire hematopoietic system of the host.

1.2 Cellular Microenvironment

The complexity of the signals in the cellular microenvironment and its effects on cell behavior are a fascinating area of biology. The cellular microenvironment is comprised of biochemical, biomechanical, and bioelectrical signals derived from surrounding cells, extracellular matrix (ECM), and soluble factors. These components work in synergistic and antagonistic manners in regulating cellular behavior.

With respect to biochemical signals, cells signal each other through small molecules such as hormones as well as larger molecules, which are mostly made of proteins. These larger molecules include signaling proteins such as chemokines, cytokines, and growth factors. Signaling molecules typically interact with cells through surface receptors. Different cells exhibit unique types of cell surface receptors, thereby interacting with specific signaling molecules. The interactions between the surface receptors and signaling molecules result in a signal transduction cascade that transfers the signal from the cell's surface to its nucleus, which results in gene expression.

Another class of molecules in the cellular microenvironment that is important in signaling cells is the ECM. The ECM provides a 3D scaffold, or a 2D surface with which cells interact. Through these interactions, cells are provided with an environment to which they can anchor and generate tissues. The ECM is often comprised of proteins such as collagen and laminin, as well as polysaccharides, such as hyaluronic acid. Furthermore, the ECM is a particularly key constituent of the cell microenvironment because it is also a reservoir for many soluble factors that interact with the cells. For example, several molecules, such as heparin sulfate, interact specifically with polysaccharides in the ECM. Thus, signaling proteins such as fibroblast growth factor (FGF), which interacts with heparin sulfate, are regulated by the presence of signal-binding ECM [5]. Through specific interactions with soluble signals, the ECM controls the diffusion rates as well as the activity of soluble factors in the cellular milieu. Also, ECM binding sites interact with cell-surface molecules such as integrin. Integrins and other ECM binding molecules initiate the assembly of cytoskeletal elements and the formation of adhesion kinases. The change in the cytoskeletal structure influences the resulting gene expression and biomechanical features of the cells.

Direct cell-cell contact is also an important cell-microenvironment interaction (Figure 1.1). Cells sense each other by direct contact in a process in which cell surface receptors detect the cell surface receptors of neighboring cells. For example, molecules such as cadherins specifically interact with cell surface receptors of other cell types. Contact with other cells can be used to regulate cell proliferation, differentiation, and migration. For example, in many tissue culture systems, cells proliferate until they make contact with each other, at which point they stop proliferating in a process called *contact inhibition*. Within the body, cell-cell contact is a key regulator in many biological systems and mediates the formation of many types of tissues, such as epithelial and endothelial barriers. In these cases, cells contact each other to form a barrier to the external environment and regulate the transfer of molecules.

1.3 Controlling Cellular Behavior

The ability to control cell behavior is important for a wide range of applications in biotechnology and tissue engineering [6]. For example, in biotechnology, the ability to optimize bioprocess conditions to maximize the synthesis of desired biomolecules is of great interest. Furthermore, controlling cell behavior is important for developing tissue cultures and artificial tissues that can be used as predictive models of biological systems and 3D structures for tissue engineering [7]. An example of importance of controlling the cellular microenvironment is found in the drug discovery process, in which the effects of specific chemicals within a large library of chemical molecules are tested on cells, to identify candidate drugs. The ability to predict the toxicity and functionality of a candidate drug is important for classifying these drugs and eliminating costly animal experiments and failures in clinical trials. Therefore, it is important to have predictive tissue models in vitro.

Another example that demonstrates the importance of controlling cellular behavior in culture is related to generating a renewable source of cells for regenerative medicine. In most cases the cells of an adult organism cannot proliferate to the extent required to generate a renewable tissue. However, it is possible to utilize stem

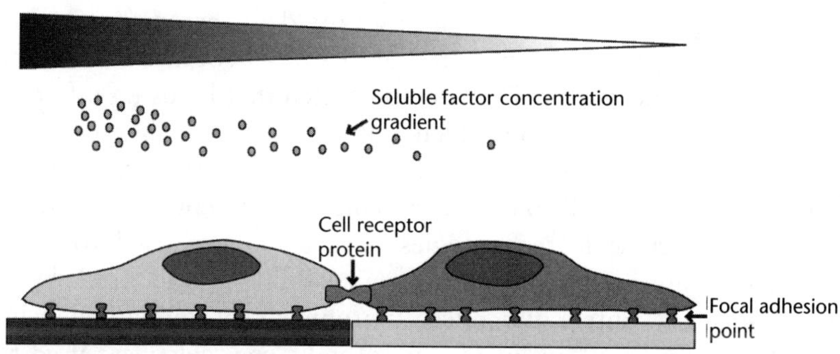

Figure 1.1 The cellular microenvironment.

cells that are either derived from the adult organism or from embryonic sources to generate a renewable source of cells. A critical challenge in stem cell–based therapy is how to uniformly direct the differentiation of stem cells into desired cell types. This is required since stem cells such as ESCs, which have the capacity to generate all the cell types of an adult organism, often differentiate in a heterogeneous manner, resulting in low yields of desired cell types. Developmental biology has shown that the microenvironment plays an important role in controlling cellular differentiation, so, to be able to utilize ESCs for regenerative medicine, cultures must be generated that mimic the features of the embryonic microenvironment.

Of particular interest in controlling the cellular microenvironment is controlling the interactions at the micro- and nano- length scales. In the body, cells interact with their surrounding tissues on a subcellular level, as well as on a tissue level. For example, many ECM fibers are on the multinanometer length scales. Therefore, to mimic features that resemble the ECM, it is important to generate structures that are on the nanometer length scale. It is also important to generate and control features of the microenvironment that are on length scales greater than individual cells, (10 μm). For example, during migration and embryonic development, gradients of soluble factors are generated within the cellular microenvironment that is often hundreds of times that of the length of a single cell. Therefore the ability to control the in vitro cellular environment on both the micro- and nanolength scales by using engineering and materials sciences approaches is of great interest in controlling cell behavior.

1.4 Micro- and Nanoengineering the Cellular Microenvironment

Recently, the development of emerging technologies at the interface of engineering and materials science has enabled a number of new methods to control the various aspects of the cellular microenvironment [6, 8]. The emergence of these technologies has mirrored the explosion of research in micro- and nanoscale applications. For microscale technologies, a number of technologies have been developed over the years to control the microenvironment of cells. Many of these technologies were adopted from the microelectronics industry techniques to fabricate miniaturized structures by using photolithographic and molding approaches. These microtechnologies can be used to control many aspects of the cellular micro- environment. For example, one of the first applications of the use of microtechnologies for controlling the cell microenvironment in which a biological phenomena was demonstrated was done by Ingber and Whitesides [9]. In their approach they used soft lithography, a technique that uses molded elastomers to fabricate microscale structures to generate micropatterned surfaces comprised of adhesive regions to which cells could attach. They demonstrated that by using different size "islands," which limited the area on which cells could grow, they could control different aspects of cell behavior. Specifically, they demonstrated that cells on larger islands proliferated, while cells on smaller islands underwent apoptosis.

Other microscale technologies have also been used to control the cellular microenvironment. For example, microfluidics was used by Whitesides and Takayama to control the microenvironment of individual cells [10]. In their approach, laminar flows inside microfluidic channels were used to create parallel streams to generate multiple environments on different sides of cells that were at the interface between these parallel streams. Using this technique, they were able to stain and study the intracellular transport of cell components.

With regards to cell-cell contact, Bhatia and Toner [11] have demonstrated that the degree of direct cell-cell contact could be controlled by using micropatterened surfaces. In their original work, they used adhesive collagen islands to pattern hepatocytes on specific regions. Subsequently, they used the adhesion of fibroblasts to the remaining substrate to generate patterned cocultures. Based on the specific shapes of the patterns, the degree of cell-cell contact between hepatocytes and fibroblasts could be mediated within these cocultures. Furthermore, they demonstrated that direct cell contact could be used to prolong the functional maintenance of hepatocytes in vitro.

Similarly, nanotechnology has been increasingly used to control the cellular microenvironment. For example, nanomaterials such as self-assembled peptides and electrospun nanofibers are emerging as powerful technologies to generate 3D structures. Self-assembled peptides are nanomaterials in which peptides assemble either by hydrophobic forces or by charge shielding to form 3D materials. The ability to modify the sequences of the peptides and to control their interactions with cells makes these materials a powerful method to control the cellular microenvironments. Furthermore, electrospun nanofibers, which can be used to generate nanofibrous materials from existing biocompatible polymers have emerged as a powerful approach to generating fibrous scaffolds with nanoscale topography. In addition, nanoscale topography has been shown to be a powerful regulator of cell behavior such as alignment, migration, and differentiation.

1.5 Book Structure

This book aims to provide a comprehensive view of the techniques available for controlling the cellular microenvironment. A number of leading authors describe their respective fields and their applications to control the cellular microenvironment. Particular emphasis is placed on the applications of micro- and nanoscale techniques to control the cellular microenvironment of mammalian cells. This, of course, is by no means an implication that other cellular systems, such as prokaryotic and nonmammalian eukaryotic systems are of less importance. There are two main reasons for the focus of this book on mammalian cells. The first reason is because of the particular importance of mammalian tissue systems for regenerative medicine as well as pharmaceutical screening applications. The second reason is that many micro- and nanoscale techniques have only been applied to control the cellular environment of these cell types. It is worth noting that most of the techniques described in this book are also applicable to other cell types.

The book is comprised of two major sections. Chapters 2 to 10 discuss the various technologies that can be applied to controlling the cellular microenvironment, while the second part of the book (Chapters 11 to 27) discuss the application of the various techniques to controlling the cellular microenvironment for different target tissues. In addition to the description of the technologies, this book will describe emerging applications areas and give specific examples of how these tools have already begun or will begin to be used in controlling and studying cell behavior in vitro and perform experiments in a more controlled and scalable manner.

Chapters 2 to 10 discuss the techniques used to manipulate the cellular microenvironment, which can be categorized into the following broad areas: 1) controlling the soluble cellular microenvironment, 2) controlling the insoluble biochemical cellular microenvironment, and 3) controlling the biophysical cues of the cellular microenvironment.

1.5.1 Controlling the Soluble Cellular Microenvironment

For cells, the soluble component of a cell's environment includes all surrounding soluble molecules, such as proteins and small molecule chemicals. Proteins include molecules such as cytokines, chemokines, and growth factors, while small soluble molecules in the surrounding environment include molecules such as glucose, oxygen, hormones and ions. Recently, microfluidic devices have been used to control the presentation of soluble molecules to cells. Microfluidic devices provide a number of advantages in comparison to macroscale systems. Microfluidic devices require fewer reagents to operate, which minimizes cost, and the space restrictions associated with macroscale systems. Also, microfluidic channels can be used to perform high-throughput experimentation and to control cell-soluble interactions in a parallel manner. Furthermore, microfluidic devices enable the presentation of soluble factors in a highly controllable manner. For example, in Chapter 2, Dr. Jeon et al. describe the formation of concentration gradients generated by microfluidic devices for controlling the cellular microenvironment.

1.5.2 Controlling the Insoluble Biochemical Cellular Microenvironment

The insoluble biochemical microenvironment includes the nondissolvable agents that affect cell behavior. These factors interact with cell surface receptors to direct cell function. Generally, the insoluble biochemical components are comprised of ECM molecules that form 2D substrates as well as 3D environments within which cells reside. Surface micropatterns can be used to control cell-substrate and cell-cell interactions. To control interaction with substrates, techniques such as lithography and surface modification have been merged. Through these modifications, properties like cell adhesion can be controlled by depositing cells on adhesive islands or controlling the nanotopography of surfaces. Furthermore, to control the interactions of cells with surrounding cells, such as in cocultures, micropatterning technologies have been utilized. These techniques have been successful at controlling the homotypic and heterotypic cell-cell contact in vitro. In Chapter 3, Dr. Jiang will describe surface patterning approaches for controlling cell-substrate interactions, while in Chapter 4, Dr. Khademhosseini and colleagues describe patterned coculture approaches for controlling cell-cell interactions.

Another method of controlling the insoluble biochemical microenvironment is through the fabrication of 3D scaffolds. 3D scaffolds enable better control of generating cellular microenvironments, since the native tissue ECM is 3D. By engineering different properties into scaffolds, the morphology and function of cells within 3D geometries can be altered. This enables greater possibilities in controlling tissue development and cellular function in culture. In Chapter 5, Dr. Desai et al. describe the application of micro- and nanofabrication technologies for generating 3D tissue constructs.

Hydrogels are a common material for fabricating 3D scaffolds since they are biocompatible and have similar properties to native tissues. In Chapter 6, Dr. West and colleagues describe methods of engineering photocrosslinkable hydrogels for 3D cell culture. Another microscale method to form 3D structures is through cell printing. This technique works by using a modified computer printer to print cells. Cell printing and its application for tissue engineering are further described in chapter 7, by Dr. Demirci et al.

1.5.3 Controlling the Biophysical Cues of the Cellular Microenvironment

Biophysical cues of the cellular microenvironment are any physical properties of the surrounding microenvironments, such as physical forces, as well as electrical and topographical features of the surrounding environment. This book describes the various techniques to engineer the mechanical properties and topographical features of the surrounding cellular microenvironment.

A common technique to control the biophysical surrounding of a cell is through engineering the mechanical properties of the surrounding biomaterial. It is also possible to control the cell mechanical environment interactions by engineering the substrate at the microscale. In Chapter 8, Dr. Chen et al. describe the use of surface patterning approaches to engineer cellular and multicellular architecture. By generating surface patterns, they demonstrate approaches for controlling the cell shape and the resulting cytoskeletal structures. In Chapter 9, Dr. Simmons et al. describe approaches of engineering surface substrate mechanics for regulating the cell microenvironment. They discuss the effects of controlling the stiffness of the scaffolding, on cell behavior such as viability and differentiation. Biophysical cues can also be modified through changes in surface topography. Classically, cell microenvironments have been developed without regards to the nanotopography of the underlying substrates. Novel patterning technology can now be used to fabricate specific micro- and nanoscale features. This change in microenvironment has been shown to affect various aspects of the cells, such as adhesion and proliferation, as discussed by Profs. Suh and Levchenko in Chapter 10.

Chapters 11 to 27 discuss the tissue specific applications of the various micro- and nanoscale technologies for controlling the cellular micro- environment. It is divided into a number of application areas that range from embryonic cell development and fertilization to neural, vascular, muscle, cartilage, and liver engineering. Each of these areas receives unique benefits from the discussed technologies. For instance, microfluidics technologies can be used to isolate motile sperm as described by Beebe, Smith and Takayama et al. in Chapter 11. Also, the ability to engineer the cellular microenvironment offers great potential in directing the differentiation of ESCs as described by Dr. McDevitt et al. in Chapter 12.

Micro- and nanofabrication can also be useful for neural applications. In Chapters 13 and 14, Drs. Folch et al. and Hosseinkhani et al., describe the application of microfabrication techniques and self-assembled nanofibers for engineering neural tissue microenvironments. Microfabrication techniques can be used to generate neural networks and to control cell biology for neurobiological studies. Furthermore, the ability to engineer the biochemical properties of self-assembled peptides can be used to induce differentiation of neural stem cells and regenerate neural tissues.

A promising area in the use of microengineering techniques in tissue engineering is to design artificial vascular structures. In Chapter 15, Dr. Borenstein and colleagues describe the use of microengineered scaffolds for generating vascularized networks within biocompatible polymers. This microscale approach is complemented with a discussion of the use of nanotechnology for inducing angiogenesis in Chapter 16, by Dr. Laurencin et al. Nanotechnology can be used to control the release rate of angiogenic molecules within tissue engineered scaffolds and to generate conditions that induce endothelial cell migration and differentiation to generate functional vasculature.

Another area of significant promise is engineering artificial muscle tissues. These tissues can be useful for tissue engineering applications as well as artificial biological actuators. In Chapter 17, Dr. Parker and colleagues describe the use of micro- engineering techniques for studying muscle cells and generating muscle actuators using microengineering techniques. While in Chapter 18, Dr. Radisic et al. describe the application of microengineering techniques for cardiac tissue engineering by providing examples of the use of separation techniques and micropatterning for isolating cardiac cells and generating myofibers.

Biomechanically relevant tissues such as bone and cartilage are two other tissue types that have been influenced by the research in micro- and nanoengineering techniques. In Chapter 19, Dr. Ellisseeff and colleagues describe the use of engineered hydrogels for generating functional cartilage tissue. In addition, microfabrication techniques offer great potential in generating the complex microstructure of bone tissues. In Chapter 20, Drs. Karp, and Langer and colleagues discuss the emerging applications of microfabrication in bone tissue engineering. In addition, Chapters 21 and 22, Drs. Webster, Jabbari and colleagues describe the use of nanoengineered materials for bone tissue engineering applications.

Chapters 23–27 provide examples of a number of other tissues that have been influenced by the emergence of micro- and nanoscale technologies. In Chapter 23 Drs. Mofrad and Borenstein and colleagues describe the use of microengineered technological devices for renal replacement therapies. In Chapter 24, Dr. Takayama et al. describes the use of microfabrication for mimicking the structure and function of pulmonary epithelial tissue for disease diagnosis and modeling. In Chapter 25, Dr. Irvine et al. discusses the use of micropatterning approaches for controlling the interaction of immune cells with each other and their microenvironment with fine spatial and temporal resolution. These studies have demonstrated new insight into the biology of the immune system and provide a foundation for future studies in immune modulation and disease therapy. In Chapter 26, Drs. Yarmush, Toner, Tilles and colleagues discuss the role of microfabricated culture for liver tissue engineering. Finally, in Chapter 27, Dr. Pins et al. discusses the application of novel micro- and nanotechniqudes for the development of skin substitutes.

In summary, this book aims to provide an in-depth overview of the technologies available to control the cellular microenvironment and their applications towards tissue engineering with mammalian cells. Clearly the field of controlling the cellular microenvironment is a large field, encompassing many aspects of biology and bioengineering. By providing this comprehensive view, our aim is to bring together the world's leading experts and to generate a volume that is of use and significance for the novice as well as the expert scientists.

References

[1] Czyz, J., and A. Wobus, "Embryonic stem cell differentiation: the role of extracellular factors," *Differentiation,* Vol. 68, No. 4–5, 2001, pp. 167–174.

[2] Watt, F. M., and B. L. Hogan, "Out of Eden: stem cells and their niches," *Science*, Vol. 287, No. 5457, 2000, pp. 1427–1430.

[3] Michalopoulos, G. K., and M. C. DeFrances, "Liver regeneration," *Science*, Vol. 276, No. 5309, 1997, pp. 60–66.

[4] Lemischka, I. R., "Microenvironmental regulation of hematopoietic stem cells," *Stem Cells,* Vol. 15, No. Suppl. 1, 1997, pp. 63–68.

[5] Matsumoto, K., H. Tajima, H. Okazaki and T. Nakamura, "Heparin as an inducer of hepatocyte growth factor," *J. Biochem. (Tokyo)*, Vol. 114, No. 6, 1993, pp. 820–826.

[6] Khademhosseini, A., R. Langer, J. Borenstein and J.P. Vacanti, "Microscale technologies for tissue engineering and biology," *Proc. Natl. Acad. Sci. USA.*, Vol. 103, No. 8, 2006, pp. 2480–2487.

[7] Langer, R., and J. P. Vacanti, "Tissue engineering," *Science,* Vol. 260, No. 5110, 1993, pp. 920–926.

[8] Bhatia, S. N., and S. C. Chen, "Tissue engineering at the micro-scale," *Biomedical Microdevices*, Vol. 2, No. 2, 1999, pp. 131–144.

[9] Chen, C. S., M. Mrksich, S. Huang, G. M. Whitesides and D.E. Ingber, "Geometric control of cell life and death," *Science*, Vol. 276, No. 5317, 1997, pp. 1425–1428.

[10] Takayama, S., E. Ostuni, P. LeDuc, K. Naruse, D. E. Ingber and G. M. Whitesides, "Subcellular positioning of small molecules," *Nature,* Vol. 411, No. 6841, 2001, p. 1016.

[11] Bhatia, S. N., U. J. Balis, M. L. Yarmush and M. Toner, "Effect of cell-cell interactions in preservation of cellular phenotype: cocultivation of hepatocytes and nonparenchymal cells," *Faseb. J.*, Vol. 13, No. 14, 1999, pp. 1883–1900.

CHAPTER 2

Gradient-Generating Microfluidic Devices for Cell Biology Research

Francis Lin, Bong Geun Chung, Wajeeh Saadi, and Noo Li Jeon

2.1 Introduction

Gradients of soluble factors play an important role in a number of physiological processes, such as chemotaxis, axon guidance, and embryogenesis [1]. Traditional devices (e.g., the Boyden chamber, the Zigmond chamber, the Dunn chamber, under-agarose assay, and micropipette-based assays) generally lack the ability to maintain and manipulate stable gradients [2–6]. Although widely used in the biology community, these macroscale devices (e.g., the Boyden chamber) have unstable gradients, poor reproducibility, and require a large number of cells. In addition, incompatibility with optical imaging methods further limit their utility in modern biology experiments

However, microfluidic devices for gradient generation hold the potential to overcome these limitations imposed by conventional devices. Microfluidic devices can generate stable gradient profiles, precisely manipulate fluid flow and allow real-time monitoring of cells. Increasingly, gradient-generating microfluidic devices are being utilized in biological applications. The majority of these applications have been in the area of chemotaxis, proliferation, differentiation, and axon guidance. Rapid development of the gradient-generating microfluidic device and its use in biological research is evidenced by the noteworthy increase in the number of total publications since 1997 (Figure 2.1). Moreover, the number of studies that utilize microfluidic gradient devices in biological applications has also followed this remarkable growth (Figure 2.1), partly explaining the added advantages of these approaches in biological research.

This chapter reviews recent developments in gradient-generating microfluidic devices for cell biology research. Microfluidic devices for gradient generation can be divided into two main classes. The first class of microfluidic devices produces stable gradients by mixing laminar flow streams in a simple channel or a microchannel network. Gradient generation typically yields close agreement with theoretical predictions, allowing gradients with defined simple or complex shapes to be produced and manipulated in the microfluidic devices. The second class of devices generates time-evolving or stable gradients by free-diffusion of chemicals in a flow-free environment such as membrane-based or hydrogel-based devices. Various off-chip or on-chip controls can also be incorporated to manipulate the gradients in space and time. The first part of the chapter reviews conventional macroscale methods for

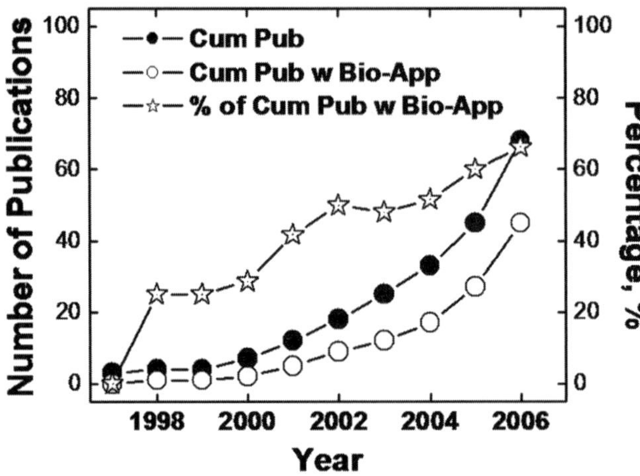

Figure 2.1 Publications related to microfluidic gradients since 1997. Solid circle: Cumulative number of publications related to microfluidic gradients from 1997 to 2006. Empty circle: Cumulative number of publications related to microfluidic gradients with biological applications from 1997 to 2006. Star: Percentage of cumulative number of publications related to microfluidic gradients with biological applications from 1997 to 2006.

generating biochemical gradients. The second part of the chapter describes the design and operating principles behind a variety of gradient-generating microfluidic devices, followed by a detailed review of their applications in biological experiments.

2.2 Conventional Devices for Soluble Gradient Generation

2.2.1 The Boyden Chamber

The Boyden chamber is one of the most commonly used assays for cell motility and chemotaxis (also called transfilter or transwell assay). Chemotaxis is a process by which cellular organisms direct their movement in response to concentration gradients of certain chemicals in their environment. The Boyden chamber measures motility and chemotaxis according to the number of cells that migrate across a filter between two compartments containing soluble factors [7]. Cells are loaded in the top compartment while chemoattractants are placed in the bottom compartment. In spite of its popularity, the Boyden chamber suffers from a number of limitations. Being an endpoint assay, it does not allow visualization of migrating cells. Chemotaxis thus cannot be directly observed, and various confounding variables can obscure the cell response. For example, the ability of cells to cross the filter depends not only on their motility but also on their size, ability to deform, and initial position relative to the pores [8, 9]. Cells that are small and easily deformable can migrate across the pores more efficiently than large and less deformable cells. Similarly, cells that happen to be seeded close to a pore can cross the filter more easily than cells that are further away. Depending on the adhesion of the cells to the filter, cells may also fall off the filter and escape the final count [10], further distorting the measured response. The response itself, irrespective of these limitations, is not necessarily a result of chemotaxis, as chemokinesis, which generally refers to random motility induced by chemicals, can also cause the cells to cross the filter [11]. The two processes cannot be distinguished from each other directly, and a checkerboard analysis is required to evaluate the relative contributions of each [8]. Even if the chemotactic response can be quantified, analysis of the results is complicated by the unstable gradient profiles, which continuously change until the concentrations in the top and bottom compartments equilibrate [12].

2.2.2 Under-Agarose Assay

Under-agarose assay (also called the agarose gel assay) overcomes some of the problems of filter-based assays (i.e., the Boyden chamber and its variants) by allowing cells to migrate over a flat surface. The assay chamber consists of an agarose gel cast on a glass coverslip, with separate wells for cells and chemoattractants punched into the agarose [13]. Diffusion of the chemoattractant through the gel produces a concentration gradient, and cells responding to the gradient can migrate under the agarose toward the chemoattractant wells. The assay is attractive due to its simplicity and ability to superimpose multiple gradients simultaneously [14]. The main disadvantage is that the gradient never reaches a steady state [12]. This is inherent in he design of the assay, as the chemoattractants diffuse radially in a relatively infinite plain [12].

2.2.3 The Zigmond Chamber

The Zigmond chamber (also known as the orientation assay or visual assay) allows direct observation of chemotaxis in gradients. The chamber consists of a Plexiglas slide containing two wells separated by a bridge [15]. A coverslip with cells attached is inverted over the bridge and wells, and the wells are filled with different concentrations of chemoattractants. Chemoattractants diffuse across the bridge, where the bridge behaves like a thin film in a source-and-sink diffusion system, establishing a linear gradient at the steady state [16, 17]. Gradients eventually decay due to the finite volume of the wells but are stable long enough for studies of leukocyte chemotaxis (~90 min. for low-MW molecules such as fMLP) [15]. The chamber design allows the migration to be visualized with phase microscopy. The gradient stability and optical characteristics of this assay represent dramatic improvements over its predecessors, namely the Boyden chamber and under-agarose assay.

2.2.4 The Dunn Chamber

The Dunn chamber improved on the Zigmond chamber with the direct-viewing chamber (now known simply as the Dunn chamber) and consists of two concentric circular wells ground into a glass slide and separated by an annular bridge [18]. As in the Zigmond chamber, cells are placed on a glass coverslip and inverted over the wells. Chemoattractants diffuse between the wells to establish a steady-state gradient across the bridge. Although the principle for generating gradient in the Dunn chamber is similar to that of the Zigmond chamber, their designs are slightly different. While the bridge region in the Zigmond chamber can be modeled as a thin film, the annular bridge region in the Dunn chamber can be modeled as a cylindrical wall

[18]. As a result, the steady-state gradient profile across the annular bridge is nonlinear [16]. Relative to the chamber dimensions, however, the gradient is very close to linear [18]. Compared to the Zigmond chamber, the Dunn chamber has a more precise construction and a fully enclosed design, allowing more stable gradients to be generated [19]. Moreover, the entire chamber is made of glass, providing better optical properties than the Zigmond chamber.

2.2.5 Micropipette Assay

Perhaps one of the most definitive approaches for studying chemotaxis is to apply the chemoattractant with a micropipette in the vicinity of the cell and to observe the movement directly [20]. A powerful feature of this assay is the ability to move the pipette and observe the cell as it reorients toward the source of chemoattractant. This allows direct observation of chemotactic behavior and provides a wealth of information about the dynamics of the process, such as the reorientation of lamellipods and distribution of receptors [21, 22]. Dynamic studies of this nature are difficult, if not impossible, with other methods. Different variants of this assay have been employed, and it has been possible to obtain stable gradients using pulsatile ejections [23, 24] or by applying a steady flow over a glass dam [25]. The main drawback of this approach is its low throughput, making statistical assessments difficult and labor intensive. The assay is thus best suited for qualitative, single-cell studies.

2.3 Microfluidic-Based Devices for Gradient Generation

2.3.1 Flow-Based Gradient Devices

A simple method for generating gradients is to continuously infuse two or more streams of different species into a simple microfluidic channel. Diffusion between the streams generated a gradient across the channel, perpendicular to the flow direction, having a diffusive profile that was stable at any fixed position in the channel but evolved along the channel. The upstream gradient had a steplike profile with a clear concentration interface, while the downstream gradient yielded a more continuous profile resulting from sufficient diffusion. Thus, different locations along a given channel could be used to carry out different experiments. The mixing process of laminar streams in microfluidic channels was governed by the diffusion model and has been well characterized by theory and experiments. Kamholz and Yager theoretically analyzed molecular diffusion in pressure-driven laminar flow in a "T"-shaped microfluidic channel [26]. Holdena et al. demonstrated gradient generation using pressure-driven flow in a "Y"-shaped microfluidic channel fabricated in glass by photolithography [27]. The end of the "Y"-shaped channel was divided into multiple subchannels, yielding a range of chemical concentrations, allowing cell sorting or screening of concentration-dependent cell responses. Using a more complicated design, Yang et al. generated concentration gradients in a microfluidic device by controlling the fluid distribution at each intersection of the channels and subsequently the diffusion between the streams in the downstream microchannels [28]. As an alternative to pressure-driven systems, gradients could be generated by electrokinetic or electrochemical means as well. Biddiss and Li reported a method of generating gradients using electrokinetic flow in microfluidic channels [29]. By varying the relative strength of the electric fields used to drive the flow streams, the interface between the streams could be shifted to yield different gradient configurations. Moreover, the gradient profile was further modulated by surface charge modification in the selected regions of the channel. Xie et al. developed an electrochemical pumping system based on electrolysis to generate gradients in a microfluidic device [30]. These systems, although limited by the electrokinetic properties of the liquid, showed several advantages such as on-chip dynamic, reversible chemical handling, and gradient generation over pressure-driven systems.

Based on the well-characterized mixing process of continuous flow streams in simple microfluidic channels, the chemical gradients could be more flexibly manipulated through microfluidic microchannel networks. Starting from a small number of chemical inputs, the microchannel network generated multiple output streams with different defined chemical concentrations, which subsequently flowed into a common channel forming a gradient across the device. The gradient was regulated by the inlet configuration and the structure of the microchannel network, yielding simple or complex shapes. Instead of requiring an increased number of inlets to achieve finer gradient resolution, the network approach improved the resolution and spatial stability of the gradient by increasing the size of the microchannel network. Gradients of soluble factors and immobilized molecules were produced in two-dimensional and potentially three-dimensional environments using the network approach. Several strategies were also developed to improve the flexibility of the network approach for dynamically varying the gradient environments. Over a relatively short period, the network method and its derivatives have been successfully applied to a number of cell-based studies [31, 32]. Jeon et al. reported the microchannel network method for generating defined chemical gradients in a microfluidic device [33]. The chemical solutions with different concentrations were introduced to the microfluidic device by syringe pumps at separate inlets and repeatedly mixed and split through the microchannel network, producing multiple diluted streams with predictable concentrations. These streams flowed side by side in a common channel and generated a soluble or surface gradient across the channel. Dertinger et al. further developed the network method to create concentration gradient with complex profiles [34]. By configuring the inlet concentrations, gradients with different shapes, such as linear or parabolic gradients, were produced. If multiple microchannel networks are arranged side by side in a single chip and the output streams are connected to a common gradient channel, individual gradients produced by each network can be aligned to configure complex gradients. Superimposed gradients of two or more species could be also formed using the same approach. Lin et al. improved the network approach for generating dynamically controlled temporal and spatial gradients [35]. A "two-inlet" mixing module was developed to control the resulting concentration by varying the relative input flow rates. When the two-inlet mixing modules were incorporated to the microchannel network device, linear gradients with different slopes, baselines, and directions, as well as nonlinear gradients with different nonlinearities, could be produced and altered dynamically. Using the microchannel network device, Jiang et al. described a general method to pattern surface molecule gradients utilizing the specific recognition between avidin and biotin [36]. In this method, a gradient of avidin was patterned on the glass substrate using the microchannel network, followed by binding reactions with uniformly applied biotinylated molecules. Thus, a gradient of any biotinylated molecule, such as proteins and DNA, was created on the glass surface. In this study, a previously reported chaotic mixing method was applied to the microchannel network to enhance the mixing [37]. The authors also demonstrated the formation of complex and superimposed surface gradients of fibronectin and laminin using this method. When used together with microcontact printing, the patterned surface gradient was confined to selected regions of the substrate. Irimia, Geba, and Toner reported the generation of arbitrary monotonic gradients by creating flow "dividers" to build the microchannel network, allowing controlled mixing of flow streams [38]. By configuring the positions of the flow "dividers," any monotonic gradient was generated using this method. The authors demonstrated gradient generation of four representative profiles, including power, exponential, error, and cubic-root function. An important feature of the microchannel network approach was that the microfluidic gradients were formulated by the input configurations and the design of the microchannel network, and the resolution, as well as the complexity, of the gradients was scalable with the input number and the size of the network. Furthermore, Campbell et al. modified a microfluidic network device developed by Jeon et al. and generated complex gradient profiles (e.g., linear, exponential, and double-parabolic shape) in planar microfluidic networks. The device consisted of horizontal and serpentine channels that created various complex gradients guided by theoretical calculations of fluidic resistances.

2.3.2 Free-Diffusion-Based Gradient Devices

In contrast to generating microfluidic gradients by continuous flow mixing, another method for gradient generation was the free-diffusion of molecules in miniaturized flow-free microenvironments. These devices improved conventional assays such as the Boyden chamber, the Zigmond chamber and micropipette-based assays [2, 3, 5]. This approach was less amenable to flexible manipulation of the gradients compared to the continuous flow mixing approach but generated gradients with reduced requirements for external control, allowing easier designs of high-throughput experimentations. The molecules diffused across the microfluidic channel, membranes, and hydrogel for forming steady gradients at the equilibrium state. The membrane and hydrogel approach minimized the effect of flow on gradients and cells even more and allowed the steady-state gradient to be maintained for a long period with a small amount of source chemicals [39–41]. If on-chip valves are incorporated into the device, the temporal delivery of the gradients can be flexibly regulated, allowing two-dimensional or three-dimensional gradients to be generated [42–45]. Flow-free gradients did not induce shear stress and thus minimized mechanical disturbance to the cells. Because of these advantages, these devices form a unique class of microfluidic gradient generators and have been applied to studies of cell migration, drug delivery, and intracellular responses. When valves and micro-/nanofabricated apertures were used to control the release of chemicals on chip, a local chemical gradient near the source was generated and controlled. This approach allowed precise manipulation of the local stimulation.

2.4 Biological Applications of Gradient-Generating Microfluidic Devices

2.4.1 Biological Applications of Flow-Based Gradient Devices

Flow-based microfluidic gradient devices allowed miniaturized real-time optical assays for cell biology research. This system allowed continuous perfusion of medium and removed autocrine, paracrine signals in a microchannel. This was a unique feature compared to conventional devices. Using flow-based gradient devices, Mao, Cremer, and Manson studied the migration of a number of strains of *E. coli* cells in the absence or presence of the chemoattractant gradients [46]. A cell inlet was located between the chemoeffector and the medium inlets. The end of the "T" channel was divided into multiple subchannels toward separate outlets to collect the cells [Figure 2.2(a)]. The attractant/repellent solution, the medium, and the cells were injected into the channel by separate syringe pumps. The authors showed that the cells distributed symmetrically in the outlets in the absence of a gradient but exhibited different motility depending on the strains. When a gradient was applied, the wild-type cells exhibited increased distribution toward the attractant gradients [Figure 2.2(b)] or away from the repellent gradients in a dose-dependent manner. A surprising finding was that the gradients of a particular chemoeffector (L-leucine) induced biphasic responses in the wild-type cells. It worked as an attractant at low concentrations but as a repellent at high concentrations. A similar device and experimental setup was used by Koyama et al. to generate gradients of mouse ovary extracts for studying mouse sperm chemotaxis [47]. The extract, buffer, and cells were infused to the channel by a pressure gradient caused by the solution level difference between the inlet and outlet reservoirs. The cells that migrated to the extract side or buffer side of the channel were counted and compared to evaluate

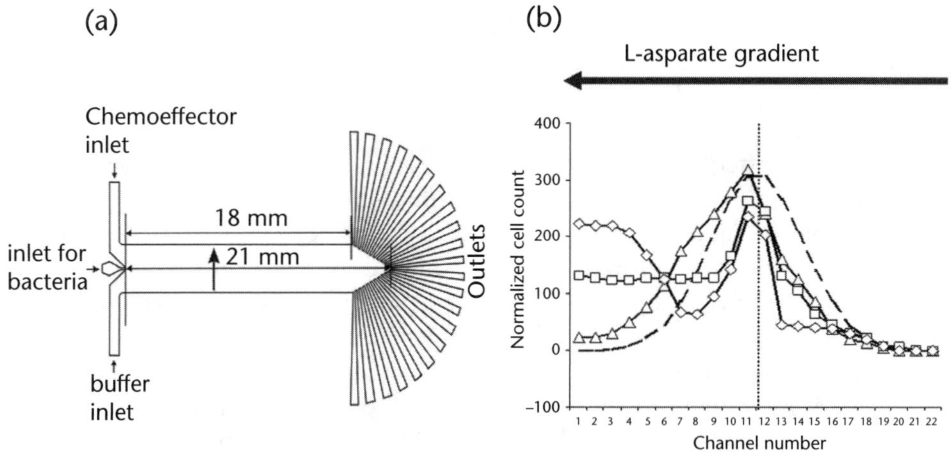

Figure 2.2 Generation of microfluidic gradients by continuous flow mixing in simple channels and its application in studying bacterial chemotaxis: (a) design of the microfluidic gradient-generating device, and (b) responses of wild-type *E. coli* cells in the microfluidic device as illustrated in part (a) to gradients of L-asparate of different concentrations: 3.2 nM L-aspartate (triangle), 10 nM L-aspartate (square), and 1 mM L-aspartate (diamond). The curve with long dashes shows the normal distribution in the presence of buffer. The vertical lines with short dashes mark the middle of the device. (*Source:* [46]. © 2003 National Academy of Sciences, USA, reprinted with permission.)

chemotaxis. Lin and Butcher reported chemotaxis of human peripheral blood T cells, a subset of white blood cells, in a simple "Y"-shaped microfluidic channel [48]. Using this device, they demonstrated robust chemotaxis of human T cells in response to single and competing gradients of chemokine CCL19 and CXCL12.

Similar to the device developed by Holdena et al. [27], Fujii et al. used a "Y"-shaped channel to test anticancer agents [49]. A gradient of an anti-stomach-cancer agent, 5-Fluorouracil (5-FU), was generated by continuous flow mixing in the microfluidic channel, resulting in a range of 5-FU concentrations in the downstream channels. The dose-dependent effect of 5-FU on preseeded SH-10-TC cells, a stomach cancer cell line, was evaluated by measuring the viability of the cells. The viability of the cells decreased with increased 5-FU concentration, suggesting that 5-FU might find potential use as an anticancer agent. In a separate study, Zhu et al. used a simple microfluidic channel with multiple inlets to produce steplike or continuous concentration gradients of a microtubule disruption drug, Colchicine [50]. The gradients were applied to preseeded C2C12 myoblasts, and the concentration-dependent effect of Colchicine on the cells was evaluated. A high percentage of the cells became round, as a sign of microtubule disruption, resulting from the drug treatment with increased drug concentration. This study utilized a unique solution-delivery system that consisted of arrays of horizontally oriented reservoirs. In contrast to vertically oriented reservoir-based systems, the pressure difference between the horizontally oriented inlet reservoirs and the outlet reservoir was fixed, maintaining the desired flow rate for a long period. The authors also showed cell growth in the channel with continuous medium perfusion compared to that in static environments. Wittig, Ryan, and Asbeck demonstrated the use of the "Y"-shaped microfluidic mixing channel in studying the growth preference of spiral ganglion neurons [51]. A gradient of neurotrophin-3 (NT-3) in cell culture medium was produced in the channel by mixing continuous flow streams driven by syringe pumps. The neurites first grew in a gradient of NT-3 and were forced to grow into either the NT-3-containing branch or the NT-3-free branch at the "Y" junction of the channel. The authors showed that the neurites preferentially grew in the medium containing NT-3, consistent with the predicted role of NT-3 in directing neutrite growth.

To pattern cells in a simple three-inlet device, Takayama et al. reported the use of multiple laminar streams in a microfluidic channel to deliver membrane-permeable molecules for partial treatment of cells [52]. Subcellular positioning of small molecules allowed mitochondrial movement and changes in cytoskeletal structure of cells to be observed. Yang, Li, and Yang used a microfluidic gradient-generating device to study the ATP-dependent calcium uptake reaction of HL-60 cells, a promyelocytic cell line [53]. The cells were loaded with calcium binding dye and were immobilized along the edge of the flow channel using a multiheight structure, in which a thin "cell-resist" channel is sandwiched between two thick flow channels. The calcium response over the gradient of ATP was monitored, and the threshold ATP concentration for inducing calcium uptake was determined. In a separate paper, Li, Yang, and Yang patterned HL-60 cells along the middle of a "T"-shaped or "V"-shaped microfluidic channel [54]. The authors showed that the "V"-shaped channel can generate a more continuous gradient along the channel by mixing continuous flows compared to the "T"-shaped channel. Using the

"V"-shaped device, patterned HL-60 cells were exposed to single or superimposed gradients of intracellular staining dyes. These studies demonstrated the advantage of using microfluidics-based gradients and cell patterning techniques for measuring dose-dependent intracellular responses.

Microchannel network-based devices for gradient generation created complex arbitrary gradients compared to simple "Y"- or "T"-shaped devices. Jeon et al. demonstrated the use of the microchannel network gradient generator in studying neutrophil chemotaxis [55]. Human peripheral blood neutrophils, an important cell type for the body's innate immune responses, were tested for their chemotactic responses to different configurations of chemokine IL-8 gradients in the microfluidic device. The authors showed that the cells migrate toward linear IL-8 gradients of a broad range of input concentrations [Figure 2.3(b)]. Moreover, when the cells were exposed to a "cliff"-type IL-8 gradient, the cells accumulated in the "cliff" region of the gradient due to the sharp change in the IL-8 concentration. In

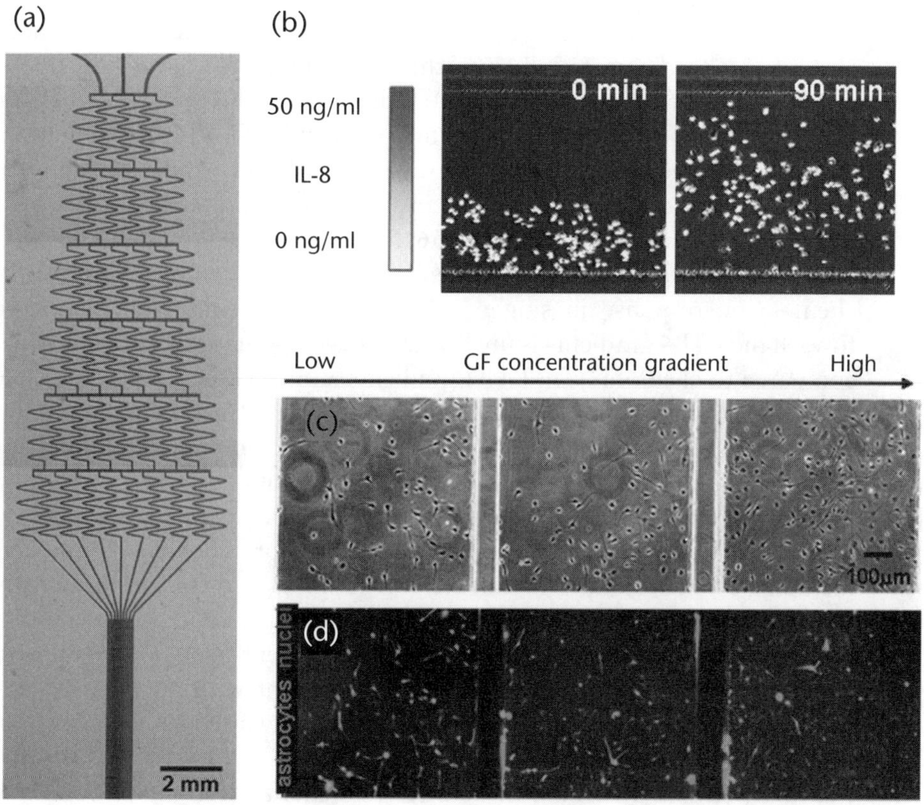

Figure 2.3 Generation of microfluidic gradients by the microchannel network approach and its applications in studying neutrophil chemotaxis and stem cell growth and differentiation: (a) illustration of microfluidic gradient generation using the network design (*Source:* [34] © 2001 American Chemical Society, reprinted with permission), (b) images showing neutrophils chemotax toward a linear IL-8 gradient in the microchannel network device (*Source:* [55] © 2002 Macmillan Publishers Ltd., [Nature Biotechnology], reprinted by permission), and (c, d) proliferation and differentiation of human neural stem cells after seven days' culture in a gradient of the growth factor produced by the microchannel network device. (c) Phase contrast images and (d) fluorescence images of the cells stained by antibodies against an astrocytes marker and nuclei. (*Source:* [58], reproduced with permission of the Royal Society of Chemistry.)

contrast, in a "hill"-type IL-8 gradient, the cells were able to migrate over the "hill" before they returned to the peak-concentration region of the gradient. These results were consistent with previous studies of neutrophil migration in under-agarose assays, showing that the cells might have a memory of their previous ligand exposure [56, 57]. Chung et al. used the network device to study the proliferation and differentiation of human neural stem cells in stable gradients of growth factors and showed that increasing growth factor concentrations resulted in increased growth rate but decreased differentiation of stem cells [Figure 2.3(c, d)] [58]. These results also demonstrated the potential of culturing stem cells for long periods in controlled gradient microenvironments. To study neutrophil chemotaxis in a microchannel network device, Lin et al. investigated the influence of the steepness and the mean concentration of linear IL-8 gradients [59]. The authors found that the optimal IL-8 concentration for random motility of cells in uniform IL-8 fields and the optimal mean concentration for effective neutrophil chemotaxis in linear IL-8 gradients coincide. When the linear gradient was configured to have a lower steepness while keeping the mean concentration unchanged, the cells yielded a similar magnitude of chemotaxis. These results suggest that the mean IL-8 concentration, not the steepness of the gradients tested, was the determining factor of effective neutrophil chemotaxis. The same group further studied neutrophil migration in competing linear gradients of two chemoattractants, IL-8 and LTB4 [60]. The authors showed that the cells preferentially migrated toward the distant gradient in competing gradients of IL-8 and LTB4, which is consistent with previous studies of neutrophil migration in under-agarose assays [56, 57]. Moreover, the cells exhibited decreased chemotaxis toward the IL-8 gradients in competing gradients compared to their chemotactic response in single IL-8 gradients. In contrast, chemotaxis of cells toward the LTB4 gradients is unchanged or even potentiated in competing gradients compared to chemotaxis in LTB4 gradients alone. These results suggested interesting cellular interactions between the two chemoattractant gradients in directing neutrophil migration. Tharp et al. examined a much broader concentration range of linear IL-8 gradients in inducing neutrophil chemotaxis and reported the surprising finding that the cells exhibit chemorepulsive migratory response in the gradients of high input IL-8 concentrations [61]. The authors showed comprehensive data including calcium flux, immunochemical staining, and tracking analysis to characterize the cell response and investigated the dependence of chemorepulsion on pertuciss toxin, wortmannin, and a Rho-dependent protein kinase, p160 ROCK. The chemorepulsion of neutrophils was consistent with previous studies showing that lymphocytes, as well as bacteria, migrated away from the attractants at high concentrations [46, 62, 63]. Lastly, Irimia et al. developed a sophisticated method that allowed fast switching of defined gradients to study neutrophil migration in response to dynamically altered linear IL-8 gradients [64]. This method constructed different gradients in parallel by separate microchannel networks and selected the gradient for the cell-containing channel using the on-chip valves. This method significantly improved the gradient switching time, compared to the previously described strategy that relies on the changes of the inlet concentrations [35]. These results were consistent with previous studies of adaptive orientation of neutrophils in dynamically changing uniform attractant fields, showing that neutrophils were more sensitive to the decrease of chemoattractant concentrations [65, 66].

The microchannel network gradient generator was also applied to studying breast cancer cell chemotaxis. Wang et al. compared the migration of MDA-MB-231 cells, a breast cancer cell line, in linear and nonlinear EGF gradients [67]. The authors showed that the cells migrate randomly in the linear EGF gradients but move chemotactically toward the nonlinear EGF gradients. Subregion analysis revealed that the cells undergo chemotaxis only in the steep region but not the flat region of the nonlinear gradients. These results suggested that cancer cells were sensitive to gradient profiles and required a larger concentration difference across the cell body to induce a significant chemotactic movement. Saadi et al. used a modified device to generate parallel EGF gradients and studied chemotaxis of MDA-MB-231 cells in different gradients simultaneously [68]. The two gradients and the cells were separated by either the laminar flow interface or a microfabricated physical barrier. This approach allowed parallel measurements of cell migration in different gradient conditions in a single chip. The authors showed diminished chemotaxis of the cells to the EGF gradient in the presence of anti-EGFR antibodies, confirming that the EGF-induced cancer cell chemotaxis was mediated by the cell surface receptor EGFR. These two studies demonstrated the effectiveness of the network device in examining the migration of slowly moving cells such as tumor cells in response to highly controlled gradients. Walker et al. performed a detailed analysis of the influence of flow speed used in the network device on the gradient profile as well as cell migration [69]. Theoretical and experimental analysis showed that the flow speed affected the linearity of the gradient. Examining the chemotaxis of differentiated HL-60 cells expressing transfected CXCR2 (the chemokine receptor for neutrophil chemoattractant IL-8) in linear IL-8 gradients under different flow speeds showed that the cell trajectory along the flow direction increased with the increasing flow speed. Sai et al. reported that the IL sequence in the LLKIL motif in CXCR2 was required for full ligand-induced activation of Erk, Akt, and chemotaxis in HL-60 cells expressing transfected CXCR2 using the microfluidic chemotaxis device [70]. Song et al. used the network device to study the migration of Dictyostelium discoideum in response to linear cyclic adenosine 3′,5′-monophosphate (cAMP) gradients and found a biphasic dependence of the chemotactic directionality of the cells on the gradient steepness [71]. Because both HL-60 cells and Dictyostelium discoideum were well-characterized model cell systems for studying the molecular mechanism of chemotaxis, the use of microfluidic devices might facilitate such studies using readily generated strains of different cell genotypes.

To generate logarithmic gradients of flow rate and chemical concentration, Kim et al. developed a microfluidic network with different flow resistances [72]. The authors demonstrated the flow effect on 3T3 murine fibroblasts culture and murine embryonic stem cells culture. This approach could be useful for optimizing culture conditions of different cell types in microfluidic chips. Wei, Cheng, and Young utilized the microfluidic network gradient generator to study cell-cell interaction by culturing cells in the microfluidic network [73]. The microfluidic device was fabricated by CO_2 laser cutting in polymethyl methacrylate (PMMA), and two different cell types were seeded in the upstream and downstream wells of the microchannel network respectively. U937 cells (macrophages) in the upstream well were stimulated by PMMA debris to express tumor necrosis factor-alpha (TNF-alpha), which subsequently flowed down the microchannel network, producing a gradient of

TNF-alpha. MG-63 cells (osteoblasts) in the downstream wells were activated by TNF-alpha to release prostaglandin E2 (PGE2), a well-known bone resorption marker. Using the chaotic mixing technique [37], Jiang et al. developed a microfluidic device that could serially dilute the input chemical with reduced channel length for mixing. It resulted in a concentration gradient in the downstream channels [74]. The authors showed that a dynamic concentration range on the order of 10^3 could be achieved using 1:1 dilution in each mixing channel. The use of this device in evaluating immunochemical reactions was illustrated by applying serially diluted solutions of human HIV+ serum that contained antibodies for HIV ENV proteins (e.g., anti-gp41 and anti-pg120) to perpendicularly patterned parallel strips of HIV ENV proteins (antigens, gp41, and pg120) on a polycarbonate membrane. The results showed increasing immunochemical detection of the HIV ENV proteins with the increased concentration of antibody-containing HIV+ serum.

Microfluidic devices can generate surface gradients for cell biology study such as axon guidance. Dertinger et al. patterned a gradient of laminin in BSA on a Poly-L-Lysine precoated surface by continuous-flow-based deposition using the microchannel network gradient generator. They used this device to investigate the growth of axons of hippocampal neurons from neonatal rats [75]. Axons oriented toward the laminin gradients, and the threshold laminin gradient for directing axon orientation was determined. Burdick, Khademhosseini, and Langer developed a method to pattern surface gradients of immobilized molecules using photocrosslinked hydrogels [76]. The ligand-containing monomer solution and the ligand-free monomer solution were introduced into the microchannel network to produce a prepolymer gradient. These solutions were subsequently polymerized into a water-swollen hydrogel by ultraviolet light exposure to create an immobilized gradient of the ligand. The authors successfully applied this method to generate a gradient of adhesive ligands, Acr-PEG-RGDS, and demonstrated increased number of attached endothelial cells along the RGDS gradient. Gunawan et al. used the microchannel network device to pattern competing surface gradients of laminin and collagen I on gold substrates and studied protein expression of several cell types via immunofluorescence [77]. The authors argued that in contrast to physisorbed gradients, covalently immobilized protein patterns preserved the gradient fidelity, making long-term cell studies feasible. In studying IEC-6 rat small intestine epithelial cells, the expression of PCNA, a protein that correlates with cell cycle progression, increased with increased density of collagen. The expression of p27, a protein that correlates with the exit of the cell cycle, increased with increased laminin density. In a later study, the migration and polarity of IEC-6 cells in response to surface gradients of laminin was investigated [78]. The results showed that the cells can migrate in response to a broad range of laminin gradients regardless of the steepness.

2.4.2 Biological Applications of Free-Diffusion-Based Gradient Devices

Similar to the principle of generating gradients in the Zigmond or Dunn chambers [3, 6], Kanegasaki et al. developed a gradient device consisting of two outer microcompartments, containing chemoattractant or medium, and an inner gradient channel [79]. A near-linear chemoattractant gradient could be generated in the gradient channel by free-diffusion and maintained at the equilibrium state for up to one hour. By creating small barriers in the gradient channel, cells could be aligned to the edge of the gradient channel at the beginning of the migration experiment. Neutrophil chemotaxis to IL-8 gradients was successfully demonstrated in this system. This system showed the potential for parallel cell-migration experiments in a single chip.

Compared to conventional membrane-based assays such as the Boyden chamber [5], microfluidic devices allowed tracking of individual cells in miniaturized membrane-free environments. Abhyankar et al. reported generation of stable linear gradient using the membrane-based microfluidic device for neutrophil chemotaxis [39]. This device consisted of a membrane-covered source region and a large-volume sink region connected by a microchannel. The chemical-containing reservoir and the buffer reservoir allowed the molecules to diffuse through a membrane into a bottom gradient channel. The stable chemical gradient was created in the gradient channel by free-diffusion and maintained in linear form at the equilibrium state. Diao et al. developed a membrane-based three-channel microfluidic gradient generator for studying bacterial chemotaxis [40]. In this device, the microchannels were cut in a nitrocellulose membrane, which was sandwiched between the upper reagent-reservoir layer and the bottom glass substrate to form sealed channels [Figure 2.4(a)]. The chemicals flowed through the two outer channels and diffused horizontally through the membrane into the inner gradient channel. The chemical gradient could be created in the inner gradient channel by free-diffusion and maintained in linear form at the equilibrium state. Using this device, the authors demonstrated the migration of a wild-type strain of *E. coli* cells RP437 toward the attractant gradient of L-asparate or against the repellent gradient of Glycerol [Figure 2.4(b)]. Mutant strains derived from RP437 were shown to be incapable of chemotaxis, except for the mutant strains KX1485 and KX1486, which interestingly exhibited similar chemotactic behaviors to the wild-type cells. Finally, migration experiments could be performed in multiplex format using this

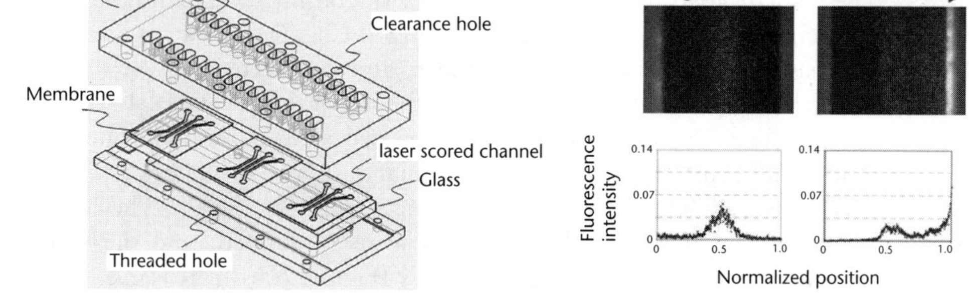

Figure 2.4 Generation of microfluidic gradients by free-diffusion through a membrane and its application in studying bacterial chemotaxis: (a) design of the microfluidic device for generating gradient, and (b) responses of bacteria strain RP437 cells in the microfluidic device with or without a 1×10^{-4}M L-aspartate gradient. (*Source:* [40], reproduced with permission of the Royal Society of Chemistry.)

device, suggesting the potential of the membrane approach for high-throughput experimentations. Wu, Huang, and Zare reported an interesting approach to generating complex gradients based on diffusion through a hydrogel with a variety of microfluidic channel designs [41]. The device consisted of an upper PDMS layer containing chemical reservoirs, a middle hydrogel layer, and a bottom PDMS layer containing gradient channels. The chemicals in the upper layer diffused through the hydrogel and formed a stable linear gradient. This gradient was further modulated by the shape of different gradient channels in the bottom layer. Thus, arbitrary gradient shapes could be generated using this method. In contrast to the previously described devices of generating microfuidic gradients that mostly used the microfluidic channels as an operator to modulate the chemical inputs, the current method presents a way to regulate the gradient profile using the microfluidic channel itself.

Gradients could be also generated by solutions released through small apertures. Chung, Lin, and Jeon developed a PDMS microfluidic device to control fluid injection by on-chip valves [42], which were fabricated using multilayer soft lithography [80]. The fluid was driven by a pressure source and could be injected to a relatively large medium-containing gradient device through a small aperture by opening the valve for a controlled period. The amount of the injected chemical was controlled by the valve opening time, and a single injection yielded a very small volume (i.e., 30 to 120 pL). When the valve was continuously operated at a controlled frequency to repeatedly inject the chemical, a stable gradient could be generated and maintained at the equilibrium state of diffusion. The authors demonstrated the generation of stable complex gradient using two side-by-side, separately controlled injectors. This method simulated the conventional micropipette-based gradient generator [2] but provided improvement in miniaturization and gradient control. Similar to the device reported by Chung et al., Frevert et al. developed a method to generate time-evolving chemical gradients in a microfluidic device [43]. In this device, the chemical and the medium were confined in separate channels. Opening the on-chip valve allowed the release of the chemical to the medium channel, resulting in the formation of a gradient by diffusion. Using this method, the authors generated an IL-8 gradient and demonstrated neutrophil chemotaxis. Peterman et al. developed a microfluidic device that allowed local chemical release through small apertures [44]. These small apertures were etched on a silicon-nitride membrane and aligned to flow channels made of SU-8 [Figure 2.5(a)]. Chemical flow was driven by electroosmosis, induced by on-chip electrodes, and delivered to the cell side of the chip through the apertures. To demonstrate the use of the device, PC12 cells were seeded on the silicon nitride membrane on the cell side of the chip and stimulated by bradykinin using this method. The cellular response to the bradykinin stimulation was measured by monitoring the intracellular Ca^{2+} levels. The authors showed that only the cells near the apertures were stimulated, and the system allowed sequential stimulations to the same cells [Figure 2.5(b)]. Kosar et al. developed a nanofabricated planar aperture as a mimic of nerve and muscle contacts [45]. Myotubes were cultured on top of nanofabricated planar apertures (2–8 μm diameter) to focally stimulate the muscle cell membrane with neural agrin, a synaptogenic factor. Focal agrin delivery through the apertures resulted in local aggregation of acetylcholine receptors (AChRs). The same group further developed

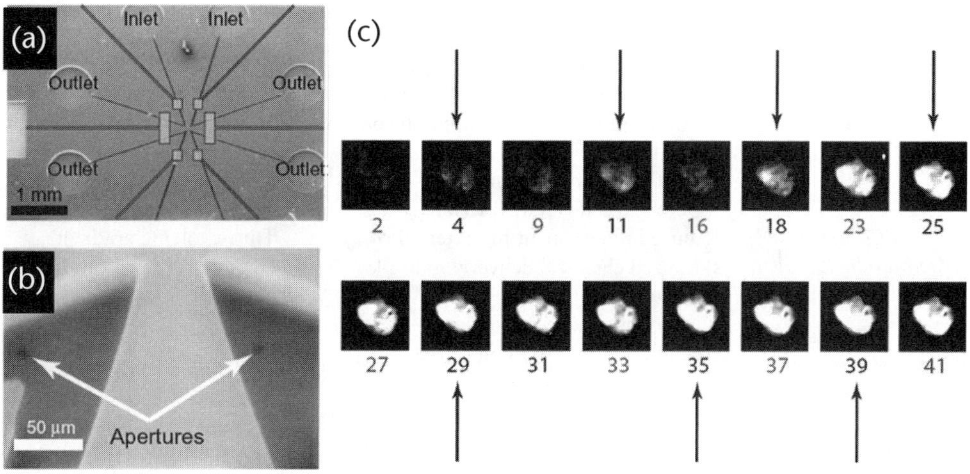

Figure 2.5 Microfluidic chemical delivery by microinjection through small apertures and its application in measuring intracellular responses of cells to stimulations: (a) scanning electron micrograph of the microfluidic device, (b) higher-magnification scanning electron micrograph of the channel corners in which the apertures are visible inside the channels as indicated by the arrows, and (c) Ca^{2+} response of two PC12 cells by repeated bradykinin stimulation via microinjections through the small apertures. The frame numbers are indicated, with red indicating the frame at which stimulation is activated. The arrows indicate the frames at which the stimulation is a maximum, always two frames after stimulation. (*Source:* [44] © 2004 National Academy of Sciences, USA, reprinted with permission.)

a microfluidic jet device for generating steady-state linear gradients of soluble molecules on open surfaces [81]. This device was fabricated by two-layer photolithography and electron beam lithography. The sink and source chambers were connected with 1.5-μm–thick microchannels. The microfluidic jet device could inject picoliter volume of fluid into an open reservoir with negligible exposure of the surface to flow.

2.5 Summary and Future Directions

The gradient-generating microfluidic device is an enabling experimental platform for fundamental research in a number of physiological and pathological processes that range from immune response, cancer cell migration, and axon guidance. Compared to conventional macroscale devices, microfluidic devices allow precise control and manipulation of cellular microenvironments. Generation of stable and well-defined gradient profiles affords more accurate and reproducible quantitative analysis of cellular behavior. The two main classes of microfluidic gradient-generating approaches have distinct desirable features, as well as limitations, in gradient configuration and cell-based studies (Table 2.1). Furthermore, there are a number of shortcomings that make the existing microfluidic gradient-generating devices difficult to adopt for routine use in a biological laboratory setting. In order for microfluidics to gain broader acceptance by the biology community and thus realize its full potential in helping to make fundamental advances, several key challenges must be addressed.

Table 2.1 Comparison of Flow-Based and Free-Diffusion-Based Microfluidic Gradient-Generating Devices

Methods	*Main Advantages*	*Main Disadvantages*
Flow-Based Microfluidic Gradient Device	Stable gradient with defined shapes Flexible gradient configuration in space and time Optical visualization; single cell analysis Minimized cell-cell interactions	Gradient limited in 1D Flow-induced shear stress on cells
Free-Diffusion-Based Microfluidic Gradient Device	Reduced requirement for external control Localized chemical delivery; multiple gradients in 2D Optical visualization; single cell ananlysis Negligible shear stress on cells	Time-evolving gradient Uncontrolled cell/cell interactions

First, device assembly and implementation need to be simplified. Most current microfluidic devices require specialized equipments (e.g., plasma chamber) and procedures for device assembly and implementation. For larger biology community end users, these additional steps represent extra hurdles for using microfluidic devices for their experiments. The microfluidics community needs to develop a simplified and shared platform for "microplumbing" that can be used to connect the microfluidic devices to external devices or to other microfluidic chips. We will also need to develop a flexible interface that can couple microfluidic devices to conventional laboratory assays or platforms (i.e., that can be used directly with cells growing in a petri dish).

Second, approaches to generating stable gradients in three-dimensional extracellular matrix will be necessary as classical methods involving cell culture and migration on two-dimensional surfaces are moving toward three-dimensional conditions. In order to investigate the regulation of cellular behavior under physiologically relevant microenvironments, it is necessary to develop three-dimensional culture conditions that incorporate the necessary cell-cell and cell-extracellular matrix interactions. Although the first generation of microfluidic devices was successful in simulating two-dimensional gradient conditions, it is much more challenging to generate and maintain gradients across small dimensions in three dimensions. Advances in the tissue-engineering field for generating three-dimensional models need to be incorporated for future devices to recreate the three-dimensional tissue microenvironment in vitro.

Finally, functional microfluidic devices that integrate cell sorting, clonal cell expansion, and biochemical analysis will be desirable for applications involving stem cells. The gradient-generating function will play an important part of an integrated platform in which cells are exposed to combinatorial mixtures of growth factors or culture conditions that can be subjected to phenotypic and/or genotypic analysis. While it is expected that new designs for generating gradients will continue to be explored, microfluidic-based devices have started a rapidly growing interest in the biology community. They will continue to play an increasingly visible role in biological research. Close collaborations between engineers and cell biologists are critical to the success of these efforts and will enable important innovations in investigating complex biological systems.

References

[1] Gurdon, J., and Bourillot, P., "Morphogen gradient interpretation," *Nature*, Vol. 413, No. 6858, 2001, pp. 797–803.

[2] Lohof, A., Quillan, M., Dan, Y., and Poo, M., "Asymmetric modulation of cytosolic cAMP activity induces growth cone turning," *J. Neurosci.*, Vol. 12, No. 4, 1992, pp. 1253–1261.

[3] Zigmond, S., "Ability of polymorphonuclear leukocytes to orient in gradients of chemotactic factors," *J. Cell. Biol.*, Vol. 75, No. 2, 1977, pp. 606–616.

[4] Nelson, R. D., Quie, P. G., and Simmons, R. L., "Chemotaxis under agarose: A new and simple method for measuring chemotaxis and spontaneous migration of human polymorphonuclear leukocytes and monocytes," *J. Immunol.*, Vol. 115, 1975, pp. 1650–1656.

[5] Boyden, S. V., "The chemotactic effect of mixtures of antibody and antigen on polymorphonuclear leukocytes," *J. Exp. Med.*, Vol. 115, 1962, p. 453.

[6] Zicha, D., Dunn, G., and Brown, A., "A new direct-viewing chemotaxis chamber," *J. Cell. Sci.*, Vol. 99, 1991, pp. 769–775.

[7] Boyden, S., "The chemotactic effect of mixtures of antibody and antigen on polymorphonuclear leucocytes," *J. Exp. Med.*, Vol. 115, No. 3, 1962, pp. 453–466.

[8] Wilkinson, P. C., "Assays of leukocyte locomotion and chemotaxis," *J. Immunol. Methods*, Vol. 216, Nos. 1–2, 1998, pp. 139–153.

[9] Wells, A., "Tumor invasion: Role of growth factor–induced cell motiltiy," *Adv. Cancer Res.*, Vol. 78, 2000, pp. 31–101.

[10] Keller, H. U., Borel, J. F., Wilkinson, P. C., Hess, M. W., and Cottier, H., "Re-assessment of Boyden's technique for measuring chemotaxis," *J. Immunol. Methods*, Vol. 1, No. 2, 1972, pp. 165–168.

[11] Wei, S. H., Parker, I., Miller, M. J., and Cahalan, M. D., "A stochastic view of lymphocyte motility and trafficking within the lymph node," *Immunol. Rev.*, Vol. 195, 2003, pp. 136–159.

[12] Lauffenburger, D. A., and Zigmond, S. H., "Chemotactic factor concentration gradients in chemotaxis assay systems," *J. Immunol. Methods*, Vol. 40, No. 1, 1981, pp. 45–60.

[13] Nelson, R. D., Quie, P. G., and Simmons, R. L., "Chemotaxis under agarose: A new and simple method for measuring chemotaxis and spontaneous migration of human polymorphonuclear leukocytes and monocytes," *J. Immunol.*, Vol. 115, No. 6, 1975, pp. 1650–1656.

[14] Entschladen, F., Drell, T. L. T., Lang, K., Masur, K., Palm, D., Bastian, P., Niggemann, B., and Zaenker, K. S., "Analysis methods of human cell migration," *Exp. Cell Res.*, Vol. 307, No. 2, 2005, pp. 418–426.

[15] Zigmond, S. H., "Ability of polymorphonuclear leukocytes to orient in gradients of chemotactic factors," *J. Cell. Biol.*, Vol. 75, No. 2, Pt. 1, 1977, pp. 606–616.

[16] Crank, J., *The Mathematics of Diffusion*, Oxford: Clarendon Press, 1964.

[17] Cussler, E. L., *Diffusion: Mass Transfer in Fluid Systems*, New York: Cambridge University Press, 1997.

[18] Zicha, D., Dunn, G., and Brown, A., "A new direct-viewing chemotaxis chamber," *J. Cell. Sci.*, Vol. 99, No. 4, 1991, pp. 769–775.

[19] Zicha, D., Dunn, G., and Jones, G., "Analyzing chemotaxis using the Dunn direct-viewing chamber," *Methods Mol. Biol.*, Vol. 75, 1997, pp. 449–457.

[20] Gerisch, G., and Keller, H. U., "Chemotactic reorientation of granulocytes stimulated with micropipettes containing fMet-Leu-Phe," *J. Cell. Sci.*, Vol. 52, 1981, pp. 1–10.

[21] Bailly, M., Yan, L., Whitesides, G. M., Condeelis, J. S., and Segall, J. E., "Regulation of protrusion shape and adhesion to the substratum during chemotactic responses of mammalian carcinoma cells," *Exp. Cell Res.*, Vol. 241, No. 2, 1998, pp. 285–299.

[22] Xiao, Z., Zhang, N., Murphy, D. B., and Devreotes, P. N., "Dynamic distribution of chemoattractant receptors in living cells during chemotaxis and persistent stimulation," *J. Cell. Biol.*, Vol. 139, No. 2, 1997, pp. 365–374.

[23] Lohof, A. M., Quillan, M., Dan, Y., and Poo, M. M., "Asymmetric modulation of cytosolic cAMP activity induces growth cone turning," *J. Neurosci.*, Vol. 12, No. 4, 1992, pp. 1253–1261.

[24] Zheng, J. Q., Felder, M., Connor, J. A., and Poo, M. M., "Turning of nerve growth cones induced by neurotransmitters," *Nature*, Vol. 368, No. 6467, 1994, pp. 140–144.

[25] Soon, L., Mouneimne, G., Segall, J., Wyckoff, J., and Condeelis, J., "Description and characterization of a chamber for viewing and quantifying cancer cell chemotaxis," *Cell Motil. Cytoskeleton*, Vol. 62, No. 1, 2005, pp. 27–34.

[26] Kamholz, A. E., and Yager, P., "Theoretical analysis of molecular diffusion in pressure-driven laminar flow in microfluidic channels," *Biophys. J.*, Vol. 80, No. 1, 2001, pp. 155–160.

[27] Holdena, M. A., Kumarb, S., Castellanaa, E. T., Beskok, A., and Cremer, P. S., "Generating fixed concentration arrays in a microfluidic device," *Sensors and Actuators B: Chemical*, Vol. 92, 2003, pp. 199–207

[28] Yang, M., Yang, J., Li, C. W., and Zhao, J., "Generation of concentration gradient by controlled flow distribution and diffusive mixing in a microfluidic chip," *Lab Chip*, Vol. 2, No. 3, 2002, pp. 158–163.

[29] Biddiss, E., and Li, D., "Electrokinetic generation of temporally and spatially stable concentration gradients in microchannels," *J. Colloid Interface Sci.*, Vol. 288, No. 2, 2005, pp. 606–615.

[30] Xie, J., Miao, Y., Shih, J., He, Q., Liu, J., Tai, Y. C., and Lee, T. D., "An electrochemical pumping system for on-chip gradient generation," *Anal. Chem.*, Vol. 76, No. 13, 2004, pp. 3756–3763.

[31] Taylor, A. M., Rhee, S. W., and Jeon, N. L., "Microfluidic chambers for cell migration and neuroscience research," *Methods Mol. Biol.*, Vol. 321, 2006, pp. 167–177.

[32] Rhoads, D. S., Nadkarni, S. M., Song, L., Voeltz, C., Bodenschatz, E., and Guan, J. L., "Using microfluidic channel networks to generate gradients for studying cell migration," *Methods Mol. Biol.*, Vol. 294, 2005, pp. 347–357.

[33] Jeon, N. L., Dertinger, S. K. W., Chiu, D. T., Choi, I. S., Stroock, A. D., and Whitesides, G. M., "Generation of solution and surface gradients using microfluidic systems," *Langmuir*, Vol. 16, No. 22, 2000, pp. 8311–8316.

[34] Dertinger, S. K. W., Chiu, D. T., Jeon, N. L., and Whitesides, G. M., "Generation of gradients having complex shapes using microfluidic networks," *Anal. Chem.*, Vol. 73, No. 6, 2001, pp. 1240–1246.

[35] Lin, F., Saadi, W., Rhee, S. W., Wang, S. J., Mittal, S., and Jeon, N. L., "Generation of dynamic temporal and spatial concentration gradients using microfluidic devices," *Lab Chip*, Vol. 4, No. 3, 2004, pp. 164–167.

[36] Jiang, X., Xu, Q., Dertinger, S. K., Stroock, A. D., Fu, T. M., and Whitesides, G. M., "A general method for patterning gradients of biomolecules on surfaces using microfluidic networks," *Anal. Chem.*, Vol. 77, No. 8, 2005, pp. 2338–2347.

[37] Stroock, A. D., Dertinger, S. K. W., Ajdari, A., Mezic, I., Stone, H. A., and Whitesides, G. M., "Chaotic mixer for microchannels," *Science*, Vol. 295, No. 5555, 2002, pp. 647–651.

[38] Irimia, D., Geba, D. A., and Toner, M., "Universal microfluidic gradient generator," *Anal. Chem.*, Vol. 78, No. 10, 2006, pp. 3472–3477.

[39] Abhyankar, V. V., Lokuta, M. A., Huttenlocher, A., and Beebe, D. J., "Characterization of a membrane-based gradient generator for use in cell-signaling studies," *Lab Chip*, Vol. 6, No. 3, 2006, pp. 389–393.

[40] Diao, J., Young, L., Kim, S., Fogarty, E. A., Heilman, S. M., Zhou, P., Shuler, M. L., Wu, M., and DeLisa, M. P., "A three-channel microfluidic device for generating static linear

gradients and its application to the quantitative analysis of bacterial chemotaxis," *Lab Chip*, Vol. 6, No. 3, 2006, pp. 381–388.

[41] Wu, H., Huang, B., and Zare, R. N., "Generation of complex, static solution gradients in microfluidic channels," *J. Am. Chem. Soc.*, Vol. 128, No. 13, 2006, pp. 4194–4195.

[42] Chung, B. G., Lin, F., and Jeon, N. L., "A microfluidic multi-injector for gradient generation," *Lab Chip*, Vol. 6, No. 6, 2006, pp. 764–768.

[43] Frevert, C. W., Boggy, G., Keenan, T. M., and Folch, A., "Measurement of cell migration in response to an evolving radial chemokine gradient triggered by a microvalve," *Lab Chip*, Vol. 6, No. 7, 2006, pp. 849–856.

[44] Peterman, M. C., Noolandi, J., Blumenkranz, M. S., and Fishman, H. A., "Localized chemical release from an artificial synapse chip," *Proc. Natl. Acad. Sci. USA*, Vol. 101, No. 27, 2004, pp. 9951–9954.

[45] Kosar, T. F., Tourovskaia, A., Figueroa-Masot, X., Adams, M. E., and Folch, A., "A nanofabricated planar aperture as a mimic of the nerve-muscle contact during synaptogenesis," *Lab Chip*, Vol. 6, No. 5, 2006, pp. 632–638.

[46] Mao, H., Cremer, P. S., and Manson, M. D., "A sensitive, versatile microfluidic assay for bacterial chemotaxis," *Proc. Natl. Acad. Sci. USA*, Vol. 100, No. 9, 2003, pp. 5449–5454.

[47] Koyama, S., Amarie, D., Soini, H. A., Novotny, M. V., and Jacobson, S. C., "Chemotaxis assays of mouse sperm on microfluidic devices," *Anal. Chem.*, Vol. 78, No. 10, 2006, pp. 3354–3359.

[48] Lin, F., and Butcher, E. C., "T cell chemotaxis in a simple microfluidic device," *Lab Chip*, Vol. 6, No. 11, 2006, pp. 1462–1469.

[49] Fujii, S., Uematsu, M., Yabuki, S., Abo, M., Yoshimura, E., and Sato, K., "Microbioassay system for an anti-cancer agent test using animal cells on a microfluidic gradient mixer," *Anal. Sci.*, Vol. 22, No. 1, 2006, pp. 87–90.

[50] Zhu, X., Yi Chu, L., Chueh, B. H., Shen, M., Hazarika, B., Phadke, N., and Takayama, S., "Arrays of horizontally-oriented mini-reservoirs generate steady microfluidic flows for continuous perfusion cell culture and gradient generation," *Analyst*, Vol. 129, No. 11, 2004, pp. 1026–1031.

[51] Wittig, J. H., Jr., Ryan, A. F., and Asbeck, P. M., "A reusable microfluidic plate with alternate-choice architecture for assessing growth preference in tissue culture," *J. Neurosci. Methods*, Vol. 144, No. 1, 2005, pp. 79–89.

[52] Takayama, S., Ostuni, E., LeDuc, P., Naruse, K., Ingber, D. E., and Whitesides, G. M., "Laminar flows subcellular positioning of small molecules," *Nature*, Vol. 411, No. 6841, 2001, pp. 1016–1016.

[53] Yang, M., Li, C. W., and Yang, J., "Cell docking and on-chip monitoring of cellular reactions with a controlled concentration gradient on a microfluidic device," *Anal. Chem.*, Vol. 74, No. 16, 2002, pp. 3991–4001.

[54] Li, C. W., Yang, J., and Yang, M., "Dose-dependent cell-based assays in V-shaped microfluidic channels," *Lab Chip*, Vol. 6, No. 7, 2006, pp. 921–929.

[55] Jeon, N. L., Baskaran, H., Dertinger, S. K., Whitesides, G. M., Van de Water, L., and Toner, M., "Neutrophil chemotaxis in linear and complex gradients of interleukin-8 formed in a microfabricated device," *Nat. Biotechnol.*, Vol. 20, No. 8, 2002, pp. 826–830.

[56] Foxman, E. F., Kunkel, E. J., and Butcher, E. C., "Integrating conflicting chemotactic signals: The role of memory in leukocyte navigation," *J. Cell. Biol.*, Vol. 147, No. 3, 1999, pp. 577–588.

[57] Foxman, E. F., Campbell, J. J., and Butcher, E. C., "Multistep navigation and the combinatorial control of leukocyte chemotaxis," *J. Cell. Biol.*, Vol. 139, No. 5, 1997, pp. 1349–1360.

[58] Chung, B. G., Flanagan, L. A., Rhee, S. W., Schwartz, P. H., Lee, A. P., Monuki, E. S., and Jeon, N. L., "Human neural stem cell growth and differentiation in a gradient-generating microfluidic device," *Lab Chip*, Vol. 5, No. 4, 2005, pp. 401–406.

[59] Lin, F., Nguyen, C. M., Wang, S. J., Saadi, W., Gross, S. P., and Jeon, N. L., "Effective neutrophil chemotaxis is strongly influenced by mean IL-8 concentration," *Biochem. Biophys. Res. Commun.*, Vol. 319, No. 2, 2004, pp. 576–581.

[60] Lin, F., Nguyen, C. M., Wang, S. J., Saadi, W., Gross, S. P., and Jeon, N. L., "Neutrophil migration in opposing chemoattractant gradients using microfluidic chemotaxis devices," *Ann. Biomed. Eng.*, Vol. 33, No. 4, 2005, pp. 475–482.

[61] Tharp, W. G., Yadav, R., Irimia, D., Upadhyaya, A., Samadani, A., Hurtado, O., Liu, S. Y., Munisamy, S., Brainard, D. M., Mahon, M. J., Nourshargh, S., van Oudenaarden, A., Toner, M. G., and Poznansky, M. C., "Neutrophil chemorepulsion in defined interleukin-8 gradients in vitro and in vivo," *J. Leukoc. Biol.*, Vol. 79, No. 3, 2006, pp. 539–554.

[62] Vianello, F., Kraft, P., Mok, Y. T., Hart, W. K., White, N., and Poznansky, M. C., "A CXCR4-dependent chemorepellent signal contributes to the emigration of mature single-positive cd4 cells from the fetal thymus," *J. Immunol.*, Vol. 8, 2005, pp. 5115–5125.

[63] Poznansky, M. C., Olszak, I. T., Evans, R. H., Wang, Z., Foxall, R. B., Olson, D. P., Weibrecht, K., Luster, A. D., and Scadden, D. T., "Thymocyte emigration is mediated by active movement away from stroma-derived factors," *J. Clin. Invest.*, Vol. 109, 2002, pp. 1101–1110.

[64] Irimia, D., Liu, S.-Y., Tharp, W. G., Samadani, A., Toner, M., and Poznansky, M. C., "Microfluidic system for measuring neutrophil migratory responses to fast switches of chemical gradients," *Lab Chip*, Vol. 6, No. 2, 2006, pp. 191–198.

[65] Zigmond, S. H., and Sullivan, S. J., "Sensory adaptation of leukocytes to chemotactic peptides," *J. Cell. Biol.*, Vol. 82, No. 2, 1979, pp. 517–527.

[66] Albrecht, E., and Petty, H. R., "Cellular memory: Neutrophil orientation reverses during temporally decreasing chemoattractant concentrations," *Proc. Natl. Acad. Sci. USA*, Vol. 95, No. 9, 1998, pp. 5039–5044.

[67] Wang, S.-J., Saadi, W., Lin, F., Minh-Canh Nguyen, C., and Li Jeon, N., "Differential effects of EGF gradient profiles on MDA-MB-231 breast cancer cell chemotaxis," *Exp. Cell Res.*, Vol. 300, No. 1, 2004, pp. 180–189.

[68] Saadi, W., Wang, S. J., Lin, F., and Jeon, N. L., "A parallel-gradient microfluidic chamber for quantitative analysis of breast cancer cell chemotaxis," *Biomed Microdevices*, Vol. 8, No. 2, 2006, pp. 109–118.

[69] Walker, G. M., Sai, J., Richmond, A., Stremler, M., Chung, C. Y., and Wikswo, J. P., "Effects of flow and diffusion on chemotaxis studies in a microfabricated gradient generator," *Lab Chip*, Vol. 5, No. 6, 2005, pp. 611–618.

[70] Sai, J., Walker, G., Wikswo, J., and Richmond, A., "The IL sequence in the LLKIL motif in CXCR2 is required for full ligand-induced activation of Erk, Akt, and chemotaxis in HL60 cells," *J. Biol. Chem.*, Vol. 281, No. 47, 2006, pp. 35931–35941.

[71] Song, L., Nadkarni, S. M., Bodeker, H. U., Beta, C., Bae, A., Franck, C., Rappel, W. J., Loomis, W. F., and Bodenschatz, E., "Dictyostelium discoideum chemotaxis: Threshold for directed motion," *Eur. J. Cell. Biol.*, Vol. 2006.

[72] Kim, L., Vahey, M. D., Lee, H. Y., and Voldman, J., "Microfluidic arrays for logarithmically perfused embryonic stem cell culture," *Lab Chip*, Vol. 6, No. 3, 2006, pp. 394–406.

[73] Wei, C. W., Cheng, J. Y., and Young, T. H., "Elucidating in vitro cell-cell interaction using a microfluidic coculture system," *Biomed Microdevices*, Vol. 8, No. 1, 2006, pp. 65–71.

[74] Jiang, X., Ng, J. M. K., Stroock, A. D., Dertinger, S. K. W., and Whitesides, G. M., "A miniaturized, parallel, serially diluted immunoassay for analyzing multiple antigens," *J. Am. Chem. Soc.*, Vol. 125, 2003, pp. 5294–5295.

[75] Dertinger, S. K., Jiang, X., Li, Z., Murthy, V. N., and Whitesides, G. M., "Gradients of substrate-bound laminin orient axonal specification of neurons," *Proc. Natl. Acad. Sci. USA*, Vol. 99, No. 20, 2002, pp. 12542–12547.

[76] Burdick, J. A., Khademhosseini, A., and Langer, R., "Fabrication of gradient hydrogels using a microfluidics/photopolymerization process," *Langmuir*, Vol. 20, No. 13, 2004, pp. 5153–5156.

[77] Gunawan, R. C., Choban, E. R., Conour, J. E., Silvestre, J., Schook, L. B., Gaskins, H. R., Leckband, D. E., and Kenis, P. J., "Regiospecific control of protein expression in cells cultured on two-component counter gradients of extracellular matrix proteins," *Langmuir*, Vol. 21, No. 7, 2005, pp. 3061–3068.

[78] Gunawan, R. C., Silvestre, J., Gaskins, H. R., Kenis, P. J., and Leckband, D. E., "Cell migration and polarity on microfabricated gradients of extracellular matrix proteins," *Langmuir*, Vol. 22, No. 9, 2006, pp. 4250–4258.

[79] Kanegasaki, S., Nomura, Y., Nitta, N., Akiyama, S., Tamatani, T., Goshoh, Y., Yoshida, T., Sato, T., and Kikuchi, Y., "A novel optical assay system for the quantitative measurement of chemotaxis," *J. Immunol. Methods*, Vol. 282, Nos. 1–2, 2003, pp. 1–11.

[80] Unger, M., Chou, H., Thorsen, T., Scherer, A., and Quake, S., "Monolithic microfabricated valves and pumps by multilayer soft lithography," *Science*, Vol. 288, 2000, pp. 113–116.

[81] Keenan, T. M., Hsu, C.-H., and Folch, A., "Microfluidic 'jets' for generating steady-state gradients of soluble molecules on open surfaces," *Applied Physics Lett.*, Vol. 89, No. 11, 2006, p. 114103.

CHAPTER 3

Surface Patterning for Controlling Cell-Substrate Interactions

Xingyu Jiang

3.1 Introduction

This chapter describes a set of tools that allow the patterning of the interactions between mammalian cells and solid substrates. The adhesion of mammalian cells requires the presence of surface-immobilized ligands, such as proteins, that bind to receptors in the cell membrane. The binding between these ligands and receptors promotes the adhesion of cells. The tools described in this chapter allow the control of the amount of adsorbed proteins and the adhesion of cells on a two-dimensional surface. The development of surface chemistry, synthetic chemistry, and lithography has made it possible to control the density of ligands on the surface, the amount of adsorbed molecules, and the geometry of cell adhesion on surfaces. Recent advances in these fields have also made it possible to modulate cell adhesion on surfaces in real time. The main contents of this chapter include: (1) an introduction to tools generally useful for surface patterning (Section 3.2); (2) methods that control the adsorption of proteins on surfaces (Section 3.3); (3) patterning of proteins and cells (Section 3.4); (4) dynamic systems (Section 3.5); and (5) other patterning methods and their applications (Section 3.6).

3.2 Self-Assembled Monolayers, Lithography, and Other Important Tools

This section introduces the basic tools commonly used to control surface properties and generate patterns. These tools include self-assembled monolayers (SAMs), tools in lithography such as molding, printing, and methods based on membranes and microchannels.

3.2.1 Introduction to SAMs

SAMs are a class of model systems, typically based on molecules that spontaneously form a single layer with medium- to long-range order on a solid substrate [1]. SAMs discussed in this chapter only refer to those that form tightly packed single layers from small molecules, such as alkanethiols on the surface of gold (Figure 3.1). Alkanethiols having the general structures of $HS(CH_2)_nX$ form an ordered layer on

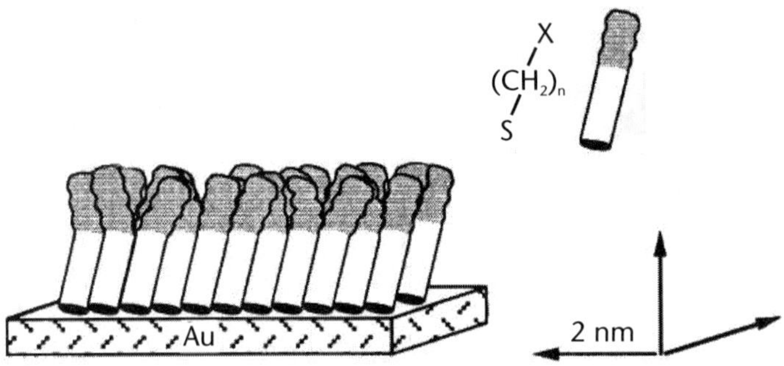

Figure 3.1 A schematic depiction of SAMs of alkanethiols on gold. $HS(CH_2)_nX$ is the general structure of the alkanethiols used in SAMs. Because of the tightly packed layer, controlling the chemical properties of the terminal group "X" allows full control of the properties of surfaces.

the surface of gold. Because of the closely packed structures of the alkanethiols, the physicochemical properties of the surface are dominated by the "X" group. It is therefore possible to modulate surface properties by simply altering the terminal group X via chemical synthesis [2]. Recent advances in the technology of SAMs make it an extraordinarily versatile technology for controlling the adhesion, migration, and differentiation of many types of cells [3].

SAMs of alkanethiols can also form on other metals, such as silver, palladium, and platinum [1]. Many of these SAMs are also compatible with biological systems, the most notable exception being silver, which leaches silver ion, toxic to most living cells [4]. Silanes can form SAMs on the surface of silicon dioxide [5, 6]. Compared with SAMs formed by alkanethiols, silanes have the advantage of being applicable to a variety of commonly accessible surfaces whose major component is SiO_x (i.e., glass, silicon wafers, and silicones). SAMs of alkanethiols on metals, however, are less accessible than SAMs of silanes due to the requirement of a layer of metal, typically in the form of a thin film on a substrate. But the formation of SAMs of silane on glass is not as rapid as thiols on metals, and the formation of a single layer of SAM of silane can be difficult due to the potential for polymerization of silanes on surfaces [6]. Many other types of SAMs do exist, but they are less used in biological applications [1].

3.2.2 Lithography and Patterning

Photolithography is one of the primary methods through which small patterns can be generated [1, 7]. Soft lithography includes a set of tools that are straightforward extensions of photolithography using soft polymers, such as polydimethylsiloxane (PDMS), to generate small features obtained in photolithography [1] (Figure 3.2). Compared to photolithography, soft lithography has several attractive features: (1) it is relatively simple in its instrument requirements and does not need the use of a clean room once the master is fabricated; (2) it allows the use of biologically compatible materials (i.e., there is no need for organic solvents and reagents toxic to cells), (3) it is compatible with optical microscopy conventionally used in cell biology due to the transparency of PDMS to visible light and its low levels of

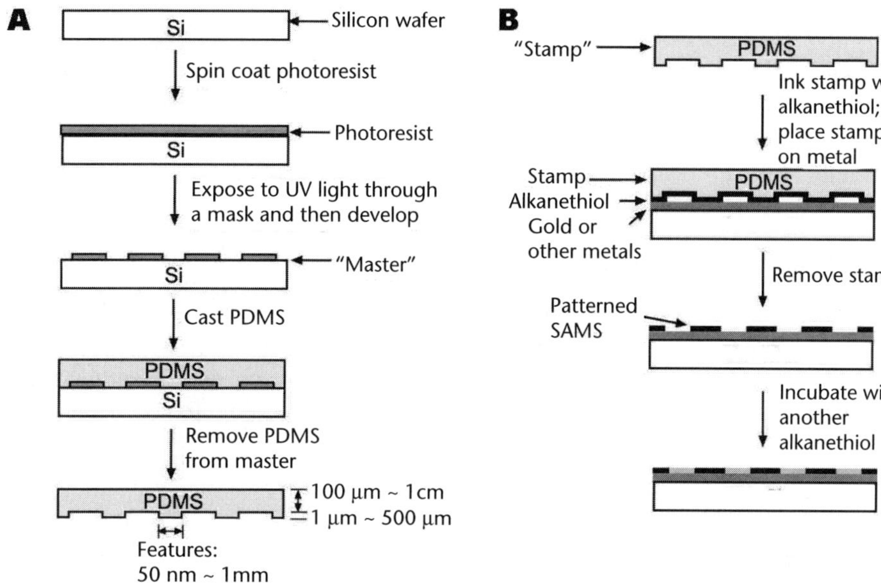

Figure 3.2 Photolithography and microcontact printing: (a) the process of photolithography, generation of small features on surfaces, and the generation of PDMS stamps with replica molding, and (b) the process of microcontact printing.

autofluorescence; (4) it allows patterning on curved and flexible substrates; and (5) PDMS itself exhibits low toxicity to most types of cells and tissues [1, 8, 9].

3.2.3 Patterning with SAMs, Stamping, Membranes, and Microfluidic Channels

Methods of patterning that employ SAMs and some tools in soft lithography are widely used in patterning the interactions between biological materials and solid substrates.

3.2.3.1 Microcontact Printing

Microcontact printing (μCP) is a method that allows the transfer of materials with high spatial resolution (Figure 3.2), resulting in a surface with spatially defined chemistries [10]. The stamp for μCP is typically fabricated by replica molding from features made with photolithography. After soaking the stamp with one type of alkanethiol as an "ink," the stamp is brought into contact with the substrate. The ink is selectively transferred onto parts of the substrate wherever the stamp comes into direct contact [Figure 3.2(b)]. At this point, SAMs only form on parts of the substrate where the ink of alkanethiol is transferred via the direct contact between the protruding features on the stamp and the substrate. The rest of the substrate is still bare. Incubation of the substrate with a second thiol fills the rest of the surface. This procedure results in a surface derivatized with two different types of SAMs; the two different terminal groups will yield a substrate with two different surface properties, each confined to precisely defined geometries.

Printing of SAMs allows the surface to have patterns of essentially any shape with different surface properties. Researchers have extended μCP on SAMs to many other applications, for example, the direct printing of other small molecules and polymers [3]. (See Section 3.4.1 for using SAMs and related tools for patterning cells.)

3.2.3.2 Membranes and Grids

Polymeric membranes or metallic plates that carry holes can also be used to generate patterns on surfaces. Metallic grids used for transmission electron microscopy are one of the first microstructures used to generate patterns in biological systems [11]. (See Section 3.4.5 for using such tools for patterning cells.)

3.2.3.3 Microfluidic Channels

Microfluidic channels guide small volumes of solutions over surfaces. Soft lithography has greatly simplified the generation of microfluidic channels [12–14]. The previous chapter has introduced this technology for patterning the fluidic environment for cells. This chapter addresses the issue of using microfluidic channels to generate patterned surfaces (see Section 3.4.6).

3.3 Controlling the Adsorption of Proteins on Surface

Most mammalian cells require the presence of an immobilized protein or an attachment factor [such as a peptide from an adhesive protein, for example, an extracellular matrix (ECM) protein, or the mimic thereof] to adhere to solid surfaces [15]. Most cells use cell-surface receptor molecules, the integrins, to bind to these ECM proteins to adhere to solid surfaces. Controlling the adsorption of proteins on surfaces is essential for controlling cell adhesion. Controlling the adsorption of biomolecules is also important in high-throughput biological screenings, such as microarrays for DNAs and proteins [16, 17]. This section examines studies that aim to understand and control the adsorption of biomolecules, particularly proteins, on solid surfaces. A variety of surface chemistries allow the introduction of chemical groups for attachment on surfaces; this chapter emphasizes surfaces that resist the nonspecific adsorption of proteins on surfaces.

3.3.1 Amount of Protein Adsorption Versus the Hydrophobicity of the Surface

The attempt to fully understand the adsorption of proteins on solid surfaces is a long-standing, but unresolved, issue [18]. SAMs provide a way to systematically investigate the relationship between the amount of protein adsorbed and the wettability of a smooth surface. On a smooth surface covered by SAMs of alkanethiols on gold, among SAMs terminated in various groups, more wettable (hydrophilic) surfaces tend to adsorb more proteins [19]. The only type of surface that resists all adsorption of proteins in this study was the SAM terminated in polyethyleneglycol (PEG), a chemical group intermediate in hydrophobicity. (This molecule is also called polyethyleneoxide, or PPO, or oligoethyleneglycol, OEG for oligomers that are low in molecular weight. In this chapter, I will use "PEG" to refer to this entire class of molecules.)

3.3.2 Inert Surfaces

A few types of surfaces completely resist the adsorption of proteins (and cells). For convenience, we call these surfaces "inert surfaces." Alternative phrases for this type of surface include "nonfouling surfaces" and "bioresistant surfaces." These surfaces include SAMs that present PEG terminal groups, polymeric PEG grafted on surfaces, natural polymers such as bovine serum albumin (BSA), and agarose. Because most proteins can nonspecifically and irreversibly adsorb onto most solid surfaces, inert surfaces often play the most important role in modulating the properties of the surface to control protein adsorption and cell adhesion.

3.3.2.1 Inert SAMs Based on PEG

Short chains of PEG (where there are three to six repeats of the ethyleneglycol units) result in inert surfaces when grafted on SAMs [20, 21]. This type of SAM has become one of the most widely used in biological applications.

It has long been known that long chains of PEG are inert. The mechanism of its inertness has not been satisfactorily established, despite a series of experimental and theoretical efforts [22–24]. The mechanism of why short oligomers of PEG, when grafted on the surface of SAMs, can be inert is not well understood either. There have been some efforts to explore this mechanism and some hypotheses developed about the mechanism, but no conclusive explanation exists at the moment [25–32]. The lack of a full understanding of the mechanism of the inertness of PEG has not hindered its application to a variety of biological systems.

3.3.2.2 Inert Polymers Based on PEG

The first application of PEG is to graft it onto solid surfaces to reduce thrombosis [33]. There are many different approaches to linking polymers of PEG to surfaces. Here, I introduce three commonly used types: direct coupling, coupling via an anchor, and coupling through layer-by-layer (LbL) adsorption of polyelectrolytes.

There are many examples of chemical transformations that directly couple PEG to the surface of the polymers. One example is to use an electrochemical-based method to directly tether PEG to surfaces of solid substrates via a chemical reaction between a nucleophile on the surface and a PEG chain terminated in acrylate [34].

Another approach relies on a copolymer containing PEG and another component that anchors the copolymer to the surface. Pluroic surfactants, PEG-polypropyleneoxide (PPO)–PEG, is an example of such a type of copolymer [35, 36]. The PPO moiety of the polymer can noncovalently anchor onto hydrophobic surfaces, while the PEG covers the surface to enable the surface inert. Similarly, poly-L-lysine (PLL)–PEG has the ability to make negatively charged surfaces inert since positively charged PLL adsorbs onto surfaces that carry negative charges [37]. Another example is the copolymer that contains side chains of both trimethylsilane and PEG, where the anchoring part (trimethoxysilane) tethers to the surface of SiO_2 and the functional part (PEG) resists protein adsorption [38].

Another approach that allows the generation of inert surfaces is to covalently link PEG to a surface with multiple layers of electrolytes fabricated by LbL [39].

3.3.2.3 Inert Natural Polymers

Natural polymers, such as polysaccharides and proteins, have been employed in the fabrication of inert surfaces. Polysaccharides, such as cellulose, dextran, and agarose, are inert [40–43]. BSA and casein are frequently used to block nonspecific protein adsorption in immunoassays and cell adhesion on solid surfaces [44].

3.3.2.4 Other Inert Systems

In order to understand what type of molecular features would give rise to inert surfaces, Whitesides et al. screened a small library presented on the surface of SAMs and found that a number of other terminal groups could also make SAMs inert [45, 46]. The common features of these groups, along with PEG, include: (1) polarity, (2) electrical neutrality, and (3) the absence of hydrogen bond donors, though they can have hydrogen bond acceptors. SAMs that present electrically charged terminal groups can also be inert as long as the overall charge on the surface is neutral, by either mixing thiols terminated in oppositely charged groups at equal molar ratios or using thiols terminated in groups that contain both positive and negative charges in the same molecule [47–49].

Based on some of the principles derived from studies on SAMs, polymers can be designed to be inert. One example is that a designed polymer that presents oppositely charged moieties but is overall electrically neutral, such as sulfobetaine, is inert [50].

3.3.2.5 Duration of Inertness

SAMs of PEG-terminated thiols on gold cannot keep the surface inert for more than twenty days in the presence of regular medium for culturing mammalian cells [51, 52]. Several approaches have been employed to improve the duration of inert surfaces. For example, using palladium, instead of gold, as the substrate to form SAMs can lengthen the inertness of the surface to at least thirty days [51]. A comparison between agarose, PEG, polyacrylamide, pluronics, and BSA in the presence of cells in regular culture medium shows that acrylamide was inert for the longest period of time (twenty-eight days) [53]. The duration of each type of surface is probably different under various testing conditions; the few examples mentioned here provide a starting point from which to experiment for those who wish to find the best-performing surface at a particular condition.

3.4 Patterning of Proteins and Cells

Central to many applications and systems in the rest of this book is a series of tools that allow the patterning of biomolecules, particularly proteins, and cells. This section introduces a few of the most important tools commonly used to achieve surface patterning for cell adhesion.

3.4.1 Patterning with SAMs

Patterning the surface with SAMs is one of the simplest and most used technologies for confining the adsorption of proteins and adhesion of cells on solid substrates. It allows the control of the density of immobilized ligands and the geometrical shapes of patches of adsorbed proteins and adhered cells.

3.4.1.1 Geometrical Confinement

Using μCP of SAMs on gold, it is possible to confine proteins, hence cells, into essentially any geometrical shape (Figures 3.2 and 3.3) [6, 54–57]. SAMs of silanes can also pattern proteins and cells [5, 58]. In most of these examples, inert SAMs prevent protein adsorption and cell adhesion, and the rest of the surface acts as the substrate for protein adsorption and cell adhesion, hence the generation of patterns.

An extension of μCP and SAMs to confine proteins on surfaces is the direct printing of proteins to form micropatterns on surfaces [59, 60]. This method is simpler than μCP of SAMs, but it involves a step that dries the printed proteins and is therefore not generally applicable to all proteins.

3.4.1.2 Control of Ligand Density

SAMs formed by different ratios of two thiols, one terminated in PEG, the other terminated in the sequence of peptide arginine-glycine-aspartate (RGD), the minimum peptide fragment isolated from ECM proteins that supports cell adhesion, can result in a surface that allows the control of ligand density on the surface, therefore the extent of the adhesion and spreading of bovine capillary endothelial cells [61]. The

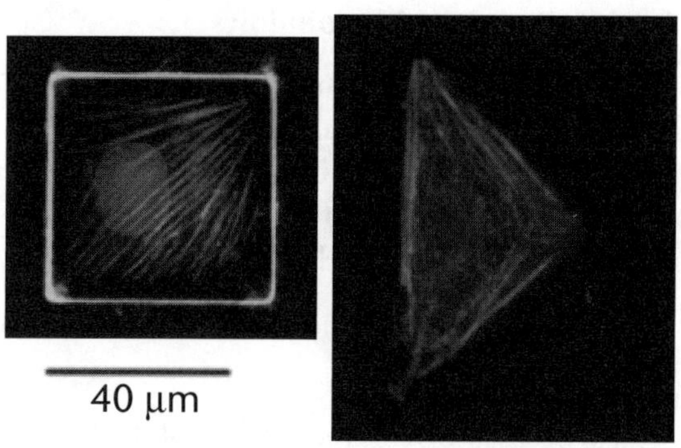

Figure 3.3 Examples of the confinement of single mammalian cells to geometrical patterns. Bright lines outline the distribution of actin filaments.

amount of adsorbed proteins can also be precisely controlled in a number of other SAMs formed by mixtures of thiols on gold [62].

3.4.2 Patterning with Other Polymers of PEG

There are many examples of polymeric systems based on PEG employed to confine proteins and cells. Poly(PEG-methyl ether methacrylate)-based copolymers have been employed to pattern smooth muscle cells [63]. Copolymers of poly(acrylamide) and PEG can confine cells on the surface of glass [64]. Copolymers that contain PEG and trimethylsilane are effective in patterning cells (compare Section 3.3.2.2) [65].

3.4.3 Patterning with Other Polymers

BSA and many types of hydrogels are useful in the fabrication of surfaces that allow patterning of proteins and cells [44, 53]. Agarose [53, 66–68], dextran and its derivatives [69, 70], polyvinyl alcohol [71], and polyacrylamide [53, 72, 73] have all been used for confining cells into geometrically defined areas. Printing on polymeric materials fabricated via LbL assembly of poly(acrylic acid) and polyacrylamide represents yet another method to pattern cells [74].

3.4.4 Patterning with Lipids

Supported lipid membranes can be used to control cell adhesion [75]. Patterned lipid membranes are applicable to high-throughput screening, such as the screen of ligands for G-protein coupled receptors [76, 77].

3.4.5 Patterning with Holes on Thin Membranes

When PDMS membranes that contain holes are brought into contact with a flat substrate, parts of the substrate are exposed through these holes. Cells can be made to grow selectively on the exposed substrate; when the membrane is removed, patterned cells can form since the areas on the substrate that were in contact with the membrane had no cells [44, 78].

3.4.6 Patterning Based on Microfluidics

Individually addressable microfluidic channels allow the patterning of small amounts of solutions of biomolecules in relatively large areas of solid surfaces with high spatial resolution [79]. Laminar flows within a single microfluidic channel allow the generation of patterned proteins and cells on surfaces, either in the form of stripes or in the form of gradients [80, 81].

3.4.7 Light-Based Methods

Photochemical processes offer modulation of surface properties with well-defined spatial resolution when combined with masks. I discuss two classes of photochemical reactions, both on SAMs, one on SAMs of thiols, the other on SAMs of silanes.

3.4.7.1 Photochemistry of Thiols on Gold

Activation of chemical groups on SAMs using spatially confined photochemistry is an effective means to achieve patterning of cells [82, 83]. It is possible to pattern multiple aligned features of proteins on SAMs using photochemistry [83]. This method relies on an alkanethiol sensitive to light both at 220 and 365 nm. After a single exposure to light (of multiple wavelengths) through a photomask having different regions that allow different wavelengths of light to pass through, three kinds of SAMs on one single gold substrate could form [83].

3.4.7.2 Photochemistry of Silanes

Photochemistry on silanes is also useful for pattering cells. For example, as BSA initially adsorbed on a SAM of silanes that carries photocleavable groups, the entire surface is inert. When light cleaves the terminal groups of SAMs, BSA desorbs from the surface. The newly exposed terminal group of the SAMs supports adsorption of the ECM protein fibronectin, and cells only attach to areas where the light has shone [84].

3.4.8 Patterning that Combines Two or More Tools

Sometimes it is necessary to use the combination of two or more of the above techniques to confine proteins and cells to a desired geometry. To generate patterns within patterns, the strategy is to initially pattern on a surface with either molding or printing, then to use the printed surface as the substrate for a microfluidic channel that brings laminar flows to the patterned surface. This approach results in stripes or gradients of proteins within geometrical patterns [51, 85].

Surface topography and microfluidic channels can result in arrays of different types of cells [86]. Details on methods for studying multiple types of cells will be covered in chapter 4.

3.5 Dynamic Patterning of Cells

Recent advances in the ability to engineer surface properties of solid substrates have allowed researchers to dynamically modulate the inertness of the substrate, hence to control the interactions between mammalian cells and the solid substrate in real time. Examples of dynamic patterning include the control of the adhesion and motility of cells, the ability to attach or detach cells at will, and the sequential patterning of two types of cells.

3.5.1 Dynamic Patterning Using SAMs

The wide range of chemical transformations available on SAMs has given rise to a number of methods for *in situ* transformation of the surface that allow dynamic control of cell adhesion and migration [87]. Some of these transformations take place between the gold substrate and the thiols; some of them take place at the terminal groups of the thiols on the SAMs.

3.5.1.1 Reactions That Take Place at the Interface Between Thiols and the Substrate

Cathodic potentials on the gold substrate desorb thiols from the surface. Desorption of EG-terminated thiols enables initially confined cells to move out of their patterns of confinement (Figure 3.4) [88]. We have used this technique to show that asymmetry in cell shape alone could bias the direction of cell migration [89]. We have also used it, along with microfluidic channels, to pattern two types of cells [90].

3.5.1.2 Reactions That Take Place at the Termini of the Thiols

Transformations at the terminal groups of SAMs also enable the dynamic control of cell adhesion [91–96]. In addition to achieving the same basic control of turning on cell motility as described in Section 3.5.1.1, one of these methods realized the initial immobilization, then the selective detachment, and finally the release of the confinement of cells [95].

3.5.2 Dynamic Patterning Using Polymers

Polymeric materials also allow the dynamic patterning of cells [97, 98]. These methods mainly rely on stopping the inertness by adsorption of polymers that support cell adhesion on the inert part of the surface to enable cells to adhere on these initially inert parts. The details of these techniques will be covered in Chapter 4.

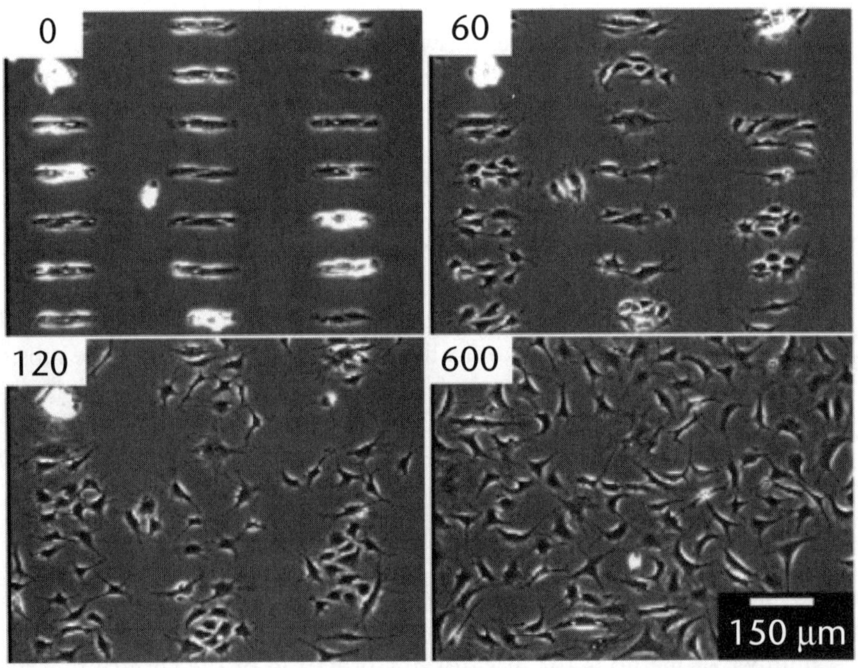

Figure 3.4 An example of dynamic patterning. BCE cells were first confined to a pattern of arrays by SAMs. Application of a cathodic voltage pulse desorbed the SAMs that confined the cells, and the cells resumed their motility. The numbers indicate the time elapsed (in minutes) after the voltage pulse.

3.5.3 Dynamic Patterning Based on Light

When agarose is used as the inert material to confine cells, it can be "etched" with an infrared laser locally. This method can stop the inertness of local areas in the substrate to allow patterned cells to invade into areas that do not support cell adhesion initially [67]. Cardiac myocytes patterned in this method could be coaxed to interact with each other, establishing an in vitro model of cardiac synchronization [68].Another example of light-based dynamic patterning uses the same chemistry as described in Section 3.4.7.2; it allows the sequentially patterning of two types of cells [84].

3.5.4 Dynamic Patterning Using Oxidation by Microelectrodes

Microelectrodes can be used to spatially alter the chemistry of small regions on surfaces. For example, when BSA is used as the inert surface for confining cells, the pattern of confinement can be altered by using microelectrodes to oxidize the BSA on selected areas of surfaces [99–102]. A recent report demonstrated patterning with the same principle but with a surface initially rendered inert by PEG-terminated SAMs [103].

3.5.5 Dynamic Patterning Using Thermally Responsive Materials

Some polymeric materials change their physical properties when the temperature is altered. This section gives a few examples of their applications in controlling cell adhesion.

3.5.5.1 Patterning Based on Poly(isopropylacrylamide)

Poly(isopropylacrylamide) is inert at 32°C and does not support cell adhesion. At 37°C, however, it does support cell adhesion [104]. When this polymer is grafted on polystyrene culture dishes at 37°C, the surface promotes the adhesion and spreading of endothelial cells. Sheets of cells formed on this substrate could detach from the substrate when temperature is changed to 32°C (at this point, the surface becomes inert) [104]. This system thus makes possible the fabrication of free-standing tissues, such as periodontal ligaments, epithelia of trachea, and myocardial tissues [105–107].

3.5.5.2 Patterning Based on Elastinlike Peptide

Whether elastinlike peptide (ELP) adsorbs on a hydrophobic surface is a function of temperature. When the temperature is high, ELP adsorbs onto the surface; when the temperature is low, it desorbs from the surface and is solubilized. The reversible adsorption of ELP allows the surface to be patterned with a protein that is covalently bound to ELP [108, 109].

3.6 Other Systems for Patterning Cells

This section sums up a few examples that do not belong to the general toolbox that the preceding sections of the chapter have introduced.

3.6.1 Patterning Cell Adhesion with Cracks

Plasma oxidation generates cracks and ripples on the surface of PDMS [110]. Cracks on PDMS and mechanical stretching have allowed the reversible control of cell adhesion and spreading (Figure 3.5) [111].

3.6.2 Other Self-Assembled Systems

Self-assembled systems besides SAMs have found uses in the fabrication of materials of many scales and levels [112]. Patterning of the cell-substrate interactions has benefited from recent advancements in self-assembled systems.

Self-assembled arrays of ligands for cell adhesion can be used to understand mechanistic details of cell adhesion, such as how far apart integrin molecules can be when cells can still adhere to surfaces. Integrins cannot function when each of them stands alone; they have to come together and form a bundle to support cell adhesion. It was not clear what the maximum separation distance between these integrins is when cells can still adhere and spread on a surface. Using hexagonal arrays of nanodots that have RGD groups (each of which is large enough to bind only one integrin molecule), it is possible to determine conclusively that the maximum distance between each integrin, when cells can still adhere and spread, is 73 nm [113].

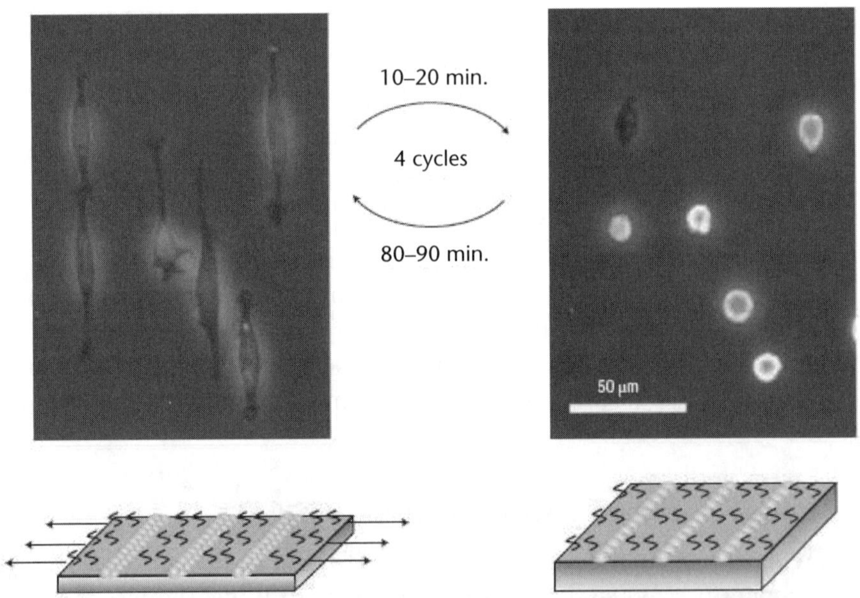

Figure 3.5 Fabrication of a substrate that allows the reversible attachment of cells on PDMS. There are cracks on the surface of the PDMS. The cracks can be reversibly opened or closed simply by stretching or releasing the PDMS in the horizontal direction. Copolymers of PEG-PPO are adsorbed on the PDMS surface when it is released to make the exposed surface of PDMS inert (the cracks are not open; the polymers are indicated by the tildes). When the cracks are stretched open, fibronectin (Fn) is deposited only in the cracks. At this time, myoblasts can adhere to and spread on these open cracks that present Fn. When the PDMS is released, the cracks close, thereby masking the fibronectin, and cells lose their attachment on the surface and retract. When the PDMS is again stretched, cracks open up to present the Fn again to allow cell spreading. This process can be repeated.

Using a technique called dip-pen nanolithography, it is also possible to confine points of cell-substrate interactions to nanoscale patterns [114].

3.7 Conclusion

Emerging tools in fabrication and chemical synthesis have enabled the development of a series of tools that allow the control of the microenvironment of mammalian cells. These tools have allowed researchers to control many aspects of the physiology of the cell, answer new questions about the cell, and achieve new applications with cells. Many aspects of the microenvironment of cell physiology are still not well understood; nor are they controllable. Both the invention of new tools and their application to various biological problems will be very important for engineering the cellular microenvironment in the decade to come.

Acknowledgments

Funding for this work was provided by the Chinese Academy of Sciences, the National Science Foundation of China, the Ministry of Science and Technology, and the Human Frontier Science Program.

References

[1] Xia, Y., and Whitesides, G. M., "Soft lithography," *Angewandte Chemie—International Edition in English*, Vol. 37, No. 5, 1998, pp. 550–575.

[2] Love, J. C., Estroff, L. A., Kriebel, J. K., Nuzzo, R. G., and Whitesides, G. M., "Self-assembled monolayers of thiolates on metals as a form of nanotechnology," *Chem. Rev.*, Vol. 105, No. 4, 2005, pp. 1103–1169.

[3] Whitesides, G. M., Ostuni, E., Takayama, S., Jiang, X., and Ingber, D. E., "Soft lithography in biology and biochemistry," *Ann. Rev. Biomed. Eng.*, Vol. 3, 2001, pp. 335–373.

[4] Ostuni, E., Yan, L., and Whitesides, G. M., "The interaction of proteins and cells with self-assembled monolayers of alkanethiols on gold and silver," *Colloids and Surfaces B: Biointerfaces*, Vol. 15, 1999, pp. 3–30.

[5] Kleinfeld, D., Kahler, K. H., and Hockberger, P. E., "Controlled outgrowth of dissociated neurons on patterned substrates," *J. Neurosci.*, Vol. 8, No. 11, 1988, pp. 4098–4120.

[6] Mrksich, M., and Whitesides, G. M., "Using self-assembled monolayers to understand the interactions of man-made surfaces with proteins and cells," *Ann. Rev. Biophysics and Biomolecular Structure*, Vol. 25, 1996, pp. 55–78.

[7] Geissler, M., and Xia, Y. N., "Patterning: Principles and some new developments," *Adv. Mater.*, Vol. 16, No. 15, 2004, pp. 1249–1269.

[8] Lee, J. N., Jiang, X., Ryan, D., and Whitesides, G. M., "Compatibility of mammalian cells on surfaces of poly(dimethylsiloxane)," *Langmuir*, Vol. 20, No. 26, 2004, pp. 11684–11691.

[9] Lugowski, S. J., Smith, D. C., Bonek, H., Lugowski, J., Peters, W., and Semple, J., "Analysis of silicon in human tissues with special reference to silicone breast implants," *J. Trace Element in Medicine and Biology*, Vol. 14, No. 1, 2000, pp. 31–42.

[10] Kumar, A., and Whitesides, G. M., "Features of gold having micrometer to centimeter dimensions can be formed through a combination of stamping with an elastomeric stamp

and an alkanethiol ink followed by chemical etching," *Applied Physics Lett.*, Vol. 63, No. 14, 1993, pp. 2002–2004.

[11] Letourneau, P. C., "Cell-to-substratum adhesion and guidance of axonal elongation," *Dev. Biol.*, Vol. 44, No. 1, 1975, pp. 92–101.

[12] McDonald, J. C., and Whitesides, G. M., "Poly(dimethylsiloxane) as a material for fabricating microfluidic devices," *Accounts of Chem. Res.*, Vol. 35, No. 7, 2002, pp. 491.

[13] Lahann, J., Balcells, M., Rodon, T., Lee, J., Choi, I. S., Jensen, K. F., and Langer, R., "Reactive polymer coatings: A platform for patterning proteins and mammalian cells onto a broad range of materials," *Langmuir*, Vol. 18, No. 9, 2002, pp. 3632–3638.

[14] Thorsen, T., Maerkl, S. J., and Quake, S. R., "Microfluidic large-scale integration," *Science*, Vol. 298, No. 5593, 2002, pp. 580–584.

[15] Giancotti, F. G., and Ruoslahti, E., "Integrin signaling," *Science*, Vol. 285, No. 5430, 1999, pp. 1028–1032.

[16] Pollack, J. R., Perou, C. M., Alizadeh, A. A., Eisen, M. B., Pergamenschikov, A., Williams, C. F., Jeffrey, S. S., Botstein, D., and Brown, P. O., "Genome-wide analysis of DNA copy-number changes using cDNA microarrays," *Nature Genetics*, Vol. 23, No. 1, 1999, pp. 41–46.

[17] Phizicky, E., Bastiaens, P. I. H., Zhu, H., Snyder, M., and Fields, S., "Protein analysis on a proteomic scale," *Nature*, Vol. 422, No. 6928, 2003, pp. 208–215.

[18] Andrade, J. D., and Hlady, V., "Protein adsorption and materials biocompatibility: A tutorial review and suggested hypotheses," *Advances in Polymer Science*, Vol. 79, No. Biopolym./Non-Exclusion HPLC, 1986, pp. 1–63.

[19] Sigal, G. B., Mrksich, M., and Whitesides, G. M., "Effect of surface wettability on the adsorption of proteins and detergents," *J. Am. Chem. Soc.*, Vol. 120, No. 14, 1998, pp. 3464–3473.

[20] Prime, K. L., and Whitesides, G. M., "Self-assembled organic monolayers: Model systems for studying adsorption of proteins at surfaces," *Science*, Vol. 252, 1991, pp. 1164–1167.

[21] Prime, K. L., and Whitesides, G. M., "Adsorption of proteins onto surfaces containing end-attached oligo(ethylene oxide)—a model system using self-assembled monolayers," *J. Am. Chem. Soc.*, Vol. 115, No. 23, 1993, pp. 10714–10721.

[22] Jeon, S. I., Lee, J. H., Andrade, J. D., and De Gennes, P. G., "Protein-surface interactions in the presence of polyethylene oxide: I. Simplified theory," *J. Colloid and Interface Science*, Vol. 142, No. 1, 1991, pp. 149–158.

[23] Kidane, A., McPherson, T. B., Szleifer, I., and Park, K., "Protein adsorption on PEO-grafted surfaces: Theoretical analysis and experimental observations," *Polymeric Materials Science and Engineering*, Vol. 77, 1997, pp. 570–571.

[24] Kingshott, P., and Griesser, H. J., "Surfaces that resist bioadhesion," *Current Opinions in Solid State and Materials Science*, Vol. 4, 1999, pp. 403–412.

[25] Harder, P., Bierbaum, K., Woell, C., Grunze, M., Heid, S., and Effenberger, F., "Induced orientational order in long alkyl chain aminosilane molecules by preadsorbed octadecyltrichlorosilane on hydroxylated si(100)," *Langmuir*, Vol. 13, No. 3, 1997, pp. 445–454.

[26] Harder, P., Grunze, M., Dahint, R., Whitesides, G. M., and Laibinis, P. E., "Molecular conformation in oligo(ethylene glycol)-terminated self-assembled monolayers on gold and silver surfaces determines their ability to resist protein adsorption," *J. Phys. Chem. B*, Vol. 102, No. 2, 1998, pp. 426–436.

[27] Zolk, M., Eisert, F., Pipper, J., Herrwerth, S., Eck, W., Buck, M., and Grunze, M., "Solvation of oligo(ethylene glycol)-terminated self-assembled monolayers studied by vibrational sum frequency spectroscopy," *Langmuir*, Vol. 16, No. 14, 2000, pp. 5849–5852.

[28] Pertsin, A. J., Hayashi, T., and Grunze, M., "The interaction of oligo(ethylene glycol) with water: Testing an atomistic force field for transferability," *Physical Chemistry Chemical Physics*, Vol. 3, No. 9, 2001, pp. 1598–1601.

[29] Pertsin, A. J., Hayashi, T., and Grunze, M., "Grand canonical monte carlo simulations of the hydration interaction between oligo(ethylene glycol)-terminated alkanethiol self-assembled monolayers," *J. Phys. Chem. B*, Vol. 106, No. 47, 2002, pp. 12274–12281.

[30] Herrwerth, S., Rosendahl, T., Feng, C., Fick, J., Eck, W., Himmelhaus, M., Dahint, R., and Grunze, M., "Covalent coupling of antibodies to self-assembled monolayers of carboxy-functionalized poly(ethylene glycol): Protein resistance and specific binding of biomolecules," *Langmuir*, Vol. 19, No. 5, 2003, pp. 1880–1887.

[31] Herrwerth, S., Eck, W., Reinhardt, S., and Grunze, M., "Factors that determine the protein resistance of oligoether self-assembled monolayers—internal hydrophilicity, terminal hydrophilicity, and lateral packing density," *J. Am. Chem. Soc.*, Vol. 125, No. 31, 2003, pp. 9359–9366.

[32] Fick, J., Steitz, R., Leiner, V., Tokumitsu, S., Himmelhaus, M., and Grunze, M., "Swelling behavior of self-assembled monolayers of alkanethiol-terminated poly(ethylene glycol): A neutron reflectometry study," *Langmuir*, Vol. 20, No. 10, 2004, pp. 3848–3853.

[33] Mori, Y., Nagaoka, S., Takiuchi, H., Kikuchi, T., Noguchi, N., Tanzawa, H., and Noishiki, Y., "A new antithrombogenic material with long polyethyleneoxide chains," *Transactions of the American Society on Artificial Internal Organs*, Vol. 28, 1982, pp. 459–463.

[34] Gabriel, S., Dubruel, P., Schacht, E., Jonas, A. M., Gilbert, B., Jerome, R., and Jerome, C., "Electrografting of poly(ethylene glycol) acrylate: A one-step strategy for the synthesis of protein-repellent surfaces," *Angewandte Chemie—International Edition in English*, Vol. 44, No. 34, 2005, pp. 5505–5509.

[35] Caldwell, K. D., "Surface modifications with adsorbed poly(ethylene oxide)-based block copolymers. Physical characteristics and biological use," *ACS Symposium Series*, Vol. 680, No. Poly(ethylene glycol), 1997, pp. 400–419.

[36] Green, R. J., Davies, M. C., Roberts, C. J., and Tendler, S. J. B., "A surface plasmon resonance study of albumin adsorption to PEO-PPO-PEO triblock copolymers," *J. Biomed. Mater. Res.*, Vol. 42, No. 2, 1998, pp. 165–171.

[37] Huang, N.-P., Michel, R., Voros, J., Textor, M., Hofer, R., Rossi, A., Elbert, D. L., Hubbell, J. A., and Spencer, N. D., "Poly(L-lysine)-g-poly(ethylene glycol) layers on metal oxide surfaces: Surface-analytical characterization and resistance to serum and fibrinogen adsorption," *Langmuir*, Vol. 17, No. 2, 2001, pp. 489–498.

[38] Jon, S. Y., Seong, J. H., Khademhosseini, A., Tran, P. E., Laibinis, T. N. T., and Langer, R., "Construction of nonbiofouling surfaces by polymeric self-assembled monolayers," *Langmuir*, Vol. 19, No. 24, 2003, pp. 9989–9993.

[39] Makamba, H., Hsieh, Y. Y., Sung, W. C., and Chen, S. H., "Stable permanently hydrophilic protein-resistant thin-film coatings on poly(dimethylsiloxane) substrates by electrostatic self-assembly and chemical cross-linking," *Analytical Chemistry*, Vol. 77, No. 13, 2005, pp. 3971–3978.

[40] Lindon, J., Rosenberg, R., Merrill, E., and Salzman, E., "Interaction of human platelets with heparinized agarose gel," *J. Laboratory and Clinical Medicine*, Vol. 91, No. 1, 1978, pp. 47–59.

[41] Merrill, E. W., Salzman, E. W., Lipps, B. J., Jr., Gilliland, E. R., Austen, W. G., and Joison, J., "Antithrombogenic cellulose membranes for blood dialysis," *Transactions of the American Society of Artificial Internal Organs*, Vol. 12, 1966, pp. 139–150.

[42] Frazier, R. A., Matthijs, G., Davies, M. C., Roberts, C. J., Schacht, E., and Tendler, S. J. B., "Characterization of protein-resistant dextran monolayers," *Biomaterials*, Vol. 21, No. 9, 2000, pp. 957–966.

[43] Elam, J. H., Nygren, H., and Stenberg, M., "Covalent coupling of polysaccharides to silicon and silicone rubber surfaces," *J. Biomed. Mater. Res.*, Vol. 18, No. 8, 1984, pp. 953–959.

[44] Ostuni, E., Kane, R., Chen, C. S., Ingber, D. E., and Whitesides, G. M., "Patterning mammalian cells using elastomeric membranes," *Langmuir*, Vol. 16, No. 20, 2000, pp. 7811–7819.

[45] Ostuni, E., Chapman, R. G., Holmlin, R. E., Takayama, S., and Whitesides, G. M., "A survey of structure-property relationships of surfaces that resist the adsorption of protein," *Langmuir*, Vol. 17, No. 18, 2001, pp. 5605–5620.

[46] Chapman, R. G., Ostuni, E., Takayama, S., Holmlin, R. E., Yan, L., and Whitesides, G. M., "Surveying for surfaces that resist the adsorption of proteins," *J. Am. Chem. Soc.*, Vol. 122, No. 34, 2000, pp. 8303–8304.

[47] Holmlin, R. E., Chen, X., Chapman, R. G., Takayama, S., and Whitesides, G. M., "Zwitterionic SAMs that resist nonspecific adsorption of protein from aqueous buffer," *Langmuir*, Vol. 17, No. 9, 2001, pp. 2841–2850.

[48] Kane, R. S., Deschatelets, P., and Whitesides, G. M., "Kosmotropes form the basis of protein-resistant surfaces," *Langmuir*, Vol. 19, No. 6, 2003, pp. 2388–2391.

[49] Chen, S., Liu, L., and Jiang, S., "Strong resistance of oligo(phosphorylcholine) self-assembled monolayers to protein adsorption," *Langmuir*, Vol. 22, No. 6, 2006, pp. 2418–2421.

[50] Chang, Y., Chen, S., Zhang, Z., and Jiang, S., "Highly protein-resistant coatings from well-defined diblock copolymers containing sulfobetaines," *Langmuir*, Vol. 22, No. 5, 2006, pp. 2222–2226.

[51] Jiang, X., Bruzewicz, D. A., Thant, M. M., and Whitesides, G. M., "Palladium as a substrate for self-assembled monolayers used in biotechnology," *Anal. Chem.*, Vol. 76, No. 20, 2004, pp. 6116–6121.

[52] Flynn, N. T., Tran, T. N. T., Cima, M. J., and Langer, R., "Long-term stability of self-assembled monolayers in biological media," *Langmuir*, Vol. 19, No. 26, 2003, pp. 10909–10915.

[53] Nelson, C. M., Raghavan, S., Tan, J. L., and Chen, C. S., "Degradation of micropatterned surfaces by cell-dependent and -independent processes," *Langmuir*, Vol. 19, No. 5, 2003, pp. 1493–1499.

[54] Singhvi, R., Kumar, A., Lopez, G. P., Stephanopoulos, G. N., Wang, D. I. C., Whitesides, G. M., and Ingber, D. E., "Engineering cell-shape and function," *Science*, Vol. 264, No. 5159, 1994, pp. 696–698.

[55] Chen, C. S., Mrksich, M., Huang, S., Whitesides, G. M., and Ingber, D. E., "Geometric control of cell life and death," *Science*, Vol. 276, No. 5317, 1997, pp. 1425–1428.

[56] Brock, A., Chang, E., Ho, C.-C., LeDuc, P., Jiang, X., Whitesides, G. M., and Ingber, D. E., "Geometric determinants of directional cell motility revealed using microcontact printing," *Langmuir*, Vol. 19, No. 5, 2003, pp. 1611–1617.

[57] Parker, K. K., Brock, A. L., Brangwynne, C., Mannix, R. J., Wang, N., Ostuni, E., Geisse, N. A., Adams, J. C., Whitesides, G. M., and Ingber, D. E., "Directional control of lamellipodia extension by constraining cell shape and orienting cell tractional forces," *Faseb J.*, Vol. 16, No. 10, 2002, pp. 1195–1204.

[58] Stenger, D. A., Georger, J. H., Dulcey, C. S., Hickman, J. J., Rudolph, A. S., Nielsen, T. B., McCort, S. M., and Calvert, J. M., "Coplanar molecular assemblies of aminoalkylsilane and perfluorinated alkylsilane—characterization and geometric definition of mammalian-cell adhesion and growth," *J. Am. Chem. Soc.*, Vol. 114, No. 22, 1992, pp. 8435–8442.

[59] Bernard, A., Fitzli, D., Sonderegger, P., Delamarche, E., Michel, B., Bosshard, H. R., and Biebuyck, H., "Affinity capture of proteins from solution and their dissociation by contact printing," *Nat. Biotechnol.*, Vol. 19, No. 9, 2001, pp. 866–869.

[60] Bernard, A., Delamarche, E., Schmid, H., Michel, B., Bosshard, H. R., and Biebuyck, H., "Printing patterns of proteins," *Langmuir*, Vol. 14, No. 9, 1998, pp. 2225–2229.

[61] Roberts, C., Chen, C. S., Mrksich, M., Martichonok, V., Ingber, D. E., and Whitesides, G. M., "Using mixed self-assembled monolayers presenting RGD and (EG)(3)OH groups to

characterize long-term attachment of bovine capillary endothelial cells to surfaces," *J. Am. Chem. Soc.*, Vol. 120, No. 26, 1998, pp. 6548–6555.

[62] Ostuni, E., Grzybowski, B. A., Mrksich, M., Roberts, C. S., and Whitesides, G. M., "Adsorption of proteins to hydrophobic sites on mixed self-assembled monolayers," *Langmuir*, Vol. 19, No. 5, 2003, pp. 1861–1872.

[63] Popescu, D. C., Lems, R., Rossi, N. A. A., Yeoh, C. T., Loos, J., Holder, S. J., Bouten, C. V. C., and Sommerdijk, N., "The patterning and alignment of muscle cells using the selective adhesion of poly(oligoethylene glycol methyl ether methacrylate)-based ABA block copolymers," *Adv. Mater.*, Vol. 17, No. 19, 2005, pp. 2324–2329.

[64] Tourovskaia, A., Barber, T., Wickes, B. T., Hirdes, D., Grin, B., Castner, D. G., Healy, K. E., and Folch, A., "Micropatterns of chemisorbed cell adhesion-repellent films using oxygen plasma etching and elastomeric masks," *Langmuir*, Vol. 19, No. 11, 2003, pp. 4754–4764.

[65] Khademhosseini, A., Jon, S., Suh, K. Y., Tran, T. N. T., Eng, G., Yeh, J., Seong, J., and Langer, R., "Direct patterning of protein- and cell-resistant polymeric monolayers and microstructures," *Adv. Mater.*, Vol. 15, No. 23, 2003, pp. 1995–2000.

[66] Nelson, C. M., and Chen, C. S., "Cell-cell signaling by direct contact increases cell proliferation via a PI3K-dependent signal," *FEBS Letters*, Vol. 514, Nos. 2–3, 2002, pp. 238–242.

[67] Moriguchi, H., Wakamoto, Y., Sugio, Y., Takahashi, K., Inoue, I., and Yasuda, K., "An agar-microchamber cell-cultivation system: Flexible change of microchamber shapes during cultivation by photo-thermal etching," *Lab Chip*, Vol. 2, No. 2, 2002, pp. 125–132.

[68] Kojima, K., Moriguchi, H., Hattori, A., Kaneko, T., and Yasuda, K., "Two-dimensional network formation of cardiac myocytes in agar microculture chip with 1480 nm infrared laser photo-thermal etching," *Lab Chip*, Vol. 3, No. 4, 2003, pp. 292–296.

[69] Zhou, Y., Andersson, O., Lindberg, P., and Liedberg, B., "Protein microarrays on carboxymethylated dextran hydrogels: Immobilization, characterization and application," *Microchimica Acta*, Vol. 147, Nos. 1–2, 2004, pp. 21–30.

[70] Zhou, Y., Andersson, O., Lindberg, P., and Liedberg, B., "Reversible hydrophobic barriers introduced by microcontact printing: Application to protein microarrays," *Microchimica Acta*, Vol. 146, Nos. 3–4, 2004, pp. 193–205.

[71] Peterbauer, T., Heitz, J., Olbrich, M., and Hering, S., "Simple and versatile methods for the fabrication of arrays of live mammalian cells," *Lab Chip*, Vol. 6, No. 7, 2006, pp. 857–863.

[72] Thomas, C. H., Lhoest, J. B., Castner, D. G., McFarland, C. D., and Healy, K. E., "Surfaces designed to control the projected area and shape of individual cells," *J. Biomechanical Engineering*, Vol. 121, No. 1, 1999, pp. 40–48.

[73] Raghavan, S., and Chen, C. S., "Micropatterned environments in cell biology," *Adv. Mater.*, Vol. 16, No. 15, 2004, pp. 1303–1313.

[74] Berg, M. C., Yang, S. Y., Hammond, P. T., and Rubner, M. F., "Controlling mammalian cell interactions on patterned polyelectrolyte multilayer surfaces," *Langmuir*, Vol. 20, No. 4, 2004, pp. 1362–1368.

[75] Groves, J. T., Mahal, L. K., and Bertozzi, C. R., "Control of cell adhesion and growth with micropatterned supported lipid membranes," *Langmuir*, Vol. 17, No. 17, 2001, pp. 5129–5133.

[76] Hong, Y., Webb, B. L., Su, H., Mozdy, E. J., Fang, Y., Wu, Q., Liu, L., Beck, J., Ferrie, A. M., Raghavan, S., Mauro, J., Carre, A., Mueller, D., Lai, F., Rasnow, B., Johnson, M., Min, H., Salon, J., and Lahiri, J., "Functional GPCR microarrays," *J. Am. Chem. Soc.*, Vol. 127, No. 44, 2005, pp. 15350–15351.

[77] Fang, Y., Frutos, A. G., and Lahiri, J., "Membrane protein microarrays," *J. Am. Chem. Soc.*, Vol. 124, No. 11, 2002, pp. 2394–2395.

[78] Folch, A., Jo, B. H., Hurtado, O., Beebe, D. J., and Toner, M., "Microfabricated elastomeric stencils for micropatterning cell cultures," *J. Biomed. Mater. Res.*, Vol. 52, No. 2, 2000, pp. 346–353.

[79] Delamarche, E., Bernard, A., Schmid, H., Michel, B., and Biebuyck, H., "Patterned delivery of immunoglobulins to surfaces using microfluidic networks," *Science*, Vol. 276, No. 5313, 1997, pp. 779–781.

[80] Takayama, S., McDonald, J. C., Ostuni, E., Liang, M. N., Kenis, P. J., Ismagilov, R. F., and Whitesides, G. M., "Patterning cells and their environments using multiple laminar fluid flows in capillary networks," *Proc. Natl. Acad. Sci. USA*, Vol. 96, No. 10, 1999, pp. 5545–5548.

[81] Jiang, X., Xu, Q., Dertinger, S. K., Stroock, A. D., Fu, T. M., and Whitesides, G. M., "A general method for patterning gradients of biomolecules on surfaces using microfluidic networks," *Analytical Chemistry*, Vol. 77, No. 8, 2005, pp. 2338–2347.

[82] Dillmore, W. S., Yousaf, M. N., and Mrksich, M., "A photochemical method for patterning the immobilization of ligands and cells to self-assembled monolayers," *Langmuir*, Vol. 20, No. 17, 2004, pp. 7223–7231.

[83] Ryan, D., Parviz, B. A., Linder, V., Semetey, V., Sia, S. K., Su, J., Mrksich, M., and Whitesides, G. M., "Patterning multiple aligned self-assembled monolayers using light," *Langmuir*, Vol. 20, No. 21, 2004, pp. 9080–9088.

[84] Nakanishi, J., Kikuchi, Y., Takarada, T., Nakayama, H., Yamaguchi, K., and Maeda, M., "Photoactivation of a substrate for cell adhesion under standard fluorescence microscopes," *J. Am. Chem. Soc.*, Vol. 126, No. 50, 2004, pp. 16314–16315.

[85] Khademhosseini, A., Suh, K. Y., Jon, S., Eng, G., Yeh, J., Chen, G.-J., and Langer, R., "A soft lithographic approach to fabricate patterned microfluidic channels," *Analytical Chemistry*, Vol. 76, No. 13, 2004, pp. 3675–3681.

[86] Khademhosseini, A., Yeh, J., Eng, G., Karp, J., Kaji, H., Borenstein, J., Farokhzad, O. C., and Langer, R., "Cell docking inside microwells within reversibly sealed microfluidic channels for fabricating multiphenotype cell arrays," *Lab Chip*, Vol. 5, No. 12, 2005, pp. 1380–1386.

[87] Sullivan, T. P., and Huck, W. T. S., "Reactions on monolayers: Organic synthesis in two dimensions," *Eur. J. Organic Chem.*, No. 1, 2003, pp. 17–29.

[88] Jiang, X., Ferrigno, R., Mrksich, M., and Whitesides, G. M., "Electrochemical desorption of self-assembled monolayers noninvasively releases patterned cells from geometrical confinements," *J. Am. Chem. Soc.*, Vol. 125, No. 9, 2003, pp. 2366–2367.

[89] Jiang, X., Bruzewicz, D. A., Wong, A. P., Piel, M., and Whitesides, G. M., "Directing cell migration with asymmetric micropatterns," *Proc. Natl. Acad. Sci. USA*, Vol. 102, No. 4, 2005, pp. 975–978.

[90] Li, Y., Bo, Y., Ji, H., Han, D., Chen, S., Tian, F., and Jiang, X. Y., "A method for patterning multiple types of cells by using electrochemical desorption of self-assembled monolayers within microfluidic channels," *Angewandte Chemie—International Edition in English*, Vol. 2006, pp. 10.1002/anie.200603844.

[91] Yousaf, M. N., Houseman, B. T., and Mrksich, M., "Turning on cell migration with electroactive substrates," *Angewandte Chemie—International Edition in English*, Vol. 40, No. 6, 2001, pp. 1093–1096.

[92] Yeo, W.-S., Hodneland, C. D., and Mrksich, M., "Electroactive monolayer substrates that selectively release adherent cells," *ChemBioChem*, Vol. 2, Nos. 7–8, 2001, pp. 590–593.

[93] Yousaf, M. N., Houseman, B. T., and Mrksich, M., "Using electroactive substrates to pattern the attachment of two different cell populations," *Proc. Natl. Acad. Sci. USA*, Vol. 98, No. 11, 2001, pp. 5992–5996.

[94] Mrksich, M., "What can surface chemistry do for cell biology?" *Current Opinions in Chemical Biology*, Vol. 6, 2002, pp. 794–797.

[95] Yeo, W.-S., Yousaf, M. N., and Mrksich, M., "Dynamic interfaces between cells and surfaces: Electroactive substrates that sequentially release and attach cells," *J. Am. Chem. Soc.*, Vol. 125, No. 49, 2003, pp. 14994–14995.

[96] Chan, E. W., and Yousaf, M. N., "Immobilization of ligands with precise control of density to electroactive surfaces," *J. Am. Chem. Soc.*, Vol. 128, No. 48, 2006, pp. 15542–15546.

[97] Kumar, G., Meng, J. J., Ip, W., Co, C. C., and Ho, C. C., "Cell motility assays on tissue culture dishes via non-invasive confinement and release of cells," *Langmuir*, Vol. 21, No. 20, 2005, pp. 9267–9273.

[98] Fukuda, J., Khademhosseini, A., Yeh, J., Eng, G., Cheng, J. J., Farokhzad, O. C., and Langer, R., "Micropatterned cell co-cultures using layer-by-layer deposition of extracellular matrix components," *Biomaterials*, Vol. 27, No. 8, 2006, pp. 1479–1486.

[99] Nishizawa, M., Takoh, K., and Matsue, T., "Micropatterning of HeLa cells on glass substrates and evaluation of respiratory activity using microelectrodes," *Langmuir*, Vol. 18, No. 9, 2002, pp. 3645–3649.

[100] Kaji, H., Kanada, M., Oyamatsu, D., Matsue, T., and Nishizawa, M., "Microelectrochemical approach to induce local cell adhesion and growth on substrates," *Langmuir*, Vol. 20, No. 1, 2004, pp. 16–19.

[101] Kaji, H., Tsukidate, K., Hashimoto, M., Matsue, T., and Nishizawa, M., "Patterning the surface cytophobicity of an albumin-physisorbed substrate by electrochemical means," *Langmuir*, Vol. 21, No. 15, 2005, pp. 6966–6969.

[102] Kaji, H., Tsukidate, K., Matsue, T., and Nishizawa, M., "In situ control of cellular growth and migration on substrates using microelectrodes," *J. Am. Chem. Soc.*, Vol. 126, No. 46, 2004, pp. 15026–15027.

[103] Zhao, C., Witte, I., and Wittstock, G., "Switching on cell adhesion with microelectrodes," *Angewandte Chemie—International Edition in English*, Vol. 45, No. 33, 2006, pp. 5469–5471.

[104] Hirose, M., Kwon, O. H., Yamato, M., Kikuchi, A., and Okano, T., "Creation of designed shape cell sheets that are noninvasively harvested and moved onto another surface," *Biomacromolecules*, Vol. 1, No. 3, 2000, pp. 377–381.

[105] Hasegawa, M., Yamato, M., Kikuchi, A., Okano, T., and Ishikawa, I., "Human periodontal ligament cell sheets can regenerate periodontal ligament tissue in an athymic rat model," *Tissue Eng.*, Vol. 11, Nos. 3–4, 2005, pp. 469–478.

[106] Kanzaki, M., Yamato, M., Hatakeyama, H., Kohno, C., Yang, J., Umemoto, T., Kikuchi, A., Okano, T., and Onuki, T., "Tissue engineered epithelial cell sheets for the creation of a bioartificial trachea," *Tissue Eng.*, Vol. 12, No. 5, 2006, pp. 1275–1283.

[107] Shimizu, T., Sekine, H., Isoi, Y., Yamato, M., Kikuchi, A., and Okano, T., "Long-term survival and growth of pulsatile myocardial tissue grafts engineered by the layering of cardiomyocyte sheets," *Tissue Eng.*, Vol. 12, No. 3, 2006, pp. 499–507.

[108] Frey, W., Meyer, D. E., and Chilkoti, A., "Dynamic addressing of a surface pattern by a stimuli-responsive fusion protein," *Adv. Mater.*, Vol. 15, No. 3, 2003, pp. 248–251.

[109] Frey, W., Meyer, D. E., and Chilkoti, A., "Thermodynamically reversible addressing of a stimuli responsive fusion protein onto a patterned surfacetemplate," *Langmuir*, Vol. 19, No. 5, 2003, pp. 1641–1653.

[110] Jiang, X., Takayama, S., Qian, X., Ostuni, E., Wu, H., Bowden, N., LeDuc, P., Ingber, D. E., and Whitesides, G. M., "Controlling mammalian cell spreading and cytoskeletal arrangement with conveniently fabricated continuous wavy features on poly(dimethylsiloxane)," *Langmuir*, Vol. 18, No. 8, 2002, pp. 3273–3280.

[111] Zhu, X., Mills, K. L., Peters, P. R., Bahng, J. H., Liu, E. H., Shim, J., Naruse, K., Csete, M. E., Thouless, M. D., and Takayama, S., "Fabrication of reconfigurable protein matrices by cracking," *Nature Materials*, Vol. 4, No. 5, 2005, pp. 403–406.

[112] Whitesides, G. M., and Grzybowski, B., "Self-assembly at all scales," *Science*, Vol. 295, No. 5564, 2002, pp. 2418–2421.

[113] Arnold, M., Cavalcanti-Adam, E. A., Glass, R., Blummel, J., Eck, W., Kantlehner, M., Kessler, H., and Spatz, J. P., "Activation of integrin function by nanopatterned adhesive interfaces," *ChemPhysChem*, Vol. 5, No. 3, 2004, pp. 383–388.

[114] Lee, K.-B., Park, S.-J., Mirkin, C. A., Smith, J. C., and Mrksich, M., "Protein nanoarrays generated by dip-pen nanolithography," *Science*, Vol. 295, No. 5560, 2002, pp. 1702–1705.

CHAPTER 4

Patterned Cocultures for Controlling Cell-Cell Interactions

Sun Min Kim, Junji Fukuda, and Ali Khademhosseini

4.1 Introduction

Cell-cell interactions play an important role in the function of many organ systems. In addition to homeostatic function, cell-cell interactions are also vital for regenerative processes, as well as for in vitro reconstruction of tissues for tissue replacement [1–8]. In the body, the dynamic interplay between various cell types regulates the function of the cells. The lack of such cell-cell interactions is one potential reason for the loss of functional capability of cell types such as hepatocytes outside the body. In tissue culture, much of the native cell-cell interactions present in the body are lost due to tissue isolation and digestion, as well as processes such as purification of specific cell populations. To overcome this limitation, cocultures of two (or more) cell types have been used to better mimic the organization and complexity of the in vivo microenvironment [9–16].

Traditionally, in vitro cellular interactions were investigated by random seeding of multiple cell types on a tissue culture substrate; however, using this approach, it is difficult to control the degree of homotypic (i.e., contact between same cell types) and heterotypic (i.e., contact between different cell types) cell-cell interactions [17]. To overcome these limitations, patterned coculture techniques have been developed based on advances in microengineering technology, material science, and chemistry [17–25].

Patterned coculture techniques enhance the spatial control of cells in culture, the precise manipulation of homotypic/heterotypic cell-cell interactions, and the function of cell types through introduction of support cells that provide the signals to maintain these cells in culture [17, 18, 20, 22, 24, 26]. Many different types of patterned coculture techniques have been investigated, most of which could be placed into one of four categories. The first approach uses the selective adhesion of cells to micropatterned substrates or islands [18–20, 27, 28]; the second approach uses the liquid flow in microfluidic channels to pattern cells and proteins on a substrate [29–31]; the third approach uses stencil-based patterning for localizing cells to specific regions on a substrate [17, 21–23, 32, 33]; and, finally, the fourth approach utilizes micropatterned surfaces that can be switched from cell repulsive to adhesive by specific stimuli (electrical potential [34–37], temperature [38–44], or light exposure [45]).

In this chapter, we review each patterned coculture method in detail based on the above-mentioned categories. Also, emerging patterned coculture methods that facilitate cell culturing in three-dimensional environments, as well as with controlled temporal control, are discussed. Finally, we analyze the future directions of the use of patterned cocultures for in vitro systems.

4.2 Random Coculture Systems

Traditionally, cocultures of two or more cell types were generated by randomly seeding the cells on a substrate [9, 46–50]. These cocultures have enabled the study of cell-cell interactions. For example, it has been shown that the coculturing of adult rat hepatocytes with liver epithelial cells improved hepatocyte function in vitro as demonstrated by the expression of liver-specific markers in culture. In these cocultures, hepatocytes showed distinct morphology and function in vitro. Cocultures have also been applied to many other biological systems to recapture the complexity of the cellular microenvironment. Random coculture systems have presented insight into homotypic and heterotypic cell-cell interactions but have been limited by the inability to vary local cell seeding density and the degree of cell-cell contact. To overcome these limitations, micropatterned coculture systems have been used to enhance the control of spatial localization of multiple cell types relative to each other and to enable detailed mechanistic studies of the processes that regulate cell-cell interactions.

4.3 Patterned Coculture Systems

The majority of current patterned coculture systems can be generated based on one of the following techniques: (1) selective adhesion to micropatterned substrates; (2) microfluidic patterning; (3) stencil-based patterning; and (4) micropatterned switchable surfaces.

4.3.1 Selective Adhesion of Cells to Micropatterned Substrate

In this approach, the cells of interest selectively adhere to specific regions of a micropatterned substrate by the surface characteristics of substrate, including variations in surface charge, chemistry, hydrophilicity, and topology [18, 51–57]. Different cell types have different levels of adhesiveness to various substrates based on their level of expression of various adhesion molecules, such as integrins and cadherins. Based on these differences, it is possible to localize specific cell types to micropatterned regions on a substrate for culturing multiple cell types. One of the first examples of the use of this approach was demonstrated by Bhatia et al. [18, 19, 28] through the use of a micropatterning technique for generating two-dimensional, anisotropic surfaces used to localize different cell types in specific region. Figure 4.1(a) shows the schematic of the process of generating such micropattern cocultures. In this approach, micropatterned cell-adhesive biomolecules (e.g.,

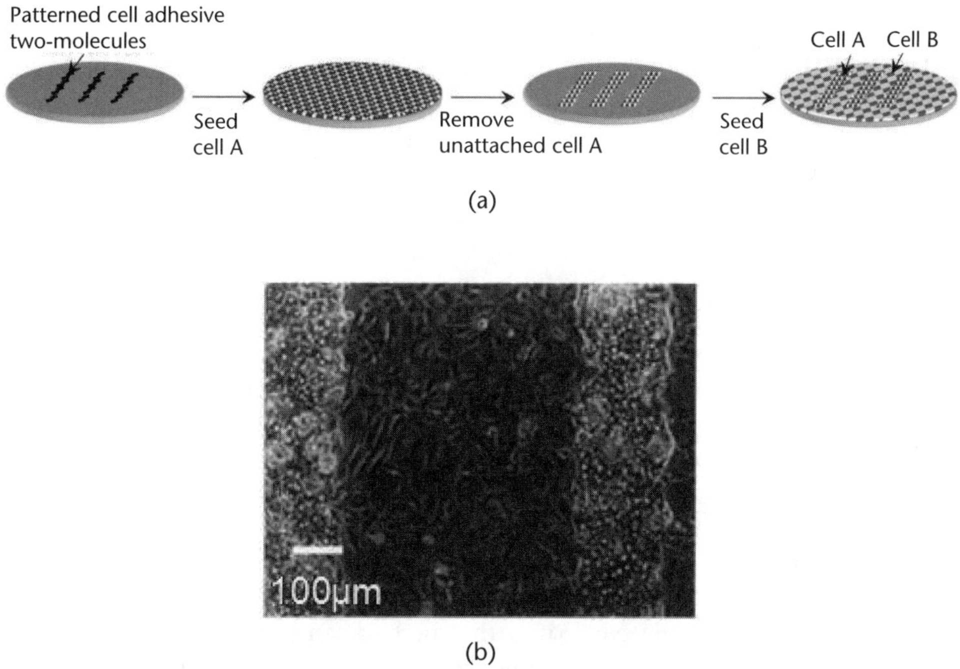

Figure 4.1 Micropatterning using selective adhesion of cells to patterned substrate: (a) A schematic diagram of the procedure to generate micropattern cocultures. Cell-adhesive biomolecules were patterned by typical photolithography process. Then, type A cells were spread and attached to the wafer, and unattached cells were washed out. Finally, type B cell were attached to an unmodified region of the wafer for coculturing. (b) A micrograph of cocultures of hepatocytes (200 μm lanes of collagen-derivatized glass) with fibroblasts (500 μm glass lanes). The photograph was taken forty-eight hours after hepatocyte seeding and twenty-four hours after fibroblast seeding. (*Source:* [18], reprinted with permission from John Wiley and Sons.)

collagen) mediated the adhesion of the first-type cell. For example, collagen micropatterns mediated the selective adhesion of hepatocytes compared to the noncoated regions. Finally, secondary cells were seeded and adhered to the unmodified region of the substrate by nonspecific, serum-mediated attachment. Using this technique, homotypic and heterotypic cell-cell interactions of rat hepatocytes and fibroblasts in maintaining liver functions were examined to yield important biological insight into the role of cell-cell contact [Figure 4.1(b)].

This micropatterning technique can efficiently localize cells within patterned cocultures. Furthermore, this approach can be easily implemented since it requires only access to existing techniques, including photolithography, soft lithography, and microcontact printing [58–67]. However, despite these advantages, the technique is limited by several issues [20]. First, this technique depends on the relative cell-cell and cell-substrate adhesiveness of each cell type. For example, first seeded cells must weakly adhere to the unmodified region but strongly attach to the patterned region. Also, secondary cells should adhere to the unmodified region of the substrate rather than the primary cells. Furthermore, the seeding order of cell, as well as the choice of the matrix materials, is limited using these cultures.

4.3.2 Microfluidic Patterning for Cocultures

Fluid flow in microfluidic channels can also be used for patterning the substrate with cells and biomolecules [29–31]. In this approach microfluidic channels can be used to deliver cells and soluble factors to specific regions of a substrate. For example, Takayama et al. [30] presented the patterning of two different cell types by using the multiple flow streams in capillary channels fabricated with poly(dimethylsiloxane) (PDMS). Within these microchannels, the fluid flow is in the low Reynolds number regime; thus, two or more laminar flow streams flow parallel to each other due to low convective mixing [25, 68–71]. This feature of multiple laminar flows enables the spatial patterning of cells and their microenvironments within microchannels. In this process, the width of each cell pattern can easily be controlled by adjusting the relative volumetric flow rate of cell suspensions and sheath flow.

This patterning technique has distinctive advantages for versatile applications, which include a straightforward fabrication process and the ability to generate patterns inside microfluidic channels [30]. However, this method patterns cells only on parallel stripes, not on the various shapes that can be achieved by other patterning methods. Moreover, this method can be applied only to a few metabolically slow cell types because the delivery of oxygen and nutrient is arrested, while the energy-consuming processes of cell anchorage and spreading occur [23, 43].

An alternate approach is to patterning multiple cell types within individual reversibly sealed microchannels can be used to localize cells to specific regions on a substrate. In this approach, an array channel containing different cell types can be used to localize the cells in specific regions [26, 72]. For example, Khademhosseini et al. [26] presented multiphenotype cell patterning within an array of reversibly sealed microfluidic channels. Microfluidic channels deliver various fluids or cells onto specific locations on a substrate, and microwells on the substrate capture and immobilize cells within low shear stress regions. Both of the microfluidic channels and microwells were fabricated by the soft lithography of PDMS polymer. This approach utilizes three distinct concepts: (1) the capture of cells inside the microwells that contain low shear stress regions; (2) reversible sealing between the top layer of microchannels and microwell-patterned substrate; and (3) alternative orthogonal placing of microchannel arrays to deliver a unique set of fluids or cell types (see Figure 4.2).

This approach can reduce the time for cell attachment on the substrate and allow the top layer of microchannel to be realigned and moved without disturbing the cells because cell are docked inside microwells. This approach also enables coculture of multiple cell types by sequences of fluid flow of different cell types and has potential applications in high-throughput screening of drug and optimization of cell-soluble signal interactions for biological research.

4.3.3 Stencil-Based Patterning

In this approach, elastomeric stencil can be used to pattern cells to specific regions of a substrate [17, 21–23]. Elastomeric stencils with microengineered holes can be reversibly sealed on a substrate to promote patterned deposition of biomolecules or cells directly on a substrate.

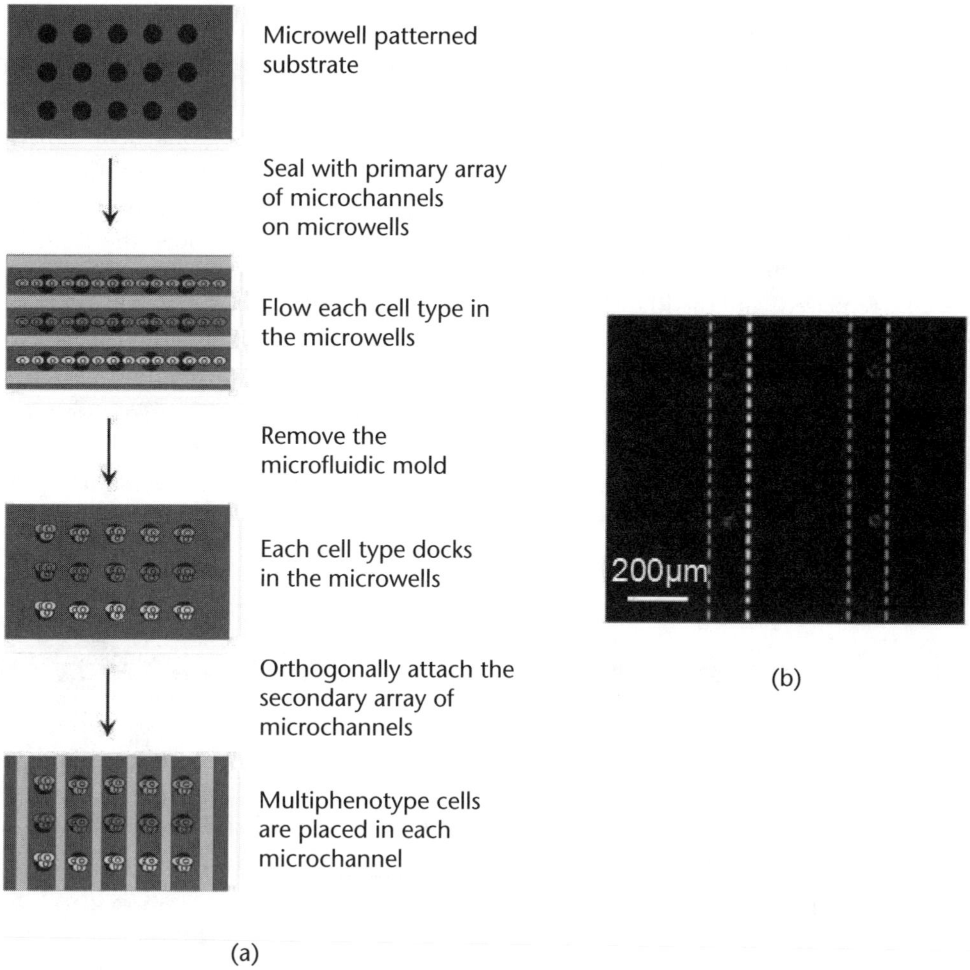

Figure 4.2 Micropatterned cell patterning using microfluidic channels: (a) A schematic of multiphenotype cell patterning by utilizing reversible sealing of microfluidic channel arrays onto microwell-patterned substrate. Each cell type flows through an independent microchannel and docks inside microwells. The PDMS microfluidic mold can be removed and replaced with another mold, which is placed orthogonally to generate multiphenotype cell arrays inside each microchannel. (b) A fluorescent image of patterned AML12 and NIH-3T3 cells inside microwells. (*Source:* [26], reprinted with permission from the Royal Society of Chemistry.)

For example, Ostuni et al. [21] presented a stencil-based lift-off approach to generate patterned cell arrays [73, 74]. In this process, PDMS polymer was spin-coated onto a microstructured silicon wafer containing photoresist posts to generate a PDMS membrane. Subsequently, the cured membrane was removed from the master and brought into conformal contact with a substrate. To generate cell patterns, cell-adhesion-promoting molecules (fibronectin or gelatin) were absorbed into the patterned holes in the membranes, and then membrane was removed from the substrate to generate a pattern of adhesion-promoting molecules. After coating the rest of the substrate with cell-repulsive molecules (bovine serum albumin) to inhibit cell adhesion, cells were only patterned on the adhesive regions. In an alternative approach, adhesive cells such as fibroblasts can be directly patterned by adhering to the substrate through the holes in the membranes. Patterning of multiple cell types can also be accomplished by sequentially patterning cell-adhesion promoters and cells using this technique (see Figure 4.3).

Folch et al. [22, 23] presented two techniques for the fabrication of elastomeric membranes for cell patterning. First, elastomeric membranes were fabricated by injecting or suctioning uncured PDMS prepolymer into the microchamber, which was composed with photoresist posts on a wafer and a thin adhesive film as a top roof. After the PDMS prepolymer was cured, the adhesive film was carefully removed, and the PDMS membrane was peeled off from the wafer. Then, the PDMS membrane was sealed onto the substrate for cell culturing. Second, the elastomeric membrane was fabricated by using a compression-molding process. PDMS prepolymer was poured on a photoresist mold, and pressure was uniformly applied by plate stack. After the PDMS curing was complete, the next steps were performed likewise. Cell patterning and cocultures of multiple cell types were performed following similar procedures to those previously discussed.

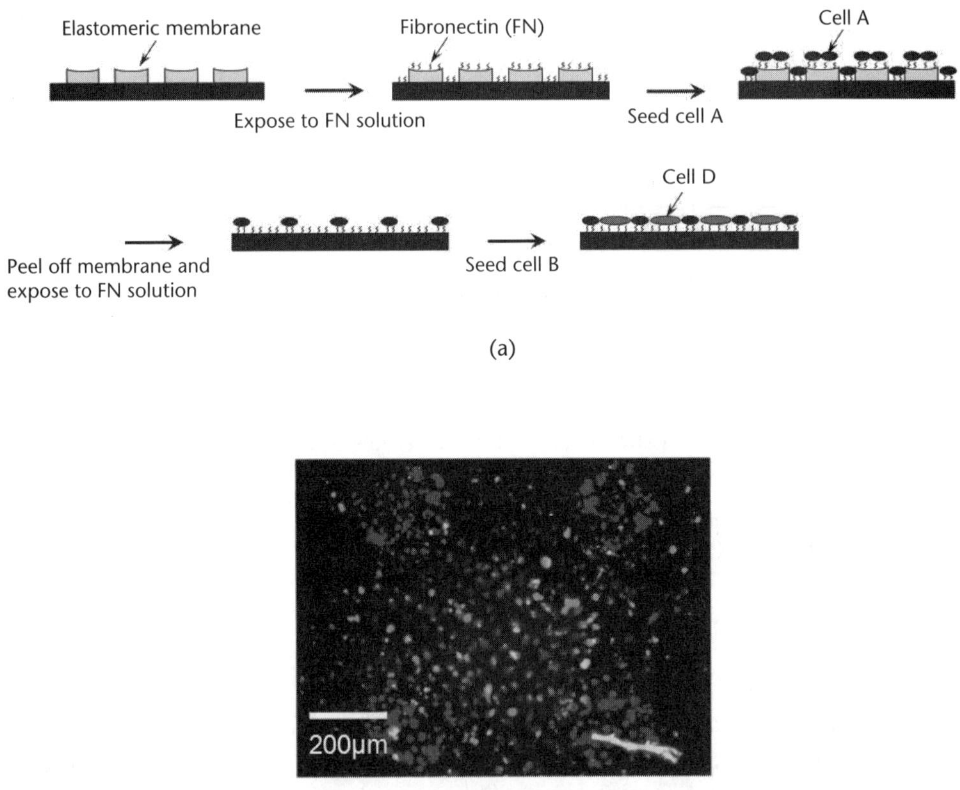

Figure 4.3 Micropatterned cell cocultures using elastomeric membrane: (a) A schematic of generic procedure of patterned cocultures using elastomeric membrane. At first, cell-adhesion-promoting molecules (fibronectin, or FN) adhered to the substrate through patterned holes in the membrane, and then type A cells adhered to FN. Subsequently, the membrane was removed from the substrate to generate a pattern, and the substrate was exposed to FN solution again for seeding second-type cells. Finally, type B cells were seeded on the substrate and cells were only attached to the unpatterned area. (b) A fluorescent image of cocultured mES cells with AML12 cells using parylene membrane. (*Source:* [32].)

These patterning methods using PDMS membranes are applicable to a broad range of substrates that can absorb adhesion-promoting proteins and can make conformal contact with PDMS membranes [21]. PDMS membranes also can be replicated many times from the same master mold because the replication procedure does not break the mold [23]. Furthermore, three-dimensional structures formed by PDMS could make it possible to generate a three-dimensional cell-culture microenvironment [21, 72]. However, the attachment of PDMS membranes over the substrates and the peeling off of membranes after cell seeding can cause trouble in large-scale applications [43].

Recently, microfabricated parylene membranes, which have been used for micropatterning methods to pattern cells, proteins, and antibodies [33, 75, 76], were used for static and dynamic cocultures of multiple cell types, which can manipulate the spatial and temporal cell-cell interactions in tissue culture by changing the cell adhesiveness to parylene membrane surfaces [32]. In this coculture system, the top surface of the parylene membrane was pretreated with hyaluronic acid (HA) to lessen nonspecific cell adhesion and then placed onto a PDMS substrate. First-type cells were seeded and only adhered to the substrate through the holes in the membrane. Collagen was then deposited on the parylene membrane to change the surface properties of parylene to cell adhesive. Subsequently, second-type cells were seeded on the membrane to form a patterned coculture. To seed the third-type cells, second-type cells were removed by peeling off the parylene membrane from the substrate. This method enables the spatial and temporal regulation of multiple cell cultures for studying the complexity of cell-cell interactions in in vitro cell cultures.

Parylene membranes have several advantages over PDMS membranes. Parylene membranes can be easily removed or attached to a surface without tearing due to their mechanical robustness compared to PDMS [77, 78] and can form a reversible binding with hydrophobic surfaces. Thus, parylene membranes could be used for multiple patterning processes. Moreover, more cells adhere to parylene compared to PDMS [32]. However, the parylene membrane fabrication procedure requires many steps and much special equipment compared to the PDMS membrane fabrication procedure [33].

4.3.4 Micropatterning Using Surfaces Switchable from Cell Repulsive to Adhesive

This approach is based on using the surfaces that can be turned from being cell repulsive to adhesive by specific stimuli: electrical potential, temperature, and light exposure. Recently, layer-by-layer assembly of biopolymers [79–82] has also been employed for cell patterning by switching the cell adhesion of the substrate [24, 67, 83].

Yousaf et al. [35] developed an electroactive mask that enables two different cell types to be patterned on a single substrate. This mask was fabricated with a self-assembled monolayer of alkanethiolates on gold, which can be switched from cell adhesive to repulsive.

Figure 4.4(a) shows the schematic strategy of patterning two different cell types using switchable monolayers. At first, the monolayer was patterned by microcontact printing [56, 63, 84], and extracellular matrix (ECM) proteins

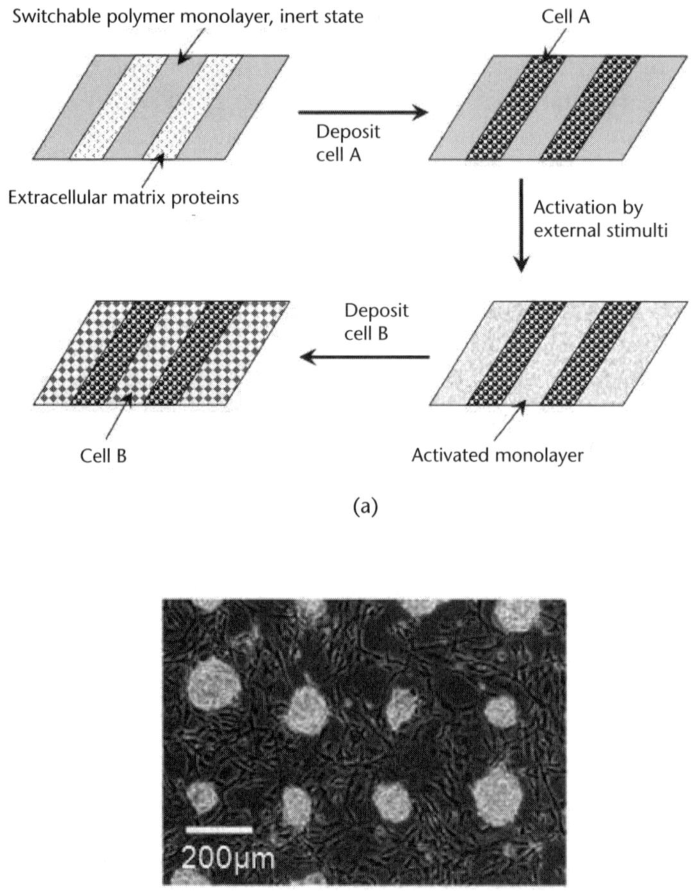

(a)

(b)

Figure 4.4 Micropatterned cell cocultures using switchable substrates via external stimuli: (a) A generic procedure for patterning two different cell types using switchable substrates. At first, the monolayer was patterned, and fibronectins were attached to the regions where no monolayer was patterned. This fibronectin pattern resulted in the attachment of first-type cells to the fibronectin-coated region. Then, external stimulus (electrical, thermal, or ultraviolet light) was applied to the entire substrate, and the monolayer was changed to cell adhesive. Finally, second-type cells were attached to this activated layer. (b) A micrograph of cocultured ES cells (bright area) with fibroblasts (dark area) using layer-by-layer deposition. (*Source:* [24], reprinted with permission from Elsevier.)

(fibronectin) were attached to the regions where no monolayer was patterned. This fibronectin pattern resulted in the attachment of first-type cells to the fibronectin-coated region. Then, external stimulus (electric potential) was applied to the entire substrate, and the monolayer was changed to cell adhesive. Finally, second-type cells were attached to this activated layer. This technique does not require extensive or invasive manipulation of the substrate, so this technique is suited for cell-culturing systems where physical manipulations are not applicable [35]. This technique can control the receptor-ligand interactions between cell and substrates, as well as the properties of substrate [85–87].

Yamato et al. [42, 43, 88] presented a coculture technique utilizing a thermo-responsive polymer whose cell-polymer adhesion is changed by temperature. A thermoresponsive poly(N-isopropylacrylamide) (PIPAAm) was covalently grafted as a thin layer onto tissue culture grade polystyrene (TCPS) dishes by electron beam radiation through a pattering mask [38–41, 88, 89]. Above the lower critical solution temperature (LCST, 32°C) of PIPAAm polymer, PIPAAm is dehydrated and cell adhesive. Under its LCST of 32°C, PIPAAm polymer is hydrated, and cell attachment is highly suppressed [40, 90]. The cell-patterning procedure using thermoresponsive PIPAAm polymer is quite simple and versatile. The first cell type was seeded over the PIPAAm-patterned TCPS dishes at 20°C and attached only to the non-PIPPAm region of TCPS dishes. Then, the second cell type was seeded above LCST and only attached to the PIPAAm-patterned region. This patterning technique possibly can be utilized for coculturing three or more cell types by varying the LCST of the PIPAAm polymer, both over successive temperature regimes and spatially across the culture surface, simultaneously using sequential masks and copolymerization. PIPAAm copolymerized with other monomers permits wide access to copolymer-grafted culture surfaces with variable hydration temperatures [91, 92]. Recently, Edahiro et. al. [45] improved this method with thermo- and photoresponsive surfaces. Culture surfaces were coated with a polymer material composed of PIPAAm having spiropyran chromophores as side chains. These surfaces can be switched from being cell repulsive to adhesive by being exposed to different wave lengths of light, as well as by temperature changing. Since light exposure can be applied to the small area of culture surfaces, this approach can improve the regional control of cell adhesion by reducing the size of patterns compared to the previous thermoresponsive method.

Khademhosseini et al. [24, 67] developed a method for patterning cell cocultures using layer-by-layer deposition of ionic biomolecules. Hyaluronic acid (HA), which is a biocompatible and biodegradable material [93, 94], was patterned on glass slides by capillary force lithography [95, 96], followed by fibronectin adsorption onto the non-HA-patterned region. Then, the first-type cells were seeded and only attached to the fibronectin-coated region. Subsequent ionic adsorption of poly-L-lysine (PLL) to the HA pattern was used to change HA surfaces from cell repulsive to cell adhesive. Finally, the second-type cells were seeded and attached to the PLL pattern. Coculturing of ES cells with fibroblasts and hepatocytes with fibroblasts was successfully performed by utilizing this method [Figure 4.4(b)] [24]. This coculturing method utilizes the switchable but does not require electroactive or thermoresponsive surfaces that may be difficult to fabricate. Both HA and PLL are commercially available and do not require any chemistry or complex techniques for immobilization.

4.3.5 Other Approaches

Recently, various micropatterned coculture approaches not in one of four categories have been developed to better mimic the organization and complexity of in vivo microenvironments using microfabrication technique. Although the patterned coculture methods discussed above provide the spatial control of cells and the manipulation of homotypic/heterotypic cell-cell interactions in culture, they do not replicate dynamic and three-dimensional aspects of the in vivo environment.

The dynamics of cell-cell interactions as regulated by embryonic morphogenesis and mechanical factors are a key regulator of cell fate decisions. Khademhosseini et. al. [32] presented the dynamic coculture system using microfabricated parylene membranes as discussed previously (see Section 4.3.4) This system enables the cocultures of multiple cell types, which can manipulate the spatial and temporal cell-cell interactions in tissue cultures by changing the cell adhesiveness to parylene membrane surfaces. Hui and Bhatia [97] have demonstrated the dynamic coculture of various cell types by using a microfabricated interdigitating system. This system utilized a silicon platform to bring cells in close proximity to each other in a dynamic manner and manipulate the complexity of cell-cell interactions in a spatially and temporally regulated way.

A three-dimensional structure containing various cell types is required to reconstruct an organ system to function as it does in vivo. Ito et. al. [98] utilized magnetic force to precisely place magnetically labeled cells onto target cells and promote heterotypic cell-cell adhesion to form a three-dimensional structure. Magnetite cationic liposomes carrying a positive surface charge accumulated in endothelial cells to improve adsorption. Subsequently, endothelial cells specifically accumulated onto hepatocyte monolayers at sites where a magnet was positioned, then adhered to form a heterotypic layered construct with tight and close contact. Tsang et. al. [99] fabricated a three-dimensional hepatic tissue construct embedding hepatocytes in poly(ethylene glycol) (PEG) hydrogel structures using a multilayer photo- patterning platform [Figure 4.5(a)]. They perfused these three-dimensional tissue structures in a continuous flow bioreactor and demonstrated improved viability and liver-specific function of the encapsulated hepatocytes over unpatterned controls. Albrecht et al. [100] also presented a method for the rapid formation of three-dimensional cellular structures within a photopolymerizable PEG hydrogel using dielectrophoretic forces. In this method, cells were micropatterned via dielectrophoretic forces, and each single hydrogel layer was incorporated into multilayer constructs for cocultures. Mikos et al. [101] suggested light-based process for creating bilayered hydrogel structures. At first, one layer of partially gelled liquid polymer solution is placed in the mold, and then a second layer of liquid polymer solution is added to the first layer. The whole construct is then fully crosslinked by light to form a bilayered hydrogel. Different cell types can be embedded in each layer for cocultures. This method also can be extended to create a multilayered hydrogel construct. Yeh et al. [102] also presented a hydrogel-based cell-coculturing method in three-dimensional constructions.

Figure 4.5(b) shows a fluorescence image of hydrogel arrangement and assembly. Cells were suspended in a HA hydrogel prepolymer solution and molded using a PDMS stamp. Hydrogels were then formed via exposing the prepolymer to ultraviolet light. Two different cell types encapsulated in separate HA hydrogels were subsequently arranged in a checkerboard pattern for cocultures.

These coculture systems have been developed to better mimic the organization and complexity of in vivo tissue structure in dynamic and three-dimensional aspects. However, these approaches are still under active investigation in the area of materials, chemistry, and pattering techniques.

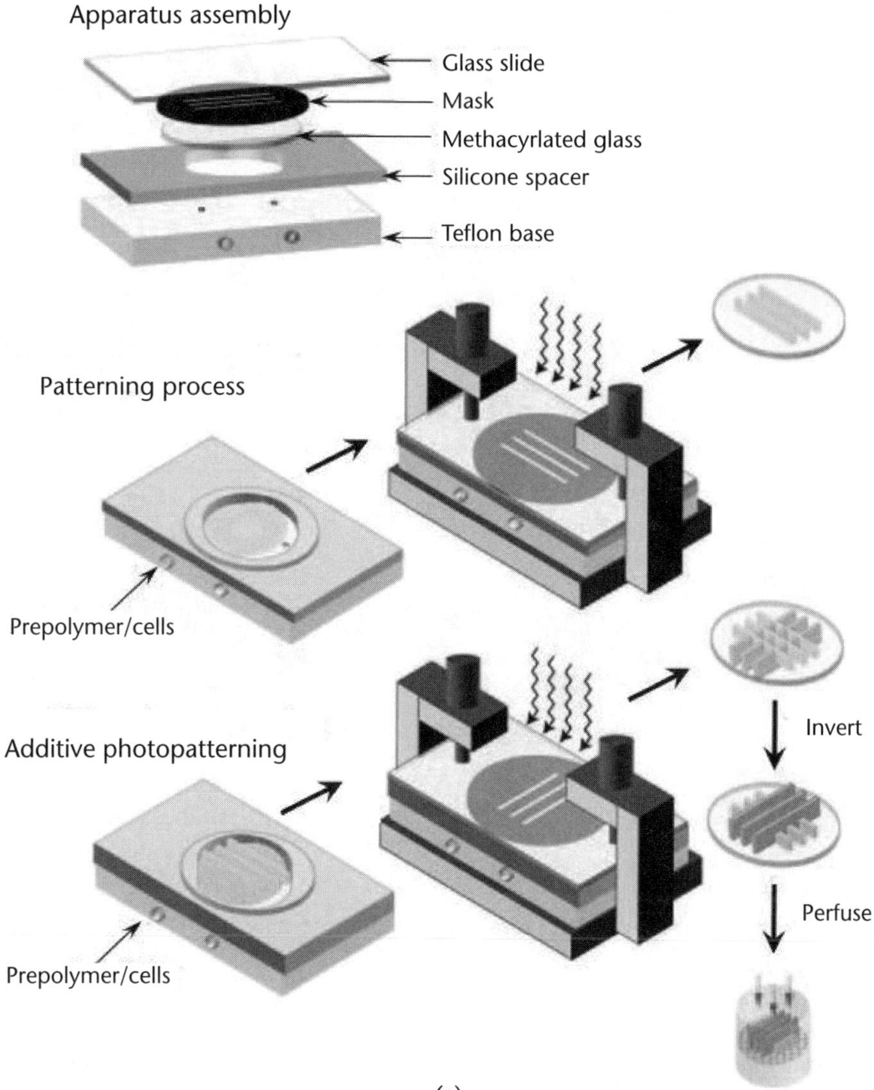

(a)

Figure 4.5 (a) Schematic procedure for fabricating three-dimensional living tissues by additive photopatterning of cellular hydrogels. The chamber was filled with cell-laden prepolymer solution, and exposure to light through a photomask causes photocrosslinking in pattern. The height of the chamber is then increased, and the process repeated, resulting in a multilayer cell-laden three-dimensional tissue construct with microscale features. (*Source:* [99], reprinted with permission from The Federation of American Societies for Experimental Biology.) (b) A fluorescence image of hydrogel arrangement and assembly. Two different cell types encapsulated in separate HA hydrogels were arranged in a checkerboard pattern for cocultures. (*Source:* [102], reprinted with permission from Elsevier.)

4.4 Conclusion

Understanding cell-cell interactions is important for tissue engineering, which studies in vitro reconstructing of tissue architecture for tissue replacement. The

(b)

Figure 4.5 (continued)

coculturing of two or more cell types has been used to establish a more biomimetic environment. In particular, micropatterning coculture methods have been developed and challenged to enhance coculture environment control through spatial localization of multiple cells. Each method has its own advantages, as well as limitations due to the lack of suitable materials and the complex fabrication procedure, as discussed. Micropatterning methods have usually been performed on static and two-dimensional culture systems; however, cells in tissue and natural organs are in dynamic and three-dimensional environments. Recently, some methods have been developed to fabricate dynamic coculture systems using specific patterning method and three-dimensional coculture constructions using hydrogel, but these systems are still under investigation in the area of biomaterials and patterning techniques.

References

[1] Michalopoulos, G. K., and DeFrances, M. C., "Liver regeneration," *Science*, Vol. 276, No. 5309, 1997, pp. 60–66.

[2] Slavkin, H. C., Snead, M. L., Zeichnerdavid, M., Jaskoll, T. F., and Smith, B. T., "Concepts of epithelial-mesenchymal interactions during development—tooth and lung organogenesis," *J. Cell. Biochem.*, Vol. 26, No. 2, 1984, pp. 117–125.

[3] Davies, P. F., "Vascular Cell-Interactions with Special Reference to the Pathogenesis of Atherosclerosis," *Laboratory Investigation*, Vol. 55, No. 1, 1986, pp. 5–24.

[4] Aufderheide, E., Chiquetehrismann, R., and Ekblom, P., "Epithelial mesenchymal interactions in the developing kidney lead to expression of tenascin in the mesenchyme," *J. Cell. Biol.*, Vol. 105, No. 1, 1987, pp. 599–608.

[5] Simonassmann, P., Bouziges, F., Arnold, C., Haffen, K., and Kedinger, M., "Epithelial mesenchymal interactions in the production of basement-membrane components in the gut," *Development*, Vol. 102, No. 2, 1988, pp. 339–347.

[6] Brown, L. F., Yeo, K. T., Berse, B., Yeo, T. K., Senger, D. R., Dvorak, H. F., and Vandewater, L., "Expression of vascular-permeability factor (vascular endothelial growth-factor) by epidermal-keratinocytes during wound-healing," *J. Exp. Med.*, Vol. 176, No. 5, 1992, pp. 1375–1379.

[7] Nishida, M., Springhorn, J. P., Kelly, R. A., and Smith, T. W., "Cell-cell signaling between adult-rat ventricular myocytes and cardiac microvascular endothelial-cells in heterotypic primary culture," *J. Clin. Invest.*, Vol. 91, No. 5, 1993, pp. 1934–1941.

[8] Grinnell, A. D., "Dynamics of nerve-muscle interaction in developing and mature neuromuscular-junctions," *Physiological Reviews*, Vol. 75, No. 4, 1995, pp. 789–834.

[9] Guguenguillouzo, C., Clement, B., Baffet, G., Beaumont, C., Morelchany, E., Glaise, D., and Guillouzo, A., "Maintenance and reversibility of active albumin secretion by adult-rat hepatocytes co-cultured with another liver epithelial-cell type," *Exp. Cell Res.*, Vol. 143, No. 1, 1983, pp. 47–54.

[10] Guillouzo, A., Delers, F., Clement, B., Bernard, N., and Engler, R., "Long-term production of acute-phase proteins by adult-rat hepatocytes co-cultured with another liver-cell type in serum-free medium," *Biochem. Biophys. Res. Comm.*, Vol. 120, No. 2, 1984, pp. 311–317.

[11] Lebreton, J. P., Daveau, M., Hiron, M., Fontaine, M., Biou, D., Gilbert, D., and Guguenguillouzo, C., "Long-term biosynthesis of complement component C-3 and alpha-1 acid glycoprotein by adult-rat hepatocytes in a coculture system with an epithelial liver cell-type," *Biochem. J.*, Vol. 235, No. 2, 1986, pp. 421–427.

[12] Jongen, W. M. F., Sijtsma, S. R., Zwijsen, R. M. L., and Temmink, J. H. M., "A cocultivation system consisting of primary chick-embryo hepatocytes and V79 chinese-hamster cells as a model for metabolic cooperation studies," *Carcinogenesis*, Vol. 8, No. 6, 1987, pp. 767–772.

[13] Conner, J., Valletcollom, I., Daveau, M., Delers, F., Hiron, M., Lebreton, J. P., and Guillouzo, A., "Acute-phase-response induction in rat hepatocytes cocultured with rat-liver epithelial-cells," *Biochem. J.*, Vol. 266, No. 3, 1990, pp. 683–688.

[14] Donato, M. T., Gomezlechon, M. J., and Castell, J. V., "Drug-metabolizing-enzymes in rat hepatocytes cocultured with cell-lines," *In Vitro Cellular and Developmental Biology*, Vol. 26, No. 11, 1990, pp. 1057–1062.

[15] Matsuo, R., Ukida, M., Nishikawa, Y., Omori, N., and Tsuji, T., "The role of Kupffer cells in complement activation in D-galactosamine lipopolysaccharide-induced hepatic-injury of rats," *Acta Medica Okayama*, Vol. 46, No. 5, 1992, pp. 345–354.

[16] Rojkind, M., Novikoff, P. M., Greenwel, P., Rubin, J., Rojasvalencia, L., Decarvalho, A. C., Stockert, R., Spray, D., Hertzberg, E. L., and Wolkoff, A. W., "Characterization and functional-studies on rat-liver fat-storing cell-line and freshly isolated hepatocyte coculture system," *Am. J. Path.*, Vol. 146, No. 6, 1995, pp. 1508–1520.

[17] Folch, A., and Toner, M., "Microengineering of cellular interactions," *Ann. Rev. Biomed. Eng.*, Vol. 2, 2000, pp. 227–256.

[18] Bhatia, S. N., Yarmush, M. L., and Toner, M., "Controlling cell interactions by micropatterning in co-cultures: Hepatocytes and 3T3 fibroblasts," *J. Biomed. Mater. Res.*, Vol. 34, No. 2, 1997, pp. 189–199.

[19] Bhatia, S. N., Balis, U. J., Yarmush, M. L., and Toner, M., "Probing heterotypic cell interactions: Hepatocyte function in microfabricated co-cultures," *J. Biomaterials Science-Polymer Edition*, Vol. 9, No. 11, 1998, pp. 1137–1160.

[20] Bhatia, S. N., Balis, U. J., Yarmush, M. L., and Toner, M., "Effect of cell-cell interactions in preservation of cellular phenotype: Cocultivation of hepatocytes and nonparenchymal cells," *Faseb J.*, Vol. 13, No. 14, 1999, pp. 1883–1900.

[21] Ostuni, E., Kane, R., Chen, C. S., Ingber, D. E., and Whitesides, G. M., "Patterning mammalian cells using elastomeric membranes," *Langmuir*, Vol. 16, No. 20, 2000, pp. 7811–7819.

[22] Folch, A., and Toner, M., "Cellular micropatterns on biocompatible materials," *Biotech. Progress*, Vol. 14, No. 3, 1998, pp. 388–392.

[23] Folch, A., Jo, B. H., Hurtado, O., Beebe, D. J., and Toner, M., "Microfabricated elastomeric stencils for micropatterning cell cultures," *J. Biomed. Mater. Res.*, Vol. 52, No. 2, 2000, pp. 346–353.

[24] Khademhosseini, A., Suh, K. Y., Yang, J. M., Eng, G., Yeh, J., Levenberg, S., and Langer, R., "Layer-by-layer deposition of hyaluronic acid and poly-L-lysine for patterned cell co-cultures," *Biomaterials*, Vol. 25, No. 17, 2004, pp. 3583–3592.

[25] Khademhosseini, A., Langer, R., Borenstein, J., and Vacanti, J. P., "Microscale technologies for tissue engineering and biology," *Proc. Natl. Acad. Sci. USA*, Vol. 103, No. 8, 2006, pp. 2480–2487.

[26] Khademhosseini, A., Yeh, J., Eng, G., Karp, J., Kaji, H., Borenstein, J., Farokhzad, O. C., and Langer, R., "Cell docking inside microwells within reversibly sealed microfluidic channels for fabricating multiphenotype cell arrays," *Lab Chip*, Vol. 5, No. 12, 2005, pp. 1380–1386.

[27] Bhatia, S. N., Toner, M., Tompkins, R. G., and Yarmush, M. L., "Selective adhesion of hepatocytes on patterned surfaces," *Annals NY Acad. Sci.*, Vol. 745, 1994, pp. 187–209.

[28] Bhatia, S. N., Balis, U. J., Yarmush, M. L., and Toner, M., "Microfabrication of hepatocyte/fibroblast co-cultures: Role of homotypic cell interactions," *Biotech. Progress*, Vol. 14, No. 3, 1998, pp. 378–387.

[29] Delamarche, E., Bernard, A., Schmid, H., Michel, B., and Biebuyck, H., "Patterned delivery of immunoglobulins to surfaces using microfluidic networks," *Science*, Vol. 276, No. 5313, 1997, pp. 779–781.

[30] Takayama, S., McDonald, J. C., Ostuni, E., Liang, M. N., Kenis, P. J. A., Ismagilov, R. F., and Whitesides, G. M., "Patterning cells and their environments using multiple laminar fluid flows in capillary networks," *Proc. Natl. Acad. Sci. USA*, Vol. 96, No. 10, 1999, pp. 5545–5548.

[31] Takayama, S., Ostuni, E., LeDuc, P., Naruse, K., Ingber, D. E., and Whitesides, G. M., "Laminar flows—subcellular positioning of small molecules," *Nature*, Vol. 411, No. 6841, 2001, pp. 1016–1016.

[32] Wright, D., Rajalingam, B., Selvarasah, S., Dokmeci, M. R., and Khademhosseini, A., "Generation of static and dynamic patterned co-cultures using microfabricated parylene-C stencils," *Lab on a Chip*, Vol. 7, No. 10, 2007, pp. 1272–1279.

[33] Wright, D., Rajalingam, B., Karp, J. M., Selvarasah, S., Ling, Y., Yeh, J., Langer, R., Dokmeci, M. R., and Khademhosseini, A., "Reusuable, reversibly sealable parylene membranes for cell and protein patterning," *Journal of Biomedical Materials Research*, Part A, 2007, DOI: 10.1002/jbm.a.31281.

[34] Mrksich, M., "A surface chemistry approach to studying cell adhesion," *Chem. Soc. Rev.*, Vol. 29, No. 4, 2000, pp. 267–273.

[35] Yousaf, M. N., Houseman, B. T., and Mrksich, M., "Using electroactive substrates to pattern the attachment of two different cell populations," *Proc. Natl. Acad. Sci. USA*, Vol. 98, No. 11, 2001, pp. 5992–5996.

[36] Jiang, X. Y., Ferrigno, R., Mrksich, M., and Whitesides, G. M., "Electrochemical desorption of self-assembled monolayers noninvasively releases patterned cells from geometrical confinements," *J. Am. Chem. Soc.*, Vol. 125, No. 9, 2003, pp. 2366–2367.

[37] Yeo, W. S., Yousaf, M. N., and Mrksich, M., "Dynamic interfaces between cells and surfaces: Electroactive substrates that sequentially release and attach cells," *J. Am. Chem. Soc.*, Vol. 125, No. 49, 2003, pp. 14994–14995.

[38] Yamada, N., Okano, T., Sakai, H., Karikusa, F., Sawasaki, Y., and Sakurai, Y., "Thermoresponsive polymeric surfaces—control of attachment and detachment of cultured-cells," *Makromolekulare Chemie—Rapid Communications*, Vol. 11, No. 11, 1990, pp. 571–576.

[39] Okano, T., Yamada, N., Sakai, H., and Sakurai, Y., "A novel recovery-system for cultured-cells using plasma-treated polystyrene dishes grafted with poly(N-isopropylacrylamide)," *J. Biomed. Mater. Res.*, Vol. 27, No. 10, 1993, pp. 1243–1251.

[40] Yamato, M., Okuhara, M., Karikusa, F., Kikuchi, A., Sakurai, Y., and Okano, T., "Signal transduction and cytoskeletal reorganization are required for cell detachment from cell culture surfaces grafted with a temperature-responsive polymer," *J. Biomed. Mater. Res.*, Vol. 44, No. 1, 1999, pp. 44–52.

[41] Yamato, M., Konno, C., Kushida, A., Hirose, M., Utsumi, M., Kikuchi, A., and Okano, T., "Release of adsorbed fibronectin from temperature-responsive culture surfaces requires cellular activity," *Biomaterials*, Vol. 21, No. 10, 2000, pp. 981–986.

[42] Hirose, M., Yamato, M., Kwon, O. H., Harimoto, M., Kushida, A., Shimizu, T., Kikuchi, A., and Okano, T., "Temperature-responsive surface for novel co-culture systems of hepatocytes with endothelial cells: 2-D patterned and double layered co-cultures," *Yonsei Med. J.*, Vol. 41, No. 6, 2000, pp. 803–813.

[43] Yamato, M., Konno, C., Utsumi, M., Kikuchi, A., and Okano, T., "Thermally responsive polymer-grafted surfaces facilitate patterned cell seeding and co-culture," *Biomaterials*, Vol. 23, No. 2, 2002, pp. 561–567.

[44] Tsuda, Y., Kikuchi, A., Yamato, M., Nakao, A., Sakurai, Y., Umezu, M., and Okano, T., "The use of patterned dual thermoresponsive surfaces for the collective recovery as co-cultured cell sheets," *Biomaterials*, Vol. 26, No. 14, 2005, pp. 1885–1893.

[45] Edahiro, J., Sumaru, K., Tada, Y., Ohi, K., Takagi, T., Kameda, M., Shinbo, T., Kanamori, T., and Yoshimi, Y., "In situ control of cell adhesion using photoresponsive culture surface," *Biomacromolecules*, Vol. 6, No. 2, 2005, pp. 970–974.

[46] Shimaoka, S., Nakamura, T., and Ichihara, A., "Stimulation of growth of primary cultured adult-rat hepatocytes without growth-factors by coculture with nonparenchymal liver-cells," *Exp. Cell Res.*, Vol. 172, No. 1, 1987, pp. 228–242.

[47] Goulet, F., Normand, C., and Morin, O., "Cellular interactions promote tissue-specific function, biomatrix deposition and junctional communication of primary cultured-hepatocytes," *Hepatology*, Vol. 8, No. 5, 1988, pp. 1010–1018.

[48] Lawrence, M. B., McIntire, L. V., and Eskin, S. G., "Effect of flow on polymorphonuclear leukocyte endothelial-cell adhesion," *Blood*, Vol. 70, No. 5, 1987, pp. 1284–1290.

[49] Lawrence, M. B., Smith, C. W., Eskin, S. G., and McIntire, L. V., "Effect of venous shear-stress on Cd18-mediated neutrophil adhesion to cultured endothelium," *Blood*, Vol. 75, No. 1, 1990, pp. 227–237.

[50] Schrode, W., Mecke, D., and Gebhardt, R., "Induction of glutamine-synthetase in periportal hepatocytes by cocultivation with a liver epithelial-cell line," *Eur. J. Cell. Biol.*, Vol. 53, No. 1, 1990, pp. 35–41.

[51] Hammarback, J. A., McCarthy, J. B., Palm, S. L., Furcht, L. T., and Letourneau, P. C., "Growth cone guidance by substrate-bound laminin pathways is correlated with neuron-to-pathway adhesivity," *Dev. Biol.*, Vol. 126, No. 1, 1988, pp. 29–39.

[52] Stenger, D. A., Georger, J. H., Dulcey, C. S., Hickman, J. J., Rudolph, A. S., Nielsen, T. B., McCort, S. M., and Calvert, J. M., "Coplanar molecular assemblies of aminoalkylsilane and perfluorinated alkylsilane—characterization and geometric definition of mammalian-cell adhesion and growth," *J. Am. Chem. Soc.*, Vol. 114, No. 22, 1992, pp. 8435–8442.

[53] Lee, J.-S., Kaibara, M., Iwaki, M., Sasabe, H., Suzuki, Y., and Kusakabe, M., "Selective adhesion and proliferation of cells on ion-implanted polymer domains," *Biomaterials*, Vol. 14, No. 12, 1993, pp. 958–960.

[54] Soekarno, A., Lom, B., and Hockberger, P. E., "Pathfinding by neuroblastoma cells in culture is directed by preferential adhesion to positively charged surfaces," *NeuroImage*, Vol. 1, No. 2, 1993, pp. 129–144.

[55] Singhvi, R., Stephanopoulos, G., and Wang, D. I. C., "Effects of substratum morphology on cell physiology—review," *Biotechnol. Bioeng.*, Vol. 43, No. 8, 1994, pp. 764–771.

[56] Singhvi, R., Kumar, A., Lopez, G. P., Stephanopoulos, G. N., Wang, D. I. C., Whitesides, G. M., and Ingber, D. E., "Engineering cell-shape and function," *Science*, Vol. 264, No. 5159, 1994, pp. 696–698.

[57] Oakley, C., and Brunette, D. M., "Topographic compensation—guidance and directed locomotion of fibroblasts on grooved micromachined substrata in the absence of microtubules," *Cell Motility and the Cytoskeleton*, Vol. 31, No. 1, 1995, pp. 45–58.

[58] Kumar, A., and Whitesides, G. M., "Features of gold having micrometer to centimeter dimensions can be formed through a combination of stamping with an elastomeric stamp and an alkanethiol ink followed by chemical etching," *Appl. Phys. Lett.*, Vol. 63, No. 14, 1993, pp. 2002–2004.

[59] Duffy, D. C., McDonald, J. C., Schueller, O. J. A., and Whitesides, G. M., "Rapid prototyping of microfluidic systems in poly(dimethylsiloxane)," *Analytical Chemistry*, Vol. 70, No. 23, 1998, pp. 4974–4984.

[60] Anderson, J. R., Chiu, D. T., Jackman, R. J., Cherniavskaya, O., McDonald, J. C., Wu, H. K., Whitesides, S. H., and Whitesides, G. M., "Fabrication of topologically complex three-dimensional microfluidic systems in PDMS by rapid prototyping," *Analytical Chemistry*, Vol. 72, No. 14, 2000, pp. 3158–3164.

[61] McDonald, J. C., Duffy, D. C., Anderson, J. R., Chiu, D. T., Wu, H. K., Schueller, O. J. A., and Whitesides, G. M., "Fabrication of microfluidic systems in poly(dimethylsiloxane)," *Electrophoresis*, Vol. 21, No. 1, 2000, pp. 27–40.

[62] McDonald, J. C., Metallo, S. J., and Whitesides, G. M., "Fabrication of a configurable, single-use microfluidic device," *Analytical Chemistry*, Vol. 73, No. 23, 2001, pp. 5645–5650.

[63] Whitesides, G. M., Ostuni, E., Takayama, S., Jiang, X. Y., and Ingber, D. E., "Soft lithography in biology and biochemistry," *Ann. Rev. Biomed. Eng.*, Vol. 3, 2001, pp. 335–373.

[64] McDonald, J. C., Chabinyc, M. L., Metallo, S. J., Anderson, J. R., Stroock, A. D., and Whitesides, G. M., "Prototyping of microfluidic devices in poly(dimethylsiloxane) using solid-object printing," *Analytical Chemistry*, Vol. 74, No. 7, 2002, pp. 1537–1545.

[65] Khademhosseini, A., Yeh, J., Jon, S., Eng, G., Suh, K. Y., Burdick, J. A., and Langer, R., "Molded polyethylene glycol microstructures for capturing cells within microfluidic channels," *Lab Chip*, Vol. 4, No. 5, 2004, pp. 425–430.

[66] Khademhosseini, A., Suh, K. Y., Jon, S., Eng, G., Yeh, J., Chen, G. J., and Langer, R., "A soft lithographic approach to fabricate patterned microfluidic channels," *Analytical Chemistry*, Vol. 76, No. 13, 2004, pp. 3675–3681.

[67] Fukuda, J., Khademhosseini, A., Yeh, J., Eng, G., Cheng, J. J., Farokhzad, O. C., and Langer, R., "Micropatterned cell co-cultures using layer-by-layer deposition of extracellular matrix components," *Biomaterials*, Vol. 27, No. 8, 2006, pp. 1479–1486.

[68] Brody, J. P., Yager, P., Goldstein, R. E., and Austin, R. H., "Biotechnology at low Reynolds numbers," *Biophys. J.*, Vol. 71, No. 6, 1996, pp. 3430–3441.

[69] Weigl, B. H., and Yager, P., "Microfluidics—microfluidic diffusion-based separation and detection," *Science*, Vol. 283, No. 5400, 1999, pp. 346–347.

[70] Kenis, P. J. A., Ismagilov, R. F., and Whitesides, G. M., "Microfabrication inside capillaries using multiphase laminar flow patterning," *Science*, Vol. 285, No. 5424, 1999, pp. 83–85.

[71] Takayama, S., Ostuni, E., Qian, X. P., McDonald, J. C., Jiang, X. Y., LeDuc, P., Wu, M. H., Ingber, D. E., and Whitesides, G. M., "Topographical micropatterning of

poly(dimethylsiloxane) using laminar flows of liquids in capillaries," *Adv. Mater.*, Vol. 13, No. 8, 2001, pp. 570–574.

[72] Chiu, D. T., Jeon, N. L., Huang, S., Kane, R. S., Wargo, C. J., Choi, I. S., Ingber, D. E., and Whitesides, G. M., "Patterned deposition of cells and proteins onto surfaces by using three-dimensional microfluidic systems," *Proc. Natl. Acad. Sci. USA*, Vol. 97, No. 6, 2000, pp. 2408–2413.

[73] Jackman, R. J., Duffy, D. C., Cherniavskaya, O., and Whitesides, G. M., "Using elastomeric membranes as dry resists and for dry lift-off," *Langmuir*, Vol. 15, No. 8, 1999, pp. 2973–2984.

[74] Duffy, D. C., Jackman, R. J., Vaeth, K. M., Jensen, K. F., and Whitesides, G. M., "Patterning electroluminescent materials with feature sizes as small as 5 mu m using elastomeric membranes as masks for dry lift-off," *Adv. Mater.*, Vol. 11, No. 7, 1999, pp. 546–552.

[75] Ilic, B., and Craighead, H. G., "Topographical patterning of chemically sensitive biological materials using a polymer-based dry lift-off," *Biomed. Microdevices*, Vol. 2, No. 4, 2000, pp. 317–322.

[76] Pal, R., Sung, K. E., and Burns, M. A., "Microstencils for the patterning of nontraditional materials," *Langmuir*, Vol. 22, No. 12, 2006, pp. 5392–5397.

[77] Noh, H. S., Moon, K. S., Cannon, A., Hesketh, P. J., and Wong, C. P., "Wafer bonding using microwave heating of parylene intermediate layers," *J. Micromechanics and Microengineering*, Vol. 14, No. 4, 2004, pp. 625–631.

[78] Lotters, J. C., Olthuis, W., Veltink, P. H., and Bergveld, P., "The mechanical properties of the rubber elastic polymer polydimethylsiloxane for sensor applications," *J. Micromechanics and Microengineering*, Vol. 7, No. 3, 1997, pp. 145–147.

[79] Decher, G., "Fuzzy nanoassemblies: Toward layered polymeric multicomposites," *Science*, Vol. 277, No. 5330, 1997, pp. 1232–1237.

[80] Vazquez, E., Dewitt, D. M., Hammond, P. T., and Lynn, D. M., "Construction of hydrolytically-degradable thin films via layer-by-layer deposition of degradable polyelectrolytes," *J. Am. Chem. Soc.*, Vol. 124, No. 47, 2002, pp. 13992–13993.

[81] Thierry, B., Winnik, F. M., Merhi, Y., and Tabrizian, M., "Nanocoatings onto arteries via layer-by-layer deposition: Toward the in vivo repair of damaged blood vessels," *J. Am. Chem. Soc.*, Vol. 125, No. 25, 2003, pp. 7494–7495.

[82] Kidambi, S., Chan, C., and Lee, I. S., "Selective depositions on polyelectrolyte multilayers: Self-assembled monolayers of m-dPEG acid as molecular template," *J. Am. Chem. Soc.*, Vol. 126, No. 14, 2004, pp. 4697–4703.

[83] Kidambi, S., Lee, I., and Chan, C., "Controlling primary hepatocyte adhesion and spreading on protein-free polyelectrolyte multilayer films," *J. Am. Chem. Soc.*, Vol. 126, No. 50, 2004, pp. 16286–16287.

[84] Mrksich, M., Dike, L. E., Tien, J., Ingber, D. E., and Whitesides, G. M., "Using microcontact printing to pattern the attachment of mammalian cells to self-assembled monolayers of alkanethiolates on transparent films of gold and silver," *Exp. Cell Res.*, Vol. 235, No. 2, 1997, pp. 305–313.

[85] Roberts, C., Chen, C. S., Mrksich, M., Martichonok, V., Ingber, D. E., and Whitesides, G. M., "Using mixed self-assembled monolayers presenting RGD and (EG)(3)OH groups to characterize long-term attachment of bovine capillary endothelial cells to surfaces," *J. Am. Chem. Soc.*, Vol. 120, No. 26, 1998, pp. 6548–6555.

[86] Houseman, B. T., and Mrksich, M., "Efficient solid-phase synthesis of peptide-substituted alkanethiols for the preparation of substrates that support the adhesion of cells," *J. Organic Chem.*, Vol. 63, No. 21, 1998, pp. 7552–7555.

[87] Houseman, B. T., and Mrksich, M., "The microenvironment of immobilized Arg-Gly-Asp peptides is an important determinant of cell adhesion," *Biomaterials*, Vol. 22, No. 9, 2001, pp. 943–955.

[88] Yamato, M., Kwon, O. H., Hirose, M., Kikuchi, A., and Okano, T., "Novel patterned cell coculture utilizing thermally responsive grafted polymer surfaces," *J. Biomed. Mater. Res.*, Vol. 55, No. 1, 2001, pp. 137–140.

[89] Hirose, M., Kwon, O. H., Yamato, M., Kikuchi, A., and Okano, T., "Creation of designed shape cell sheets that are noninvasively harvested and moved onto another surface," *Biomacromolecules*, Vol. 1, No. 3, 2000, pp. 377–381.

[90] Uchida, K., Sakai, K., Ito, E., Kwon, O. H., Kikuchi, A., Yamato, M., and Okano, T., "Temperature-dependent modulation of blood platelet movement and morphology on poly(N-isopropylacrylamide)-grafted surfaces," *Biomaterials*, Vol. 21, No. 9, 2000, pp. 923–929.

[91] Feil, H., Bae, Y. H., Jan, F. J., and Kim, S. W., "Effect of comonomer hydrophilicity and ionization on the lower critical solution temperature of N-isopropylacrylamide copolymers," *Macromolecules*, Vol. 26, No. 10, 1993, pp. 2496–2500.

[92] Takei, Y. G., Aoki, T., Sanui, K., Ogata, N., Okano, T., and Sakurai, Y., "Temperature-responsive bioconjugates .2. molecular design for temperature-modulated bioseparations," *Bioconjugate Chem.*, Vol. 4, No. 5, 1993, pp. 341–346.

[93] Vercruysse, K. P., and Prestwich, G. D., "Hyaluronate derivatives in drug delivery," *Critical Reviews in Therapeutic Drug Carrier Systems*, Vol. 15, No. 5, 1998, pp. 513–555.

[94] Luo, Y., Kirker, K. R., and Prestwich, G. D., "Cross-linked hyaluronic acid hydrogel films: New biomaterials for drug delivery," *J. Controlled Release*, Vol. 69, No. 1, 2000, pp. 169–184.

[95] Suh, K. Y., Seong, J., Khademhosseini, A., Laibinis, P. E., and Langer, R., "A simple soft lithographic route to fabrication of poly(ethylene glycol) microstructures for protein and cell patterning," *Biomaterials*, Vol. 25, No. 3, 2004, pp. 557–563.

[96] Suh, K. Y., Khademhosseini, A., Yang, J. M., Eng, G., and Langer, R., "Soft lithographic patterning of hyaluronic acid on hydrophilic substrates using molding and printing," *Adv. Mater.*, Vol. 16, No. 7, 2004, pp. 584–588.

[97] Hui, E. E., and Bhatia, S. N., "Micromechanical control of cell-cell interactions," *Proc. Natl. Acad. Sci. USA*, Vol. 2007, pp. 0608660104.

[98] Ito, A., Takizawa, Y., Honda, H., Hata, K. I., Kagami, H., Ueda, M., and Kobayashi, T., "Tissue engineering using magnetite nanoparticles and magnetic force: Heterotypic layers of cocultured hepatocytes and endothelial cells," *Tissue Eng.*, Vol. 10, Nos. 5–6, 2004, pp. 833–840.

[99] Tsang, V. L., Chen, A. A., Cho, L. M., Jadin, K. D., Sah, R. L., DeLong, S., West, J. L., and Bhatia, S. N., "Fabrication of 3-D hepatic tissues by additive photopatterning of cellular hydrogels," *Faseb J.*, Vol. 21, No. 3, 2007, pp. 790–801.

[100] Albrecht, D. R., Underhill, G. H., Wassermann, T. B., Sah, R. L., and Bhatia, S. N., "Probing the role of multicellular organization in three-dimensional microenvironments," *Nature Methods*, Vol. 3, No. 5, 2006, pp. 369–375.

[101] Mikos, A. G., Herring, S. W., Ochareon, P., Elisseeff, J., Lu, H. H., Kandel, R., Schoen, F. J., Toner, M., Mooney, D., Atala, A., Dyke, M. E. V., Kaplan, D., and Vunjak-Novakovic, G., "Engineering complex tissues," *Tissue Eng.*, Vol. 12, No. 12, 2006, pp. 3307–3339.

[102] Yeh, J., Ling, Y. B., Karp, J. M., Gantz, J., Chandawarkar, A., Eng, G., Blumling, J., Langer, R., and Khademhosseini, A., "Micromolding of shape-controlled, harvestable cell-laden hydrogels," *Biomaterials*, Vol. 27, No. 31, 2006, pp. 5391–5398.

CHAPTER 5

Micro- and Nanofabricated Scaffolds for Three-Dimensional Tissue Recapitulation

James Norman, Sarah Tao, Ketul Popat, Carlos Lopez, Kristen La Flamme, Rahul Thakar, and Tejal Desai

5.1 Introduction

When cells are cultured outside the body in an artificial environment, they inevitably lose many of their in vivo characteristics. In the human body, cells are in contact with a complex combination of proteins, proteoglycans, and glycoproteins that make up the basement membrane and the extracellular matrix of a tissue. These acellular portions of a cell's natural environment vary from tissue to tissue in both composition and structure.

Moreover, cells are organized with precise geometric complexity such that structure often dictates function. In most tissues, there are also multiple cell types present, which each play their own important role in regulating the function of the other cells present. Then, finally, a cell lives in a dynamic four-dimensional environment: three spatial dimensions and time. Even trying to capture the in vivo aspects of a cell in one moment in time (thereby reducing the complexity to a three-dimensional problem) becomes a monumentally complex undertaking. This presents a huge challenge for scientists studying these cells or for bioengineers who want to build artificial tissues. Given the multilevel hierarchical structure of tissues, researchers look for ways to better recapitulate tissue structure and structure.

When cultured on flat surfaces, such as the standard petri dish, cells will often lose their native morphology. Cells typically flatten and spread out on flat surfaces, and as a population, there is little cohesive organization. The problems run deeper than just the appearance of the cells, however. By altering the cell's microenvironment, one also alters the cell's genetic expression. Genes that are normally expressed can be upregulated or downregulated, while genes that normally are not expressed at all may be turned on. This may lead to very different functionality in cultured cells compared to their in vivo counterparts.

5.2 Microfabricated Interfaces

One method that researchers have used to control the way cells grow in vitro is to build a physical environment at the scale of individual cells. Many cells are known to respond to physical cues that can guide their migration and organization. In order to build at the size scale of cells, researchers and engineers turned to the semiconductor industry for inspiration. Photolithography, used for patterning the complex layers of integrated circuits, allows for building structures on flat substrates in photoresist. Researchers working with cells have since learned to transfer this pattern into elastomeric polymers, such as silicone, for use as a cell-culture substrate that now plays an active role in modulating cell behavior [1]. By using these techniques, the complex patterns of cells found in the body have been able to be recreated for a number of tissues, and often, by recreating the spatial complexity, the cells begin to express genes in a manner in keeping with their natural functionality.

While photolithography of flat substrates (typically silicon wafers) has revolutionized the way biologists and bioengineers work with cells, there is one major problem: these techniques are mainly planar ones. These surfaces still lack the third dimension in which all cells live in the body. It has been shown that this third dimension is extremely important to regulating cell function and the can in itself bring cells closer to their natural functionality [2–5]. The most common way of creating three-dimensional cultures in the lab is to mix the cells into a biological polymer such as collagen or, to create an even more complex milieu, a mixture of extracellular matrix proteins that match the target tissue may be used. By doing this, the cells have now moved from a two-dimensional to a three-dimensional culture. However, these amorphous gels provide little in the way of organizational structure to the cells. Cells still have little cohesive ultrastructural organization in the gel, and while they come closer to their native counterparts, there is still a lack of control in regulating the structural and chemical microenvironment. This chapter provides several examples of ways in which micro-/nanofabrication techniques coupled with materials engineering can be used to create organized three-dimensional culture environments (Figure 5.1).

5.2.1 Microtopography for Cardiac Tissue Engineering

The heart attack, or myocardial infarction, is the culmination of ischemia due to an upstream blockage of the blood vessel. The necrotic damage that results from the lack of oxygen severely stresses and damages the heart muscles. In severe myocardial

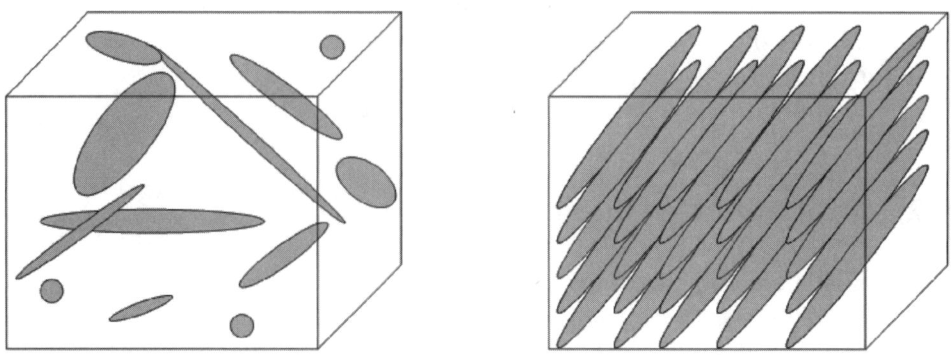

Figure 5.1 (a)Schematic of cells in a typical three-dimensional culture. Cells are randomly organized and take on a variety of morphologies. (b) Schematic of a proposed three-dimensional culture system that organizes the cells into a cohesive pattern and guides all cells to a common morphology.

infarctions, the initial damage ultimately yields heart failure and death. In these severe cases, lifestyle changes and medications alone are insufficient to overcome this damage. To combat this malady, cell-replacement therapy has not only become the vogue but has proven to be effective. The critical elements in the efficacy of this therapy are the ability for the transplanted cells to survive and for the transplanted cells to functionally couple and align for force transmission and cardiac function. To aid in the development of a therapy, microfabrication techniques have been developed to control the structural niches of cells. This work complements other research elucidating the importance of chemical induction via growth factors and cell-matrix adhesion for control of cell proliferation, differentiation, and apoptosis [6].

Further investigations with controlled niches that promote cardiac myocyte functionality have led to the development of microtopographies that increase cardiac myocyte alignment, decrease fibroblast proliferation, and up-regulate connexin 43 expression [7–10]. Microtopography has been shown previously to modulate cell shape, which in turn modulated vascular smooth muscle cell shape and organization, as well as proliferation [11, 12]. For cardiac muscle, however, a more three-dimensional microarchitecture is required for efficacious therapies. Initial efforts utilized a series of microchannels with microposts protruding out of the channels. The microchannels give topographical alignment cues that help minimize fibroblast proliferation, and the microposts give topographical mechanical cues for cardiac myocyte force transduction.

One way to extend two-dimensional micropatterning into three dimensions is the fusion of an internal microfabricated scaffold with a precisely defined architecture with a matrix that can support cell growth. This composite structure leads to an engineered construct that takes on a three-dimensional organization not possible with standard two- or three-dimensional culture techniques. To control and guide cells, a spatial differential inside the matrix is needed in the environment local to a cell. By using composite systems for cell culture, the internal stresses the cells apply to the matrix can be resisted, causing the cells to migrate and organize in a pattern laid out by the internal structure of the filler material. Photolithography allows for the construction of geometrically precise features. Unlike in two-dimensional cultures, these features can be removed from the substrate and put into the interior of a biological matrix along with cells. If the features provide enough of a physical change from the surrounding matrix, in porosity or stiffness for example, over time, the cells will interact with these internal features and be guided to grow according to their geometry. A common pattern of cells found in the body is a parallel arrangement of cells. This can be seen in tissues such as muscle or layers of skin. This pattern has been easily replicated in two-dimensional culture through the use of photolithographically created microgrooves on a surface [13]. A similar pattern can be used to extend this into a three-dimensional culture. High-aspect ratio walls created using a combination of photolithography and soft-lithography can be embedded into a collagen matrix containing fibroblasts (Figure 5.2) [14]. Since the walls are made of a material that is impermeable to cell migration, the cells are restricted from migrating and grow in only certain areas of the matrix. The width of the walls (less than the length of the cell) constrains the entire fibroblast population to extend in parallel. The walls themselves run the entire height of the matrix, allowing for

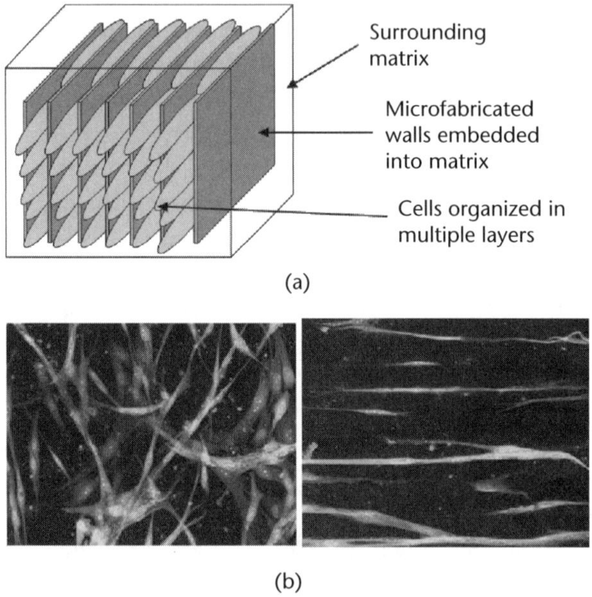

Figure 5.2 (a) Schematic of a composite three-dimensional scaffold containing microfabricated wall structures (dark gray) that guide the cells (represented by light-gray ovals) to organize into parallel layers throughout the entire matrix. (b) Left: Control collagen gel containing fibroblasts, stained with propidium ioded (nucleus) and FITC (cell body). Cells are randomly organized. Right: Gel containing microfabricated walls. Fibroblasts are all organized in parallel with each other in multiple layers. Images are stacks of confocal images taken through the height of the matrix (100 μm).

many layers of fibroblasts to align according to the pattern. This composite system provides the biological milieu in three dimensions that is more natural to the cells and provides the contact guidance of patterned features so effective in two-dimensional cultures into one system.

More recently, we have developed (in collaboration with the Russell Lab at the University of Illinois, Chicago) injectable microrod scaffolds that can provide topographical cues in three dimensions to modulate fibroblast proliferation and direct cardiac myocyte organization and contractility. Microfabricated rods, microrods, can be fabricated to the dimensions of an individual cell (approximately 100 microns long with a 15×15 micron cross section). When incorporated into a biological matrix along with the cells, the cells will come in contact with the microrods (Figure 5.3). The microrods are stiff compared to the surrounding matrix. Because of this, the cells cannot migrate through the microrods, and they attach and remain attached in a manner similar to that of micropegged surfaces.

This interaction of fibroblasts with microrods appears to halt the proliferation of the fibroblasts. Fibroblasts are cells that are important in the body for maintaining the proper extracellular matrix environment by producing collagens. This is also the reason they produce scar tissue. In vitro, fibroblasts will proliferate excessively. In a biological matrix, fibroblasts migrate through the matrix and will proliferate and increase in number over time [15]. In bioengineered tissues where fibroblasts need to be present, but only in a limited number, this can be a challenging issue. How can their proliferation be controlled so as not to overtake other cells in the cul-

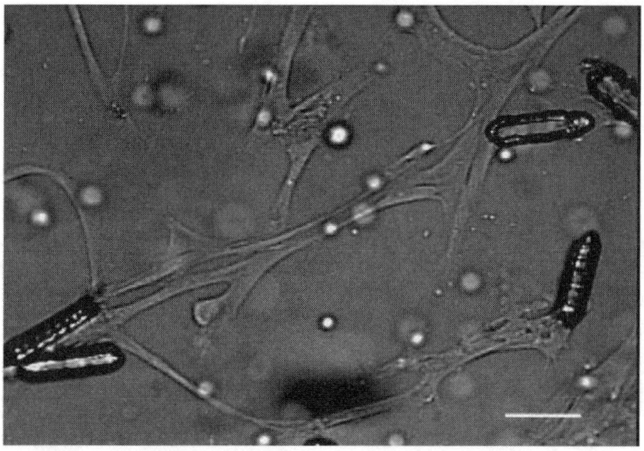

Figure 5.3 Fibroblasts in gel attaching to microrods that are 100 μm in length. Cells are seen bridging between rods and orienting in response to rod microstructure.

ture? It turns out that microstructural cues could be one way. In matrices seeded with microrods and fibroblasts, the population of fibroblasts after one week does not increase significantly over that of the initial seeding density. Without the incorporation of the microrods, however, fibroblasts increased in number over this period. This modulation of fibroblast phenotype can be accomplished with a very small number of microrods present in the culture (100:1 ratio of cells to microrods). Due to their extremely low volume, the microrods make up a very small volume fraction of the entire matrix (less than 10^{-3}). Therefore, the bulk properties of the matrix do not change. Using this method, bioengineers can control the cell population while still tailoring the mechanics of the bulk artificial tissue to the necessary standards of the native tissue. Cell modulation and material properties have been decoupled through the use of a microfabricated composite tissue scaffold.

5.2.2 Blood Vessel Microengineering

During the course of one's lifetime, blood vessels critical to life can become damaged. Damage can manifest itself as a loss of elasticity, occlusion of the lumen, high blood pressure, and many other means. A close relationship exists among the levels in the body of blood cholesterol, triglycerides, and other lipids and the development of atherosclerosis, a disorder caused by plaques containing cholesterol deposited on artery walls [16]. Atherosclerosis leads to hardening and roughening of the arteries, as well as to their occlusion. The cholesterol coats the blood vessels much like the fouling of pipes in a chemical plant. If this situation continues over a period of time, cardiovascular disease (CVD) will claim another victim, whether it is from a stroke, heart attack, or any other form of heart disease [17, 18]. A dislodged plaque occluding an artery can shunt the blood supply to the brain causing a stroke. A stenosis, a constriction of the blood vessel, can also cut off the blood supply to the heart causing a heart attack. The description of heart disease pales next to the number of victims it claims. As of 2006, approximately 70.1 million Americans have a form of CVD, which is approximately one in every five Americans.

The fabrication of a functional small-diameter blood vessel analogue will have implications in the treatment of vascular disease, as well as facilitate in drug testing and elucidating the physiological characteristics of blood vessels. One of the most difficult aspects of the vasculature to recreate in biological constructs has been the strength of native vessels to withstand the physiological pressures. The medial layer of blood vessel is composed of multiple layers of smooth muscle cells arranged in alternating spiral layers [19]. The organization of extracellular matrix (ECM) proteins, such as collagen, fibronectin, and elastin, also plays an important role in vessel strength and integrity. The medial layer provides the strength, elasticity, and contractility to the vessel. It is believed that this distinct architecture plays a role in the function of the medial layer of the vessel.

Since the medial layer has a distinct three-dimensional architecture that changes from cell layer to cell layer, a different method of building up the multilayer tissue needs to be used from that described previously. Each cell layer has its own arrangement in relation to the others. Within each layer, however, the cells are organized in parallel rows so that they are all running in the same direction (Figure 5.4). This is a common organizational motif that has been studied using microfabricated channels in substrates for cell monolayers. Cell patterning using topographical cues in the form of grooves on poly(dimethylsiloxane) (PDMS) membranes has shown promise in achieving VSMC alignment, eliciting in vivo–like VSMC morphology and promoting increased ECM remodeling (Figure 5.5) [12, 20]. By aligning the network of actin filaments throughout the population of cells, the entire unit will act to contract in a single direction, thereby providing tissue-appropriate functionality. Studies have shown, though, that with increasing pattern size, control over VSMC behavior

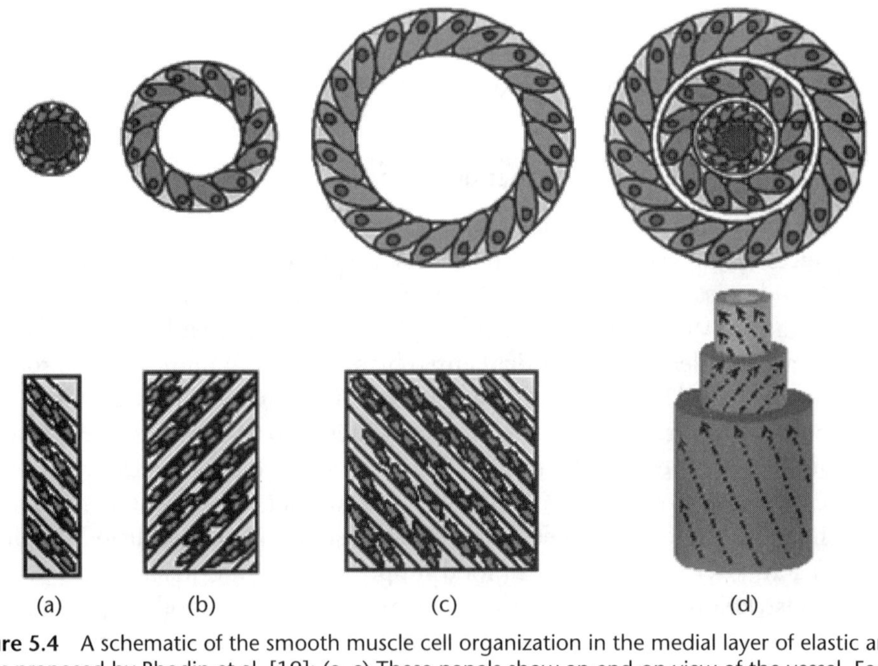

(a) (b) (c) (d)

Figure 5.4 A schematic of the smooth muscle cell organization in the medial layer of elastic arteries as proposed by Rhodin et al. [19]: (a–c) These panels show an end-on view of the vessel. Each layer is arranged in a spiral pattern (top) that, if unrolled, would show the cells organized in parallel strips (bottom). (d) By layering VSMC seeded, micropatterned membranes and rolling into a cylinder, the herringbone pattern of VSMC in the vessel can be recreated. (Image reproduced with permission from [12].)

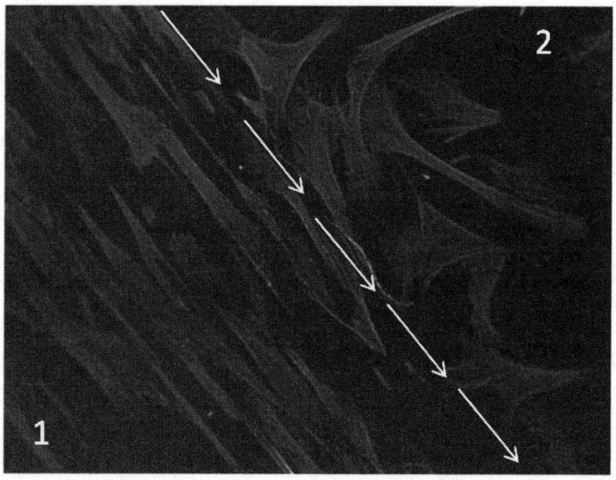

Figure 5.5 Rhodamine-phalloidin staining of the F-actin of VSMCs on a PDMS substrate. Area 1 is patterned with 48 μm-wide channels. Area 2 is unpatterned. The white arrows indicate the border between the two surfaces. The actin in the cells in area 1 have a high order of alignment, while those in area 2 have a random network of actin filaments that are not organized as a population. (Image reproduced with permission from [12].)

is lost [11]. It is therefore important to be able to incorporate precise small-scale microcues into tissue-engineered structures when tight control over cell behavior is necessary.

In order to build up the three-dimensional architecture that could lead to an engineered small-diameter blood vessel with proper strength and functionality, individual patterned VSMC layers can be stacked on top of each other to build a medial layer with the proper number of layers and helical pitch between layers, as described in Figure 5.4. The ability to control the number of layers, as well as the pitch between layers, is important because it is known that the pitch and thickness can change between vessels found in different tissues.

An important part of these stacked membranes will be the ability for nutrients to diffuse in through the layers and for cellular waste to diffuse out so that the cells in interior layers can survive. In traditional scaffolds where the volume of the scaffold is made from one piece of material instead of being built up layer by layer, diffusion of nutrients to the center of the scaffold is accomplished through a porous network in the material. Without creating a porous network between the layers in the stacked membranes, the patterned cells in the interior of the construct will die due to inadequate nutrient supply. Approaches to the introduction of porous architectures into three-dimensional scaffolds include particulate leaching of salts and sugars, phase separation, and freeze drying [21]. Current porous techniques, however, produce relatively large pores on the order of tens of microns or larger. At these size scales, cells conform to the porous microarchitecture presented [22], which, especially in the case of highly organized tissues such as blood vessels, does not provide the precise control needed to organize cells into in vivo–like formations. By creating nano-size pores, the feature size of the pores is reduced several orders of magnitude below that of the surface pattern.

In order to accomplish this, a novel particulate leaching technique was developed where poly(lactic-co-glycolic acid) (PLGA) micro/nanospheres were used as a porogen, and the degradation properties of both the porogen and the bulk scaffold material polycaprolactone (PCL) were utilized to leach the embedded particles [23]. The microchannels necessary for VSMC alignment can be created in PCL that is loaded with the PLGA spheres through a hot embossing method [24]. The PLGA spheres embedded in thin micropatterned PCL scaffolds were shown to be well distributed and numerous, indicating potential for a well-connected porous network (Figure 5.6). Pore shape and size relative to grooved micropatterning was investigated for both PLGA-leached and traditional salt-leached scaffolds where PLGA-leached scaffolds had small micron-/submicron scale circular pores, while salt-leached scaffolds had larger, irregularly shaped, disruptive pores. The presence of interconnected porous networks that span the width of the scaffolds was demonstrated by an increase in media diffusion through porous scaffolds. It was also confirmed that any changes in scaffold topography caused by the introduction of pores using the novel PLGA-leaching technique does not significantly alter VSMC alignment. The ability to pattern cells on porous scaffolds will facilitate in transferring microtechnology from simple two-dimensional substrates to complex three-dimensional architectures.

5.2.3 Microtextured Thin-Film Scaffolds for Retinal Tissue Engineering

The retina is composed of a highly polarized, multilayered architecture. Light must travel through each retinal layer before reaching the outer segments of photoreceptors where phototransduction occurs. The pathophysiology of retinitis pigmentosa and age-related macular degeneration both result in degeneration of the photoreceptor cell layer and permanent visual loss in the corresponding region of the visual field. Currently, photoreceptor loss is untreatable, although a variety of approaches to preserve or restore vision are under investigation. Research efforts to advance treatments and cures for retinal degenerative diseases are closely focused on developing therapies that can ameliorate the genetic dysfunction or slow and halt

Figure 5.6 Three-dimensional confocal laser-scanning microscopy image of a PCL-patterned scaffold containing PCL nanospheres mixed in at a ratio of 4:1 PCL to PLGA spheres. Texas Red was mixed into the PLGA during the emulsion process in order to visualize them. Image at 20×.

the progress of the disease. These methods include gene therapy [25–27], delivery of pharmaceutical agents [28], and retinal transplantation [29–31]. One of the most promising therapies for late-stage retinal degenerations involves the delivery of stem or progenitor cells to the outer retina.

Retinal progenitor cells (RPCs), self-renewing cells capable of differentiating into several retinal cell types, have shown promise for replacing photoreceptor cells in experimental models of retinal degeneration. The retina is composed of seven different cell types that include ganglion cells, amacrine cells, bipolar cells, horizontal cells, rod and cone photoreceptor cells, and Muller glial cells. Thymidine birth-dating studies have shown that each cell type emerges from RPCs in an invariant temporal sequence [32]. Progenitor cells in the developing retina are multipotential, and their choice of cell fate is governed by intrinsic, as well as by extrinsic, signals from the microenvironment [32].

RPCs have recently been successfully isolated from embryonic and adult eyes in rodents and humans [33–36]. Maintenance of stem cells in an undifferentiated, multipotent state during in vitro expansion is challenging. However, it has been previously shown that RPC proliferation and long-term expansion can be achieved in vitro. Upon removal of mitogens and serum, RPCs have an excellent survival rate and can be differentiated along neuronal or glial pathways [37].

RPCs have shown promise for replacing photoreceptor cells in experimental models of retinal degeneration. Numerous studies have been performed to characterize the interactions between precursor cells and the retina after donor cell implantation into the eye. It has been shown that an actively degenerating retina may release factors that induce migration of donor cells into the retina, where in response to environmental cues and factors not present in normal retinas or from retinas in which the degeneration has progressed to completion, they undergo terminal differentiation [38].

However, a number of obstacles to the development of a practical cell-based therapy for the retina remain, including immediate reflux at the time of injection and massive death of donor cells following the standard bolus injection method. Several studies have demonstrated improvements in stem and progenitor survival when the cells are delivered to the subretinal space on polymer scaffolds [39–41]. Compared to cell-injection methods, RPCs cultured on polymer scaffolds prior to subretinal transplantation in rho-/- mice showed a ten- and fourteenfold increase in survival and cell delivery, respectively. However, due to the physical constraint of the subretinal space, the use of thick scaffolds (> 100 μm) increased the incidence of trauma during the transplantation procedure, implicating the need for an alternate approach. Although RPCs have been shown to differentiate into retinal-specific neurons after in vivo transplantation, they have not been able to maintain morphologic development, lamination, or extensive integration with the host retina. In addition, typical bolus injection of RPCs into the subretinal space has been shown to lead to a large degree of cell death rather than differentiation into functional photoreceptors. Furthermore, injection alone is not sufficient to recreate the complex architecture of the retina, particularly when multiple retinal layers have been lost or disrupted.

Surface-modified biomaterials may provide a new means for delivering retinal progenitor and stem cells to the outer retina. The highly ordered physical characteristics of the retinal microenvironment can be approximated by utilizing microelectromechanical systems (MEMS)–based technology to provide topographical cues that influence RPC survival, attachment, migration, and differentiation. Traditional MEMS techniques have allowed for the fabrication of ultrathin poly(methyl methacrylate) (PMMA) films that contain specific topographies [42]. The choice of PMMA for use as a scaffold in the present study was based on its use as a resist in MEMS fabrication, incorporation of a functional methyl ester group for potential modification, and, most importantly, its biocompatibility with ocular tissue, which has been established through its long use as a material for contact and intraocular lenses [43]. Micropatterned PMMA thin-film scaffolds were fabricated to contain through pores using a dual process of photolithography and reactive ion etching [Figure 5.7(a)]. After the etching process, the diameter of the pores was found to be approximately 11 μm with an interpore distance of 63 μm [Figure 5.7(b)]. The thickness of the ultra-thin-film scaffold was measured to be approximately 6 μm by profilometry (Dektak 8, Veeco, Tucson, AZ). During the transplantation process, a focal retinal detachment is produced at the transplantation site. The ultrathin (6 μm) PMMA is well suited for incorporation into this restricted microenvironment. Due to their thinness, these film scaffolds provide a means to increase the ease of delivery and reduce the risk of trauma, while allowing the scaffold to rest against the retina, thereby enhancing potential integration with the host. The goal of this study was to determine the survival, migration, and differentiation of RPCs [isolated from postnatal day-one enhanced green fluorescent protein positive (GFP+) transgenic C57BL/6 mice] adherent to unpatterned (nonporous) and patterned (porous) micromachined PMMA scaffolds transplanted to the subretinal space of C57BL/6 mice.

It was demonstrated that RPCs show similar survival and proliferation rates on micromachined PMMA and polystryene culture well surfaces in vitro. Both nonporous and porous PMMA demonstrated full biocompatibility with RPCs in

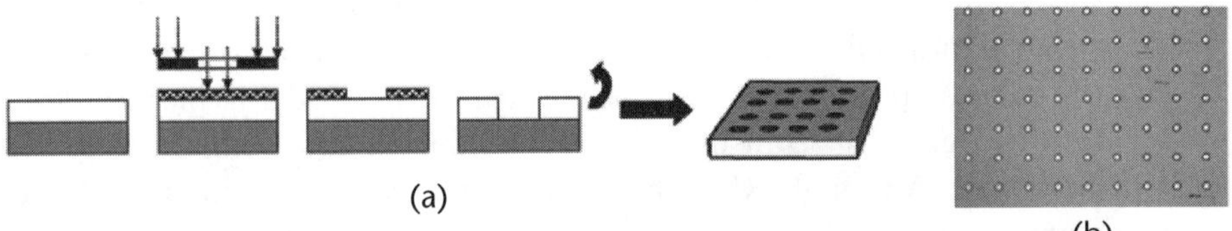

(a) (b)

Figure 5.7 (a) Schematic of ultra-thin-film PMMA scaffold fabrication. PMMA and positive photoresist are first spun on a wafer. The photoresist is exposed to ultraviolet light through a mask and developed. Areas of PMMA unmasked by photoresist are then dry-etched. The thin-film PMMA is then lifted off the wafer in a single sheet. (b) PMMA scaffold, 6 μm thick, containing pores approximately 11 μm in diameter and spaced 63 μm apart. (Reproduced by permission of the Royal Society of Chemistry.)

culture with no observable difference in the survival or proliferation of RPCs cultured on tissue-culture polystyrene. Although RPC adherence and survival were nearly identical in culture on both types of scaffolds, transplantation with nonporous scaffolds showed limited RPC retention. During the transplantation process, nonporous scaffolds lost the majority of their RPCs. Porous PMMA scaffolds demonstrated consistently higher retention of GFP+ RPCs (Figure 5.8). The porous topography allowed for RPC adherence through transplantation to the posterior eye for up to four weeks. Under microscopic examination, many RPCs appeared closely bound to the porous scaffold and exhibited signs of survival across the entire surface. Enhanced RPC attachment to porous scaffolds further provided a cytoarchtectural microenvironment permissive for eventual cell migration into host retinal layers. Adhesion of RPCs to porous PMMA during the process of subretinal transplantation demonstrated the utility of a surface topography comprising pores 11 μm in diameter. The average RPC diameter is slightly less than 10 μm. Across the surface of a 1 mm graft containing approximately two hundred pores, it seems probable that a mechanism of cell anchorage would involve the insertion of individual cells or their processes into pores. RPCs embedded into pores could remain attached to the scaffold during transplantation while also serving as an anchorage point for surrounding RPCs through cell-to-cell contacts. It is probable that the majority of RPCs lost from polymers in vivo actually lose adherence during the mechanical stressors of the transplantation process.

Nonporous grafts were associated with RPC integration into the host retina in only one out of five transplant recipients. A very limited number of RPCs were visibly integrated into host retinal layers in the GFP+ RPC retaining nonporous transplantation [Figure 5.9(a)]. The limited integrated RPCs from the nonporous membrane extended relatively short processes and failed to exhibit immunohistochemical markers of retinal differentiation [Figure 5.9(d)]. In the remaining four nonporous transplantations there were no GFP+ RPCs visible in any

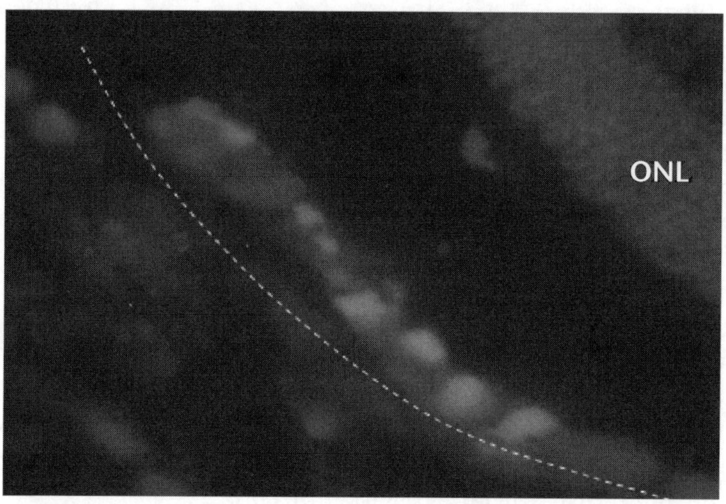

Figure 5.8 Micromachined porous PMMA with GFP+ RPCs attached to its surface inserted into the subretinal space of a C57bl/6 host. The dashed line traces the approximate contour of the membrane. ONL = outer nuclear layer. Bar = 50 μm. (Reproduced by permission of the Royal Society of Chemistry.)

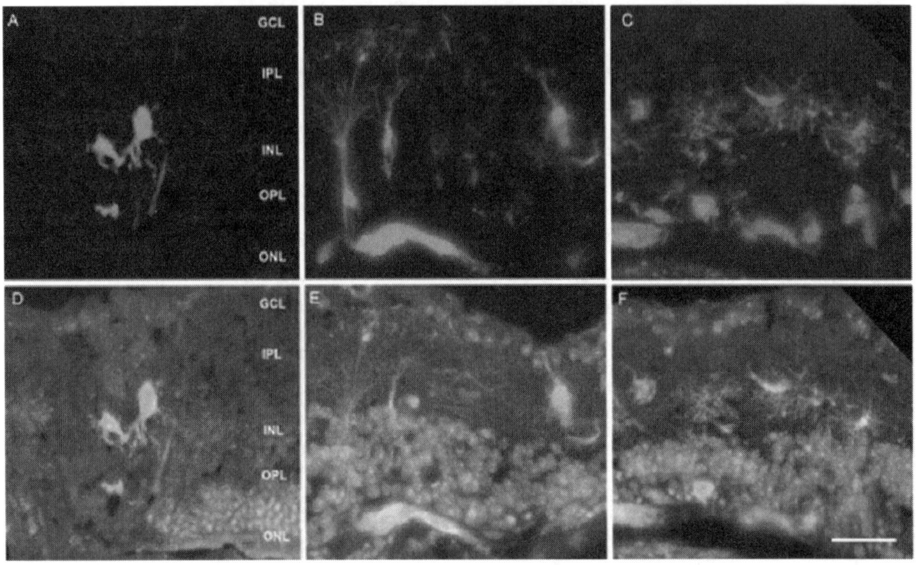

Figure 5.9 Porous PMMA scaffold RPC retention leads to enhanced integration and differentiation in host retina. (a) After four weeks in vivo, a nonporous PMMA RPC graft is analyzed for host retina RPC integration. Few, (~3) per 12 μm section, GFP+ RPCs appear integrated into the INL and ONL region from the nonporous graft. (b, c) A significantly higher number (~45) of GFP+ RPCs integrate into all host retinal layers from porous grafts. RPCs integrated from porous grafts exhibit a range of retinal neural morphologic differentiation. (d) Immunohistochemical analysis revealed that RPCs integrated from nonporous grafts failed to express GFAP. (e, f) RPCs integrated from porous grafts with morphologies that either spanned all retinal layers or branched radially in the inner plexiform layer expressed GFAP. Bar = 50 μm. (Reproduced by permission of the Royal Society of Chemistry.)

region closely adjacent to or incorporated into host retinal layers. Porous grafts allowed for RPC integration in four out of five transplantations, a fourfold increase when compared to nonporous grafts. The most robust porous graft transplants exhibited a greater number of integrated GFP+ RPCs (n = 80) across host retinal layers [Figure 5.9(b, c)]. Initial fluorescent analysis of GFP+ RPCs originating from porous grafts and integrated into host retinal layers revealed morphologic differentiation consistent with retinal neurons. Differentiated RPCs in the inner retina developed radial processes similar to those of astrocytes or amacrine cells. Some RPC-derived cells exhibited morphology that spanned the radial extent of the retina, similar to Mueller cells, and these profiles labeled positively for the glial cell marker GFAP [Figure 5.9(e, f)]. In GFP+ RPCs, expression of GFAP is likely an indication of differentiation toward either a retinal astrocytre or Mueller glial cell fate [44]. These cell types have many functions in the eye and work directly to facilitate retinal protection and homeostasis. In the outer retina, integrated RPCs localized to the region of the outer limiting membrane and expressed the retinal-specific protein recoverin. Recoverin is normally expressed only by photoreceptors and a subset of bipolar cells. RPCs in the earlier stages of migration from porous scaffolds into the outer retina extended processes from scaffolds to retina and expressed the progenitor marker nestin and the early neuronal marker nf-200.

Although further analysis is required to determine the ultimate level of integration achieved by the grafted cells, it is important to note that when compared to bolus injections, RPC differentiation toward a photoreceptor fate appears to be increased significantly through the use of these thin-film polymer scaffolds. RPCs delivered using these micromachined ultra-thin-film PMMA scaffolds send projections into the host retina and express at least three markers appropriate to this tissue. By observing the progenitor cell response to a precisely defined architectural environment, a better understanding as to how physical cues control either the plasticity or lineage commitment of the cell can be achieved. This may ultimately lead to the combined use of RPCs and polymer substrates for the generation of tissue equivalents in culture.

5.3 Nanofabricated Interfaces

5.3.1 Nanostructures for Osseointegration

Total joint replacement is an effective treatment for relieving pain and restoring function for patients with damaged or degenerative joints. Approximately five hundred thousand total hip and knee replacements are performed each year in the United States. These numbers will increase as the population continues to age and as the indications for joint arthroplasty extend to younger patients [45]. Although many of the outcomes are successful, there are still significant problems with implant loosening and failure. In fact, 25 percent of hip-replacement surgeries were revisions due to previous implant failure [46]. Surgery to replace these failures is more difficult and costly to perform and has a poorer outcome than the original joint-replacement surgery. For example, for a hip implant to be successful, it is critical that the femoral stem remains adequately fixed over a period of several years without any signs of an immune reaction by the natural bone. If fixation is not sufficient, loosening and osteolysis of the implant can occur [42]. To overcome this problem, it is thought that bone implant materials need to stimulate rapid bone regeneration in order to fill in deficient bone and fix the implant firmly with the adjacent bone. The material surface must be able to recruit bone-forming cells (i.e., osteoblasts) such that they can colonize and synthesize new bone tissue. For successful implants, sufficiently regenerated bone fills the gap between an implant and the adjacent bone, thus firmly attaching the implant. However, some implant materials promote inflammation and the formation of undesirable soft connective tissue rather than bone. In this case, the tissue may improperly fix the implant into the surrounding bone, leading to periprosthetic osteolysis and eventual implant failure [42]. Thus, there a need to design a more robust implant, one that can promote long-term in vivo osseointegration. This would have significant implications not only for older patients but also for young patients who would otherwise need many revision surgeries throughout their lives.

In an attempt to enhance the stability of endosseous implants, a large number of implant materials and designs have been used. In addition to cement-based prosthetics, much attention in recent years has turned to microinterlocked implants, which have microporous surfaces to allow for the ingrowth of bone. Early work by Hulbert, Morrison, and Klawitter using oxide ceramics showed that a minimum interconnected pore diameter of approximately 100 μm was needed for adequate bone ingrowth [47]. It was thought that smaller pore sizes allowed incomplete mineralization of the infiltrating tissue. Subsequently, work by Bobyn et al. using metallic implants showed bone ingrowth with pore sizes between 50 and 400 μm [48]. Recently, however, studies by our group and several others have revealed the possibility that much smaller pores may allow bone ingrowth when presented at high density within metal-oxide substrates. For example, nanoporous Ca-P coatings on implants have shown apposition of human bone growth within two to three weeks after surgery [49]. Osteoblasts cultured on ceramics of different nm-scale textures also exhibit altered morphologies and growth rates [45, 50–57]. Nonetheless, there are several problems related to dissolution of nanoscale coatings over time, as well as cracking and separation from the metallic substrate [58, 59]. These studies point to the importance of developing more robust and flexible nanoscale architectures to enhance the apposition of bone from existing bone surfaces and stimulate new bone formation.

We have fabricated novel nanotubular titania surfaces using a simple anodization process. Titanium, or rather titania (native oxide TiO_2 on the surface of Ti), has been used in prosthetic devices since the 1970s. As a biocompatible material, titanium and its alloys, particularly Ti-6Al-4V, are extensively used in orthopedic and dental implants. The biocompatibility of metal oxides has already been proven as the materials have current clinical applications in orthopedic prostheses and dental implants [50]. Our technique of producing well-controlled nanostructured materials using anodization provides ease of control over the size and configuration of structure, with maintenance of mechanical properties that is otherwise not possible [60]. Bone marrow stromal cells (MSCs) are used to investigate the ability of nanotubular surfaces to enhance differentiation. The ability of MSCs to induce bone formation in vivo is believed to be due to the interaction of osteoprogenitors present within the cell populations with osteoinductive factors, such as bone morphogenetic proteins and various growth factors and cytokines, which cause them to differentiate into bone-forming cells (i.e., osteoblasts) [61–65], which will then eventually form bone matrix.

Current orthopedic implant technology uses dense alumina, titanium, or titanium-based alloys such as Ti-6Al-4V as implant materials. Thus, in this work we have developed nanotubular surfaces that can be applied to existing implants, pursuing a strategy that can be applied quite readily in the clinical environment. Our approach to achieving an optimal material nanoarchitecture uses a simple anodization technique to fabricate vertically oriented, immobilized, high-aspect-ratio titania nanotube arrays. Titania nanotubes with a pore size of approximately 80 nm and a length of 400 nm can be prepared using an anodization voltage of 20V for twenty minutes. It can be seen that the nanotube array is uniform over the substrate. There is a precise correlation between the anodization voltage and pore size; thus, by varying the voltage and anodization time, substrates with different size scales can be fabricated [66, 67]. The large surface area of the nanotube-array structure and the ability to precisely tune pore size, wall thickness, and nanotube length to optimize biotemplating properties are among the many desirable properties of this architecture for use in orthopedic applications. Studies have shown a 40 percent increase in the number of cells present on nanotubular titania surfaces compared to flat titania surfaces ($p < 0.05$) after seven days of culture. Moreover, there is approximately a 50 percent increase in alkaline phosphatase levels on nanotubular surfaces after three weeks of culture ($p < 0.05$). This suggests that topographical cues at the nanoscale level promote cell adhesion and proliferation. These results are strongly supported by results obtained by several other researchers with different types of nanostructured surfaces [45, 51, 55, 57, 68, 69].

As the cells differentiate, they begin to deposit bone matrix on the surface. The bone matrix predominantly consists of calcium phosphate. Figure 5.10 shows scanning electron microscope (SEM) images of MSCs on nanotubular surfaces for up to three weeks of culture. The cells show a spreading morphology and network formation on the surface after one week of culture [Figure 5.10(a)]. High magnification SEM images show the presence of granular material on the nanotubular surface [Figure 5.10(b)]. After two weeks, the SEM images show that the whole surface is covered with a network of well-spread cells [Figure 5.10(c)]. A close look at the areas surrounding the cells confirms that the nanopores are being filled in with matrix [Figure 5.10(d)]. After three weeks of culture, the SEM images show that the whole surface is completely covered with both cells and mineralized matrix compo-

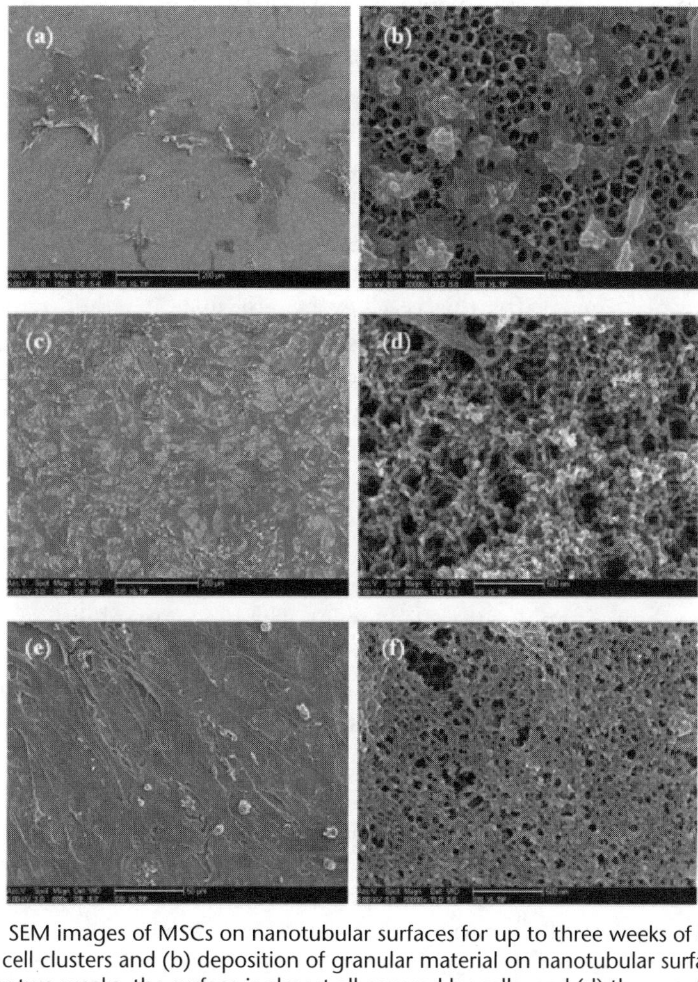

Figure 5.10 SEM images of MSCs on nanotubular surfaces for up to three weeks of culture: (a) Formation of cell clusters and (b) deposition of granular material on nanotubular surfaces after one week. (c) After two weeks, the surface is almost all covered by cells, and (d) the nanotubes are further filled with matrix. (e) After three weeks, the entire surface is covered with well-spread cells, and (f) the nanotubes are almost completely filled with matrix constituents. (Image reproduced with permission from [94].)

nents [Figure 5.10(e)]. Again, a high-magnification SEM image of the area around the cells shows that the nanotubular structures are completely filled with a porous material. X-ray photoelecton spectroscopy (XPS) analysis suggested that this deposited material predominantly consists of calcium and phosphorous, important bone matrix constituents. Thus, the development of nanostructured platforms based on novel metal-oxide films can provide insight into cell-material interactions for the development of improved implant surfaces.

5.3.2 Nanoporous Interfaces for Cellular Delivery

Cellular encapsulation, or immunoisolation, has emerged over the past several decades as a method of treating a number of diseases, including Alzheimer's, Parkinson's, liver failure, chronic pain, or Type 1 diabetes [70]. Therapeutic cells are isolated from the host immune system by a selectively permeable membrane. Low-molecular-weight substances such as oxygen, nutrients, therapeutic proteins, and cellular waste are exchanged freely across the membrane, while immune components that can jeopardize the integrity of the graft are excluded. Thus, transplanted cells can function without being rejected by the host. These features are extremely important for a medical device since they help facilitate the replacement of physiological function of the damaged tissue or organ without the need for an immunosuppressive regime that is required for full organ transplantation; this is an important advantage since generalized immunosupression is associated with a number of problems, including increased risk of infection, mouth ulcers, diarrhea, acne, edema, and ovarian cysts. Immunoisolation therapy can also theoretically permit the use of xenografts; this is a key benefit since it would circumvent the lack of human cadaveric donors, which is a major issue for full organ transplantation.

When designing a cellular encapsulation device, an optimal balance has to be maintained among the various capsule properties in order to support cell survival and function in the context of a particular application. Generally, the mass-transport properties of the membrane are critical since the influx rate of molecules essential for cell survival and the outflow rate of therapeutic agents, as well as the exclusion of large-molecular-weight molecules such as immune components, will ultimately determine the success of the device. The molecular weight cutoff for permeability is application dependent. The biocompatibility of the capsule material is crucial since this will determine the nature and extent of the host reaction; initiation of the foreign-body response and subsequent fibrous encapsulation of the device would drastically limit its ability to allow the diffusion of therapeutic molecules.

Typically, polymeric microcapsules are used for immunoisolation applications, with alginate-poly L-lysine beads being the most commonly used [71–73]. While polymeric membranes are biocompatible and easy to fabricate and have a long history of use in immunoisolation procedures, there are several disadvantages associated with them. Polymeric membranes have demonstrated poor chemical resistance and mechanical stability, as well as broad pore-size distributions, all of which lead to the eventual destruction of the transplanted cells. Also, polymeric microcapsules have poor retrievability and can form aggregates in vivo which reduce their effectiveness [74–80].

The advent of nanotechnology has led to the development of more robust immunoisolation membranes, such as silicon or metal oxides. These inorganic membranes are much more chemically and mechanically stable and the pore size can be easily controlled by varying the fabrication parameters. Furthermore, a tight pore-size distribution can be achieved [74–80]. Finally, the surface of inorganic membranes can be chemically tailored to resist the adhesion of blood proteins and immune cells that may block the pores and reduce the diffusion of therapeutic agents [81, 82]. In the following sections, two types of technologies for creating robust nanoporous membranes, both for cell encapsulation applications, are introduced. Silicon- and alumina-based membranes will be discussed.

5.3.2.1 Silicon-Based Nanoporous Membranes

Nanoporous silicon immunoisolating biocapsules were originally described by Desai et al. [76]. These capsules are fabricated out of silicon and its oxides, which offer several advantages including biochemical inertness and mechanical durability, as well as availability, a well-defined machining processes, and ease of surface modification. Using traditional photolithography and other MEMS fabrication techniques, a highly reproducible device can be created with tight control over several important features, such as pore size, pore density, and membrane thickness. A tapered pore structure can be created such that the narrow end still allows the diffusion of relevant molecules and maintains immunoisolation, while the wide end promotes neovascularization at the implant site. The surface of a silicon device can easily be chemically modified to improve biocompatibility and promote site-specific delivery.

The fabrication process for nanoporous silicon membranes is outlined in Figure 5.11(a–i). In short, beginning with 4″ boron-doped (p^+) single crystal silicon (100) wafers, a low-stress silicon nitride layer, which acts as an etch-stop layer, is deposited and then buried under a p^+ polysilicon film [Figure 5.11(a)]. This polysilicon acts as a structural layer for later steps. After the growth of an oxide mask, holes, which will define the overall shape and location of the pores, are etched into the base layer using chlorine plasma [Figure 5.11(b)]. The next and most critical step is the growth of a sacrificial oxide layer on the polysilicon [Figure 5.11(c)]. The thickness of the sacrificial oxide will determine the pore size for the completed membrane. Thermal oxidation with dry oxygen allows control of the sacrificial oxide layer thickness down to 5 nm. Next, anchor points are defined in the sacrificial oxide layer, and a second polysilicon layer, or plug layer, is deposited [Figure 5.11(d)]. This plug layer is then planarized to the base layer using chemical mechanical polishing [Figure 5.11(e)]. A protective nitride layer is then deposited on both sides of the original wafer [Figure 5.11(f)], and the backside (nonmembrane side) is etched to form a window exposing the single-crystal silicon underneath [Figure 5.11(g)]. The silicon is then etched to the etch-stop layer (buried silicon nitride) using an 80°C KOH bath [Figure 5.11(h)]. Finally, the remaining protective, sacrificial, and etch-stop layers are removed using hydrofluoric acid (HF) to release the membrane and its pores [Figure 5.11(i)]. These procedures create channel-/trench-shaped pores in the membrane instead of the cylindrical "holes" normally expected. SEM images of the final membrane are shown in Figure 5.11(j, k).

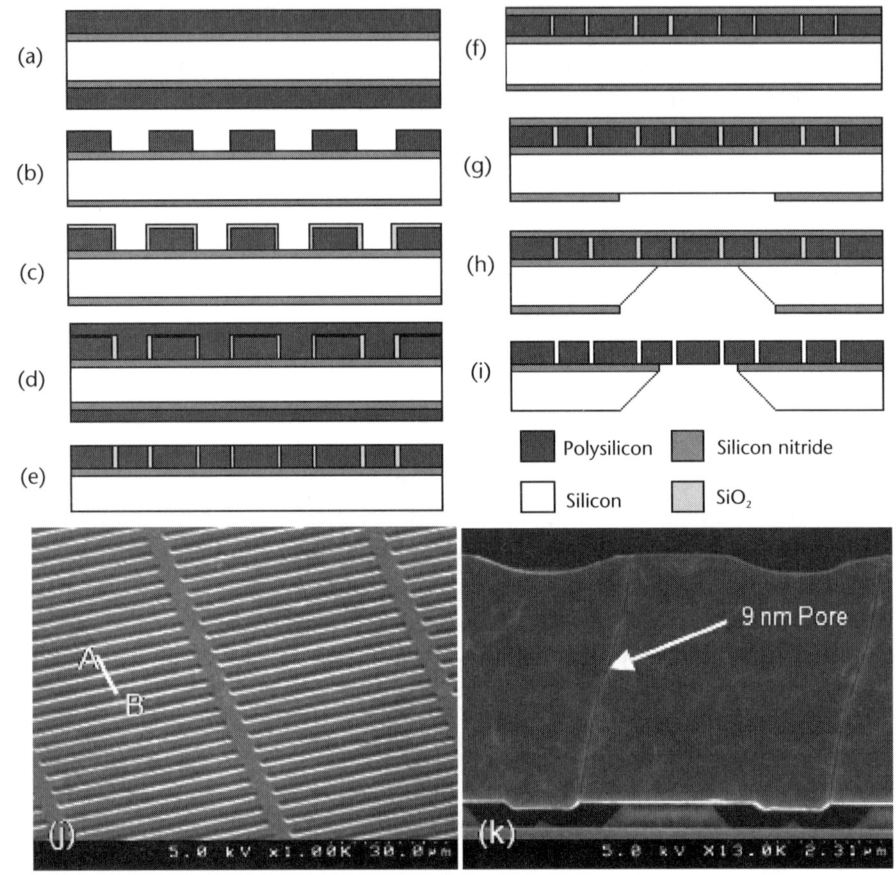

Figure 5.11 Cross-sectional schematic depiction of the microfabrication process (not to scale) and SEM images of the nanoporous membranes: (a) A silicon nitride layer is deposited below a polysilcon layer on both sides of a 4″ boron-doped (p^+) single crystal silicon (100) wafers. (b) After creation of an oxide mask, holes are etched into the polysilicon layer using chlorine plasma. (c) A sacrificial oxide layer is grown. (d) Anchor points are defined, and a polysilicon plug layer is deposited. (e) The plug layer is planarized. (f) A nitride layer is deposited on both sides of the wafer. (g) A window is formed in the nitride layer via a backside etch. (h) The silicon wafer is etched through to the nitride layer (etch-stop). (i) The structure is released using HF. The critical submicron pore size is defined by the controlled growth of a thin layer of SiO_2 in (c), which is etched away in (i). Membrane sets with pore sizes ranging from 5 to 50 nm can be routinely fabricated. (j, k) SEM images of a microfabricated silicon membrane with 9 nm pores: (j) the top-side surface, and (k) the line A–B corresponds to the cross-sectional image. (Image reporoduced with permission from [83].)

Final membrane and pore dimensions can be precisely controlled to ±2 percent of the desired values. The pores currently fabricated are 45 μm in channel length, and the membranes are about 5 to 10 μm thick.

Silicon biocapsules are assembled by bonding two membranes directly together, by inserting two membranes into a plastic housing with latex gaskets between the membrane and plastic body, or by directly bonding the membranes to either side of a PMMA spacer. Therapeutic cells can be loaded in a reservoir created by the space between the membranes, which measures about 20 to 40 μL. A number of investigations done with these capsules have shown that they are capable of allowing transport of small molecules such as glucose, while at the same time excluding larger immune components [79]. They are also biocompatible and do not induce excessive inflammation when implanted into the omental pouch of rats [78]; they can also reverse diabetes in mice short term [77].

5.3.2.2 Nanoporous Microenvironments for Neurons

Researchers have shown that pheochromocytoma cells can attach, proliferate, and differentiate on the surface of microfabricated silicon nanoporous membranes as easily as they can on traditional tissue culture polystyrene (TCPS) surfaces [83]. An acceptable level of phenotypic "function" on these surfaces opens many avenues for their use; neuroengineering and neurointerfacing applications include neuronal network modeling, design of neuroprostheses (e.g., neurosecretory biohybrid macrocapsules), and design of novel microelectrodes for data collection and other unique therapies [84]. We compared the viability of PC12 cells on the silicon nanoporous membranes to those grown on TCPS. The cell clusters shown in Figure 5.12(a, b) are located directly on the nanoporous membrane surface. Qualitatively these image suggest that the cells do not perceive the polysilicon membrane differently than they do TCPS for the purpose of differentiation. Figure 5.12(c–j) shows a three-dimensional microenvironment composed of a 20 nm silicon membrane and PC12 cells embedded within a collagen type I matrix. The cell differentiation imaged in this figure, combined with supporting data for the functionality of PC12 cells in these types of hybrid environments, demonstrates that an "in vivo–like" milieu can be established. This would be beneficial for replicating xenogenic tissue with various cell lines and has also been regarded as an extremely important parameter for proper functioning of an encapsulation device [85].

Silicon nanoporous membranes have been shown to provide substantial immunoisolation capabilities, as well as a high level of transport for small soluble factors, such as glucose, dopamine (DA), DOPAC, and HVA. The diffusive transport and cell-derived release of some of these substances through the membranes have been shown to be highly linear in many cases, for limited regimes, and in other cases highly nonlinear (due to molecular instability and strict metabolic control over secretion). Figure 5.13 shows the cell-derived release of DA from encapsulated PC12 cells under conditions of basal and activated secretion through these nanoporous membranes with pore size of 18 nm [cells were seeded on tissue culture plastic (TCPS) as a control]. DA release from the neurons was monitored under both basal (i.e., unactivated) and activated conditions. Activation was achieved using a high 5: salt solution to depolarize the cells and stimulate the exocytosis of vesicular DA. Activation of the neurons results in increased levels of DA to be released. The amount of DA that diffuses through the membrane is greater in the activated case and demonstrates the ability of the membrane to pass increasing levels of DA. Under both basal and activated conditions, the level of measured DA was lower than the control conditions where there is no barrier (i.e., membrane) for the DA to diffuse through; therefore, the concentration builds up more rapidly in the control conditions. This data demonstrates the high level of transport and control that can be achieved for small molecule release for an in vitro biohybrid encapsulation device. However, in this respect, much research still needs to be completed to

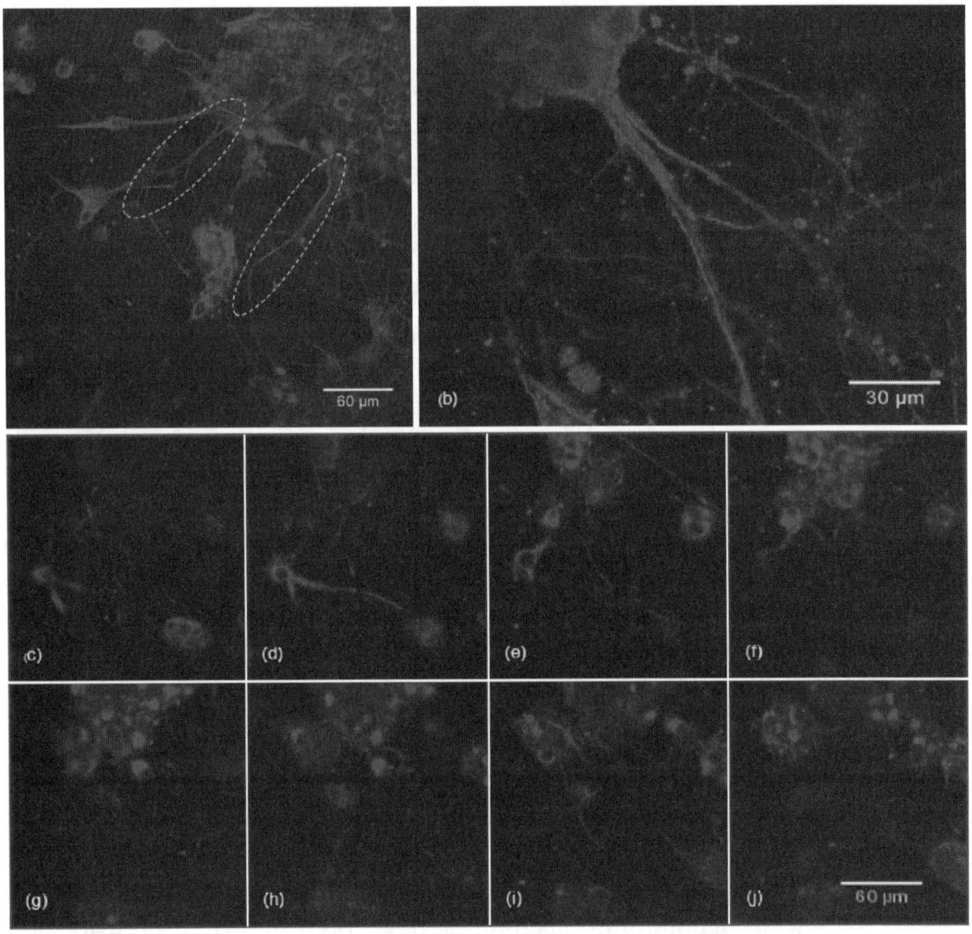

Figure 5.12 PC12 cells cultured on (a) a 20 nm silicon membrane and (b) a TCPS surface showing significant neurite extension after seven [(dashed ovals in (a)] and nineteen days (b). (c–j) represent a confocal microscope z-scan montage of PC12 cells embedded in a collagen Type I gel cultured on the same membrane surface imaged in (a) and showing neurite extension throughout the collagen gel matrix after seven days in culture. Imaging toward the membrane surface, frame (j) is 35 μm below frame (c). All images were obtained using a 40× objective. (Image reproduced with permission from [83].)

understand and design a reliable therapeutic device for the delivery of a substance as ubiquitous as a neurotransmitter.

5.3.2.3 Aluminum Oxide–Based Nanoporous Membranes

Aluminum oxide is a bioinert material that has been utilized in numerous biomedical applications, such as bone prostheses, dental implants, and artificial eye sockets [52, 53, 86, 87]. Through a simple two-step anodization process, physically strong membranes with an organized array of pores can be fabricated. Compared with polymeric encapsulation devices, it is possible to achieve a high pore density (~10^{10}/cm^2) and a narrow pore size distribution [88]. The pore size can range from 10 to 100 nm and is controlled by the anodization voltage. The surface of the mem-

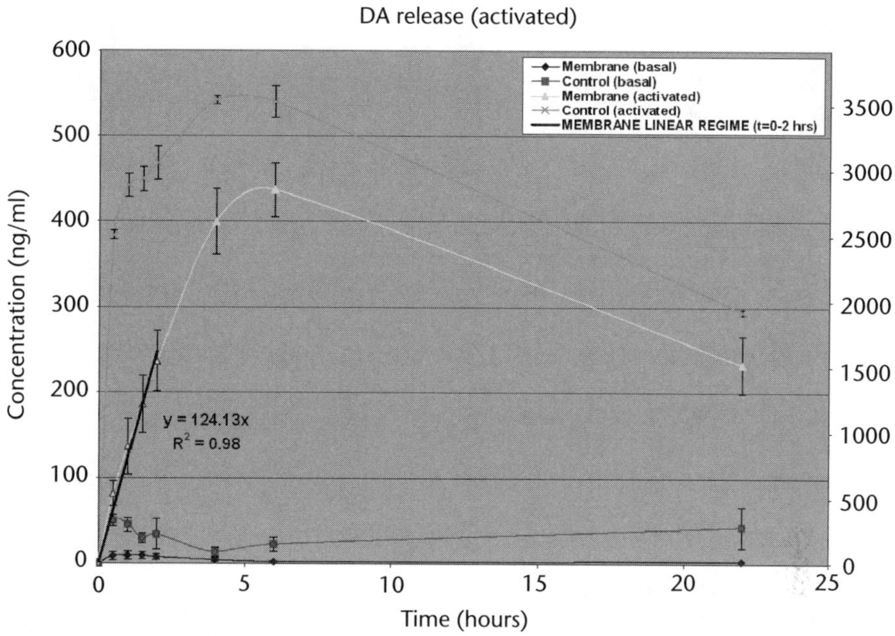

Figure 5.13 Activated and basal cell-derived DA release through microfabricated silicon membranes with 18 nm pores. Collagen-embedded PC12 cells were loaded in an in vitro "biocapsule" with a nanoporous membrane (and also seeded on TCPS for the control condition). Activation of the neurons results in a greater release of DA over time. The membrane itself slows the diffusion of DA in comparison to cells seeded on TCPS with no diffusion barrier. Note the high degree of linearity in the activated concentrations achieved over the first two hours for the membrane condition.

brane can be modified, with PEG for example, to improve the biocompatibility in vivo. Alumina also offers certain advantages over its silicon-based counterparts. It is easier and cheaper to fabricate than silicon, requiring only the setting up of a simple anodization cell, which can be done on any lab bench top [88]. Also, while microfabricated silicon membranes are always flat, the anodization technology used to create alumina membranes can be adapted to any desired geometry, including curved surfaces such as a cylindrical capsule. As with microfabricated membranes, it is possible to create a tapered pore structure that immunoisolates encapsulated cells while promoting neovascularization.

The fabrication method for nanoporous alumina capsules is demonstrated in Figure 5.14. Briefly, an aluminum tube with starting length, outer diameter, and thickness of 2.5 cm, 6.35 mm, and 710 μm, respectively, is cleaned and precoated with a resist on the outer surface to protect it from being anodized during the subsequent fabrication steps [Figure 5.14(a)]. The inner surface of the tube is then anodized in an acid, usually sulfuric or oxalic, for several hours. The resulting thin film of aluminum oxide (AO) is etched away using a solution of chromic acid and phosphoric acid, leaving a pretextured aluminum surface necessary for the formation of an organized nanoarray of pores during the second anodization step [Figure 5.14(b)]. The inner side of the tube is then anodized a second time under the same conditions as the first, resulting in a layer of nanoporous alumina that serves as the final mem-

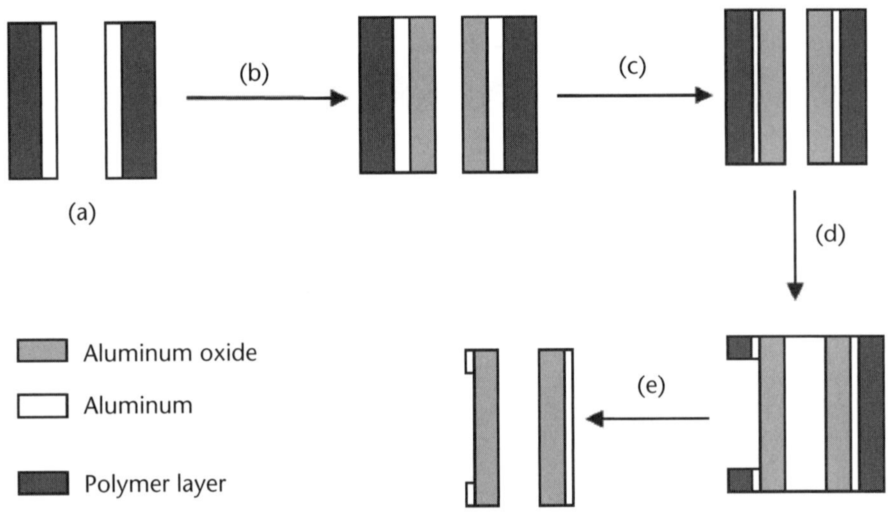

Figure 5.14 Schematic of the two-step anodization process for the fabrication of nanoporous alumina membranes. (a) The outer surface of an aluminum tube is coated with a protective resist. (b) The inner surface of the aluminum tube is anodized, forming a thin layer of aluminum oxide, which is subsequently etched away, leaving a pretextured surface necessary for the formation of an organized pore array. (c) The inner surface of the tube is anodized a second time. (d) A small area of the resist is removed and the underlying aluminum and barrier layer are etched away, exposing the aluminum oxide. (e) The remaining protective resist is removed. Pore diameter is controlled via anodization voltage, with a pore-size-to-anodizing-voltage correlation of 1.29 nm/V. (Image reproduced with permission from [95].)

brane [Figure 5.14(c)]. A small area of the resist on the outer surface of the tube is then removed, and the aluminum layer is etched away with a solution of HCl and $CuCl_2$, and a superficial amorphous barrier layer of alumina is etched away using phosphoric acid exposing the nanoporous alumina membrane [Figure 5.14(d)]. Finally, the remaining protective resist is removed [Figure 5.14(e)].

Since the membranes are actually incorporated into the tube, they are strong enough for easy handling and use, withstanding up to 32.6 mPa before failure [89]. Pore size is determined by anodization voltage, with a correlation of 1.29 nm/V. This fabrication process can easily be manipulated for any geometry or membrane pattern desired by the user. The inset in Figure 5.15 shows an SEM image of a membrane with a pore size of ~75 nm; plainly, a very organized array of hexagonally shaped pores is achieved.

Figure 5.16 shows data that provides some insight into the diffusion behavior of some relevant molecules across nanoporous alumina membranes. Clearly, these membranes permit the rapid transport of glucose (MW = 180), a cellular nutrient and insulin secretagogue, while at the same time markedly impeding the diffusion of IgG (MW = 150,000), an immune component, suggesting that these capsules would be effective in immunoisolation applications. Figure 5.17 shows the profile of glucose-stimulated insulin secretion from encapsulated MIN-6 β cells after seven days in culture. The cells begin to release insulin within minutes of glucose stimulation, indicating that these capsules are able to maintain viable, functional cells.

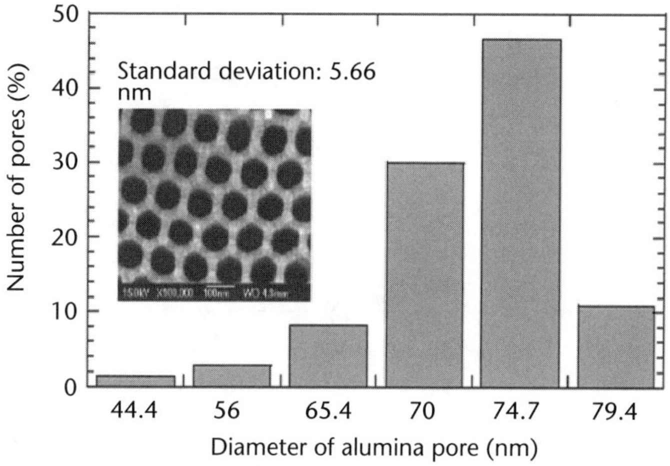

Figure 5.15 The pore size distribution for alumina membranes anodized at 40V. Insets show SEM images of each membrane, demonstrating the highly organized arrangement of hexagonal pores. (Image reproduced with permission from [95].)

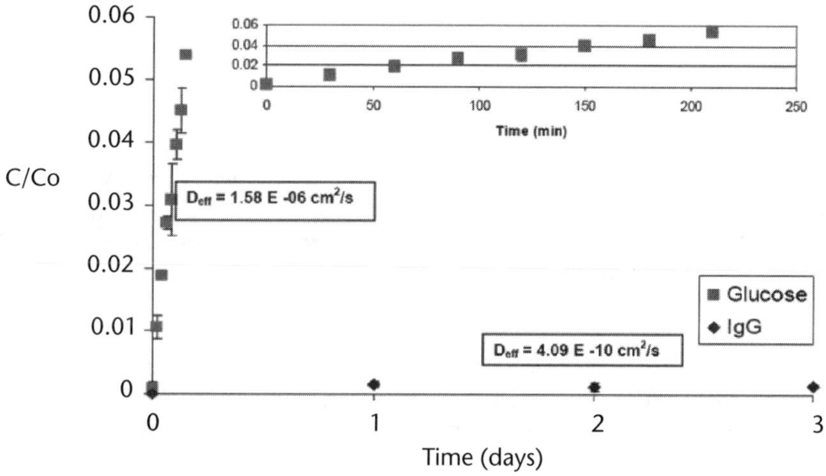

Figure 5.16 Normalized release of glucose (squares, Deff = 1.58E–06 cm^2/s) and IgG (diamonds, Deff = 4.09E–10 cm^2/s) through a nanoporous alumina membrane with a nominal pore size of 75 nm. *C* = concentration at time *t*, *Co* = loading concentration. Inset: glucose release is shown on a 210 minute time scale. (Image reproduced with permission from [96].)

5.4 Conclusion

As studies continue to uncover the inner workings of cells, it is becoming apparent that cells must be affected in their nano-/microenvironments. The ability to utilize cells for therapeutic applications relies on the ability to manipulate these niches. Aside from microfabrication processes, techniques such as polymer demixing [90], nanoprecipitation [91], and electrospun nanofibers [92] are available for use for in vitro studies that shed further light on the interaction between cells and their

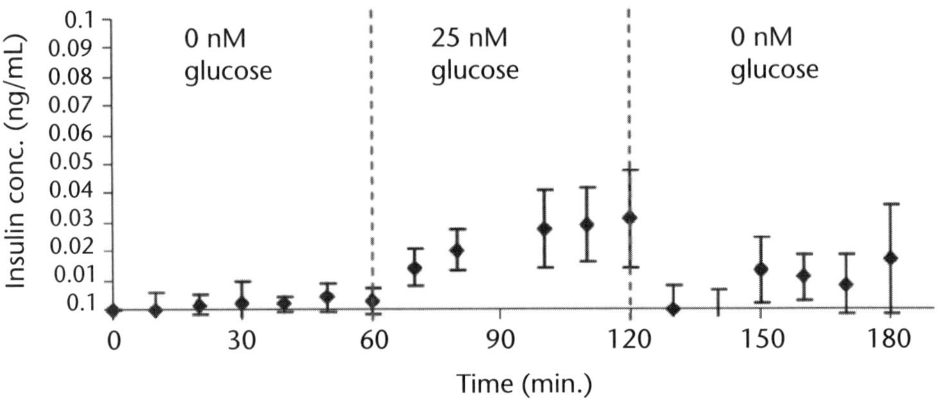

Figure 5.17 The response of encapsulated MIN6 cells to alternating high (25 mM) and low (0 mM) glucose signals. Encapsulated MIN6 cells seem to regulate insulin secretion according to changes in glucose levels. (Image reproduced with permission from [96].)

micro-/nanoenvironments or for in vivo studies that will advance the boundaries of the therapies currently available for patients needing tissue replacements.

References

[1] Whitesides, G. M., Ostuni, E., Takayama, S., Jiang, X., and Ingber, D. E., "Soft lithography in biology and biochemistry," *Annu. Rev. Biomed. Eng.*, Vol. 3, 2001, pp. 335–373.

[2] Bissell, M. J., Radisky, D. C., Rizki, A., Weaver, V. M., and Petersen, O. W., "The organizing principle: Microenvironmental influences in the normal and malignant breast," *Differentiation*, Vol. 70, Nos. 9–10, 2002, pp. 537–546.

[3] Bissell, M. J., Rizki, A., and Mian, I. S., "Tissue architecture: The ultimate regulator of breast epithelial function," *Curr. Opin. Cell. Biol.*, Vol. 15, No. 6, 2003, pp. 753–762.

[4] Gudjonsson, T., Ronnov-Jessen, L., Villadsen, R., Bissell, M. J., and Petersen, O. W., "To create the correct microenvironment: Three-dimensional heterotypic collagen assays for human breast epithelial morphogenesis and neoplasia," *Methods*, Vol. 30, No. 3, 2003, pp. 247–255.

[5] Weaver, V. M., Fischer, A. H., Peterson, O. W., and Bissell, M. J., "The importance of the microenvironment in breast cancer progression: Recapitulation of mammary tumorigenesis using a unique human mammary epithelial cell model and a three-dimensional culture assay," *Biochem. Cell. Biol.*, Vol. 74, No. 6, 1996, pp. 833–851.

[6] Yang, S., Leong, K. F., Du, Z., and Chua, C. K., "The design of scaffolds for use in tissue engineering. Part I. Traditional factors," *Tissue Eng.*, Vol. 7, No. 6, 2001, pp. 679–689.

[7] Boateng, S. Y., Hartman, T. J., Ahluwalia, N., Vidula, H., Desai, T. A., and Russell, B., "Inhibition of fibroblast proliferation in cardiac myocyte cultures by surface microtopography," *Am. J. Physiol. Cell. Physiol.*, Vol. 285, No. 1, 2003, pp. C171–C182.

[8] Deutsch, J., Motlagh, D., Russell, B., and Desai, T. A., "Fabrication of microtextured membranes for cardiac myocyte attachment and orientation," *J. Biomed. Mater. Res.*, Vol. 53, No. 3, 2000, pp. 267–275.

[9] Motlagh, D., Hartman, T. J., Desai, T. A., and Russell, B., "Microfabricated grooves recapitulate neonatal myocyte connexin43 and N-cadherin expression and localization," *J. Biomed. Mater. Res. A*, Vol. 67, No. 1, 2003, pp. 148–157.

[10] Motlagh, D., Senyo, S. E., Desai, T. A., and Russell, B., "Microtextured substrata alter gene expression, protein localization and the shape of cardiac myocytes," *Biomaterials*, Vol. 24, No. 14, 2003, pp. 2463–2476.

[11] Thakar, R. G., Ho, F., Huang, N. F., Liepmann, D., and Li, S., "Regulation of vascular smooth muscle cells by micropatterning," *Biochem. Biophys. Res. Commun.*, Vol. 307, No. 4, 2003, pp. 883–890.

[12] Sarkar, S., Dadhania, M., Rourke, P., Desai, T. A., and Wong, J. Y., "Vascular tissue engineering: Microtextured scaffold templates to control organization of vascular smooth muscle cells and extracellular matrix," *Acta Biomaterialia*, Vol. 1, No. 1, 2005, pp. 93.

[13] Walboomers, X. F., Croes, H. J., Ginsel, L. A., and Jansen, J. A., "Growth behavior of fibroblasts on microgrooved polystyrene," *Biomaterials*, Vol. 19, No. 20, 1998, pp. 1861–1868.

[14] Norman, J. J., and Desai, T. A., "Control of cellular organization in three dimensions using a microfabricated polydimethylsiloxane-collagen composite tissue scaffold," *Tissue Eng.*, Vol. 11, Nos. 3–4, 2005, pp. 378–386.

[15] Tan, W., Oldenburg, A. L., Norman, J. J., Desai, T. A., and Boppart, S. A., "Optical coherence tomography of cell dynamics in three-dimensional tissue models," *Optics Express*, Vol. 14, No. 16, 2006, pp. 7159–7171.

[16] Ross, R., "Atherosclerosis is an inflammatory disease," *Am. Heart J.*, Vol. 138, No. 5, Pt. 2, 1999, pp. S419–S420.

[17] Nerem, R. M., and Cornhill, J. F., "Hemodynamics and atherogenesis," *Atherosclerosis*, Vol. 36, No. 2, 1980, pp. 151–157.

[18] Wakhloo, A. K., Lieber, B. B., Seong, J., Sadasivan, C., Gounis, M. J., Miskolczi, L., and Sandhu, J. S., "Hemodynamics of carotid artery atherosclerotic occlusive disease," *J. Vasc. Interv. Radiol.*, Vol. 15, No. 1, Pt. 2, 2004, pp. S111–S121.

[19] Rhodin, J., *Handbook of Physiology*, ed. R. Berne, Baltimore: Waverly Press, Inc., 1980.

[20] L'Heureux, N., Paquet, S., Labbe, R., Germain, L., and Auger, F. A., "A completely biological tissue-engineered human blood vessel," *Faseb J.*, Vol. 12, No. 1, 1998, pp. 47–56.

[21] Oh, S. H., Kang, S. G., Kim, E. S., Cho, S. H., and Lee, J. H., "Fabrication and characterization of hydrophilic poly(lactic-co-glycolic acid)/poly(vinyl alcohol) blend cell scaffolds by melt-molding particulate-leaching method," *Biomaterials*, Vol. 24, No. 22, 2003, pp. 4011–4021.

[22] Hou, Q., Grijpma, D. W., and Feijen, J., "Porous polymeric structures for tissue engineering prepared by a coagulation, compression moulding and salt leaching technique," *Biomaterials*, Vol. 24, No. 11, 2003, pp. 1937–1947.

[23] Sarkar, S., Lee, G. Y., Wong, J. Y., and Desai, T. A., "Development and characterization of a porous micro-patterned scaffold for vascular tissue engineering applications," *Biomaterials*, Vol. 27, No. 27, 2006, pp. 4775–4782.

[24] Sarkar, S., Lee, G. Y., Wong, J. Y., and Desai, T. A., "Development and characterization of a porous micro-patterned scaffold for vascular tissue engineering applications," *Biomaterials*, Vol. 27, 2006, pp. 4775–4782.

[25] Bessant, D. A., Ali, R. R., and Bhattacharya, S. S., "Molecular genetics and prospects for therapy of the inherited retinal dystrophies," *Curr. Opin. Genet. Dev.*, Vol. 11, No. 3, 2001, pp. 307–316.

[26] Ali, R. R., Reichel, M. B., Hunt, D. M., and Bhattacharya, S. S., "Gene therapy for inherited retinal degeneration," *Br. J. Ophthalmol.*, Vol. 81, No. 9, 1997, pp. 795–801.

[27] Reichel, M. B., Ali, R. R., Hunt, D. M., and Bhattacharya, S. S., "Gene therapy for retinal degeneration," *Ophthalmic Res.*, Vol. 29, No. 5, 1997, pp. 261–268.

[28] Thanos, C., and Emerich, D., "Delivery of neurotrophic factors and therapeutic proteins for retinal diseases," *Expert Opin. Biol. Ther.*, Vol. 5, No. 11, 2005, pp. 1443–1452.

[29] Aramant, R. B., and Seiler, M. J., "Progress in retinal sheet transplantation," *Prog. Retin. Eye Res.*, Vol. 23, No. 5, 2004, pp. 475–494.

[30] Aramant, R. B., and Seiler, M. J., "Retinal transplantation—advantages of intact fetal sheets," *Prog. Retin. Eye Res.*, Vol. 21, No. 1, 2002, pp. 57–73.

[31] Aramant, R., and Seiler, M., "Cryopreservation and transplantation of immature rat retina into adult rat retina," *Brain Res. Dev. Brain Res.*, Vol. 61, No. 2, 1991, pp. 151–159.

[32] Zahir, T., Klassen, H., and Young, M. J., "Effects of ciliary neurotrophic factor on differentiation of late retinal progenitor cells," *Stem Cells*, Vol. 23, No. 3, 2005, pp. 424–432.

[33] Ahmad, I., Dooley, C. M., Thoreson, W. B., Rogers, J. A., and Afiat, S., "In vitro analysis of a mammalian retinal progenitor that gives rise to neurons and glia," *Brain Res.*, Vol. 831, Nos. 1–2, 1999, pp. 1–10.

[34] Klassen, H. J., Ng, T. F., Kurimoto, Y., Kirov, I., Shatos, M., Coffey, P., and Young, M. J., "Multipotent retinal progenitors express developmental markers, differentiate into retinal neurons, and preserve light-mediated behavior," *Invest. Ophthalmol. Vis. Sci.*, Vol. 45, No. 11, 2004, pp. 4167–4173.

[35] Tropepe, V., Coles, B. L., Chiasson, B. J., Horsford, D. J., Elia, A. J., McInnes, R. R., and van der Kooy, D., "Retinal stem cells in the adult mammalian eye," *Science*, Vol. 287, No. 5460, 2000, pp. 2032–2036.

[36] Yang, P., Seiler, M. J., Aramant, R. B., and Whittemore, S. R., "In vitro isolation and expansion of human retinal progenitor cells," *Exp. Neurol.*, Vol. 177, No. 1, 2002, pp. 326–331.

[37] Qiu, G., Seiler, M. J., Arai, S., Aramant, R. B., and Sadda, S. R., "Alternative culture conditions for isolation and expansion of retinal progenitor cells," *Curr. Eye Res.*, Vol. 28, No. 5, 2004, pp. 327–336.

[38] Meyer, J. S., Katz, M. L., and Kirk, M. D., "Stem cells for retinal degenerative disorders," *Annals NY Acad. Sci.*, Vol. 1049, 2005, pp. 135–145.

[39] Klassen, H., Sakaguchi, D. S., and Young, M. J., "Stem cells and retinal repair," *Prog. Retin. Eye Res.*, Vol. 23, No. 2, 2004, pp. 149–181.

[40] Tomita, M., Lavik, E., Klassen, H., Zahir, T., Langer, R., and Young, M. J., "Biodegradable polymer composite grafts promote the survival and differentiation of retinal progenitor cells," *Stem Cells*, Vol. 23, No. 10, 2005, pp. 1579–1588.

[41] Warfvinge, K., Kiilgaard, J. F., Lavik, E. B., Scherfig, E., Langer, R., Klassen, H. J., and Young, M. J., "Retinal progenitor cell xenografts to the pig retina: Morphologic integration and cytochemical differentiation," *Arch. Ophthalmol.*, Vol. 123, No. 10, 2005, pp. 1385–1393.

[42] Takagi, M., "Bone-implant interface biology—foreign body reaction and periprosthetic osteolysis in artificial hip joints," *J. Clin. Exp. Hematopathol.*, Vol. 41, No. 2, 2001, pp. 81–87.

[43] Tao, S. L., Lubeley, M. W., and Desai, T. A., "Bioadhesive poly(methyl methacrylate) microdevices for controlled drug delivery," *J. Control. Release*, Vol. 88, No. 2, 2003, pp. 215–228.

[44] Zahir, T., Klassen, H., Tomita, M., and Young, M. J., "Sorbitol causes preferential selection of Muller glial precursors from late retinal progenitor cells in vitro," *Mol. Vis.*, Vol. 12, 2006, pp. 1606–1614.

[45] "Improving medical implant performance through retrieval information: Challenges and opportunities," Paper delivered at the *NIH Technology Assessment Conference*, January 10–12, 2000.

[46] Webster, T. J., "Nanophase ceramics as improved bone tissue engineering materials," *Am. Ceramic Soc. Bull.*, Vol. 82, No. 6, 2003, pp. 1–8.

[47] Hulbert, S. F., Morrison, S. J., and Klawitter, J. J., "Tissue reaction to three ceramics of porous and non-porous structures," *J. Biomed. Mater. Res.*, Vol. 6, No. 5, 1972, pp. 347–374.

[48] Bobyn, J. D., Pilliar, R. M., Cameron, H. U., and Weatherly, G. C., "The optimum pore size for the fixation of porous-surfaced metal implants by the ingrowth of bone," *Clin. Orthop. Relat. Res.*, No. 150, 1980, pp. 263–270.

[49] Lee, T. M., Wang, B. C., Yang, Y. C., Chang, E., and Yang, C. Y., "Comparison of plasma-sprayed hydroxyapatite coatings and hydroxyapatite/tricalcium phosphate composite coatings: In vivo study," *J. Biomed. Mater. Res.*, Vol. 55, No. 3, 2001, pp. 360–367.

[50] Boyan, B. D., Hummert, T. W., Dean, D. D., and Schwartz, Z., "Role of material surfaces in regulating bone and cartilage cell response," *Biomaterials*, Vol. 17, No. 2, 1996, pp. 137–146.

[51] Popat, K. C., Daniels, R. H., Dubrow, R. S., Hardev, V., and Desai, T. A., "Nanostructured surfaces for bone biotemplating applications," *J. Orthop. Res.*, Vol. 24, No. 4, 2006, pp. 619–627.

[52] Popat, K. C., Leary Swan, E. E., Mukhatyar, V., Chatvanichkul, K. I., Mor, G. K., Grimes, C. A., and Desai, T. A., "Influence of nanoporous alumina membranes on long-term osteoblast response," *Biomaterials*, Vol. 26, No. 22, 2005, pp. 4516–4522.

[53] Swan, E. E., Popat, K. C., and Desai, T. A., "Peptide-immobilized nanoporous alumina membranes for enhanced osteoblast adhesion," *Biomaterials*, Vol. 26, No. 14, 2005, pp. 1969–1976.

[54] Swan, E. E., Popat, K. C., Grimes, C. A., and Desai, T. A., "Fabrication and evaluation of nanoporous alumina membranes for osteoblast culture," *J. Biomed. Mater. Res. A*, Vol. 72, No. 3, 2005, pp. 288–295.

[55] Webster, T. J., and Ejiofor, J. U., "Increased osteoblast adhesion on nanophase metals: Ti, Ti6Al4V, and CoCrMo," *Biomaterials*, Vol. 25, No. 19, 2004, pp. 4731–4739.

[56] Webster, T. J., Ergun, C., Doremus, R. H., and Lanford, W. A., "Increased osteoblast adhesion on titanium-coated hydroxylapatite that forms CaTiO3," *J. Biomed. Mater. Res. A*, Vol. 67, No. 3, 2003, pp. 975–980.

[57] Webster, T. J., Ergun, C., Doremus, R. H., Siegel, R. W., and Bizios, R., "Enhanced functions of osteoblasts on nanophase ceramics," *Biomaterials*, Vol. 21, No. 17, 2000, pp. 1803–1810.

[58] Bauer, T. W., Taylor, S. K., Jiang, M., and Medendorp, S. V., "An indirect comparison of third-body wear in retrieved hydroxyapatite-coated, porous, and cemented femoral components," *Clin. Orthop. Relat. Res.*, Vol. 298, 1994, pp. 11–18.

[59] Bloebaum, R. D., Beeks, D., Dorr, L. D., Savory, C. G., DuPont, J. A., and Hofmann, A. A., "Complications with hydroxyapatite particulate separation in total hip arthroplasty," *Clin. Orthop. Relat. Res.*, No. 298, 1994, pp. 19–26.

[60] Karlsson, M., Palsgard, E., Wilshaw, P. R., and Di Silvio, L., "Initial in vitro interaction of osteoblasts with nano-porous alumina," *Biomaterials*, Vol. 24, No. 18, 2003, pp. 3039–3046.

[61] Harakas, N. K., "Demineralized bone-matrix-induced osteogenesis," *Clin. Orthop. Relat. Res.*, Vol. 188, 1984, pp. 239–251.

[62] Urist, M. R., "Bone: Formation by autoinduction," *Science*, Vol. 150, No. 698, 1965, pp. 893–899.

[63] Urist, M. R., DeLange, R. J., and Finerman, G. A., "Bone cell differentiation and growth factors," *Science*, Vol. 220, No. 4598, 1983, pp. 680–686.

[64] Wang, J., and Glimcher, M. J., "Characterization of matrix-induced osteogenesis in rat calvarial bone defects: II. Origins of bone-forming cells," *Calcif. Tissue Int.*, Vol. 65, No. 6, 1999, pp. 486–493.

[65] Wang, J., Yang, R., Gerstenfeld, L. C., and Glimcher, M. J., "Characterization of demineralized bone matrix-induced osteogenesis in rat calvarial bone defects: III. Gene and protein expression," *Calcif. Tissue Int.*, Vol. 67, No. 4, 2000, pp. 314–320.

[66] Gong, D., Grimes, C. A., Varghese, O. K., Hu, W., Singh, R. S., Chen, Z., and Dickey, E. C., "Titanium oxide nanotube arrays prepared by anodic oxidation," *J. Mater. Res.*, Vol. 16, No. 12, 2001, pp. 3331–3334.

[67] Mor, G. K., Varghese, O. K., Paulose, M., Mukherjee, N., and Grimes, C. A., "Fabricaton of tapered, conical-shaped titania nanotubes," *J. Mater. Res.*, Vol. 18, No. 11, 2003, pp. 2588–2593.

[68] Dalby, M. J., McCloy, D., Robertson, M., Agheli, H., Sutherland, D., Affrossman, S., and Oreffo, R. O., "Osteoprogenitor response to semi-ordered and random nanotopographies," *Biomaterials*, Vol. 27, No. 15, 2006, pp. 2980–2987.

[69] Dalby, M. J., McCloy, D., Robertson, M., Wilkinson, C. D., and Oreffo, R. O., "Osteoprogenitor response to defined topographies with nanoscale depths," *Biomaterials*, Vol. 27, No. 8, 2006, pp. 1306–1315.

[70] Lanza, R. P., Hayes, J. L., and Chick, W. L., "Encapsulated cell technology," *Nat. Biotechnol.*, Vol. 14, No. 9, 1996, pp. 1107–1111.

[71] Chen, J., Chu, I., Shio, M., Hsu, B., and Fu, S., "Microencapsulation of iselts in PEG-amine modified alginate-poly-(L-lysine)-alginate microcapsules for constructing bioartifical pancreas," *J. Ferment. Bioeng.*, Vol. 86, No. 2, 1998, pp. 185–190.

[72] Orive, G., Tam, S. K., Pedraz, J. L., and Halle, J. P., "Biocompatibility of alginate-poly-L-lysine microcapsules for cell therapy," *Biomaterials*, Vol. 27, No. 20, 2006, pp. 3691–3700.

[73] Thu, B., Bruheim, P., Espevik, T., Smidsrod, O., Soon-Shiong, P., and Skjak-Braek, G., "Alginate polycation microcapsules: II. Some functional properties," *Biomaterials*, Vol. 17, No. 11, 1996, pp. 1069–1079.

[74] Desai, T. A., Hansford, D., and Ferrari, M., "Characterization of micromachined silicon membranes for immunoisolation and bioseparation applications," *J. Memb. Sci.*, Vol. 159, 1999, pp. 221–231.

[75] Desai, T. A., Hansford, D. J., and Ferrari, M., "Micromachined interfaces: New approaches in cell immunoisolation and biomolecular separation," *Biomol. Eng.*, Vol. 17, No. 1, 2000, pp. 23–36.

[76] Desai, T. A., Hansford, D. J., Kulinsky, L., Nashat, A. H., Rasi, G., Tu, J., Wang, Y., Zhang, M., and Ferrari, M., "Nanopore technology for biomedical applications," *Biomed. Microdevices*, Vol. 2, No. 1, 1999 pp. 11–40.

[77] Desai, T. A., West, T., Cohen, M., Boiarski, T., and Rampersaud, A., "Nanoporous microsystems for islet cell replacement," *Adv. Drug Deliv. Rev.*, Vol. 56, No. 11, 2004, pp. 1661–1673.

[78] Leoni, L., Boiarski, A., and Desai, T. A., "Characterization of nanoporous membranes for immunoisolation: diffusion properties and tissue effects," *Biomed. Microdevices*, Vol. 4, No. 2, 2002, pp. 131–139.

[79] Leoni, L., and Desai, T. A., "Micromachined biocapsules for cell-based sensing and delivery," *Adv. Drug Deliv. Rev.*, Vol. 56, No. 2, 2004, pp. 211–229.

[80] Tao, S. L., and Desai, T. A., "Microfabricated drug delivery systems: From particles to pores," *Adv. Drug Deliv. Rev.*, Vol. 55, No. 3, 2003, pp. 315–328.

[81] Popat, K. C., Mor, G., Grimes, C. A., and Desai, T. A., "Surface modification of nanoporous alumina surfaces with poly(ethylene glycol)," *Langmuir*, Vol. 20, No. 19, 2004, pp. 8035–8041.

[82] Sharma, S., Johnson, R. W., and Desai, T. A., "Ultrathin poly(ethylene glycol) films for silicon-based microdevices," *Appl. Surface Sci.*, Vol. 206, 2003, pp. 218–229.

[83] Lopez, C. A., Fleischman, A. J., Roy, S., and Desai, T. A., "Evaluation of silicon nanoporous membranes and ECM-based microenvironments on neurosecretory cells," *Biomaterials*, Vol. 27, No. 16, 2006, pp. 3075–3083.

[84] Rutten, W. L., "Selective electrical interfaces with the nervous system," *Annu. Rev. Biomed. Eng.*, Vol. 4, 2002, pp. 407–452.

[85] Uludag, H., De Vos, P., and Tresco, P. A., "Technology of mammalian cell encapsulation," *Adv. Drug Deliv. Rev.*, Vol. 42, Nos. 1–2, 2000, pp. 29–64.

[86] Galindo, M. L., Hagmann, E., Marinello, C. P., and Zitzmann, N. U., "Long-term clinical results with Procera AllCeram full-ceramic crowns," *Schweiz Monatsschr Zahnmed*, Vol. 116, No. 8, 2006, pp. 804–809.

[87] Morel, X., Rias, A., Briat, B., el Aouni, A., D'Hermies, F., Adenis, J. P., Legeais, J. M., and Renard, G., "Biocompatibility of a porous alumina orbital implant. Preliminary results of an animal experiment," *J. Fr. Ophtalmol.*, Vol. 21, No. 3, 1998, pp. 163–169.

[88] Gong, D., Yadavalli, V., Paulose, M., Pishko, M., and Grimes, C. A., "Controlled molecular release using nanoporous alumina capsules," *Biomed. Microdevices*, Vol. 5, 2003, pp. 75–80.

[89] Itoh, N., Tomura, N., Tsuji, T., and Hongo, M., "Strengthened porous alumina membrane tube prepared by means of internal anodic oxidation," *Microporous Mesoporous Mater.*, Vol. 20, Nos. 4–6, 1998, pp. 333–337.

[90] Dalby, M. J., Riehle, M. O., Johnstone, H. J., Affrossman, S., and Curtis, A. S., "Polymer-demixed nanotopography: Control of fibroblast spreading and proliferation," *Tissue Eng.*, Vol. 8, No. 6, 2002, pp. 1099–1108.

[91] Lanza, G., Winter, P., Cyrus, T., Caruthers, S., Marsh, J., Hughes, M., and Wickline, S., "Nanomedicine opportunities in cardiology," *Annals NY Acad. Sci.*, Vol. 1080, 2006, pp. 451–465.

[92] Li, M., Guo, Y., Wei, Y., MacDiarmid, A. G., and Lelkes, P. I., "Electrospinning polyaniline-contained gelatin nanofibers for tissue engineering applications," *Biomaterials*, Vol. 27, No. 13, 2006, pp. 2705–2715.

[93] Tao, S., Young, C., Redenti, S., Zhang, Y., Klassen, H., Desai, T., Young, M., "Survival, migration and differentiation of retinal progenitor cells transplanted on micro-machined poly(methyl methacrylate) scaffolds to the subretinal space," *Lab Chip*, Vol. 7, 2007, pp 695–701.

[94] Popat, K., Leoni, L., Grimes, C. A., Desai, T. A., "Influence of engineered titania nanotubular surfaces on bone cells," *Biomaterials*, Vol. 28, No. 21, 2007, pp. 3188–3917.

[95] La Flamme, K. E., Popat, K. C., Leoni, L., Markiewicz, E., La Tempa, T. J., Roman, B. B., Grimes, C. A., Desai, T. A., "Biocompatibility of nanoporous alumina membranes for immunoisolation," *Biomaterials*, Vol. 28, No. 16, 2007, pp. 684–694.

[96] La Flamme, K. E., Mor, G., Gong, D., La Tempa, T. J., Fusaro, V. A., Grimes, C. A., Desai, T. A., "Nanoporous Alumina Capsules or Cellular Macroencapsulation: Transport and Biocompaitbility," *Diabetes. Technol. Ther.*, Vol. 7, No. 5, 2005, pp 684–694.

CHAPTER 6

Biomimetic Hydrogels to Support and Guide Tissue Formation

Jordan S. Miller and Jennifer L. West

6.1 Introduction

Each year, tens of thousands of patients die worldwide simply waiting for donor organs to become available for transplantation. While patient-specific, whole-organ regeneration has not yet been realized in the laboratory, research is ongoing to create three-dimensional, vascularized structures for tissue-engineered organ replacements. Most tissue-engineering approaches in current use involve random seeding of cells within porous polymer structures. While this has yielded success in some cases, generation of complex tissue structures may require control over the localization and behavior of multiple cell types in three dimensions. Hydrogels are a particularly attractive biomaterial for use in tissue engineering and regenerative medicine. As discussed below, they are highly biocompatible, they have mechanical properties that can approximate those of various soft tissues, and they can be crosslinked under mild reaction conditions that can be tolerated by encapsulated cells. Furthermore, it is often possible to modify bioinert hydrogel materials with bioactive moieties to control cell processes such as adhesion, proliferation, migration, and differentiation. Current research in the patterning of hydrogels has involved directing their formation in both two and three dimensions, as well as spatially controlling the covalent immobilization of bioactive factors such as peptides and proteins to generate controlled cellular microenvironments.

6.2 Hydrogels and Their Synthesis

Hydrogels, are an exciting class of materials for tissue-engineering applications because of their excellent biocompatibility and the ease with which their mechanical properties can be tuned to match those of soft tissues. They have been utilized over the past several decades for a broad range of medical applications, including contact lenses [1]; implant coatings [2–4]; tissue coatings and wound dressings [5–8]; cell transplantation [9]; microfluidic valves, actuators, and sensors [10–15]; and drug delivery [16–19]. In addition to their roles as medical devices and therapeutic vehicles, hydrogels have been continuously refined to exhibit many of the characteristics required for use as a scaffold for tissue engineering. These crosslinked, polymer-based networks are hydrophilic, three-dimensional scaffolds that can swell to many times their weight and size in water or biologic fluids. Their tunable chemical, mechanical, and biological properties make synthetic hydrogels an excellent system for approaching complex tissue-engineering challenges such as the construction of patient-specific whole organ replacements.

These scaffolds can be formed by physically or covalently crosslinking a liquid prepolymer solution into a solid hydrogel. A variety of material compositions can be used to make hydrogels, including agarose [20], alginate [21, 22], chitosan [23], polyvinyl alcohol [24], and even peptide amphiphiles [25, 26]. Collagen and fibrin gels are not considered hydrogels because of the insoluble, hydrophobic nature of the material compositions (these mechanically weak gels are formed from soluble materials that are precipitated to form gels by pH neutralization for the former, or enzymatic activity for the latter). Gelatin is more appropriately referred to as a colloidal gel, while hyaluronic acid gels are considered true hydrogels. One particular system of hydrogels based on polyethylene glycol (PEG) [also known as polyethylene oxide (PEO)] and its derivatives has proven extremely versatile for tissue-engineering applications.

PEG is Food and Drug Administration approved in a variety of applications and intrinsically exhibits high hydrophilicity, high biocompatibility, and little to no immunogenicity [27]. PEG hydrogels can be formed by first modifying individual PEG chains with two or more crosslinkable moieties, such as acrylate groups (Figure 6.1). Acrylates are susceptible to free-radical polymerization, which leads to crosslinking of the prepolymer solution and the formation of a solid hydrogel, as shown in Figure 6.1. Furthermore, the hydrophobic nature of the acrylate moieties with the hydrophilic PEG chains results in formation of micellelike structures in water, clustering the crosslinkable groups, resulting in very fast crosslinking. By utilizing photoinitiators, polymerization can be spatially and temporally controlled by exposure to ultraviolet or visible light [9, 28, 29]. This degree of control over the polymerization process can enable a variety of patterning technologies.

The photoinitiators typically used for hydrogel formation, as well as the photopolymerization process itself, exhibit low cytotoxicity, making it possible to polymerize a cell suspension and thus encapsulate viable cells within hydrogels [9, 30–32]. However, some cell types are dramatically less tolerant of photopolymerization, owing to a substantial cell-type dependence on the biocompatibility of different photoinitiators [33–35]. The concentrations of photoinitiator used for polymerization as well as exposure time to UV light are known to play a role as well. Cells that have been viably encapsulated within hydrogels include fibroblasts [36, 37], hepatocytes [33, 35], chondrocytes [38, 39], smooth muscle cells [29, 30, 40], osteochondrocytes and their progenitors [41–43], adipocytes [44], and neuronal cells [45]. Numerous photoinitiators are available, each with different characteristics of water solubility, biocompatibility, and means of decomposition. A selection of biocompatible photoinitiators is given in Table 6.1 [46–55].

6.3 Incorporating Bioactive Factors into Hydrogels

Most materials implanted in the body become coated with a monolayer of adsorbed proteins from the host almost immediately upon implantation. Inflammatory and interstitial cells interact with implanted materials through activation and clustering

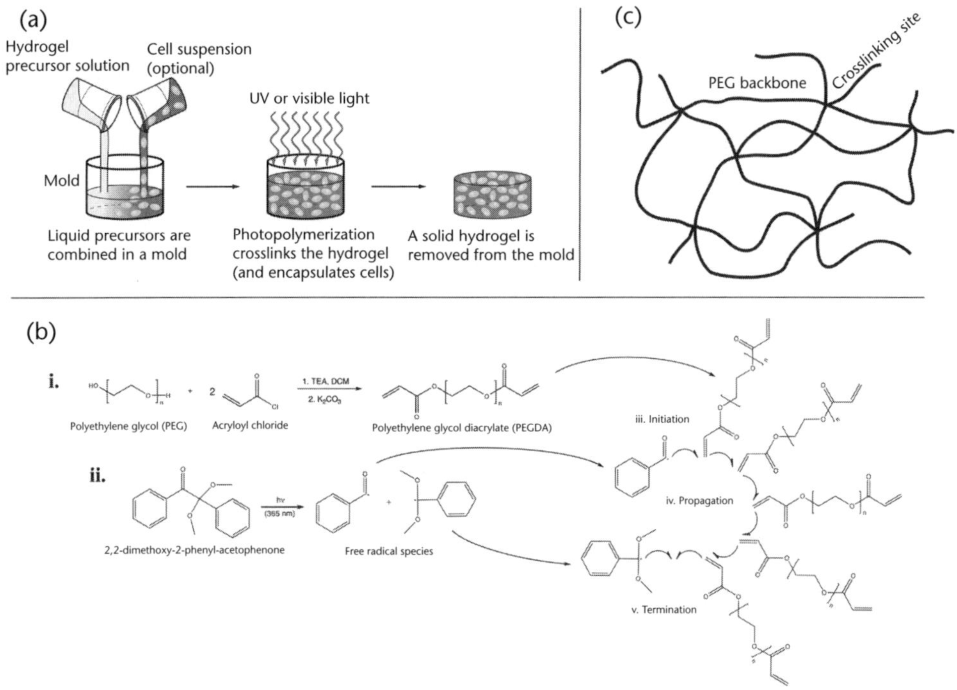

Figure 6.1 PEG hydrogel formation can be mediated by free-radical polymerization: (a) Macroscopic view of PEG hydrogel photopolymerization. A liquid precursor solution is poured into a mold, optionally along with a cell suspension. Exposure to light induces radical formation and crosslinking of the hydrogel network. The mold can be removed to reveal a solid hydrogel. (b) Synthetic scheme of PEG hydrogel formation: (i) Acrylated derivatives of PEG can be synthesized in a dry organic solvent such as dichloromethane (DCM) [30, 31]. (ii) Decomposition: the photoinitiator 2,2-dimethoxy-2-phenyl acetophenone is decomposed by long-wave ultraviolet light (365 nm) into two free-radical species. (iii) Initiation: a free radical attacks an acrylate group on a modified PEG chain. (iv) Propagation: preorganization of acrylate groups into micellelike structures by the hydrophobic effect enhances free-radical polymerization. Each modified PEG chain can participate in a polymerization event at each of its acrylate groups, leading to crosslinking of the prepolymer network and the formation of a solid hydrogel. (v) Termination: annihilation of two radical species terminates polymerization. (c) Schematic of hydrogel mesh network after photopolymerization. Each "crosslinking site" represents a micelle-like polymerization site as depicted in b.

of proteins found at the surface of these cells when they interact with adsorbed proteins on the implant. Thus, protein attachment to biomaterials is thought to be the key mediator of the deleterious effects, such as acute or chronic inflammation and the foreign-body response, seen after many device implantations [56, 57]. PEG exhibits excellent resistance to protein adsorption in an aqueous environment. While oligomers of ethylene glycol as short as three units have been shown to prevent protein adsorption to some degree [58], the best prevention of protein adsorption is typically seen in much longer PEG chains (with molecular weights greater than ~2 kDa) that are densely packed [59]. Crosslinked hydrogels composed of PEG chains are similarly able to inhibit protein adsorption and cell adhesion and elicit a minimal inflammatory response in vivo.

However, this inability of proteins to adsorb to PEG has another ramification that must be addressed when using PEG hydrogels as scaffolds for tissue engineer-

Table 6.1 Selection of Biocompatible Photoinitiators for Hydrogel Photopolymerization

Trade Name	*Structure (Chemical Name)*	*Abs. Max. (nm)*	*Ref(s)*
Irgacure-2959	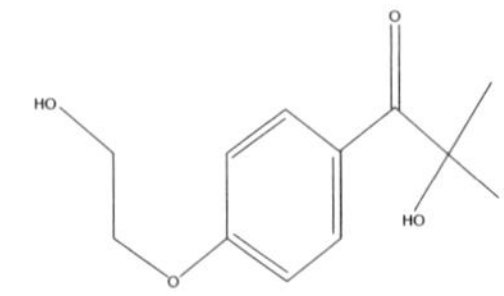 (2-hydroxy-1-[4-(hydroxyethoxy)phenyl]-2-methyl-1-propanone)	< 313	[39, 46, 47]
Darocur-2959	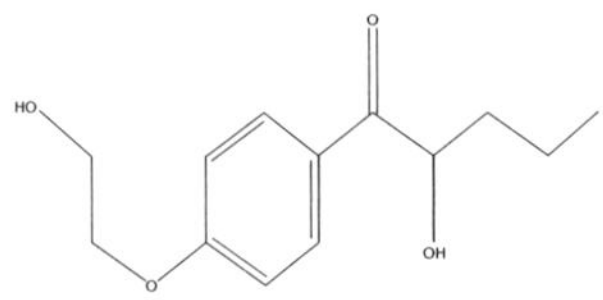 (2-hydroxy-1-[4-(hydroxyethoxy)phenyl]-1-propanone)	< 313	[48]
Irgacure-184	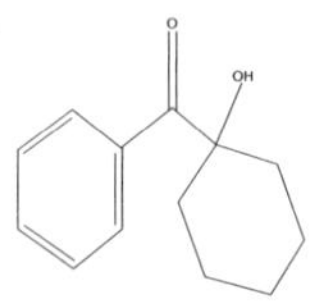 (1-hydroxycyclohexane acetophenone)	326	[47, 48]
Irgacure-651	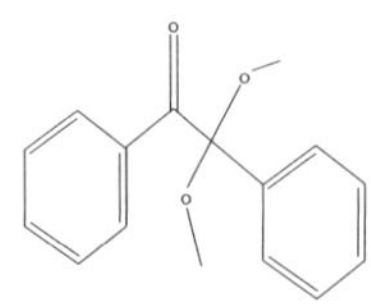 (2,2-dymethoxy-2-phenyl acetophenone)	335	[28–30, 47–50]
THX	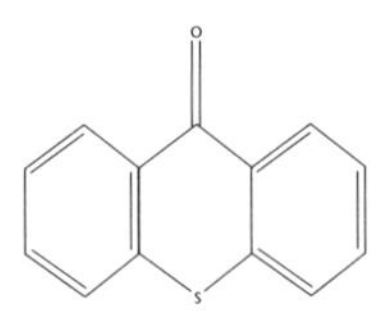 (thioxanthone)	378	[47, 48]
Eosin Y	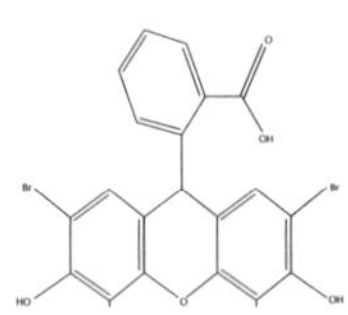	514	[51–55]

ing. We have mentioned that cells can be encapsulated inside of PEG hydrogels and can tolerate the photopolymerization process, but a corollary to low protein adsorption to PEG is that encapsulated cells cannot then recognize or bind to their surrounding PEG microenvironment. The solution to maintaining the high biocompatibility and low protein adsorption of PEG hydrogels, while simultaneously allowing cells to thrive in a largely PEG environment, is to add very specific moieties back into the hydrogel to mediate specific receptor-ligand interactions. The unmodified PEG hydrogel we have described until now serves as a "blank slate" to which bioactive factors can be added to mediate and modulate cell behavior. Rather than having to use whole adhesive proteins commonly found in the extracellular matrix (ECM), it was found in the 1980s that some short fragments of these proteins can have similar binding potential as the entire protein itself [60]. For example, the tripeptide arginine-glycine-aspartic acid (abbreviated as RGD) serves as the recognition site in a multitude of ECM proteins such as fibronectin, vitronectin, collagens, and fibrin, interacting with cell-adhesion proteins called integrins. Immobilizing this tripeptide to a surface enables many mammalian cell types to attach and proliferate at faster rates than if the peptide were not present [60–63]. Furthermore, it has been shown that soluble versions of the peptide have the opposite effect: soluble adhesive peptides bind to the same cellular proteins responsible for substrate adhesion and thus outcompete the substrate. Therefore, peptides such as RGD and their corresponding proteins cannot simply be mixed into the hydrogel prepolymer solution before polymerization; they must be covalently crosslinked and immobilized into the hydrogel network in order to stimulate cell adhesion rather than being soluble and inhibiting adhesion. RGD is presented here as an illustrative example. There are many cell-adhesive peptides that can be used similarly. Depending on the receptor(s) to which a specific peptide can bind, it is sometimes possible to mediate adhesion of only targeted cell types. For example, the peptides REDV and YIGSR have been shown to be selective for endothelial cells [61, 64], IKVAV for neurons [65], and VAPG for smooth muscle cells [29, 49].

Additionally, unmodified PEG hydrogels are sometimes inappropriately termed "biodegradable." We wish here to draw a distinction between "readily biodegradable" and "ultimately biodegradable" as outlined by the Organisation for Economic Cooperation and Development. Many PEGs used as biocompatible hydrogels are acrylated as described in Figure 6.1, meaning that after crosslinking, they will still have single ester functional groups at their termini that can be cleaved by slightly acidic or basic conditions in their local environment over very long periods. Thus, PEG hydrogels can be considered "ultimately biodegradable" because over long periods, the hydrogels will eventually erode and disintegrate into PEG chains and polyacrylic acid that are excreted from the body. This time scale is typically much longer (up to several years) than that generally attributed to most implant materials commonly termed "biodegradable," such as biodegradable chitin-based sutures, which degrade naturally in only a few weeks.

A way to make PEG hydrogels more biodegradable, or "readily biodegradable," is to incorporate either readily hydrolyzable groups or proteolytically degradable peptides into the backbone of individual PEG chains. Readily hydrolyzable groups include oligomers of lactic acid [66], which spontaneously and autocatalytically degrade under physiologic conditions [67–69]. A key advantage of using degradable peptides over such readily hydrolyzable groups is that the peptides can be engineered to degrade only by cellular enzymes, which are part of the normal ECM remodeling and healing process [28, 29, 36, 37, 70–72]. With control over degradation kinetics, the loss of mechanical integrity of a synthetic hydrogel during cell-mediated remodeling can be tailored to match the same kinetics as cellular deposition of natural ECM. For example, a tissue-engineered vascular graft (TEVG) must fulfill biological and mechanical functions at the site of implantation. A successful TEVG will fulfill both of these roles at the time of implantation and will also be able to maintain them throughout the regenerative process as the scaffold is replaced with native tissue [31, 73–77]. The specific enzymes utilized by cells for ECM remodeling are termed matrix-metalloproteinases (MMPs) because they proteolytically degrade ECM and require specific metallic cations to function properly (such as Zn^{2+}). These degradable peptides, often termed MMP-sensitive peptides, permit cell-mediated remodeling of, and migration through, synthetic hydrogels, which will ultimately be required for the synthesis of true three-dimensional tissue replacements.

In addition to peptides, larger proteins, such as growth factors, may also be covalently incorporated into the hydrogel network. The chemistry to effect these changes can simply involve reacting the peptide or protein of interest with a heterobifunctional PEG chain that is then incorporated into the hydrogel as it is crosslinked. A common heterobifunctional PEG used for this purpose is acrylate-PEG-NHS (3400 Da), which consists of an N-hydroxysuccinimide (NHS) active ester at one terminus for reaction with free amines on the peptide or protein of interest and an acrylate at the other terminus for incorporation into the hydrogel (Figure 6.2). The result is a complex tissue-engineering scaffold that can elicit very specific cellular responses (Figure 6.3). Now that the toolkit for hydrogel synthesis and modification has been laid out, one needs only devise ways to spatially deposit diverse hydrogel chemistries in three dimensions in order to build up large tissue analogs for tissue-engineering applications.

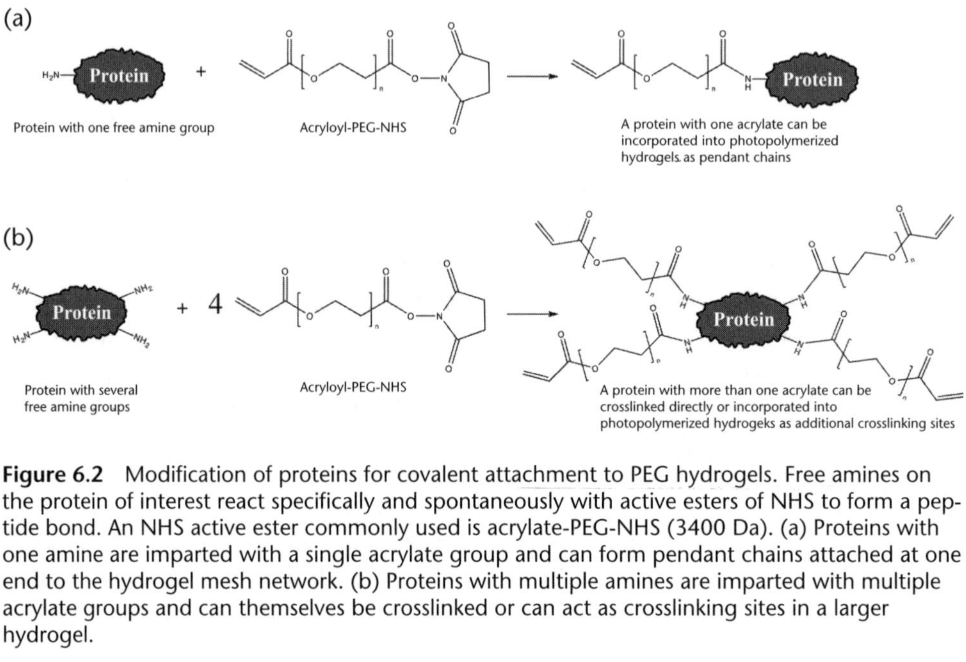

Figure 6.2 Modification of proteins for covalent attachment to PEG hydrogels. Free amines on the protein of interest react specifically and spontaneously with active esters of NHS to form a peptide bond. An NHS active ester commonly used is acrylate-PEG-NHS (3400 Da). (a) Proteins with one amine are imparted with a single acrylate group and can form pendant chains attached at one end to the hydrogel mesh network. (b) Proteins with multiple amines are imparted with multiple acrylate groups and can themselves be crosslinked or can act as crosslinking sites in a larger hydrogel.

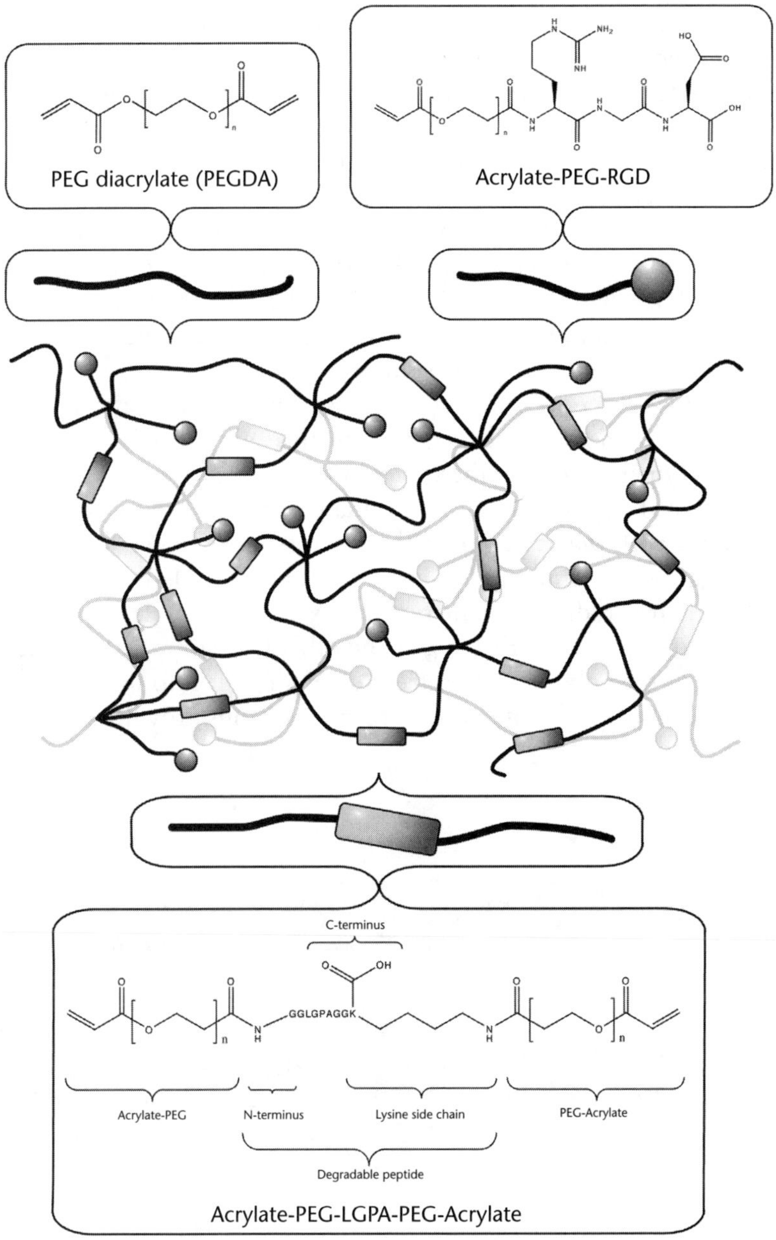

Figure 6.3 Bioactive PEG hydrogels: (a) Modification of peptides (shown here) or proteins with acrylate groups allows their covalent incorporation into PEG hydrogels during crosslinking. MMP-sensitive peptides may also be incorporated along the PEG backbone to allow proteolytic, cell-mediated scaffold degradation.

6.4 Two-Dimensional Patterning of Hydrogels

The union of hydrogel photopolymerization chemistry with photolithographic techniques from the microprocessor industry has enabled the rapid prototyping of two-dimensional hydrogel structures for in vitro studies. While these structures are not technically "two-dimensional" according to the rigorous mathematical definition, the term "two-dimensional hydrogel structure" generally refers to hydrogels that are uniform in the z-direction but vary in the x- and y-directions. These structures are typically thin hydrogel sheets; no more than a few hundred micrometers thick, they can span several centimeters in length and width.

In one method developed by Pishko and coworkers, a cell suspension is mixed with a hydrogel prepolymer solution and applied to a flat substrate or within microfluidic channels [10, 78, 79]. If the entire surface were illuminated, the prepolymer solution would crosslink everywhere and form a single hydrogel. Instead, illumination of the solution through a transparency film printed with opaque regions permits the crosslinking of individual regions of the prepolymer cell suspension. The opaque regions of this photomask block light and inhibit photopolymerization in these shadowed regions, while the transparent regions allow light to pass through and initiate hydrogel crosslinking. Simply rinsing the substrate with water removes the uncrosslinked regions, revealing free-standing hydrogel posts containing living cells (Figure 6.4). It is important to note that although free-radical polymerization is a fast chemical process, it is efficiently quenched by atmospheric oxygen and other free radicals. This quenching is believed to prevent the polymerization from extending far into the light-shielded regions on the typical time scales of hydrogel formation, thus providing high contrast at feature edges [34].

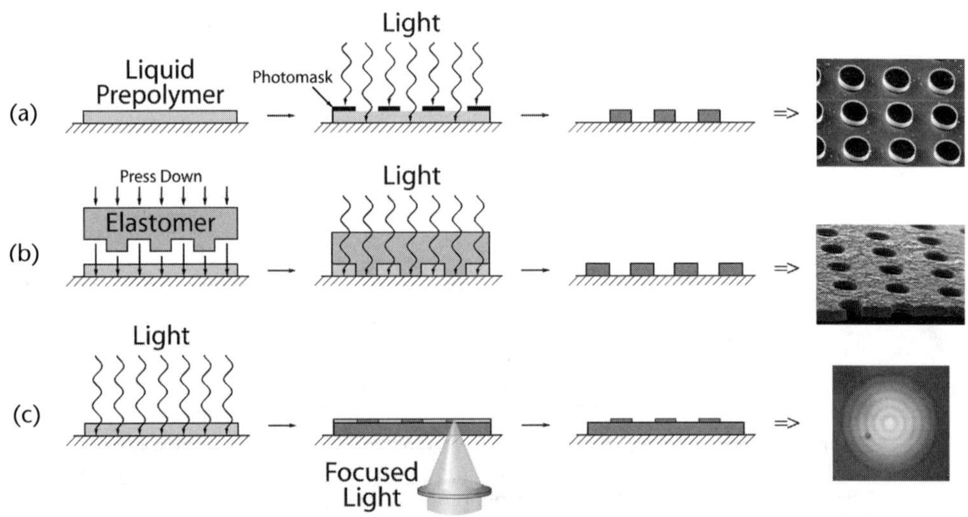

Figure 6.4 Methods for the rapid prototyping of two-dimensional hydrogels: (a) Photomasking: a photomask and light source are used to polymerize a thin layer of prepolymer solution in specific regions. The unpolymerized regions are rinsed clean, revealing free-standing hydrogel structures [78]. Resolution is largely determined by the resolution of the photomask and the nature of the light source (collimated or Gaussian). (b) Capillary force lithography: a micropatterned elastomer (typically PDMS) is applied with force to a thin layer of hydrogel prepolymer solution, squeezing out solution from patterned areas. After crosslinking, the elastomer is removed to reveal a patterned hydrogel [80]. (c) Hydrogel surface modification: the surface of preformed, incompletely crosslinked hydrogels can be further modified with additional photolithographic techniques, such as traditional photomasking or lithography with focused light (e.g., from a confocal microscope) [81].

Transparency photolithography can also be utilized to create two-dimensional patterns of bioactive factors on the surface of preformed hydrogels. This technique utilizes incomplete polymerization of a base hydrogel, leaving free acrylate groups available for further modification. When acrylated bioactive factors are applied to the surface of the hydrogel and illuminated through a photomask, the factors are covalently immobilized to the surface of the hydrogel [82]. Iterative replication of this process enables the attachment of multiple factors to the surface of the hydrogel with much lower expense and complication than if the factors were dissolved in the original prepolymer solution. Lower expense is achieved because the bioactive factors need only be applied to the surface of the hydrogel rather than the bulk prepolymer solution. Additionally, a key advantage of using a hydrogel for the immobilization of ligands is that unpatterned regions remain highly bioinert, while the entire hydrogel is both flexible and compliant.

Rather than utilizing photomasks, liquid crystal display (LCD) projection photolithography utilizes a commercially available LCD projector to cast an image from a computer directly onto the prepolymer solution to create hydrogel structures [83–86]. This technique typically has a feature resolution of 50 μm and sometimes suffers exposure artifacts from the embedded wires used to control each pixel in the LCD [83]. Nevertheless, LCD projectors are inexpensive and can serve as an excellent starting point for exploring the parameters needed for the photolithography of hydrogels.

An inverse approach involves creating free-standing hydrogel that form individual wells for the isolation of cells and cell populations. Research in this area is ongoing to create addressable templates for high-throughput screening, stimulate cell differentiation and proliferation, and investigate neural network formation and communication. Microwells can be created directly with photolithography using various masking techniques [87, 88]. Alternatively, a process named capillary force lithography can provide similar functionality by utilizing a micropatterned elastomeric polydimethylsiloxane (PDMS) stamp to squeeze out PEG prepolymer from regions of interest, thus generating the desired microwells [89, 90].

An intriguing approach for patterning thick hydrogels has involved the design of a photosensitive agarose hydrogel [20]. A confined laser caused photochemical modification of the engineered hydrogel to reveal free sulfhydryl groups within the patterned regions. The sulfhydryl groups were functionalized with derivatized RGD peptides to create columns of RGD peptide in the thick hydrogel. These patterned RGD-containing channels were able to confine the extension of primary rat dorsal root ganglia cells to the patterned regions of the scaffold. In this approach, the patterned channels went through the entire thickness of the hydrogel and came closer to approaching a three-dimensional construct than the methods discussed above. However, due to the use of a confined laser (beam width was ~200 μm) rather than a focused laser, the patterned channels were vertically uniform, and so we still classify this technique as two-dimensional patterning.

While all of these methods for two-dimensional patterning are relatively straightforward and useful for creating in vitro systems for exploratory research, they lack the resolution needed to create scaffolds with feature resolution on the order of one micron, the same scale in which cells sense and respond to their insoluble environment through focal contacts. The resolution of these techniques typically ranges in the tens to hundreds or thousands of microns. While equipment commonly found in a microprocessor clean room, such as a mask maker and contact mask aligner, could address hydrogel chemistry on this scale, they do not lend themselves well to wet polymer chemistry or cell-permissive environments.

The construction of higher-resolution two-dimensional hydrogel structures can be achieved with higher-precision instrumentssuch as a confocal microscope that are designed to work with cells in their native environment. Confocal microscopes have been optimized over the past several decades for cell imaging. These microscopes illuminate the sample with a laser focused to a diffraction-limited spot size that is then raster-scanned across the sample surface. Computer control of the scanning lasers and the laser shutter enables two-dimensional hydrogel patterning without the need for a physical photomask. Although raster scanning is slower than wide-field illumination through a photomask, the trade-off is increased precision of the patterned region, typically on the order of 1μ [81].

These patterning techniques have proven useful for simple studies of cellular behavior, such as adhesion, migration, differentiation, proliferation, and coculture. These two-dimensional systems, while lacking fundamental characteristics necessary to adequately mimic a three-dimensional tissue, aid in understanding some of the fundamental controls that influence cell behavior before attempting to move to a necessarily more complicated three-dimensional environment. While many of these studies have utilized PEG as a biocompatible and versatile foundation for hydrogel synthesis, these techniques are usually easily translated to other photocrosslinkable or photosensitive materials.

6.5 Three-Dimensional Rapid Prototyping of Hydrogels

While two-dimensional patterning of hydrogels continues to be widely investigated and has proven useful for numerous elegant in vitro studies, the most promising and versatile methods for constructing mimics of native tissue are those techniques that enable the creation of true three-dimensional constructs, that is, constructs that can chemically or structurally vary in the x-, y-, and z- directions. Additionally, although PEG hydrogels exhibit excellent permeability to water, oxygen, and nutrients compared to many other biocompatible materials, the passive permeability of these hydrogels is typically in the range of hundreds of microns to a few millimeters [10, 35, 38, 46, 79, 91]. Therefore, hydrogels thicker than a few hundred microns should be made porous or, even better, made with active vascular networks to nourish entrapped or invading cells. Current research into three-dimensional prototyping of hydrogels typically involves creating structurally heterogeneous cell-containing constructs, as well as providing means for nourishing attached and entrapped cells, thus allowing the cells to proliferate and remodel the hydrogel into something approaching native tissue.

6.5.1 Single-Photon Excitation

The simplest method to create three-dimensional constructs is iterative or layer-by-layer extension of the techniques discussed above for two-dimensional patterning (Figure 6.5). Liu and Bhatia utilized multiple applications of a

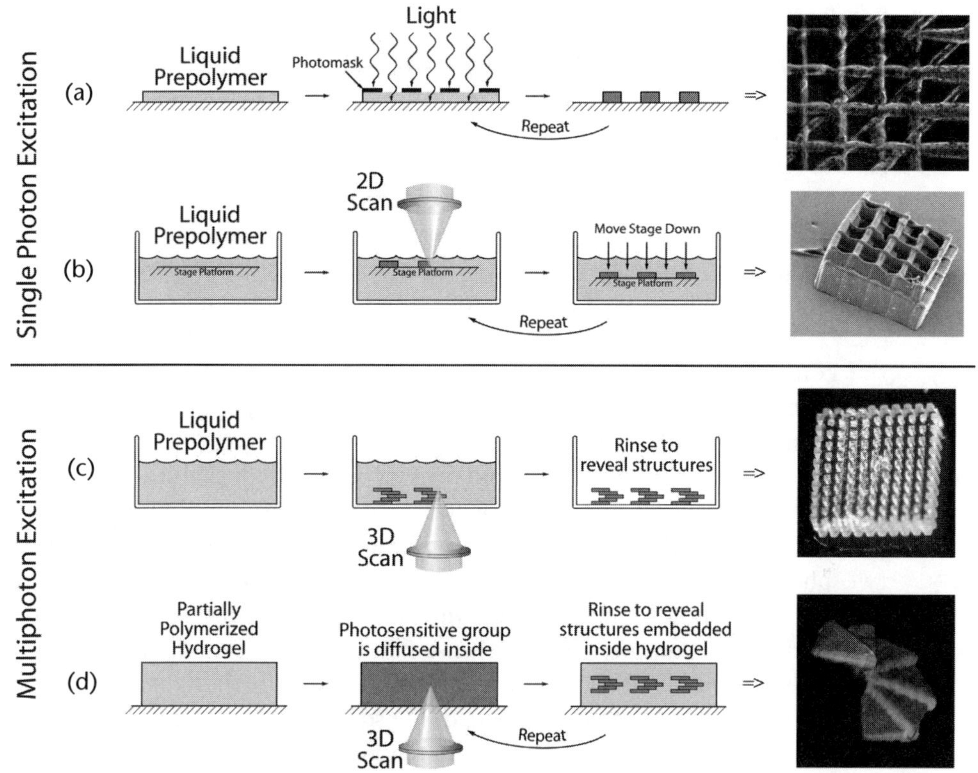

Figure 6.5 Methods for the rapid prototyping of three-dimensional hydrogels: (a) Layer-by-layer lithography: a photomask and light source are used to polymerize a thin layer of prepolymer solution in specific regions. The unpolymerized regions are rinsed clean. The supporting stage is incremented to provide room for the next layer, and the process is repeated [34]. Resolution is largely determined by the resolution of the photomask, the nature of the light source (collimated or Gaussian), and the step size of the supporting stage. (b) Laser-based stereolithography: focused lasers improve efficiency and control compared to layer-by-layer photomasking [92]. (c) Multiphoton lasers can be used to buildup arbitrary structures in three dimensions without the danger of polymerizing regions that are out of focus. (d) Multiphoton lasers can also be used to heterogeneously modify the biochemical and mechanical properties inside bulk hydrogels [93].

photomask to polymerize overlapping PEG hydrogel microstructures [34]. In this study, up to three hydrogel layers were overlaid with a resolution of hundreds of microns. More recent extension of this work has resulted in much larger structures with improved resolution. Additionally, these structures were open and interconnected, which enabled cellular maintenance by convective flow of tissue culture media [35].

Laser-based layer-by-layer polymerization has typically involved custom-built laser-scanning systems with a programmable stage. Roy and coworkers have created extended microwell scaffolds hundreds of microns tall for the study of stem cell differentiation [86, 92]. Even more recently, Arcaute, Mann, and Wicker used a similar process to create hydrogel structures tens of millimeters tall [94]. Multilumen PEG hydrogel conduits and hydrogel constructs with extended channel architecture have also been demonstrated.

6.5.2 Multiphoton Excitation

In the construction of a three-dimensional scaffold to support and guide tissue formation, one would ideally like a fast, easy-to-use method that can achieve subcellular resolution. It is well known that cells sense and respond to their surroundings on the scale of 1μ or less [61], so to gain excellent cellular control, one would like to construct the cellular microenvironment with a resolution on the order of 1μ. The methods mentioned above for three-dimensional polymerization suffer from resolution problems that will hamper the ability to precisely control this microenvironment as well as time-consuming development and alignment steps for each fabricated layer. To achieve three-dimensional subcellular resolution in material constructs, photopolymerization directed by multiphoton lasers has recently proven extremely useful [93, 95–100].

A multiphoton laser system attached to a commercially available confocal microscope can be used for the rapid prototyping of hydrogels for tissue-engineering applications. Multiphoton lasers operate according to quantum mechanical principles first theorized in the 1930s by Göppert-Mayer [101]. In the simplest case, two-photon excitation, two photons of a given wavelength can excite the electrons of a target molecule with effects equivalent to a single photon at half the wavelength. In a practical example, two photons of red light can induce fluorescence of a fluorophore, which is typically excited only by green light (Figure 6.6). This so-called two-photon effect varies with the square of the laser intensity, or the number of photons passing through a unit area per unit time [100]. For the two-photon effect to occur, two photons need to be absorbed by the same molecule nearly simultaneously (~10^{-16} s). It was found that an extremely high photon flux, from MW/cm^2 to GW/cm^2, was needed to achieve this multiphoton effect on a scale large enough to be observed. Light intensities in this range would obliterate biologic samples, so pulsed Ti:Sapphire lasers were developed. These lasers are able to achieve MW/cm^2

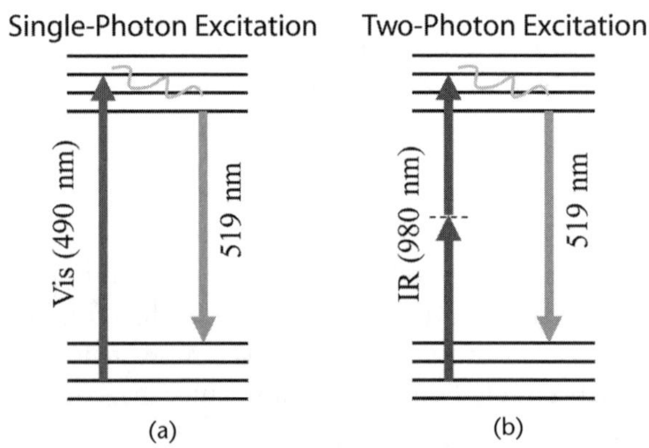

Figure 6.6 Jablonski energy diagrams of single- and two-photon excitation of a fluorophore: (a) In single-photon excitation of a fluorophore, a single photon of visible light (e.g., λ = 490 nm) excites electrons to higher energy levels. When the electrons drop down to lower energy levels, a photon of a longer wavelength (e.g., λ = 519 nm) is released. (b) In two-photon excitation of a fluorophore, two photons at twice the original wavelength (e.g., λ = 980 nm) impact the fluorophore nearly simultaneously (~10^{-16} s), promoting electrons to higher energy levels. When the electrons drop down to lower energy levels, a photon of a shorter wavelength (e.g., λ = 519 nm) is released. Two photons at a longer wavelength can have the same effect as a single photon at half the wavelength.

to GW/cm^2 but bypass these huge energy requirements by outputting this photon flux only over very short time scales (10^{-13} s) spaced very far apart (10^{-8} s); thus, their average power output is only 1W. Fluorescence occurs on much longer time scales (10^{-9} s), than the light pulses, and strong fluorescence emission can be detected.

Many groups have leveraged the two-photon effect with confocal microscopes for high-resolution fluorescence imaging and deeper tissue penetration, as first reported by Denk, Strickler, and Webb [96]. Confocal microscopes focus a laser beam to a diffraction-limited size, so the laser intensity at the focal point is exponentially higher than that outside the focal plane. By optimizing the power output of these pulsed lasers, the two-photon effect has been successfully confined to within the focal point and nowhere else (Figure 6.7) [96, 102]. These same principles used for three-dimensional fluorescence excitation can also be applied for the present purpose: the photopolymerization of hydrogels with high three-dimensional resolution. To photopolymerize three-dimensional structures, the hydrogel precursors are polymerized only at the focal point of the two-photon laser. Computer control of the laser shutter, the focal point position, and the sample stage enable buildup of complex hydrogel structures. After structure formation, unpolymerized precursors are rinsed away in a single step revealing the three-dimensional scaffold of interest. The three-dimensional resolution inherent in two-photon lasers obviates the need for layer-by-layer substrate deposition, making structure formation much more efficient and precise. A complementary method is to start with a bulk PEG hydrogel that has been incompletely polymerized. The hydrogel can be permeated with peptide-modified PEG-acrylate precursors, and the two-photon laser is raster-scanned inside the hydrogel in three dimensions. Following polymerization, residual precursors are allowed to diffuse out of the hydrogel to reveal a heterogeneously distributed peptide. Initial work in this area has demonstrated the ability to create complicated three-dimensional patterns of bioactivity within a largely unmodified PEG hydrogel [93]. Extension of this technique will be useful for creating vascular structures for nourishing cells inside thicker tissue mimics and creating migratory channels of bioactivity for three-dimensional directed cell migration. Further work will theoretically enable resolution below 1ì, finally allowing the creation of tissue-engineering scaffolds that can address individual cells on the same scale in which they perceive the world.

6.6 Summary

Hydrogels—mesh networks of crosslinked, hydrophilic polymers—are an exciting class of materials for approaching complex tissue-engineering challenges such as the construction of patient-specific whole organ replacement. A variety of material compositions and crosslinking techniques can be utilized for their formation, and we have highlighted the use of PEG as a widely investigated and promising hydrogel base material. Hydrogels can be quickly formed by crosslinking a macromer solution containing free-radical photoinitiators and can encapsulate viable cells. The hydrogel mesh network can be engineered with covalently attached bioactive factors, such as adhesive peptides and growth factors, to further direct cell behavior.

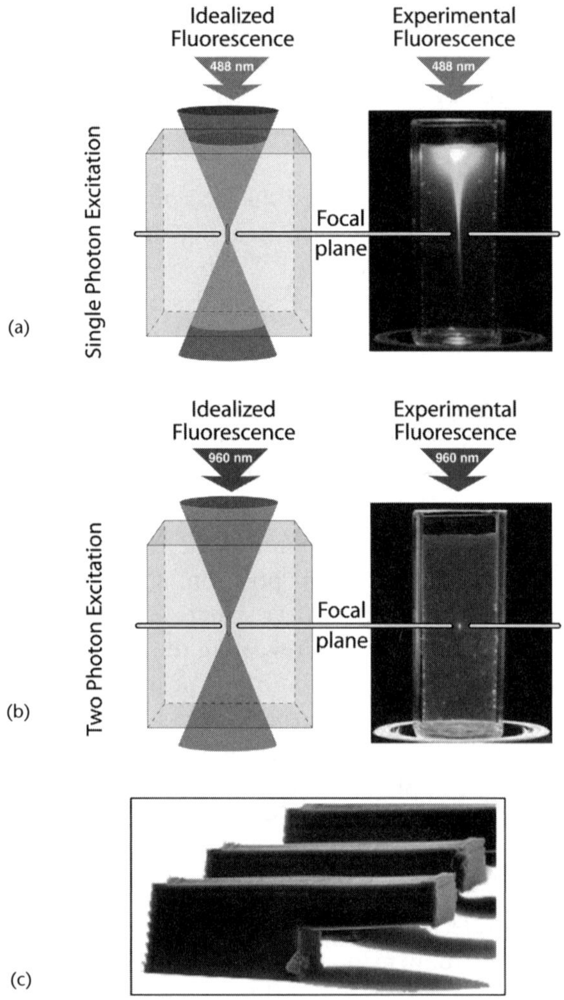

Figure 6.7 Idealized and experimental single- and two-photon excitation. A focused laser beam is projected into a cuvette filled with a solution of fluorescein, a common fluorophore. (a) In idealized single-photon excitation, all blue light entering the solution causes excitation, even outside of the focal plane. Experimental observation of this phenomenon demonstrates the difficulties in using single-photon excitation for photolithography because any precursor solution within this cone-shaped beam path will be crosslinked. (b) In idealized two-photon excitation, only red light at the focal point is at sufficient intensity to cause two-photon events; thus, excitation is confined to the focal point. Experimental observation of two-photon excitation demonstrates that much higher axial resolution is possible compared to single-photon excitation [100]. (c) The high axial confinement of two-photon excitation allows it to be used for rapid, high-resolution, three-dimensional microfabrication [95].

Additionally, readily hydrolyzable groups and MMP-sensitive degradable peptides can be incorporated along the hydrogel backbone to engineer its degradation kinetics. Conventional photolithography has enabled the construction of hydrogel microstructures, which can vary in both two and three dimensions. Two-photon photopolymerization enables even greater precision when constructing these scaffolds and can be used to add bioactive factors back inside bulk hydrogels at subcellular resolution.

The high biocompatibility of hydrogels, their tunable biochemical and mechanical properties, and the myriad methods available for patterning their structure in both two and three dimensions make hydrogels an excellent family of materials for investigating fundamental mechanisms of cell biology and engineering the cellular microenvironment. Ultimately, the vascularization of hydrogels through three-dimensional rapid prototyping and three-dimensional biochemical patterning at subcellular resolution will enable the investigation of small tissue mimics and continue progress toward true synthetic tissues and organs for human transplantation.

References

[1] Goda, T., and Ishihara, K., "Soft contact lens biomaterials from bioinspired phospholipid polymers," *Expert Rev. Med. Devices*, Vol. 3, No. 2, 2006, pp. 167–174.

[2] Harris, L. G., Patterson, L. M., Bacon, C., Gwynn, I., and Richards, R. G., "Assessment of the cytocompatibility of different coated titanium surfaces to fibroblasts and osteoblasts," *J. Biomed. Mater. Res. A*, Vol. 73, No. 1, 2005, pp. 12–20.

[3] Prokop, A., Kozlov, E., Nun Non, S., Dikov, M. M., Sephel, G. C., Whitsitt, J. S., Davidson, J. M., "Towards retrievable vascularized bioartificial pancreas: Induction and long-lasting stability of polymeric mesh implant vascularized with the help of acidic and basic fibroblast growth factors and hydrogel coating," *Diabetes Technol. Ther.*, Vol. 3, No. 2, 2001, pp. 245–261.

[4] Spargo, B. J., Rudolph, A. S., and Rollwagen, F. M., "Recruitment of tissue resident cells to hydrogel composites: In vivo response to implant materials," *Biomaterials,* Vol. 15, No. 10, 1994, pp. 853–858.

[5] Eaglstein, W. H., "Experiences with biosynthetic dressings," *J. Am. Acad. Dermatol.*, Vol. 12, No. 2, Pt. 2, 1985, pp. 434–440.

[6] Leaper, D. J., Brennan, S. S., Simpson, R. A., and Foster, M. E., "Experimental infection and hydrogel dressings," *J. Hosp. Infect.*, Vol. 5, Suppl. A, 1984, pp. 69–73.

[7] Geronemus, R. G., and Robins, P., "The effect of two new dressings on epidermal wound healing," *J. Dermatol. Surg. Oncol.*, Vol. 8, No. 10, 1982, pp. 850–852.

[8] West, J. L., Chowdhury, S. M., Sawhney, A. S., Pathak, C. P., Dunn, R. C., and Hubbell, J. A., "Efficacy of adhesion barriers: Resorbable hydrogel, oxidized regenerated cellulose and hyaluronic acid," *J. Reprod. Med.*, Vol. 41, No. 3, 1996, pp. 149–154.

[9] Pathak, C. P., Sawhney, A. S., and Hubbell, J. A., "Rapid photopolymerization of immunoprotective gels in contact with cells and tissue," *J. Am. Chem. Soc.*, Vol. 114, 1992, pp. 8311–8312.

[10] Koh, W. G., Itle, L. J., and Pishko, M. V., "Molding of hydrogel microstructures to create multiphenotype cell microarrays," *Anal. Chem.*, Vol. 75, No. 21, 2003, pp. 5783–5789.

[11] Sheppard, N. F., Jr., Lesho, M. J., McNally, P., and Francomacaro, A. S., "Microfabricated conductimetric pH sensor," *Sensors and Actuators B: Chemical*, Vol. 28, No. 2, 1995, pp. 95–102.

[12] Russell, R. J., Pishko, M. V., Gefrides, C. C., McShane, M. J., and Coté, G. L., "A fluorescence-based glucose biosensor using concanavalin A and dextran encapsulated in a poly(ethylene glycol) hydrogel," *Anal. Chem.*, Vol. 71, No. 15, 1999, pp. 3126–3132.

[13] Sudhölter, E. J. R., van der Wal, P. D., Skowronska-Ptasinska, M., van den Berg, A., Bergveld, P., and Reinhoudt, D. N., "Modification of ISFETs by covalent anchoring of poly(hyroxyethyl methacrylate) hydrogel: Introduction of a thermodynamically defined semiconductor-sensing membrane interface," *Analytica Chimica Acta*, Vol. 230, 1990, pp. 59–65.

[14] Kim, D. H., Kim, P., Song, I., Cha, J. M., Lee, S. H., Kim, B., and Suh, K. Y., "Guided three-dimensional growth of functional cardiomyocytes on polyethylene glycol nanostructures," *Langmuir*, Vol. 22, No. 12, 2006, pp. 5419–5426.

[15] Sershen, S. R., Mensing, G. A., Ng, M., Halas, N. J., Beebe, D. J., and West, J. L., "Independent optical control of microfluidic valves formed from optomechanically responsive nanocomposite hydrogels," *Adv. Mater.*, Vol. 17, No. 11, 2005, pp. 1366–1368.

[16] Eddington, D. T., and Beebe, D. J., "Flow control with hydrogels," *Adv. Drug Deliv. Rev.*, Vol. 56, No. 2, 2004, pp. 199–210.

[17] Peppas, N. A., Bures, P., Leobandung, W., and Ichikawa, H., "Hydrogels in pharmaceutical formulations," *Eur. J. Pharm. Biopharm.*, Vol. 50, No. 1, 2000, pp. 27–46.

[18] Peppas, N. A., Wood, K. M., and Blanchette, J. O., "Hydrogels for oral delivery of therapeutic proteins," *Expert Opin. Biol. Ther.*, Vol. 4, No. 6, 2004, pp. 881–887.

[19] Serra, L., Doménech, J., and Peppas, N. A., "Drug transport mechanisms and release kinetics from molecularly designed poly(acrylic acid-g-ethylene glycol) hydrogels," *Biomaterials*, Vol. 27, No. 31, 2006, pp. 5440–5451.

[20] Luo, Y., and Shoichet, M. S., "A photolabile hydrogel for guided three-dimensional cell growth and migration," *Nat. Mater*, Vol. 3, No. 4, 2004, pp. 249–253.

[21] Augst, A. D., Kong, H. J., and Mooney, D. J., "Alginate hydrogels as biomaterials," *Macromol. Biosci.*, Vol. 6, No. 8, 2006, pp. 623–633.

[22] Drury, J. L., Boontheekul, T., Boontheeku, T., and Mooney, D. J., "Cellular cross-linking of peptide modified hydrogels," *J. Biomech. Eng.*, Vol. 127, No. 2, 2005, pp. 220–228.

[23] Ishihara, M., Obara, K., Nakamura, S., Fujita, M., Masuoka, K., Kanatani, Y., Takase, B., Hattori, H., Morimoto, Y., Ishihara, M., Maehara, T., and Kikuchi, M., "Chitosan hydrogel as a drug delivery carrier to control angiogenesis," *J. Artif. Organs*, Vol. 9, No. 1, 2006, pp. 8–16.

[24] Schmedlen, R. H., Masters, K. S., and West, J. L., "Photocrosslinkable polyvinyl alcohol hydrogels that can be modified with cell adhesion peptides for use in tissue engineering," *Biomaterials*, Vol. 23, No. 22, 2002, pp. 4325–4332.

[25] Hartgerink, J. D., Beniash, E., and Stupp, S. I., "Peptide-amphiphile nanofibers: A versatile scaffold for the preparation of self-assembling materials," *Proc. Natl. Acad. Sci. USA*, Vol. 99, No. 8, 2002, pp. 5133–5138.

[26] Hartgerink, J. D., Beniash, E., and Stupp, S. I., "Self-assembly and mineralization of peptide-amphiphile nanofibers," *Science*, Vol. 294, No. 5547, 2001, pp. 1684–1688.

[27] Gombotz, W. R., Wang, G. H., Horbett, T. A., and Hoffman, A. S., "Protein adsorption to poly(ethylene oxide) surfaces," *J. Biomed. Mater. Res.*, Vol. 25, No. 12, 1991, pp. 1547–1562.

[28] Gobin, A. S., and West, J. L., "Cell migration through defined, synthetic ECM analogs," *Faseb J.*, Vol. 16, No. 7, 2002, pp. 751–753.

[29] Mann, B. K., and West, J. L., "Cell adhesion peptides alter smooth muscle cell adhesion, proliferation, migration, and matrix protein synthesis on modified surfaces and in polymer scaffolds," *J. Biomed. Mater. Res.*, Vol. 60, No. 1, 2002, pp. 86–93.

[30] Mann, B. K., Gobin, A. S., Tsai, A. T., Schmedlen, R. H., and West, J. L., "Smooth muscle cell growth in photopolymerized hydrogels with cell adhesive and proteolytically degradable domains: Synthetic ECM analogs for tissue engineering," *Biomaterials*, Vol. 22, No. 22, 2001, pp. 3045–3051.

[31] Hahn, M. S., McHale, M. K., Wang, E., Schmedlen, R. H., and West, J. L., "Physiologic pulsatile flow bioreactor conditioning of poly(ethylene glycol)-based tissue engineered vascular grafts," *Ann. Biomed. Eng.*, Vol. 35, No. 2, 2007, pp. 190–200.

[32] Hill-West, J. L., Chowdhury, S. M., Sawhney, A. S., Pathak, C. P., Dunn, R. C., and Hubbell, J. A., "Prevention of postoperative adhesions in the rat by in situ photopolymerization of bioresorbable hydrogel barriers," *Obstet. Gynecol.*, Vol. 83, No. 1, 1994, pp. 59–64.

[33] Itle, L. J., Koh, W. G., and Pishko, M. V., "Hepatocyte viability and protein expression within hydrogel microstructures," *Biotechnol. Prog.*, Vol. 21, No. 3, 2005, pp. 926–932.

[34] Liu, V. A., and Bhatia, S. N., "Three-dimensional photopatterning of hydrogels containing living cells," *Biomed. Microdevices*, Vol. 4, No. 4, 2002, pp. 257–266.

[35] Tsang, V. L., Chen, A. A., Cho, L. M., Jadin, K. D., Sah, R. L., Delong, S., West, J. L., and Bhatia, S. N., "Fabrication of 3-D hepatic tissues by additive photopatterning of cellular hydrogels," *Faseb J.*, 2006.

[36] Lutolf, M. P., and Hubbell, J. A., "Synthetic biomaterials as instructive extracellular microenvironments for morphogenesis in tissue engineering," *Nat. Biotechnol.*, Vol. 23, No. 1, 2005, pp. 47–55.

[37] Lutolf, M. P., Lauer-Fields, J. L., Schmoekel, H. G., Metters, A. T., Weber, F. E., Fields, G. B., and Hubbell, J. A., "Synthetic matrix metalloproteinase-sensitive hydrogels for the conduction of tissue regeneration: Engineering cell-invasion characteristics," *Proc. Natl. Acad. Sci. USA*, Vol. 100, No. 9, 2003, pp. 5413–5418.

[38] Bryant, S. J., Bender, R. J., Durand, K. L., and Anseth, K. S., "Encapsulating chondrocytes in degrading PEG hydrogels with high modulus: Engineering gel structural changes to facilitate cartilaginous tissue production," *Biotechnol. Bioeng.*, Vol. 86, No. 7, 2004, pp. 747–755.

[39] Cooper, J. A., Li, W. J., Bailey, L. O., Hudson, S. D., Lin-Gibson, S., Anseth, K. S., Tuan, R. S., and Washburn, N. R., "Encapsulated chondrocyte response in a pulsatile flow bioreactor," *Acta Biomater*, Vol. 3, No. 1, 2007, pp. 13–21.

[40] Peyton, S. R., Raub, C. B., Keschrumrus, V. P., and Putnam, A. J., "The use of poly(ethylene glycol) hydrogels to investigate the impact of ECM chemistry and mechanics on smooth muscle cells," *Biomaterials,* Vol. 27, No. 28, 2006, pp. 4881–4893.

[41] Elisseeff, J., Puleo, C., Yang, F., and Sharma, B., "Advances in skeletal tissue engineering with hydrogels," *Orthod. Craniofac. Res.*, Vol. 8, No. 3, 2005, pp. 150–161.

[42] Sharma, B., Williams, C. G., Khan, M., Manson, P., and Elisseeff, J. H., "In vivo chondrogenesis of mesenchymal stem cells in a photopolymerized hydrogel," *Plast. Reconstr. Surg.*, Vol. 119, No. 1, 2007, pp. 112–120.

[43] Yang, F., Williams, C. G., Wang, D. A., Lee, H., Manson, P. N., and Elisseeff, J., "The effect of incorporating RGD adhesive peptide in polyethylene glycol diacrylate hydrogel on osteogenesis of bone marrow stromal cells," *Biomaterials*, Vol. 26, No. 30, 2005, pp. 5991–5998.

[44] Patel, P. N., Gobin, A. S., West, J. L., and Patrick, C. W., "Poly(ethylene glycol) hydrogel system supports preadipocyte viability, adhesion, and proliferation," *Tissue Eng.,* Vol. 11, Nos. 9–10, 2005, pp. 1498–1505.

[45] Mahoney, M. J., and Anseth, K. S., "Three-dimensional growth and function of neural tissue in degradable polyethylene glycol hydrogels," *Biomaterials,* Vol. 27, No. 10, 2006, pp. 2265–2274.

[46] Bryant, S. J., and Anseth, K. S., "The effects of scaffold thickness on tissue engineered cartilage in photocrosslinked poly(ethylene oxide) hydrogels," *Biomaterials*, Vol. 22, No. 6, 2001, pp. 619–626.

[47] Bryant, S. J., Nuttelman, C. R., and Anseth, K. S., "Cytocompatibility of UV and visible light photoinitiating systems on cultured NIH/3T3 fibroblasts in vitro," *J. Biomater. Sci. Polym. Ed.*, Vol. 11, No. 5, 2000, pp. 439–457.

[48] Lai. Y. C., and Quinn, E. T., "The effects of initiator and diluent on the photopolymerization of 2-hydroxyethyl methacrylate and on properties of hydrogels obtained," in A. B. Scranton, C. N. Bowman, and R. W. Peiffer (eds.), *Photopolymerization: Fundamentals and Applications*, Washington, D.C.: American Chemical Society, 1997, pp. 35–50.

[49] Gobin, A. S., and West, J. L., "Val-ala-pro-gly, an elastin-derived non-integrin ligand: Smooth muscle cell adhesion and specificity," *J. Biomed. Mater. Res. A*, Vol. 67, No. 1, 2003, pp. 255–259.

[50] Mann, B. K., Schmedlen, R. H., and West, J. L., "Tethered-TGF-beta increases extracellular matrix production of vascular smooth muscle cells," *Biomaterials*, Vol. 22, No. 5, 2001, pp. 439–444.

[51] Sawhney, A. S., Pathak, C. P., and Hubbell, J. A., "Interfacial photopolymerization of poly(ethylene glycol)-based hydrogels upon alginate-poly(l-lysine) microcapsules for enhanced biocompatibility," *Biomaterials*, Vol. 14, No. 13, 1993, pp. 1008–1016.

[52] Cruise, G. M., Hegre, O. D., Scharp, D. S., and Hubbell, J. A., "A sensitivity study of the key parameters in the interfacial photopolymerization of poly(ethylene glycol) diacrylate upon porcine islets," *Biotechnol. Bioeng.*, Vol. 57, No. 6, 1998, pp. 655–665.

[53] An, Y., and Hubbell, J. A., "Intraarterial protein delivery via intimally-adherent bilayer hydrogels," *J. Control. Release*, Vol. 64, Nos. 1–3, 2000, pp. 205–215.

[54] Kizilel, S., Pérez-Luna, V. H., and Teymour, F., "Photopolymerization of poly(ethylene glycol) diacrylate on eosin-functionalized surfaces," *Langmuir*, Vol. 20, No. 20, 2004, pp. 8652–8658.

[55] Kizilel, S., Sawardecker, E., Teymour, F., and Pérez-Luna, V. H., "Sequential formation of covalently bonded hydrogel multilayers through surface initiated photopolymerization," *Biomaterials*, Vol. 27, No. 8, 2006, pp. 1209–1215.

[56] Horbett, T. A., "Protein adsorption on biomaterials," *Advances in Chemistry Series*, Vol. 199, 1982, pp. 233–244.

[57] Young, B. R., Lambrecht, L. K., Cooper, S. L., and Mosher, D. F., "Plasma proteins: Their role in initiating platelet and fibrin deposition on biomaterials," *Advances in Chemistry Series*, Vol. 199, 1982, pp. 317–350.

[58] Mrksich, M., Chen, C. S., Xia, Y., Dike, L. E., Ingber, D. E., and Whitesides, G. M., "Controlling cell attachment on contoured surfaces with self-assembled monolayers of alkanethiolates on gold," *Proc. Natl. Acad. Sci. USA*, Vol. 93, No. 20, 1996, pp. 10775–10778.

[59] Szleifer, I., "Protein adsorption on surfaces with grafted polymers: A theoretical approach," *Biophys. J.*, Vol. 72, No. 2, Pt. 1, 1997, pp. 595–612.

[60] Pierschbacher, M. D., and Ruoslahti, E., "Cell attachment activity of fibronectin can be duplicated by small synthetic fragments of the molecule," *Nature*, Vol. 309, No. 5963, 1984, pp. 30–33.

[61] Massia, S. P., and Hubbell, J. A., "An RGD spacing of 440 nm is sufficient for integrin alpha V beta 3-mediated fibroblast spreading and 140 nm for focal contact and stress fiber formation," *J. Cell. Biol.*, Vol. 114, No. 5, 1991, pp. 1089–1100.

[62] Pierschbacher, M., Hayman, E. G., and Ruoslahti, E., "Synthetic peptide with cell attachment activity of fibronectin," *Proc. Natl. Acad. Sci. USA*, Vol. 80, No. 5, 1983, pp. 1224–1227.

[63] Ruoslahti, E., and Pierschbacher, M. D., "Arg-Gly-Asp: A versatile cell recognition signal," *Cell*, Vol. 44, No. 4, 1986, pp. 517–518.

[64] Hubbell, J. A., Massia, S. P., Desai, N. P., and Drumheller, P. D., "Endothelial cell-selective materials for tissue engineering in the vascular graft via a new receptor," *Biotechnology* (NY), Vol. 9, No. 6, 1991, pp. 568–572.

[65] Tashiro, K., Sephel, G. C., Weeks, B., Sasaki, M., Martin, G. R., Kleinman, H. K., and Yamada, Y., "A synthetic peptide containing the IKVAV sequence from the A chain of laminin mediates cell attachment, migration, and neurite outgrowth," *J. Biol. Chem.*, Vol. 264, No. 27, 1989, pp. 16174–16182.

[66] Sawhney, A. S., Pathak, C. P., van Rensburg, J. J., Dunn, R. C., and Hubbell, J. A., "Optimization of photopolymerized bioerodible hydrogel properties for adhesion prevention," *J. Biomed. Mater. Res.*, Vol. 28, No. 7, 1994, pp. 831–838.

[67] Vert, M., Schwach, G., Engel, R., and Coudane, J., "Something new in the field of PLA/GA bioresorbable polymers?" *J. Control. Release*, Vol. 53, Nos. 1–3, 1998, pp. 85–92.

[68] Vert, M., Mauduit, J., and Li, S., "Biodegradation of PLA/GA polymers: Increasing complexity," *Biomaterials*, Vol. 15, No. 15, 1994, pp. 1209–1213.

[69] Athanasiou, K. A., Agrawal, C. M., Barber, F. A., and Burkhart, S. S., "Orthopaedic applications for PLA-PGA biodegradable polymers," *Arthroscopy*, Vol. 14, No. 7, 1998, pp. 726–737.

[70] Raeber, G. P., Lutolf, M. P., and Hubbell, J. A., "Molecularly engineered PEG hydrogels: A novel model system for proteolytically mediated cell migration," *Biophys. J.*, Vol. 89, No. 2, 2005, pp. 1374–1388.

[71] Seliktar, D., Zisch, A. H., Lutolf, M. P., Wrana, J. L., and Hubbell, J. A., "MMP-2 sensitive, VEGF-bearing bioactive hydrogels for promotion of vascular healing," *J. Biomed. Mater. Res. A*, Vol. 68, No. 4, 2004, pp. 704–716.

[72] West, J. L., and Hubbell, J. A., "Polymeric biomaterials with degradation sites for proteases involved in cell migration," *Macromolecules*, Vol. 32, No. 1, 1999, pp. 241–244.

[73] Dimuzio, P., Fischer, L., McIlhenny, S., Dimatteo, C., Golesorhki, N., Grabo, D., Tarola, N., Mericli, A., Shapiro, I., and Tulenko, T., "Development of a tissue-engineered bypass graft seeded with stem cells," *Vascular*, Vol. 14, No. 6, 2006, pp. 338–342.

[74] Matsuda, T., "Recent progress of vascular graft engineering in Japan," *Artif. Organs*, Vol. 28, No. 1, 2004, pp. 64–71.

[75] Riha, G. M., Lin, P. H., Lumsden, A. B., Yao, Q., and Chen, C., "Review: Application of stem cells for vascular tissue engineering," *Tissue Eng.*, Vol. 11, Nos. 9–10, 2005, pp. 1535–1552.

[76] Schmedlen, R. H., Elbjeirami, W. M., Gobin, A. S., and West, J. L., "Tissue engineered small-diameter vascular grafts," *Clin. Plast. Surg.*, Vol. 30, No. 4, 2003, pp. 507–517.

[77] Thomas, A. C., Campbell, G. R., and Campbell, J. H., "Advances in vascular tissue engineering," *Cardiovasc. Pathol.*, Vol. 12, No. 5, 2003, pp. 271–276.

[78] Revzin, A., Russell, R. J., Yadavalli, V. K., Koh, W. G., Deister, C., Hile, D. D., Mellott, M. B., and Pishko, M. V., "Fabrication of poly(ethylene glycol) hydrogel microstructures using photolithography," *Langmuir*, Vol. 17, No. 18, 2001, pp. 5440–5447.

[79] Koh, W. G., Revzin, A., and Pishko, M. V., "Poly(ethylene glycol) hydrogel microstructures encapsulating living cells," *Langmuir*, Vol. 18, No. 7, 2002, pp. 2459–2462.

[80] Khademhosseini, A., Yeh, J., Jon, S., Eng, G., Suh, K. Y., Burdick, J. A., and Langer, R., "Molded polyethylene glycol microstructures for capturing cells within microfluidic channels," *Lab Chip*, Vol. 4, No. 5, 2004, pp. 425–430.

[81] Hahn, M. S., Miller, J. S., and West, J. L., "Laser scanning lithography for surface micropatterning on hydrogels," *Adv. Mater.*, Vol. 17, No. 24, 2005, pp. 2939–2942.

[82] Hahn, M. S., Taite, L. J., Moon, J. J., Rowland, M. C., Ruffino, K. A., and West, J. L., "Photolithographic patterning of polyethylene glycol hydrogels," *Biomaterials*, Vol. 27, No. 12, 2005, pp. 2519–2524.

[83] Itoga, K., Yamato, M., Kobayashi, J., Kikuchi, A., and Okano, T., "Cell micropatterning using photopolymerization with a liquid crystal device commercial projector," *Biomaterials*, Vol. 25, No. 11, 2004, pp. 2047–2053.

[84] Itoga, K., Yamato, M., Kobayashi, J., Kikuchi, A., and Okano, T., "Micropatterned surfaces prepared using a liquid crystal projector-modified photopolymerization device and microfluidics," *J. Biomed. Mater. Res. A*, Vol. 69, No. 3, 2004, pp. 391–397.

[85] Itoga, K., Kobayashi, J., Yamato, M., Kikuchi, A., and Okano, T., "Maskless liquid-crystal-display projection photolithography for improved design flexibility of cellular micropatterns," *Biomaterials*, Vol. 27, No. 15, 2006, pp. 3005–3009.

[86] Lu, Y., Mapili, G., Suhali, G., Chen, S., and Roy, K., "A digital micro-mirror device-based system for the microfabrication of complex, spatially patterned tissue engineering scaffolds," *J. Biomed. Mater. Res. A*, Vol. 77, No. 2, 2006, pp. 396–405.

[87] Chin, V. I., Taupin, P., Sanga, S., Scheel, J., Gage, F. H., and Bhatia, S. N., "Microfabricated platform for studying stem cell fates," *Biotechnol. Bioeng.*, Vol. 88, No. 3, 2004, pp. 399–415.

[88] Yu, T., and Ober, C. K., "Methods for the topographical patterning and patterned surface modification of hydrogels based on hydroxyethyl methacrylate," *Biomacromolecules*, Vol. 4, No. 5, 2003, pp. 1126–1131.

[89] Khademhosseini, A., Langer, R., Borenstein, J., and Vacanti, J. P., "Microscale technologies for tissue engineering and biology," *Proc. Natl. Acad. Sci. USA*, Vol. 103, No. 8, 2006, pp. 2480–2487.

[90] Suh, K. Y., Seong, J., Khademhosseini, A., Laibinis, P. E., and Langer, R., "A simple soft lithographic route to fabrication of poly(ethylene glycol) microstructures for protein and cell patterning," *Biomaterials*, Vol. 25, No. 3, 2004, pp. 557–563.

[91] Nuttelman, C. R., Tripodi, M. C., and Anseth, K. S., "Synthetic hydrogel niches that promote hMSC viability," *Matrix Biol.*, Vol. 24, No. 3, 2005, pp. 208–218.

[92] Mapili, G., Lu, Y., Chen, S., and Roy, K., "Laser-layered microfabrication of spatially patterned functionalized tissue-engineering scaffolds," *J. Biomed. Mater. Res. B Appl. Biomater.*, Vol. 75, No. 2, 2005, pp. 414–424.

[93] Hahn, M. S., Miller, J. S., and West, J. L., "Three-dimensional biochemical and biomechanical patterning of hydrogels for guiding cell behavior," *Adv. Mater.*, Vol. 18, No. 20, 2006, pp. 2679–2684.

[94] Arcaute, K., Mann, B. K., and Wicker, R. B., "Stereolithography of three-dimensional bioactive poly(ethylene glycol) constructs with encapsulated cells," *Ann. Biomed. Eng.*, Vol. 34, No. 9, 2006, pp. 1429–1441.

[95] Cumpston, B. H., Ananthavel, S. P., Barlow, S., Dyer, D. L., Ehrlich, J. E., Erskine, L. L., Heikal, A. A., Kuebler, S. M., Lee, I. Y. S., McCord-Maughon, D., Qin, J., Röckel, H., Rumi, M., Wu, X. L., Marder, S. R., and Perry, J. W., "Two-photon polymerization initiators for three-dimensional optical data storage and microfabrication," *Nature*, Vol. 398, 1999, pp. 51–54.

[96] Denk, W., Strickler, J. H., and Webb, W. W., "Two-photon laser scanning fluorescence microscopy," *Science*, Vol. 248, No. 4951, 1990, pp. 73–76.

[97] Helmchen, F., and Denk, W., "New developments in multiphoton microscopy," *Curr. Opin. Neurobiol.*, Vol. 12, No. 5, 2002, pp. 593–601.

[98] Helmchen, F., and Denk, W., "Deep tissue two-photon microscopy," *Nat. Methods*, Vol. 2, No. 12, 2005, pp. 932–940.

[99] Olson, C. E., Previte, M. J., and Fourkas, J. T., "Efficient and robust multiphoton data storage in molecular glasses and highly crosslinked polymers," *Nat. Mater.*, Vol. 1, No. 4, 2002, pp. 225–228.

[100] Zipfel, W. R., Williams, R. M., and Webb, W. W., "Nonlinear magic: Multiphoton microscopy in the biosciences," *Nat. Biotechnol.*, Vol. 21, No. 11, 2003, pp. 1369–1377.

[101] Göppert-Mayer, M., "Über elementarakte mit zwei quantensprüngen," *Annalen der Physik*, Vol. 401, No. 3, 1931, pp. 273–294.

[102] Xu, C., Zipfel, W., Shear, J. B., Williams, R. M., and Webb, W. W., "Multiphoton fluorescence excitation: New spectral windows for biological nonlinear microscopy," *Proc. Natl. Acad. Sci. USA*, Vol. 93, No. 20, 1996, pp. 10763–10768.

CHAPTER 7

Three-Dimensional Cell-Printing Technologies for Tissue Engineering

N. Gözde Durmus, Richard L. Lin, Grace Lee, SangJun Moon, Gokhan Percin, Emre Kayaalp, Edward Hæggstrom, and Utkan Demirci

7.1 Overview

Loss or failure of organs and tissues is a frequent, devastating, and costly problem in clinical medicine. Each year, millions suffer from organ dysfunction and rely on biomedical technologies, such as transplantation, surgical reconstruction, and mechanical devices, to maintain their health and well-being. The interdisciplinary field of tissue engineering strives to meet the needs of these patients. The ultimate goal of tissue engineering is to fabricate functional three-dimensional tissues and organs in order to maintain, restore, or enhance native tissue and organ function [1, 2].

The three-dimensional tissue architecture arises from the coordination of multicellular processes, emergent mechanical properties, and integration with other organ systems via the microcirculation [3]. Individual cells integrate external cues arising from various extracellular matrix (ECM) components, mechanical stimulation, and soluble signals from both adjacent and distant cells. This allows the generation of a basal phenotype and response to perturbations in their environment. Tissues require a three-dimensional environment to provide cell adhesion, cell contraction, and cell-cell signaling in vivo. Only a proper three-dimensional environment yields the physiological structure necessary to promote the coordinated cellular function that defines living tissue. Further, the ability to model a three-dimensional tissue environment in vitro might prove instrumental in both understanding and promoting morphogenetic and remodeling events that occur over larger-length scales. These events include epithelial acinar formation, epithelial-cell behavior, and morphogenesis. Figure 7.1 details the importance of the three-dimensional environment for normal cell function [4].

Recent advances in fabrication methods have redirected the field of tissue engineering. Using soft lithography, some researchers fabricate scaffolds, whereas others use printing technology to build three-dimensional microarchitectures. Tissue-printing technology involves designing printers, cellular "bioink," and three-dimensional hydrogel "biopaper." The technology creates functional, living tissues using modified desktop computer printers filled with cell suspensions, rather than ink, which are deposited on three-dimensional hydrogels rather than flat paper. Cell printing has the potential to be faster and more precise than other methods to

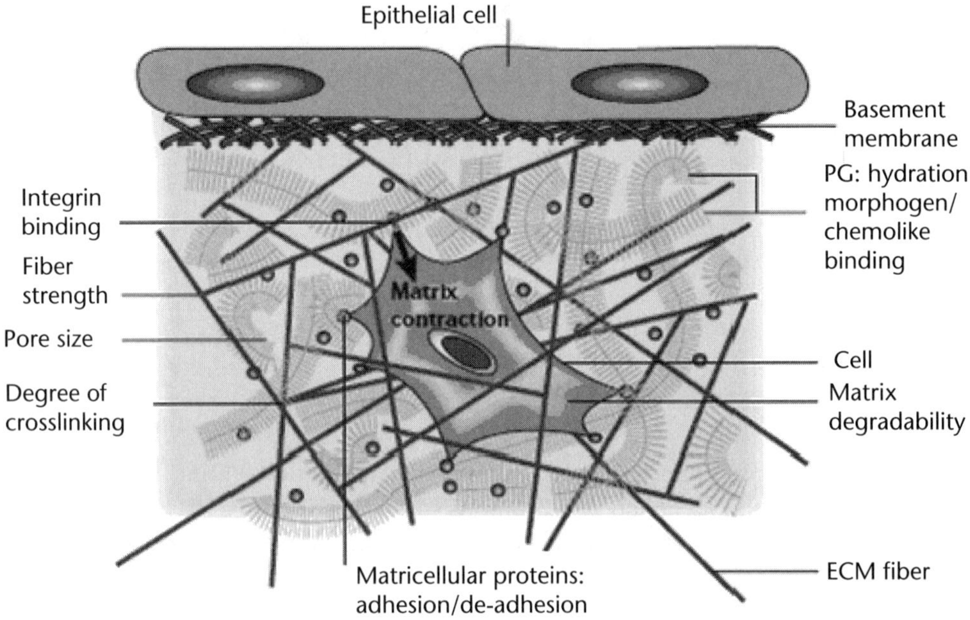

Figure 7.1 Schematic representation of a normal three-dimensional cell environment. The ECM supports the cell's three-dimensional shape and controls cell adhesion, migration, and transmission of mechanical forces. (*Source:* [4], reprinted with permission.)

deposit living cells into ordered arrays. Hydrogel biopaper consists of the same molecules found in the ECM of biological tissue. This new technology has the potential to overcome obstacles facing tissue engineers struggling to build functional, dimensionally correct organs. For example, promoting vascularization of engineered tissues might be done by printing branching vascular trees directly into engineered tissues.

This chapter provides information about three-dimensional cell-printing technologies for tissue engineering applications. It first discusses the development of cell-printing technologies by reviewing conventional three-dimensional printing methods. It then discusses "organ printing," an exciting application of three-dimensional cell-printing technologies. Next, it examines self-assembly of tubular structures, bioink, biopaper, and potential applications of printed cells in biology and medicine. Finally, it introduces an emerging technology for three-dimensional cell printing, acoustic picoliter droplet generators.

7.2 Development of Cell-Printing Technologies

Cell printing has recently developed due to the simultaneous advancement of inkjet printers and the growing need for transplantable organs. Given the difficulty of controlling structured three-dimensional cell growth using only scaffold materials, printing technology is a promising means to overcome this challenge via rapid, layer-by-layer deposition of cells or ECM to create the desired assembly. Although the term "organ printing" was introduced only in 1989, the groundwork was laid earlier by developmental biologists studying tissue affinity and cell-to-cell adhesion [5]. Their work facilitated the development of cell-printing technologies [6–8].

One classical tissue-engineering approach relied on solid polyglycolic acid (PGA) scaffolds to maintain the desired structure. Isolated cells were deposited on this biodegradable material [9]. However, problems arise with such a methodology: (1) it is difficult to promote uniform cell growth and to attain adequate cell penetration; (2) organs comprise different cells that require placement at specific locations, which is difficult to achieve with a solid scaffold method; (3) the rigid, solid scaffolds are ill suited for culturing contractile tissues, such as cardiac muscle and vascular structures; and (4) vascularization is hard to achieve through this approach [10].

These problems motivated scientists to combine printing technology with developmental biology, which led to the idea of a cell printer [11]. Thus, the bulk of development on cell/organ printing has occurred in the past few years. Tissue engineers have used custom-designed printers or modified off-the-shelf models to print tissues using bioink comprising proteins or individual cells [12].

Rapid-prototyping technologies were incorporated to enhance the precision and complexibility of printable tissue patterns [13]. These technologies, subcategorized into contact and noncontact deposition, allow layer-by-layer growth of solids, liquids, or powders. Common rapid prototyping techniques are fused deposition modeling, stereolithography, selective laser sintering, laser direct writing, microcontact printing, and inkjet printing [14]. Moreover, using computer-aided design (CAD) software, tissue engineers can achieve a specificity that has eluded the traditional solid scaffolding method. Such "desktop organ printers" have been built and have generated promising results [15].

7.3 Conventional Three-Dimensional Cell-Printing Methods

Tissue engineering, particularly organ printing, has expanded widely as researchers strive both to create viable transplantable organs and to develop alternative methods to repair tissue damage. A challenge in three-dimensional cell printing is the assembly of cells into healthy, vascularized tissues. Traditional tissue-engineering methods involve forming homogenous, porous scaffolds, usually hydrogels or polymers, and then seeding these scaffolds with cells. However, simply seeding cells onto prefabricated scaffolds produces a mass of disorganized, randomly attached cells that lack vasculature. Functional tissues are generally heterogeneous, with certain cells placed strategically in order to optimize their function in the tissue [16].

Two conventional methods have been developed to create such heterogeneous, multicell structures. Both methods involve depositing individual cells in a controlled pattern on a thin layer of the scaffold material, then stacking scaffold layers. This layer-by-layer approach allows three-dimensional spatial control of cell placement, as well as embedding vascular constructs within the engineered tissue during formation. The two mainstream printing methods are laser printing and inkjet printing.

7.3.1 Laser Printing

Cell printing began in 1999, when Odde and Renn developed a laser-based method to direct cells to form a pattern on a target surface [17]. When a laser beam hits a dielectric particle, it is reflected and refracted, causing its photons to be redirected. This redirection results in a change in the photons' momentum, which is in turn transferred to the particle (Figure 7.2). Due to this momentum transfer, the particle is pulled radially inward toward the center of the laser light and pushed axially in the direction of the light.

Odde and Renn expanded the work of Tudor Buican, who used a weak laser light to direct cells through an aqueous media. Odde and Renn created their "laser-guided direct-writing" process by focusing a weak laser beam on a suspension of cells, which were then propelled onto a surface. This process can place particles with a precision of below 1 μm. They further refined their process by placing a hollow optical fiber between the cell suspension and the target surface, using the fiber to direct each cell to its desired position with greater precision. Adding an optical fiber provided several significant advantages. One problem with the original method was that currents and other fluid movement in the cell suspension solution frequently overcame the optical forces provided by the laser, causing particles to veer off course and miss their targets. The fiber solved this problem by insulating the particles from convective fluid motion. Also, the fiber allowed the laser light to be directed over a longer distance. Using an intensity gradient within the hollow

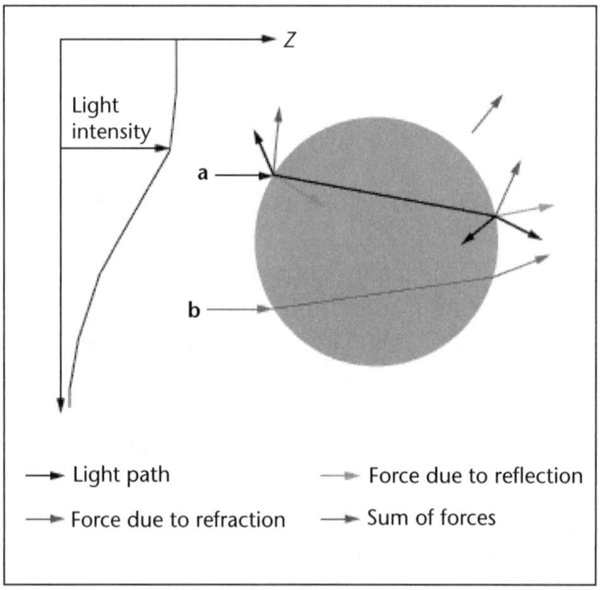

Figure 7.2 "Optical forces on a dielectric particle. Laser light [whose intensity (Z) varies across space (r) in the graph] is reflected and refracted at each interface, resulting in a redirection of the light. Photons have momentum and so their redirection by interacting with the particle results in a corresponding momentum transfer to the particle. The net result of the interactions from ray *a* is to push the particle along the beam axis and to pull the particle radially inwards. By symmetry, ray *b* pushes the particle axially and pushes the particle radially outward. However, ray *a* is stronger than ray *b*, so it overcomes the radial force directed outward. In the absence of other forces, the particle is simultaneously pulled radially inward and pushed axially in the direction of the laser beam." (*Source:* [17], reprinted with permission.)

interior of the fiber, particles could be drawn to the center of the fiber, preventing them from adhering to the fiber walls. Further, the fiber presented a buffer between the cells in suspension and the target surface, which prevented the cells from randomly attaching to the target surface. Finally, the fiber opened the door for printing multiple types of cells or materials alongside one another, as fibers attached to different reservoirs could all print simultaneously to the same surface [17]. This laser printing was tried with various kinds of particles, including embryonic chick spinal cord cells. The researchers deposited the 9 μm–diameter spinal cord cells on a glass target surface using a 450 mW laser beam. The cells remained viable and appeared to develop the characteristics of normal nerve cells. Odde and Renn predicted that they would be able to use their procedure to print various cell types alongside each other and on top of each other to create a three-dimensional tissue block. However, because this process takes some seconds to deposit each cell, it is rather inefficient in handling large cell volumes [17]. In 2005, Barron, Krizman, and Ringeisen modified the existing process to develop a "laser forward transfer technique" [18]. In this process, the cells are suspended in extremely shallow wells that are mounted on a "laser absorption layer," a piece of optically transparent, metal-coated quartz. The laser beam is focused on the laser absorption layer with a 10× microscope objective lens, causing the layer to thermally deform. Vaporization due to this sudden heat creates a gas bubble, which causes the directional propulsion of a droplet of the cell solution toward a receiving substrate. The wells containing cell solution are placed on an XY stage and the receiving substrate is placed on a separate XY stage, allowing precise cell patterning. The device prints up to one hundred cell solution droplets into a 75 μm line every second [18].

The laser-printing technique developed by Barron and his colleagues does not appear to harm the printed cells. In a study using osteosarcoma cells, cell viability was nearly 100 percent one week after being printing onto a hydrogel [18]. Since only the portion of the cell solution in contact with the laser absorption layer was volatilized, the majority of the cell solution did not undergo heating. In addition, the vast majority of the printed cells did not express the HSP60 or HAP70 heat shock proteins, suggesting that these cells did not undergo thermal stress. Barron and his colleagues also used their technique with genetically engineered *Escherichia coli* cells, which were also printed with high cell viability [19]. In addition, the biological laser method (BioLP) has printed large and reproducible arrays of both active proteins and cells, with spot sizes ranging from 30 to 70 μm and with resolution ranging from a single cell to tens of cells per spot [20, 21]. Twenty-seven spots of human osteosarcoma cells were printed on a 200 μm–thick hydrogel layer. After live/dead staining on the substrate (twenty-four hours after printing), high viability of the osteosarcoma cells was demonstrated.

The process of laser-guided direct writing (LGDW) offers a possible means to print vascular structures in tissue prior to implantation. The LGDW technique patterns human umbilical vein endothelial cells (HUVECs) in two and three dimensions by inducing self-assembly along a printed pattern. Coculture studies have shown the formation of aggregates of tubular structures, which can mimic the hepatic sinusoid [22]. In another study, Ringeisen printed pluripotent embryonal carcinoma cells on hydrogel layers and reported greater than 95 percent cell survival [16]. He found that laser-transferred cells express Microtubular Associated

Protein 2 after retinoic acid stimulation and Myosin Heavy Chain Protein after dimethyl sulfoxide stimulus, which indicate successful neural and muscular pathway differentiation, respectively.

7.3.2 Current Issues with Laser-Printing Approaches

Laser printing of cells has advantages. This technology allows accurate cell printing with preserved high cell viability. The major advantage of this method is directionality [5]. In addition, it does not require an orifice, thus eliminating the nozzle clogging risk. This increases the number of materials that can be printed. Laser cell printing features high reproducibility and ability to perform continuous and accurate single-cell printing. However, laser printing cannot attain the high throughput rates needed to create tissues consisting of thousands of cells [10, 23]. Moreover, these technologies all heat the suspended cells to some extent due to the absorption of laser light by the biolayer around the cell. Thus, since the cells are transferred via heating, vaporization, and expansion, thermal trauma jeopardizes proper cell growth after placement and could represent a disadvantage for organ and three-dimensional tissue printing [23].

7.3.3 Inkjet Printing

Inkjet printing and CAD techniques, borrowed from the field of electronics and mechanics, are widely used in the field of biotechnology for purposes ranging from biochips to DNA arrays. Microscale patterning of biological materials, including proteins, monofunctional acrylate ester, sinapinic acid, DNA, and DNA scaffolds, has been done using piezoelectric inkjet printing [14]. When introduced to the field of tissue engineering, inkjet printing techniques were used to create layer-by-layer scaffolds by depositing a chemical binding agent with a "drop-on-demand" mechanism into a layer of powder polymeric material. However, this process requires complex equipment and toxic chemicals, both of which render it impractical for printing viable tissues [24].

Thermal inkjet printing was developed by Hewlett-Packard and Canon in 1979. In 2003, Mironov and Boland improved this process and created their "drop-on-demand" technique to deliver small volumes of liquid to locations in a three-dimensional scaffold [25]. They modified Canon Bubble Jet and HP Deskjet printers for biomedical applications. In practice, they disabled the paper-feed mechanism, placed it under a hood, sterilized it with ultraviolet radiation, and put hydrogel under the print head to serve as "paper." They cleaned the print cartridges multiple times with 70 percent ethanol to sterilize the containers and prepared them for cell suspensions. In one set of experiments, they printed primary neurons from chicken embryos and quail mesodermal stem cells into 1.5 mg/mL collagen hydrogels. Initially, few mesodermal cells were observed, but after six days of incubation, the cells began to proliferate and fluoresce, indicating viability. Most of the neurons appeared stable during the first few days of culture, but after five days, many had differentiated and exhibited neural processes [25].

Mironov, Borland, and colleagues demonstrated that fragile mammalian cells, though extremely sensitive to heat and shear stresses, could be printed with slightly modified conventional desktop printers. By limiting the maximum temperatures, they increased the cell viability significantly. Mironov et al. also conducted printing experiments with Chinese hamster ovary (CHO) cells, showing that less than 8 percent of the cells were lysed during printing [26].

The drop-on-demand inkjet mechanism opens up opportunities for organ printing in addition to constructing simple two-dimensional arrays of cells. Three-dimensional alginate-calcium gels are created by constructing a chamber from a sterile 50 mL conical tube and placing a 2.5 cm round coverslip attached to a metal rod inside the chamber [10]. This glass coverslip serves as the elevator platform and is controlled by a stepper motor, which exerts z-axis control. A chamber is filled with 2 percent alginic acid solution, and the ink cartridge is filled with 0.25M $CaCl_2$, that crosslinks alginate into a biodegradable hydrogel. Thus, when $CaCl_2$ is printed onto the elevator platform, and the platform moves down into the alginate solution, the alginate crosslinks into a hydrogel. The process of printing a layer of $CaCl_2$ and moving the elevator platform down is continued until the desired gel shape has been achieved. This study also confirms that endothelial cells can migrate through the alginate-calcium hydrogel. This three-dimensional hydrogel has potential to create favorable scaffolds for cell printing and tissue engineering. Researchers have created nontoxic, biodegradable, and thermoreversible gels by printing layer upon layer of thermosensitive gels (K-70). Cells could then be printed on the gel layers in an organized pattern. They also printed cell aggregates on very thin gel layers, ensuring direct contact between cells in sequential hydrogel layers. In some cases, closely placed cells in different gel layers fused together. Relying on this mechanism, tubes were constructed by stacking cell aggregate ring layers [13].

Nakamura et al. investigated the use of a biocompatible inkjet head and the feasibility of microseeding with living cells [27]. They prepared a bovine vascular endothelial cell suspension as bioink and ejected it onto culture disks. The bioink contained 1.0 to 1.5 million cells/mL, and each printed dot contained up to four endothelial cells. The number of cells in each dot depended on the cell suspension concentration and on the ejection frequency. Ejected cells were viable after the printing procedure and the cell-printing system did not cause significant mechanical damage to the ejected cells. This living cell microseeding appears to have potential to print three-dimensional tissues. Moreover, bacterial colony arrays and complex patterns were fabricated using inkjet printers. Bacterial colonies, 100 colonies/cm^2, were obtained by directly ejecting *E. coli* onto agar-coated substrates at an arraying speed of 880 spots per second. Single colonies of viable bacteria could be obtained by adjusting the printer speed. This noncontact inkjet printer can deliver single or multiple colonies onto targeted positions. Printed colonies can be used to screen colonies, to construct genomic and expression libraries, or to fabricate bacterial biosensors. They can also be used to generate cell-density gradients for pharmacological screening for those instances in which drug-effectiveness and toxicity measurements require testing at various cell densities [28].

Boland et al. created complex cellular patterns and structures by automated and direct inkjet printing of primary embryonic hippocampal and cortical neurons.

Usually, the neural cells of interest formed cellular patterns through preferential attachment to cell-adhesive molecules. Although these two-step methods are high throughput, flexible, and cost-effective, they are limited to two-dimensional neuronal patterning due to their reliance on a thin layer of cell-adhesive molecules. This reliance makes it difficult to adapt these approaches to building three-dimensional complex neural tissuelike constructs. To overcome these limitations, researchers expanded the capability of the inkjet printer to eject living cells. In addition, a new method was developed to create three-dimensional cellular structures. Sheets of neural cells were stacked by inkjet printing of NT2 cells and fibrin gels. Since fibrin gelling is rapid, it may present an alternative material for inkjet printing. Printing thrombin onto fibrinogen causes geometry-specific crosslinking, allowing rapid construction of three-dimensional fibrin scaffolds of specific structure and form. Neurons are anchorage dependent, and their functionality and differentiation depend on their attachment to extracellular scaffolds. To demonstrate that printed embryonic hippocampal and cortical neurons maintain their phenotypes and electrophysiological characteristics, whole-cell patch-clamp experiments and immunostaining analysis were performed. After one day of culture, the printed neurons began to differentiate and to develop their axonal processes. In this study, a viability of 74 percent was obtained, and these cells exhibited normal neuronal appearance and good neurite growth on collagen-based biopaper [29]. Inkjet printing techniques are also used to create scaffolds for tissue-engineering purposes. Limpanuphap and Derby invented a novel printing process using the lost mold technique for biomaterial manufacture [30]. They prepared porous TCP (tricalcium phosphate) and polymer/TCP composites using inkjet printing to fabricate the molds. Scaffolds fabricated by inkjet process are porous, which allows them to be used for cell adhesion. Weiss et al. studied cellular responses to inkjet printed patterns of hormones. They used preosteoblastic cells and explored their response to fibroblast growth factor 2 (FGF-2) printed on fibrin films. Concentration gradients were made by overprinting twenty times in patterned arrays on fibrin films using a bioink composed of 50 μg/ml FGF-2 labeled with 50 μg/ml cyanine 3 dye. Experimental results showed that immobilized FGF-2 was biologically active and that the printed patterns persisted for ten days during cell culture. This study indicates that inkjet printing allows engineering persistent hormone patterns that can be used in developmental-biology and tissue-regeneration applications. In addition, hormone inkjet printing methodology can create complex three-dimensional structures and has potential utility in tissue-engineering therapeutics [31].

An emerging technology in the field of jet-based printing applications is electric field driven jetting, also known as electrohydrodynamic jetting (EHDJ). In this method, a multiphase liquid is exposed to a high-intensity electric field as it comes out at a needle with a controlled flow rate. The liquid flowing in the needle is charged by the high-intensity electric field, which promotes jet formation. Using the EHDJ method, Jurkat cells were jetted using 0.67 to 2 kV/mm electric field strength, corresponding to an applied voltage of 10 to 30 kV. The printed cells were unharmed and retained their integrity. The emerging electrohydrodynamic jetting technique has the potential to compete with conventional inkjet cell-printing methods [32].

7.3.4 Current Problems with Inkjet-Printing Approaches

The inkjet printing technology allows building three-dimensional scaffolds and engineering tissues. The process is not only extremely precise and inexpensive, but it is also capable of generating high-throughput cell patterning. Commercial inkjet printers allow printing biomolecules onto target substrates without reducing their biological activities [19]. However, inkjet printing technologies were not specifically created for cell printing; hence, they are compromised in dealing with living cells. Inkjet printing may harm ejected cells during droplet generation by way of exerting high heat, pressure, and shear stress on the cells. In addition, unreliable ejection, nonuniform droplet size, and nozzle clogging hamper efficient inkjet cell printing. The ejection system inefficiency is a major problem when printing millions of cells in a continuous manner, as is necessary for printing complex tissues and organs.

7.4 Current Applications of Cell-Printing Technology: Organ Printing

As outlined at the start of this chapter, the loss or failure of organs and tissues is a central concern in medicine. Tissue engineers have applied the principles of engineering, materials science, and cell and molecular biology to develop viable structures that restore, maintain, or improve the function of human organs and tissues. Conventional tissue-engineering approaches generally use solid, rigid, porous, biodegradable scaffolds onto which cells are seeded. Four major drawbacks of these conventional methods were described at the start of this chapter. The major challenge in tissue engineering is to mimic the complex organization of native tissues. Reproducing the complexity of native tissue would be a critical step toward developing new methods for tissue repair and the production of transplantable organs. The fundamental problem remains how to arrange cells in precise three-dimensional patterns. It is also challenging to achieve efficient vascularization of complex tissues. Three-dimensional tissue and organ printing is evolving into a promising approach to tackle these considerable technical challenges [2].

Organ printing is a rapid prototyping, computer-aided, three-dimensional printing technology that uses layer-by-layer deposition of cells and/or cell aggregates into a three-dimensional gel. Sequential maturation of the printed construct forms a vascularized living tissue or organ. Organ printing applies microfluidic constructs to cells and cell aggregates to trigger biologically relevant phenomena. This technology offers promising solutions for organ and tissue development by exploiting cellular self-assembly into tissues (similar to the way embryonic tissues sort and fuse into functional forms). There are three sequential steps to organ printing: preprocessing, processing, and postprocessing. Preprocessing primarily includes developing a CAD blueprint of the specific organ [10, 13]. CAD-based modeling includes contour-based surface extraction and solid approaches [33]. Weiss et al. used a Bayesian CAD-aided experimental design to create heterogeneous scaffolds. They developed the solid free-form fabrication (SFF) method—computer-aided design/computer- aided manufacturing—to fabricate complex shapes automatically. Scaffolds with controlled microstructures and complex interior architectures can be built. Growth factors loaded into these heterogeneous scaffolds are released as the scaffolds degrade. With SFF, the distribution of growth factors can be controlled. Researchers used this computer model to fabricate fibrin and hydroxyapatite scaffolds for bone tissue generation [34]. The dataset for the fabrication process originates from medical imaging sources or CAD programs. During the process, a three-dimensional dataset is mathematically converted to a two-dimensional dataset. The two-dimensional dataset is used as a 1:1 copy of the computer model to generate the physical model. Cells attach well to the scaffolds fabricated by three-dimensional printing and proliferate on HA scaffolds in both static and dynamic cultures. As a result, three-dimensional printing with the help of CAD systems seems well suited to tissue-engineering applications [35].

Organ printing utilizes a bioprinter to create precisely arranged living cells. Living cells are spun in a centrifuge and assemble to form tiny spheres. These bioink spheres are then loaded into the syringelike nozzle of the bioprinter. Bioink is arranged into layered rings and tubular structures, respectively. Different cell types can be loaded into different bioprinter nozzles to yield specific cellular patterns.

Biomolecules interact with each other and self-organize to form structures with well-established functionality. Self-assembly is a fundamental process that drives structural organization in all living systems. In the early stages of development, the self-assembly of cells and tissues occurs, guided by cell-cell and cell-ECM interactions. These interactions direct the ultimate morphology of the organism. In organ-printing applications, cellular aggregates serve as building blocks of tissues. Aggregates allow faster organ formation as compared to individual cells [36–38]. Whereas the fusion of cellular aggregates is a random process, these aggregates can be delivered into the gels as a specific geometry (Figure 7.3). The propensity of cell aggregates to fuse promotes the assembly of printed droplets to form three-dimensional organ structures. In order to demonstrate this property of cell aggregates, Jakab et al. created rings of tissue using genetically altered cells with adhesive properties [38]. They also hypothesized that closely placed aggregates in appropriately chosen three-dimensional gels could fuse to form tissue constructs of desired geometry. Figure 7.3 shows the schematic diagram of the vascularized structures formed by Jakab et al. If these aggregates were printed in multiple layers, they would be expected to fuse in both the horizontal and vertical directions, thus forming a lumenous organlike module. This study demonstrates that cells form aggregates by self-assembly and that these spherical cell aggregates can be used as building blocks in tissue-engineering applications. In addition, these aggregates can be used as bioink in cell printers. Thus, bioink may contain self-assembled spherical cellular aggregates rather than just individual cells. The optimum size for cell aggregates is defined as a diameter small enough to allow centrally positioned aggregate cells to survive (usually less than 1 mm), yet it must be large enough to yield reasonable thickness to the printed cell layers. These restrictions place well-defined limits on the smallest acceptable diameter [10].

One challenge in the tissue engineering of tubular structures is the rapid seeding of porous scaffolds with a desired cell type, density, and sustained cell viability. Soletti et al. developed a novel seeding device using the effects of vacuum, centrifugal force, and flow [39]. This device allows porous tubular scaffolds to be uniformly

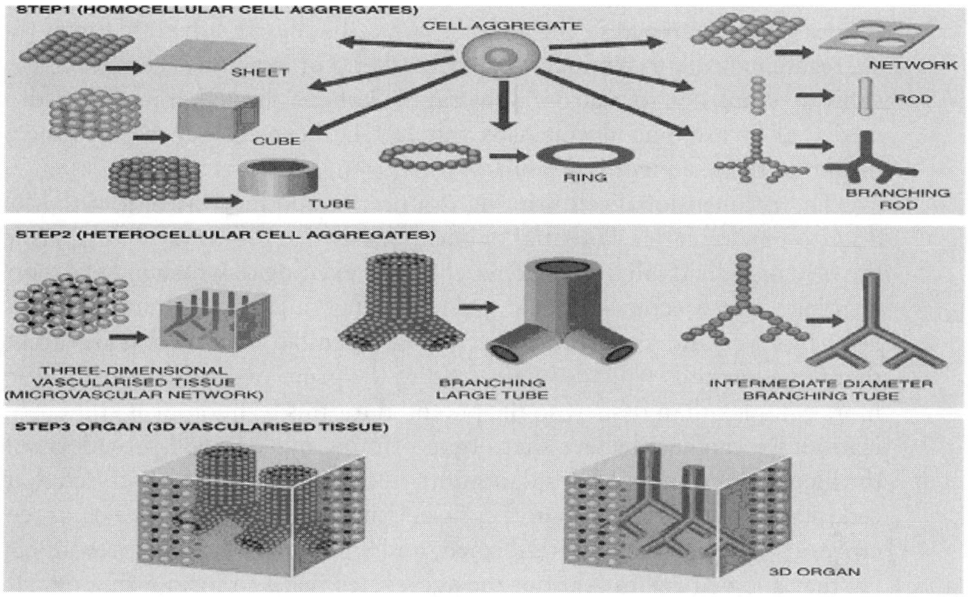

Figure 7.3 Schematic representation of constructing three-dimensional vascularized tissues using aggregates of multiple cell types. (*Source:* [38], reprinted with permission.)

bulk- and surface-seeded with cells. This study showed that the scaffolds were uniformly seeded along their longitudinal and circumferential dimensions within the tube wall without affecting cell viability or exposing cells to excessive shear stress. The use of this device and the seeding of tubular structures have the potential to improve tissue-engineering applications.

Another component of bioprinting is biopaper, a biocompatible gel. Bioink may contain cells of one or several cell types, depending on the specific requirements of the printed organ or tissue. The gel layer between the cell aggregates provides mechanical strength and stability to the construct and simultaneously acts as a drug-delivery system [10]. However, there are problems with seeding cell aggregates onto the gel. If the cell aggregate is too cohesive, cells cannot migrate along the gel. If, instead, the aggregate is not sufficiently cohesive, cells disperse. In either case, the cell fusion is disrupted. To achieve fusion and appropriate structure formation, biopaper has to be designed with optimal properties for the employed cell types and aggregates. The biopaper must be nontoxic, biocompatible, and biodegradable. The gel should be thermosensitive, photosensitive, pH sensitive, and even molecule sensitive. In addition, it must be fluid enough to pass through the relatively thin printer nozzles. Cell-gel adhesion must be controlled to avoid structure formation problems [23, 40]. The biopaper thickness is an important factor for organ printing; the gel thickness should match the aggregates size. The gel layer thickness depends on gel kinetics and on drop size, which is governed by surface tension and needle diameter. Boland et al. used thermoreversible gel as a new biopaper, which seems to offer an opportunity to mimic cell positions within organs [10]. This thermoreversible gel seems to affect both the aggregation of the cell aggregates and their positioning within the gel. If the cell-substrate force is stronger than the cell-cell force, a monolayer is created. If the force of cell-cell contact is predominant, the cells form aggregates. The absence of a cell-adhesive substrate forces the cells to aggregate in order to interact with each other. For example, cells attach to collagen substrates but not to agarose substrates. Perhaps the tethering effect of collagen mediated by arginine-glycine-aspartate (RGD) keeps cell aggregates together and promotes their sequential fusion.

Three-dimensional cell printing is a precise and innovative tool that opens the door to new research. Potential applications of this tool carry enormous impact in the field of individualized medicine, the most exciting of which may be whole-organ printing. Since conventional cell printing allows printing cells precisely layer-by-layer and since printed cells self-assemble into tissues, researchers hope that these methods will eventually lead to the printing of entire organs. Organs are more complex than tissues, so the potential for this technology is still far from realization. Researchers believe that organ printing one day will provide a solution to the limitations of organ transplantation. Initial work has demonstrated that rings of cells placed adjacent to one another fuse. Using cardiac cells from a chick, concentric circles of cell aggregates were formed, and when these aggregates were placed in a gel that allowed cell migration, the aggregates fused to form a ring or a thick disc [13]. These rings and discs could be the beginning of a blood vessel or a piece of gut. Though this is an exciting development, several challenges remain. A difficult challenge is the necessity to create vascularized tissue. Organs require complex vascular trees that branch into three main types of vasculature: large vessels (arteries and veins), smaller vessels (arterioles and venules), and microvasculature (capillaries and sinusoids) [23, 41]. The main difficulty is to create intermediate-sized vasculature, which is too small for direct cell printing but too large for microvasculature methods. Researchers hope that combining technologies such as laser-based ablation-induced apoptosis, cell sorting, and the formation of more complex aggregates will lead to in vitro formation of intermediate-sized blood vessels.

7.5 Other Applications of Cell Printing

One application of cell printing is to create tissue composed of endothelial cells and angioblasts [13]. These tissues could be used to test medications on actual human tissue, a vast improvement over using animal tissue for this purpose. In addition, testing medications on a patient's own tissue might become possible. For example, one could print tissue from a patient's actual tumor and use that tissue to test the efficacy of chemotherapeutic agents on that particular tumor tissue [13].

A vascular tissue can potentially be printed. The ability to create new skin or cartilaginous appendages, such as noses and ears, could improve the life quality of burn victims and trauma patients. An artificial nose could be generated using a patient's own cells. The ability to model any desired three-dimensional structure would mean that the artificial nose could be dimensionally identical to the original [23].

Another important application of cell printing is to use two-dimensional patterned cells as a research tool to explore factors that control cell adhesion, cell proliferation, and cell differentiation [42]. By precise cell positioning, researchers can carefully control their microenvironments.

7.6 Technologies for Three-Dimensional Cell Printing: Single Cell Epitaxy by Acoustic Picoliter Droplets

Existing three-dimensional tissue-formation methods include scaffolding, laser jet, and inkjet printing. As mentioned, scaffolding can be used to create complex shapes. However, placing various cell types next to each other in precise arrays and monitoring their interactions in a controllable and repeatable way is difficult. In addition, there is the persistent problem of adequately perfusing cells near the scaffold center [43]. Current laser- and inkjet-printing approaches have been only marginally successful in printing scalable tissues because these technologies have not been specifically developed for cell printing. In addition, the ejection directionality of traditional inkjet devices is limited by the nozzle geometry [44]. Therefore, there is a need for droplet-generation technologies specifically designed for cell printing. Recently, an acoustic technology was developed to generate picoliter droplets of cell suspensions for use with cell types that are sensitive to heat, pressure, and/or shear [52]. This technology offers significant advantages over existing printing technologies. Droplets are generated from an open-pool without a nozzle. The gentle acoustic wave does not harm cells. In general, cells (~5 to 20 μm in diameter) are smaller than the acoustic wavelength. Therefore, the cells will not experience the acoustic waves traveling through the medium. Instead, they move in a circular pattern as though exposed to a large fluid flow. Ejectors can be packed uniformly into arrays, yet still be addressed on an individual basis. Such devices could eject hydrogels, print biodegradable microfluidic channels directly, and manage different fluids or cell types simultaneously from the same array. Finally, they offer tightly controlled droplet directionality, a feature that allows producing cocultures. Acoustic picoliter droplet generators designed specifically for cell printing have great potential to impact tissue engineering.

In 1997, Percin et al. proposed a novel, piezoelectrically actuated droplet ejector to overcome the limitations of previous ejectors [49]. This novel ejector does not harm sensitive fluids and can work at high speed (megahertz range) and at high flow rates (up to 100 ml/s). Although this method allows the user to individually address each array element, the fabrication process is complex. Therefore, Demirci et al. developed the fabrication of two-dimensional micromachined ejector arrays using silicon-rich nitride (Si_XN_Y) or, preferably, single-crystal silicon membranes [50]. Water ejection was demonstrated at 470 kHz, 1.24 MHz, and 2.26 MHz, producing droplet diameters of 6.5 μm, 5.0 μm, and 3.5 μm, respectively. The device can deposit a specific number of droplets at any desired location. The directionality of the device depends on the initial shape of the fluid surface at the orifice, the uniformity of the orifice shape, and the pressure distribution in the fluid reservoir.

Most recently, a micromachined, acoustically actuated, two-dimensional microdroplet ejector array was developed. The ejector operated at 36.2 MHz and generated 37 μm–diameter droplets (Figure 7.4). These ejectors have several advantages. Since ejection takes place from an open-pool, nozzleless reservoir, droplet directionality does not depend on nozzle geometry. Directionality is thus easy to control. In addition, they can be packed uniformly and addressed individually. They can be loaded with various fluids, polymers, and biological samples, so two-dimensional micromachined microdroplet ejectors have applications such as drug testing, cell sorting, and cell printing [44].

The unit cell of an acoustic picoliter droplet generator is an interdigitated circular transducer on a piezoelectric substrate. This unit cell can be repeated periodically to build two-dimensional arrays. Fluidic medium is placed on the intedigital ring, and the interdigital ring transducers, which launch surface acoustic waves, which leak into the medium. These waves travel toward the medium surface on which the waves interfere to form a focus. When the acoustic pressure at the focal point exceeds the surface tension of the fluid, a droplet is generated [45].

Acoustic picoliter droplet ejectors hold promise for cell printing since they can precisely dispense chemicals and biological samples to desired locations [44, 46]. This novel, nozzle-free, clog-free, directional, picoliter droplet technology prints single live cells encapsulated in droplets for cell-by-cell tissue printing. By using such devices, droplets can be generated from an open pool, and various cell and fluid types can be ejected simultaneously from the same array in multiple directions (e.g., laterally, downwards). The focus location can be controlled without being affected by evaporation or tilt of the device. Silicon microfluidic channel spacers can be designed so that the fluid can continuously fill the open pool through microfluidic channels, thus maintaining a steady fluid level. In this way, single hydrogel and polymer droplets may be patterned on desired locations of a surface. Various cell types can be positioned according to a CAD template. The ejector and the substrate are parallel and facing each other, with the substrate placed on top of a precise submicron resolution controlled *xyz* stage that allows exact placement of cells and other fluids at desired locations on the surface. The necessary coordinates are automatically computed, and cell printing takes place as the micrometer stage moves. As a result, cells can be printed by droplets onto any surface in their own media or in hydrogels to ensure viability over long periods as described in three-dimensional settings. The results show high cell viability after ejection for various cell types. The ratio of the ejected cell viability to that in the non-ejected control solution shows the effect of the acoustic picoliter ejector. For all cell types the viabilities exceed 89.8% [51].

7.7 Conclusion

Cell printing has the potential to revolutionize the field of tissue engineering. This technology, combining our understanding of printing and tissue engineering, brings us one step closer to artificially building organs that possess the functionality, specificity, and complexity of endogenous organs. Many challenges lie ahead. Continued extensive research is required to attain the goal of in vitro organ development.

Acknowledgments

The authors would like to thank Professor Martin Yarmush, Professor Farhang Shadman, and Professor Krishna Saraswat for their invaluable support. Dr. Demirci

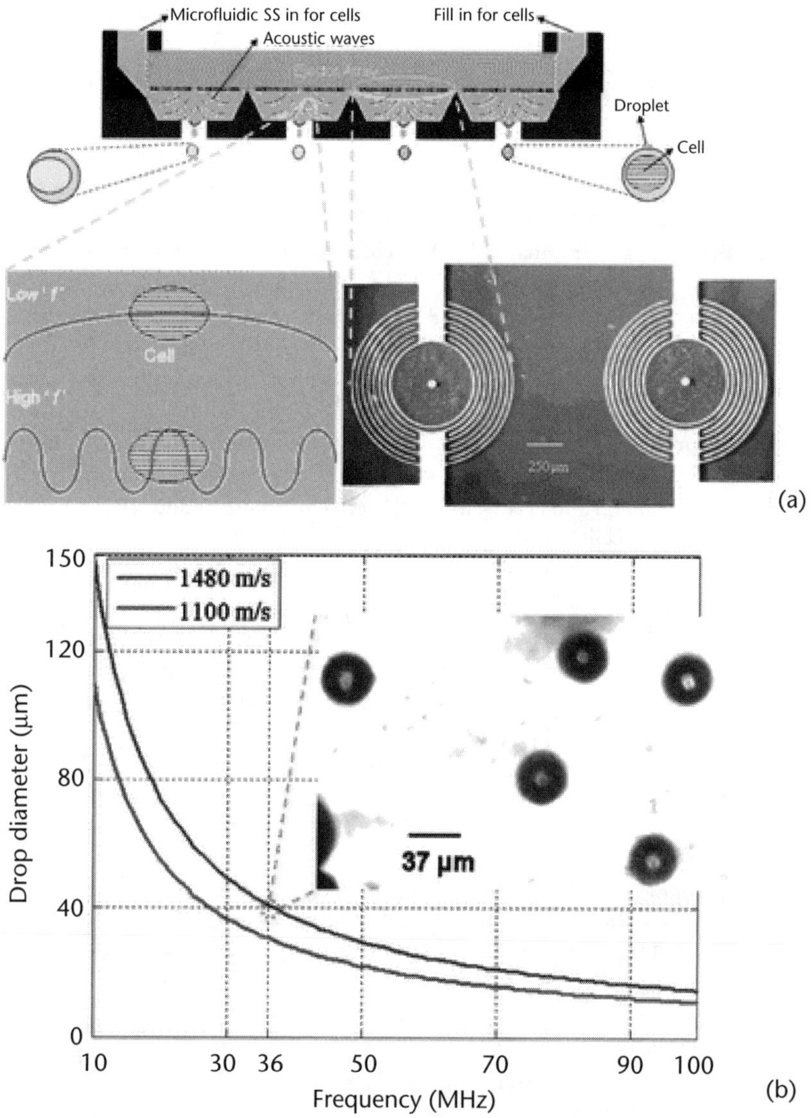

Figure 7.4 (a) Setup for acoustic picoliter droplet generation is shown. Droplets can be deposited drop-on-demand at predetermined locations. Periodically spaced interdigitated gold rings of an acoustic picoliter droplet ejector with the innermost 500 μm–diameter ring are demonstrated. The acoustic wavelength (low *f*) is much larger than the cell size, resulting in harmless ejection of cells. (b) Theoretical droplet size as a function of frequency is plotted for two sound speeds. Ejected droplets at 36.2 MHz were 37 μm in diameter and uniform in size. Theory predicts the droplet size to be 40 μm. Droplet size decreases with increasing frequency. Fluids with low sound speed generate smaller droplets than fluids with large sound speed. (*Source:* [51], reprinted with permission.)

thanks Professor Greg Kovacs for introducing him to the field of microelectromechanical systems, Professor Pierre Khuri-Yakub for introducing him to the field of acoustics, and Professor Mehmet Toner for introducing him to the field of biotechnology.

References

[1] Khademhosseini, A., Langer, R., Borenstein, J., and Vacanti, J. P., "Microscale technologies for tissue engineering and biology," *Proc. Natl. Acad. Sci. USA,* Vol. 103, No. 8, 2006, pp. 2480–2487.

[2] Lalan, S., Pomerantseva, I., and Vacanti, J. P., "Tissue engineering and its potential impact on surgery," *World J. Surg.,*Vol 25, 2001, pp. 1458–1466.

[3] Tsang, V. L., and Bhatia, S. N., "Three-dimensional tissue fabrication,"*Adv. Drug Deliv. Rev.*, Vol. 56, 2004, pp. 1635–1647.

[4] Griffith, L. G., and Swart, M. A., "Capturing complex 3-D tissue physiology in vitro," *Nat. Rev. Molec. Cell Biol.*, Vol 7, 2006, pp. 211–224.

[5] Mironov, V., et al. "Organ printing: Computer-aided jet-based 3-D tissue engineering," *Trends in Biotechnol.*, Vol. 21, No. 4, 2003, pp. 157–61.

[6] Mironov, V., and Markwald, R. R. "Anatomy of tissue engineering," *Anat. Rec.*, 2001, pp. 263, 335.

[7] Townes, P. L., and Holtfreter, J. "Directed movements and selective adhesion of embryonic amphibian cells," *J. Exp. Zool.*, Vol. 128, 1955, pp. 53–120.

[8] Steinberg, M. S. "Adhesion in development: An historical overview," *Dev. Biol.*, Vol. 180, 1996, pp. 377–388.

[9] Langer, R., and Vacanti, J. P., "Tissue engineering," *Science*, Vol. 260, 1993, pp. 920–926.

[10] Boland, T., Mironov, V., Gutowska, A., Roth, E. A., and Markwald, R. R., "Cell and organ printing 2: Fusion of cell aggregates in three-dimensional gels," *Anatomical Record*, Vol. 272A, 2003, pp. 497–502.

[11] Wilson, W.C., Jr, and Boland, T., "Cell and organ printing 1: Protein and cell printers," *Anat. Rec. A Discov. Mol. Cell. Evol. Biol.,* Vol. 272, No. 2, 2003, pp. 491–496

[12] Jennifer, C., "Printing muscle and bone," *Tech. Rev.*, at www.technologyreview.com/BioTech/17913/.

[13] Mironov, V., et al., "Organ printing: Computer-aided jet-based 3-D tissue engineering," *Trends in Biotechnol.*, Vol. 21, 2003, pp. 157–161.

[14] Sumerel, J., et al., "Piezoelectric ink jet processing of materials for medical and biological applications," *Biotechnol. J.,* Vol. 1, 2006, pp. 976–987.

[15] Markwald, R., "Desktop organ printing,"*Anat. Rec. B New Anat.,* Vol. 273, No. 1, 2003, pp. 120–121.

[16] Ringeisen, B. R. "Laser printing of pluripotent embryonal carcinoma cells," *Tissue Eng.*, Vol. 10, 2004, pp. 483–491.

[17] Odde, D. J., and Renn, M. J., "Laser-guided direct writing for applications in biotechnology," *Trends in Biotechnol.*, Vol. 17, 1999, pp. 385–389.

[18] Barron, J. A., Krizman, D. B., and Ringeisen, B. R., "Laser printing of single cells: Statistical analysis, cell viability, and stress," *Annals Biomed. Eng.*, Vol. 33, 2005, pp. 121–130.

[19] Barron, J. A., et al., "Biological laser printing of genetically modified Escherichia coli for biosensor applications," *Biosensors and Bioelectronics,* Vol. 20, 2004, pp. 246–252.

[20] Barron J. A., Wu, P., Ladoceur, H. D., and Ringeisen, B. R., "Biological laser printing: A novel technique for creating heterogenous 3-dimensional cell patterns," *Biomed. Microdevices*, Vol. 6, No. 2, 2004, pp. 139–147.

[21] Ringeisen, B. R., et al., "Jet-based methods to print living cells," *Biotechnol. J.,* Vol. 1, 2006, pp. 930–948.

[22] Nahmias, Y., Schwartz, R. E., Verfaillie, C. M., and Odde, D. J., "Laser-guided direct writing for three-dimensional tissue engineering." *Biotechnol. Bioeng.*, Vol. 92, No. 2, 2005, pp. 129–136.

[23] Mironov, V., and Forgacs, G., "Organ printing: Self-assembling cell aggregates as bioink," *Science and Medicine*, Vol. 9, No. 2, 2003, pp. 69–71.

[24] Roth, E. A., et al., "Inkjet printing for high-throughput cell patterning," *Biomaterials,* Vol. 25, No. 3, 2004, pp. 707–3715.

[25] Boland, T., Xu, T., Damon, B., and Cui, X., "Application of inkjet printing to tissue engineering," *Biotechnol. J.*, Vol. 1, 2006, pp. 1–8.

[26] Xu, T., et al., "Inkjet printing of viable mammalian cells," *Biomaterials,* Vol. 26, 2005, pp. 93–99.

[27] Nakamura, M., et al., "Biocompatible inkjet printing technique for designed seeding of individual living cells," *Tissue Eng.,* Vol. 11, No. 2, 2005, pp. 1658–1666.

[28] Xu, T., et al., "Construction of high-density bacterial colony arrays and patterns by the ink-jet method." *Biotechnol. Bioeng.,* Vol. 85, No. 1, 2003, pp. 29–33.

[29] Xu, T., et al., "Viability and electrophysiology of neural cell structures generated by the inkjet printing method," *Biomaterials*, Vol. 27, 2006, pp. 3580–3588

[30] Limpanuphap, S. S., and Derby, B., "Manufacture of biomaterials by a novel printing process," *J. Mater. Sci.: Mater. in Med.*, Vol. 13, 2002, pp. 1163–1166.

[31] Campbell, P. G., et al., "Engineered spatial patterns of FGF-2 immobilized on fibrin direct cell organization." *Biomaterials*, Vol. 26, No. 33, 2005, pp. 6762–70.

[32] Jayasinghe, N., and P. A. E., Quresh, A. N., "Electric field driven jetting: An emerging approach for processing living cells," *Biotechnol. J.,* Vol. 1, 2006, pp. 86–94.

[33] Sun, W., and Lal, P. "Recent development on computer aided tissue engineering—a review," *Computer Methods and Programs in Biomedicine*, Vol. 67, 2002, pp. 85–103.

[34] Weiss, L. E., et al., "Bayesian computer-aided experimental design of heterogeneous scaffolds for tissue engineering," *Computer-Aided Design*, Vol. 37, 2005, pp. 1127–1139.

[35] Leukers, B., et al., "Hydroxyapatite scaffolds for bone tissue engineering made by 3-D printing." *J. Mater. Sci.: Mater. in Med.*, Vol. 16, 2005, pp. 1121–1124.

[36] Martin, I., et al., *Cytometry,* Vol. 28, 1997, pp. 2141–2146.

[37] Layer, P. G., Robitzki, A., Rothermel, A., and Willbold, E., "Of layers and spheres: The reaggregate approach in tissue engineering," *Trends Neurosc.*, Vol. 3, 2002, pp. 131–134.

[38] Jakab, K., et al., "Engineering biological structures of prescribed shape using self-assembling multicellular systems," *PNAS,* Vol. 101, No. 9, 2004, pp. 2864–2869.

[39] Soletti, L., et al., "A seeding device for tissue engineered tubular structures,"*Biomaterials*, Vol. 27, 2006, pp. 4863–4870.

[40] Jakab, K., Neagu, A., Mironov, V., and Forgacs, G., "Organ printing: Fiction or science," *Biorheology*, Vol. 41, Nos. 3–4, 2004, pp. 371–375.

[41] Barron, J. A., et al., "Application of laser printing to mammalian cells," *Thin Solid Films*, Vol. 453, 2004, pp. 383–387.

[42] Falconnet, D., Csucs, G., Grandin, H. M., and Textor, M., "Surface engineering approaches to micropattern surfaces for cell-based assays," *Biomaterials*, Vol. 27, 2006, pp. 3044–3063.

[43] Langer, R. S., and Vacanti, J. P., "Tissue engineering: the challenges ahead," *Sci. Am.*, Vol. 280, No. 4, 1999, pp. 86–89.

[44] Demirci, U. "Acoustic picoliter droplets for emerging applications in semiconductor industry and biotechnology," *J. Microelectromechanical Syst.*, Vol. 15, No. 4, 2006, pp. 957–966.

[45] Demirci, U., " Droplet-based photoresist deposition," *Appl. Phys. Lett.*, Vol. 88, 2006, pp. 144104–144106.

[46] Demirci, U., "Picoliter droplets for spinless photoresist deposition," *Rev. Sci. Instrum.*, Vol. 76, No. 6, 2005, pp. 5103–5107.

[47] Percin, G., and Khuri-Yakup, L. T. S. B., "Controlled ink-jet printing and deposition of organic polymers and solid particles," *Appl. Phys. Lett.*, Vol. 73, No. 16, 1998, pp. 2375–2377.

[48] Percin, G., Levin, L., and Khuri-Yakup, T., "Piezoelectrically actuated droplet ejector." *Rev. Sci. Instrum.*, Vol. 68, No. 12, 1997, pp. 4561–4563.

[49] Perçin, G., "Micromachined piezoelectrically actuated flextensional transducers for high resolution printing and imaging," Ph.D. dissertation, Stanford University, Stanford, CA, 2002.

[50] Demirci, U., et al., "Acoustically actuated flextensional SixNy and single-crystal silicon 2-D micromachined ejector arrays," *IEEE Trans. Semiconductor Manufacturing*, Vol. 17, No. 4, 2004, pp. 517–524.

[51] Entcheva, E., and Bein, H., "Acoustic micromachining of three-dimensional surfaces for biological applications," *Lab Chip*, Vol. 5, 2004, pp. 179–183.

[52] Demirci, U., and Montesano, G., "Single cell epitaxy by acoustic picolitre droplets," *Lab Chip*, 2007, Vol. 7, pp. 1139–1145.

CHAPTER 8

Using Microfabrication to Engineer Cellular and Multicellular Architecture

Sami Alom Ruiz and Christopher S. Chen

8.1 Introduction

Invented by the German bacteriologist Julius Petri in 1877, the petri dish was initially used to culture microbial organisms but was then extended to fungus, plant, and animal cells. Today, it remains the standard instrument for cell culture. While the petri dish has been instrumental to countless biological discoveries, a brief comparison to the physiologic microenvironment reveals its inadequacies. Cells in vivo are exposed to gradients of cytokines, are surrounded by extracellular matrix (ECM), make contacts with other cells in three dimensions, and interact with other cell types. These microenvironmental cues have important roles in regulating cell function, yet they are all altered in culture dishes. As a result, cells extracted from the body and placed in such simplified culture conditions lose many of their normal behaviors. Replicating physiologic characteristics in vitro is necessary if cells are to be studied in a state that more closely resembles their native one.

One major recent improvement has been to produce spatially defined environments in which cells and their scaffolds can be organized on a surface in a specific arrangement. In vivo, such spatial organization determines the arrangement of many extracellular cues that are sensed by cells via transmembrane receptors and the cytoskeleton and transduced into intracellular signals that drive function. These more sophisticated approaches to culturing cells are developed using technologies adapted from the semiconductor industry, collectively referred to as microfabrication [1]. Photolithography, the technology used to fabricate the initial template of electronic circuits, has been used extensively to generate micrometer-scale structure to the cell substrate. The technique uses ultraviolet light and a photomask to pattern a layer of photosensitive resist on top of a substrate [1]. The patterned photoresist has regions covered by polymerized photoresist, as well as bare regions where photoresist has been removed. This photoresist pattern can, in turn, be used as a sacrificial layer to control the spatial distribution of other materials. One type of molecule can be patterned on the bare regions, and then, after stripping the photoresist, a second molecule of different chemistry can be deposited on the previously protected areas [Figure 8.1(a)]. Alternatively, the patterned photoresist can be used as a template for soft lithography. Soft lithography is the set of techniques that use elastomeric stamps with patterned relief features to generate micro- and nanoscale patterns [2, 3]. The stamps are made by casting prepolymer elastomer, usually polydimethylsiloxane (PDMS), onto the template prepared by photolithography, then curing and peeling off the crosslinked elastomer [Figure 8.1(b)].

Microfabrication has been used to provide the experimenter with exquisite control over the deposition of cells and proteins in spatially defined patterns. Microcontact printing, for example, has been used to regulate cell shape with microscale resolution [4–6], and microfluidic channels have been used to control the spatial and temporal distribution of cytokines [7–9]. Most of these tools have been developed to culture cells on two-dimensional surfaces as a natural transition from petri dishes. This chapter describes these and other techniques in the context of such two-dimensional patterning, as well as how recent work has extended this into patterning three-dimensional culture environments.

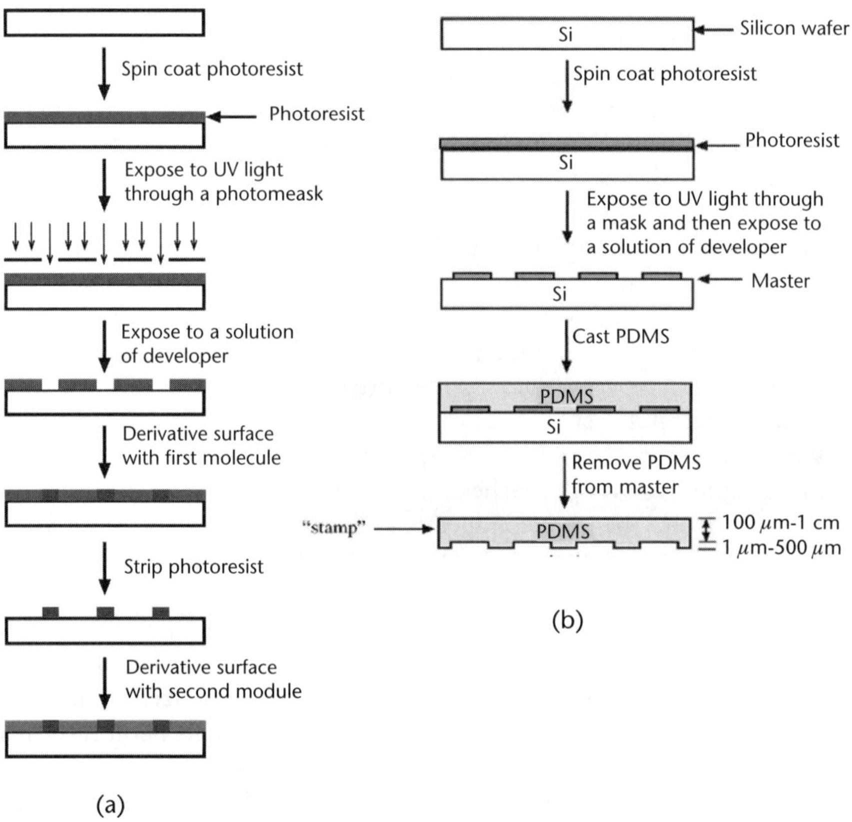

Figure 8.1 Illustrations of two patterning methods: (a) Diagram illustrating the patterning of surface chemistry using photolithography: (1) a quartz coverslip is coated with photoresist, (2) the photoresist is exposed to ultraviolet light through a photomask, (3) unpolymerized photoresist is developed leaving a lithographically defined surface, (4) a molecule with a particular chemistry is bound to the bare regions, (5) the photoresist is removed, and (6) a second molecule with a different chemistry is deposited onto the previously protected regions creating a chemically defined surface. (b) Diagram illustrating the fabrication of a PDMS stamp: after creation of a lithographically defined master, prepolymer PDMS is cast onto the master, cured, and peeled to create a stamp with relief features. (Adapted from [3].)

8.2 Patterning Adhesion

Defined organization of cells on a culture substrate allows one to begin studying how the structure of cells and tissues contributes to their function. The key to achieving control over cellular positioning is the placement of cell-adhesive islands on substrates, surrounded by nonadhesive regions. Cells seeded onto such substrates adhere only to the islands and spread to conform to their size and shape if the islands are of the same length scale as the cells. Several fabrication-based approaches have been used to pattern such surfaces. Early techniques used photolithography to directly pattern these regions. Ultraviolet light and a photomask were used to spatially control the photocleavage of head groups on self-assembled monolayer (SAM) surfaces [10]. The photocleaved regions were amenable to further chemical modification. These two-component SAM films were used to pattern cell attachment. In another study, photolithography was used first to pattern photoresist and, subsequently, a SAM-forming molecule was added to derivatize the unprotected regions. The photoresist was then stripped, and the newly uncovered regions were derivatized with another SAM to create a pattern of adhesive strips surrounded by a nonadhesive background to spatially regulate cell placement [11]. Another method for patterning cells entailed scanning a laser beam across a SAM surface to pattern the photoimmobilization of arginine-glycine-aspartic acid (RGD), a moiety that cells adhere to [12]. Using these techniques to constrain cells to shapes that were more similar to those found in vivo allowed investigators to make novel observations about the role of cell shape in regulating cell function.

A simpler approach for patterning substrate adhesivity known as microcontact printing (μCP) was developed [13]. One major advantage of μCP is that, unlike photolithography, this process requires a clean room only for the fabrication of the initial template, making the patterning of adhesivity more accessible. μCP is a technique in which SAMs or proteins are printed onto a substrate using an elastomeric stamp with relief features cast from a template (reviewed in [2, 14, 15]). After pressing the stamp against the substrate, an imprint of the SAM or protein is left in the configuration of the stamp features. A template can be used to make multiple stamps, and a single stamp can be used to print many substrates, allowing for rapid production. To pattern SAMs, the stamp is first loaded by applying a solution of alkanethiol to the surface with raised features, then drying it gently with a stream of air. When the stamp is then brought into contact with a gold or silver surface, the alkanethiol molecules coordinate with the surface through the thiol groups at the end of their tails. The head groups of the alkanethiols, which determine the properties of the SAM, are presented at the surface. To create adhesive islands surrounded by nonadhesive regions, islands of methyl-terminated alkanethiol were μCP-ed to form an adsorption-promoting SAM [16]. The substrate was then immersed in poly(ethylene glycol)-terminated alkanethiol to form an adsorption-resistant SAM surrounding the islands. To make such a substrate amenable to cell attachment, it is immersed in a solution of ECM protein that adsorbs onto the methyl-terminated regions and promotes cell attachment. Alternatively, the protein can be patterned directly. The stamp is loaded by allowing the protein molecules to adsorb to the stamp through nonspecific interactions [17, 18]. Protein printing differs from printing SAMs in that it is dependent on the hydrophilicity of the substrate relative to the stamp [19]. Following protein transfer, the surrounding areas were rendered nonadhesive with a detergent, pluronic F127 [20]. Once these substrates are derivatized with adhesive and nonadhesive regions, seeding cells onto them would result in their patterning.

8.3 Patterning Single Cells

Cultured cells exhibit differences in behavior from cells in vivo, such as decreased differentiation [21, 22] and increased proliferation [23, 24]. Because cells spread and flatten against substrates such as glass or plastic much more so than often observed in vivo, it was postulated that this increased spreading was in part responsible for the transformation in cell phenotype. Also, cells of the same type may have distinct geometries in vivo when they perform different functions, such as the leading endothelial cell in a sprouting vessel [25], cardiomyocytes in the different regions of the ventricle [26], or neurons in the different cerebral cortex layers [27]. To give cultured cells more physiologic shapes and to examine whether such manipulations would alter cell function, micropatterning techniques to control the area of cell adhesion were utilized. μCP was used to pattern isolated single cells on adhesive islands of different shapes and sizes [4–6]. In hepatocytes, as adhesion area (hence, the area of cell spreading and flattening) increased, proliferation rose, and albumin production, an indicator of liver differentiation, decreased [4]. In endothelial cells, cell area was found to regulate a similar switch, except that large islands supported proliferation, while small ones triggered apoptosis, or programmed cell death [6] [Figure 8.2(a)]. Mesenchymal stem cells grown on islands of printed ECM differentiated into adipocytes on small islands and osteoblasts on large ones [28]. This agrees with previous studies where increased osteoblast spreading, regulated by ligand density, correlates with increased matrix mineralization [29, 30]. These studies demonstrate that cell geometry is an important regulator of apoptosis, differentiation, and growth. Further manipulation of cell shape may help uncover the importance of geometry in other cell behaviors.

When cells attach to a substrate, they adhere through discrete, punctate structures called focal adhesions (FA) at the interface with the ECM [31–33]. These structures serve to anchor the cell to the matrix and act as signaling hubs that regulate cell shape, signaling pathways, and function. Since the ECM to which cells adhere in vivo is discontinuous, patterning the distribution of ECM as discrete subcellular patches to control FA localization is complementary to controlling overall cell shape. μCP of subcellular adhesive patches has been the primary method for controlling FA distribution [6, 34–38]. Stamps with features in the range of 3 to 10 μm were used to print dots of ECM [34]. Cells spread over the dots forming FAs with the ECM [Figure 8.2(b)]. This technique made it possible to uncouple cell area from adhesion area and demonstrate that the former, not the latter, dictated proliferation [6]. The amount of FA per cell also depended on the cell area [34]. Using a much wider range of feature sizes (0.3 to 3 μm) and spacings (1 to 30 μm), cell spreading on arrays of squares was shown to depend not on the size of the features but on the percentage of substrate covered by ECM [35]. μCP was also used to vary the spatial distribution of adhesion while maintaining the shape constant by printing ECM in

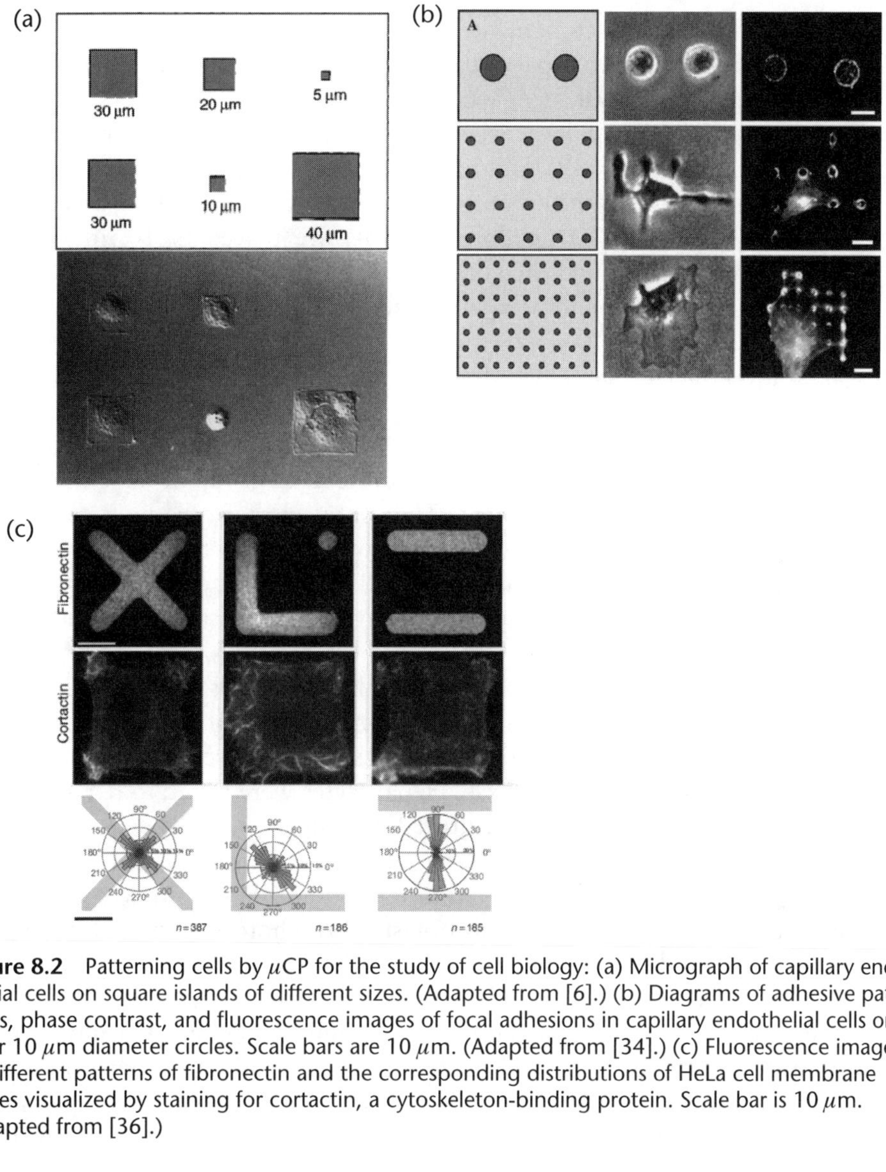

Figure 8.2 Patterning cells by μCP for the study of cell biology: (a) Micrograph of capillary endothelial cells on square islands of different sizes. (Adapted from [6].) (b) Diagrams of adhesive patterns, phase contrast, and fluorescence images of focal adhesions in capillary endothelial cells on 3, 5, or 10 μm diameter circles. Scale bars are 10 μm. (Adapted from [34].) (c) Fluorescence images of different patterns of fibronectin and the corresponding distributions of HeLa cell membrane ruffles visualized by staining for cortactin, a cytoskeleton-binding protein. Scale bar is 10 μm. (Adapted from [36].)

different geometries [36–38] [Figure 8.2(c)]. Cell-division axis orientation [36], focal adhesion and actin stress fiber patterns [37], and polarity [38] were determined by the adhesion pattern rather than cell shape. These studies prove that the distribution of adhesion, not just its absence or presence, is important in regulating function. The μCP of ECM to pattern subcellular adhesion may prove useful in furthering our understanding of how in vivo matrices regulate cell behavior.

8.4 Multicellular Patterning

In most tissues, cells are in contact with other cells in their environment. Adjacent cells can influence behavior through cell-cell contacts or paracrine signaling. To replicate such architecture in vitro, μCP has been used not only to control the geometry of single cells but also to create multicellular patterns by stamping ECM shapes greater than the size of a single cell. Endothelial monolayers of different shapes and sizes prepared by μCP large islands of ECM exhibited higher proliferation rates at the monolayer edges [39] [Figure 8.3(a)]. The variation in growth was due to differences in stress across the monolayer caused by the cells pulling on each other. Interestingly, when capillary endothelial cells are placed in 10 μm–wide lines, they cooperate to roll around each other, possibly through these intercellular forces, and form three-dimensional cylinders, whereas the cells remain as monolayers on 30 μm–wide lines [40]. Monolayer patterns of different shapes were used to observe cell migration out of them [41]. Mrksich and colleagues developed a method to switch the surrounding nonadhesive regions to adhesive regions after the cells are patterned, allowing cells to migrate out of multicellular patterns. This was achieved with a brief electrical pulse that altered the head group of the initially nonadhesive SAMs to permit cell attachment. They showed that cells migrated out of circular patterns more quickly than they did out of patterns with straight edges. The organization of cells into large patterns by μCP has made it easier to begin to understand how structure affects multicellular behavior in ways not possible by conventional culture methods.

Besides participating in interactions between cells of the same type, it is necessary for cells in many tissues to make contacts with other cell types to perform certain functions, such as endothelial and smooth muscle cells in blood vessels and neurons and myocytes in muscle. Hepatocytes were found to remain more differentiated in vitro if they were cocultured with liver epithelial cells [42, 43]. Such randomly seeded cocultures, however, did not permit control over the coculture interaction. To overcome this limitation, photolithography was used to pattern the coculture of hepatocytes and fibroblasts [44]. Photoresist was patterned first, collagen was bound to the bare regions, and the photoresist was then removed [Figure 8.3(b)]. Hepatocytes were added first and attached only to collagen; fibroblasts were added next and adhered to the unmodified regions. Coculture of hepatocyte islands surrounded by fibroblasts resulted in improved liver function as measured by urea synthesis and albumin secretion. Patterning allowed the comparison of hepatic function in different hepatocyte colony sizes [44]. Such patterned hepatocyte cocultures where hepatic function is retained may be of use in liver-toxicity studies.

Soft lithography methods have also been used to generate patterned coculture. One method that was used for patterning the organization of two different cell types exploited the previously described electroswitchable SAMs [41]. Two cell populations were patterned by seeding one population that occupied the stamped adhesive areas first, then switching the nonadhesive regions to become adhesive and seeding a second population to occupy the surrounding switched regions [Figure 8.3(c)]. Another technique for patterned coculture utilized a multilevel stamp to seed two different cell types in juxtaposition [45] [Figure 8.3(d)]. ECM was first adsorbed onto regions of the substrate not blocked by the stamp. When the stamp was pushed further, another level collapsed against the substrate shielding part of the adhesive regions. The first cell type was then seeded through holes in the stamp and allowed to adhere to the unprotected regions. After removing the stamp and blocking regions not coated with ECM, the second cell type was added and attached to the adhesive,

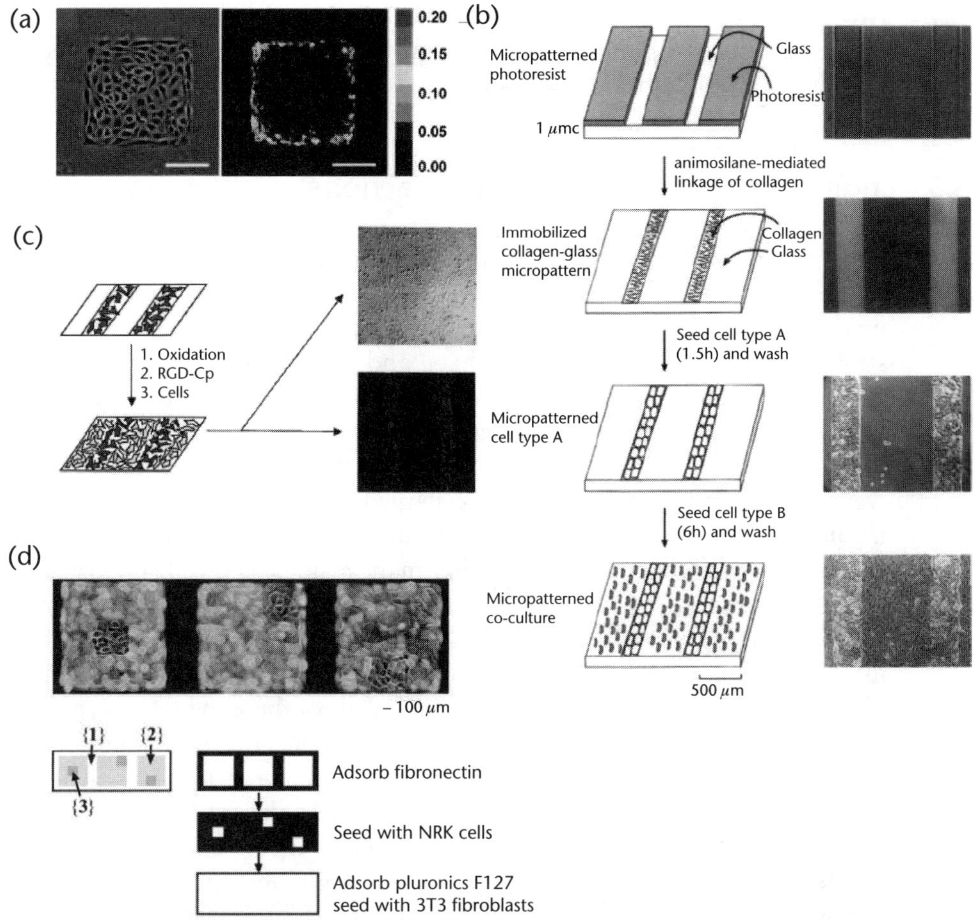

Figure 8.3 Patterning multicellular aggregates: (a) A monolayer of bovine pulmonary artery endothelial cells cultured on 250 μm squares of fibronectin. The heat map represents average proliferation localization in fifty monolayers. Scale bars are 100 μm. (Adapted from [39].) (b) Schematic and micrographs of method to generate patterned cocultures using photolithography. Collagen was immobilized onto the unprotected areas of a substrate with patterned photoresist, then the photoresist was stripped. Hepatocytes localized to the collagen-covered regions. Fibroblasts added subsequently localized to the rest of the substrate. (Adapted from [44].) (c) Schematic of method to generate patterned cocultures using μCP. Fibroblasts are seeded initially to the areas with ECM. Electrochemical oxidation of the monolayer in the presence of RGD resulted in the tethering of this cell-adhesive moiety to the nonadhesive regions to render them adhesive. Addition of another population of cells led to their localization to these switched areas. (Adapted from [41].) (d) Coculture of NRK cells and fibroblasts using a multilevel PDMS membrane. The membrane was placed against the dish covering region {1}. {2} and {3} were then coated with fibronectin. When the membrane was pushed against the substrate, the middle level collapsed, shielding {2}. After NRK cells were seeded onto {3}, the membrane was removed, the substrate was immersed in pluronics F127 to render {1} nonadhesive, and fibroblasts were seeded onto {2}. (Adapted from [45].)

but previously obstructed, regions. Even though fabricating a multilevel template involves an extra level of complexity, this technique facilitates the rapid preparation of patterned cocultures without the need for SAMs or a clean room. These approaches made possible the controlled placement of heterotypic contacts between two cell populations. Because cells in many tissues interact with multiple cell types, future coculture techniques to explore multiple heterotypic interactions will create even more physiologically relevant settings in which to investigate cell function.

8.5 Engineering Single Cell-Cell Interactions

Even though cells in vivo normally make many contacts with their neighbors, reducing the interaction to a sole connection has proven to be a useful strategy to investigate the effects of a cell-cell contact on cell behavior. A novel method for constraining pairs of cells to a fixed shape and a single contact was developed using elastomeric stamps similar to those used in μCP [46]. Stamps with bowtie-shaped features were brought into contact with glass slides [Figure 8.4(a)]. Agarose was then wicked in between the PDMS and glass and dried, such that when the stamp was lifted off, glass-bottomed bowtie-shaped wells were formed, surrounded by the nonadhesive agarose. When these substrates were seeded with cells, each half of the bowtie was occupied with one cell that made a single cell-cell contact of controlled dimension with its neighbor at the bowtie center [Figure 8.4(b)]. By controlling the shape of cells, potentially confounding shape changes brought about by contact are inhibited, keeping cell area constant. Pairs of endothelial [46, 47] and epithelial cells [48] in bowties proliferate more than single cells of the same shape. This system was also useful in studying the effect of cell-cell contacts on focal adhesions [49]. This tool made possible observations regarding cell-cell contacts that may not have been realized using standard cell-culture techniques. In the future, this approach could be extended to study single heterotypic cell contacts such as those formed by T-cell-B-cell and pericyte-endothelial cell interactions, in addition to other functions that may be influenced by homotypic cell-cell adhesion, such as polarity and differentiation.

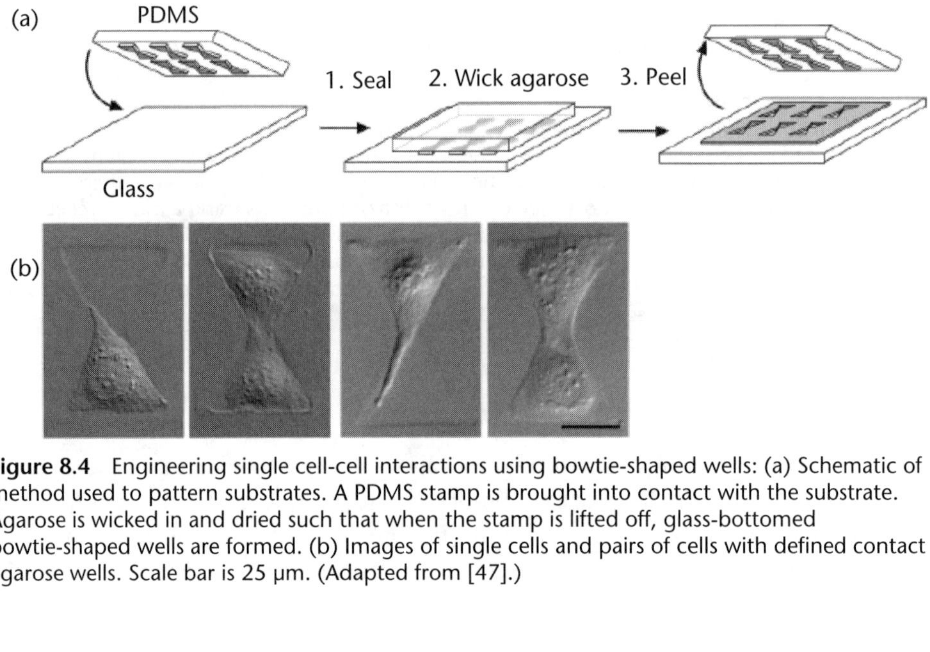

Figure 8.4 Engineering single cell-cell interactions using bowtie-shaped wells: (a) Schematic of method used to pattern substrates. A PDMS stamp is brought into contact with the substrate. Agarose is wicked in and dried such that when the stamp is lifted off, glass-bottomed bowtie-shaped wells are formed. (b) Images of single cells and pairs of cells with defined contact in agarose wells. Scale bar is 25 μm. (Adapted from [47].)

8.6 Cell Patterning by Active Positioning: Dielectrophoresis and Microfluidics

One challenge in organizing cell architecture is achieving the precise positioning of cells on patterned surfaces. Cells that land on adhesive areas generated by μCP attach to the substrate, whereas those that settle on nonadhesive areas are washed away. Because cells are randomly distributed onto the substrate, the placement of individual cells cannot be controlled. This problem was overcome using two "active-positioning" techniques that directly position cells on specified regions of the substrate. The first, dielectrophoresis (DEP), allows the user to trap cells in specific locations with single-cell resolution [50, 51]. DEP force is the interaction of a nonuniform electric field with the dipole moment it induces in an object [Figure 8.5(a)]. Photolithography was used to fabricate an array of electrodes onto a glass slide. The electrodes functioned as an array of traps and were used to trap beads [52] and cells [50] in suspension against destabilizing flows. By controlling the timing of trap activation, the device functioned as a cell sorter [50]. A similar device was used to improve cell attachment on ECM patterns [51]. In this device, the electrodes were fabricated under the surface and not inside the culture chamber, and ECM adsorption was patterned using a membrane of arrayed holes that was aligned to the electrode array such that the holes and electrodes coincided [Figure 8.5(b)]. After removing the membrane, the regions surrounding the ECM were rendered nonadhesive using pluronic F127. When a cell suspension was flowed over the array and current was applied, cells were trapped above the electrodes. When the flow was stopped and the current turned off, the trapped cells spread and filled up the patterns [Figure 8.5(c)]. DEP greatly enhanced the efficiency of cell patterning: in a device with two electrodes per 30×60 μm ECM pattern, the yield of correct patterns (with two cells) attained was 70 percent, a great improvement over the 17 percent achieved by random seeding [51]. This demonstrates that DEP could be a useful technique for applications where high pattern fidelity is required. Even though DEP is presently not as widely used as μCP because of its complexity, future advances in microfabrication, as well as the need for finer control over cell placement, will propel the use of such active trapping methods.

The second approach used to deliver cells to specified locations utilizes microfluidics. Microfluidic channels have been used extensively to control the spatial and temporal profiles of the soluble microenvironment in order to better mimic the biochemical signals that cells sense in vivo [7–9, 53]. Microfluidic chips have also been used to organize cellular architecture. PDMS stamps containing microfluidic channels with multiple inlets were cast from templates made by photolithography, sealed against a petri dish, and used to pattern bacteria and erythrocytes [7] [Figure 8.6(a)]. When multiple inlet streams merge, they undergo minimal mixing in the channel due to low Reynolds number laminar flow, and their contents remain relatively segregated. These parallel streams containing different solutions are the key to patterning using microfluidic devices. The advantage of using such a system, besides being able to pattern cells and their soluble environment in the same substrate, is that cells can be dispensed to the regions on the substrate that are accessed by the different streams. This additional level of control is not available using μCP or photolithography, where ECM is patterned and cells are seeded onto

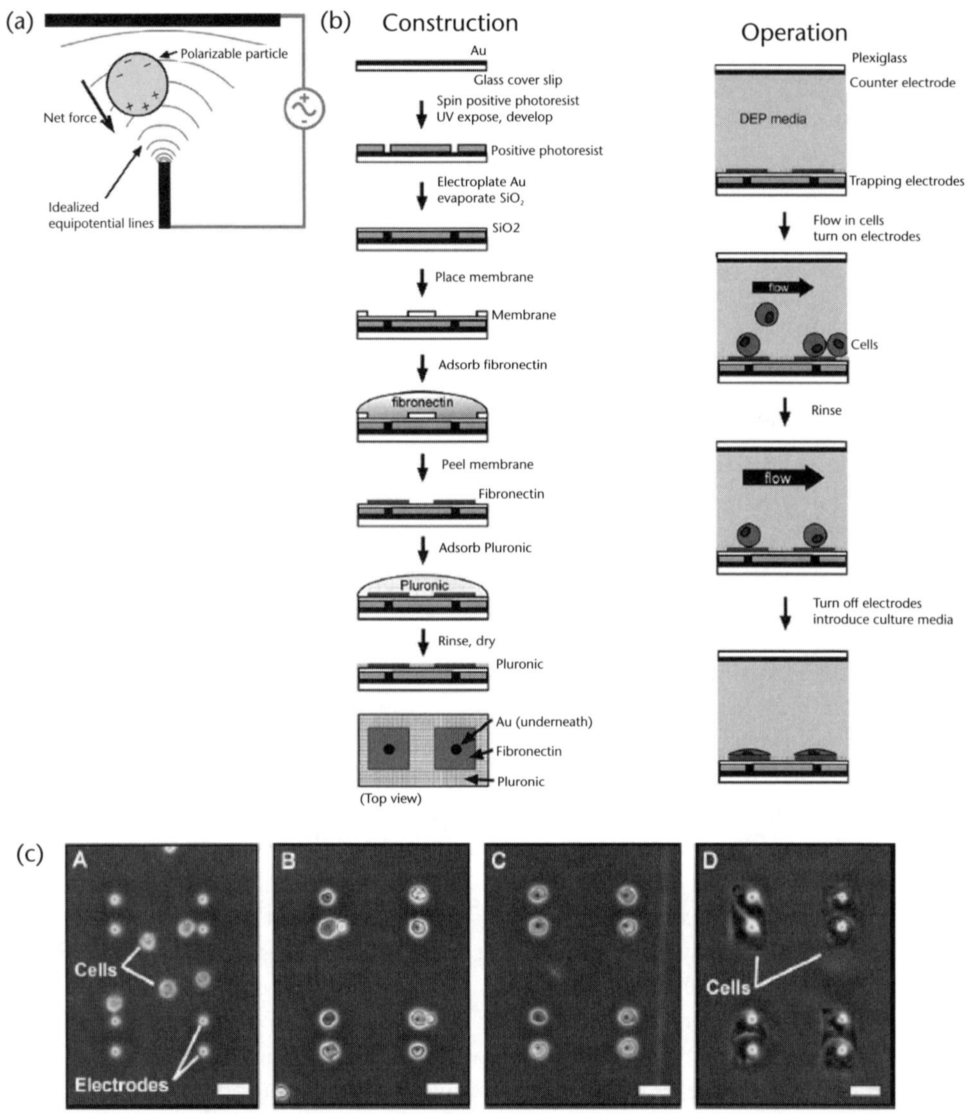

Figure 8.5 Cell patterning using DEP: (a) A polarizable object in a nonuniform electric field experiences a net dielectrophoretic force. (b) Schematics showing the construction and operation of DEP substrates. (c) Images of cells patterned by DEP in various stages. Image of randomly distributed cells with electrodes turned off (A). Energized electrodes trap single and multiple cells with low flow rates (B) but only single cells with high flow rates (C). One hour after DEP, cells spread on the adhesive areas (d). Scale bars are 30 μm. (Adapted from [51].)

the substrate randomly. Another method for patterning cells in microfluidic chips utilized an array of U-shaped hydrodynamic structures within the channel to trap single cells [54] [Figure 8.6(b)]. As cells were flowed into the channel, they became trapped in the structures, which were large enough to contain one cell. This method eliminated the need for parallel streams normally used to pattern ECM or cells, thus reducing fluidic connections and requiring a simpler flow system.

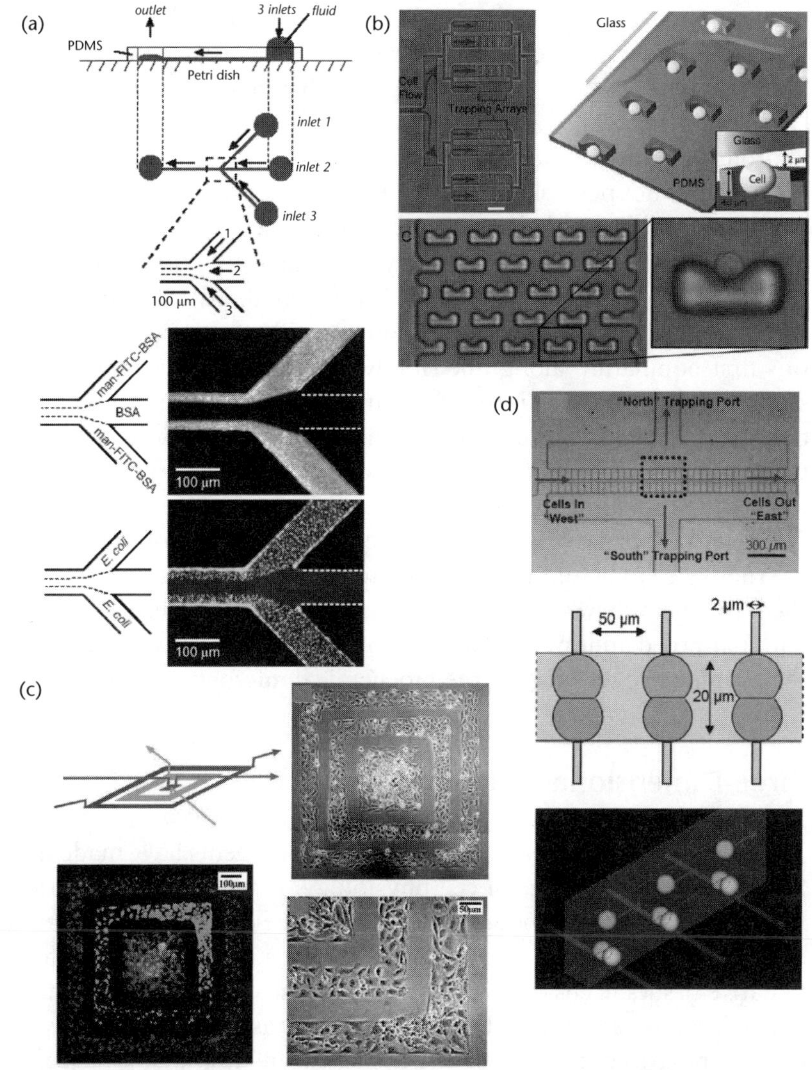

Figure 8.6 Patterning cells using microfluidic channels: (a) Side- and top-view schematics of a microfluidic channel with three laminar streams. This system was used to pattern fluorescent protein and bacteria. (Adapted from [7].) (b) Microfluidic arrays for trapping single cells. Photograph showing the branching architecture and trapping chambers. Schematic of a trapping array with an inset showing the geometry of a single trap. Micrograph of a trapping array with cells in the traps and a magnification of an individual trap. Scale bar is 500 μm. (Adapted from [54].) (c) Trajectories of three channels within a multilevel stamp used to coculture different cell types onto one substrate. Fluorescence and phase-contrast images of bladder cancer cells and endothelial cells arranged in concentric squares. (Adapted from [55].) (d) Layout of microfluidic trapping device used to generate an array of cell pairs. Image of the arrayed device showing four inlets. Cells are flowed through the channel from "west" to "east." The four valves that regulate flow through the inlets are used to control trapping. Two independently regulated channels are used to trap different cell populations to either the "north" or "south" wall of the main channel. Top-view and three-dimensional schematics showing trapped cells and channel dimensions. (Adapted from [56].)

Microfluidics also offers another approach to control the coculture of multiple cell types. Using multilayered soft lithography, three-dimensional microfluidic systems were used to pattern endothelial and bladder cancer cells in complex, discontinuous structures, such as concentric squares or chessboard patterns [55] [Figure 8.6(c)]. Because many segregated channels can be contained in the multilayered stamp, multiple cell types can be patterned more easily than using μCP or photolithography, although the placement of the cell populations in contact with each other is not possible because of the presence of stamp walls separating the compartments. Individual heterotypic cell-cell contacts were produced inside a microfluidic chamber using an array of smaller 2×2 μm channels within the main channel side walls that act as traps [56] [Figure 8.6(d)]. The first cell population was flowed in while trapping channels on one side were activated, selectively trapping this first population along one side wall. After flushing away untrapped cells, the second population was introduced and captured by the trapping channels on the opposite side wall. Because the channel width was slightly smaller than the sum of the two cell diameters, contacts were formed between opposite cells. While this technique offers a higher yield of paired cells than the bowtie system described above, it affords no control over cell shape or contact size. Microfluidic channels provide an alternative to μCP and photolithography for patterning cells, physically delivering them to where they are desired rather than randomly seeding them onto substrates with patterned adhesivity. Though not as simple to use as μCP, microfluidics offers the advantage of regulating the biochemical microenvironment.

8.7 Three-Dimensional Patterning

Traditional two-dimensional cell-culture experiments have made the greatest contribution to our knowledge of cell physiology. However, as our understanding of cell biology deepens, it has become clear that three-dimensional substrates are also necessary for studying cells. Cells cultured on flat surfaces behave differently from those in native tissues. Cells on two-dimensional substrates can undergo dedifferentiation and exhibit increased growth [21–24]. One reason may be because cells on surfaces become polarized, flatten against the substrate attaining a pancakelike shape, and make focal adhesions only on their ventral side because ECM is only available on one planar surface. Because gels are three-dimensional and have some properties that are more similar to tissues than two-dimensional substrates, biologists have started adopting them as platforms for studying cells.

As these gels become more commonly used, the need for patterning them becomes more apparent. In the last few years, several microfabrication techniques have been used to pattern cells in gels. Laser-scanning lithography (LSL) [57] and photolithography [58] were used to pattern the adhesiveness of the surface of a poly(ethylene glycol) (PEG)–based hydrogel [Figure 8.7(a)]. Using a confocal laser-scanning microscope and software to dictate which locations became irradiated, the linkage of monomeric acryloyl-PEG-RGDS to the hydrogel by photopolymerization was patterned, and the attachment of fibroblasts was spatially regulated [57]. LSL and conventional photolithography, however, cannot be used to pattern the gel bulk because they utilize single-photon light, which cannot penetrate into the gel. To overcome this limitation, two-photon absorption (TPA) photolithography was used in conjunction with LSL to create three-dimensional patterns in

(a)

(b)

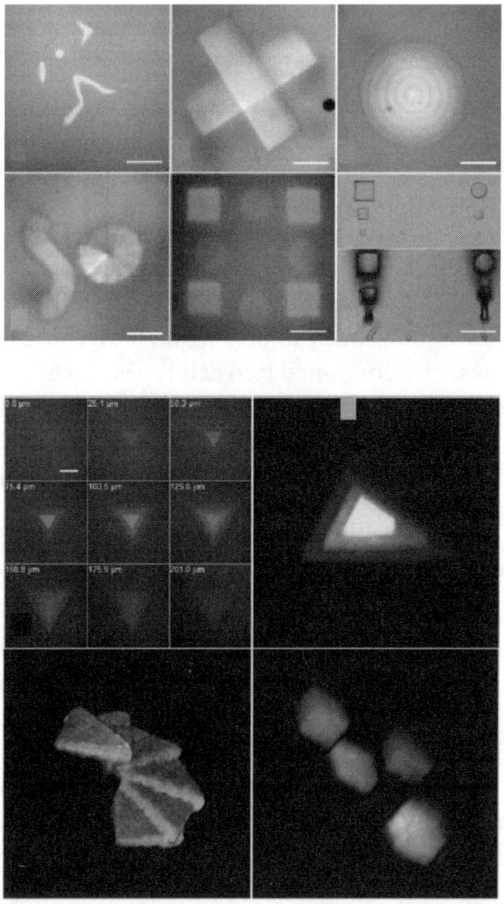

Figure 8.7 Patterning the surface and bulk of PEG-based hydrogels: (a) Patterning ligand onto hydrogel surfaces using LSL. Grayscale fluorescence images of free-form shapes made by using LSL to pattern single density and two-dimensional gradients (made by varying exposure times) of fluorescently-labeled ACRL-PEG-RGDS onto hydrogel surfaces. Fluorescence image of two different adhesive peptides. DIC images of 4 μm–tall and 20 μm–tall three-dimensional structures formed on hydrogel surfaces by using diacrylated monomers and varying the exposure focal plane. (Adapted from [57].) (b) Patterning ligand density in the interior of hydrogels using TPA photolithography. Confocal images showing a *z*-series of a hydrogel patterned with fluorescently-labeled ACRL-PEG-RGDS. Each triangular structure received a different exposure time, with the large base triangle and the intermediate triangle receiving one-quarter and one-half of the exposure time per micrometer squared, respectively, as the top triangle. Three-dimensional rendering of the images in with each color representing a different density of captured fluorescent ligand. Three-dimensional rendering of images taken from a spiral and parallelograms. Scale bar is 200 μm. (Adapted from [59].)

the interior of the gel [59]. By tightly focusing the beam, the region of TPA was confined to a focal volume, and the locations of photoinduced conjugation were precisely regulated. This technique provided control over ligand (monoacrylate) and crosslinker (diacrylate) incorporation to regulate moiety density and stiffness, respectively. Free-form patterns with varying ligand density within each structure were formed in the gel bulk [Figure 8.7(b)]. Cells migrated into and remained restricted to the regions patterned with RGD ligand. While this technique requires equipment that is more sophisticated than that used in standard photolithography, it provides the user with fine spatial control over chemical and mechanical properties in three dimensions.

Other methods have sought to control the placement of gels and the cells within them. One approach used conventional photolithography, DEP, or a combination thereof to pattern the polymerization of PEG gels and the distribution of the cells within them, respectively [60] [Figure 8.8(a, b)]. Two domains within a gel were patterned with different cell types by using a photomask to control the polymerization of the first gel/cell solution inside a chamber, flushing out the unpolymerized gel, flowing in a second gel/cell solution, and patterning again by photolithography. The resulting construct contained two cell types, each randomly dispersed within one of two gel regions. DEP was used to pattern the cells into clusters before photopolymerizing the gel [60, 61]. By increasing the chamber height after polymerization of the first layer, another layer of gel with a second cell type patterned in a different conformation was created [61] [Figure 8.8(c, d)]. The combination of photopolymerization and DEP afforded exquisite macro- and microscale control of the spatial organization of gels and the cells within them [60].

As with patterning surfaces, alternatives to photolithography for patterning gels have been developed that take advantage of soft lithography. Such approaches require a clean room only for the fabrication of the initial template and do not require sophisticated equipment. Elastomeric stamps were pressed against unpolymerized collagen loaded with cells to mold the gel [62]. After the gel polymerized, the stamp was removed, and the molded gel was overlayed with collagen precursor containing another cell type to generate a patterned gel domain with one cell population surrounded by another domain containing a second [Figure 8.9(a)]. PDMS stamps also served to construct microvascular tubes in collagen [63]. A chamber within a stamp was filled with collagen precursor, and a needle was inserted parallel to the chamber [Figure 8.9(b)]. After the gel polymerized, the needle was removed, leaving a hollow cylinder within the collagen. Endothelial cells seeded in the cylinder adhered to the collagen, forming tubes that had good barrier function and were resistant to leukocyte adhesion. The inflammatory agonists thrombin, histamine, and TNF-α increased permeability and promoted leukocyte adhesion to the endothelial wall. These functional microvessels through which soluble factors and cells could be flowed could not have been engineered without a microfabrication technique to organize cell architecture. Elastomeric stamps were also used to construct wells in collagen to study the branching in epithelial tubules [64]. Different shape wells were filled with cells and overlayed with collagen to encase the multicellular aggregates [Figure 8.9(c)]. The geometry of the tubules dictated branching, which was regulated by inhibitory autocrine signals. These observations were facilitated by the ability to organize the aggregates into different geometries. Elastomeric stamps present a quick and easy method for patterning gels, and the cells within them and are expected to gain widespread use in the near future.

Even though the patterning of three-dimensional substrates is still in its infancy, using microfabrication to control the spatial seeding of cells in gels has already been useful in controlling cell cluster size, coculture, engineering vascular capillaries, and studying epithelial tube migration. The field is expected to blossom in the coming years as engineers adapt more microfabrication techniques for patterning gels in

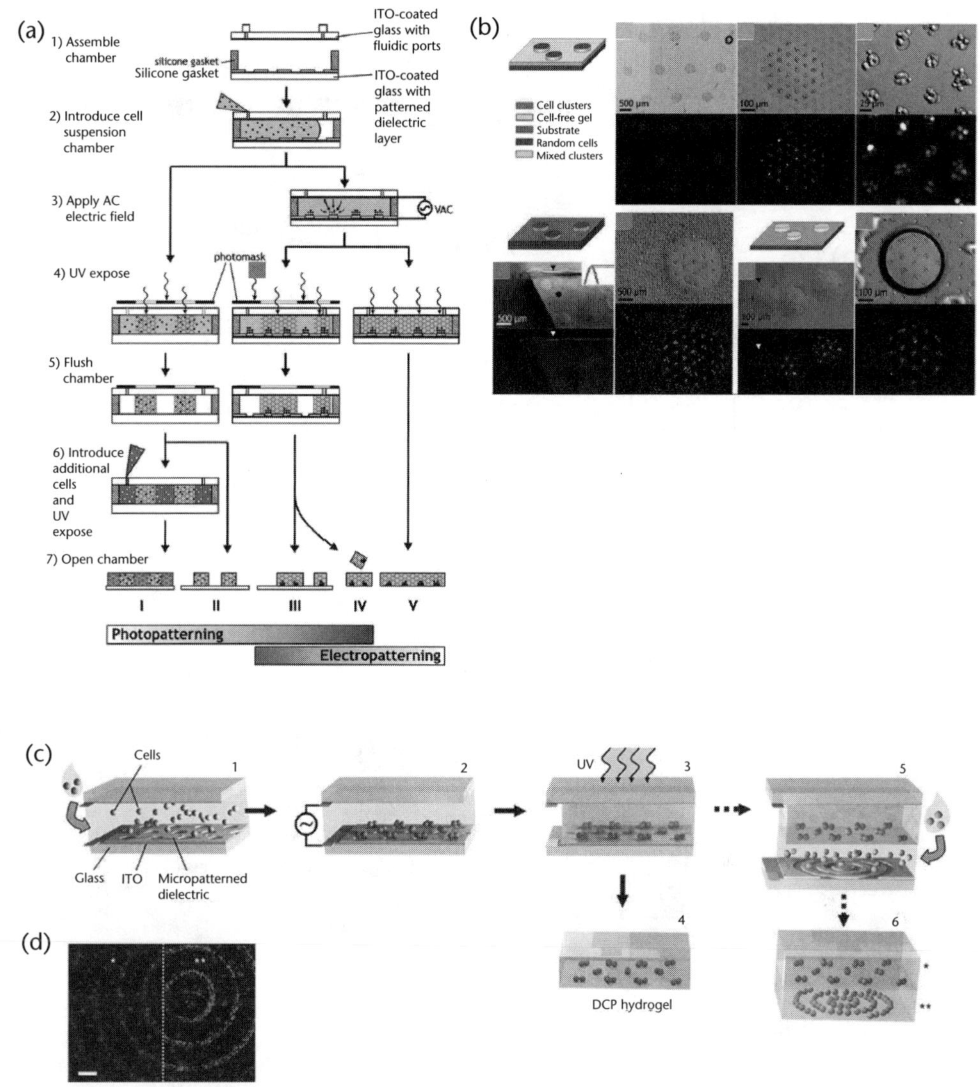

Figure 8.8 The patterning of gels and the cells within them using photolithography and DEP: (a) Schematic showing protocols for the patterning of hydrogels and the cells within them by photolithography (i, ii) and DEP (v), respectively. The combination of both techniques yields patterned cells within pattern gel microstructures (iii, iv). (Adapted from [60].) (b) Structures with patterned gel domains containing cells in an organized arrangement made by combining photo- and electropatterning. Clustered cells were patterned in gel domains surrounded by gel without cells (grey). Micrographs and fluorescence images of the gels at different magnifications. Gel domains with clustered cells surrounded by gel with randomly distributed cells. Gel domains with mixed cell clusters surrounded by air. (Adapted from [60].) (c) Schematic showing the method to fabricate two-layered gels with cells arranged in different conformations in each layer. Prepolymer gel containing cells is flowed into a chamber (1) containing a DEP electrode array that is used to cluster the cells (2). After photopolymerization (3), the patterned hydrogel can either be removed and cultured (4), or incorporated into a bilayered gel by raising the chamber height, flowing in another cell/gel solution and repeating steps 2 to 3. (Adapted from [61].) (d) A bilayered gel containing cells in clusters above and cells in rings below. (Adapted from [61].)

order to more accurately simulate in vivo microenvironments for use in basic science research and tissue engineering.

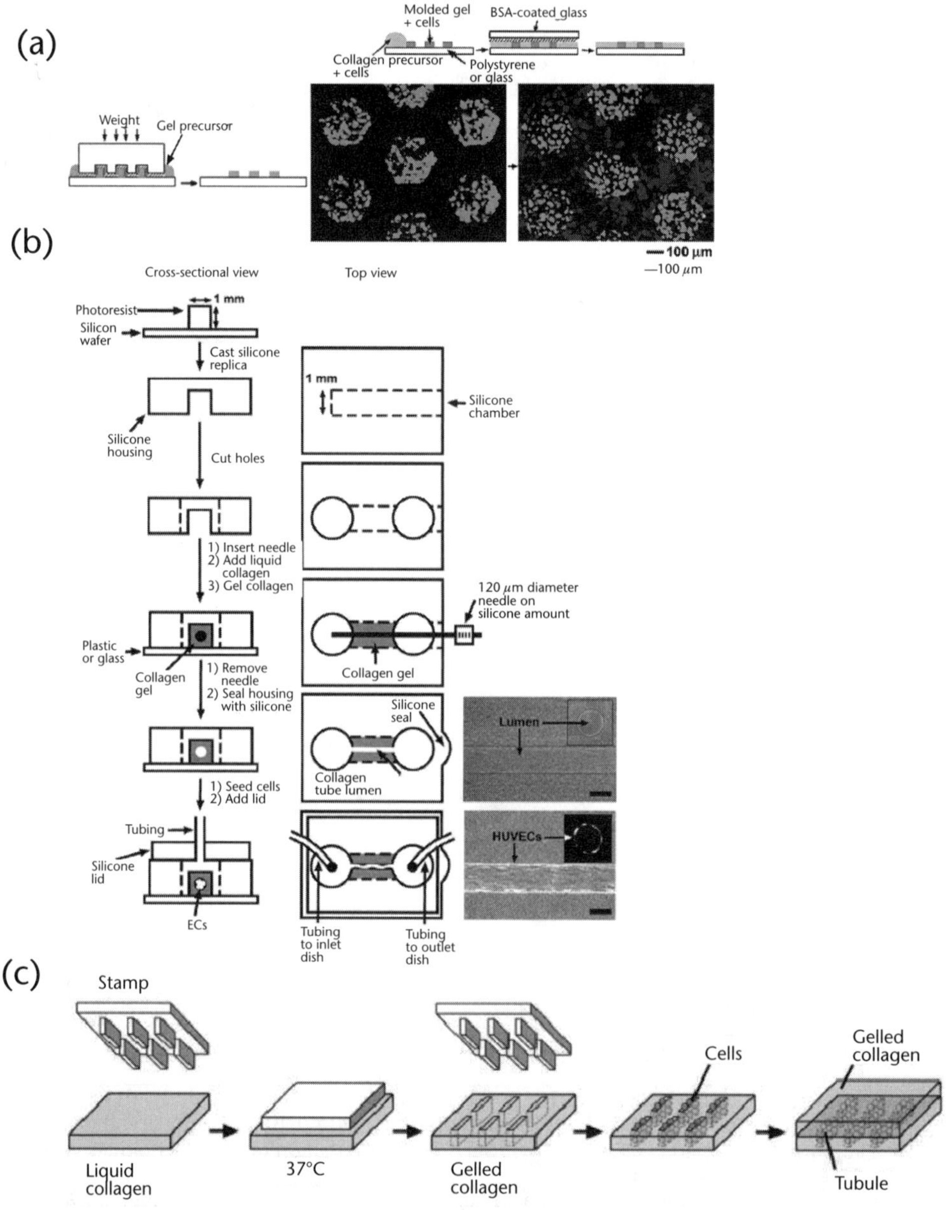

Figure 8.9 Patterning of cells in gels using elastomeric stamps: (a) Schematics of methods used to pattern gels and fluorescence images of the cells within them. The first schematic shows the molding of gel by pushing a stamp against a surface containing precursor gel. The second schematic shows how hexagonal gel microstructures containing cells are surrounded by gel containing another cell population. (Adapted from [62].) (b) Schematic showing the preparation of an endothelial tube using a stamp. Two holes cut into the silicon chamber serve as the inlet and outlet. A needle is inserted into the chamber with liquid collagen. After the collagen gels, the needle is removed, and endothelial cells are flowed through the cylindrical cavity to form an endothelial tube. The micrographs show top and cross-section views of the lumen before and after seeding cells. The fluorescence image shows the cross-section of an endothelial tube. (Adapted from [63].) (c) Schematic of method using a stamp to engineer epithelial tubules. The stamp is placed on top of liquid collagen. After raising the temperature to gel the collagen and removing the stamp, cells are seeded on the substrate and overlayed with another layer of collagen. (Adapted from [64].)

8.8 Future Directions

Because tissues are more complex than the substrates generated with the technologies that have thus far been developed, future efforts will focus on integrating these complexities into new substrates. Printing nanoscale features can provide molecular- level control over cell adhesion. Stamps with submicron features cast from templates fabricated with higher-resolution lithography technologies, such as extreme ultraviolet, electron-beam, and X-ray lithography have been used to print arrays of ECM squares with sides as small as 300 nm [35] and to facilitate single protein molecule transfer [65]. Besides enhancing resolution, the capture of matrix proteins on surfaces can also be improved, for example, by taking advantage of the high affinity of the biotin-avidin bond [66] or by covalently linking protein to the surface by reactive μCP [67, 68]. Recent studies have demonstrated that the composition of ECM is an important determinant of cell function [69], and it is likely that complex mixtures of matrices will be explored in coming years. In vivo matrices also have complex topologies [70] that can be replicated in vitro to regulate function [71, 72].

Current reports have shown that it is possible to integrate many spatially controlled cues into three-dimensional matrices as well. Growth factors have been trapped in hydrogels [73] for sustained release [74] and for dual delivery [75] and could be used to dynamically modify the biochemical environment. Degradable gels may provide a means for developing substrates with changing mechanical properties [76]. New platforms will offer the user control over adhesion, mechanical properties, the soluble microenvironment, and coculture, as well as the dynamics of the aforementioned facets. Integration of microfabrication techniques to engineer complex microenvironments will yield more promising systems for research and therapeutic purposes.

References

[1] Madou, M. J., *Fundamentals of microfabrication: The science of miniaturization*, 2nd ed., Boca Raton, FL: CRC Press, 2002, p. 723, [716] of plates.

[2] Xia, Y. N., and Whitesides, G. M., "Soft lithography," *Ann. Rev. Mater. Sci.*, Vol. 28, 1998, pp. 153–184.

[3] Kane, R. S., Takayama, S., Ostuni, E., Ingber, D. E., and Whitesides, G. M., "Patterning proteins and cells using soft lithography," *Biomaterials*, Vol. 20, Nos. 23–24, 1999, pp. 2363–2376.

[4] Singhvi, R., Kumar, A., Lopez, G. P., Stephanopoulos, G. N., Wang, D. I. C., Whitesides, G. M., and Ingber, D. E., "Engineering cell-shape and function," *Science*, Vol. 264, No. 5159, 1994, pp. 696–698.

[5] Mrksich, M., Chen, C. S., Xia, Y. N., Dike, L. E., Ingber, D. E., and Whitesides, G. M., "Controlling cell attachment on contoured surfaces with self-assembled monolayers of alkanethiolates on gold," *Proc. Natl. Acad. Sci. USA*, Vol. 93, No. 20, 1996, pp. 10775–10778.

[6] Chen, C. S., Mrksich, M., Huang, S., Whitesides, G. M., and Ingber, D. E., "Geometric control of cell life and death," *Science*, Vol. 276, No. 5317, 1997, pp. 1425–1428.

[7] Takayama, S., McDonald, J. C., Ostuni, E., Liang, M. N., Kenis, P. J. A., Ismagilov, R. F., and Whitesides, G. M., "Patterning cells and their environments using multiple laminar fluid flows in capillary networks," *Proc. Natl. Acad. Sci. USA*, Vol. 96, No. 10, 1999, pp. 5545–5548.

[8] Jeon, N. L., Baskaran, H., Dertinger, S. K. W., Whitesides, G. M., Van de Water, L., and Toner, M., "Neutrophil chemotaxis in linear and complex gradients of interleukin-8 formed in a microfabricated device," *Nat. Biotechnol.*, Vol. 20, No. 8, 2002, pp. 826–830.

[9] Takayama, S., Ostuni, E., LeDuc, P., Naruse, K., Ingber, D. E., and Whitesides, G. M., "Selective chemical treatment of cellular microdomains using multiple laminar streams," *Chem. Biol.*, Vol. 10, No. 2, 2003, pp. 123–130.

[10] Dulcey, C. S., Georger, J. H., Krauthamer, V., Stenger, D. A., Fare, T. L., and Calvert, J. M., "Deep UV photochemistry of chemisorbed monolayers—patterned coplanar molecular assemblies," *Science*, Vol. 252, No. 5005, 1991, pp. 551–554.

[11] Healy, K. E., Thomas, C. H., Rezania, A., Kim, J. E., McKeown, P. J., Lom, B., and Hockberger, P. E., "Kinetics of bone cell organization and mineralization on materials with patterned surface chemistry," *Biomaterials*, Vol. 17, No. 2, 1996, pp. 195–208.

[12] Herbert, C. B., McLernon, T. L., Hypolite, C. L., Adams, D. N., Pikus, L., Huang, C. C., Fields, G. B., Letourneau, P. C., Distefano, M. D., and Hu, W. S., "Micropatterning gradients and controlling surface densities of photoactivatable biomolecules on self-assembled monolayers of oligo(ethylene glycol) alkanethiolates," *Chem. Biol.*, Vol. 4, No. 10, 1997, pp. 731–737.

[13] Kumar, A., and Whitesides, G. M., "Features of gold having micrometer to centimeter dimensions can be formed through a combination of stamping with an elastomeric stamp and an alkanethiol ink followed by chemical etching," *Appl. Phys. Lett.*, Vol. 63, No. 14, 1993, pp. 2002–2004.

[14] Quist, A. P., Pavlovic, E., and Oscarsson, S., "Recent advances in microcontact printing," *Analytical and Bioanalytical Chem.*, Vol. 381, No. 3, 2005, pp. 591–600.

[15] Alom Ruiz, S., and Chen, C. S., "Microcontact printing: A tool to pattern," *Soft Matter*, Vol. 3, No. 2, 2007, pp. 168–177.

[16] Kumar, A., Biebuyck, H. A., and Whitesides, G. M., "Patterning self-assembled monolayers—applications in materials science," *Langmuir*, Vol. 10, No. 5, 1994, pp. 1498–1511.

[17] James, C. D., Davis, R. C., Kam, L., Craighead, H. G., Isaacson, M., Turner, J. N., and Shain, W., "Patterned protein layers on solid substrates by thin stamp microcontact printing," *Langmuir*, Vol. 14, No. 4, 1998, pp. 741–744.

[18] Bernard, A., Delamarche, E., Schmid, H., Michel, B., Bosshard, H. R., and Biebuyck, H., "Printing patterns of proteins," *Langmuir*, Vol. 14, No. 9, 1998, pp. 2225–2229.

[19] Tan, J. L., Tien, J., and Chen, C. S., "Microcontact printing of proteins on mixed self-assembled monolayers," *Langmuir*, Vol. 18, No. 2, 2002, pp. 519–523.

[20] Tan, J. L., Liu, W., Nelson, C. M., Raghavan, S., and Chen, C. S., "Simple approach to micropattern cells on common culture substrates by tuning substrate wettability," *Tissue Eng.*, Vol. 10, Nos. 5–6, 2004, pp. 865–872.

[21] Bissell, M. J., "The differentiated state of normal and malignant cells or how to define a 'normal' cell in culture," *Int. Rev. Cytology*, Vol. 70, 1981, pp. 27–100.

[22] Ingber, D. E., and Folkman, J., "Mechanochemical switching between growth and differentiation during fibroblast growth factor–stimulated angiogenesis invitro—role of extracellular-matrix," *J. Cell. Biol.*, Vol. 109, No. 1, 1989, pp. 317–330.

[23] Ausprunk, D. H., and Folkman, J., "Migration and proliferation of endothelial cells in preformed and newly formed blood vessels during tumor angiogenesis," *Microvascular Res.*, Vol. 14, No. 1, 1977, pp. 53–65.

[24] Folkman, J., and Moscona, A., "Role of cell-shape in growth-control," *Nature*, Vol. 273, No. 5661, 1978, pp. 345–349.

[25] Gerhardt, H., Golding, M., Fruttiger, M., Ruhrberg, C., Lundkvist, A., Abramsson, A., Jeltsch, M., Mitchell, C., Alitalo, K., Shima, D., and Betsholtz, C., "VEGF guides angiogenic sprouting utilizing endothelial tip cell filopodia," *J. Cell. Biol.*, Vol. 161, No. 6, 2003, pp. 1163–1177.

[26] Auman, H. J., Coleman, H., Riley, H. E., Olale, F., Huai-Jen, T., and Yelon, D., "Functional modulation of cardiac form through regionally confined cell shape changes," *Plos. Biol.*, Vol. 5, No. 3, 2007, pp. 604–615.

[27] Kandel, E. R., Schwartz, J. H., and Jessell, T. M. (eds.), *Principles of Neural Science*, 4th ed., New York: McGraw-Hill Health Professions Division, 2000, xli, p. 1414.

[28] McBeath, R., Pirone, D. M., Nelson, C. M., Bhadriraju, K., and Chen, C. S., "Cell shape, cytoskeletal tension, and RhoA regulate stem cell lineage commitment," *Developmental Cell*, Vol. 6, No. 4, 2004, pp. 483–495.

[29] Rezania, A., and Healy, K. E., "Biomimetic peptide surfaces that regulate adhesion, spreading, cytoskeletal organization, and mineralization of the matrix deposited by osteoblast-like cells," *Biotech. Progress*, Vol. 15, No. 1, 1999, pp. 19–32.

[30] Rezania, A., and Healy, K. E., "The effect of peptide surface density on mineralization of a matrix deposited by osteogenic cells," *J. Biomed. Mater. Res.*, Vol. 52, No. 4, 2000, pp. 595–600.

[31] Geiger, B., and Bershadsky, A., "Assembly and mechanosensory function of focal contacts," *Curr. Opin. Cell. Biol.*, Vol. 13, No. 5, 2001, pp. 584–592.

[32] Geiger, B., Bershadsky, A., Pankov, R., and Yamada, K. M., "Transmembrane extracellular matrix-cytoskeleton crosstalk," *Nat. Rev. Molec. Cell Biol.*, Vol. 2, No. 11, 2001, pp. 793–805.

[33] Zamir, E., and Geiger, B., "Molecular complexity and dynamics of cell-matrix adhesions," *J. Cell Sci.*, Vol. 114, No. 20, 2001, pp. 3583–3590.

[34] Chen, C. S., Alonso, J. L., Ostuni, E., Whitesides, G. M., and Ingber, D. E., "Cell shape provides global control of focal adhesion assembly," *Biochem. Biophys. Res. Comm.*, Vol. 307, No. 2, 2003, pp. 355–361.

[35] Lehnert, D., Wehrle-Haller, B., David, C., Weiland, U., Ballestrem, C., Imhof, B. A., and Bastmeyer, M., "Cell behaviour on micropatterned substrata: Limits of extracellular matrix geometry for spreading and adhesion," *J. Cell Sci.*, Vol. 117, No. 1, 2004, pp. 41–52.

[36] Thery, M., Racine, V., Pepin, A., Piel, M., Chen, Y., Sibarita, J. B., and Bornens, M., "The extracellular matrix guides the orientation of the cell division axis," *Nat. Cell Biol.*, Vol. 7, No. 10, 2005, pp. 947–U929.

[37] Thery, M., Pepin, A., Dressaire, E., Chen, Y., and Bornens, M., "Cell distribution of stress fibres in response to the geometry of the adhesive environment," *Cell Motility and the Cytoskeleton*, Vol. 63, No. 6, 2006, pp. 341–355.

[38] Thery, M., Racine, V., Piel, M., Pepin, A., Dimitrov, A., Chen, Y., Sibarita, J. B., and Bornens, M., "Anisotropy of cell adhesive microenvironment governs cell internal organization and orientation of polarity," *Proc. Natl. Acad. Sci. USA*, Vol. 103, No. 52, 2006, pp. 19771–19776.

[39] Nelson, C. M., Jean, R. P., Tan, J. L., Liu, W. F., Sniadecki, N. J., Spector, A. A., and Chen, C. S., "Emergent patterns of growth controlled by multicellular form and mechanics," *Proc. Natl. Acad. Sci. USA*, Vol. 102, No. 33, 2005, pp. 11594–11599.

[40] Dike, L. E., Chen, C. S., Mrksich, M., Tien, J., Whitesides, G. M., and Ingber, D. E., "Geometric control of switching between growth, apoptosis, and differentiation during angiogenesis using micropatterned substrates," *In Vitro Cell. Dev. Biol. Anim.*, Vol. 35, No. 8, 1999, pp. 441–448.

[41] Yousaf, M. N., Houseman, B. T., and Mrksich, M., "Using electroactive substrates to pattern the attachment of two different cell populations," *Proc. Natl. Acad. Sci. USA*, Vol. 98, No. 11, 2001, pp. 5992–5996.

[42] Guguenguillouzo, C., Clement, B., Baffet, G., Beaumont, C., Morelchany, E., Glaise, D., and Guillouzo, A., "Maintenance and reversibility of active albumin secretion by adult-rat hepatocytes co-cultured with another liver epithelial-cell type," *Exp. Cell Res.*, Vol. 143, No. 1, 1983, pp. 47–54.

[43] Clement, B., Guguenguillouzo, C., Grimaud, J. A., Rissel, M., and Guillouzo, A., "Effect of hydrocortisone on deposition of type-1 and type-4 collagen in primary culture of rat hepatocytes," *Cell. Molec. Biol.*, Vol. 34, No. 5, 1988, pp. 449–460.

[44] Bhatia, S. N., Balis, U. J., Yarmush, M. L., and Toner, M., "Effect of cell-cell interactions in preservation of cellular phenotype: Cocultivation of hepatocytes and nonparenchymal cells," *Faseb J.*, Vol. 13, No. 14, 1999, pp. 1883–1900.

[45] Tien, J., Nelson, C. M., and Chen, C. S., "Fabrication of aligned microstructures with a single elastomeric stamp," *Proc. Natl. Acad. Sci. USA*, Vol. 99, No. 4, 2002, pp. 1758–1762.

[46] Nelson, C. M., and Chen, C. S., "Cell-cell signaling by direct contact increases cell proliferation via a PI3K-dependent signal," *Febs Letters*, Vol. 514, Nos. 2–3, 2002, pp. 238–242.

[47] Nelson, C. M., and Chen, C. S., "VE-cadherin simultaneously stimulates and inhibits cell proliferation by altering cytoskeletal structure and tension," *J. Cell Sci.*, Vol. 116, No. 17, 2003, pp. 3571–3581.

[48] Liu, W. F., Nelson, C. M., Pirone, D. M., and Chen, C. S., "E-cadherin engagement stimulates proliferation via Rac1," *J. Cell. Biol.*, Vol. 173, No. 3, 2006, pp. 431–441.

[49] Nelson, C. M., Pirone, D. M., Tan, J. L., and Chen, C. S., "Vascular endothelial-cadherin regulates cytoskeletal tension, cell spreading, and focal adhesions stimulating RhoA," *Molec. Biol. Cell*, Vol. 15, No. 6, 2004, pp. 2943–2953.

[50] Voldman, J., Gray, M. L., Toner, M., and Schmidt, M. A., "A microfabrication-based dynamic array cytometer," *Analytical Chemistry*, Vol. 74, No. 16, 2002, pp. 3984–3990.

[51] Gray, D. S., Tan, J. L., Voldman, J., and Chen, C. S., "Dielectrophoretic registration of living cells to a microelectrode array," *Biosensors and Bioelectronics*, Vol. 19, No. 7, 2004, pp. 771–780.

[52] Voldman, J., Braff, R. A., Toner, M., Gray, M. L., and Schmidt, M. A., "Holding forces of single-particle dielectrophoretic traps," *Biophys. J.*, Vol. 80, No. 1, 2001, pp. 531–541.

[53] Chung, B. G., Flanagan, L. A., Rhee, S. W., Schwartz, P. H., Lee, A. P., Monuki, E. S., and Jeon, N. L., "Human neural stem cell growth and differentiation in a gradient-generating microfluidic device," *Lab Chip*, Vol. 5, No. 4, 2005, pp. 401–406.

[54] Di Carlo, D., Wu, L. Y., and Lee, L. P., "Dynamic single cell culture array," *Lab Chip*, Vol. 6, No. 11, 2006, pp. 1445–1449.

[55] Chiu, D. T., Jeon, N. L., Huang, S., Kane, R. S., Wargo, C. J., Choi, I. S., Ingber, D. E., and Whitesides, G. M., "Patterned deposition of cells and proteins onto surfaces by using three-dimensional microfluidic systems," *Proc. Natl. Acad. Sci. USA*, Vol. 97, No. 6, 2000, pp. 2408–2413.

[56] Lee, P. J., Hung, P. J., Shaw, R., Jan, L., and Lee, L. P., "Microfluidic application-specific integrated device for monitoring direct cell-cell communication via gap junctions between individual cell pairs," *Appl. Phys. Lett.*, Vol. 86, No. 22, 2005, Art. No. 223902.

[57] Hahn, M. S., Miller, J. S., and West, J. L., "Laser scanning lithography for surface micropatterning on hydrogels," *Adv. Mater.*, Vol. 17, No. 24, 2005, pp. 2939.

[58] Hahn, M. S., Taite, L. J., Moon, J. J., Rowland, M. C., Ruffino, K. A., and West, J. L., "Photolithographic patterning of polyethylene glycol hydrogels," *Biomaterials*, Vol. 27, No. 12, 2006, pp. 2519–2524.

[59] Hahn, M. S., Miller, J. S., and West, J. L., "Three-dimensional biochemical and biomechanical patterning of hydrogels for guiding cell behavior," *Adv. Mater.*, Vol. 18, No. 20, 2006, pp. 2679.

[60] Albrecht, D. R., Tsang, V. L., Sah, R. L., and Bhatia, S. N., "Photo- and electropatterning of hydrogel-encapsulated living cell arrays," *Lab Chip*, Vol. 5, No. 1, 2005, pp. 111–118.

[61] Albrecht, D. R., Underhill, G. H., Wassermann, T. B., Sah, R. L., and Bhatia, S. N., "Probing the role of multicellular organization in three-dimensional microenvironments," *Nature Methods*, Vol. 3, No. 5, 2006, pp. 369–375.

[62] Tang, M. D., Golden, A. P., and Tien, J., "Molding of three-dimensional microstructures of gels," *J. Am. Chem. Soc.*, Vol. 125, No. 43, 2003, pp. 12988–12989.

[63] Chrobak, K. M., Potter, D. R., and Tien, J., "Formation of perfused, functional microvascular tubes in vitro," *Microvascular Res.*, Vol. 71, No. 3, 2006, pp. 185–196.

[64] Nelson, C. M., VanDuijn, M. M., Inman, J. L., Fletcher, D. A., and Bissell, M. J., "Tissue geometry determines sites of mammary branching morphogenesis in organotypic cultures," *Science*, Vol. 314, No. 5797, 2006, pp. 298–300.

[65] Renault, J. P., Bernard, A., Bietsch, A., Michel, B., Bosshard, H. R., Delamarche, E., Kreiter, M., Hecht, B., and Wild, U. P., "Fabricating arrays of single protein molecules on glass using microcontact printing," *J. Phys. Chem. B*, Vol. 107, No. 3, 2003, pp. 703–711.

[66] Patel, N., Bhandari, R., Shakesheff, K. M., Cannizzaro, S. M., Davies, M. C., Langer, R., Roberts, C. J., Tendler, S. J. B., and Williams, P. M., "Printing patterns of biospecifically-adsorbed protein," *J. Biomater. Sci.—Polymer Ed.*, Vol. 11, No. 3, 2000, pp. 319–331.

[67] Feng, J., Gao, C. Y., Wang, B., and Shen, J. C., "A novel process for inking the stamp with biomacromolecule solution used in reactive microcontact printing," *Colloids and Surfaces B—Biointerfaces*, Vol. 36, Nos. 3–4, 2004, pp. 177–180.

[68] Feng, C. L., Vancso, G. J., and Schonherr, H., "Fabrication of robust biomolecular patterns by reactive microcontact printing on N-hydroxysuccinimide ester-containing polymer films," *Adv. Functional Mater.*, Vol. 16, No. 10, 2006, pp. 1306–1312.

[69] Flaim, C. J., Chien, S., and Bhatia, S. N., "An extracellular matrix microarray for probing cellular differentiation," *Nature Methods*, Vol. 2, No. 2, 2005, pp. 119–125.

[70] Flemming, R. G., Murphy, C. J., Abrams, G. A., Goodman, S. L., and Nealey, P. F., "Effects of synthetic micro- and nano-structured surfaces on cell behavior," *Biomaterials*, Vol. 20, No. 6, 1999, pp. 573–588.

[71] Elias, K. L., Price, R. L., and Webster, T. J., "Enhanced functions of osteoblasts on nanometer diameter carbon fibers," *Biomaterials*, Vol. 23, No. 15, 2002, pp. 3279–3287.

[72] Schindler, M., Ahmed, I., Kamal, J., Nur-E-Kamal, A., Grafe, T. H., Young Chung, H., and Meiners, S., "A synthetic nanofibrillar matrix promotes in vivo–like organization and morphogenesis for cells in culture," *Biomaterials*, Vol. 26, No. 28, 2005, pp. 5624–5631.

[73] Burdick, J. A., Ward, M., Liang, E., Young, M. J., and Langer, R., "Stimulation of neurite outgrowth by neurotrophins delivered from degradable hydrogels," *Biomaterials*, Vol. 27, No. 3, 2006, pp. 452–459.

[74] Ennett, A. B., Kaigler, D., and Mooney, D. J., "Temporally regulated delivery of VEGF in vitro and in vivo," *J. Biomed. Mater. Res. A*, Vol. 79A, No. 1, 2006, pp. 176–184.

[75] Richardson, T. P., Peters, M. C., Ennett, A. B., and Mooney, D. J., "Polymeric system for dual growth factor delivery," *Nat. Biotechnol.*, Vol. 19, No. 11, 2001, pp. 1029–1034.

[76] Burdick, J. A., Chung, C., Jia, X. Q., Randolph, M. A., and Langer, R., "Controlled degradation and mechanical behavior of photopolymerized hyaluronic acid networks," *Biomacromolecules*, Vol. 6, No. 1, 2005, pp. 386–391.

CHAPTER 9

Technologies and Applications for Engineering Substrate Mechanics to Regulate Cell Response

Cindy Y. Y. Yip, Jan-Hung Chen, and Craig A. Simmons

9.1 Introduction

Many cells in the body adhere to an extracellular matrix (ECM) and must do so to remain viable. In conjunction with soluble chemical signals and neighboring cells, the ECM provides important cues to adherent cells to regulate their function and ultimately define tissue phenotype. While the surface chemistry of the ECM (e.g., cell-adhesion ligands) regulates cell function, it is increasingly apparent that cells also sense and respond to the physical properties of the substrate to which they are adhered.

Cells are able to sense, at the nano- to microscale, both the geometric and inherent mechanical properties of the adhesion substrate. This chapter focuses on the mechanical properties of the substrate and, in particular, its stiffness or rigidity. We use stiffness here to describe the intrinsic resistance of a solid material to deformation under an applied stress, as described by the elastic (E) or shear (G′) modulus. Most cells and extracellular matrices have elastic moduli on the order of 10^1 to 10^6 pascals (Pa) [1, 2], which is much more compliant than glass or plastic, the substrates typically used to study cell behavior in vitro. As reviewed here, many cellular responses and behaviors are significantly different on compliant versus stiff substrates. Substrate stiffness can be as potent a regulator of cell phenotype as soluble chemical cues, and in many cases substrate stiffness modulates the response of a cell to other stimuli from its microenvironment. Thus, understanding how substrate stiffness regulates cell function is a prerequisite to understanding general cell mechanobiological responses. It also has important implications for understanding developmental and pathological processes that involve changes in the physical properties of the ECM [3]. Further, the responsiveness of cells to mechanical cues suggests that cell function can be regulated through appropriate engineering of physical aspects of the microenvironment, with important implications for rational design of biomaterials [4].

In this chapter, we first describe the mechanisms by which cells sense the stiffness of their substrate. We then describe several approaches that have been used to engineer the mechanical properties of cell-adhesion substrates. Finally, we review a number of cellular responses that are regulated by substrate stiffness. Further details of these and other emerging topics related to cellular response to substrate mechanics can be found in several excellent reviews [5–8].

9.2 How Cells Sense the Stiffness of Their Substrate

Cells adhere to and interact with their substrate through surface receptors, most notably integrins. Integrins are a family of transmembrane heterodimeric glycoproteins that physically link the ECM, the cell surface, and the intracellular cytoskeleton [9, 10]. Integrins consist of α and β subunits, and specific combinations of subunits interact with specific adhesion ligands on matrix proteins. The interactions are promiscuous, however, with an individual integrin often binding several distinct ligands and individual ligands often recognized by more than one integrin [9].

Integrin binding to the ECM induces clustering and recruitment of scaffolding proteins inside the cell to physically connect integrins to the actin cytoskeleton at sites of focal adhesion [11]. Forces applied to ECM-integrin-cytoskeleton connections induce the adhesions to mature, strengthen, and couple to actin bundles called stress fibers [11, 12]. Conversely, loss of force triggers disassembly of stress fibers and focal adhesions [13]. Force may be applied to focal adhesions through movement of the ECM induced by exogenous forces or may be generated by the cell pulling on its matrix. Endogenous cell-based forces are generated by myosin II, a motor protein that interacts with actin to generate contractile forces in the cell. Myosin-generated contractile forces are in part resisted by the cytoskeleton (for example, microtubules), but the majority are transmitted to the substrate via integrins at focal adhesions [14]. Cells respond to the stiffness of the substrate by altering integrin expression, focal adhesions, and cytoskeletal organization to establish a force balance between the cell-generated traction force and the resistance provided by the substrate [15]. As contact and cytoskeletal proteins turn over on a time scale of minutes or less [7], the cell is continually monitoring its environment as new contacts are formed.

The mechanotransductory mechanisms by which the cell senses substrate stiffness and alters its cytoskeleton in response are the subject of much investigation. Forces applied to integrins can be transferred directly to the nuclear membrane through the cytoskeleton [16] and may regulate gene expression through as of yet unknown mechanisms. However, biochemical signaling pathways traditionally associated with cell response to soluble signals (e.g., cytokines and growth factors) are also differentially activated on substrates of different stiffness. In particular, tyrosine kinases and phosphatases appear to have crucial roles in stiffness sensing [7, 17]. There are also important roles of the mitogen-activated protein kinases, including extracellular signal-regulated kinase [18] and p38 [19, 20], as well as proteins of the Ras superfamily, particularly Rho subfamily members that regulate cytoskeletal assembly, cell growth, and transcription [21–23]. Notably, the signaling pathways involved in stiffness responses depend on the type of ECM protein [7]; presumably, this results from specific ligand-integrin interactions and the activation of specific pathways by particular integrins [9]. Therefore, substrate stiffness and surface chemistry responses are coupled, which can confound interpretation of experimental data obtained with some model systems (see Section 9.3).

A position-dependent mechanism may explain how substrate stiffness can alter biochemical signaling pathways [7, 8, 24] (Figure 9.1). The crucial feature of this molecular model is that there are two complexes in the cell, one that is linked to the inner cell membrane or another adhesion site and one that is linked to the cytoskeleton and the external ligand. Under force application, the first complex remains stationary, whereas the second complex translocates with actin filaments as they move rearward due to myosin contraction. Thus, there is relative motion between the two complexes. The rigidity of the substrate will determine how far the cytoskeletal-linked complex is displaced in a given period: on a stiff substrate, the two complexes would remain in close contact, whereas on compliant substrates, the distance between the complexes would be greater. If the first complex contains an enzyme (e.g., a kinase) and the second complex contains its target (e.g., a signaling molecule), then the activation of the target can only occur if the two complexes remain close enough for modification to occur. Furthermore, forces applied to enzymes or their targets may alter protein conformations to affect reaction kinetics [25, 26]. Signaling molecules or transcription factors that are activated in this manner at focal complexes may initiate additional signaling pathways or may translocate directly to the nucleus to alter transcription [27].

Therefore, cells appear to use an integrated mechanotransductory pathway that links substrate stiffness, through integrin clustering, to cytoskeletal tension and bio-

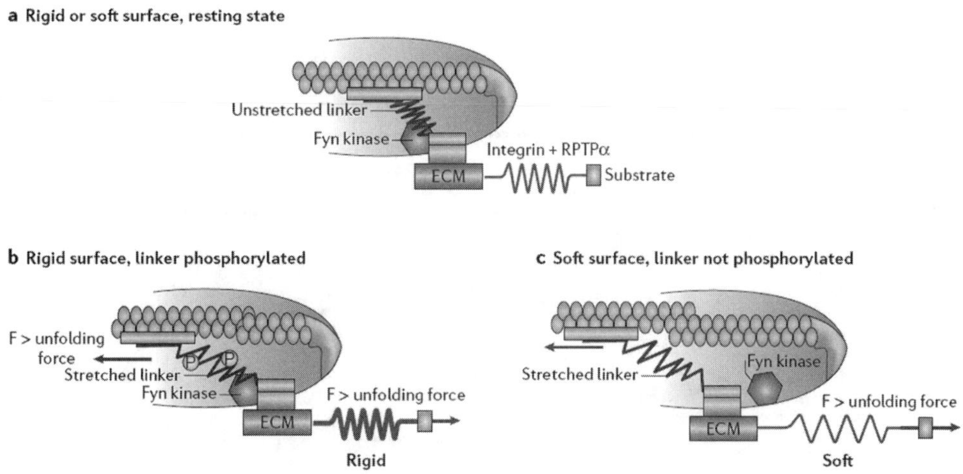

Figure 9.1 Proposed position-dependent mechanism by which substrate stiffness can activate biochemical signaling pathways. Mechanosensing may involve relative displacement of two complexes due to cytoskeletal contraction (force F in the diagram). In the case shown in the diagram, one complex may be an enzyme (Fyn kinase) that is stationary. The other complex is the enzyme target ("linker" in the diagram), which is grouped as part of the integrin-focal adhesion-actin cytoskeleton complex and therefore displaces with rearward actin transport. In the case of a stiff substrate (panel b), the substrate resists the cell-traction force, and the complexes remain close enough for modification of one complex by the other to occur (phosphorylation of the linker molecule in this case). If the substrate is compliant and, therefore, does not resist the cell-generated contraction, the two complexes may separate before an activation reaction can occur (panel c). RPTPα = receptorlike protein tyorinse phosphatase-α. (*Source:* [7] after [24].)

chemical signaling pathways (Figure 9.2). Sensing and transduction events occur on subsecond to minute time scales but ultimately may result in alterations in protein expression and cell function that persist for days or longer. Several model systems have been developed to study these responses in vitro. These systems and the changes in cell function and phenotype induced in response to substrate stiffness are summarized in Sections 9.3 and 9.4, respectively.

9.3 Technologies to Engineer the Mechanical Properties of the Substrate

Several systems have been developed to study the effects of substrate mechanical properties (particularly stiffness) on cell function, with hydrogels being the most used. Hydrogels are networks of natural and/or synthetic polymer chains that are swollen with water. Their significant water content contributes to their rubberlike consistency, low interfacial tension [28], and physical properties that are similar to those of living tissues. Cells can be cultured on the surface of gels (two-dimensional systems) or, in some cases, within the gels (three-dimensional systems). A common feature of these hydrogel systems is the ability to tune the substrate stiffness sensed by cells. This is accomplished either by altering the structural stiffness of the gels (e.g., by altering the geometry of the gel or its physical constraints) or by altering the elastic modulus of the material that makes up the gel (e.g., by altering the polymer concentration or extent of crosslinking).

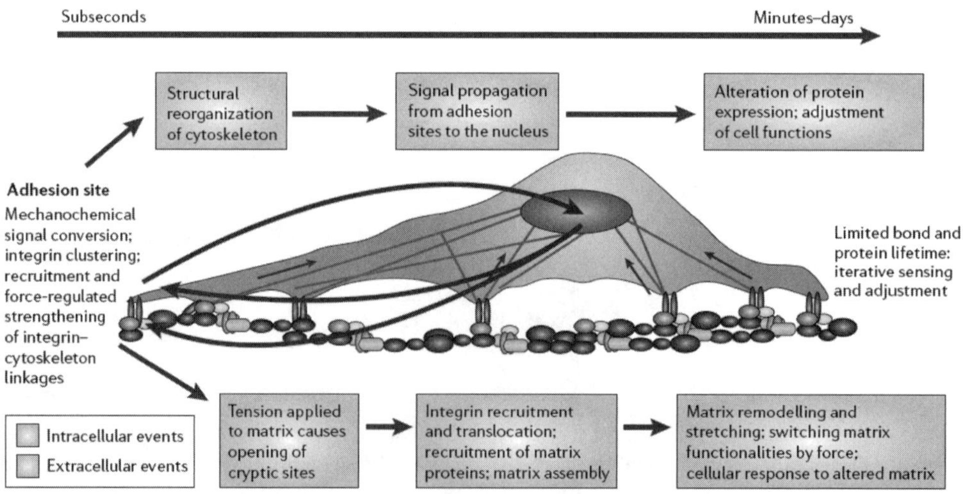

Figure 9.2 Time sequence of processes associated with mechanosensing and cellular response to substrate stiffness. Cells continually probe the mechanical properties of their substrate, initiating signaling responses that rapidly impact the cytoskeleton and matrix and ultimately (over minutes to days) alter a wide range of cell functions and the structure and composition of the ECM. (*Source:* [7].)

9.3.1 Natural Biopolymer-Based Systems

Hydrogels made from natural biopolymers, such as ECM proteins or polysaccharides, are readily available, biocompatible, and physiologically relevant, contributing to their popularity for use in cell-substrate studies. As described below, a variety of strategies have been developed to alter the stiffness of these hydrogels to study how cellular behavior is influenced by matrix mechanics.

9.3.1.1 ECM Protein-Based Hydrogels

Hydrogels made of natural ECM proteins, such as collagen and fibrin, are frequently used in cell studies because of their physiological relevance and their potential use as implantable biomaterials. Furthermore, ECM proteins can undergo condensation polymerization without the addition of crosslinking agents that can be toxic [29]. Therefore, gelation of solubilized ECM proteins is a gentle, biocompatible process, allowing cells to be mixed with the protein solution prior to polymerization to produce cell-seeded three-dimensional gels. Alternatively, cells can be grown on the surface of gels for two-dimensional studies.

9.3.1.1.1 Collagen

Collagen is a fibrous structural protein found in connective tissues and is the most abundant protein in the body, making up about 25 percent of the total protein content. The collagen molecule, called tropocollagen, consists of three polypeptide strands that form a helical subunit. Tropocollagen subunits aggregate to form fibrils, which in turn bundle to form fibers. While there are at least twenty-eight types of collagen, type I collagen is the most abundant in soft tissues and is the one used most frequently in cell-substrate studies.

Collagen that is solubilized in an acidic solution can be gelled by inducing polymerization through neutralization, followed by incubation at 37°C. Collagen can be combined with other macromolecules, including fibronectin [30], glycosaminoglycans [31], and basement membrane proteins (i.e., Matrigel) [22], to customize the chemical and physical properties of the hydrogel for specific applications.

The stiffness of collagen gels can be tuned using a variety of approaches. The simplest method is to change the total concentration of the collagen solution, as the elastic shear modulus of the polymerized network is approximately proportional to the square of the protein concentration [32]. The drawback to this approach is that the density and distribution of ECM adhesion ligands also depend on the amount of ECM protein present. Therefore, changes in protein concentration simultaneously alter the stiffness and adhesive properties of the substrate. This complicates interpretation of cellular responses as many cell processes are regulated by both stiffness and ligand density (see Section 9.4).

Alternative methods to modify gel stiffness have been developed to circumvent this limitation. One approach is to manipulate the structural stiffness of a gel by changing its geometry or physical constraints (boundary conditions). For example, thin (<10 μm) collagen films polymerized on rigid glass or plastic surfaces are stiffer than thicker (>1 mm) gels [33], and gels that are anchored to a rigid surface provide more resistance to contraction than do unconstrained, free-floating gels [33]. The limitation of these structural approaches is that they do not permit precise tuning of the gel stiffness.

The stiffness of a collagen substrate can also be adjusted by varying the extent of crosslinking between collagen polypeptide chains or by modifying the fibril nanostructure. Increased crosslinking increases the gel stiffness and can be achieved by dehydrothermal, irradiation, or chemical (e.g., glutaraldehyde) treatment [34]. However, these treatments can denature collagen, causing loss of its native triple helical structure. Independently of their effect on the mechanical properties of the substrate, alterations in the supramolecular structure of collagen fibrils impact cell response (e.g., [35]), possibly through topographical cues. Even in native fibrillar collagen gels, there is significant variability in local stiffness due to microstructural heterogeneity in individual fibers and fiber aggregates [36].

More homogenous and reproducible collagen substrates have been formed by self-assembly of monomeric collagen from solution into fibrils at an alkanethiol-coated surface [37–39]. The resulting thin film consists of a submicron-thick bed of collagen fibrils, the properties of which can be manipulated systematically to change the nanoscale structure or mechanical properties of the fibrils. Stiffening of the fibrils is achieved by dehydration of the film [39]. This treatment does not alter substrate nanotopography or integrin recognition by cells, but it does impact cell morphology and phenotype, suggesting that the stiffness of individual fibrils influences cell functions. Therefore, while this approach does not allow the substrate stiffness to be tuned precisely, it does provide a method to study the influence of the nanoscale mechanical properties of a substrate on cell behavior while maintaining many physical and chemical characteristics of the native ECM.

9.3.1.1.2 Fibrin

Fibrin is a fibrillar protein and the major structural component in blood clots. Fibrin gels are formed by the addition of thrombin to fibrinogen to effect polymerization. The modulus of fibrin gels is determined by the total protein (fibrinogen) concentration. Further tuning of the bulk properties of fibrin gels can be achieved by altering the concentration of activated factor XIII (a plasma transglutaminase) [40] or calcium and zinc cations [41] in the prepolymerized solution. These treatments alter the size and density of fibrin fibers, thereby impacting not only gel modulus but also the density and distribution of adhesion ligands for cell attachment. As with collagen, this coupling of surface chemistry and mechanics complicates interpretation of cell responses to these materials.

9.3.1.1.3 Matrigel

Matrigel is a solubilized mixture of basement membrane proteins extracted from EHS mouse sarcoma, a tumor that is rich in ECM proteins. It contains laminin, collagen IV, heparan sulfate proteoglycans, entactin, and a variety of growth factors. It gels rapidly in temperature ranges of 22°C to 35°C and is frequently used as a substrate for cell culture studies. The modulus of Matrigel gels can be tuned by altering the matrigel concentration or by crosslinking the gels using glutaraldehyde, for example, [42–44]. However, because of the heterogeneity of its composition and the potential for batch-to-batch variability, Matrigel is not an ideal substrate for systematic investigations of the independent roles of substrate stiffness and chemistry in regulating cell response.

9.3.1.2 Polysaccharide-Based Hydrogels

As an alternative to ECM-protein-based gels, substrate regulation of cell response has been studied with gels made from natural polysaccharides, such as alginate, agarose, and chitosan. In their native state, polysaccharide polymers are hydrophilic and therefore tend to discourage protein adsorption and subsequent cell adhesion. Thus, the nonadhesive surface of polysaccharide hydrogels provides a "blank slate" that can be engineered to provide specific cell-substrate interactions by covalently coupling ECM proteins or adhesion peptides to activated hydroxyl groups on the polymer. As described below, the gel elastic modulus can also be prescribed by changing the polymer mass, the polymer composition, or the extent of crosslinking. Finally, polysaccharide hydrogels are processed under relatively mild conditions, have mechanical and structural properties similar to many tissues and the ECM, and can be delivered in a minimally invasive manner, making them amenable to both in vitro and in vivo cell-based applications.

9.3.1.2.1 Alginate

Alginates are naturally derived linear polysaccharides derived from brown seaweed and bacteria [45]. They have been used extensively as synthetic matrices for cell immobilization, cell transplantation, and tissue engineering. Alginates are linear block copolymers of (1-4)-linked β-D-mannuronic acid (M) and α-L-guluronic acid (G) [45]. Like other polysaccharide polymers, alginate can be modified by covalent coupling of cell-adhesion ligands using aqueous carbodiimide chemistry with reaction efficiencies greater than 80 percent [46]. Furthermore, the ligand density can be varied by several orders of magnitude, affording excellent control of cell-matrix interactions.

The mechanical properties of alginate hydrogels are readily manipulated. The compressive and equilibrium shear moduli of alginate gels can be increased most simply by increasing the concentration of the alginate solution [47]. The composition of the polymer is also important: a high ratio of G to M subunits and longer G blocks result in alginate gels with higher compressive elastic moduli [45]. The type and density of crosslinkers used to form the gels also influence the mechanical properties. Typically alginate hydrogels are ionically crosslinked with divalent cations, such as Ca^{2+}, Ba^{2+}, or Sr^{2+}. Gel stiffness increases with increasing ion concentration and with ions that have high affinity for the alginate [45]. However, the stiffness of ionically crosslinked alginate gels decreases uncontrollably with time in solution due to ion exchange and polymer chain dissociation. To address this limitation, alginate hydrogels can be covalently crosslinked with various macromolecules, including poly(ethylene glycol)-diamine, adipic dihydrazide, and lysine [48, 49]. The moduli and swelling properties of covalently crosslinked alginate hydrogels can be controlled tightly by varying the type of crosslinker, the chain length of the crosslinking molecule, or the crosslinker density [48, 49].

9.3.1.2.2 Agarose

Agarose is a naturally derived thermosensitive polysaccharide consisting of alternating copolymers of (1-4)-linked 3,6-anhydro-α-L-galactopyranosyl and (1-3)-linked-β-D-galactopyranose. Agarose gels through chain entanglement and hydrogen bonding at temperatures in the range of 17°C to 40°C, depending on the degree of hydroxylethyl substitutions [50]. Like other polysaccharide polymers, agarose is hydrophilic and generally nonadhesive to mammalian cells but can be modified by immobilizing adhesion ligands on the surface of the gel through activation of hydroxyl groups [51]. Gel stiffness is controlled by changing the agarose concentration [50].

9.3.1.2.3 Chitosan

Chitosan is a linear polysaccharide consisting of β-(1-4) linked D-glucosamine and N-acetyl-glucosamine residues [52] produced by deacetylation of chitin, a structural component of arthropod exoskeletons [53]. Chitosan is soluble in dilute acids and can be gelled in alkaline solutions [52, 53] through chain entanglement and hydrogen bonding [54]. The stiffness of chitosan-based hydrogels can be controlled by changing the chitosan concentration. Chitosan and its derivatives and blends have also been gelled by covalent or ionic crosslinking, ultraviolet irradiation, and thermal variations [54], providing means to tune the hydrogel properties, including the elastic modulus [55]. Chitosan can be chemically derivatized through a variety of modification reactions [53] to engineer specific cell-substrate interactions [52].

9.3.2 Synthetic Polymer-Based Systems

Synthetic polymeric gels offer several advantages over natural polymer-based systems as tools to study matrix regulation of cell function. First, synthetic polymers are more homogenous and less susceptible to biodegradation than naturally derived materials, thereby providing more consistent and predictable control over the cell microenvironment. Second, similar to polysaccharide gels, synthetic gels do not adhere cells or other macromolecules, thereby providing the opportunity to engineer specific cell-matrix interactions through covalent coupling of adhesion ligands. Third, the elastic moduli of synthetic gels can often be varied over a wide range by minor changes in the crosslinker concentration; this avoids the changes in ligand density and microstructure that occur with ECM-protein-based gels when the protein concentration is varied to modify matrix stiffness. Finally, synthetic hydrogels are optically clear, making them convenient for microscopy. For some synthetic polymers (e.g., polyacrylamide), these advantages are somewhat offset by the poor biocompatibility of the materials, limiting their use to date as model systems and not implantable biomaterials.

9.3.2.1 Polyacrylamide

Polyacrylamide (PA) gels were first used to investigate the sensitivity of cell function to substrate stiffness in 1997 by Pelham and Wang [17]. Since then, PA has emerged as the material of choice for the majority of similar studies. PA is composed of acrylamide monomers and bis-acrylamide, which crosslinks the PA chains. The elastic moduli of PA gels can be tuned over three orders of magnitude (shear moduli of 10 Pa to 5×10^3 Pa) by adjusting the acrylamide concentration or the ratio of the monomer to crosslinker [56]. PA gels stiffness is particularly sensitive to the bis-acrylamide concentration, meaning a wide range of gel stiffness can be achieved by changing the crosslinker concentration minimally while keeping the monomer concentration constant [56]. Because bis-acrylamide represents only a small fraction of the total polymer (typically less than 1 percent weight/volume (w/v) versus up to 12 percent w/v acrylamide), minor modifications in the crosslinker concentration have negligible impact on the total polymer concentration and surface texture [6]. The extent of crosslinking, and therefore the stiffness of the gel, is also affected by the addition of ammonium persulfate and tetramethylethylene, which initiate free-radical reactions between the crosslinker and monomer, thereby accelerating polymerization [57]. PA gels can also be formed through photopolymerization, with the resulting gel stiffness dependent on the concentration of the photoinitiator (typically an ultraviolet-light-sensitive compound that initiates free-radical polymerization) and the intensity of irradiation. An advantage of photopolymerization is that one can create single gels with spatially defined stiffness or gradients in stiffness by simply irradiating the gel through a mask with varying opacity, with the mask imparting spatial control of the light intensity and, therefore, the extent of crosslinking [58]. Microscale spatial control of the stiffness of PA substrates has also been achieved using microfluidic gradient generators to define a gradation in crosslinker concentration prior to photopolymerization [59].

PA is inert, and cells only bind adhesion ligands that are covalently grafted to the gel surface. Thus, the adhesive properties of PA gels can be engineered independently of the mechanical properties, a distinct advantage over ECM-protein- based gels. However, covalent coupling of ECM proteins and other ligands to PA is a challenge because of the inertness of the polymer. Conjugation requires surface functionalization, which is most commonly accomplished using N-Sulfosuccinimidyl-6-[4´-azido-2´-nitrophenylamino] hexanoate (sulfo-SANPAH), a photoactivatable crosslinker [60]. Other conjugation chemistries have been developed [61–63], including the use of hydrazine hydrate, a reducing agent that modifies nonreactive amide groups in PA to reactive hydrazide groups that can form covalent bonds with adehyde or ketone groups in oxidized proteins [64].

The limitations of PA include that it is applicable only to two-dimensional studies, it may not support long-term cell culture [65], and has not been demonstrated to be a practical biomaterial for applications in vivo. Nonetheless, PA gels have proven to be important model systems for dissecting the mechanisms by which substrate stiffness regulates cell function.

9.3.2.2 Poly(ethylene glycol)

Poly(ethylene glycol) (PEG) is used widely as a biomaterial for tissue-engineering applications because it is biocompatible, has physical characteristic similar to those of soft tissues, and is highly permeable to oxygen, nutrients, and other water-soluble metabolites [66]. The characteristics of PEG also make it a useful system to study substrate-stiffness regulation of cell function, although few studies to date have done so [65]. Notably, the excellent biocompatibility of PEG allows it to be used for long-term, three-dimensional culture, both in vitro or in vivo, providing several advantages over other synthetic systems.

PEG hydrogels are crosslinked and formed by ultraviolet photopolymerization of a mixture of PEG with acrylate end groups (PEG-diacrylate, or PEGDA) and nonacrylated PEG, with the extent of crosslinking being dependent on the ratio of PEGDA to PEG. Therefore, the elastic moduli of PEG hydrogels can be prescribed over a relatively wide range by altering the total polymer concentration or the PEGDA:PEG ratio [65].

Similar to PA, PEG is hydrophilic and is not adhesive to cells but can be functionalized with adhesion ligands or ECM proteins. Thus, the adhesive properties of PEG gels can be prescribed independently of their mechanical properties. PEG can be functionalized for two-dimensional studies by activation of the gel surface with sulfo-SANPAH and conjugation with full-length ECM molecules [65]. For three-dimensional systems, cell-binding peptides can be incorporated directly into the polymer backbone [67]. Three-dimensional PEG systems have also been generated by combining fibrinogen fragments into the PEG gel mixture prior to photopolymerization [68]. Cell adhesion to PEGylated fibrinogen occurs through binding with ligands on the fibrinogen backbone. However, because the mechanical properties of PEGylated fibrinogen gels depend on the fibrinogen concentration [69], substrate chemistry and mechanics are not decoupled in this particular system.

9.3.2.3 Polydimethylsiloxane

Polydimethylsiloxane (PDMS) is a silicone elastomer used frequently in soft lithography and microfabrication for biomedical applications. The most widely used, commercially available PDMS (Sylgard 184; Dow Corning Corporation, Midland, MI) is heat curable and supplied as a two-part kit consisting of prepolymer (base) and crosslinker (curing agent) components. The properties of PDMS, including its modulus, can be altered by manipulating the base:curing agent ratio without changing the surface chemistry [70]. This strategy has been used to investigate how cells migrate on substrates micropatterned with stiff regions surrounded by compliant regions [71]. Functionalization of PDMS-based substrates for cell adhesion is challenging and even more difficult than with PEG-based hydrogels as PDMS is relatively chemically inert and hydrophobic, and has high chain mobility [1]. Nonetheless, a variety of approaches for immobilizing proteins to PDMS surfaces exist [72], including simple protein adsorption on the native PDMS surface [71]. PDMS can therefore provide for independent control of substrate chemistry and mechanics, at least over relatively short culture periods of a few days.

9.4 Effects of Substrate Mechanics on Cell Response

The effects of substrate mechanics on cell phenotype and function have been demonstrated in a number of culture models. For the most part, cells have been studied in two dimensions by growing them on top of natural or synthetic hydrogels. In general, cells express tissue-specific functions and fail to proliferate when attached to compliant substrates, but they grow without differentiation on more rigid substrates

[73]. However, as reviewed below, several recent studies indicate that multiple cell functions are regulated by substrate mechanics and that the relationships between substrate stiffness and cell migration, growth, differentiation, and organization are more complex than originally believed, with important dependencies on cell type and surface chemistry.

9.4.1 Cell Spreading and Motility

The ability of a cell to adhere, spread, and migrate on its substrate is regulated by the stiffness of the substrate. This was first demonstrated by Pelham and Wang [17], who used type I collagen–coated polyacrylamide gels to study the movement of epithelial cells and fibroblasts on compliant (E~5kPa) versus relatively stiff (E~70kPa) substrates. Cells on compliant substrates exhibited reduced spreading, greater migration rates, and elevated lamellipodial protrusion and retraction activity compared with cells on more rigid substrates. Increased motility and lamellipodial activity on compliant substrates was associated with more dynamic focal adhesions, whereas cells on rigid substrates had more regularly shaped, stable adhesions. Because of more stable focal adhesions, fibroblasts on stiff substrates generate greater traction forces [74] and are more strongly adhered to the substrate [75]. When presented with two-dimensional substrates with discrete changes in stiffness [58, 74] or gradations in stiffness [58], cells migrate toward stiffer regions of the substrate, a phenomenon called durotaxis or mechanotaxis. Durotaxis has also been observed in three-dimensional matrices [76, 77].

The inverse relationship between two-dimensional substrate stiffness and migration speed has been observed for many cell types, including vascular smooth muscle cells [23, 58], fibroblasts [17, 75, 78], and preosteoblast cells [18]. The ability of neurites to extend their processes is also dependent on substrate stiffness, with higher extension rates on compliant substrates [50]. While this trend is seemingly independent of the type of adhesion ligand, it is modulated by the density of adhesion ligands on the substrate surface. Vascular smooth muscle cells on glass or stiff (E ~ 8 kPa) polyacrylamide gels coated with type I collagen spread to a greater extent, with higher expression of actin stress fibers and focal adhesions than cells on more compliant (E ~ 1 kPa) gels [79]; this response is also seen in endothelial cells and fibroblasts [56] (Figure 9.3). However, the dependency of spreading on substrate stiffness and ligand density for smooth muscle cells is biphasic, with maximal spreading on intermediate collagen densities. Migration speeds of both vascular smooth muscle cells [23] and preosteoblasts [18] are also dependent on both stiffness and ligand density. With high ECM ligand densities, migration rate is greatest on compliant substrates, whereas with lower ligand densities, migration rate is greatest on stiffer substrates (Figure 9.4). Furthermore, the stiffness at which migration rate is maximal is different for smooth muscle cells and preosteoblasts, indicating that the effects of substrate stiffness on migration and other cell functions (as discussed below) are cell-type specific. Notably, neutrophil spreading is unaffected by substrate stiffness (ranging from gels with elastic modulus of 2 Pa to rigid glass) [56], further supporting the hypothesis that different cell types respond in fundamentally different ways to their local mechanical environment.

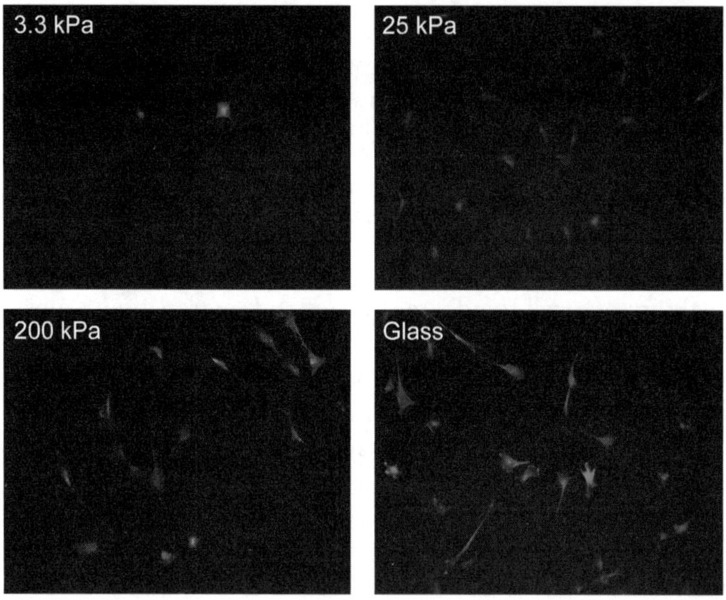

Figure 9.3 Aortic valve fibroblast morphological response to type I collagen–coated polyacrylamide gels of increasing elastic modulus and on glass. On stiffer substrates, fibroblasts are more spread, express more α-smooth muscle actin, and assemble the actin into more pronounced stress fibers. Vascular smooth muscle cells and other fibroblasts show similar responses. (*Source:* Courtesy of Stephanie Ting, University of Toronto.)

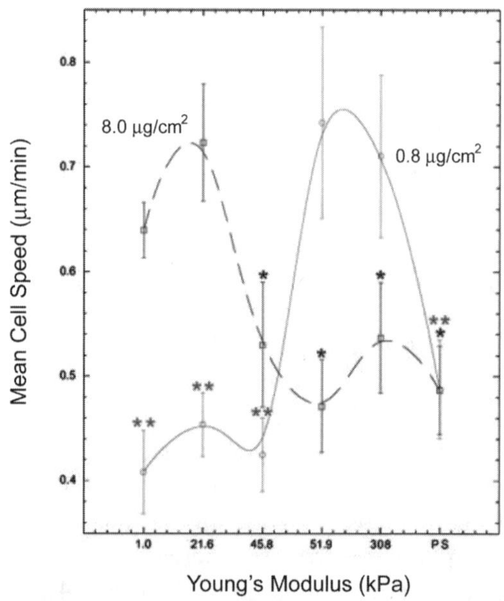

Figure 9.4 Dependency of vascular smooth muscle cell migration speed on substrate stiffness and ligand density. Smooth muscle cells grown on fibronectin-coated polyacrylamide or polystyrene (PS) exhibit a biphasic dependence of migration speed on substrate stiffness. On high fibronectin ligand density (dashed line), the maximum cell speed occurs at a lower modulus than with a low fibronectin ligand density (solid line). Data are presented as mean ± SE. *$P < 0.05$ compared to 21.6 kPa substrate for high fibronectin density. **$P < 0.05$ compared to 51.9 kPa substrate for low fibronectin density. (*Source:* [23].)

9.4.2 Cell Growth and Death

Regardless of cell type or matrix composition, cells grown on stiffer two-dimensional substrates tend to proliferate more rapidly than those on compliant substrates. This has been demonstrated for fibroblasts on type I collagen–coated polyacrylamide gels [17, 80] or on a hyaluronan/fibronectin composite gels [78], bone cells and skeletal myoblasts on alginate hydrogels modified with arginine-glycine-aspartate (RGD)–containing peptides [81, 82], bone cells on type I collagen–coated polyacrylamide gels [18], and chondrocytes on chitosan gels [55]. Apoptosis of preosteoblasts has also been shown to be reduced on stiffer substrates [81].

Although cells proliferate more rapidly on stiffer substrates, the optimal stiffness for proliferation is cell-type-dependent. This was demonstrated for mixed cultures of prenatal rat embryo neuronal cells and astrocytes grown on laminin-coated polyacrylamide gels and on fibrin [83]. Although both cell types proliferated to a greater extent on more rigid substrates, compliant substrates selected for neuronal over astrocyte adhesion and growth. Notably, the modulus of the substrate that was optimal for neuronal growth was similar to that of intact brain (G' ~ 300 Pa). Interestingly, F-actin polymerization in neurons was comparable on the compliant (G' ~ 200Pa) and stiff (G′ ~ 9 kPa) gels, suggesting that actin-dependent contraction of neurons may not be the motor that drives neurites forward.

Exploiting the fact that proliferating cells are more likely to incorporate plasmid DNA, Kong et al. [81] used substrate stiffness to regulate cell growth and improve the efficiency of nonviral gene delivery. In preosteoblasts cultured on alginate hydrogels, uptake of DNA condensates by the cells increased proportionately with substrate stiffness, suggesting a strategy by which one could engineer the cell microenvironment for effective gene delivery via matrix mechanics.

9.4.3 Cell Differentiation

Early studies on the effects of substrate stiffness focused on cell differentiation, particularly of epithelial cells, which express the fully differentiated phenotype on compliant substrates [84–86]. Recent studies demonstrate that differentiation of a wide variety of cell types also depends on substrate stiffness. Notably, many cells express their differentiated phenotype in vitro on substrates that match the stiffness of their native ECM. Thus, the optimal stiffness for differentiation is cell-type dependent and, in many cases, physiological. This clearly has implications for engineering the cell micromechanical environment and designing biomaterials to guide cell fate. It may also be a mechanism by which the ECM regulates cell phenotype in development and in diseases involving matrix alterations.

The sensitivity of cell differentiation to substrate stiffness was elegantly demonstrated by Engler et al. [87] for marrow-derived mesenchymal stem cells (MSCs). By manipulating the stiffness of type I collagen–coated two-dimensional polyacrylamide substrates from compliant (E ~ 0.1–1 kPa) to intermediate stiffness (E ~ 8–17 kPa) to relatively stiff (E ~ 25–40 kPa), they were able to direct MSC differentiation to neuronal, myogenic, and osteogenic lineages, respectively (Figure 9.5). Remarkably, these moduli mimic those of brain tissue, muscle tissue, and collagenous osteoid, respectively, suggesting that multipotent progenitor cells

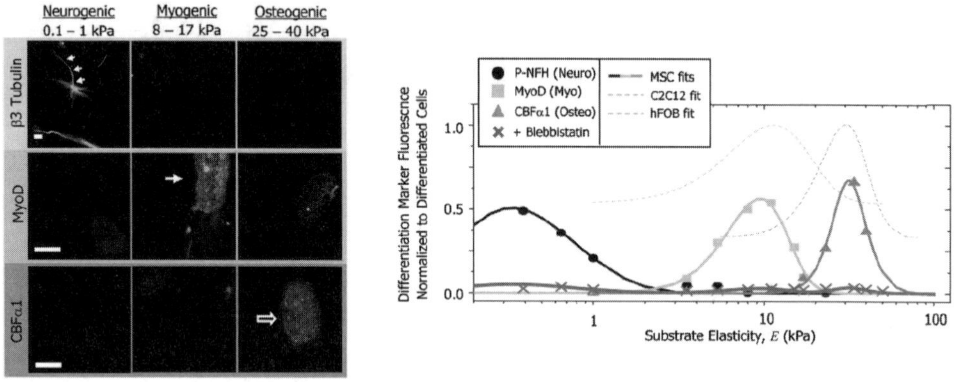

Figure 9.5 Effect of substrate stiffness on the differentiation of mesenchymal stem cells. Initially undifferentiated MSCs express neuronal (β3 tubulin), myogenic (myoD1), and osteogenic (Cbfa-1) differentiation markers only in specific stiffness ranges (left panel), which remarkably mimic those of native brain, muscle, and osteoid tissue. In the right panel, fluorescent intensity measurements [normalized to peak expression of fully differentiated C2C12 (muscle) and hFOB (osteoblast) control cells] demonstrate maximal lineage specification at various substrate stiffnesses. Treatment of MSCs with blebbistatin (a myosin inhibitor) blocked all differentiation marker expression. In these experiments, MSCs were grown on compliant type I collagen–coated polyacrylamide substrates under identical media conditions. (*Source:* [87].)

are able to recognize physiologically relevant ECM stiffness and differentiate accordingly. MSC lineage specification in response to matrix mechanics was mediated by pathways involving nonmuscle myosin II and the cytoskeleton, as pharmacological inhibition of myosin abrogated the stiffness-dependent differentiation of MSCs.

Substrate mechanics regulation of differentiation has also been observed in committed cells. Skeletal muscle differentiation, evidenced by fusion of myoblasts into striated myotubes, is dependent on matrix stiffness [88]. Myoblasts plated on type I collagen–coated polyacrylamide gels with elastic moduli ranging from 1 to 17 kPa only formed striated myotubes on gels of physiologically relevant stiffness (E = 8 to 11 kPa). Myoblasts plated on either very compliant (E = 1 kPa) or stiff (E = 17 kPa) gels showed poor striation.

Modulating matrix stiffness can also selectively induce or repress differentiated functions of hepatocytes [43, 44]. In particular, hepatocytes express a more differentiated phenotype (measured by growth factor–induced albumin secretion) on stiff, slightly crosslinked Matrigel (G′ ~ 118 Pa) than on compliant Matrigel (G′ ~ 34 Pa). Note that moduli of the Matrigel used in these studies was significantly lower than that of liver (G′ ~ 5 kPa) [89]. It is possible that there exists an optimal matrix stiffness (perhaps in the physiologically relevant stiffness range) for maintaining differentiated hepatocyte functions, but this has yet to be determined.

The differentiation of MC3T3-E1 cells, a preosteoblast cell line, to bone-forming osteoblasts is substrate-stiffness-dependent. However, two recent studies reported different dependencies. On two-dimensional alginate gels modified with RGD-containing peptides, MC3T3-E1 cells expressed osteocalcin (a protein expressed by mature osteoblasts) and deposited bone mineral more readily on compliant substrates (E = 20 kPa) than on stiff substrates (E = 110 kPa) [90]. In contrast, MC3T3-E1 cells grown on rigid collagen-coated polystyrene deposited more bone mineral than cells on type I collagen–coated two-dimensional polyacrylamide gels [18]. These conflicting results may be explained by the differences in adhesion ligands (short RGD-containing peptides versus full-length collagen) in the two studies: cells typically interact with RGD adhesion sequences through $\alpha_5\beta_3$ and $\alpha_5\beta_1$ integrins, whereas interaction with type I collagen is via $\alpha_2\beta_1$. While osteoblast differentiation is known to depend on activation of particular sets of integrins [91, 92], the studies of Kong et al. [90] and Khatiwala et al. [18] suggest that osteogenic differentiation is also regulated by substrate mechanics. Furthermore, the differentiation response to surface chemistry may be modulated by substrate mechanics and vice versa.

Substrate stiffness can affect cell differentiation by modulating the cell response to soluble cues. An example is fibroblast differentiation to contractile myofibroblasts, which express α-smooth muscle actin (α-SMA) filaments. Myofibroblast differentiation is regulated by growth factors, such as transforming growth factor–β (TGF-β), and by substrate stiffness, as has been demonstrated in several studies that used anchored (stiff) versus free-floating (compliant) collagen gels (reviewed in [93, 94]). However, TGF-β-induced myofibroblast differentiation depends on a stiff substrate that resists cell-mediated deformation, thereby promoting cytoskeletal tension [33, 95]. Substrate-stiffness modulation of other cell responses to soluble cues is a largely unexplored phenomenon but one that may be critical in predicting and controlling the cellular response to environmental cues.

9.4.4 Cellular Organization

It has been suggested that stiffness gradients and alterations in tissue stiffness during embryological development are responsible for tissue patterning and organization in vivo [3]. As reviewed below, cell-culture studies show that cell-cell interactions, cellular organization, and tissue morphogenesis are in fact regulated by substrate mechanics.

On compliant type I collagen–coated polyacrylamide substrates, fibroblasts aggregate and establish cell-cell contacts [75]. Migration to form aggregates in this study was facilitated by the compliant substrate, upon which fibroblasts are particularly motile (at least at the ligand densities considered). Aggregation appears to occur when dynamic, filopodialike extensions from cells come in contact, bind, and then contract, pulling the cells together. Cells within the aggregate are unable to migrate out, but those on the periphery maintain dynamic filopodia to establish additional cell-cell contacts and draw cells into the aggregate. This process is dependent on myosin II and is regulated by Rho signaling, as pharmacological inhibition of myosin II ATPase or Rho-dependent kinase prevents cell aggregation and even disperses existing aggregates. Substrate-stiffness-dependent aggregation is also observed with hepatocytes. Hepatocytes are more responsive to growth factor–induced aggregation on slightly crosslinked Matrigel ($G' \sim 118$ Pa) than on basal Matrigel ($G' \sim 34$ Pa) [43, 44] but are less motile and do not form aggregates with more extensive substrate crosslinking [42].

Tissue-specific functions that depend on cell-cell contact and proper organization are also regulated by substrate stiffness. Neonatal rat cardiomyocytes cultured within compliant three-dimensional PEGylated fibrinogen gels organize and beat synchronously, resulting in a significantly larger region of contraction compared to stiffer gels [69]. Cultured human breast epithelial cells undergo normal tissue morphogenesis only in type I collagen or basement membrane three-dimensional matrices with compliances typical of mammary stroma (E ~ 167 Pa) [22]. Stiffer matrices, typical of tumors, increase cytoskeletal tension, enhance growth, and perturb tissue organization, promoting malignant transformation of the cells and tissue. Endothelial cells grown in three-dimensional collagen [96, 97] or fibronectin [98] gels form capillary-like tube structures and networks more readily in compliant gels than stiff gels. However, when endothelial cells are grown to confluence on two-dimensional substrates, their morphologies are indistinguishable on compliant versus stiff gels [56]. This suggests that, in some cases, cadherin-mediated cell-cell junctions may transmit mechanical or biochemical signals that override those of the substrate.

9.5 Summary and Future Challenges

Adherent cells continually probe and respond to the physical properties of their substrate through mechanosensory pathways that involve integrin clustering, focal adhesion maturation, myosin-mediated cytoskeletal contraction, and activation of intracellular signaling pathways. Ultimately, this response impacts many diverse cell functions, including spreading, migration, growth, death, differentiation, and organization. The relationship between substrate stiffness and cell response is complex, with strong dependencies on cell type, ligand density, the nature of the ligand, and other microenvironmental factors. Aspects of these relationships are beginning to be defined using a variety of natural and synthetic polymer systems, many of which enable some decoupling of interacting stimuli (e.g., substrate stiffness and ligand density).

Despite these advances, several issues remain to be resolved before the intricacies of substrate-stiffness regulation are well enough understood to predictably control cell response by engineering substrate mechanics. Some emerging issues include:

- *Mechanosensory mechanisms and time-dependent effects:* A fundamental challenge is to determine how matrix mechanical properties are sensed by cells and what governs the cytoskeletal tension generated by the cell. Tractions applied to gels by cells differ very little (3 to 4 percent) between gels that differ by twofold in elastic modulus, suggesting that substrate strain may be sensed by cells as a "tactile set-point" [5]. The position-dependent model (Figure 9.1) provides some insight into the molecular mechanisms, but more investigation is required. Whatever the mechanosensory mechanisms, it is likely that they will be time-dependent [7]. Stiffness sensing is a dynamic process, involving assembly, stabilization, movement, and disassembly of adhesions sites and the actin cytoskeleton on time scales of subseconds to minutes [8]. Similarly, the ECM and many biomaterial model systems exhibit time-dependent viscoelastic behavior, meaning that their deformation in response to a constant cell-traction force will increase with time (i.e., creep). The relative time constants for stiffness-dependent cytoskeletal organization versus viscoelastic substrate deformation may impact the dynamics of mechanosensing, but these issues are largely unexplored. Similarly, many tissues are mechanically loaded, which superimposes dynamic, time-varying forces on the resistance forces exerted by the substrate. It will be important to determine how cell response to dynamic exogenous forces is modulated by the rheological properties of the substrate. Finally, cells alter their matrix on both short (e.g., cell traction on ECM fibers in seconds) and long (e.g., matrix deposition and remodeling over hours to days) time scales. Thus, cell response and ultimate tissue phenotype will likely be impacted by dynamic cell-mediated changes in substrate composition, chemistry, and mechanics.

- *Stiffness responses in two dimensions versus three dimensions:* To date, most studies of stiffness response have only considered cells grown on top of two-dimensional gels in culture. It is well recognized that cells grown in three-dimensional matrices exhibit different cell-matrix interactions, signaling, and other cellular functions from cells grown on two-dimensional surfaces in vitro [76, 77, 99]. Presumably, three-dimensional systems better mimic the microenvironment in vivo. Therefore, it will be important to define stiffness responses in three-dimensional matrices if these studies are to guide biomaterial design or improve our understanding of the role of native ECM mechanics in cell regulation.
- *Multivariate analyses of cell mechanoresponse:* Existing data clearly demonstrate that the responses to substrate stiffness, surface chemistry, soluble signals, and other microenvironmental cues are coupled and cell-type-dependent. It will be important to determine the independent effects of these parameters and their interactions if these data are to be used to rationally design biomaterials or better understand the molecular regulators of developmental and pathological processes. This may be accomplished through combinatorial experimentation and multivariate analyses (perhaps facilitated by high-throughput technologies, as suggested below) or by modeling efforts that build on advances in our understanding of mechanosensing and mechanotransduction mechanisms.
- *Advances in the design and characterization of model systems:* To address the issues described above, it will be necessary to use model systems with well-characterized and tunable material properties. This may be accomplished through advances in biomaterial design or through improved characterization of material properties at the cellular and molecular scale. In particular, the nano- and microscale mechanical properties may be characterized with tools like atomic force microscopes [79, 100, 101] and micropipette aspirators [102, 103]. To facilitate multivariate analyses, cell-based arrays [104] that integrate materials with tunable stiffness will permit efficient, high-throughput combinatorial investigations of the complex interplay between substrate stiffness, surface chemistry, soluble signals, and cell response.

Fundamentally, an improved understanding of the integrated response of cells to microenvironmental cues is required to predict and control these processes for therapeutic applications. Substrate mechanics plays a critical role in regulating and modulating many cellular responses and, therefore, likely has important implications for development, disease, and regeneration. Therefore, further investigation and delineation of the response of cells to their mechanical environment will likely reveal mechanobiological regulators of developmental and pathological processes. Furthermore, these studies may define important design criteria and design strategies for biomaterial systems that predictably guide cell function and tissue regeneration.

References

[1] Wong, J.Y., Leach, J.B., and Brown, X.Q., "Balance of chemistry, topography, and mechanics at the cell-biomaterial interface: Issues and challenges for assessing the role of substrate mechanics on cell response," *Surface Science*, Vol. 570, Nos. 1–2, 2004, pp. 119–133.

[2] Ethier, C. R., and Simmons, C. A., *Introductory Biomechanics: From Cells to Organisms*, New York: Cambridge University Press, 2007.

[3] Huang, S., and Ingber, D. E., "The structural and mechanical complexity of cell-growth control," *Nat. Cell Biol.*, Vol. 1, 1999, pp. E131–E138.

[4] Simmons, C. A., and Mooney, D. J., "Regulation of cellular response to mechanical signals by matrix design," in Guilak, F. (ed.), *Functional Tissue Engineering*, New York: Springer-Verlag, 2003.

[5] Discher, D. E., Janmey, P., and Wang, Y. L., "Tissue cells feel and respond to the stiffness of their substrate," *Science*, Vol. 310, No. 5751, 2005, pp. 1139–1143.

[6] Georges, P. C., and Janmey, P. A., "Cell type-specific response to growth on soft materials," *J. Appl. Physiol.*, Vol. 98, No. 4, 2005, pp. 1547–1553.

[7] Vogel, V., and Sheetz, M., "Local force and geometry sensing regulate cell functions," *Nat. Rev. Mol. Cell. Biol.*, Vol. 7, No. 4, 2006, pp. 265–275.

[8] Giannone, G., and Sheetz, M. P., "Substrate rigidity and force define form through tyrosine phosphatase and kinase pathways," *Trends Cell Biol.*, Vol. 16, No. 4, 2006, pp. 213–223.

[9] Hynes, R. O., "Integrins: Versatility, modulation, and signaling in cell adhesion," *Cell*, Vol. 69, 1992, pp. 11–25.

[10] Juliano, R. L., and Haskill, S., "Signal transduction from the extracellular matrix," *J. Cell. Biol.*, Vol. 120, No. 3, 1993, pp. 577–585.

[11] Choquet, D., Felsenfeld, D. P., and Sheetz, M. P., "Extracellular matrix rigidity causes strengthening of integrin-cytoskeleton linkages," *Cell*, Vol. 88, 1997, pp. 39–48.

[12] Riveline, D., et al., "Focal contacts as mechanosensors: Externally applied local mechanical force induces growth of focal contacts by an mDia1-dependent and ROCK-independent mechanism," *J. Cell. Biol.*, Vol. 153, No. 6, 2001, pp. 1175–1186.

[13] Balaban, N. Q., et al., "Force and focal adhesion assembly: A close relationship studied using elastic micropatterned substrates," *Nat. Cell Biol.*, Vol. 3, No. 5, 2001, pp. 466–472.

[14] Wang, N., et al., "Mechanical behavior in living cells consistent with the tensegrity model," *Proc. Natl. Acad. Sci. USA*, Vol. 98, No. 14, 2001, pp. 7765–7770.

[15] Ingber, D. E., "Cellular mechanotransduction: Putting all the pieces together again," *FASEB J.*, Vol. 20, No. 7, 2006, pp. 811–827.

[16] Maniotis, A. J., Chen, C. S., and Ingber, D. E., "Demonstration of mechanical connections between integrins, cytoskeletal filaments, and nucleoplasm that stabilize nuclear structure," *Proc. Natl. Acad. Sci. USA*, Vol. 94, No. 3, 1997, pp. 849–854.

[17] Pelham, R. J., Jr., and Wang, Y.-L., "Cell locomotion and focal adhesions are regulated by substrate flexibility," *Proc. Natl. Acad. Sci. USA*, Vol. 94, 1997, pp. 13661–13665.

[18] Khatiwala, C. B., Peyton, S. R., and Putnam, A. J., "Intrinsic mechanical properties of the extracellular matrix affect the behavior of preosteoblastic MC3T3-E1 cells," *Am. J. Physiology Cell Physiology*, Vol. 290, No. 6, 2006, pp. C1640–1650.

[19] Wang, J., et al., "Smooth muscle actin determines mechanical force-induced p38 activation," *J. Biol. Chem.*, Vol. 280, No. 8, 2005, pp. 7273–7284.

[20] Wang, J., Zohar, R., and McCulloch, C. A., "Multiple roles of alpha-smooth muscle actin in mechanotransduction," *Exp. Cell Res.*, Vol. 312, No. 3, 2006, pp. 205–214.

[21] McBeath, R., Pirone, D. M., Nelson, C. M., Bhadriraju, K., and Chen, C. S., "Cell shape, cytoskeletal tension, and RhoA regulate stem cell lineage commitment," *Dev. Cell*, Vol. 6, No. 4, 2004, pp. 483–495.

[22] Paszek, M. J., et al., "Tensional homeostasis and the malignant phenotype," *Cancer Cell*, Vol. 8, No. 3, 2005, pp. 241–254.

[23] Peyton, S. R., and Putnam, A. J., "Extracellular matrix rigidity governs smooth muscle cell motility in a biphasic fashion," *J. Cell. Physiol.*, Vol. 204, No. 1, 2005, pp. 198–209.

[24] Jiang, G., Huang, A. H., Cai, Y., Tanase, M., and Sheetz, M. P., "Rigidity sensing at the leading edge through alphavbeta3 integrins and RPTPalpha," *Biophys. J.*, Vol. 90, No. 5, 2006, pp. 1804–1809.

[25] Khan, S., and Sheetz, M. P., "Force effects on biochemical kinetics," *Ann. Rev. Biochem.*, Vol. 66, 1997, pp. 785–805.

[26] Bustamante, C., "Of torques, forces, and protein machines," *Protein Sci.*, Vol. 13, No. 11, 2004, pp. 3061–3065.

[27] Woods, A. J., Kantidakis, T., Sabe, H., Critchley, D. R., and Norman, J. C., "Interaction of paxillin with poly(A)-binding protein 1 and its role in focal adhesion turnover and cell migration," *Mol. Cell Biol.*, Vol. 25, No. 9, 2005, pp. 3763–3773.

[28] Kudela, V., "Hydrogels," in I. Jacqueline (ed.), *Encyclopedia of Polymer Science and Engineering,. New York: Wiley Interscience, 1976, pp. 783–807.*

[29] Sannino, A., Pappada, S., Madaghiele, M., Maffezzoli, A., Ambrosio, L., and Nicolais, L., "Crosslinking of cellulose derivatives and hyaluronic acid with water-soluble carbodiimide," *Polymer*, Vol. 46, No. 25, 2005, pp. 11206–11212.

[30] Gildner, C. D., Lerner, A. L., and Hocking, D. C., "Fibronectin matrix polymerization increases tensile strength of model tissue," *Am. J. Physiol. Heart Circulatory Physiol.*, Vol. 287, No. 1, 2004, pp. H46–H53.

[31] Yannas, I. V., Lee, E., Orgill, D. P., Skrabut, E. M., and Murphy, G. F., "Synthesis and characterization of a model extracellular matrix that induces partial regeneration of adult mammalian skin," *Proc. Natl. Acad. Sci. USA*, Vol. 86, No. 3, 1989, pp. 933–937.

[32] MacKintosh, F. C., Kas, J., and Janmey, P. A., "Elasticity of semiflexible biopolymer networks," *Phys. Rev. Lett.*, Vol. 75, No. 24, 1995, pp. 4425–4428.

[33] Arora, P. D., Narani, N., and McCulloch, C. A., "The compliance of collagen gels regulates transforming growth factor–beta induction of alpha-smooth muscle actin in fibroblasts," *Am. J. Path.*, Vol. 154, No. 3, 1999, pp. 871–882.

[34] Khor, E., "Methods for the treatment of collagenous tissues for bioprostheses," *Biomaterials*, Vol. 18, No. 2, 1997, pp. 95–105.

[35] Mercier, I., Lechaire, J. P., Desmouliere, A., Gaill, F., and Aumailley, M., "Interactions of human skin fibroblasts with monomeric or fibrillar collagens induce different organization of the cytoskeleton," *Exp. Cell Res.*, Vol. 225, No. 2, 1996, pp. 245–256.

[36] Mizutani, T., Haga, H., Kato, K., Matsuda, K., and Kawabata, K., "Observation of stiff domain structure on collagen gels by wide-range scanning probe microscopy," *Japanese J. Appl. Phys.*, Vol. 45, No. 3B, 2006, pp. 2353–2356.

[37] Elliott, J. T., Woodward, J. T., Langenbach, K. J., Tona, A., Jones, P. L., and Plant, A. L., "Vascular smooth muscle cell response on thin films of collagen," *Matrix Biol.*, Vol. 24, No. 7, 2005, pp. 489–502.

[38] Langenbach, K. J., Elliott, J. T., Tona, A., McDaniel, D., and Plant, A. L., "Thin films of Type 1 collagen for cell by cell analysis of morphology and tenascin-C promoter activity," *BMC Biotechnol.*, Vol. 6, 2006, p. 14.

[39] McDaniel, D. P., et al., "The stiffness of collagen fibrils influences vascular smooth muscle cell phenotype," *Biophys. J.*, Vol. 92, No. 5, 2007, pp. 1759–1769.

[40] Carr, M. E., Jr., Gabriel, D. A., and McDonagh, J., "Influence of factor XIII and fibronectin on fiber size and density in thrombin-induced fibrin gels," *J. Lab. Clin. Med.*, Vol. 110, No. 6, 1987, pp. 747–752.

[41] Marx, G., "Elasticity of fibrin and protofibrin gels is differentially modulated by calcium and zinc," *Thromb Haemost*, Vol. 59, No. 3, 1988, pp. 500–503.

[42] Coger, R., Toner, M., Moghe, P., Ezzell, R., and Yarmush, M. L., "Hepatocyte aggregation and reorganization of EHS matrix gel," *Tissue Eng.*, Vol. 3, No. 4, 1997, pp. 375–390.

[43] Semler, E. J., and Moghe, P. V., "Engineering hepatocyte functional fate through growth factor dynamics: The role of cell morphologic priming," *Biotechnol. Bioeng.*, Vol. 75, No. 5, 2001, pp. 510–520.

[44] Semler, E. J., Ranucci, C. S., and Moghe, P. V., "Mechanochemical manipulation of hepatocyte aggregation can selectively induce or repress liver-specific function," *Biotechnol. Bioeng.*, Vol. 69, No. 4, 2000, pp. 359–369.

[45] Smidsrod, O., and Skjak-Braek, G., "Alginate as immobilization matrix for cells," *Trends Biotechnol.*, Vol. 8, No. 3, 1990, pp. 71–78.

[46] Rowley, J. A., Madlambayan, G., and Mooney, D. J., "Alginate hydrogels as synthetic extracellular matrix materials," *Biomaterials*, Vol. 20, No. 1, 1999, pp. 45–53.

[47] LeRoux, M. A., Guilak, F., and Setton, L. A., "Compressive and shear properties of alginate gel: Effects of sodium ions and alginate concentration," *J. Biomed. Mater. Res.*, Vol. 47, No. 1, 1999, pp. 46–53.

[48] Eiselt, P., Lee, K. Y., and Mooney, D. J., "Rigidity of two-component hydrogels prepared from alginate and poly(ethylene glycol)-diamines," *Macromolecules*, Vol. 32, 1999, pp. 5561–5566.

[49] Lee, K. Y., Rowley, J. A., Eislet, P., Moy, E. M., Bouhadir, K. H., and Mooney, D. J., "Controlling mechanical and swelling properties of alginate hydrogels independently by cross-linker type and cross-linking density," *Macromolecules*, Vol. 33, 2000, pp. 4291–4294.

[50] Balgude, A. P., Yu., X., Szymanski, A., and Bellamkonda, R. V., "Agarose gel stiffness determines rate of DRG neurite extension in 3-D cultures," *Biomaterials*, Vol. 22, No. 10, 2001, pp. 1077–1084.

[51] Luo, Y., and Shoichet, M. S., "Light-activated immobilization of biomolecules to agarose hydrogels for controlled cellular response," *Biomacromolecules*, Vol. 5, No. 6, 2004, pp. 2315–2323.

[52] Suh, J. K., and Matthew, H. W., "Application of chitosan-based polysaccharide biomaterials in cartilage tissue engineering: A review," *Biomaterials*, Vol. 21, No. 24, 2000, pp. 2589–2598.

[53] Kurita, K., "Chitin and chitosan: Functional biopolymers from marine crustaceans," *Mar. Biotechnol.* (NY), Vol. 8, No. 3, 2006, pp. 203–226.

[54] Berger, J., Reist, M., Mayer, J. M., Felt, O., Peppas, N. A., and Gurny, R., "Structure and interactions in covalently and ionically crosslinked chitosan hydrogels for biomedical applications," *Eur. J. Pharm. Biopharm.*, Vol. 57, No. 1, 2004, pp. 19–34.

[55] Subramanian, A., and Lin, H. Y., "Crosslinked chitosan: Its physical properties and the effects of matrix stiffness on chondrocyte cell morphology and proliferation," *J. Biomed. Mater. Res. A*, Vol. 75, No. 3, 2005, pp. 742–753.

[56] Yeung, T., et al., "Effects of substrate stiffness on cell morphology, cytoskeletal structure, and adhesion," *Cell Motil. Cytoskeleton*, Vol. 60, No. 1, 2005, pp. 24–34.

[57] Janney, M. A., et al., "Gelcasting," in *The Handbook of Ceramic Engineering*, ed. Rahman, M. N., Marcel Dekker, 1998, New York.

[58] Wong, J. Y., Velasco, A., Rajagopalan, P., and Pham, Q., "Directed movement of vascular smooth muscle cells on gradient-compliant hydrogels," *Langmuir*, Vol. 19, No. 5, 2003, pp. 1908–1913.

[59] Zaari, N., Rajagopalan, P., Kim, S. K., Engler, A. J., and Wong, J. Y., "Photopolymerization in microfluidic gradient generators: Microscale control of substrate compliance to manipulate cell response," *Adv. Mater.*, Vol. 16, Nos. 23–24, 2004, pp. 2133–2137.

[60] Beningo, K. A., Lo, C. M., and Wang, Y. L., "Flexible polyacrylamide substrata for the analysis of mechanical interactions at cell-substratum adhesions," *Methods Cell Biol.*, Vol. 69, 2002, pp. 325–339.

[61] Reinhart-King, C. A., Dembo, M., and Hammer, D. A., "Endothelial cell traction forces on RGD-derivatized polyacrylamide substrata," *Langmuir*, Vol. 19, No. 5, 2003, pp. 1573–1579.

[62] Wang, Y. L., Pelham, R. J., Jr., "Preparation of a flexible, porous polyacrylamide substrate for mechanical studies of cultured cells," *Methods Enzymol.*, Vol. 298, 1998, pp. 489–496.

[63] Burnham, M. R., Turner, J. N., Szarowski, D., and Martin, D. L., "Biological functionalization and surface micropatterning of polyacrylamide hydrogels," *Biomaterials*, Vol. 27, No. 35, 2006, pp. 5883–5891.

[64] Damljanovic, V., Lagerholm, B. C., and Jacobson, K., "Bulk and micropatterned conjugation of extracellular matrix proteins to characterized polyacrylamide substrates for cell mechanotransduction assays," *BioTechniques*, Vol. 39, No. 6, 2005, pp. 847–851.

[65] Peyton, S. R., Raub, C. B., Keschrumrus, V. P., and Putnam, A. J., "The use of poly(ethylene glycol) hydrogels to investigate the impact of ECM chemistry and mechanics on smooth muscle cells," *Biomaterials*, Vol. 27, No. 28, 2006, pp. 4881–4893.

[66] Nguyen, K. T., and West, J. L., "Photopolymerizable hydrogels for tissue-engineering applications," *Biomaterials*, Vol. 23, No. 22, 2002, pp. 4307–4314.

[67] Hern, D. L., and Hubbell, J. A., "Incorporation of adhesion peptides into nonadhesive hydrogels useful for tissue resurfacing," *J. Biomed. Mater. Res.*, Vol. 39, No. 2, 1998, pp. 266–276.

[68] Almany, L., and Seliktar, D., "Biosynthetic hydrogel scaffolds made from fibrinogen and polyethylene glycol for 3-D cell cultures," *Biomaterials*, Vol. 26, No. 15, 2005, pp. 2467–2477.

[69] Shapira-Schweitzer, K., and Seliktar, D., "Matrix stiffness affects spontaneous contraction of cardiomyocytes cultured within a PEGylated fibrinogen biomaterial," *Acta Biomater*, Vol. 3, No. 1, 2007, pp. 33–41.

[70] Mata, A., Fleischman, A. J., and Roy, S., "Characterization of polydimethylsiloxane (PDMS) properties for biomedical micro/nanosystems," *Biomed. Microdevices*, Vol. 7, No. 4, 2005, pp. 281–293.

[71] Gray, D. S., Tien, J., and Chen, C. S., "Repositioning of cells by mechanotaxis on surfaces with micropatterned Young's modulus," *J. Biomed. Mater. Res. A*, Vol. 66, No. 3, 2003, pp. 605–614.

[72] Huang, B., Wu, H. K., Kim, S., Kobilka, B. K., and Zare, R. N., "Phospholipid biotinylation of polydimethylsiloxane (PDMS) for protein immobilization," *Lab Chip*, Vol. 6, No. 3, 2006, pp. 369–373.

[73] Ingber, D. E., et al., "Cellular tensegrity: Exploring how mechanical changes in the cytoskeleton regulate cell growth, migration, and tissue pattern during morphogenesis," *Int. Rev. Cytol.*, Vol. 150, 1994, pp. 173–224.

[74] Lo, C.-M., Wang, H.-B., Dembo, M., and Wang, Y.-L., "Cell movement is guided by the rigidity of the substrate," *Biophys. J.*, Vol. 79, No. 1, 2000, pp. 144–152.

[75] Guo, W.-H., Frey, M. T., Burnham, N. A., and Wang, Y.-L. "Substrate rigidity regulates the formation and maintenance of tissues," *Biophys. J.*, Vol. 90, No. 6, 2006, pp. 2213–2220.

[76] Cukierman, E., Pankov, R., Stevens, D. R., and Yamada, K. M., "Taking cell-matrix adhesions to the third dimension," *Science*, Vol. 294, No. 5547, 2001, pp. 1708–1712.

[77] Beningo, K. A., Dembo, M., and Wang, Y. L., "Responses of fibroblasts to anchorage of dorsal extracellular matrix receptors," *Proc. Natl. Acad. Sci. USA*, Vol. 101, No. 52, 2004, pp. 18024–18029.

[78] Ghosh, K., et al., "Cell adaptation to a physiologically relevant ECM mimic with different viscoelastic properties," *Biomaterials*, Vol. 28, No. 4, 2007, pp. 671–679.

[79] Engler, A. J., Richert, L., Wong, J. Y., Picart, C., and Discher, D. E., "Surface probe measurements of the elasticity of sectioned tissue, thin gels and polyelectrolyte multilayer films: Correlations between substrate stiffness and cell adhesion," *Surface Science*, Vol. 570, Nos. 1–2, 2004, pp. 142–154.

[80] Wang, H. B., Dembo, M., and Wang, Y. L., "Substrate flexibility regulates growth and apoptosis of normal but not transformed cells," *Am. J. Physiology Cell Physiology*, Vol. 279, No. 5, 2000, pp. C1345–C1350.

[81] Kong, H. J., Liu, J., Riddle, K., Matsumoto, T., Leach, K., and Mooney, D. J., "Non-viral gene delivery regulated by stiffness of cell adhesion substrates," *Nat. Mater.*, Vol. 4, No. 6, 2005, pp. 460–464.

[82] Rowley, J. A., and Mooney, D. J., "Alginate type and RGD density control myoblast phenotype," *J. Biomed. Mater. Res.*, Vol. 60, No. 2, 2002, pp. 217–223.

[83] Georges, P. C., Miller, W. J., Meaney, D. F., Sawyer, E. S., and Janmey, P. A., "Matrices with compliance comparable to that of brain tissue select neuronal over glial growth in mixed cortical cultures," *Biophys. J.*, Vol. 90, No. 8, 2006, pp. 3012–3018.

[84] Opas, M., "Substratum mechanics and cell differentiation," *Int. Rev. Cytol.*, Vol. 150, 1994, pp. 119–137.

[85] Opas, M., "Expression of the differentiated phenotype by epithelial cells in vitro is regulated by both biochemistry and mechanics of the substratum," *Dev. Biol.*, Vol. 131, 1989, pp. 281–293.

[86] Streuli, C. H., and Bissell, M. J., "Expression of extracellular matrix components is regulated by substratum," *J. Cell. Biol.*, Vol. 110, 1990, pp. 1405–1415.

[87] Engler, A. J., Sen, S., Sweeney, H. L., and Discher, D. E., "Matrix elasticity directs stem cell lineage specification," *Cell*, Vol. 126, No. 4, 2006, pp. 677–689.

[88] Engler, A. J., Griffin, M. A., Sen, S., Bonnemann, C. G., Sweeney, H. L., and Discher, D. E., "Myotubes differentiate optimally on substrates with tissuelike stiffness: Pathological implications for soft or stiff microenvironments," *J. Cell. Biol.*, Vol. 166, No. 6, 2004, pp. 877–887.

[89] Kruse, S. A., et al., "Tissue characterization using magnetic resonance elastography: Preliminary results," *Phys. Med. Biol.*, Vol. 45, No. 6, 2000, pp. 1579–1590.

[90] Kong, H. J., Polte, T. R., Alsberg, E., and Mooney, D. J., "FRET measurements of cell-traction forces and nanoscale clustering of adhesion ligands varied by substrate stiffness," *Proc. Natl. Acad. Sci. USA*, Vol. 102, No. 12, 2005, pp. 4300–4305.

[91] Keselowsky, B. G., Collard, D. M., and Garcia, A. J., "Integrin binding specificity regulates biomaterial surface chemistry effects on cell differentiation," *Proc. Natl. Acad. Sci. USA*, Vol. 102, No. 17, 2005, pp. 5953–5957.

[92] Xiao, G., Wang, D., Benson, M. D., Karsenty, G., and Franceschi, R. T., "Role of the alpha2-integrin in osteoblast-specific gene expression and activation of the Osf2 transcription factor," *J. Biol. Chem.*, Vol. 273, No. 49, 1998, pp. 32988–32994.

[93] Tomasek, J. J., Gabbiani, G., Hinz, B., Chaponnier, C., and Brown, R. A., "Myofibroblasts and mechano-regulation of connective tissue remodelling," *Nat. Rev. Mol. Cell Biol.*, Vol. 3, No. 5, 2002, pp. 349–363.

[94] Grinnell, F., "Fibroblasts, myofibroblasts, and wound contraction," *J. Cell. Biol.*, Vol. 124, No. 4, 1994, pp. 401–404.

[95] Grinnell, F., "Fibroblast-collagen-matrix contraction: growth-factor signalling and mechanical loading," *Trends Cell. Biol.*, Vol. 10, No. 9, 2000, pp. 362–365.

[96] Sieminski, A. L., Hebbel, R. P., and Gooch, K. J., "The relative magnitudes of endothelial force generation and matrix stiffness modulate capillary morphogenesis in vitro," *Exp. Cell Res.,* Vol. 297, No. 2, 2004, pp. 574–584.

[97] Vernon, R. B., Angello, J. C., Iruela-Arispe, M. L., Lane, T. F., and Sage, E. H., "Reorganization of basement membrane matrices by cellular traction promotes the formation of cellular networks in vitro," *Lab. Invest.*, Vol. 66, No. 5, 1992, pp. 536–547.

[98] Vailhe, B., Ronot, X., Tracqui, P., Usson, Y., and Tranqui, L., "In vitro angiogenesis is modulated by the mechanical properties of fibrin gels and is related to alpha(v)beta3 integrin localization," *In Vitro Cell. Dev. Biol. Anim.*, Vol. 33, No. 10, 1997, pp. 763–773.

[99] Cukierman, E., Pankov, R., and Yamada, K. M., "Cell interactions with three-dimensional matrices," *Curr. Opin. Cell. Biol.*, Vol. 14, No. 5, 2002, pp. 633–639.

[100] Mahaffy, R. E., Shih, C. K., MacKintosh, F. C., and Kas, J., "Scanning probe-based frequency-dependent microrheology of polymer gels and biological cells," *Phys. Rev. Lett.*, Vol. 85, No. 4, 2000, pp. 880–883.

[101] Dimitriadis, E. K., Horkay, F., Maresca, J., Kachar, B., and Chadwick, R. S., "Determination of elastic moduli of thin layers of soft material using the atomic force microscope," *Biophys. J.*, Vol. 82, No. 5, 2002, pp. 2798–2810.

[102] Boudou, T., Ohayon, J., Arntz, Y., Finet, G., Picart, C., and Tracqui, P., "An extended modeling of the micropipette aspiration experiment for the characterization of the Young's modulus and Poisson's ratio of adherent thin biological samples: Numerical and experimental studies," *J. Biomech.*, Vol. 39, No. 9, 2006, pp. 1677–1685.

[103] Boudou, T., Ohayon, J., Picart, C., and Tracqui, P., "An extended relationship for the characterization of Young's modulus and Poisson's ratio of tunable polyacrylamide gels," *Biorheology*, Vol. 43, No. 6, 2006, pp. 721–728.

[104] Flaim, C. J., Chien, S., and Bhatia, S. N., "An extracellular matrix microarray for probing cellular differentiation," *Nat. Methods*, Vol. 2, No. 2, 2005, pp. 119–125.

CHAPTER 10

Engineered Surface Nanotopography for Controlling Cell-Substrate Interactions

Deok-Ho Kim, Andre Levchenko, and Kahp Y. Suh

10.1 Introduction

Living tissues are intricate ensembles of different cell types embedded in complex and well-defined structures of extracellular matrix (ECM) with nanoscale topographical features. The organization of ECM is also frequently hierarchical, with many proteins capable of forming large-scale structures with feature size up to several hundred microns. For example, hierarchical packing of collagen proteins can lead to formation of fibrils, long (tens of microns) cylindrical structures, whose diameter may vary from 20 to 200 nm range. Furthermore, in connective tissues, it is common to find bundles of collagen microfibrils running parallel to each other, with cells of various origins attached to them. These cells might both affect and be affected by the superstructures of collagen and other ECM components. For instance, fibroblasts can reorganize collagen fibrils by secreting collagen, pulling on the fibrils, thus deforming them, but can also polarize and migrate along the fibrils, using them as guidance cues [1]. Epithelial cells affect the composition of, and are affected by, the embedded factors present in the basement membrane—a complex ECM superstructure with feature sizes of tens and hundreds nanometers. Cells contain nanoscale physical features whose size is also compatible with embedded ECM, including the intracellular organelles such as cytoskeletal elements (e.g., actin filaments and microtubules) and adhesive structures (e.g., focal adhesions) and the surface structures with fine processes such as microspikes, filopodia, and fimbriae. It is therefore quite reasonable to expect that functioning of many cell types can be significantly affected in vivo by the nanoscale features of the surrounding ECM. Unfortunately, the extent and the importance of nanotopography of ECM in defining cell behavior are currently poorly understood, in part due to an almost complete neglect of this factor in most in vitro experimentation.

Most of the previous in vitro studies examine cells cultured on a two-dimensional flat and rigid substrate although most cells in vivo are exposed to a three-dimensional environment with complex topographical features. Recent ultrastructural analysis using electron microscopy has shown that the basement membrane of epithelium tissues exhibit three-dimensional features of pores, fiber, and ridges in the nanometer range [2, 3], and collagen fibers in connective tissues are composed of tropocollagen molecules associating to form microfibrils with an apparent nanometer periodicity [4, 5]. Additionally, experimental evidence by cell culture in three-dimensional matrices suggested that signaling and resulting cell functions may differ significantly in three-dimensional, compared with two-dimensional, culture systems made of an artificially flat and rigid substrate [6, 7]. This difference is obvious in that cell-adhesion structures in three dimensions can evolve in vitro toward in vivo–like adhesions with distinct biological activities.

For the last decade, microfabrication and surface-chemistry-based approaches have provide a versatile set of tools to control the spatial organization and temporal presentation of cellular cues on a microscale [8]. To better understand the mechanisms involved in cellular behavior and function in a microenvironment, cells have often been attached to various microengineered substrata such as bioactive substrates patterned with natural biomaterials like collagen or fibronectin with tissue-specific properties and/or artificial substrates fabricated with synthetic biomaterials [9]. In micropatterned cell culture, topographical and physicochemical modification of material surface enable selective localization and phenotypic and genotypic control of living cells [10]. The effects of microscale ECM features on cellular behavior have been well documented so far, suggesting that control of cell mechanics and shape by microtopography is crucial for many cellular functions, including gene expression, adhesion, migration, proliferation, and differentiation [11].

Recent advancement of nanofabrication techniques has led to a rapid accumulation of evidence on the importance of ECM nanotopography in controlling cellular processes [12]. Using some of the advanced nanofabrication methods, nanoscale topographic features can now be incorporated into the in vitro experimental platform to mimic various in vivo three-dimensional ECM environments with structural and mechanical similarity. Preliminary experiments with various nanotopographic features, such as grooves, ridges, pores, wells, and pillars, have shown that nanotopography can strongly influence cell morphology, adhesion, proliferation, and gene regulation, but the mechanism mediating this cell response still remains unclear [13].

Here, we describe the nanofabrication techniques for generating nanotopography of cell substrata and the current studies investigating the role of ECM nanotopography in defining cell polarity, migration, and differentiation. We also discuss topical issues that could be addressed by this approach to probe cellular process and ultimately engineer cellular functions with desired phenotypic responses. This research is of high significance due to its potential to yield both fundamental knowledge of mechanisms of cell motility and the resulting control of cellular function that can be utilized in advanced tissue engineering.

10.2 Methods for Generating Nanotopography

Nanolithography generally refers to a set of lithographic technologies used to fabricate very small features (at scales ranging from 1 to 100 nm) that can be integrated into more complex nano- and microsystems. Important biological phenomena are regulated in this nanoscale regime, but their analysis is hampered as the feature size approaches a fundamental length scale of physical and chemical processes (e.g., it becomes comparable to the wavelengths of visible light). Therefore, nanolithographically defined substrates could offer unique opportunities for in vitro cell culture experimentation by allowing the presentation of controlled biophysical stimuli. To investigate the role of nanotopography in cell functions, it is of great importance to fabricate well-defined nanotopography or adhesive nanopatterns mimicking ECM environments with similar structural and biochemical properties. In this section, we describe a number of currently available nanolithographic methods, including optical lithography and unconventional contact-based lithographies (nanoimprint, soft, capillary-force, and scanning-probe lithographies). Some techniques based on self-assembly and self-organization (e.g., colloidal and block copolymer lithographies) are also included.

10.2.1 Conventional Nanolithography

Optical lithography or photolithography, which has been and will be the mainstay of lithography for the near future, is currently used for manufacturing microelectronic structures and typically involves a projection-printing system (usually called a stepper). An optical lithographic process consists of three successive steps: (1) coating a substrate with irradiation-sensitive polymer layer (resist); (2) exposing the resist with light, electron, or ion beams, and (3) developing the resist image with a suitable chemical. Exposures can be done by either scanning a focused beam pixel by pixel from a designed pattern or exposing through a mask for parallel replication (Figure 10.1). Serial beam scanning is used for mask fabrication and single component fabrication, but it does not supply adequate throughput for manufacturing. Industrial techniques must be fast, reliable, and cost-effective. Optical projection lithography with deep ultraviolet light is now used for large-scale fabrication. As next-generation lithographies, several nonoptical techniques have been developed, including extreme ultraviolet lithography, X-ray lithography, and projection lithography with either electrons or ions [14]. These methods, generally referred to as top-down methods, are very expensive but sufficiently flexible for new development out of the traditional microelectronic industry.

Conventional lithographic exposures involve the interaction of an incident beam with a solid substrate. Absorption of light or inelastic scattering of particles can affect the chemical structure of the resist by changing its solubility. One can choose the resist so that the response of the resist to a patterned light exposure can be either positive or negative, depending on whether the exposed or unexposed portions will be removed from the substrate after development. The next step after lithography is the pattern transfer from the resist to the substrate. There are a number of pattern-transfer techniques: selective growth of materials in the trenches of

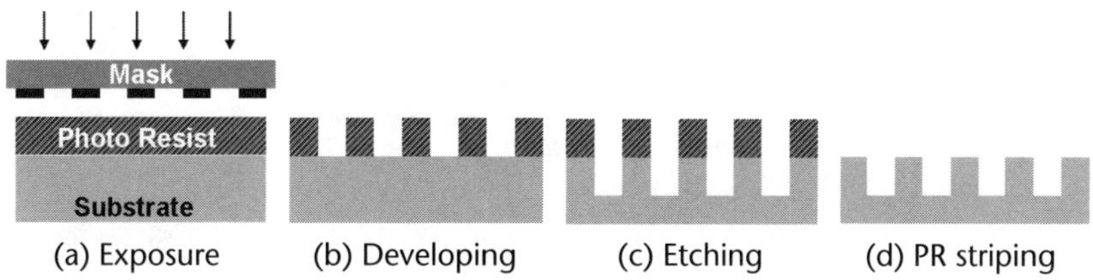

Figure 10.1 (a–d) Schematic representation of optical lithography.

the resist, etching of the unprotected areas, and doping through the open areas of the resist by diffusion or implantation [15]. Both wet chemical etching and dry plasma etching can be used. For a high-resolution pattern transfer, dry etching is more suitable and often requires a metallic layer as a mask. This metal mask is obtained by lift-off (i.e., by first depositing a thin metallic layer over the developed resist pattern and then dissolving the resist in order to leave only the metal portions that are directly in contact with the substrate).

In the process of optical lithography, the image of a reticle (Figure 10.2) is reduced and projected onto a thin film of photoresist that is spin-coated on a wafer through a high numerical aperture lens system. The resolution of the stepper (R) is subject to the limitations set by optical diffraction according to the Rayleigh equation [16] that is given by

$$R = k_1 \lambda / NA \tag{10.1}$$

where λ is the wavelength of the illuminating light, NA is the numerical aperture of the lens system, and k_1 is a constant determined by the photoresist. The theoretical minimum feature size imposed by the relationship is usually the wavelength of the light used. Therefore, extensive efforts have been made to introduce illuminating sources with shorter wavelengths to generate structures with smaller feature sizes [17]. Fabrication of nanostructures using optical lithography is a well-established technology. More information on optical nanolithography can be found elsewhere [18, 19].

10.2.2 Unconventional Nanolithography

As structures get smaller than 100 nm, it becomes increasingly difficult and expensive to fabricate using the conventional optical lithography. Therefore, there have

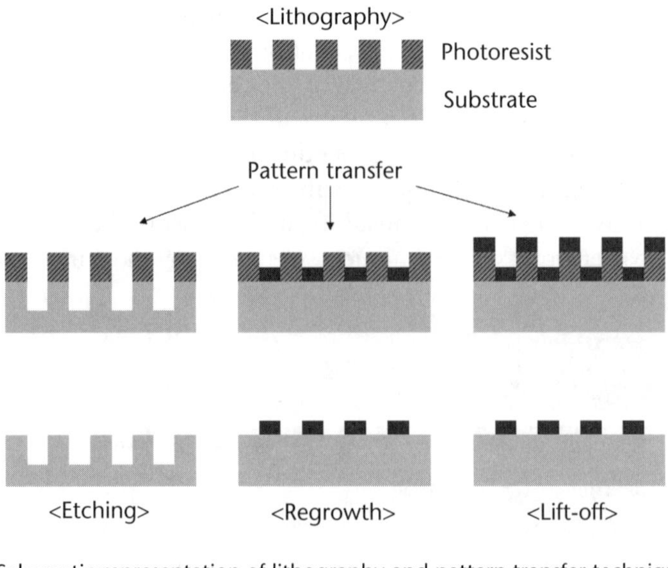

Figure 10.2 Schematic representation of lithography and pattern transfer techniques.

been extensive efforts made since mid-1990s to develop methods that can replace photolithography. In contrast to the conventional optical lithography, a contact-based unconventional lithographic process generally consists of three consecutive steps: (1) coating a substrate with a mobile polymer layer (resist); (2) treating the resist with physical force such as pressure, capillary force, solvation force, or electric force; and (3) transferring the pattern on the resist into the underlying substrate. These methods are relatively cheap and compatible with optical lithography in that they conserve the lithographic printing strategy. Here, we overview several notable unconventional nonphotolithographic methods that can be applicable to patterning sub-100 nm features on large areas in a relatively simple and compatible manner.

10.2.2.1 Nanoimprint Lithography

Embossing with a rigid master is an unconventional, low-cost technique for high-resolution pattern replication and is the standard for manufacturing replicas of holograms, diffraction grating, and compact disks [20]. Recently, this technique has been reexamined as a method for producing nanometer-sized structures of semiconductors, metals, and other materials commonly used in microelectronic circuitry or information storage, which is now called nanoimprint lithography (NIL) (Figure 10.3) [21].

NIL generates resist relief patterns in a thermoplastic layer such as polystyrene by physically compressing the resist that has been thermally softened, rather than by

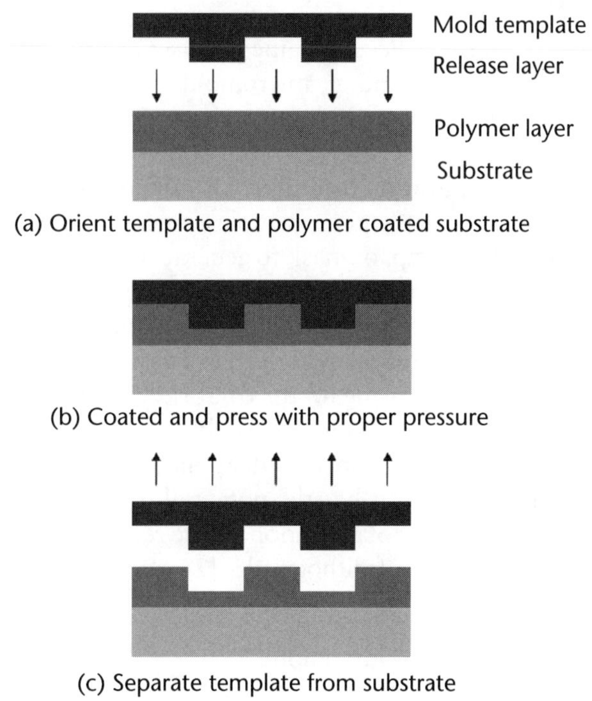

Figure 10.3 (a–c) Schematic representation of nanoimprint lithography.

modifying the resist's chemical structure by irradiation. A rigid mold is used to physically deform the polymer layer, which can be made of metal or thermal silicon dioxide produced on a silicon substrate. An anisotropic etching process, such as reactive ion etching, is used to remove the residual resist in the compressed region for the subsequent hard material pattern transfer. The mold can be reused many times without damage. In typical NIL (i.e., hot-embossing lithography), temperature is raised above the glass transition temperature of the polymer [typically 170°C for polymethylmethacrylate (PMMA)], and the resist thickness can vary from 50 to 200 nm. Structures as small as 6 nm could be produced [22]. However, there are a few problems with the associated high-temperature processes, primarily due to the clean release of the resist from the mold being one of the critical requirements. Usually, antiadhesion agents have been used to achieve clean mold release [21]. Distortion of the imprinted structures or mold features due to the thermal cycle of heating followed by cooling is another problem to contend with.

To overcome these shortcomings, room-temperature imprint methods have been proposed, which gave rise to a series of successful low-temperature processes. The easiest method is to use a low-molecular-weight polymer for its low viscosity or high mobility even at low temperatures. Examples are standard optical resist S1805 [23] and newly developed hybrane, as well as hydrogen-bonded polymers. In terms of processing conditions in order to achieve room-temperature process, there are two useful techniques; one is room-temperature (RT)–NIL by solvent treatment [24], and the other uses plastic deformation along with free-volume contraction [25].

10.2.2.2 Soft Lithography

Soft lithography is the collective name for a set of lithographic techniques—replica molding, microcontact printing, micromolding in capillaries, microtransfer molding, solvent-assisted micromolding, and near-field conformal photolithography using an elastomeric phase-shifting mask—developed as an alternative to photolithography and as a replication technology for micro- and nanofabrication [26–28].

These techniques use a patterned elastomer (usually polydimethylsiloxane, or PDMS) as the mold, stamp, or mask to generate or transfer the pattern (Figure 10.4). Soft lithography offers immediate advantages over photolithography and other conventional microfabrication techniques for applications where patterning of nonplanar substrates, unusual materials, or large-area patterning are the major concerns. It is especially promising for microfabrication of relatively simple, single-layer structures for uses such as in cell culture, sensors, microanalytical systems, microelectromechanical systems, and applied optics. The initial success of soft lithography indicates that it has the potential to become an important addition to the field of micro- and nanofabrication. Some recent review articles are available for detailed information on soft lithography [26–28].

10.2.2.3 Capillary-Force Lithography

The imprinting method requires very high pressure, typically on the order of 10^9 pascals (N/m^2), which could cause breakage of the substrate. In addition, a recent

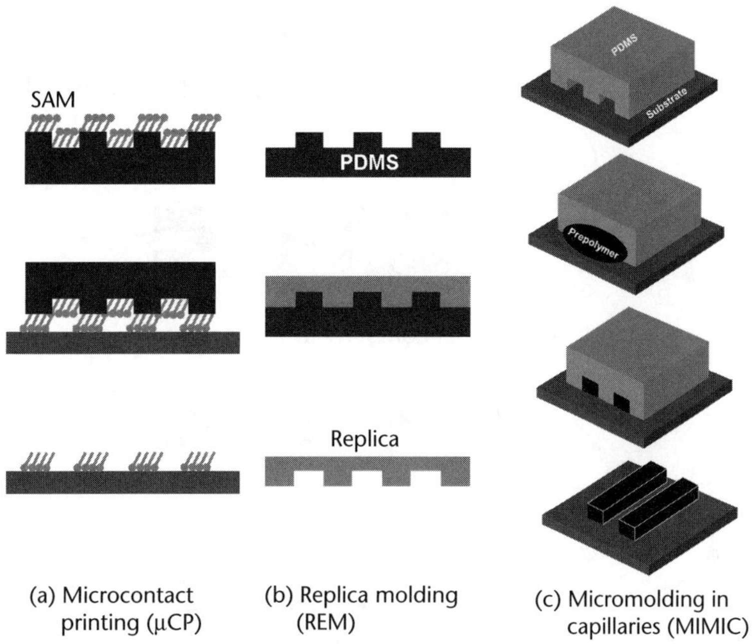

Figure 10.4 (a–c) Schematic representation of soft lithography.

study [29] reported that material transport limits the performance of NIL in general, especially when the mold is negative or has recessed features within a large elevated surface level. In this case, the pattern transfer turned out to be unsatisfactory, and bubblelike defects were observed, possibly due to lack of conformal contact.

In order to solve these problems coming from high pressure, a new lithographic technique called capillary-force lithography (CFL) has been developed by simply combining nanoimprint and soft lithographies [30, 31]. When a patterned (positive or negative) PDMS mold is placed on a polymer surface and heated above the polymer's glass transition temperature (T_g), capillarity forces the polymer melt into the void space of the channels formed between the mold and the polymer, thereby generating a negative replica of the mold. The pattern formation is also possible with a solvent-laden polymer or an ultraviolet-curable (Figure 10.5) resin followed by solvent evaporation or exposure to ultraviolet light. This experimental procedure is similar to the nanoimprint lithography except that a hard mold is replaced by a soft mold, and no pressure is applied.

In CFL, the essential feature of nanoimprint lithography, molding a polymer melt, is combined with the prime element of soft lithography, the use of a PDMS mold. As a result, the advantage of imprint lithography over soft lithography is retained in meeting stringent pattern-fidelity requirements, as in the fabrication of integrated circuits, while eliminating the need to use the extremely high pressure needed in nanoimprint lithography. More recently, an ultraviolet curable mold made from polyurethane functionalized with acrylate groups has been introduced to replace PDMS mold for sub-100 nm lithography, thus expanding the use of CFL to studies of cell biology [32, 33].

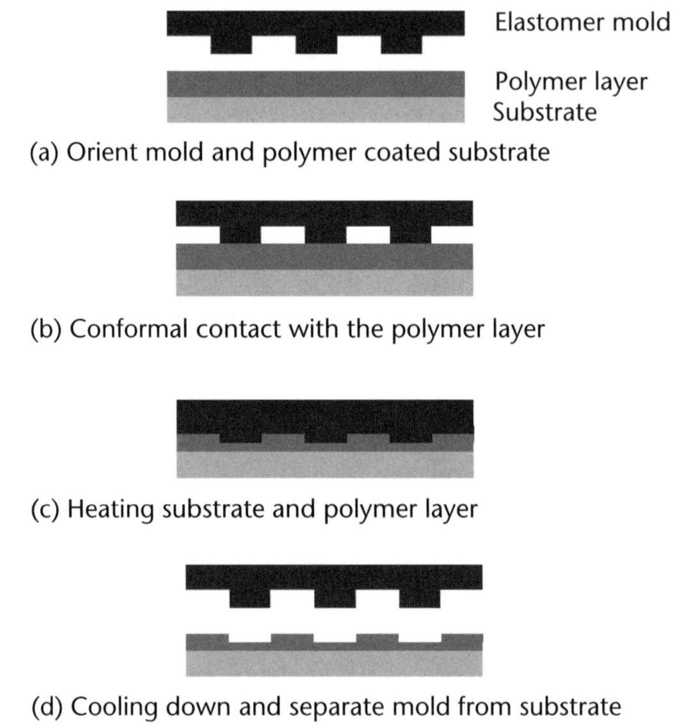

Figure 10.5 (a–d) Schematic representation of capillary-force lithography.

10.2.2.4 Scanning Probe Lithography

If the economics is not a major concern, there are technologies already available for sub-100 nm features, such as electron-beam lithography [15, 34], X-ray lithography [35, 36], and scanning-probe lithography [37–39], etc. These methods all share the same basic operational principle and only have different exposure methods. The principle is that exposure of an appropriate material to electromagnetic radiation (ultraviolet, deep ultraviolet, extreme ultraviolet, or X-ray) (Figure 10.6) introduces a latent modification (usually a difference in solubility) into the material as a result of a set of chemical changes in its molecular structure. Then, this latent modification is subsequently developed into relief structures through etching. When probe tips do the writing, the lithography can be termed scanning-probe lithography (SPL). SPL uses small (<50 nm) tips scanned near the surface of a sample in the configuration of scanning tunneling microscopes (STMs) [40], atomic force microscopes (AFMs) [41], scanning electrochemical microscopes (SECMs) [42], or near-field scanning optical microscopes (NSOMs) [43]. Since the first demonstration that STMs can be used as a lithographic tool, STMs, AFMs, SECMs, and NSOMs have been used in many different ways to perform sophisticated lithography. These tip-based lithographies have the advantages of resolving atomic level, realizing features with nearly arbitrary geometries, and patterning over surface topography that deviates significantly from planarity [39]. However, these methods are serial techniques and have writing speeds that are typically limited by the mechanical resonance of the tips

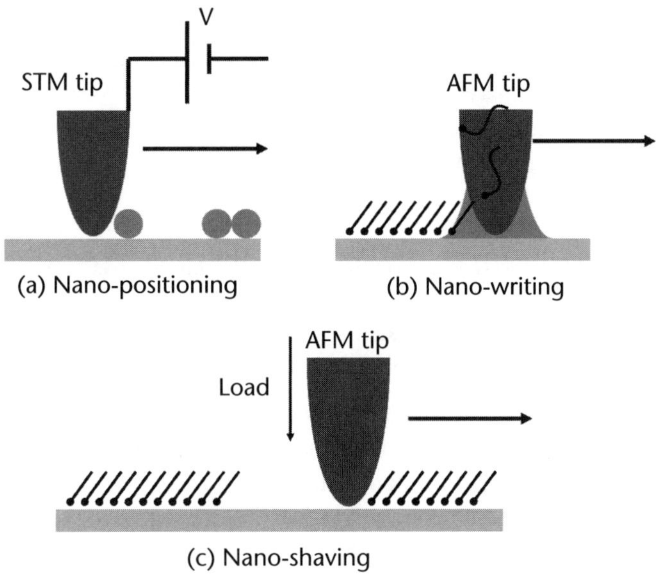

Figure 10.6 (a–c) Schematic representation of scanning-probe lithography.

and piezoelectric elements that maintain constant separation between the tips and the sample.

10.2.2.5 Other Lithographic Methods Based on Self-Assembly and Self-Organization

Numerous possibilities of nanostructure formation exist based on chemical and biological approaches relying on self-assembly and self-organization. The advantage of these techniques is the realization of specific nanostructures without having to construct them atom by atom. Chemical processing, in which self-assembly takes place in organic or inorganic materials, can be used for forming aggregates and colloidal assemblies. For example, the electrostatically self-assembled periodic structure from colloids can be used as templates. These nanocolloids can be used as an etching mask to produce large areas of nanotopography with short range order (so called colloidal lithography). The colloids are dispersed as a monolayer and electrostatically self-assembled on a surface [44, 45]. The colloids are then etched away, resulting in a nanopatterned substrate with controlled dimension. If the etch is directly from above, this technique produces nanocolumns, nanospheres, or nanocones with different etching conditions.

Another example is the self-organized growth of nanostructures on a strained substrate. By molecular beam epitaxy, InAs quantum dots were produced by depositing a fraction of InAs monolayer on a GaAs substrate, followed by another GaAs layer [46]. Metallic quantum dots were also obtained by depositing a Cu bilayer on a strain-relief pattern of the (111) Pt surfaces [47]. If structure regularity is not a major concern, there are a number of other methods for constructing a nanopatterned substrate using templates of organic nanostructure patterns. For example, block copolymers can be used to create a range of highly ordered morphologies in thin films where the size scale of the features is only limited by the size of the polymer chains (block copolymer lithography) [48]. Also, supramolecular networks could be created from nanoscale rods in binary, phase-separating mixtures [49]. These ordered structures can act as a template to transfer nanostructures into an underlying substrate.

10.3 Topical Issues in Controlling Cell-Substrate Interactions

10.3.1 Spatial Control of Cell-Adhesive Proteins on Nanostructured Surface for Cell-Matrix and Cell-Cell Interactions

It would greatly improve our understanding of how cell-ECM interactions affect cellular processes if one could fabricate nanostructured ECM on various two-dimensional and three-dimensional substrates. A few nanopatterning approaches have been proposed to produce ECM protein patterns at the nanoscale level and to control the spatial distribution of ECM proteins. For example, an SPL-based method was used to print collagens and collagenlike peptides (30 to 50 nm line width) without disrupting the triple-helical structures and biological activities of collagens [Figure 10.7(a)] [50, 51]. Nanopatterning of self-assembled monolayers by electron beam lithography has been developed to produce organic templates for creating high-density protein nanoarrays [52, 53]. Selective molecular assembly patterning in combination with lithographically prepatterned substrate also has proven

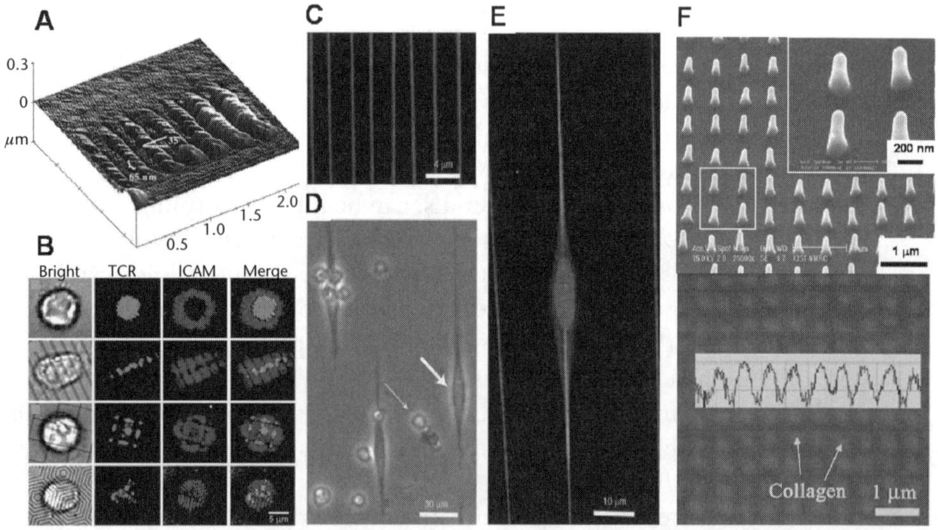

Figure 10.7 Nanopatterned proteins for cell-matrix adhesion and cell-cell interactions: (a) SPL-based nanopatterning of collagen molecules at varying forces. (*Source:* [50].) (b) Immunological synapse formation altered by geometrical constraints of the nanostructured substrate. (*Source:* [61].) (c) Parallel crack pattern formation using fluorescent protein (FITC-BSA). (d) A micrograph of C2C12 myoblasts spreading on wide cracks and elongated along collagen-coated cracks (thick arrow). (e) A C2C12 myoblast cell stained with Phalloidin-TRITC for actin and Hoechst 33342 for the nucleus. The cell is attached to a crack coated with collagen. (*Source:* [60].) (f) Control over wettability by surface nanotopography modulating spatial distribution of absorbed collagens. (*Source:* [32].)

to be a fast and reliable method to create protein arrays over large areas with feature sizes comparable to scanning-probe based techniques [33, 54–58]. Mechanically mediated direct deposition of functional proteins to create ordered nanoarrays was also studied [59]. The technique has been also introduced to fabricate reconfigurable protein matrices by cracking that can span the 100–3,000 nm scale [Figure 10.7(c–e)] [60]. Such reconfigurable nanoscale-to-microscale ECM protein patterns are desired because the sizes of adhesion plaques fall within this range, and the dynamics of these plaques are known to vary according to the state of the cell and its surrounding matrix.

Nanopatterned biomaterials, such as ECM proteins or biocompatible polymers, have shown a great promise when it comes to interfacing directly with living cells. For example, surface nanopatterning techniques can be also used to produce biomimetic display of cellular ligands and receptors in cell-cell interactions, which is relatively less investigated than cell-ECM adhesion. Particularly, cell- membrane-like supported lipid bilayer (SLB) offers a fluid lipid bilayer on a glass surface, allowing high-resolution analysis of membrane receptor-mediated signaling dynamic events in a more precisely controlled manner [62, 63]. This bilayer can be copatterned with ECM and soluble ligands for tailoring more complex, in vivo–like cellular environments for live cell assays [61, 64]. Most recently, it was shown that T cell receptor signaling can be altered by SLB and patterned protein surfaces that modulate immunological synapse structure formations [Figure 10.7(b)] [61, 65]. Similar benefits are expected for the study of the formation and function of neuronal synapses [66].

Recent studies have showed that surface nanotopography enhances the cell-substratum adhesion of human corneal epithelial cells [67] and primary rat cardiomyocytes [32]. It was also reported that proteins (collagen, fibronectin, immunoglobulin, and albumin) and cells (CHO, fibroblasts, and P19 EC cells) preferentially adhere on nanostructured poly(ethylene glyocol) (PEG) surfaces in comparison to nonpatterned PEG films [Figure 10.7(f)] [32, 33]. Other studies using CFL have revealed that the wetting property of the nanostructured surface of a given chemical composition can be systematically controlled by rendering nanoscale roughness [68]. These findings suggest that the control over wettability by surface nanotopography made of polymeric nanostructure can also modulate spatial distribution of absorbed proteins and cell-adhesion properties. However, the role that adhesive ECM proteins coated on the nanostructured polymeric surface play in cell adhesion has not been extensively addressed. Further studies using the combined nanoscale mechanical topography and adhesive ECM nanopatterns on polymer cell-culture substrates will possibly provide more insight into how cells synergically interact with surface nanotopography and surface adhesive nanopatterns [69].

10.3.2 Nanotopography-Induced Changes in Cell Morphology and Motility

It has been found that nanotopography induces change in the morphology and motility of many different cell types, which is also called contact guidance. Contact guidance refers to the reactions of cells with the topography of their substratum [70, 71]. Although the phenomenon of contact guidance has been known for decades, the mechanism of cellular responses to substrate topography remains poorly understood [70, 71]. Current hypotheses on the mechanism of contact guidance rely on experimental observation of the dynamic behavior of cellular cytoskeletons. Arrays of parallel nanogroove (alternatively, nanoridge) patterns have been used extensively to provide an in vitro experimental model for investigating how nanotopography of the in vivo ECM can affect cell motility. Previous studies with smooth muscle cells (SMCs) [72], keratocytes [73], epithelial cells [67, 74], and neuronal cells [70, 75] showed that cells elongated and aligned along nanogrooves. Further quantitative analysis with corneal epithelial cells showed that the percentage of aligned cells was constant on the substrate topographies with lateral dimensions ranging from 70 nm to 2 μm and increased with groove depth ranging from 150 to 600 nm [74]. Interestingly, recent ultrastructural analysis showed that primary human corneal epithelial cells oriented perpendicular to 400 nm pitch patterns (70 nm ridges) extended filopodia in a crownlike fashion with individual filopodia having orientations that were perpendicular, oblique, or parallel to the underlying pattern [Figure 10.8(a, b)] [76]. In contrast, in cells that were aligned parallel to larger-scale patterns (4,000 nm pitch, 1,900 nm ridges), filopodia were generally more guided by the substrate topographies, resulting in most filopodia extending parallel to the underlying features [76].

The effect of soluble factors on cell morphology in the presence of physical stimuli by nanotopography has also been investigated. For example, the use of DMEM/F12 medium resulted in a larger percentage of epithelial cells aligning along the topographic patterns than when no serum was added to the basal medium regardless of feature size [74]. In contrast, in Epilife medium, cells aligned preferentially in the direction perpendicular to nanogrooves and ridges [76]. PC 12 cells cultured with suboptimal concentration of nerve growth factor extended more neurites on nanogroove than on flat substrates [77]. These findings suggest that alignment responses guided by nanoscale topographic anisotropies (i.e., contact guidance) may be regulated by soluble factors in the culture medium as well [73]. More specifically,

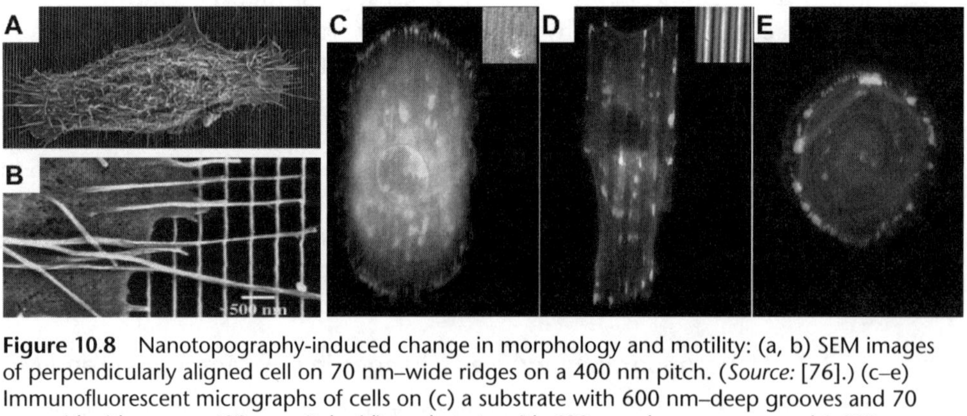

Figure 10.8 Nanotopography-induced change in morphology and motility: (a, b) SEM images of perpendicularly aligned cell on 70 nm–wide ridges on a 400 nm pitch. (*Source:* [76].) (c–e) Immunofluorescent micrographs of cells on (c) a substrate with 600 nm–deep grooves and 70 nm–wide ridges on a 400 nm pitch, (d) a substrate with 600 nm–deep grooves and 1,900 nm ridges on a 4,000 nm pitch, and (e) a smooth silicon oxide substrate stained for actin, vinculin, and the nucleus. Inset: a reflection image of the substrates. (*Source:* [74].)

the mechanical signal by contact guidance may cooperate with soluble-factor-mediated signaling pathways to guide cellular functions.

Primary neuronal cells cultured on quartz etched with a series of parallel nanogrooves showed that Xenopus neurites grew in parallel to the grooves as shallow as 14 nm and as narrow as 1 μm, while hippocampal neurites grew in parallel to deeper, wider grooves but perpendicularly to shallower and narrower ones [70, 75]. The frequency of instances of perpendicular alignment of hippocampal neurites depended on the age of the embryos from which neurons were isolated [70]. These findings suggest that substratum topography is a potent morphogenetic factor for developing central nervous system neurons, and ECM topography may play a critical role in establishing neuronal polarity and pathfinding [70]. Collectively, one may conclude that topography-induced morphogenesis (i.e., contact guidance) should be carefully interpreted in a more tissue-developmental and cell-physiological context.

Nanotopography has influence on cell migration as well as morphological change. For instance, individual SV40-transformed human corneal epithelial cells frequently migrate along grooves and ridges of a variety of pitch values, whereas the direction of migrating cells on a flat substratum is significantly more random [78]. On the biochemical level, unlike in cells on nonpatterned control surfaces, cells on the grooved substrata have their actin cytoskeletons and focal adhesions aligned with the nanogrooves supporting the strong cell polarization along the nanogroove direction [Figure 10.8(c–e)] [74, 76]. The width of the focal adhesions is determined by the width of the ridges in the underlying substrate once cells align along the nanotopographic cues [Figure 10.8(c–d)] [74].

It is also of importance to examine the spatiotemporal sequence of dynamic events involved in cell adhesion on nanogrooved substrates. Immunocytochemical analysis has provided initial insights into the temporal sequence in which microtubules, focal contacts, and actin filaments become aligned to the substratum topography as fibroblasts spread on grooved substrata [71]. According to these observations, microtubules were the first element to become oriented parallel to the direction of the grooves. Aligned actin microfilament bundles were not observed until forty to sixty minutes and were observed first at the wall-ridge edges. At early times, focal contacts were distributed radially. However, the majority of cells demonstrated aligned focal contacts after three hours. The polarization of microtubule organizing centers of SMC cultured on nanopatterned surfaces, which are associated with cell migration, also showed a preferential orientation along the axis of cell alignment [72]. Based on the assumption that the first aligned cytoskeletal component is the major determinant of cell orientation, these data suggest that microtubules in fibroblasts may determine cell orientation on grooved titanium surfaces.

10.3.3 Focal Adhesion and Integrin-Dependent Signal Transduction

Recent evidence suggests that nanopatterned surfaces with various nanotopography and nanoscale ECM ligand density can regulate focal adhesion assembly and influence cell growth and differentiation [13, 74, 79–84]. It is obvious that nanoscale biomechanical and biochemical adhesive patterns can offer a set of versatile experimental platforms with more accuracy and better controllability for the study of cell adhesion and signaling.

Integrins play a major role in adhesion-dependent signal transduction (i.e., mechanotransduction) as the major cell-ECM adhesion receptors and signal transducers for regulating cell motility, growth, and differentiation. As force transducers, integrins serve as a link between mechanical stimuli and biochemical signals. The capacity of integrins to transduce extracellular forces is mediated by the proteins localized at the interface between the receptor and the cytoskeleton, called the focal-adhesion complex. In vitro cellular studies on adhesion-dependent signal transduction and the development of new biomaterials would benefit from engineered matrices that allow modulation of the cell-adhesive signals at the scale that is commensurate with individual adhesion complexes [60, 85]. Many cell responses induced by integrins require both receptor occupancy and receptor aggregation and appear to be regulated by both biochemical and biomechanical stimuli. Several key biomechanical natures of integrin-mediated cell-matrix interactions have been observed. First, cell motility can be regulated by modulating the magnitude of cell-substratum adhesion [86, 87], cell integrin expression level [88], integrin-ECM binding affinity [88, 89], or ECM surface rigidity [90–93]. Second, integrin clustering enables the assembly of multiple cytoplasmic regulatory and structural proteins at the sites of aggregated integrin cytoplasmic domains, activating certain intracellular signaling pathways [94].

Cell biologists and bioengineers are beginning to use surface micro- and nanotopography as a noninvasive tool to investigate the dynamics of the force-dependent modulation of focal adhesions and the related adhesion-dependent signal transduction [13, 95]. Among various topographical features, elastic pillar substrates have been extensively used to probe focal-adhesion-dependent cell behavior. First, these allow migrating cells to continuously encounter localized topographic stimuli and facilitate the direct comparison of cell behavior on surfaces with or without topographic features. Second, pillars can be used to transmit mechanical forces into cellular cytoskeletons and form focal contacts locally around them. The arrays of silicone pillars (down to 2 μm in diameter) were used to measure subcellular traction force [95, 96] and modulate the focal-adhesion size that controls the tension-dependent recruitment of alpha-smooth muscle actin to stress fibers [97]. By using microcontact printing, threshold adhesive area and adhesion strength were quantitatively measured in terms of integrin binding and focal-adhesion assembly [89]. Colloidal lithography was used to quantify filopodia response to cylindrical nanocolumns (100 nm diameter, 160 nm high), showing that fibroblasts produced more filopodia per cell perimeter by interacting with surface nanotopography [13, 98]. The increased surface contact provided by topographic features guided cell migration by regulating the strength of local adhesions and contractions through a focal-adhesion kinase– and myosin II–dependent mechanism [99]. Highly defined nanopatterns arranged with cyclic arginine-glycine-aspartate (RGD) peptides were created to investigate the dynamics of cell spreading and the molecular composition of adhesion sites in relation to the fixed spacing between the peptides on the surface [83, 100]. This approach indicates that in addition to the biochemical composition, lateral spacing of integrin ligands influences cell spreading and focal-adhesion assembly [83, 100].

10.3.4 Nanotopography-Integrated Tissue Scaffolds

Tissue engineering aims at restoring or replacing damaged tissues or organs by combining functional cells, supporting scaffolds, and appropriate environmental cues, such as biochemical and biomechanical stimuli. During the past decade, considerable effort has been devoted to the development of scaffolds for tissue engineering using biodegradable or biocompatible synthetic and natural polymers. In principle, the design of the scaffolds should mimic the structure and biological function of native ECM proteins, which provide mechanical support and regulate cell activities. To construct mechanically and functionally stable tissue grafts, scaffolds must provide environmental cues that are properly recognized by cells, similarly to naturally occurring interaction between cells and ECM molecules.

Incorporation of common integrin-adhesive peptide sequence, RGD, into biomaterial scaffolds with topographic features could be beneficial for tissue engineering. Conjugated biotinylated RGD to biotinylated PEG and poly lactic acid copolymers promoted endothelial cell spreading in the polymer [101]. In addition to the simple availability of adhesion legends, the spacing of ligands is also important. By varying the nonsocial organization of RGD in alginate gel, it was found that changing the spacing between the RGD molecules alters cell spreading and proliferation [79]. A self-assembling nanofiber scaffold enabled selective differentiation of neural progenitor cells by incorporating the pentapeptide epitope iso-

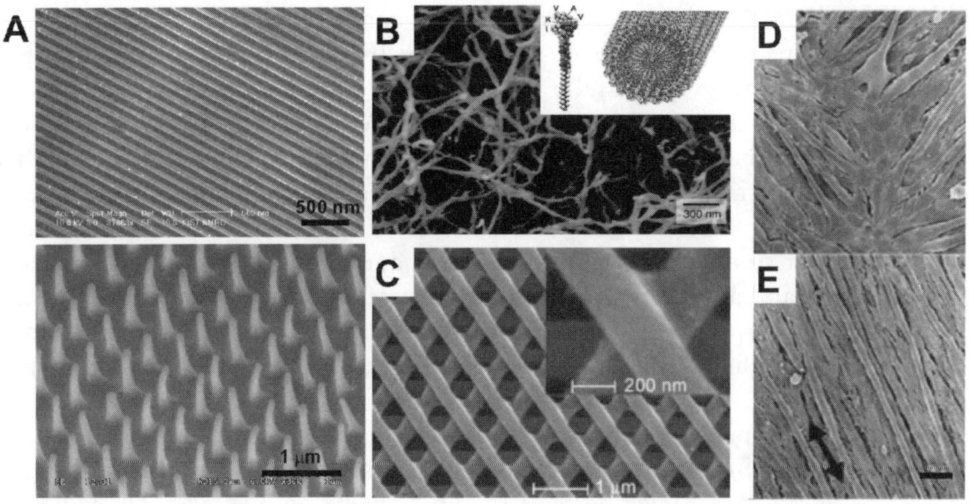

Figure 10.9 Nanofabricated scaffolds for mimicking the nanotopology of the ECM: (a) Scanning electron microscope (SEM) images of various topographically nanopatterned substrates of PEG hydrogels. (b) SEM image of a self-assembling scaffold incorporating peptide sequences (IKVAV) known to direct cell differentiation. Inset: the pentapeptide epitope IKVAV. (*Source:* [102]). (c) A three-layer polymer nanostructure created by nanoimprint lithography. (*Source:* [103].) (d, e) Myotube alignment after seven days in differentiation media on (d) randomly oriented and (e) aligned poly(L-lactic acid) nanofibrous scaffolds [104].

lucinelysine-valine-alanine-valine (IKVAV), which is known to promote neurite sprouting and to direct neurite growth [Figure 10.9(b)] [102].

Along with the biochemical modification of tissue scaffolds, physical manipulation of scaffolds allows better control over microenvironmental factors. A cardiac graft using the degradable scaffold with an ECM topography was presented for in vitro tissue engineering [105]. Microstructured ECM proteins have been shown to assist the formation of the microvessels, which is required for proper oxygen and nutrient supply for newly generated tissues [106]. Furthermore, it has been reported that microfabricated ECM proteins lead to enhanced differentiation of engineered epidermis [107, 108]. These recent findings suggest that deliberate patterning of scaffolds in microscale or even in nanoscale may facilitate tissue regeneration with appropriate functions. Nanotopographical synthetic biomaterials could be used as instructive extracellular microenvironments for morphogenesis in tissue engineering [109]. For example, nanofibrous polymers fabricated by electrospinning were employed to regulate cell and cytoskeleton alignment, myotube assembly, myotube striation, and myoblast proliferation [Figure 10.9(d–e)] [104, 110]. Structurally organized cardiomyocytes by substrate anisotropy with uniformity and continuity demonstrated functional anisotropy, which may enable systematic two-dimensional in vitro studies of macroscopic structure-related mechanisms of reentrant arrhythmias [111].

For tissue-engineering scaffolds, a simple and highly reproducible method to fabricate well-defined nanostructures with aligned nanoembosses or nanofibers of controllable aspect ratios on a large area is required to mimic the ECM environment with remarkable structural and mechanical similarity. Recent advances in nanofabrication techniques show their potential to generate tailored nanostructures in tissue scaffolds. For example, CFL-based nanofabrication techniques would be suitable for fabricating well-defined nanostructures with various shapes of biocompatible hydrogels on a large area [Figure 10.9(a)] [33]. A nanoimprinting technique was successfully employed to generate three-dimensional, multilayered, well-defined nanostructures [Figure 10.9(c)] [103]. Three-dimensional collagen fibers comparable with ECM can be also created through a technique called electrospinning, which has also been used for fabricating tissue scaffolds [112–114]. Electrospinning uses a strong electric current to force a charged polymer solution through a tiny needle and allow the resulting polymer fibers to whip back and forth in the shifting electric field and accumulate on a rotating drum. The process yields well-ordered collagen fibers about 100 nm in diameter that allow living cells, such as those from the muscle tissue, to grow properly on the matrix.

Developing tissue scaffolds is also of benefit if, in addition to varying the geometrical design of nanostructures in tissue scaffolds, mechanical properties of these structures, such as their rigidity, can be modulated. This could be simply achieved by fabricating nanostructured scaffolds using ultraviolet-curable hydrogels with different amounts of curing agent. The rigidity increases with the larger amount of curing agent. Alternatively, polymers of different molecular weight can be used. This approach is especially promising in light of the recent finding that naive mesenchymal stem cells specify lineage and commit to phenotypes with extreme sensitivity to tissue-level elasticity [115]. The results have significant implications for understanding physical effects of the in vivo microenvironment and also for therapeutic uses of nanostructured polymeric scaffolds with tissuelike rigidity [115].

10.4 Conclusion

In this chapter, we have summarized currently available methods for generating nanotopography and its potential as an experimental platform for better understanding cell-substrate interactions in cell biology and tissue-engineering applications. Previous studies with nano- and microengineered substrata demonstrated the control of cell shape and differentiation by biophysical cues. It can be now envisioned that the nanotopography-based approaches capable of guiding tissue morphogenesis have the potential to create biomimetic cell and tissue structures. Taking advantage of the strong and well-defined cellular responses to nanopatterned polymeric surfaces, microfluidic chambers could be integrated with topographically nanopatterned substrate in the near future and provide engineered niche microenviroments for embryonic stem cell proliferation and guidance of cell differentiation. Better understanding of signaling and cellular functions through the nanoscale control of cell-ECM interaction could open up novel strategies to manipulate cell locomotion, matrix assembly, and cell interactions in tissue-engineering applications.

References

[1] Meshel, A., Wei, Q., Adelstein, R., and Sheetz, M., "Basic mechanism of three-dimensional collagen fibre transport by fibroblasts," *Nat. Cell Biol.*, Vol. 7, No. 2, 2005, pp. 157–U170.

[2] Abrams, G. A., Murphy, C. J., Wang, Z. Y., Nealey, P. F., and Bjorling, D. E., "Ultrastructural basement membrane topography of the bladder epithelium," *Urol. Res.*, Vol. 31, No. 5, 2003, pp. 341–346.

[3] Abrams, G. A., Bentley, E., Nealey, P. F., and Murphy, C. J., "Electron microscopy of the canine corneal basement membranes," *Cells Tissues Organs*, Vol. 170, No. 4, 2002, pp. 251–257.

[4] Baldock, C., Koster, A., Ziese, U., Rock, M., Sherratt, M., Kadler, K., Shuttleworth, C., and Kielty, C., "The supramolecular organization of fibrillin-rich microfibrils," *J. Cell. Biol.*, Vol. 152, No. 5, 2001, pp. 1045–1056.

[5] Kielty, C., Wess, T., Haston, L., Ashworth, J., Sherratt, M., and Shuttleworth, C., "Fibrillin-rich microfibrils: Elastic biopolymers of the extracellular matrix," *J. Muscle Res. Cell Motility*, Vol. 23, Nos. 5–6, 2002, pp. 581–596.

[6] Cukierman, E., Pankov, R., Stevens, D., and Yamada, K., "Taking cell-matrix adhesions to the third dimension," *Science*, Vol. 294, No. 5547, 2001, pp. 1708–1712.

[7] Grinnell, F., "Fibroblast biology in three-dimensional collagen matrices," *Trends in Cell Biol.*, Vol. 13, No. 5, 2003, pp. 264–269.

[8] Raghavan, S., and Chen, C., "Micropatterned environments in cell biology," *Adv. Mater.*, Vol. 16, No. 15, 2004, pp. 1303–1313.

[9] Wu, N., Pease, L. F., III, and Russel, W. B., "Electric-field-induced patterns in thin polymer films: Weakly nonlinear and fully nonlinear evolution," *Langmuir*, Vol. 21, No. 26, 2005, pp. 12290–12302.

[10] Jung, D. R., Kapur, R., Adams, T., Giuliano, K. A., Mrksich, M., Craighead, H. G., and Taylor, D. L., "Topographical and physicochemical modification of material surface to enable patterning of living cells," *Crit. Rev. Biotechnol.*, Vol. 21, No. 2, 2001, pp. 111–154.

[11] **"Mechanobiology of cells and tissues. Selected papers of the Euromech Colloquium No. 420. Nancy, France, 24–26 April 2001,"** *Biorheology*, Vol. 40, Nos. 1–3, 2003, pp. 1–430.

[12] Curtis, A., "Tutorial on the biology of nanotopography," *IEEE Trans. Nanobiosci.*, Vol. 3, No. 4, 2004, pp. 293–295.

[13] Dalby, M. J., Riehle, M. O., Sutherland, D. S., Agheli, H., and Curtis, A. S., "Use of nanotopography to study mechanotransduction in fibroblasts—methods and perspectives," *Eur. J. Cell. Biol.*, Vol. 83, No. 4, 2004, pp. 159–169.

[14] Chen, Y., and Pepin, A., "Nanofabrication: Conventional and nonconventional methods," *Electrophoresis*, Vol. 22, No. 2, 2001, pp. 187–207.

[15] Smith, H., and Craighead, H., "Nanofabrication," *Physics Today*, Vol. 43, No. 2, 1990, pp. 24–30.

[16] Okazaki, S., "Resolution limits of optical lithography," *J. Vacuum Sci. Tech. B*, Vol. 9, No. 6, 1991, pp. 2829–2833.

[17] Levenson, M., "Extending optical lithography," *Solid State Tech.*, Vol. 38, No. 2, 1995, p. 57.

[18] Timp, G., *Nanotechnology*. Springer-Verlag, 1998.

[19] Ball, P., "Designing the molecular world " Vol. 1994. **[[AQ: need full citation info?]]**

[20] Emmelius, M., Pawlowski, G., and Vollmann, H., "Materials for optical-data storage," *Angewandte Chemie—International Edition in English*, Vol. 28, No. 11, 1989, pp. 1445–1471.

[21] Chou, S., Krauss, P., and Renstrom, P., "Imprint lithography with 25-nanometer resolution," *Science*, Vol. 272, No. 5258, 1996, pp. 85–87.

[22] Chou, S., Krauss, P., Zhang, W., Guo, L., and Zhuang, L., "Sub-10 nm imprint lithography and applications," *J. Vacuum Sci. Tech. B*, Vol. 15, No. 6, 1997, pp. 2897–2904.

[23] Lebib, A., Chen, Y., Carcenac, F., Cambril, E., Manin, L., Couraud, L., and Launois, H., "Tri-layer systems for nanoimprint lithography with an improved process latitude," *Microelectronic Eng.*, Vol. 53, Nos. 1–4, 2000, pp. 175–178.

[24] Khang, D., and Lee, H., "Room-temperature imprint lithography by solvent vapor treatment," *Appl. Phys. Lett.*, Vol. 76, No. 7, 2000, pp. 870–872.

[25] Khang, D., Yoon, H., and Lee, H., "Room-temperature imprint lithography," *Adv. Mater.*, Vol. 13, No. 10, 2001, pp. 749–752.

[26] Xia, Y., Rogers, J., Paul, K., and Whitesides, G., "Unconventional methods for fabricating and patterning nanostructures," *Chemical Rev.*, Vol. 99, No. 7, 1999, pp. 1823–1848.

[27] Xia, Y., and Whitesides, G., "Soft lithography," *Angewannte Chemie—International Edition*, Vol. 37, No. 5, 1998, pp. 551–575.

[28] Xia, Y., and Whitesides, G., "Soft lithography," *Ann. Rev. Mater. Sci.*, Vol. 28, 1998, pp. 153–184.

[29] Scheer, H., Schulz, H., Hoffmann, T., and Torres, C., "Problems of the nanoimprinting technique for nanometer scale pattern definition," *J. Vacuum Sci. Tech. B*, Vol. 16, No. 6, 1998, pp. 3917–3921.

[30] Suh, K., Kim, Y., and Lee, H., "Capillary force lithography," *Adv. Mater.*, Vol. 13, No. 18, 2001, pp. 1386–1389.

[31] Suh, K., and Lee, H., "Capillary force lithography: Large-area patterning, self-organization, and anisotropic dewetting," *Adv. Functional Mater.*, Vol. 12, Nos. 6–7, 2002, pp. 405–413.

[32] Kim, D., Kim, P., Song, I., Cha, J., Lee, S., Kim, B., and Suh, K., "Guided three-dimensional growth of functional cardiomyocytes on polyethylene glycol nanostructures," *Langmuir*, Vol. 22, No. 12, 2006, pp. 5419–5426.

[33] Kim, P., Kim, D., Kim, B., Choi, S., Lee, S., Khademhosseini, A., Langer, R., and Suh, K., "Fabrication of nanostructures of polyethylene glycol for applications to protein adsorption and cell adhesion," *Nanotechnology*, Vol. 16, No. 10, 2005, pp. 2420–2426.

[34] Gibson, J., "Reading and writing with electron beams," *Physics Today*, Vol. 50, No. 10, 1997, pp. 56–61.

[35] Chen, Y., Kupka, R., Rousseaux, F., Carcenac, F., Decanini, D., Ravet, M., and Launois, H., "50-nm X-ray-lithography using synchrotron-radiation," *J. Vacuum Sci. Tech. B*, Vol. 12, No. 6, 1994, pp. 3959–3964.

[36] Silverman, J., "Challenges and progress in X-ray lithography," *J. Vacuum Sci. Tech. B*, Vol. 16, No. 6, 1998, pp. 3137–3141.

[37] Piner, R., Zhu, J., Xu, F., Hong, S., and Mirkin, C., "'Dip-pen' nanolithography," *Science*, Vol. 283, No. 5402, 1999, pp. 661–663.

[38] Lee, K., Park, S., Mirkin, C., Smith, J., and Mrksich, M., "Protein nanoarrays generated by dip-pen nanolithography," *Science*, Vol. 295, No. 5560, 2002, pp. 1702–1705.

[39] Minne, S., Manalis, S., Atalar, A., and Quate, C., "Contact imaging in the atomic force microscope using a higher order flexural mode combined with a new sensor," *Appl. Phys. Lett.*, Vol. 68, No. 10, 1996, pp. 1427–1429.

[40] Kramer, N., Birk, H., Jorritsma, J., and Schonenberger, C., "Fabrication of metallic nanowires with a scanning tunneling microscope," *Appl. Phys. Lett.*, Vol. 66, No. 11, 1995, pp. 1325–1327.

[41] Dagata, J., "Device fabrication by scanned probe oxidation," *Science*, Vol. 270, No. 5242, 1995, pp. 1625–1626.

[42] Bard, A., Denuault, G., Lee, C., Mandler, D., and Wipf, D., "Scanning electrochemical microscopy—a new technique for the characterization and modification of surfaces," *Accounts Chem. Res.*, Vol. 23, No. 11, 1990, pp. 357–363.

[43] Betzig, E., and Trautman, J., "Near-field optics—microscopy, spectroscopy, and surface modification beyond the diffraction limit," *Science*, Vol. 257, No. 5067, 1992, pp. 189–195.

[44] Hanarp, P., Sutherland, D., Gold, J., and Kasemo, B., "Control of nanoparticle film structure for colloidal lithography," *Colloids and Surfaces A—Physicochemical and Engineering Aspects*, Vol. 214, Nos. 1–3, 2003, pp. 23–36.

[45] Denis, F., Hanarp, P., Sutherland, D., and Dufrene, Y., "Fabrication of nanostructured polymer surfaces using colloidal lithography and spin-coating," *Nano Lett.*, Vol. 2, No. 12, 2002, pp. 1419–1425.

[46] Marzin, J., Gerard, J., Izrael, A., Barrier, D., and Bastard, G., "Photoluminescence of single InAs quantum dots obtained by self-organized growth on GaAs," *Phys. Rev. Lett.*, Vol. 73, No. 5, 1994, pp. 716–719.

[47] Brune, H., Giovannini, M., Bromann, K., and Kern, K., "Self-organized growth of nanostructure arrays on strain-relief patterns," *Nature*, Vol. 394, No. 6692, 1998, pp. 451–453.

[48] Hawker, C., and Russell, T., "Block copolymer lithography: Merging 'bottom-up' with 'top-down' processes," *MRS Bull.*, Vol. 30, No. 12, 2005, pp. 952–966.

[49] Peng, G., Qiu, F., Ginzburg, V., Jasnow, D., and Balazs, A., "Forming supramolecular networks from nanoscale rods in binary, phase-separating mixtures," *Science*, Vol. 288, No. 5472, 2000, pp. 1802–1804.

[50] Wilson, D. L., Martin, R., Hong, S., Cronin-Golomb, M., Mirkin, C. A., and Kaplan, D. L., "Surface organization and nanopatterning of collagen by dip-pen nanolithography," *Proc. Natl. Acad. Sci. USA*, Vol. 98, No. 24, 2001, pp. 13660–13664.

[51] Jiang, F., Khairy, K., Poole, K., Howard, J., and Muller, D., "Creating nanoscopic collagen matrices using atomic force microscopy," *Microscopy Res. Technique*, Vol. 64, Nos. 5–6, 2004, pp. 435–440.

[52] Cherniavskaya, O., Chen, C., Heller, E., Sun, E., Provezano, J., Kam, L., Hone, J., Sheetz, M., and Wind, S., "Fabrication and surface chemistry of nanoscale bioarrays designed for

the study of cytoskeletal protein binding interactions and their effect on cell motility," *J. Vacuum Sci. Tech. B*, Vol. 23, No. 6, 2005, pp. 2972–2978.

[53] Zhang, G., Tanii, T., Zako, T., Hosaka, T., Miyake, T., Kanari, Y., Funatsu, T., and Ohdomari, I., "Nanoscale patterning of protein using electron beam lithography of organosilane self-assembled monolayers," *Small*, Vol. 1, Nos. 8–9, 2005, pp. 833–837.

[54] Michel, R., Reviakine, I., Sutherland, D., Fokas, C., Csucs, G., Danuser, G., Spencer, N., and Textor, M., "A novel approach to produce biologically relevant chemical patterns at the nanometer scale: Selective molecular assembly patterning combined with colloidal lithography," *Langmuir*, Vol. 18, No. 22, 2002, pp. 8580–8586.

[55] Lussi, J., Tang, C., Kuenzi, P., Staufer, U., Csucs, G., Voros, J., Danuser, G., Hubbell, J., and Textor, M., "Selective molecular assembly patterning at the nanoscale: A novel platform for producing protein patterns by electron-beam lithography on SiO2/indium tin oxide-coated glass substrates," *Nanotechnology*, Vol. 16, No. 9, 2005, pp. 1781–1786.

[56] Cheng, L., Kao, M., Meyhofer, E., and Guo, L., "Highly efficient guiding of microtubule transport with imprinted CYTOP nanotracks," *Small*, Vol. 1, No. 4, 2005, pp. 409–414.

[57] Cai, Y., and Ocko, B., "Large-scale fabrication of protein nanoarrays based on nanosphere lithography," *Langmuir*, Vol. 21, No. 20, 2005, pp. 9274–9279.

[58] Michel, R., Lussi, J., Csucs, G., Reviakine, I., Danuser, G., Ketterer, B., Hubbell, J., Textor, M., and Spencer, N., "Selective molecular assembly patterning: A new approach to micro- and nanochemical patterning of surfaces for biological applications," *Langmuir*, Vol. 18, No. 8, 2002, pp. 3281–3287.

[59] Lynch, M., Mosher, C., Huff, J., Nettikadan, S., Johnson, J., and Henderson, E., "Functional protein nanoarrays for biomarker profiling," *Proteomics*, Vol. 4, No. 6, 2004, pp. 1695–1702.

[60] Zhu, X., Mills, K. L., Peters, P. R., Bahng, J. H., Liu, E. H., Shim, J., Naruse, K., Csete, M. E., Thouless, M. D., and Takayama, S., "Fabrication of reconfigurable protein matrices by cracking," *Nat. Mater*, Vol. 4, No. 5, 2005, pp. 403–406.

[61] Mossman, K. D., Campi, G., Groves, J. T., and Dustin, M. L., "Altered TCR signaling from geometrically repatterned immunological synapses," *Science*, Vol. 310, No. 5751, 2005, pp. 1191–1193.

[62] Groves, J. T., Ulman, N., and Boxer, S. G., "Micropatterning fluid lipid bilayers on solid supports," *Science*, Vol. 275, No. 5300, 1997, pp. 651–653.

[63] Groves, J., "Molecular organization and signal transduction at intermembrane junctions," *Angewandte Chemie—International Edition*, Vol. 44, No. 23, 2005, pp. 3524–3538.

[64] Nam, J., Nair, P., Neve, R., Gray, J., and Groves, J., "A fluid membrane-based soluble ligand-display system for live-cell assays," *ChemBioChem*, Vol. 7, No. 3, 2006, pp. 436–440.

[65] Doh, J., and Irvine, D. J., "Immunological synapse arrays: Patterned protein surfaces that modulate immunological synapse structure formation in T cells," *Proc. Natl. Acad. Sci. USA*, Vol. 103, No. 15, 2006, pp. 5700–5705.

[66] Pautot, S., Lee, H., Isacoff, E., and Groves, J., "Neuronal synapse interaction reconstituted between live cells and supported lipid bilayers," *Nat. Chem. Biol.*, Vol. 1, No. 5, 2005, pp. 283–289.

[67] Karuri, N., Liliensiek, S., Teixeira, A., Abrams, G., Campbell, S., Nealey, P., and Murphy, C., "Biological length scale topography enhances cell-substratum adhesion of human corneal epithelial cells," *J. Cell Sci.*, Vol. 117, No. 15, 2004, pp. 3153–3164.

[68] Suh, K., and Jon, S., "Control over wettability of polyethylene glycol surfaces using capillary lithography," *Langmuir*, Vol. 21, No. 15, 2005, pp. 6836–6841.

[69] Charest, J., Eliason, M., Garcia, A., and King, W., "Combined microscale mechanical topography and chemical patterns on polymer cell culture substrates," *Biomaterials*, Vol. 27, No. 11, 2006, pp. 2487–2494.

[70] Rajnicek, A., Britland, S., and McCaig, C., "Contact guidance of CNS neurites on grooved quartz: Influence of groove dimensions, neuronal age and cell type," *J. Cell Sci.*, Vol. 110, 1997, pp. 2905–2913.

[71] Oakley, C., and Brunette, D. M., "The sequence of alignment of microtubules, focal contacts and actin filaments in fibroblasts spreading on smooth and grooved titanium substrata," *J. Cell. Sci.*, Vol. 106, Pt. 1, 1993, pp. 343–354.

[72] Yim, E. K., Reano, R. M., Pang, S. W., Yee, A. F., Chen, C. S., and Leong, K. W., "Nanopattern-induced changes in morphology and motility of smooth muscle cells," *Biomaterials*, Vol. 26, No. 26, 2005, pp. 5405–5413.

[73] Teixeira, A. I., Nealey, P. F., and Murphy, C. J., "Responses of human keratocytes to micro- and nanostructured substrates," *J. Biomed. Mater. Res. A*, Vol. 71, No. 3, 2004, pp. 369–376.

[74] Teixeira, A. I., Abrams, G. A., Bertics, P. J., Murphy, C. J., and Nealey, P. F., "Epithelial contact guidance on well-defined micro- and nanostructured substrates," *J. Cell. Sci.*, Vol. 116, Pt. 10, 2003, pp. 1881–1892.

[75] Rajnicek, A., and McCaig, C., "Guidance of CNS growth cones by substratum grooves and ridges: Effects of inhibitors of the cytoskeleton, calcium channels and signal transduction pathways," *J. Cell Sci.*, Vol. 110, 1997, pp. 2915–2924.

[76] Teixeira, A. I., McKie, G. A., Foley, J. D., Bertics, P. J., Nealey, P. F., and Murphy, C. J., "The effect of environmental factors on the response of human corneal epithelial cells to nanoscale substrate topography," *Biomaterials*, Vol. 27, No. 21, 2006, pp. 3945–3954.

[77] Foley, J., Grunwald, E., Nealey, P., and Murphy, C., "Cooperative modulation of neuritogenesis by PC12 cells by topography and nerve growth factor," *Biomaterials*, Vol. 26, No. 17, 2005, pp. 3639–3644.

[78] Diehl, K., Foley, J., Nealey, P., and Murphy, C., "Nanoscale topography modulates corneal epithelial cell migration," *J. Biomed. Mater. Res. Part A*, Vol. 75A, No. 3, 2005, pp. 603–611.

[79] Lee, K., Alsberg, E., Hsiong, S., Comisar, W., Linderman, J., Ziff, R., and Mooney, D., "Nanoscale adhesion ligand organization regulates osteoblast proliferation and differentiation," *Nano Lett.*, Vol. 4, No. 8, 2004, pp. 1501–1506.

[80] Maheshwari, G., Brown, G., Lauffenburger, D. A., Wells, A., and Griffith, L. G., "Cell adhesion and motility depend on nanoscale RGD clustering," *J. Cell. Sci.*, Vol. 113, Pt. 10, 2000, pp. 1677–1686.

[81] Koo, L. Y., Irvine, D. J., Mayes, A. M., Lauffenburger, D. A., and Griffith, L. G., "Co-regulation of cell adhesion by nanoscale RGD organization and mechanical stimulus," *J. Cell. Sci.*, Vol. 115, Pt. 7, 2002, pp. 1423–1433.

[82] Irvine, D. J., Hue, K. A., Mayes, A. M., and Griffith, L. G., "Simulations of cell-surface integrin binding to nanoscale-clustered adhesion ligands," *Biophys. J.*, Vol. 82, No. 1, Pt. 1, 2002, pp. 120–132.

[83] Arnold, M., Cavalcanti-Adam, E. A., Glass, R., Blummel, J., Eck, W., Kantlehner, M., Kessler, H., and Spatz, J. P., "Activation of integrin function by nanopatterned adhesive interfaces," *ChemPhysChem*, Vol. 5, No. 3, 2004, pp. 383–388.

[84] Teixeira, A., Abrams, G., Murphy, C., and Nealey, P., "Cell behavior on lithographically defined nanostructured substrates," *J. Vacuum Sci. Tech. B*, Vol. 21, No. 2, 2003, pp. 683–687.

[85] Chen, C. S., Tan, J., and Tien, J., "Mechanotransduction at cell-matrix and cell-cell contacts," *Annu. Rev. Biomed. Eng.*, Vol. 6, 2004, pp. 275–302.

[86] Powers, M., Rodriguez, R., and Griffith, L., "Cell-substratum adhesion strength as a determinant of hepatocyte aggregate morphology," *Biotechnol. Bioeng.*, Vol. 53, No. 4, 1997, pp. 415–426.

[87] Palecek, S. P., Horwitz, A. F., and Lauffenburger, D. A., "Kinetic model for integrin-mediated adhesion release during cell migration," *Ann. Biomed. Eng.*, Vol. 27, No. 2, 1999, pp. 219–235.

[88] Palecek, S. P., Loftus, J. C., Ginsberg, M. H., Lauffenburger, D. A., and Horwitz, A. F., "Integrin-ligand binding properties govern cell migration speed through cell-substratum adhesiveness," *Nature*, Vol. 385, No. 6616, 1997, pp. 537–540.

[89] Gallant, N. D., Michael, K. E., and Garcia, A. J., "Cell adhesion strengthening: Contributions of adhesive area, integrin binding, and focal adhesion assembly," *Mol. Biol. Cell*, Vol. 16, No. 9, 2005, pp. 4329–4340.

[90] Lo, C., Wang, H., Dembo, M., and Wang, Y., "Cell movement is guided by the rigidity of the substrate," *Biophys. J.*, Vol. 79, No. 1, 2000, pp. 144–152.

[91] Choquet, D., Felsenfeld, D. P., and Sheetz, M. P., "Extracellular matrix rigidity causes strengthening of integrin-cytoskeleton linkages," *Cell*, Vol. 88, No. 1, 1997, pp. 39–48.

[92] Peyton, S. R., and Putnam, A. J., "Extracellular matrix rigidity governs smooth muscle cell motility in a biphasic fashion," *J. Cell. Physiol*, Vol. 204, No. 1, 2005, pp. 198–209.

[93] Wang, Y. K., Wang, Y. H., Wang, C. Z., Sung, J. M., Chiu, W. T., Lin, S. H., Chang, Y. H., and Tang, M. J., "Rigidity of collagen fibrils controls collagen gel-induced down-regulation of focal adhesion complex proteins mediated by alpha2beta1 integrin," *J. Biol Chem*, Vol. 278, No. 24, 2003, pp. 21886–21892.

[94] Kumar, C. C., "Signaling by integrin receptors," *Oncogene*, Vol. 17, No. 11 Reviews, 1998, pp. 1365–1373.

[95] Balaban, N. Q., Schwarz, U. S., Riveline, D., Goichberg, P., Tzur, G., Sabanay, I., Mahalu, D., Safran, S., Bershadsky, A., Addadi, L., and Geiger, B., "Force and focal adhesion assembly: A close relationship studied using elastic micropatterned substrates," *Nat. Cell Biol.*, Vol. 3, No. 5, 2001, pp. 466–472.

[96] Tan, J. L., Tien, J., Pirone, D. M., Gray, D. S., Bhadriraju, K., and Chen, C. S., "Cells lying on a bed of microneedles: An approach to isolate mechanical force," *Proc. Natl. Acad. Sci. USA*, Vol. 100, No. 4, 2003, pp. 1484–1489.

[97] Goffin, J. M., Pittet, P., Csucs, G., Lussi, J. W., Meister, J. J., and Hinz, B., "Focal adhesion size controls tension-dependent recruitment of alpha-smooth muscle actin to stress fibers," *J. Cell. Biol.*, Vol. 172, No. 2, 2006, pp. 259–268.

[98] Dalby, M., Riehle, M., Sutherland, D., Agheli, H., and Curtis, A., "Changes in fibroblast morphology in response to nano-columns produced by colloidal lithography," *Biomaterials*, Vol. 25, No. 23, 2004, pp. 5415–5422.

[99] Frey, M. T., Tsai, I. Y., Russell, T. P., Hanks, S. K., and Wang, Y. L., "Cellular responses to substrate topography: Role of myosin II and focal adhesion kinase," *Biophys. J.*, Vol. 90, No. 10, 2006, pp. 3774–3782.

[100] Cavalcanti-Adam, E. A., Micoulet, A., Blummel, J., Auernheimer, J., Kessler, H., and Spatz, J. P., "Lateral spacing of integrin ligands influences cell spreading and focal adhesion assembly," *Eur. J. Cell. Biol.*, Vol. 85, Nos. 3–4, 2006, pp. 219–224.

[101] Cannizzaro, S., Padera, R., Langer, R., Rogers, R., Black, F., Davies, M., Tendler, S., and Shakesheff, K., "A novel biotinylated degradable polymer for cell-interactive applications," *Biotechnol. Bioeng.*, Vol. 58, No. 5, 1998, pp. 529–535.

[102] Silva, G. A., Czeisler, C., Niece, K. L., Beniash, E., Harrington, D. A., Kessler, J. A., and Stupp, S. I., "Selective differentiation of neural progenitor cells by high-epitope density nanofibers," *Science*, Vol. 303, No. 5662, 2004, pp. 1352–1355.

[103] Guo, L., "Recent progress in nanoimprint technology and its applications," *J. Phys. D—Appl. Phys.*, Vol. 37, No. 11, 2004, pp. R123–R141.

[104] Huang, N., Patel, S., Thakar, R., Wu, J., Hsiao, B., Chu, B., Lee, R., and Li, S., "Myotube assembly on nanofibrous and micropatterned polymers," *Nano Lett.*, Vol. 6, No. 3, 2006, pp. 537–542.

[105] Ishii, O., Shin, M., Sueda, T., and Vacanti, J. P., "In vitro tissue engineering of a cardiac graft using a degradable scaffold with an extracellular matrix-like topography," *J. Thorac. Cardiovasc. Surg.*, Vol. 130, No. 5, 2005, pp. 1358–1363.

[106] Kaihara, S., Borenstein, J., Koka, R., Lalan, S., Ochoa, E., Ravens, M., Pien, H., Cunningham, B., and Vacanti, J., "Silicon micromachining to tissue engineer branched vascular channels for liver fabrication," *Tissue Eng.*, Vol. 6, No. 2, 2000, pp. 105–117.

[107] Downing, B., Cornwell, K., Toner, M., and Pins, G., "The influence of microtextured basal lamina analog topography on keratinocyte function and epidermal organization," *J. Biomed. Mater. Res. Part A*, Vol. 72A, No. 1, 2005, pp. 47–56.

[108] Pins, G., Toner, M., and Morgan, J., "Microfabrication of an analog of the basal lamina: Biocompatible membranes with complex topographies," *Faseb J.*, Vol. 14, No. 3, 2000, pp. 593–602.

[109] Lutolf, M. P., and Hubbell, J. A., "Synthetic biomaterials as instructive extracellular microenvironments for morphogenesis in tissue engineering," *Nat. Biotechnol.*, Vol. 23, No. 1, 2005, pp. 47–55.

[110] Deutsch, J., Motlagh, D., Russell, B., and Desai, T. A., "Fabrication of microtextured membranes for cardiac myocyte attachment and orientation," *J. Biomed. Mater. Res.*, Vol. 53, No. 3, 2000, pp. 267–275.

[111] Bursac, N., Parker, K. K., Iravanian, S., and Tung, L., "Cardiomyocyte cultures with controlled macroscopic anisotropy: A model for functional electrophysiological studies of cardiac muscle," *Circ. Res.*, Vol. 91, No. 12, 2002, pp. e45–54.

[112] Matthews, J. A., Wnek, G. E., Simpson, D. G., and Bowlin, G. L., "Electrospinning of collagen nanofibers," *Biomacromolecules*, Vol. 3, No. 2, 2002, pp. 232–238.

[113] Rho, K. S., Jeong, L., Lee, G., Seo, B. M., Park, Y. J., Hong, S. D., Roh, S., Cho, J. J., Park, W. H., and Min, B. M., "Electrospinning of collagen nanofibers: Effects on the behavior of normal human keratinocytes and early-stage wound healing," *Biomaterials*, Vol. 27, No. 8, 2006, pp. 1452–1461.

[114] Chua, K., Chai, C., Lee, P., Tang, Y., Ramakrishna, S., Leong, K., and Mao, H., "Surface-aminated electrospun nanofibers enhance adhesion and expansion of human umbilical cord blood hematopoietic stem/progenitor cells," *Biomaterials*, Vol. 27, No. 36, 2006, pp. 6043–6051.

[115] Engler, A. J., Sen, S., Sweeney, H. L., and Discher, D. E., "Matrix elasticity directs stem cell lineage specification," *Cell*, Vol. 126, No. 4, 2006, pp. 677–689.

CHAPTER 11

Microfluidics for Assisted Reproductive Technologies

C. L. Bormann, M. B. Wheeler, D. J. Beebe, S. Takayama, and G. D. Smith

11.1 Introduction

Assisted reproductive technologies (ARTs) have been evolving for several decades with the development of new technologies such as global gene-expression patterning utilizing microarray technology, gene silencing using interfering ribonucleic acid (RNA) sequences, the improving efficiency of nuclear transfer/cloning, and the continuous refinement of in vitro production (IVP) of mammalian embryos, which is central to several other areas of biotechnology. IVP of mammalian embryos is utilized for treatment of human infertility [1], derivation of genetically modified animals [2], cryopreservation of gametes, and nuclear transfer [3].

Continuous refinement of IVP over the past decade has been achieved through modification of medium components, their concentrations, and supplementation of new components at specific time points. While there have been many advances in IVP in regard to efficiency and embryo quality, these embryos are still suboptimal compared to their in vivo counterparts. Differences in the quality of embryos produced in vitro versus in vivo are believed to result from stresses imposed on oocytes/embryos during the IVP procedure. These stresses begin as early as oocyte retrieval, as even a small reduction in handling temperature affects the ability of oocytes to continue through the IVP procedure [4]. Additionally, IVP procedures entail as many as twenty washes and transfers to different culture drops that potentially impose stress (change in pH, osmolarity, and temperature) on gametes and embryos [5]. Furthermore, human factors associated with IVP have the potential to introduce oocyte/embryo loss at each handling step. In the treatment of human infertility, where costs of embryos are high, loss of an oocyte or embryo at any stage could be devastating to the patient and thus damaging to the infertility clinic. In the past, scientists have worked to develop tools that will allow gamete/embryo manipulation under more in vivo–like conditions. This would reduce human error and stress imposed during IVP. Options available to formulate such tools were limited. Fortunately, over the past twenty-five years, significant developments in fabrication of micro-/nanotechnology devices have greatly impacted biology and medicine.

The integration of miniaturized devices into medicine and biology in the research arena has progressed rapidly [6–9]. The development of micro-/nanodevices has begun to enhance the study of cells at single-cell and subcellular levels in ways not previously possible. These devices may allow us to track protein production, metabolism, and additional viability markers in living cells versus current fixed-cell systems. This type of science becomes especially important when the availability of cells is limited, such as when dealing with gametes and embryos in treating human infertility. In addition, microfluidic devices could allow the development of more in vivo–like systems for cell culture [10] and embryo development [11] by precise environmental control at the microscale.

11.2 Micro-/Nanotechnology

Advancements in nanotechnology and the miniaturization of bioanalytical procedures have allowed scientists a powerful platform to study biological systems. Deoxyribonucleic acid (DNA) sequencing, polymerase chain reaction (PCR), electrophoresis, DNA separation, and cell counting and sorting are just a small list of biological systems that have begun the transformation from the macro- to the micro-/nanolevels [9]. Microfluidics is one area of advancement with great promise. Microfluidics deals with the behavior, specific control, and manipulation of microliter and nanoliter volumes of fluid. This technology first emerged in the 1990s and has enabled molecular biology procedures and single-cell-based bioassays. One example of how this technology has been applied was in development of a polydimethylsiloxane (PDMS)–based microfluidic device for the analysis of a single rat pheochromocytoma (PC12) cell. In this study, hydraulic pressure, which is believed to maintain the viability and original state of the cell, was utilized to transport the single PC12 cell to the microchamber. Once the cell was in the microchamber, nicotine was introduced into the system and microelectrodes were used to effectively monitor release of dopamine from the cell [12].

Microfluidic systems have also been designed for a unidirectional flow of oocytes, embryos, sperm, medium, and medium components [6, 13–16]. The small-scale design of these microfluidic devices permits laminar flow, characterized as parallel streams flowing without disruption between currents. When laminar flow is present, mixing between streams occurs predominately by diffusion [9]. A similar methodology could be used to expose gametes in different regions of the channels to different molecules with subcellular precision. In addition, one could develop molecular gradients to investigate the migratory events of gametes. The unique physical environment in the microchannel opens countless avenues to increase knowledge of basic embryo physiology that were not previously possible.

11.2.1 Assisted Reproductive Technologies

Currently, IVP of mammalian embryos is relatively inefficient. As mentioned, this low efficiency is due primarily to artificial stresses imposed on gametes and embryos in their in vitro environment. In a single IVP run, oocytes/embryos are moved as many as twenty times between media, requiring removal from their incubator environment. This manipulation introduces stresses such as changes in temperature, pH, osmolarity, and mechanical stress. Modification of IVP systems has reduced biochemical stress to gametes and embryos; however, most of these modifications involve manipulation of media and its composition. Relatively little attention has been paid to improving the mechanical platform on which IVP occurs.

11.2.2 In Vitro Oocyte Maturation

Oocyte maturation is the first step of IVP and arguably the most important. Oocyte maturation can be divided into two independent events: (1) nuclear maturation (progression to metaphase II (MII); and (2) cytoplasmic maturation (preparation of the cytoplasm to support development to MII, subsequent fertilization, and embryo development). Given that embryo viability closely reflects oocyte quality, it is clear that optimal oocyte maturation is essential for successful preimplantation development.

In vitro maturation (IVM) of oocytes is an emerging technology that holds great potential for the treatment of human infertility (see [17–20]). Oocyte maturation may reduce the need for ovarian hormone stimulation, limiting the risk of ovarian hyperstimulation and reducing the cost associated with in vitro fertilization (IVF) [21]. Additionally, IVM is an alternative to cancellation of the treatment cycle for patients with a poor response to ovarian stimulation [22, 23]. Furthermore, IVM may be a useful tool coupled with oocyte, follicle, or ovarian tissue cryopreservation for preserving fertility in women undergoing gonadotoxic chemo- or radiation therapy for cancers. Current human IVM is relatively inefficient, with many questions regarding the safety of this technique [24–26]. Therefore, it is important to develop a culture system that adequately supports nuclear and cytoplasmic maturation of oocytes.

It has been hypothesized that in vitro maturation of immature oocytes in a microfluidic environment may more closely resemble oocyte maturation conditions in vivo and, thus, benefit oocyte maturation and subsequent embryo development. This is especially important as it has been shown that events during oocyte maturation affect subsequent embryonic development [27–29]. Two experiments were performed to determine if microchannels supported IVM of porcine oocytes. Walters, Beebe, and Wheeler reported that, despite inferior cumulus expansion, PDMS microchannels did not inhibit oocyte nuclear maturation, (71 percent compared to control cultured oocytes at 61 percent) [30]. This study was quickly followed by a study from Hester et al., which showed a significant increase ($P < 0.05$) in embryos cleaved from total oocytes matured in PDMS/borosilicate microchannels (67 percent) compared to controls (49 percent) [31]. This improvement in embryo cleavage is thought to result from improved cytoplasmic maturation. The exact size and configuration of the microchannels used in these studies were not given [6]. Nevertheless, IVM is still an evolving technique in assisted reproduction. These initial experiments on oocyte maturation using microfluidics are promising and may provide a more supportive in vivo–like culture environment for maturing human oocytes in the future.

11.2.3 Sperm Selection and Isolation

Approximately 10 percent of all couples have infertility problems, of which nearly half are male-factor related [32]. In vitro fertilization was originally developed as a treatment for tubal infertility; however, the high rates of males with poor semen quality have led to the development of several preparation techniques to isolate motile sperm. Under in vivo conditions, potentially fertile spermatozoa are separated from immotile spermatozoa, debris, and seminal plasma in the female genital tract by active migration through the cervical mucus [33]. Several techniques have been developed to artificially isolate motile sperm from an ejaculate. The three most commonly used techniques include density gradient centrifugation, sperm migration or "swim-up," and simple dilution and washing [34]. Both sperm migration and gradient centrifugation result in recovery of motile sperm; however, several groups have raised concern as to whether these methods of sperm isolation damage sperm DNA or impair motile sperm by exposing them to reactive oxygen species [35–37]. It has been shown that increased levels of sperm DNA damage during IVF correlate with reduction in embryo morphology at early cleavage stages [38], failure to advance to the blastocyst stage in vitro [39, 40], decreased pregnancy rates [41–43], and an increase in spontaneous abortions [44]. Furthermore, these separation techniques are suboptimal for patients with severe oligozoospermia, as samples from many of these patients have large amounts of debris, and recovery rates from initial sperm samples have been reported to be as low as 0.8 percent for direct sperm migration [45, 46].

Compromised spermatozoa quality or number during IVF results in poor and sometimes failed fertilization. In these cases, fertilization by injecting a single spermatozoon into the ooplasm of a mature oocyte is possible and referred to as introcytoplasmic sperm injection (ICSI). The first successful pregnancies in humans by ICSI were achieved in the early 1990s [47], and it has since become a widely accepted procedure to treat severe male infertility. Although only a single healthy sperm is required for ICSI, processing, isolating, and selecting the most viable sperm can be challenging with current sperm-separation techniques. Importantly, one can recognize that ICSI is an invasive technique and circumvents processes of natural fertilization [48]. Because of its invasiveness, even in fully competent hands, this technique can result in lysis and death of 5 to 10 percent of injected oocytes [49, 50].

Because direct sperm injection into the oocyte bypasses all normal sperm-selection processes, it is imperative to have an efficient artificial selection process that minimizes stress and damage to sperm and maximizes the probability of successful pregnancy, birth, and healthy offspring [48]. Therefore, separation of sperm using microfluidics has been investigated as a possible alternative to traditional sperm-isolation techniques.

Development of a microfluidic device for sperm characterization was first developed by Kricka et al. using fabricated silicon and glass microfluidic devices [51]. In this study, sperm motility was assessed along the length of a network of branching microchannels (80 μm wide by 20 μm deep). The authors demonstrated that the amount of time it took for sperm to transverse the microchannel correlated with forward progression scores [52]. This method of sperm isolation provides a good measure of motility and forward progression but does not give reliable information on sperm concentration or the percentage of sperm motility.

Recent advances in microfabrication (e.g., soft lithography) have made it possible to easily construct any desired microstructures [53, 54]. Using these techniques, Cho et al. developed a novel gravity-driven pumping system to sort sperm samples

[13]. The device, termed a microscale integrated sperm sorter (MISS), contains inlet/outlet ports, fluid reservoirs, gravity-driven power sources, and converging microchannels with laminar flow, all integrated components working together to facilitate sperm sorting (see Figure 11.1) [13]. The premise for this design was based on a natural phenomenon of parallel laminar flow characteristic of fluid movement at the microscale [16]. Laminar flow occurs when the Reynolds number (Re) is below 2,300. This number is a result of a calculation based on fluid density, velocity and viscosity, and hydraulic diameter. As the Re approaches 2,300, the fluid becomes less steady, and above this point, the current becomes turbulent [55]. The orientation, geometry, and size of the MISS reservoirs were designed to balance gravitational and surface-tension forces and to provide a pumping system, free of external energy, that generates steady flow rates over extended periods [13]. This device was designed so that converging stream of semen and media would flow in parallel, in a laminar fashion within a microchannel. The two parallel streams only mix by diffusion at the interface between streams, but motile sperm are able to swim across the contacting streamline and into the media for collection. Nonmotile sperm, cellular debris, and seminal plasma do not cross this barrier and are shuttled into a waste reservoir.

Testing of this laminar-flow sorting system was performed in two steps. First, 40 μl of unprocessed human semen were processed; second, semen samples were artificially filled with debris using a stock solution of round immature germ cells to stimulate poor-quality samples. For unprocessed semen, the device consistently produced a sorted fraction with increased motility (mean 98 percent motile) and

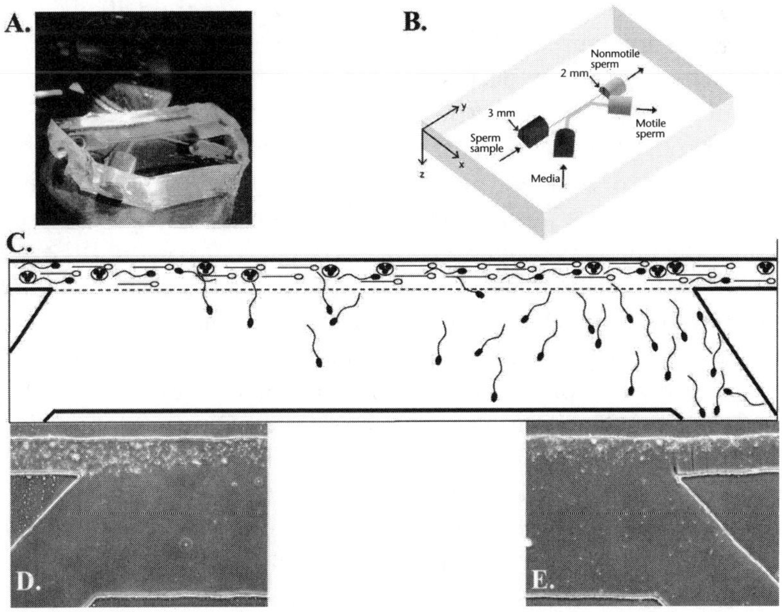

Figure 11.1 (a) Picture of microfluidic sperm sorter. (b) Three-dimensional view of microfluidic sperm sorter. (c) Theoretical vision for microfluidic sperm sorter. Media flows from left to right. (c, d) Semen sample is loaded into the upper stream inlet, and fresh media is placed in the lower stream inlet. (c, e) Motile sperm are able to deviate from the initial streamline and cross the interface of the laminar-flow streams, exiting into the lower stream outlet for recovery. Debris, nonmotile sperm, and some motile sperm are collected in the upper stream outlet. (Images reproduced with permission from [13].)

improved strict sperm morphology (mean 22 percent normal forms) versus the initial specimen (mean 44 and 10 percent, respectively). For debris-filled samples, the device not only concentrated motile sperm (mean 98 percent motile) within the collected fraction but also produced a round cell: sperm ratio of 1:33 compared with a 10:1 ratio in the starting specimen [16]. This device provides a simple, safe method of obtaining motile sperm of enriched normal morphology from both unprocessed normal semen and poor-quality specimens containing significant debris.

One of the limitations of the current microfluidic device is that it cannot process an entire sample due to its small capacity. In addition, it does not isolate every motile sperm. However, isolation of small samples of motile sperm is fast and easy and can be readily used for isolating sperm for ICSI, insemination in microdrops under oil, or for insemination in a microfluidic environment. Additional modifications of the current system are underway, which may allow for large-scale processing of semen [16].

11.2.4 In Vitro Fertilization

Once mature oocytes and quality sperm are obtained, the next step in IVP is insemination. Species differences exist in the relative difficulty of achieving IVF. This is especially apparent in animal systems, such as the pig, where polyspermy occurs at a high frequency. This is one of the reasons that pig IVP has the reputation of being difficult to perform. Furthermore, large lab-to-lab variations are seen in the blastocyst formation (25 to 30 percent) of IVP pig embryos. In general, IVF involves coincubation of oocytes with an appropriate concentration of motile and morphologically normal sperm. Media volumes may range from 20 μl microdrops to 1 ml in dishes at a minimum concentration of 500,000 sperm/ml [34]. The volume of media in which insemination occurs varies between laboratories and between total numbers of motile sperm collected from an individual. Because traditional insemination in a microdrop requires such a high number of sperm, which then deplete metabolic substrates and produce potentially harmful waste products in the IVF media [56, 57], there is a push to develop new technologies for in vitro insemination to reduce the numbers of sperm needed to achieve successful IVF.

Several expendables have been developed for insemination, such as the climbing-over-a-wall (COW) method used by Funahashi, Fujiwara, and Nagai [58] and straw IVF [59]. In the COW method, different concentrations of frozen-thawed boar sperm were placed into an outer chamber filled with modified tris-buffered medium for the fertilization of porcine oocytes. Oocytes were positioned in the inner well of the COW chamber, in which only motile sperm could swim over the wall and reach oocytes for fertilization. This experiment was designed to select the highest-quality sperm in terms of progressive motility and viability, which more mimics the in vivo system of the reproductive tract. The COW-IVF method resulted in higher penetration rates at increased sperm concentrations of 0.5, 1.0, and 5.0×10^5 cells/ml while still maintaining monospermic penetration rates comparable to the traditional IVF microdrop system [58].

The straw IVF method has also been developed to reduce the incidence of polyspermy and maintain penetration rates [59]. The straw method utilizes 0.25 ml straws with oocytes in medium (~5 cm length) at one end of the straw and an adjacent 1 cm section of medium containing spermatozoa. In this set-up, sperm must swim to the oocytes in order to initiate fertilization. The straw method reduces the number of sperm at the site of fertilization by selecting the highest-quality sperm [59]. Both the COW and straw IVF methods limit the total number of sperm reaching the egg, enriching the population for those with the best motility. Similar techniques have also been applied to human IVF with some success [60–62]. However, these techniques are time-consuming and have not gained widespread acceptance.

Following natural sperm insemination in vivo, sperm must transverse obstacles in the female reproductive tract that select for the highest-quality spermatozoa. The oviduct acts as a reservoir for sperm and regulates their passage to the site of fertilization in the ampullary-isthmus junction [63]. Advancements in microfluidics and soft-lithography have made it possible to more closely mimic this environment in vitro. Research by Clark, Beebe, and Wheeler investigated the ability of a PDMS/borosilicate microchannel to support porcine IVF compared with traditional porcine IVF methods [64]. In this study, there were no statistical differences in percentage fertilization or early embryo cleavage between the two fertilization systems. These data suggest that microchannels can support porcine IVF.

More recently, Clark et al. studied porcine fertilization in a microchannel designed to recapitulate the function and structure of the oviduct (see Figure 11.2)

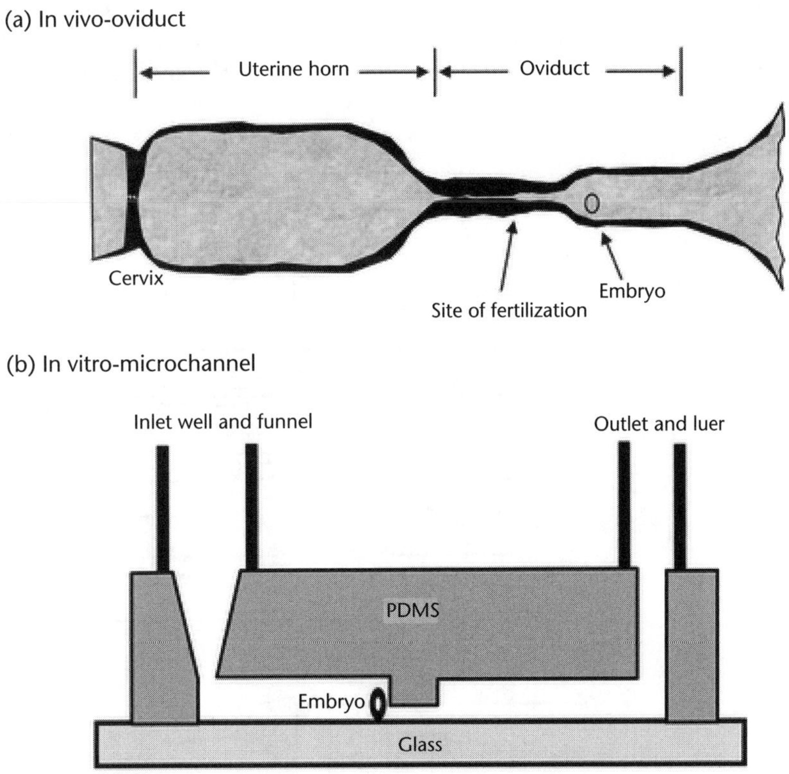

Figure 11.2 A microchannel device attempting to mimic the in vivo physioanatomical function attributed of the oviduct: (a) In vivo the sperm is guided to the oocyte via a narrowing fluid path in the oviduct. (b) The microchannel design recapitulates the in vivo function. The microchannel routes the sperm to the "parked" oocyte.

[65]. In this study, oocytes in microchannels had a significantly higher incidence of monospermic penetration ($P < 0.05$) as compared to oocytes fertilized in a traditional microdrop system, with comparable penetration and male pronuclear formation rates [65]. These results demonstrate that the biomimetic microchannel IVF system can reduce polyspermy and thereby increase the number of potentially viable embryos without reducing overall IVP efficiency.

It has been well demonstrated that only a few hundred sperm eventually reach the ampulla of the oviduct for fertilization in humans [66, 67]. Advancements in microchannel design and microfluidics may ultimately provide an alternative method of in vivo–like insemination by decreasing the volume of media and numbers of sperm needed, while increasing oocyte-sperm interaction in a dynamic environment. Suh et al. demonstrated that mouse IVF can be conducted quickly and successfully within microfluidic channels [68] as was done in Figure 11.2. In this experiment, a PDMS-based microfluidic insemination channel was fabricated consisting of a single micro channel (500 μm wide × 3,180 μm deep), separated by a barrier grate manufactured in three dimensions (see Figure 11.3). This grate permits unimpeded flow of sperm and media through the microchannel but prevents migration of the oocyte [68]. Fertilization of mouse oocytes has been achieved within this device at a sperm concentration of 1×10^{-6} (standard recommendation of mouse IVF) [69]. Furthermore, fertilization rates were significantly increased in microchannels, compared to conventional means of insemination, when the concentration of sperm used for insemination was reduced [68]. Therefore, not only are fewer sperm required for fertilization because of the small amounts of media needed

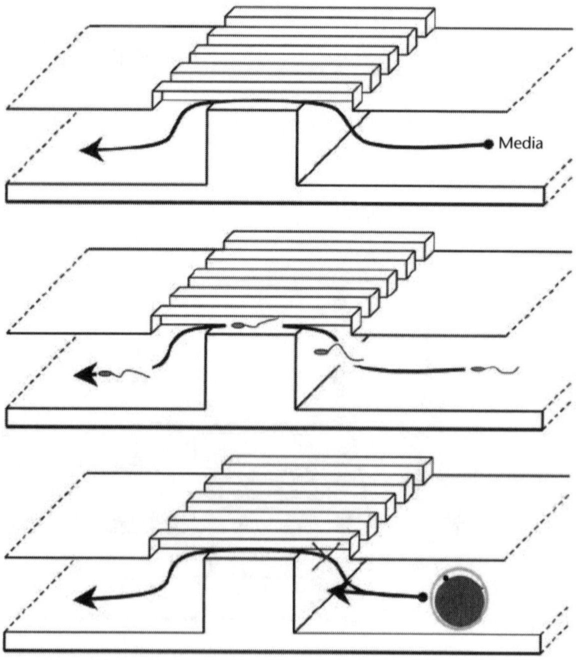

Figure 11.3 Schematic of a microfluidic device consisting of a microchannel for coincubation of sperm and oocytes for IVF. The three-dimensional construct of the channel prevents movement of the oocyte beyond the grate but allows for flow of medium and sperm through the channel. (Images were reproduced with permission from [68])

for microfluidic culture, but a lower concentration of sperm is also required, further reducing the number of sperm needed for successful fertilization [68]. This increase in fertilization is thought to be due to an increase in the local concentration and interaction of sperm and oocytes, which may result in optimization of sperm and oocyte cofactors. Furthermore, the microchannel itself may act as a guide to transport sperm directly to the oocyte. The use of microfluidics and flow results in a predictable delivery of sperm to oocytes and may reduce the randomness of sperm and oocyte interaction that takes place in larger volumes during standard IVF [68].

Advancements in the design and development of microfluidic chambers that support insemination with low numbers of sperm would not only greatly benefit individuals with oligozoospermia by reducing their need for ICSI but also serve as a replacement for standard insemination procedures, reducing the likelihood of polyspermy. Furthermore, this method may be useful when inseminating with sexed semen [47] and/or samples with compromised sperm numbers due to cryopreservation and thawing.

11.2.5 Manipulation of Oocytes/Embryos Using Microfluidics

Following insemination, cumulus-cell removal (CR) is a step in the IVP procedure that occurs at different times depending on species. Removal of the cumulus cells is important for determination of fertilization and/or embryo development. Currently, there are several efficient methods for removing cumulus cells, including vortexing, pipetting, and chemical exposure. Zeringue et al. developed a microchannel with interconnected smaller channels that can be connected with syringes, which provide a pressure-driven system, as well as a conditioning region for cumulus-oocyte-complexes (Figure 11.4) [70]. The conditioning region is only slightly larger than the oocyte, which flattens the cumulus cells, allowing for ease of cumulus-cell removal. Suction applied to the small channel forces the oocyte to move around the opening of the channel, which ultimately forces removal of cumulus cells. Zeringue et al. investigated the development of IVP-produced bovine embryos after removal of cumulus cells in microchannels compared to the standard vortexing system. In initial work, the cumulus-cell-removal device was not detrimental to development of presumptive bovine embryos [70]. Subsequent studies showed that the microfluidic CR (μFCR) device increased development on day two (20 ± 4 percent to 35 ± 6 percent, $P < 0.01$) and blastocyst formation at day eight (33 ± 1 percent to 57 ± 5 percent, $P < 0.01$) when compared to vortex CR. The effects of vortexing on embryo development were studied using 15, 30, and 120 s vortex doses. Development at day two was inversely proportional to the duration of vortexing. In order to understand the potential mechanism of the improved embryo development, an in situ transcription assay was used to assess the biochemical activity of zygotes after cumulus removal. There was a spike of RNA transcription in the vortexed zygotes at two hours after CR that was not seen in the microfluidic treatment. This spike pattern of transcription activity between μFCR and vortex CR is similar to the transcription activity patterns reported from in vivo–derived and in vitro–cultured mouse embryos [71]. In the mouse, a spike of heat-shock protein 70 (HSP70), a repair protein, was seen at the two-cell stage of in vitro–cultured embryos when compared with in vivo–derived embryos. Production of repair pro-

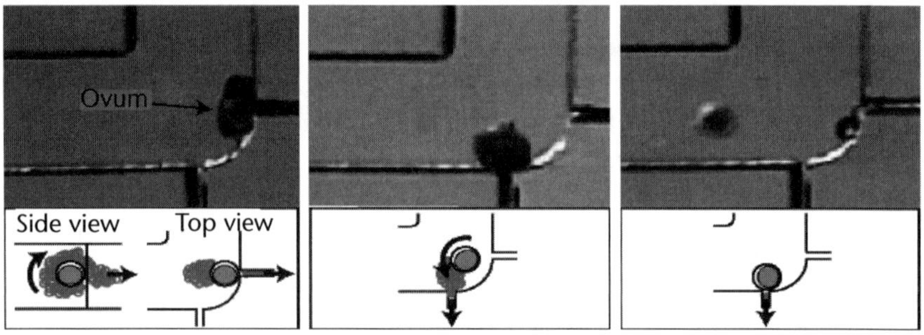

(a) (b) (c)

Figure 11.4 These pictures illustrate the three main stages of the cumulus-removal process. Insets show a schematic representation of the picture. (a) The cumulus-oocyte complex (COC) is at removal port 1. Inset shows top (left) and side (right) views of the cumulus being sucked into the removal port. (b) The cumulus rotates as it moves from port 1 to port 2. (c) Remaining cumulus is removed at port 2, leaving a clean zygote.

teins is consistent with the notion of vortex-induced damage. In a very preliminary study, with a small number of in vitro–produced cattle embryos, cumulus cells were removed with aid of hand stripping (standard method as a control) or a microfluidic device (μFD). The data have not been statistically analyzed, but there were more grade one blastocysts in the μFD (n = 58) than in the hand-stripped group (n = 37). The raw pregnancy rates for the two groups were 24.3 percent for the hand-stripped group and 36.2 percent for the μFD group (Reeder et al., unpublished). These results suggest the potential for microfluidic methods to enhance production efficiencies while providing insight into basic developmental mechanisms.

11.2.6 Embryo Culture

In vivo preimplantation embryos develop in the absence of direct cell contact with the reproductive tract for approximately five to seven days (mouse, humans, bovine) before implanting in the uterus. They are free-floating, lack a blood supply, and move continuously within the female reproductive tract through a changing fluid environment. During this time, the embryo is undergoing cell divisions and differentiation. Preimplantation embryos' nutrition is derived from maternally derived product substrates obtained from luminal secretions of the oviduct and uterus. Its cellular activities, including division, gene expression, and metabolism, are influenced by the environment and factors produced by the embryo itself and cells of the reproductive tract.

Success rates for IVF may be compromised by the limitations of current embryo-culture methods, resulting in impaired embryo development and decreased viability [34]. Improvements in embryo-culture methods have been based on simulating in vivo conditions and meeting the ever-changing needs of the embryo. Research into sequential media methods of embryo culture takes this hypothesis one step further [73]. Conventional embryo-culture systems, however, mostly employ a single medium, or double media system, from the time of insemination until embryo transfer. In vivo, the embryo is bathed in a constantly changing environment as it moves through the oviduct to the uterus. Conditions required by the embryo at one point of development may be detrimental at later stages. Special media have been developed for sequential embryo culture.

Current IVP technologies are not ideal for sequential media embryo-culture systems. Embryos must be handled during media changes, and environments are voluminous compared to in vivo conditions that may temper the presence of embryotrophic factors. Microfluidics is well suited to meet the needs of embryo culture. Changing of media is straightforward, and no manipulation of the embryo is necessary. The media can be gradually changed around the embryo, rather than subjecting it to sudden changes in the microenvironment. In theory, the microenvironment that exists within microfluidic devices might simulate physiological conditions and therefore also may benefit embryo development [6].

Thus far, there have been encouraging outcomes with microfluidic embryo-culture technology. Glasgow et al. first established that manipulation and movement of an embryo in a microfluidic environment is possible at low flow rates [74]. In this experiment, microchannels were tested and designed with constriction areas that prevented movement of the embryo down the channel while allowing for flow of media around and past the embryo. No deformation or injury to the embryo was noted at the low flow rates used [74]. Once the feasibility of embryo handling and media flow within a microchannel was established, mouse embryos were cultured and development rates compared [6]. Two-cell embryos were collected from superovulated mice. Controls were conducted in culture dishes in 30 μl microdrops, and the remainder was loaded into a microfluidic device prefilled with the same media. A significantly higher percentage of embryos within the microfluidic channels reached the morula, blastocyst, and hatched-blastocyst stages compared with controls. In addition, early embryonic development was investigated to determine if PDMS/borosilicate microchannels under static conditions could support development of in vivo–derived four-cell porcine embryos to blastocysts. There was no difference ($P > 0.05$) in the blastocyst formation in microchannels (79 percent) compared to control drops (84 percent) [75]. In addition, some microchannel-produced blastocysts were transferred to a recipient and produced five live piglets. All five animals had average birth and weaning weights and subsequently went on to reproduce normally [75]. Though preliminary, this study suggests the ability of embryo culture in microfluidic environments to produce embryos capable of transfer, implantation, and production of live births.

In 2004, experiments were conducted to determine if microchannel culture could support development of two-cell mouse embryos. Initially, three different prototypes were used: (1) one composed of entirely PDMS (complete PDMS); (2) a PDMS channel and borosilicate glass bottom; and (3) a silicon wafer/borosilicate sandwich [76]. Static embryo-culture experiments revealed that the three prototype microchannels could support early mouse embryonic development to blastocyst using a hybrid strain of mice [76]. For all stages of development and different time points, there was no significant difference between the control and the complete PDMS microchannels ($P > 0.05$). Using PDMS:borosilicate glass microchannels, it was reported that there were significantly more blastocysts ($P > 0.01$) in the microchannels than in controls following seventy-two hours of culture (73 percent and 43 percent, respectively) [76]. Using the third prototype (silicon wafer/

borosilicate sandwich), it was demonstrated that the percentage of blastocysts was also increased compared to the control, and the kinetics of development was more in vivo–like. Birth of live offspring was achieved following culture in microchannels [75].

Similarly, Cabrera et al. demonstrated an advantage of culturing mouse embryos on a microfluidic platform with dynamic media flow [77]. These experiments utilized a computer-controlled, integrated microfluidic system with on-chip pumps and valves powered by individually actuated Braille pins to pump fresh media through a series of elastomeric capillaries (see Figure 11.5) [78] as was done in Figure 11.2. This system takes advantage of the resilient, yet elastic, nature of PDMS microchannels fabricated with soft lithography, together with the movement of Braille pins to "squeeze" fluid through channels. When synchronized to various valving patterns, each stroke of a Braille pin generates forward or backward flow of media through microchannels. Results from these experiments showed embryos cultured under dynamic conditions had significantly more cells than those cultured under static conditions. Furthermore, blastocyst cell counts following microfluidic dynamic embryo culture more closely mirrored results obtained from their in vivo counterparts [77]. These results suggest that development of embryos in a dynamic microenvironment may closely mimic in vivo embryo growth conditions. Using the same culture device design, Bormann et al. showed a significant improvement in the

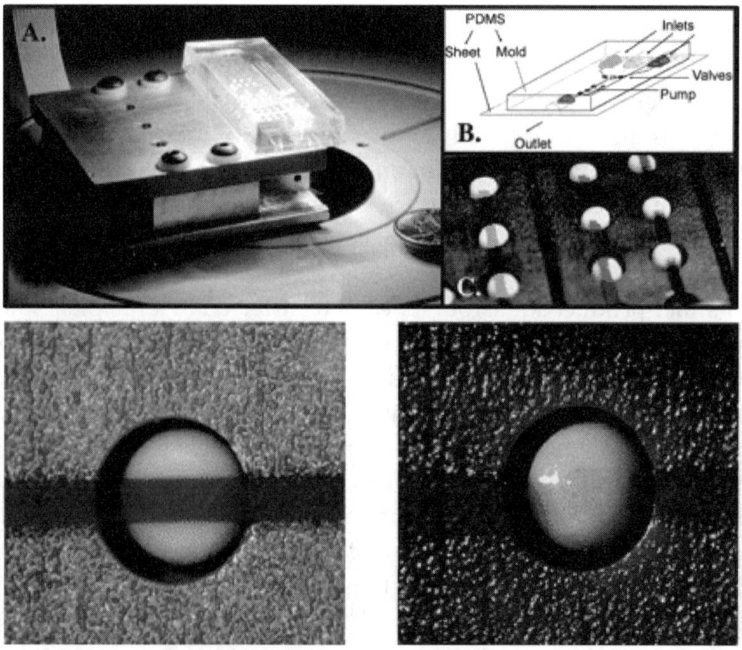

Figure 11.5 Braille displays: (a) A handheld, battery-operated, refreshable Braille display that can be used in combination with (b) elastomeric microfluidic chips for valving and actuation. (c) An array of Braille pins positioned in sequence under a microfluidic chip with media of various colors. Where a microchannel of media is interrupted (white Braille pinhead is completely visible), the pin is depressed up onto the PDMS sheet, which subsequently depresses into the channel to interrupt the colored media. (d) When the Braille pin pushed against the PDMS-parylene-PDMS hybrid membrane, the channel was fully closed. (e) When the pin was released, the membrane was restored, and channel was opened.

percentage of IVF bovine embryos reaching the blastocyst stage compared to static controls [79]. These results further demonstrate the importance of the physical and mechanical environment on embryo development, as well as the potentially wide application of this technology in IVP. These studies are currently being followed with a series of experiments looking at implantation rates, fetal development, term delivery, global gene expression, and epigenetic imprinting patterns in embryos cultured in dynamic and static microchannel culture environments.

It is clear from these early reports that microfluidic technology is beneficial in IVP of embryos. One major difference between static microfluidic devices and traditional microdrop systems is that the embryo-to-volume ratio is significantly decreased. This lower ratio means that the diffusion of medium constituents and embryo products is more tightly constrained in the microchannel as compared to in a microdrop. In the microdrop system, the volume of medium can range from 50 to 950 μL [11, 80, 81]. In some microfluidic culture systems presented in this review, there is approximately 0.125 μl of medium surrounding oocytes and embryos [6]. This small volume is comparable to the nanoliter quantity found within the lumen of the reproductive tract [82]. Because the volume is significantly reduced in the microchannel, the embryo may exert greater influence on the local environment compared to larger volumes. This embryo-derived microenvironment may prove beneficial for development under culture conditions.

11.2.7 Integrated IVP Microfluidic Devices

One of the greatest advantages of using microfluidics is the ability to integrate various aspects of IVP and cell-monitoring systems on a single chip. Clark et al. reported the integration of two different aspects of IVP (maturation and fertilization) in a single microchannel with porcine oocytes [83]. In this integrated system, oocytes were left in place at a constriction region, while the surrounding medium was removed, and fresh medium was added. The procedure was performed initially under a stereo microscope to ensure oocytes were not lost in transition from one medium to another. There was no statistical difference in the percentage of cleaved embryos between the integrated system and control microdrops [83]. This is one of the first studies to combine multiple aspects of IVP on a single chip. Additional research is needed in design and development of a platform that can meet all the needs of IVP on a single chip. This will not only more closely mirror the maternal micro- environment but will eliminate potential loss of oocytes/embryos by handling and human error.

Integration of various detection devices has already been incorporated in somatic cell culture using microfluidics. Cheng et al. demonstrated the ability to monitor Ca^{2+} transients, extracellular pH, and relative amounts of intra- and extracellular lactate production from a single cardiomyocyte [84]. Furthermore, Shackman et al. developed a novel system for high-resolution monitoring of insulin secretion from single islets of Langerhans [85], and Mehta et al. developed a method to measure oxygen content in real time in microfluidic perfusion systems [86]. Such assays, if applied to a microfluidic platform for gamete/embryo culture and analysis, could be used to develop monitoring systems for culture-condition optimization and real-time selection of embryos with the greatest implantation potential.

11.2.8 Embryo Bioanalyses

One of the greatest limiting factors in IVF is selecting the most viable embryo for transfer from a pool of visually similar embryos. It has been an accepted practice in most human clinics to transfer embryos either on day three (approximately the eight-cell stage) or on days five and six (the blastocyst stages). It is difficult to predict which day three embryos will continue to develop to the blastocyst stage. In the human, this developmental stage is also the point of maternal-to-embryonic genomic transition in which a high percentage of embryos undergo developmental arrest. Therefore, many clinicians transfer multiple embryos to obtain acceptable pregnancy rates. To reduce the number of multiple pregnancies, many clinics are beginning to move toward single-blastocyst transfer [87], making embryo selection even more critical. Currently, embryo selection is based primarily on morphological characteristics, such as cleavage rate, cell number, blastomere size, and degree of fragmentation. These methods of selection are highly subjective and lack predictive power.

One of the most reliable indicators of embryo viability is the rate at which an embryo consumes glucose or pyruvate from the culture medium and produces lactate and other metabolic waste products. Lane and Gardner showed that embryos with glycolytic activity similar to their in vivo counterparts have a fourfold increase in pregnancy rate compared to embryos selected and transferred at random [88]. This study further demonstrates the strength and reliability of using metabolism as an indicator of embryo viability. Recently, Bormann et al. developed a metabolic assay that can be readily integrated into a microfluidic platform [89]. This will allow noninvasive quantification of metabolism by individual preimplantation embryo. This assay, along with others currently being developed, may soon be used to assess embryo quality and viability.

Another measure of embryo viability may be proteins secreted by the developing embryo. The embryo secretome is defined as a class of proteins secreted from cells into extracellular media. This represents a major class of molecules involved in intercellular communication. Over the past decade, several autocrine and paracrine factors secreted by embryos essential for development and implantation have been described (reviewed by [90]).

Advancements in proteomics have made it possible to identify minute quantities of secreted peptides from preimplantation-stage embryos. Using Micro Liquid Chromatography Electrospray Ionization Mass Spectrometry and Matrix Assisted Laser Desorption Ionization Time-of-Flight Mass Spectrometry, Katz-Jaffe, Gardner, and Schoolcraft analyzed the proteome of individual human embryos and related protein expression profiles to morphology [91, 92]. These types of studies provide the basis for development of noninvasive assays of embryo viability in human IVF. One can envision that the determination of protein profiles at preimplantation stages of development may make it possible to predict which embryos are most likely to continue to develop and have the highest chance of initiating a healthy pregnancy.

Protein profiling has many benefits and potential applications in assisted reproduction. By measuring an embryo's secretome, it may be possible in the future to identify certain markers for male and female embryos and determine the sex of the developing embryo without using invasive approaches. This can be an important diagnostic tool in circumventing X-linked congenital disorders. Furthermore, another benefit to this noninvasive approach to profiling developing embryos would be to detect embryos with chromosome abnormalities or other health issues. Recently, Katz-Jaffe et al. described a relationship between the embryonic secretome and chromosomal abnormalities in human IVF [93]. Further advances in protein profiling would not just enable selection of the best-quality embryo but may also allow identification of abnormal embryos without having to perform invasive procedures, such as embryo biopsy and preimplantation genetic screening.

11.2.9 Microfluidic Platform Construction

The first step in fabrication of microfluidic devices is testing of materials to determine biocompatibility. Several potential materials have been analyzed for biocompatibility. Chan et al. reported that the rate of mouse two-cell embryo development to the blastocyst stage depended heavily on the materials on which embryos were cultured [94]. They reported that material such as silicon nitrate, silicon oxide, and borosilicate glass did not inhibit this development. Polymers PI2611 and PI2721 also did not inhibit embryonic development, suggesting that these materials could be used for development of a microfluidic system. It was further reported that embryonic development was not inhibited when embryos were cultured on chromium, gold, and titanium compared to control embryos [95]. Biocompatibility of adhesive material (RTV 118 and RTV 108) used in fabrication of microchannel devices is also important. It appears that RTV 118 is a suitable device adhesive [95]. The most commonly used elastomer in soft lithography is PDMS, a silicone elastomer that is optically transparent, gas permeable, nontoxic, and mechanically compliant [96, 97]. Furthermore, PDMS chips are cheap and easy to manipulate and can be formed by molding, then thermally sealed [98]. PDMS-based systems have specifically been shown to be compatible with mouse and pig embryo cultures, as well as nontoxic for manipulation of mouse and human sperm [6, 16].

A common problem with microchannels is the evaporation of the small volumes of liquid used. Evaporation of media at such small volumes can greatly affect the osmolarity of the culture medium. Elevation in osmolarity can seriously affect ion balance, metabolism, gene expression, and numerous other factors involved in regulating cell growth [99]. In general, gametes and embryos tolerate an osmolarity between 265 and 285 mmol/kg, and significant changes outside that range do not support oocyte maturation, fertilization, or embryo development [100]. Therefore, careful consideration must be given regarding materials and their properties used for IVP. Heo et al. demonstrated that the thickness of the PDMS used to develop microfluidic chips greatly affected osmolarity and subsequent embryo development of mouse embryos [101]. In this experiment, PDMS-based microfluidic chambers were used to culture mouse embryos with PDMS thickness ranging from 10 to 0.1 mm. Although there were no differences in embryo development or osmolarity shifts between microfluidic chips developed with PDMS thickness ranging from 1 to

10 mm, significantly fewer embryos cultured in PDMS chips of 0.1 mm thickness developed past the four-cell stage ($P < 0.01$). These developmental results were due to a significant shift in osmolarity of embryo-culture media from 265 to over 400 mmol/kg [101]. Dynamic embryo culture using Braille pin actuators requires a thin/flexible surface for controlling media movement. Therefore, thick layers of PDMS on the base of the microchannel are not suitable for this device. To overcome this problem, parylene, a conformal protective polymer, was tested to prevent evaporation. In this study Heo et al. showed a significant improvement in the percentage of blastocyst development using parylene coating when compared to an identical system with no parylene coating ($P < 0.01$); they further demonstrated that there were no significant shifts outside of the 265 to 285 mmol/kg osmolarity range when PDMS was coated with parylene [101]. This study describes development of a "hybrid membrane: PDMS-parylene-PDMS" that can prevent evaporation while maintaining a stable environment for cell growth and also providing a system that permits integration of microfluidic pumping using Braille pin technology [101]. Lastly, it is important to recognize that PDMS can also absorb small molecules [102]. What influence this might have as a supportive or nonsupportive component of microfluidic cell cultures has yet to be defined.

11.3 Conclusions and Future Directions

With the development of microscaled systems similar in size to mammalian gametes and preimplantation embryos, the birth of a new era has begun for assisted reproduction. It has been demonstrated that important functions necessary for carrying out basic IVP procedures, including oocyte maturation, insemination, and embryo culture, can be performed in microchannel devices. Integration of this "ART lab on a chip" will allow for automation and chemical and mechanical manipulation. Microfluidic platforms open many avenues in the investigation of why IVP-produced embryos are suboptimal compared to those derived in vivo. Such platforms also will provide general information to improve our knowledge of basic gamete/embryo physiology and developmental biology. Lastly, new prototypes of microfluidic devices for ARTs will be developed that enable real-time live-oocyte/-embryo metabolic and secretome analysis, without human manipulation-introduced error. Such live-cell assays will enhance our understanding of cellular physiological processes and provide additional parameters to measure oocyte/embryo health and viability.

Acknowledgments

We would like to thank Drs. Carrie Smith and Jason Swain for critically reviewing this chapter. Additionally, we would like to thank members of the Beebe, Smith, Takayama, and Wheeler labs, especially L. M. Cabrera, Y. S. Heo, E. M. Walters, S. G. Clark, and S. Raty, for the production and testing of mircrofluidic prototypes. We recognize and appreciate financial support from the National Institutes of Health (NIH; HD 049607-01—Takayama and Smith), the U.S. Department of Agriculture (USDA; 2005-35203-16148—Smith and Takayama), the Michigan Economic Development Corporation (MEDC; GR 696—Takayama and Smith), the Critical Research Initiatives at the University of Illinois, the Council for Food and Agricultural Research, the University of Wisconsin, and the USDA Multi-State Research Project W-1171. Support for Bormann was provided by NIH Training Grant in Reproductive Sciences T32-HD07048 (Smith).

References

[1] Hagmann, M., "Fertility therapy may aid gene transfer," *Science,* Vol. 284, No. 5417, 1999, pp. 1097–1098.

[2] Velander, W. H., Lubon, H., and Drohan, W. N., "Transgenic livestock as drug factories," *Sci. Am.*, Vol. 276, No. 1, 1997, pp. 70–74.

[3] Polejaeva, I. A., Chen, S. H., Vaught, T. D., Page, R. L., Mullins, J., Ball, S., Dai, Y., Boone, J., Walker, S., Ayares, D. L., Colman, A., and Campbell, K. H., "Cloned pigs produced by nuclear transfer from adult somatic cells," *Nature*, Vol. 407, No. 6800, 2000, pp. 86–90.

[4] Ravindranatha, B. M., Nandi, S., Raghu, H. M., and Reddy, S. M., "In vitro maturation and fertilization of buffalo oocytes: Effects of storage of ovaries, IVM temperatures, storage of processed sperm and fertilization media," *Reprod. Domest. Anim.*, Vol. 38, No. 1, 2003, pp. 21–26.

[5] Xie, Y., Wang, F., Zhong, W., Puscheck, E., Shen, H., and Rappolee, D. A., "Shear stress induces preimplantation embryo death that is delayed by the zona pellucida and associated with stress-activated protein kinase-mediated apoptosis," *Biol. Reprod.*, Vol. 75, No. 1, 2006, pp. 45–55.

[6] Beebe, D., Wheeler, M., Zeringue, H., Walters, E., and Raty, S., "Microfluidic technology for assisted reproduction," *Theriogenology*, Vol. 57, No. 1, 2002, pp. 125–135.

[7] Beebe, D. J., Mensing, G. A., and Walker, G. M., "Physics and applications of microfluidics in biology," *Ann. Rev. Biomed. Eng.*, Vol. 4, 2002, pp. 261–286.

[8] Quake, S. R., and Scherer, A., "From micro- to nanofabrication with soft materials," *Science,* Vol. 290, No. 5496, 2000, pp. 1536–1540.

[9] Sia, S. K., and Whitesides, G. M., "Microfluidic devices fabricated in poly(dimethylsiloxane) for biological studies," *Electrophoresis*, Vol. 24, No. 21, 2003, pp. 3563–3576.

[10] Walker, G., O. M., and Beebe, D. J., "Insect cell culture in microfluidic channels," *Biomed. Microdevices*, Vol. 4, 2002, pp. 161–166.

[11] Raty, S. D. J., Beebe, D. J., Rodriguez-Zas, S. L., and Wheeler, M. B., "Culture in microchannels enhances in vitro embryonic development of preimplantation mouse embryos," *Theriogenology*, Vol. 55, 2001, p. 241.

[12] Huang, W. H., Cheng, W., Zhang, Z., Pang, D. W., Wang, Z. L., Cheng, J. K., and Cui, D. F., "Transport, location, and quantal release monitoring of single cells on a microfluidic device," *Anal. Chem.*, Vol. 76, No. 2, 2004, pp. 483–488.

[13] Cho, B. S., Schuster, T. G., Zhu, X., Chang, D., Smith, G. D., and Takayama, S., "Passively driven integrated microfluidic system for separation of motile sperm," *Anal. Chem.*, Vol. 75, No. 7, 2003, pp. 1671–1675.

[14] Davis, J. A., Raty, S., Eddington, D. T., Glasgow, I. K., Zeringue, H. C., Wheeler, M. B., and Beebe, D. J., "Development of microfluidic channels for the culture of mammalian embryos," *1st Ann. Inter. IEEE-EMBS Spec. Topic Conf. Microtechnol. Med. Biol. Proc.*, Vol. 2000, pp. 307–310.

[15] Glasgow, I. K., Zeringue, H. C., Beebe, D. J., Choi, S. J., Lyman, J., and Wheeler, M. B., "Individual embryo transport on a chip for a total analysis system," *3rd Inter. Symp. Micro-Total Anal. Syst.*, 1998, pp. 13–16.

[16] Schuster, T. G., Cho, B., Keller, L. M., Takayama, S., and Smith, G. D., "Isolation of motile spermatozoa from semen samples using microfluidics," *Reprod. Biomed. Online*, Vol. 7, No. 1, 2003, pp. 75–81.

[17] Jurema, M. W., and Nogueira, D., "In vitro maturation of human oocytes for assisted reproduction," *Fertil. Steril.*, Vol. 86, No. 5, 2006, pp. 1277–1291.

[18] Mikkelsen, A. L., "Strategies in human in vitro maturation and their clinical outcome," *Reprod. Biomed. Online*, Vol. 10, No. 5, 2005, pp. 593–599.

[19] Papanikolaou, E. G., Platteau, P., Albano, C., Nogueira, D., Cortvrindt, R., Devroey, P., and Smitz, J., "Immature oocyte in vitro maturation: Clinical aspects," *Reprod. Biomed. Online*, Vol. 10, No. 5, 2005, pp. 587–592.

[20] Piquette, G. N., "The in vitro maturation (IVM) of human oocytes for in vitro fertilization (IVF): Is it time yet to switch to IVM-IVF?" *Fertil. Steril.*, Vol. 85, No. 4, 2006, pp. 833–835; discussion 841.

[21] Brinsden, P. R., Wada, I., Tan, S. L., Balen, A., and Jacobs, H. S., "Diagnosis, prevention and management of ovarian hyperstimulation syndrome," *Br. J. Obstet. Gynaecol.*, Vol. 102, No. 10, 1995, pp. 767–772.

[22] Check, M. L., Brittingham, D., Check, J. H., and Choe, J. K., "Pregnancy following transfer of cryopreserved-thawed embryos that had been a result of fertilization of all in vitro matured metaphase or germinal stage oocytes: Case report," *Clin. Exp. Obstet. Gynecol.*, Vol. 28, No. 2, 2001, pp. 69–70.

[23] Liu, J., Lu, G., Qian, Y., Mao, Y., and Ding, W., "Pregnancies and births achieved from in vitro matured oocytes retrieved from poor responders undergoing stimulation in in vitro fertilization cycles," *Fertil. Steril.*, Vol. 80, No. 2, 2003, pp. 447–449.

[24] Albertini, D. F., Sanfins, A., and Combelles, C. M., "Origins and manifestations of oocyte maturation competencies," *Reprod. Biomed. Online*, Vol. 6, No. 4, 2003, pp. 410–415.

[25] Fauser, B. C., Bouchard, P., Coelingh Bennink, H. J., Collins, J. A., Devroey, P., Evers, J. L., and van Steirteghem, A., "Alternative approaches in IVF," *Hum. Reprod. Update*, Vol. 8, No. 1, 2002, pp. 1–9.

[26] Hardy, K., Wright, C. S., Franks, S., and Winston, R. M., "In vitro maturation of oocytes," *Br. Med. Bull.*, Vol. 56, No. 3, 2000, pp. 588–602.

[27] Hendriksen, P. J., Vos, P. L., Steenweg, W. N., Bevers, M. M., and Dieleman, S. J., "Bovine follicular development and its effect on the in vitro competence of oocytes," *Theriogenology*, Vol. 53, No. 1, 2000, pp. 11–20.

[28] Krisher, R. L., Lane, M., and Bavister, B. D., "Developmental competence and metabolism of bovine embryos cultured in semi-defined and defined culture media," *Biol. Reprod.*, Vol. 60, No. 6, 1999, pp. 1345–1352.

[29] Spindler, R. E., Pukazhenthi, B. S., and Wildt, D. E., "Oocyte metabolism predicts the development of cat embryos to blastocyst in vitro," *Mol. Reprod. Dev.*, Vol. 56, No. 2, 2000, pp. 163–171.

[30] Walters, E. M., Beebe, D. J., and Wheeler, M.B, "In vitro maturation of pig oocytes in polydimethylsiloxane (PDMS) and silicon microfluidic devices," *Theriogenology*, Vol. 55, 2001, p. 497.

[31] Hester, P. N., Roseman, H. M., Clark, S. G., Walters, E. M., Beebe, D. J., and Wheeler, M. B., "Enhanced cleavage rates following in vitro maturation of pig oocytes within polydimethylsiloxane-borosilicate microchannels," *Theriogenology*, Vol. 57, 2002, p. 723.

[32] Mosher, W. D., and Pratt, W. F., "Fecundity and infertility in the United States: Incidence and trends," *Fertil. Steril.*, Vol. 56, No. 2, 1991, pp. 192–193.

[33] Mortimer, D., "Sperm transfer in the human female reproductive tract," in F. CA. (ed.), *Oxford Reviews of Reproductive Biology*, Vol. 5., Oxford: Oxford University Press, 1989.

[34] Trounson, A. O. a. G., D. K., *Handbook of In Vitro Fertilization*, 2nd ed., Boca Raton, FL: CRC Press LLC, 2000.

[35] Aitken, R. J., and Clarkson, J. S., "Significance of reactive oxygen species and antioxidants in defining the efficacy of sperm preparation techniques," *J. Androl.*, Vol. 9, No. 6, 1988, pp. 367–376.

[36] Zini, A., Finelli, A., Phang, D., and Jarvi, K., "Influence of semen processing technique on human sperm DNA integrity," *Urology*, Vol. 56, No. 6, 2000, pp. 1081–1084.

[37] Zini, A., Mak, V., Phang, D., and Jarvi, K., "Potential adverse effect of semen processing on human sperm deoxyribonucleic acid integrity," *Fertil. Steril.*, Vol. 72, No. 3, 1999, pp. 496–499.

[38] Virant-Klun, I., Tomazevic, T., and Meden-Vrtovec, H., "Sperm single-stranded DNA, detected by acridine orange staining, reduces fertilization and quality of ICSI-derived embryos," *J. Assist. Reprod. Genet.*, Vol. 19, No. 7, 2002, pp. 319–328.

[39] Benchaib, M., Braun, V., Lornage, J., Hadj, S., Salle, B., Lejeune, H., and Guerin, J. F., "Sperm DNA fragmentation decreases the pregnancy rate in an assisted reproductive technique," *Hum. Reprod.*, Vol. 18, No. 5, 2003, pp. 1023–1028.

[40] Seli, E., Gardner, D. K., Schoolcraft, W. B., Moffatt, O., and Sakkas, D., "Extent of nuclear DNA damage in ejaculated spermatozoa impacts on blastocyst development after in vitro fertilization," *Fertil. Steril.*, Vol. 82, No. 2, 2004, pp. 378–383.

[41] Bungum, M., Humaidan, P., Spano, M., Jepson, K., Bungum, L., and Giwercman, A., "The predictive value of sperm chromatin structure assay (SCSA) parameters for the outcome of intrauterine insemination, IVF and ICSI," *Hum. Reprod.*, Vol. 19, No. 6, 2004, pp. 1401–1408.

[42] Henkel, R., Hajimohammad, M., Stalf, T., Hoogendijk, C., Mehnert, C., Menkveld, R., Gips, H., Schill, W. B., and Kruger, T. F., "Influence of deoxyribonucleic acid damage on fertilization and pregnancy," *Fertil. Steril.*, Vol. 81, No. 4, 2004, pp. 965–972.

[43] Tesarik, J., Greco, E., and Mendoza, C., "Late, but not early, paternal effect on human embryo development is related to sperm DNA fragmentation," *Hum. Reprod.*, Vol. 19, No. 3, 2004, pp. 611–615.

[44] Carrell, D. T., Liu, L., Peterson, C. M., Jones, K. P., Hatasaka, H. H., Erickson, L., and Campbell, B., "Sperm DNA fragmentation is increased in couples with unexplained recurrent pregnancy loss," *Arch. Androl.*, Vol. 49, No. 1, 2003, pp. 49–55.

[45] Englert, Y., Van den Bergh, M., Rodesch, C., Bertrand, E., Biramane, J., and Legreve, A., "Comparative auto-controlled study between swim-up and Percoll preparation of fresh semen samples for in vitro fertilization," *Hum. Reprod.*, Vol. 7, No. 3, 1992, pp. 399–402.

[46] Smith, S., Hosid, S., and Scott, L., "Use of postseparation sperm parameters to determine the method of choice for sperm preparation for assisted reproductive technology," *Fertil. Steril.*, Vol. 63, No. 3, 1995, pp. 591–597.

[47] Palermo, G., Joris, H., Devroey, P., and Van Steirteghem, A. C., "Pregnancies after intracytoplasmic injection of single spermatozoon into an oocyte," *Lancet*, Vol. 340, No. 8810, 1992, pp. 17–18.

[48] Schultz, R. M., and Williams, C. J., "The science of ART," *Science,* Vol. 296, No. 5576, 2002, pp. 2188–2190.

[49] Palermo, G. D., Alikani, M., Bertoli, M., Colombero, L. T., Moy, F., Cohen, J., and Rosenwaks, Z., "Oolemma characteristics in relation to survival and fertilization patterns of oocytes treated by intracytoplasmic sperm injection," *Hum. Reprod.*, Vol. 11, No. 1, 1996, pp. 172–176.

[50] Nagy, Z. P., Liu, J., Joris, H., Bocken, G., Desmet, B., Van Ranst, H., Vankelecom, A., Devroey, P., and Van Steirteghem, A. C., "The influence of the site of sperm deposition and mode of oolemma breakage at intracytoplasmic sperm injection on fertilization and embryo development rates," *Hum. Reprod.*, Vol. 10, No. 12, 1995, pp. 3171–3177.

[51] Kricka, L. J., Nozaki, O., Heyner, S., Garside, W. T., and Wilding, P., "Applications of a microfabricated device for evaluating sperm function," *Clin. Chem.*, Vol. 39, No. 9, 1993, pp. 1944–1947.

[52] Kricka, L. J., Faro, I., Heyner, S., Garside, W. T., Fitzpatrick, G., McKinnon, G., Ho, J., and Wilding, P., "Micromachined analytical devices: Microchips for semen testing," *J. Pharm. Biomed. Anal.*, Vol. 15, Nos. 9–10, 1997, pp. 1443–1447.

[53] Anderson, J. R., Chiu, D. T., Jackman, R. J., Cherniavskaya, O., McDonald, J. C., Wu, H., Whitesides, S. H., and Whitesides, G. M., "Fabrication of topologically complex three-dimensional microfluidic systems in PDMS by rapid prototyping," *Anal. Chem.*, Vol. 72, No. 14, 2000, pp. 3158–3164.

[54] Unger, M. A., Chou, H. P., Thorsen, T., Scherer, A., and Quake, S. R., "Monolithic microfabricated valves and pumps by multilayer soft lithography," *Science*, Vol. 288, No. 5463, 2000, pp. 113–116.

[55] Figeys, D., P. D., "Lab-on-a-chip: A revolution in biological and medical sciences," *Anal. Chem.*, Vol. 72A, 2000, pp. 330–335.

[56] Dietl, J. A., and Rauth, G., "Molecular aspects of mammalian fertilization," *Hum. Reprod.*, Vol. 4, No. 8, 1989, pp. 869–875.

[57] Hickman, D. L., Beebe, D. J., Rodriguez-Zas, S. L., and Wheeler, M. B., "Comparison of static and dynamic medium environments for culturing of pre-implantation mouse embryos," *Comp. Med.*, Vol. 52, No. 2, 2002, pp. 122–126.

[58] Funahashi, H., Fujiwara, T., and Nagai, T., "Modulation of the function of boar spermatozoa via adenosine and fertilization promoting peptide receptors reduce the incidence of polyspermic penetration into porcine oocytes," *Biol. Reprod.*, Vol. 63, No. 4, 2000, pp. 1157–1163.

[59] Li, Y. H., Ma, W., Li, M., Hou, Y., Jiao, L. H., and Wang, W. H., "Reduced polyspermic penetration in porcine oocytes inseminated in a new in vitro fertilization (IVF) system: Straw IVF," *Biol. Reprod.*, Vol. 69, No. 5, 2003, pp. 1580–1585.

[60] Ranoux, C., Poirot, C., Foulot, H., Dubuisson, J. B., Aubriot, F. X., Chevallier, O., and Cardone, V., "Human egg fertilization in capillary tubes," *J. In Vitro Fert. Embryo. Transf.*, Vol. 5, No. 1, 1988, pp. 49–50.

[61] Ranoux, C., and Seibel, M. M., "New techniques in fertilization: intravaginal culture and microvolume straw," *J. In Vitro Fert. Embryo. Transf.*, Vol. 7, No. 1, 1990, pp. 6–8.

[62] van der Ven, H. H., Hoebbel, K., al-Hasani, S., Diedrich, K., and Krebs, D., "Fertilization of human oocytes in capillary tubes with very small numbers of spermatozoa," *Hum. Reprod.*, Vol. 4, No. 1, 1989, pp. 72–76.

[63] Topfer-Peterson, E., P. M. a. E.-H. M., *Animal Reproductive Science,* Vol. 60–61, 2000, p. 653.

[64] Clark, S. G., W. E., Beebe, D. J., and Wheeler, M. B., "In vitro fertilization of porcine oocytes in polydimethylsiloxane (PDMS)-glass microchannels," *Biol. Reprod.*, Vol. 66, 2002, p. 528.

[65] Clark, S. G., Haubert, K., Beebe, D. J., Ferguson, C. E., and Wheeler, M. B., "Reduction of polyspermic penetration using biomimetic microfluidic technology during in vitro fertilization," *Lab Chip*, Vol. 5, No. 11, 2005, pp. 1229–1232.

[66] Ahlgren, M., "Sperm transport to and survival in the human fallopian tube," *Gynecol. Invest.*, Vol. 6, Nos. 3–4, 1975, pp. 206–214.

[67] Settlage, D. S., Motoshima, M., and Tredway, D. R., "Sperm transport from the external cervical os to the fallopian tubes in women: A time and quantitation study," *Fertil. Steril.*, Vol. 24, No. 9, 1973, pp. 655–661.

[68] Suh, R. S., Zhu, X., Phadke, N., Ohl, D. A., Takayama, S., and Smith, G. D., "IVF within microfluidic channels requires lower total numbers and lower concentrations of sperm," *Hum. Reprod.*, Vol. 21, No. 2, 2006, pp. 477–483.

[69] Hogan, B. B. R., Constantini, F., and Lacy, E., *Manipulating the Mouse Embryo: A Laboratory Manual*, 2nd ed., New York: Cold Spring Harbor Laboratory Press, 1994.

[70] Zeringue, H. C., W. E., Leibfried-Rutledge, L. L., Wheeler, M. B., and Beebe, D. J., "Development of bovine in vitro–produced embryos after cumulus cell removal with a microfluidic microchannel device," *Theriogenology*, Vol. 57, 2002, p. 533.

[71] Christians, E., Campion, E., Thompson, E. M., and Renard, J. P., "Expression of the HSP 70.1 gene, a landmark of early zygotic activity in the mouse embryo, is restricted to the first burst of transcription," *Development*, Vol. 121, No. 1, 1995, pp. 113–122.

[72] Zeringue, H. C., *Applications of microfluidics in production of mammalian embryos*, Madison: University of Wisconsin: 2003, p. 123.

[73] Gardner, D. K., and Lane, M., "Culture of viable human blastocysts in defined sequential serum-free media," *Hum. Reprod.*, Vol. 13, Suppl. 3, 1998, pp. 148–159; discussion 160.

[74] Glasgow, I. K., Zeringue, H. C., Beebe, D. J., Choi, S. J., Lyman, J. T., Chan, N. G., and Wheeler, M. B., "Handling individual mammalian embryos using microfluidics," *IEEE Trans. Biomed. Eng.*, Vol. 48, No. 5, 2001, pp. 570–578.

[75] Walters, E. M., C. S., Roseman, H. M., Beebe, D. J., and Wheeler, M.B., "Production of live piglets following in vitro embryo culture in a microfluidic environment," *Theriogenology*, Vol. 59, 2003, p. 353.

[76] Raty, S., Walters, E. M., Davis, J., Zeringue, H., Beebe, D. J., Rodriguez-Zas, S. L., and Wheeler, M. B., "Embryonic development in the mouse is enhanced via microchannel culture," *Lab Chip*, Vol. 4, No. 3, 2004, pp. 186–190.

[77] Cabrera, L. M., H. Y., Ding, J., Takayama, S., and Smith, D. G., "Improved blastocyst development with microfluidics and Braille pin actuator enabled dynamic culture," *Fertil. Steril.*, Vol. 86, 2006, p. S43.

[78] Gu, W., Zhu, X., Futai, N., Cho, B. S., and Takayama, S., "Computerized microfluidic cell culture using elastomeric channels and Braille displays," *Proc. Natl. Acad. Sci. USA*, Vol. 101, No. 45, 2004, pp. 15861–15866.

[79] Bormann, C., L. M. C., Heo, Y. S., Takayama, S., and Smith, G., "Dynamic microfluidic embryo culture enhances blastocyst development of murine and bovine embryos," *Submitted to Biol. Reprod.*, Vol. 2007.

[80] Kouba, A. J., Abeydeera, L. R., Alvarez, I. M., Day, B. N., and Buhi, W. C., "Effects of the porcine oviduct-specific glycoprotein on fertilization, polyspermy, and embryonic development in vitro," *Biol. Reprod.*, Vol. 63, No. 1, 2000, pp. 242–250.

[81] Raty, S., "Development and testing of a new microfluidic system for in vitro culture of preimplantation mammalian embryos," in *Memoire d'ingenieur, Ecole Superieure d'Agriculture de Purpan, Toulouse, France*, Vol. 2000.

[82] Wheeler, M. B., Beebe, D. J., Walters, E. M., and Raty, S., "Microfluidic technology for in vitro embryo production," *Med. Biol. Proc.*, Vol. 2, 2002, pp. 104–108.

[83] Clark, S. G., Walters, E. M., Beebe, D. J., and Wheeler, M. B., "A novel integrated in vitro maturation and in vitro fertilization system for swine," *Theriogenology*, Vol. 59, 2003, p. 441.

[84] Cheng, W., Klauke, N., Sedgwick, H., Smith, G. L., and Cooper, J. M., "Metabolic monitoring of the electrically stimulated single heart cell within a microfluidic platform," *Lab Chip*, Vol. 6, No. 11, 2006, pp. 1424–1431.

[85] Shackman, J. G., Dahlgren, G. M., Peters, J. L., and Kennedy, R. T., "Perfusion and chemical monitoring of living cells on a microfluidic chip," *Lab Chip*, Vol. 5, No. 1, 2005, pp. 56–63.

[86] Mehta, G., Mehta, K., Sud, D., Song, J. W., Bersano-Begey, T., Futai, N., Heo, Y. S., Mycek, M. A., Linderman, J. J., and Takayama, S., "Quantitative measurement and control of oxygen levels in microfluidic poly(dimethylsiloxane) bioreactors during cell culture," *Biomed. Microdevices*, Vol. 2006.

[87] Papanikolaou, E. G., Camus, M., Fatemi, H. M., Tournaye, H., Verheyen, G., Van Steirteghem, A., and Devroey, P., "Early pregnancy loss is significantly higher after day 3 single embryo transfer than after day 5 single blastocyst transfer in GnRH antagonist stimulated IVF cycles," *Reprod. Biomed. Online*, Vol. 12, No. 1, 2006, pp. 60–65.

[88] Lane, M., and Gardner, D. K., "Selection of viable mouse blastocysts prior to transfer using a metabolic criterion," *Hum. Reprod.*, Vol. 11, No. 9, 1996, pp. 1975–1978.

[89] Bormann, C. L., Chisolm, C. N., Heo, Y. S., Takayama, S., and Smith, G. D., "Development of a sensitive assay to measure lactate production of preimplantation embryos," *Fertil. Steril.*, Vol. 86, 2006, p. S115.

[90] Hardy, K., and Spanos, S., "Growth factor expression and function in the human and mouse preimplantation embryo," *J. Endocrinol.*, Vol. 172, No. 2, 2002, pp. 221–236.

[91] Katz-Jaffe, M. G., Gardner, D. K., and Schoolcraft, W. B., "Proteomic analysis of individual human embryos to identify novel biomarkers of development and viability," *Fertil. Steril.*, Vol. 85, No. 1, 2006, pp. 101–107.

[92] Katz-Jaffe, M. G., Schoolcraft, W. B., and Gardner, D. K., "Analysis of protein expression (secretome) by human and mouse preimplantation embryos," *Fertil. Steril.*, Vol. 86, No. 3, 2006, pp. 678–685.

[93] Katz-Jaffe, M. G., S. J., Kearns, W. G., Gardner, D. K., and Schoolcraft, W. B., "Relationship between embryonic secretome and chromosomal abnormalities in human IVF," *Fertil. Steril.*, Vol. 86, 2006, p. S57.

[94] Chan, N. G., Lyman, J. T., Choi, S. J., Zeringue, H. C., Glasgow, I. K., Beebe, D. J., and Wheeler, M. B., "Development of an embryo transport and analysis system: Material biocompatibility," *Theriogenology*, Vol. 51, 1999, p. 234.

[95] Chan, N. G., R. S., Zeringue, H. C., Beebe, D. J., and Wheeler, M.B., "Development of microfabricated devices for embryo production: Embryo biocompatibility," *Theriogenology*, Vol. 55, 2001, p. 332.

[96] McDonald, J. C., Chabinyc, M. L., Metallo, S. J., Anderson, J. R., Stroock, A. D., and Whitesides, G. M., "Prototyping of microfluidic devices in poly(dimethylsiloxane) using solid-object printing," *Anal. Chem.*, Vol. 74, No. 7, 2002, pp. 1537–1545.

[97] McDonald, J. C., and Whitesides, G. M., "Poly(dimethylsiloxane) as a material for fabricating microfluidic devices," *Acct. Chem. Res.*, Vol. 35, No. 7, 2002, pp. 491–499.

[98] McDonald, J. C., Duffy, D. C., Anderson, J. R., Chiu, D. T., Wu, H., Schueller, O. J., and Whitesides, G. M., "Fabrication of microfluidic systems in poly(dimethylsiloxane)," *Electrophoresis*, Vol. 21, No. 1, 2000, pp. 27–40.

[99] Moor, A. N., Murtazina, R., and Fliegel, L., "Calcium and osmotic regulation of the Na+/H+ exchanger in neonatal ventricular myocytes," *J. Mol. Cell. Cardiol.*, Vol. 32, No. 6, 2000, pp. 925–936.

[100] Brinster, R. L., "Studies on the Development of Mouse Embryos In Vitro: I. The effect of osmolarity and hydrogen ion concentration," *J. Exp. Zool.*, Vol. 158, 1965, pp. 49–57.

[101] Heo, Y. S., Cabrera, L. M., Song, J. W., Futai, N., Tung, Y. C., Smith, G. D., and Takayama, S., "Characterization and resolution of evaporation-mediated osmolality shifts that constrain microfluidic cell culture in poly(dimethylsiloxane) devices," *Anal. Chem.*, Vol. 79, No. 3, 2007, pp. 1126–1134.

[102] Toepke, M. W., and Beebe, D. J., "PDMS absorption of small molecules and consequences in microfluidic applications," *Lab Chip*, Vol. 6, No. 12, 2006, pp. 1484–1486.

CHAPTER 12

Microscale Technologies for Engineering Embryonic Stem Cell Environments

Richard L. Carpenedo and Todd C. McDevitt

12.1 Embryonic Stem Cells

Embryonic stem cells (ESCs) are pluripotent cells derived from the inner cell mass of the blastocyst stage of development that are capable of infinite self-renewal and differentiating into cells from all three germ layers (ectoderm, mesoderm, and endoderm), as well as germ cells. Embryonic stem cells were first derived from mice [1–3] but have also been isolated from nonhuman primate [4, 5] and human embryos [6, 7]. The inherent plasticity of ESCs has made them useful for in vitro studies of developmental cell biology and suggests they could be effectively used as a robust cell source for regenerative cell therapies, as well as the development of in vitro test beds for pharmacological drug screening and toxicity. ESCs are exquisitely sensitive to a variety of microenvironmental cues that regulate both self-renewal and differentiation. Environmental cues can be transduced by ESCs via different classes of receptors that interact specifically with soluble factors, extracellular matrix (ECM) components, and homotypic and heterotypic cell-cell adhesions (Figure 12.1). Thus, microscale technologies capable of controlling cellular assembly and the presentation of molecular cues at the individual cellular level may elicit enhanced control of ESC growth and differentiation and an improved understanding of the signaling pathways that regulate such processes.

12.1.1 ESC Self-Renewal

Self-renewal is the process by which ESCs divide and multiply in an undifferentiated state, thereby preserving their pluripotency [8–10]. In contrast to many other cell types, mouse and human ESCs typically grow as compact, generally round colonies (~10^{1-2} μm in diameter) of cells with a high nuclei-to-cytoplasm ratio (Figure 12.2). Ultimately, to generate sufficient numbers of ESCs to be useful for cell therapeutics, diagnostics, or developmental studies, prolonged expansion of undifferentiated ESCs is necessary. Separate, yet comparable, strategies for the culture of mouse and human ESCs have been developed to stimulate growth of ESCs while inhibiting differentiation.

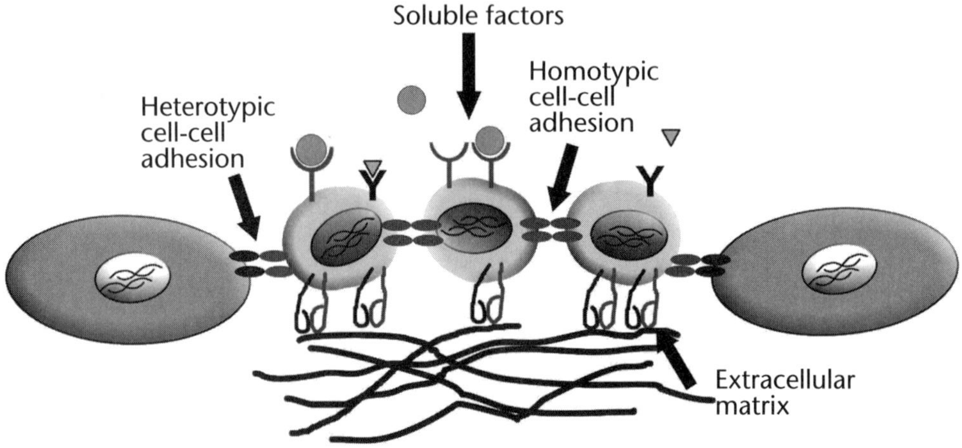

Figure 12.1 Microenvironmental factors affecting embryonic stem cell fate. Embryonic stem cell self-renewal and differentiation are influenced by homotypic and heterotypic cell-cell adhesions, soluble factors, and the ECM. ESC heterotypic interactions can be with mouse embryonic fibroblasts for self-renewal, stromal cells for differentiation coculture, or differentiating cells within embryoid bodies (EBs). Soluble factors such as cytokines [e.g., leukemia inhibitory factor (LIF)] or growth factors (TGF, NGF, VEGF) that influence stem cell fate can be added exogenously or secreted by cells endogenously. Similarly, ECM proteins that interact with ESCs may be used to coat surfaces (gelatin, laminin, or Matrigel) or may be produced endogenously by adherent ESCs or by cells within EBs.

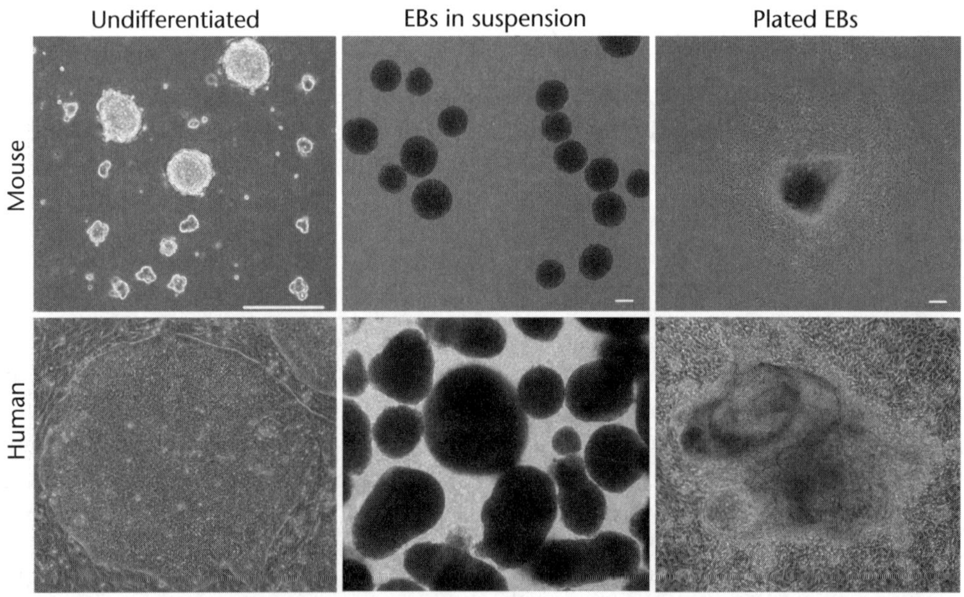

Figure 12.2 Mouse and human embryonic stem cells. Undifferentiated mouse and human ESCs both grow as compact colonies; however, when ESC colonies are dislodged from their surface and cultured in suspension conditions, ESCs spontaneously aggregate to form embryoid bodies (EBs). After some period of suspension culture (generally four to seven days), EBs are often plated on an adherent surface in order to further cell differentiation. (Mouse ESC images: scale bar = 200 μm; human ESC images acquired with 10× objective.)

12.1.1.1 Mouse ESCs

Mouse ESCs (mESCs) have traditionally been derived and propagated on mouse embryonic fibroblast (MEF) feeder layers [1, 2]. Biochemical studies discovered that LIF secreted by MEFs was the primary mitogen responsible for ESC self-renewal and that addition of LIF to serum-containing media could be used in lieu of MEFs for a feeder-free culture system [11, 12]. Bone morphogenic protein 4 has also been shown to be important in self-renewal and, in combination with LIF, can stimulate pluripotent ESC growth in serum-free media [13]. In general, mESCs are maintained on either inactivated MEFs or on gelatin-coated substrates in media supplemented with 10 to 15 percent serum and 10^3 units/ml of LIF. Mouse ESCs can be passaged by enzymatic dissociation with trypsin to produce single-cell suspensions and can be replated while remaining viable and pluripotent.

12.1.1.2 Human ESCs

Human ESCs (hESCs), like mESCs, were originally derived and maintained on MEFs [6]. However, unlike mESCs, LIF is not sufficient to inhibit differentiation of hESCs in the absence of a MEF feeder layer [6, 7], and basic fibroblast growth factor (bFGF) appears to be the primary mitogen responsible for hESC growth [6, 14]. However, the use of mouse feeder cells poses a threat to the clinical applicability of hESCs, as exposure to xenogenic cells is a potential source of hESC contamination [15, 16]. Feeder-free culture methods have been established by culturing hESCs on Matrigel- or laminin-coated surfaces with MEF-conditioned media [17] or media supplemented with bFGF [18] or activin A [19]; however, the use of MEF-conditioned media or bovine serum does not completely eliminate hESC exposure to xenogenic materials. Alternatively, human cell feeder layers may be capable of supporting hESC self-renewal [20, 21], and recent reports also suggest completely xenogenic-free methods may be possible for hESC derivation and expansion [22, 23]. Passaging of hESCs is generally performed using mechanical disruption or collagenase treatment [14], as opposed to trypsin dissociation, since reduction of hESC colonies to single cells has been shown to negatively impact cell viability. Additionally, continuous enzymatic passaging has been shown to alter the karyotype of hESCs [24]. The limitations in current hESC self-renewal methods, such as the need for defined feeder-free culture conditions and efficient, controlled passaging techniques, are potential applications for recently developed microscale technologies.

12.1.2 ESC Differentiation

Differentiation of ESCs in vitro can be induced by a variety of methods, such as adherent monolayer format (in monoculture [25], coculture with differentiated cell types [26]) or the formation of EBs in suspension culture [3]. In general, differentiation requires the removal or reduction of factors that stimulate self-renewal and preserve pluripotency, in addition to the introduction of molecular and cellular cues that instruct ESC differentiation pathways. Differentiation of ESCs is influenced by soluble factors (produced endogenously or added exogenously), interactions with

ECM components, cell-cell adhesions (both homotypic and heterotypic), and various combinations of these different types of stimuli (Figure 12.1). Microscale technologies can be applied to present each of the different classes of differentiation cues in a highly defined and precise manner in order to elicit directed control of ESC fate.

12.1.2.1 Adherent Monoculture

In adherent monoculture, ESCs are plated on an ECM-coated surface in the absence of self-renewal factors. The ECM coating used to encourage cell attachment can be chosen to promote the growth or survival of specific differentiated cell types [26, 27]. Adherent monoculture has the advantage of being simple to perform and allows all of the cells to be equally exposed to the same levels of nutrients and differentiation factors, making adherent monoculture ideal for screening assays for the discovery and development of drugs and differentiation agents. This method also has the advantage of producing a highly homogenous population of differentiated cells [25]. However, while adherent monoculture has been shown to be useful for the production of some cell types, particularly neural precursors [25, 28–30], the inherent lack of heterotypic cell-cell interactions that appear to be necessary for the development of particular cell types limits the potential utility of this method for directed differentiation of ESCs into all possible cell types.

12.1.2.2 Heterotypic Coculture

Coculture of ESCs with mature cell types is another commonly used differentiation method. Similar to coculture with MEFs for undifferentiated self-renewal of ESCs, coculture with cell types such as PA6 [31, 32] and OP9 [26, 33–35] stromal cells can induce ESC differentiation via a combination of secreted soluble factors and cell-cell adhesions. This technique has been implemented to direct differentiation of ESCs toward many mature cell types, including neurons [31, 36], cardiomyocytes [37], osteoblasts [38], hepatocytes [39], and hematopoietic precursor cells [26, 40]. However, heterotypic coculture methods are somewhat limited in the lack of precisely defined microenvironmental control they provide. The presence of both secreted factors and heterogeneous cell-cell interactions makes the mechanism of directed differentiation in coculture difficult to study directly. Additionally, coculture differentiation systems can induce heterogeneous differentiation of undesired cell types in addition to desired cell phenotypes.

12.1.2.3 Embryoid Body Culture

When ESCs are cultured under suspension conditions in the absence of self-renewal factors (e.g., LIF for mESCs), they spontaneously aggregate into spheroids referred to as EBs (Figure 12.2) [41, 42]. After some period of EB suspension culture (generally four to seven days), EBs are seeded onto adherent substrates to promote EB attachment, cell migration, and continued differentiation of the EB-derived cells (Figure 12.2). Common methods for the formation of EBs include hanging drop [43], static liquid suspension [44, 45], methylcellulose suspension [46], and the use of bioreactor systems [47–50]. In hanging-drop culture, two hundred to one thousand cells are suspended in 10 to 20 μl of drops of media on the top lid of a Petri dish to promote cell aggregation and formation of generally one EB per drop. In suspension culture, cells are inoculated at densities ranging from $1 \times 10^{4-6}$ cells/ml in either standard differentiation media or semisolid methylcellulose in a nonadherent petri dish, whereupon the ESCs begin to spontaneously aggregate and form numerous EBs varying in size and shape. In order to provide more control over the aggregation process, a variety of bioreactors, ranging from simple orbital shakers to more elaborate spinner flasks and slow-turning lateral vessels, have been employed to control the formation and proliferation of EBs [47–51].

Embryoid body culture is the most commonly applied differentiation method because the progression of cellular differentiation within EBs closely mimics patterns of genotypic and phenotypic differentiation of developing embryos [3]. Initially, a primitive endoderm layer on the periphery of EBs surrounds the undifferentiated core of ESCs and provides instructive cues that subsequently stimulate differentiation of cells comprising all three germ layers (ectoderm, mesoderm, and endoderm). However, one of the disadvantages of EBs is their inherent heterogeneous cellularity. The inability to specifically control the cell-cell interactions and local molecular cues at the microscale level in three-dimensional cell aggregates promotes a high level of cellular heterogeneity within EBs.

12.2 Microscale Technologies

Various microscale technologies have either been examined or may be applicable for enhanced control and improved study of ESC self-renewal and differentiation. Micropatterning and microfluidics can be used to spatially control the positioning and delivery of molecules to ESCs, respectively, in two dimensions, whereas microencapsulation and microparticles may be used for spatiotemporal control of ESC microenvironments in three dimensions (Figure 12.3).

12.2.1 Micropatterning

Micropatterning technologies have been broadly applied to the development of DNA and protein arrays [52, 53], as well as to cell-based sensors [54] and in vitro tissue-engineered constructs [55, 56]. Common micropatterning techniques are based on traditional photolithography practices [57] or soft lithography tools [58], such as microcontact printing, microfluidic patterning, and elastomeric stencils. Micropatterning techniques provide the ability to spatially control the distribution and organization of individual cells or groups of cells in two or three dimensions by geometrically defining surface properties to either selectively promote or inhibit regions of cell adhesion. Microscale control of surface features and chemistries has been utilized to precisely study several basic mechanisms of cell biology [59], particularly cell-cell [60], cell-matrix [61, 62], and cell-biomaterial interactions [57]. Spatial organization of a variety of cell types has been achieved with micropatterned substrates, including fibroblasts [63] and hepatocytes [64], as well as endothelial [65], cardiac [66, 67], and skeletal muscle cells [68, 69], and, most recently, embryonic stem cells [70, 71].

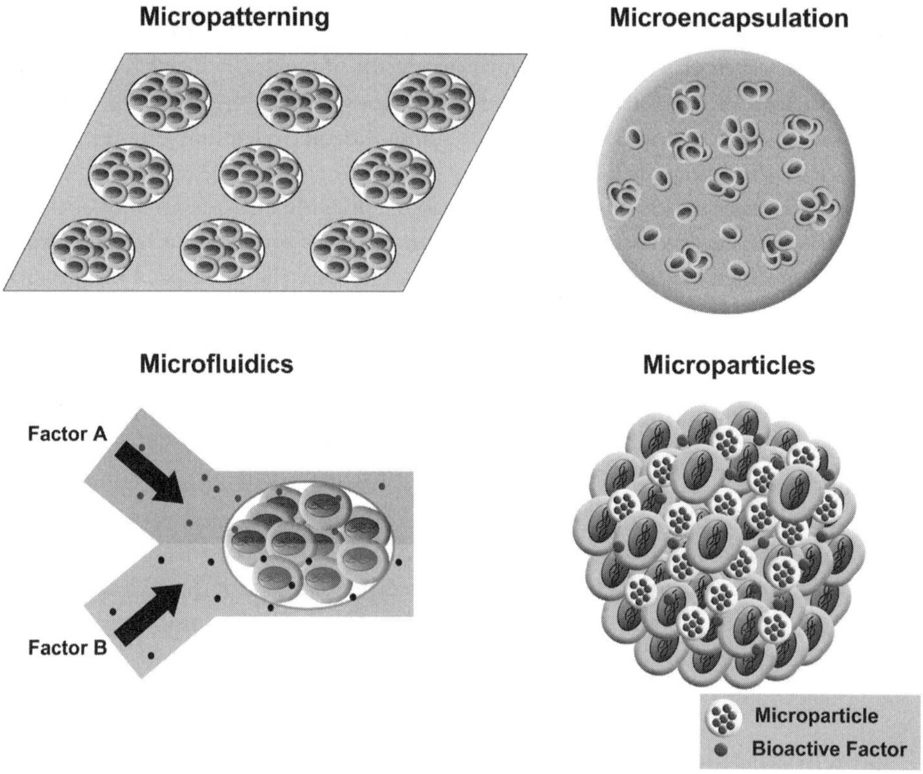

Figure 12.3 Microscale technologies for control of ESC microenvironments. Micropatterning can be used to control the size of ESC colonies, as well as the relative positions of ESCs and feeder or stromal cells. Microfluidics can be applied to control the spatial and temporal presentation of bioactive factors to ESCs, to create gradients of soluble factors, and to apply a wide range of shear stresses to cells. Microencapsulation can be used to entrap ESCs within a defined biochemical microenvironment in three dimensions, as well as to physically separate distinct ESC aggregates. Microparticles may be used to deliver chemical signals to the interior of EBs in a temporally controllable fashion.

12.2.1.1 Micropatterning and ESC Self-Renewal

Micropatterning can be used to control the size of ESC colonies in two dimensions by controlling the surface area for attachment, spreading, and cell growth. In general, control of ESC colony size is an important concern for both MEF and feeder-free cultures. When individual ESC colonies grow to become too large or individual colonies begin to merge, spontaneous differentiation can occur. Thus, it is desirable to produce colonies of controlled size in order to minimize differentiation while maximizing cell viability. ESC colony size in MEF cocultures has been controlled using layer-by-layer deposition and microwell patterning [70, 72]. Control over colony size by micropatterning has also been applied to feeder-free culture [71]. Self-assembled monolayers (SAMs) terminated with triethylene glycol, which resist protein adsorption and cell adhesion, have been used to localize Matrigel deposition and ESC attachment to the bottom of patterned microwells. Pluipotency of ESCs cultured in this manner was maintained after eighteen days, as indicated by no significant decrease in Oct-4 expression. Additionally, ESCs cultured in microwells displayed increased viability compared to those cultured on standard Matrigel-coated substrates, suggesting that controlling ESC colony size can effectively increase cell viability while maintaining pluripotency in feeder-free conditions.

In addition to directly controlling ESC colony size, micropatterning methods can spatially control the extent of cell-cell interactions within individual ESC colonies or between ESCs and feeder cell layers, such as MEFs. An important consideration for cocultures is the relative positions and ratios of the different cell types in the culture since both homotypic (ESC-ESC) and heterotypic (ESC-MEF) cell-cell interactions are important determinants of ESC self-renewal (Figure 12.1). Previous micropatterning studies with differentiated cell types have clearly demonstrated how the relative levels of homotypic and heterotypic interactions can influence cell function in vitro [63, 64]. Reproducibly controlling the degree of homotypic and heterotypic cell interactions may be important for reliable maintenance of stem cell pluripotency. One approach to control the spatial arrangement of ESCs and fibroblasts has been through layer-by-layer deposition of hyaluronic acid (HA) and poly-l-lysine (PLL) [72]. Glass substrates were patterned with HA and subsequently coated with fibronectin, such that fibronectin adsorbed only to the region's HA-free surfaces. A first cell type (ESCs or fibroblasts) was then seeded on the fibronectin surface, PLL was subsequently deposited on the HA surface, and the second seeded cell type (either fibroblasts or ESCs, respectively) adhered to the PLL surface regions. By controlling the size and shape of the fibronectin- and PLL-coated regions, the levels of heterotypic and homotypic interactions could be controlled.

Other methods have also been employed to spatially regulate attachment of two cell types. Electrical stimulation in conjunction with surface chemistry has been used to convert a nonadherent surface into an adherent surface capable of cell attachment [73]. Arginine-glycine-aspartic acid (RGD) peptides have been coupled to functionalized SAMs in the presence of electrical stimulation, allowing cell attachment on surfaces that resisted adhesion prior to stimulation. By controlling the position of the responsive SAMs using technologies such as microcontact printing, the positions of two cell types, such as MEFs and ESCs, can be controlled. Thermoresponsive materials with similar utility have also been developed. Poly(N-isopropylacrylamide) (PNIPAAm) is a polymer that goes through a hydrophobic-to-hydrophilic transition as temperature decreases below a lower critical solution temperature (LCST) of 32°C. At physiological temperature (above the LCST), cells readily attach to the polymer, but at temperatures below the LCST, protein and cell adhesion are inhibited. PNIPAAm can be patterned on a surface to form islands so that a first cell type seeded at a low temperature cannot adhere on the islands, but a second cell type seeded at 37°C can attach to PNIPAAm regions, thus allowing spatially controlled cell cocultures [74].

While layer-by-layer deposition or stimuli-responsive materials may be useful for spatial control of cocultures in two dimensions, they do not provide a means for controlling arrangement of cells in three dimensions. This is an important concern as ESC colonies are typically composed of multilayered cells. Microwell-patterned surfaces have been produced in order to control MEF interactions with aggregates of ESCs [70]. Poly(dimethyl siloxane) (PDMS) was cured on silicone masters to produce circular microwells 200 μm in diameter and 120 μm deep. After preadsorbing a layer of fibronectin from solution to promote cell attachment, MEFs were seeded over the entire PDMS surface and mitotically inactivated with mitomycin C prior to seeding hESCs. The hESCs came to reside within the microwells; therefore, by controlling the height and diameter of the microwells, ESC-MEF interactions could be controlled in three dimensions.

12.2.1.2 Micropatterning and ESC Differentiation

Micropatterning has many applications for differentiation of ESCs in addition to self-renewal. While micropatterning can be used to control spatial arrangements of MEFs and ESCs for self-renewal, it can also be applied to control the position of mature cells types used to induce differentiation of ESCs. Cocultures are commonly used in ESC differentiation protocols [26, 37–40] and are intended to mimic specific elements of stem cell niches that direct differentiation. Unlike in vitro culture conditions, these microenvironments are tightly regulated in the body. Micropatterning can be employed to increase control over the in vitro stem cell microenvironment by spatial regulation of both cell types in coculture, thereby increasing control over differentiation.

Micropatterning has also been used to study the effect of individual cell shape on stem cell differentiation [75]. Micropatterned adhesive islands were produced to either permit or inhibit spreading of individual mesenchymal stem cells (10,000 μm^2 or 1,024 μm^2, respectively). Cell spreading appeared to preferentially promote osteoblast differentiation, whereas round cells predominantly differentiated into adipocytes. Similar studies could also be performed with ESCs and ESC-derived progenitor cells to examine the effect of cell shape on self-renewal and differentiation.

Micropatterning techniques can also be applied for high-throughput analysis of individual stem cell fate. Neural progenitor cells were seeded onto a coverslip patterned with microwells, and an automated image-acquisition system was used to determine the number of cells in each well, with a majority of wells found to contain zero cells or one cell [76]. Using this format, the effect of different exogenous signals on stem cell proliferation and differentiation can be directly analyzed, and the effect of soluble signals can be distinguished from cell-cell interactions as single cells can be plated and monitored over time. While this platform has been utilized for studying neural stem cells, it may also be applicable for murine ESCs or ESC-derived cells. High-throughput analysis using microarrays has also been used to study stem cell–biomaterial interactions [77]. Copolymers were synthesized from a variety of monomers in different ratios using ultraviolet-initiated polymerization. Cells derived from human EBs were then plated on the polymer array, and the influence of cell-material interactions on cell differentiation was analyzed. Similar technology has been applied to produce protein arrays consisting of various ECM proteins in order to study the role these interactions play in ESC differentiation in a high-throughput manner [62].

Micropatterning may also be used to template ESC colony size prior to EB formation. Control of EB size may be important for the reproducibility of ESC differentiation, as cell-cell interactions play a prominent role in the differentiation process. Reliably producing EBs of similar size is challenging, especially for human ESCs, which are generally not reduced to single-cell suspensions due to the negative impact this has on cell viability. As a result, human EBs typically form from clusters of cells of variable size, making it difficult to control the input number of cells, as well as the aggregation and proliferation of clusters to form EBs. One approach to address this is to produce EBs from ESCs grown on micropatterned islands or within microwells [70–72]. The physical constraints imposed by these culture conditions reliably produce a controlled number of cells within each island or well. ESC colonies can then be dissociated by collagenase or by mechanical disruption to produce uniform cell aggregates in suspension. EBs formed from ESC aggregates grown in microwells and dislodged by gentle pipetting were shown to be more uniform in diameter than EBs formed without microwells [71]. This approach could be applied to control EB size based on microwell dimensions in order to elucidate the role of EB size in ESC differentiation.

12.2.2 Microfluidics

Microfluidics entails manipulation and processing of micro- to nanoliter volumes of fluids. Microfluidic technology generally involves micron-scale channels (tens to hundreds of micrometers), which are often fabricated using soft lithography with PDMS [78]. Applications of microfluidic technology include chemical and biomolecular analysis [79–81], molecular separations [82], protein crystallization [83, 84], and culture of cells [85–87]. Microfluidics has been used in a variety of manners to manipulate and study cells, including micropatterning of cells and proteins [88–91], subcellular localization of media components [92, 93], high-throughput drug screening [94, 95], introduction of a large range of laminar- flow rates [96], and creation of soluble factor gradients. Microfluidic technology has also been utilized for the manipulation of embryos to study development [97, 98].

Microfluidic technology can be used to subject cells to laminar flow, which can be applied for both ESC self-renewal and differentiation. Microfluidic arrays have been fabricated to produce a logarithmic scale of flow rates and a logarithmic concentration gradient [96]. The influence of flow rate on cell proliferation was studied, with high flow resulting in increased proliferation. One advantage of microfluidic studies over macroscale laminar-flow perfusion experiments is a significant reduction in the amount of media necessary. This is especially important if media is not recirculated since ESC media is typically composed of an expensive cocktail of growth factors and cytokines. As a result, microfluidic perfusion may be utilized for the screening of media components necessary for ESC self-renewal and proliferation. For example, screening for serum components and products of MEF-conditioned media necessary for self-renewal could be conducted in a microfluidic chamber to minimize consumption of expensive growth factors and cytokines.

Differentiation culture of ESCs can also benefit from microfluidic technology. Shear stress invoked by laminar flow has been shown to influence differentiation of many cell types, including mesenchymal stem cells [99], osteoblasts [100], endothelial progenitor cells [101], marrow stromal cells [102], and ESCs [103]. By combining micropatterning with controlled flow rates, in vivo environments that invoke cell differentiation can be more accurately replicated [88, 104]. Concentration gradients have been shown to be important for regulating cell differentiation. Creation of stable gradients has been achieved by ligand immobilization. Stable gradients of factors in solution in microfluidic systems have been commonly reported [86, 96,

105–108]. Temporal aspects of interactions between ESCs and soluble factors are also important in differentiation [109, 110]. In development, cells are presented with different signals in precisely controlled sequences. This level of precision cannot be easily achieved in static bulk culture where media components are likely to be more heterogeneously distributed. Microfluidic culture methods may be used to deliver different signaling molecules at specific time points, thus granting control over the temporal presentation of various factors. Microfluidic culture systems can be used to provide spatial and temporal control over cells and signaling molecules, as well as to introduce laminar-flow and concentration gradients. By combining these aspects, it may be possible to achieve a high degree of microenvironmental control of soluble factors in order to study the response of ESC fate decisions.

12.2.3 Microencapsulation

Encapsulation of cells in micron- to millimeter-size beads is commonly performed with different cell types in a variety of materials. Microencapsulation of cells can be used as a delivery vehicle for cell-transplantation therapies. One of the main advantages of this technique is that microcapsules can be designed to allow free diffusion of nutrients and metabolic waste products, yet prohibit entry of immune cells and antibodies, allowing transplantation without the need for immunosuppression. The three-dimensional environment provided by gels has been shown to enhance the function of cell types such as chrondrocytes [111] and hepatocytes [112]. Common materials for microencapsulation include agarose, alginate, and combinations of alginate, PLL, chitosan, and polyethylene glycol (PEG). Agarose beads encapsulating cells are generally produced using a water-in-oil emulsion. Alginate-based microcapsules are commonly produced using external gelation, in which cells in an alginate solution are dropped into a calcium-containing solution, which induces crosslinking of guluronic residues in the alginate. Microcapsule size can be controlled by vibrational or electrostatic disruption of alginate drops. Alginate beads are often coated with PLL, chitosan, PEG, or a combination thereof, in order to stabilize the polyanionic alginate. Additional stability is often added by a final coating of alginate.

12.2.3.1 Microencapsulation and ESC Self-Renewal

Microencapsulation may be a useful tool for maintenance of ESC pluripotency. Although traditional undifferentiated ESC culture methods rely on two-dimensional substrates, such environments do not provide the three-dimensional environment found in the blastocyst from which pluripotent ESCs are originally derived. Prolonged passaging in two-dimensional culture can lead to issues with cell pluripotency, viability, and altered karyotype, but maintenance of ESCs in three-dimensional microcapsules may be able to alleviate such concerns. Additionally, microencapsulation may be applicable to scalable culture methods for undifferentiated ESCs. Stirred tank bioreactors are commonly used for large-scale culture of cells that can be grown in suspension. The stirring in such bioreactors increases the homogeneity of media components and conditions such as cytokine and growth factor concentration, glucose concentration, and oxygen tension. Culture of ESCs in stirred bioreactors has been shown to result in massive uncontrolled agglomeration, severely reducing cell viability and limiting the utility of this approach [48, 113]. However, encapsulation of ESCs in microcapsules has been shown to inhibit agglomeration of cell clusters, effectively producing isolated cell microenvironments [113]. The use of MEF-conditioned media or addition of exogenous self-renewal factors, such as bFGF or LIF, may facilitate scaleable production of undifferentiated ESCs.

Microencapsulation may also be useful to mimic the microenvironment provided by MEFs. Feeder-free culture of hESCs is an important consideration as xenogenic contamination of hESC from MEFs limits the clinical applicability of stem cells. While replacement of MEFs with Matrigel-coated surfaces has been used to maintain undifferentiated ESCs, these methods still require MEF-conditioned media [17]. Culture of ESCs in completely MEF-free conditions has been reported [18, 19]; however, the long-term effectiveness in maintaining an undifferentiated phenotype is not known. Therefore, a system to effectively mimic the microenvironment provided by MEFs is of great interest. Hydrogels can be produced with various degrees of stiffness and elasticity, which may be an important characteristic of feeder layers to be mimicked. Matrix stiffness has been shown to affect cell fate for mesenchymal stem cells (MSCs) [114], indicating that this may be an important consideration for ESC self-renewal as well. Additionally, hydrogels can be functionalized to present peptide sequences such as RGD to cells [115, 116]. Production of hydrogels containing biomimetic peptide sequences important for ESC-MEF interactions may be useful for effectively maintaining the pluripotency of ESCs in microcapsules. Hydrogels made from a semipenetrating polymer network containing functional peptide sequences designed for cell adhesion, matrix metalloproteinase catalyzed degradation, and solubility have been produced [116]. Matrix stiffness of the polymer network can be controlled through the density of crosslinker used. Short-term studies showed that the synthetic ECM was able to maintain the pluripotency of ESCs, as indicated by Oct-4 and SSEA-4 expression. The independent control of both the mechanical and biochemical microenvironments in this system provides a unique method to study biomaterial-based regulation of stem cell fate.

12.2.3.2 Microencapsulation and ESC Differentiation

The culturing of ESCs within microcapsules has important applications for differentiation as well as self-renewal. One advantage that microencapsulation can provide over traditional differentiation methods is steric control of EB size. ESCs aggregated for one to three days have been encapsulated in agarose beads, which inhibited the ability of EBs to agglomerate into clusters of variable size and resulted in a relatively uniform population of EBs [113]. Additionally, after E-cadherin, the cell-adhesion molecule important for ESC clustering, is down-regulated, release of EBs from capsules into suspension does not result in EB agglomeration. Inhibition of EB agglomeration has been shown to increase cell viability and is therefore an important consideration for scalable differentiation processes. In addition, the ability to control EB size has important implications for developing robust protocols for directed differentiation.

The choice of materials used to encapsulate ESCs appears to be an important factor in cell differentiation. EBs encapsulated in alginate displayed different characteristics depending on alginate concentration [117]. In low alginate concentration (1.1 percent), cystic EBs were observed, as were spontaneously beating foci, indicative of cardiomyocyte differentiation. In contrast, when higher alginate concentration was used to produce microbeads, no cystic cell aggregates were found, and no spontaneous beating was observed. When these aggregates were released from the microbeads by calcium chelation and cultured in suspension, cystic EBs were shown to form, and subsequent plating on gelatin-coated dishes resulted in spontaneously beating regions. Alginate-PLL microbeads have also been investigated for ESC encapsulation [118, 119]. When single-cell suspensions of ESCs were encapsulated in alginate/poly-L-lysine microbeads, hepatocyte function, as indicated by urea secretion and intracellular albumin levels, was similar to or greater than that found in EB-derived cells [118]. This demonstrated that hepatocyte differentiation can be achieved within alginate-PLL microbeads, which may in turn be used in a scalable system to generate larger cell yields.

Functionalization of microcapsules with growth factors or peptide sequences provides an additional level of control over ESC microenvironments. Different ECM proteins have been shown to influence ESC fate [26, 27], indicating that integrin signaling may be important in differentiation. Peptide sequences that mimic ECM cell-adhesion motifs may therefore be useful for directing stem cell differentiation. Hydrogels modified with RGD sequences have been studied for chrondrogenic differentiation of hESC-derived MSC-like cells [115]. The addition of RGD-containing peptides within PEG hydrogels resulted in increased production of glycosaminoglycans and collagen, as well as cartilage-specific genes, compared to unmodified gels. Use of additional adhesion peptide sequences, such as YIGSR, may be useful for inducing different differentiation patterns. Immobilization or entrapment of growth factors into gels may also enhance control over ESC differentiation. TFG-β tethered to PEG has been shown to increase matrix production in vascular smooth muscle cells, indicating that certain growth factors can retain function after immobilization [120]. Tethering of growth factors into PEG microgels may be useful for controlling ESC differentiation. Nerve growth factor (NGF) has been tethered to polyethylene oxide segments of Pluronic surfaces for surface immobilization, and culture of EB derived cells on these surfaces demonstrated increased tyrosine hydroxylase expression, indicative of dopaminergic neuronal differentiation [121].

12.2.4 Microparticles

Biodegradable polymer microparticles are commonly employed for controlled release of bioactive factors both in vitro and in vivo [122]. Microparticles can be administered through minimally invasive injections, and the physical and chemical properties of the particles can be selected to tailor the course of factor release. The particle size, porosity, and morphology, as well as the molecular weight, rate of hydrolysis, and copolymer composition of the polymer, all influence the release profile of the particles [123, 124]. Additionally, particles that are environmentally responsive have also been investigated, including pH [125], temperature [126], and enzyme-sensitive materials [127]. Polymer microparticles encapsulating hydrophobic molecules are generally produced using an oil-in-water single emulsion, whereas hydrophilic molecules, such as proteins and growth factors, are typically encapsulated using a water-in-oil-in-water double emulsion [123]. While particle size can be varied by adjusting the energy applied to disperse the emulsion, this technique has limited effectiveness in producing a uniform size distribution. More advanced techniques, such as spraying technology utilizing an ultrasonic transducer to form uniform organic droplets, have been applied to produce uniformly sized particles [128, 129].

Polymer microspheres can be incorporated within cell aggregates in order to locally release bioactive factors. Cells derived from fetal rat brains were mixed with poly(lactic-co-glycolic acid) (PLGA) microparticles containing NGF under rotation culture to produce spheroids composed of both cells and particles termed neotissues [130]. This approach allows the production of a programmable synthetic microenvironment through the use of matrix proteins to coat the particles and the controlled release of bioactive factors from the biodegradable polymer microspheres. Such an approach may be utilized for controlled differentiation of ESCs within EBs. Microparticles containing growth factors or other instructive cues can be incorporated within the interior of EBs to locally delivery molecules in a temporally controlled fashion. This may allow for more homogeneous distribution of signaling molecules within EBs and result in more consistent differentiation patterns. Additionally, sequential addition and subtraction of differentiation factors is often practiced in vitro but cannot be easily translated into cell-transplantation therapies. The use of multiple particles with different release kinetics containing different factors may be used to provide a sequence of instructions to differentiating cells without altering media components, which may be more amenable to in vivo applications.

12.3 Conclusion

ESCs are subject to a variety of different molecular cues within their microenvironments capable of directly and indirectly influencing self-renewal and differentiation. The application of microscale technologies to ESC studies provides unique opportunities to precisely control aspects of cellular microenvironments that regulate ESC fate and function at the individual cell level. Thus, the application of microscale technologies to ESC microenvironments should facilitate the development of improved cell therapies and diagnostics.

References

[1] Evans, M. J., and Kaufman, M. H., "Establishment in culture of pluripotential cells from mouse embryos," *Nature*, Vol. 292, No. 5819, 1981, pp. 154–156.

[2] Martin, G. R., "Isolation of a pluripotent cell line from early mouse embryos cultured in medium conditioned by teratocarcinoma stem cells," *Proc. Natl. Acad. Sci. USA*, Vol. 78, No. 12, 1981, pp. 7634–7638.

[3] Doetschman, T. C., Eistetter, H., Katz, M., Schmidt, W., and Kemler, R., "The in vitro development of blastocyst-derived embryonic stem cell lines: Formation of visceral yolk sac, blood islands and myocardium," *J. Embr. Exp. Morph.*, Vol. 87, 1985, pp. 27–45.

[4] Thomson, J. A., Kalishman, J., Golos, T. G., Durning, M., Harris, C. P., Becker, R. A., and Hearn, J. P., "Isolation of a primate embryonic stem cell line," *Proc. Natl. Acad. Sci. USA*, Vol. 92, No. 17, 1995, pp. 7844–7848.

[5] Thomson, J. A., Kalishman, J., Golos, T. G., Durning, M., Harris, C. P., and Hearn, J. P., "Pluripotent cell lines derived from common marmoset (Callithrix jacchus) blastocysts," *Biol. Reprod.*, Vol. 55, No. 2, 1996, pp. 254–259.

[6] Thomson, J. A., Itskovitz-Eldor, J., Shapiro, S. S., Waknitz, M. A., Swiergiel, J. J., Marshall, V. S., and Jones, J. M., "Embryonic stem cell lines derived from human blastocysts," *Science,* Vol. 282, No. 5391, 1998, pp. 1145–1147.

[7] Reubinoff, B. E., Pera, M. F., Fong, C. Y., Trounson, A., and Bongso, A., "Embryonic stem cell lines from human blastocysts: Somatic differentiation in vitro," *Nat. Biotechnol.*, Vol. 18, No. 4, 2000, pp. 399–404.

[8] Liu, N., Lu, M., Tian, X., and Han, Z., "Molecular mechanisms involved in self-renewal and pluripotency of embryonic stem cells," *J. Cell. Physiol.*, Vol. 211, No. 2, 2007, pp. 279–286.

[9] Noggle, S. A., James, D., and Brivanlou, A. H., "A molecular basis for human embryonic stem cell pluripotency," *Stem Cell Rev.*, Vol. 1, No. 2, 2005, pp. 111–118.

[10] Avery, S., Inniss, K., and Moore, H., "The regulation of self-renewal in human embryonic stem cells," *Stem Cells Dev.*, Vol. 15, No. 5, 2006, pp. 729–740.

[11] Smith, A. G., Heath, J. K., Donaldson, D. D., Wong, G. G., Moreau, J., Stahl, M., and Rogers, D., "Inhibition of pluripotential embryonic stem cell differentiation by purified polypeptides," *Nature*, Vol. 336, No. 6200, 1988, pp. 688–690.

[12] Williams, R. L., Hilton, D. J., Pease, S., Willson, T. A., Stewart, C. L., Gearing, D. P., Wagner, E. F., Metcalf, D., Nicola, N. A., and Gough, N. M., "Myeloid leukaemia inhibitory factor maintains the developmental potential of embryonic stem cells," *Nature*, Vol. 336, No. 6200, 1988, pp. 684–687.

[13] Ying, Q. L., Nichols, J., Chambers, I., and Smith, A., "BMP induction of Id proteins suppresses differentiation and sustains embryonic stem cell self-renewal in collaboration with STAT3," *Cell*, Vol. 115, No. 3, 2003, pp. 281–292.

[14] Amit, M., Carpenter, M. K., Inokuma, M. S., Chiu, C. P., Harris, C. P., Waknitz, M. A., Itskovitz-Eldor, J., and Thomson, J. A., "Clonally derived human embryonic stem cell lines maintain pluripotency and proliferative potential for prolonged periods of culture," *Dev. Biol.*, Vol. 227, No. 2, 2000, pp. 271–278.

[15] Martin, M. J., Muotri, A., Gage, F., and Varki, A., "Human embryonic stem cells express an immunogenic nonhuman sialic acid," *Nat. Med.*, Vol. 11, No. 2, 2005, pp. 228–232.

[16] Heiskanen, A., Satomaa, T., Tiitinen, S., Laitinen, A., Mannelin, S., Impola, U., Mikkola, M., Olsson, C., Miller-Podraza, H., Blomqvist, M., Olonen, A., Salo, H., Lehenkari, P., Tuuri, T., Otonkoski, T., Natunen, J., Saarinen, J., and Laine, J., "N-glycolylneuraminic acid xenoantigen contamination of human embryonic and mesenchymal stem cells is substantially reversible," *Stem Cells*, Vol. 25, No. 1, 2007, pp. 197–202.

[17] Xu, C., Inokuma, M. S., Denham, J., Golds, K., Kundu, P., Gold, J. D., and Carpenter, M. K., "Feeder-free growth of undifferentiated human embryonic stem cells," *Nat. Biotechnol.*, Vol. 19, No. 10, 2001, pp. 971–974.

[18] Amit, M., Shariki, C., Margulets, V., and Itskovitz-Eldor, J., "Feeder layer- and serum-free culture of human embryonic stem cells," *Biol. Reprod.*, Vol. 70, No. 3, 2004, pp. 837–845.

[19] Beattie, G. M., Lopez, A. D., Bucay, N., Hinton, A., Firpo, M. T., King, C. C., and Hayek, A., "Activin A maintains pluripotency of human embryonic stem cells in the absence of feeder layers," *Stem Cells*, Vol. 23, No. 4, 2005, pp. 489–495.

[20] Xu, C., Jiang, J., Sottile, V., McWhir, J., Lebkowski, J., and Carpenter, M. K., "Immortalized fibroblast-like cells derived from human embryonic stem cells support undifferentiated cell growth," *Stem Cells*, Vol. 22, No. 6, 2004, pp. 972–980.

[21] Amit, M., Margulets, V., Segev, H., Shariki, K., Laevsky, I., Coleman, R., and Itskovitz-Eldor, J., "Human feeder layers for human embryonic stem cells," *Biol. Reprod.*, Vol. 68, No. 6, 2003, pp. 2150–2156.

[22] Li, Y., Powell, S., Brunette, E., Lebkowski, J., and Mandalam, R., "Expansion of human embryonic stem cells in defined serum-free medium devoid of animal-derived products," *Biotechnol. Bioeng.*, Vol. 91, No. 6, 2005, pp. 688–698.

[23] Ludwig, T. E., Levenstein, M. E., Jones, J. M., Berggren, W. T., Mitchen, E. R., Frane, J. L., Crandall, L. J., Daigh, C. A., Conard, K. R., Piekarczyk, M. S., Llanas, R. A., and Thomson, J. A., "Derivation of human embryonic stem cells in defined conditions," *Nat. Biotechnol.*, Vol. 24, No. 2, 2006, pp. 185–187.

[24] Mitalipova, M. M., Rao, R. R., Hoyer, D. M., Johnson, J. A., Meisner, L. F., Jones, K. L., Dalton, S., and Stice, S. L., "Preserving the genetic integrity of human embryonic stem cells," *Nat. Biotechnol.*, Vol. 23, No. 1, 2005, pp. 19–20.

[25] Ying, Q. L., Stavridis, M., Griffiths, D., Li, M., and Smith, A., "Conversion of embryonic stem cells into neuroectodermal precursors in adherent monoculture," *Nat. Biotechnol.*, Vol. 21, No. 2, 2003, pp. 183–186.

[26] Nakano, T., Kodama, H., and Honjo, T., "Generation of lymphohematopoietic cells from embryonic stem cells in culture," *Science,* Vol. 265, No. 5175, 1994, pp. 1098–1101.

[27] Goetz, A. K., Scheffler, B., Chen, H. X., Wang, S., Suslov, O., Xiang, H., Brustle, O., Roper, S. N., and Steindler, D. A., "Temporally restricted substrate interactions direct fate and specification of neural precursors derived from embryonic stem cells," *Proc. Natl. Acad. Sci. USA*, Vol. 103, No. 29, 2006, pp. 11063–11068.

[28] Tropepe, V., Hitoshi, S., Sirard, C., Mak, T. W., Rossant, J., and van der, D., Kooy, "Direct neural fate specification from embryonic stem cells: A primitive mammalian neural stem cell stage acquired through a default mechanism," *Neuron*, Vol. 30, No. 1, 2001, pp. 65–78.

[29] Gerrard, L., Rodgers, L., and Cui, W., "Differentiation of human embryonic stem cells to neural lineages in adherent culture by blocking bone morphogenetic protein signaling," *Stem Cells*, Vol. 23, No. 9, 2005, pp. 1234–1241.

[30] Yamazoe, H., Kobori, M., Murakami, Y., Yano, K., Satoh, M., Mizuseki, K., Sasai, Y., and Iwata, H., "One-step induction of neurons from mouse embryonic stem cells in serum-free media containing vitamin B12 and heparin," *Cell Transplan.*, Vol. 15, No. 2, 2006, pp. 135–145.

[31] Kawasaki, H., Mizuseki, K., Nishikawa, S., Kaneko, S., Kuwana, Y., Nakanishi, S., Nishikawa, S. I., and Sasai, Y., "Induction of midbrain dopaminergic neurons from ES cells by stromal cell-derived inducing activity," *Neuron*, Vol. 28, No. 1, 2000, pp. 31–40.

[32] Zeng, X., Cai, J., Chen, J., Luo, Y., You, Z. B., Fotter, E., Wang, Y., Harvey, B., Miura, T., Backman, C., Chen, G. J., Rao, M. S., and Freed, W. J., "Dopaminergic differentiation of human embryonic stem cells," *Stem Cells*, Vol. 22, No. 6, 2004, pp. 925–940.

[33] Vodyanik, M. A., Bork, J. A., Thomson, and Slukvin, J. A., II, "Human embryonic stem cell-derived CD34+ cells: Efficient production in the coculture with OP9 stromal cells and analysis of lymphohematopoietic potential," *Blood*, Vol. 105, No. 2, 2005, pp. 617–626.

[34] Eto, K., Murphy, R., Kerrigan, S. W., Bertoni, A., Stuhlmann, H., Nakano, T., Leavitt, A. D., and Shattil, S. J., "Megakaryocytes derived from embryonic stem cells implicate CalDAG-GEFI in integrin signaling," *Proc. Natl. Acad. Sci. USA*, Vol. 99, No. 20, 2002, pp. 12819–12824.

[35] Cho, S. K., Webber, T. D., Carlyle, J. R., Nakano, T., Lewis, S. M., and Zuniga-Pflucker, J. C., "Functional characterization of B lymphocytes generated in vitro from embryonic stem cells," *Proc. Natl. Acad. Sci. USA*, Vol. 96, No. 17, 1999, pp. 9797–9802.

[36] Kawasaki, H., Suemori, H., Mizuseki, K., Watanabe, K., Urano, F., Ichinose, H., Haruta, M., Takahashi, M., Yoshikawa, K., Nishikawa, S., Nakatsuji, N., and Sasai, Y., "Generation of dopaminergic neurons and pigmented epithelia from primate ES cells by stromal cell-derived inducing activity," *Proc. Natl. Acad. Sci. USA*, Vol. 99, No. 3, 2002, pp. 1580–1585.

[37] Mummery, C., Ward-van Oostwaard, D., Doevendans, P., Spijker, R., van den Brink, S., Hassink, R., van der Heyden, M., Opthof, T., Pera, M., de la Riviere, A. B., Passier, R., and Tertoolen, L., "Differentiation of human embryonic stem cells to cardiomyocytes: Role of coculture with visceral endoderm-like cells," *Circulation*, Vol. 107, No. 21, 2003, pp. 2733–2740.

[38] Buttery, L. D., Bourne, S., Xynos, J. D., Wood, H., Hughes, F. J., Hughes, S. P., Episkopou, V., and Polak, J. M., "Differentiation of osteoblasts and in vitro bone formation from murine embryonic stem cells," *Tissue Eng.*, Vol. 7, No. 1, 2001, pp. 89–99.

[39] Fair, J. H., Cairns, B. A., Lapaglia, M., Wang, J., Meyer, A. A., Kim, H., Hatada, S., Smithies, O., and Pevny, L., "Induction of hepatic differentiation in embryonic stem cells by co-culture with embryonic cardiac mesoderm," *Surgery*, Vol. 134, No. 2, 2003, pp. 189–196.

[40] Kaufman, D. S., Hanson, E. T., Lewis, R. L., Auerbach, R., and Thomson, J. A., "Hematopoietic colony-forming cells derived from human embryonic stem cells," *Proc. Natl. Acad. Sci. USA*, Vol. 98, No. 19, 2001, pp. 10716–10721.

[41] Desbaillets, I., Ziegler, U., Groscurth, P., and Gassmann, M., "Embryoid bodies: An in vitro model of mouse embryogenesis," *Exp. Physiol.*, Vol. 85, No. 6, 2000, pp. 645–651.

[42] Hopfl, G., Gassmann, M., and Desbaillets, I., "Differentiating embryonic stem cells into embryoid bodies," *Methods in Mol. Biol.*, Vol. 254, 2004, pp. 79–98.

[43] Maltsev, V. A., Wobus, A. M., Rohwedel, J., Bader, M., and Hescheler, J., "Cardiomyocytes differentiated in vitro from embryonic stem cells developmentally express cardiac-specific genes and ionic currents," *Circ. Res.*, Vol. 75, No. 2, 1994, pp. 233–244.

[44] Zhang, S. C., Wernig, M., Duncan, I. D., Brustle, O., and Thomson, J. A., "In vitro differentiation of transplantable neural precursors from human embryonic stem cells," *Nat. Biotechnol.*, Vol. 19, No. 12, 2001, pp. 1129–1133.

[45] Klug, M. G., Soonpaa, M. H., Koh, G. Y., and Field, L. J., "Genetically selected cardiomyocytes from differentiating embronic stem cells form stable intracardiac grafts," *J. Clin. Invest.*, Vol. 98, No. 1, 1996, pp. 216–224.

[46] Vittet, D., Prandini, M. H., Berthier, R., Schweitzer, A., Martin-Sisteron, H., Uzan, G., and Dejana, E., "Embryonic stem cells differentiate in vitro to endothelial cells through successive maturation steps," *Blood*, Vol. 88, No. 9, 1996, pp. 3424–3431.

[47] Cameron, C. M., Hu, W. S., and Kaufman, D. S., "Improved development of human embryonic stem cell-derived embryoid bodies by stirred vessel cultivation," *Biotechnol. Bioeng.*, Vol. 94, No. 5, 2006, pp. 938–948.

[48] Gerecht-Nir, S., Cohen, S., and Itskovitz-Eldor, J., "Bioreactor cultivation enhances the efficiency of human embryoid body (hEB) formation and differentiation," *Biotechnol. Bioeng.*, Vol. 86, No. 5, 2004, pp. 493–502.

[49] Fok, E. Y., and Zandstra, P. W., "Shear-controlled single-step mouse embryonic stem cell expansion and embryoid body-based differentiation," *Stem Cells*, Vol. 23, No. 9, 2005, pp. 1333–1342.

[50] Schroeder, M., Niebruegge, S., Werner, A., Willbold, E., Burg, M., Ruediger, M., Field, L. J., Lehmann, J., and Zweigerdt, R., "Differentiation and lineage selection of mouse embryonic stem cells in a stirred bench scale bioreactor with automated process control," *Biotechnol. Bioeng.*, Vol. 92, No. 7, 2005, pp. 920–933.

[51] Carpenedo, R. L., Sargent, C. Y., and McDevitt, T. C., "Rotary suspension culture enhances the efficiency, yield and homogeneity of embryoid body differentiation," *Stem Cells*, Vol. 2007.

[52] Brown, P. O., and Botstein, D., "Exploring the new world of the genome with DNA microarrays," *Nat. Genet.*, Vol. 21, No. 1, 1999, pp. 33–37.

[53] Templin, M. F., Stoll, D., Schrenk, M., Traub, P. C., Vohringer, C. F., and Joos, T. O., "Protein microarray technology," *Trends in Biotechnol.*, Vol. 20, No. 4, 2002, pp. 160–166.

[54] Jung, D. R., Kapur, R., Adams, T., Giuliano, K. A., Mrksich, M., Craighead, H. G., and Taylor, D. L., "Topographical and physicochemical modification of material surface to enable patterning of living cells," *Crit. Rev. Biotechnol.*, Vol. 21, No. 2, 2001, pp. 111–154.

[55] Desai, T. A., "Micro- and nanoscale structures for tissue engineering constructs," *Med. Eng. Phys.*, Vol. 22, No. 9, 2000, pp. 595–606.

[56] Khademhosseini, A., Langer, R., Borenstein, J., and Vacanti, J. P., "Microscale technologies for tissue engineering and biology," *Proc. Natl. Acad. Sci. USA*, Vol. 103, No. 8, 2006, pp. 2480–2487.

[57] Falconnet, D., Csucs, G., Grandin, H. M., and Textor, M., "Surface engineering approaches to micropattern surfaces for cell-based assays," *Biomaterials*, Vol. 27, No. 16, 2006, pp. 3044–3063.

[58] Kane, R. S., Takayama, S., Ostuni, E., Ingber, D. E., and Whitesides, G. M., "Patterning proteins and cells using soft lithography," *Biomaterials*, Vol. 20, Nos. 23–24, 1999, pp. 2363–2376.

[59] Li, N., Tourovskaia, A., and Folch, A., "Biology on a chip: Microfabrication for studying the behavior of cultured cells," *Crit. Rev. Biomed. Eng.*, Vol. 31, Nos. 5–6, 2003, pp. 423–488.

[60] Hui, E. E., and Bhatia, S. N., "Microscale control of cell contact and spacing via three-component surface patterning," *Langmuir*, Vol. 23, No. 8, 2007, pp. 4103–4107.

[61] Blawas, A. S., and Reichert, W. M., "Protein patterning," *Biomaterials*, Vol. 19, Nos. 7–9, 1998, pp. 595–609.

[62] Flaim, C. J., Chien, S., and Bhatia, S. N., "An extracellular matrix microarray for probing cellular differentiation," *Nat. Methods*, Vol. 2, No. 2, 2005, pp. 119–125.

[63] Bhatia, S. N., Balis, U. J., Yarmush, M. L., and Toner, M., "Microfabrication of hepatocyte/fibroblast co-cultures: Role of homotypic cell interactions," *Biotech. Prog.*, Vol. 14, No. 3, 1998, pp. 378–387.

[64] Bhatia, S. N., Balis, U. J., Yarmush, M. L., and Toner, M., "Probing heterotypic cell interactions: Hepatocyte function in microfabricated co-cultures," *J. Biomat. Sci.: Polymer Ed.*, Vol. 9, No. 11, 1998, pp. 1137–1160.

[65] Wu, C. C., Li, Y. S., Haga, J. H., Kaunas, R., Chiu, J. J., Su, F. C., Usami, S., and Chien, S., "Directional shear flow and Rho activation prevent the endothelial cell apoptosis induced by micropatterned anisotropic geometry," *Proc. Natl. Acad. Sci. USA*, Vol. 104, No. 4, 2007, pp. 1254–1259.

[66] McDevitt, T. C., Angello, J. C., Whitney, M. L., Reinecke, H., Hauschka, S. D., Murry, C. E., and Stayton, P. S., "In vitro generation of differentiated cardiac myofibers on micropatterned laminin surfaces," *J. Biomed. Mater. Res.*, Vol. 60, No. 3, 2002, pp. 472–479.

[67] McDevitt, T. C., Woodhouse, K. A., Hauschka, S. D., Murry, C. E., and Stayton, P. S., "Spatially organized layers of cardiomyocytes on biodegradable polyurethane films for myocardial repair," *J. Biomed. Mater. Res. A*, Vol. 66, No. 3, 2003, pp. 586–595.

[68] Engler, A. J., Griffin, M. A., Sen, S., Bonnemann, C. G., Sweeney, H. L., and Discher, D. E., "Myotubes differentiate optimally on substrates with tissue-like stiffness: Pathological implications for soft or stiff microenvironments," *J. Cell. Biol.*, Vol. 166, No. 6, 2004, pp. 877–887.

[69] Acarturk, T. O., Peel, M. M., Petrosko, P., LaFramboise, W., Johnson, P. C., and DiMilla, P. A., "Control of attachment, morphology, and proliferation of skeletal myoblasts on silanized glass," *J. Biomed. Mater. Res.*, Vol. 44, No. 4, 1999, pp. 355–370.

[70] Khademhosseini, A., Ferreira, L., Blumling, J., III, Yeh, J., Karp, J. M., Fukuda, J., and Langer, R., "Co-culture of human embryonic stem cells with murine embryonic fibroblasts on microwell-patterned substrates," *Biomaterials*, Vol. 27, No. 36, 2006, pp. 5968–5977.

[71] Mohr, J. C., de Pablo, J. J., and Palecek, S. P., "3-D microwell culture of human embryonic stem cells," *Biomaterials*, Vol. 27, No. 36, 2006, pp. 6032–6042.

[72] Khademhosseini, A., Suh, K. Y., Yang, J. M., Eng, G., Yeh, J., Levenberg, S., and Langer, R., "Layer-by-layer deposition of hyaluronic acid and poly-L-lysine for patterned cell co-cultures," *Biomaterials*, Vol. 25, No. 17, 2004, pp. 3583–3592.

[73] Yousaf, M. N., Houseman, B. T., and Mrksich, M., "Using electroactive substrates to pattern the attachment of two different cell populations," *Proc. Natl. Acad. Sci. USA*, Vol. 98, No. 11, 2001, pp. 5992–5996.

[74] Yamato, M., Konno, C., Utsumi, M., Kikuchi, A., and Okano, T., "Thermally responsive polymer-grafted surfaces facilitate patterned cell seeding and co-culture," *Biomaterials*, Vol. 23, No. 2, 2002, pp. 561–567.

[75] McBeath, R., Pirone, D. M., Nelson, C. M., Bhadriraju, K., and Chen, C. S., "Cell shape, cytoskeletal tension, and RhoA regulate stem cell lineage commitment," *Dev. Cell*, Vol. 6, No. 4, 2004, pp. 483–495.

[76] Chin, V. I., Taupin, P., Sanga, S., Scheel, J., Gage, F. H., and Bhatia, S. N., "Microfabricated platform for studying stem cell fates," *Biotechnol. Bioeng.*, Vol. 88, No. 3, 2004, pp. 399–415.

[77] Anderson, D. G., Levenberg, S., and Langer, R., "Nanoliter-scale synthesis of arrayed biomaterials and application to human embryonic stem cells," *Nat. Biotechnol.*, Vol. 22, No. 7, 2004, pp. 863–866.

[78] McDonald, J. C., Duffy, D. C., Anderson, J. R., Chiu, D. T., Wu, H., Schueller, O. J., and Whitesides, G. M., "Fabrication of microfluidic systems in poly(dimethylsiloxane)," *Electrophoresis*, Vol. 21, No. 1, 2000, pp. 27–40.

[79] Auroux, P. A., Koc, Y., deMello, A., Manz, A., and Day, P. J., "Miniaturised nucleic acid analysis," *Lab Chip*, Vol. 4, No. 6, 2004, pp. 534–546.

[80] Huh, D., Gu, W., Kamotani, Y., Grotberg, J. B., and Takayama, S., "Microfluidics for flow cytometric analysis of cells and particles," *Physiol. Meas.*, Vol. 26, No. 3, 2005, pp. R73–R98.

[81] Sia, S. K., and Whitesides, G. M., "Microfluidic devices fabricated in poly(dimethylsiloxane) for biological studies," *Electrophoresis*, Vol. 24, No. 21, 2003, pp. 3563–3576.

[82] Manz, A., Harrison, D. J., Verpoorte, E. M., Fettinger, J. C., Paulus, A., Ludi, H., and Widmer, H. M., "Planar chips technology for miniaturization and integration of separation techniques into monitoring systems—capillary electrophoresis on a chip," *J. Chromatography*, Vol. 593, Nos. 1–2, 1992, pp. 253–258.

[83] Zheng, B., Roach, L. S., and Ismagilov, R. F., "Screening of protein crystallization conditions on a microfluidic chip using nanoliter-size droplets," *J. Am. Chem. Soc.*, Vol. 125, No. 37, 2003, pp. 11170–11171.

[84] Hansen, C. L., Skordalakes, E., Berger, J. M., and Quake, S. R., "A robust and scalable microfluidic metering method that allows protein crystal growth by free interface diffusion," *Proc. Natl. Acad. Sci. USA*, Vol. 99, No. 26, 2002, pp. 16531–16536.

[85] Leclerc, E., Sakai, Y., and Fujii, T., "Microfluidic PDMS (polydimethylsiloxane) bioreactor for large-scale culture of hepatocytes," *Biotech. Prog.*, Vol. 20, No. 3, 2004, pp. 750–755.

[86] Hung, P. J., Lee, P. J., Sabounchi, P., Lin, R., and Lee, L. P., "Continuous perfusion microfluidic cell culture array for high-throughput cell-based assays," *Biotechnol. Bioeng.*, Vol. 89, No. 1, 2005, pp. 1–8.

[87] Wheeler, A. R., Throndset, W. R., Whelan, R. J., Leach, A. M., Zare, R. N., Liao, Y. H., Farrell, K., Manger, I. D., and Daridon, A., "Microfluidic device for single-cell analysis," *Anal. Chem.*, Vol. 75, No. 14, 2003, pp. 3581–3586.

[88] Rhee, S. W., Taylor, A. M., Tu, C. H., Cribbs, D. H., Cotman, C. W., and Jeon, N. L., "Patterned cell culture inside microfluidic devices," *Lab Chip*, Vol. 5, No. 1, 2005, pp. 102–107.

[89] Takayama, S., McDonald, J. C., Ostuni, E., Liang, M. N., Kenis, P. J., Ismagilov, R. F., and Whitesides, G. M., "Patterning cells and their environments using multiple laminar fluid flows in capillary networks," *Proc. Natl. Acad. Sci. USA*, Vol. 96, No. 10, 1999, pp. 5545–5548.

[90] Tan, W., and Desai, T. A., "Microfluidic patterning of cells in extracellular matrix biopolymers: Effects of channel size, cell type, and matrix composition on pattern integrity," *Tissue Eng.*, Vol. 9, No. 2, 2003, pp. 255–267.

[91] Chiu, D. T., Jeon, N. L., Huang, S., Kane, R. S., Wargo, C. J., Choi, I. S., Ingber, D. E., and Whitesides, G. M., "Patterned deposition of cells and proteins onto surfaces by using three-dimensional microfluidic systems," *Proc. Natl. Acad. Sci. USA*, Vol. 97, No. 6, 2000, pp. 2408–2413.

[92] Takayama, S., Ostuni, E., LeDuc, P., Naruse, K., Ingber, D. E., and Whitesides, G. M., "Subcellular positioning of small molecules," *Nature*, Vol. 411, No. 6841, 2001, p. 1016.

[93] Takayama, S., Ostuni, E., LeDuc, P., Naruse, K., Ingber, D. E., and Whitesides, G. M., "Selective chemical treatment of cellular microdomains using multiple laminar streams," *Chem. Biol.*, Vol. 10, No. 2, 2003, pp. 123–130.

[94] Pihl, J., Karlsson, M., and Chiu, D. T., "Microfluidic technologies in drug discovery," *Drug Discov. Today*, Vol. 10, No. 20, 2005, pp. 1377–1383.

[95] Dittrich, P. S., and Manz, A., "Lab-on-a-chip: Microfluidics in drug discovery," *Nat. Rev. Drug Discov.*, Vol. 5, No. 3, 2006, pp. 210–218.

[96] Kim, L., Vahey, M. D., Lee, H. Y., and Voldman, J., "Microfluidic arrays for logarithmically perfused embryonic stem cell culture," *Lab Chip*, Vol. 6, No. 3, 2006, pp. 394–406.

[97] Raty, S., Walters, E. M., Davis, J., Zeringue, H., Beebe, D. J., Rodriguez-Zas, S. L., and Wheeler, M. B., "Embryonic development in the mouse is enhanced via microchannel culture," *Lab Chip*, Vol. 4, No. 3, 2004, pp. 186–190.

[98] Glasgow, I. K., Zeringue, H. C., Beebe, D. J., Choi, S. J., Lyman, J. T., Chan, N. G., and Wheeler, M. B., "Handling individual mammalian embryos using microfluidics," *IEEE Trans. Biomed. Eng.*, Vol. 48, No. 5, 2001, pp. 570–578.

[99] Meinel, L., Karageorgiou, V., Fajardo, R., Snyder, B., Shinde-Patil, V., Zichner, L., Kaplan, D., Langer, R., and Vunjak-Novakovic, G., "Bone tissue engineering using human mesenchymal stem cells: Effects of scaffold material and medium flow," *Annals Biomed. Eng.*, Vol. 32, No. 1, 2004, pp. 112–122.

[100] Kapur, S., Baylink, D. J., and Lau, K. H., "Fluid flow shear stress stimulates human osteoblast proliferation and differentiation through multiple interacting and competing signal transduction pathways," *Bone*, Vol. 32, No. 3, 2003, pp. 241–251.

[101] Yamamoto, K., Takahashi, T., Asahara, T., Ohura, N., Sokabe, T., Kamiya, A., and Ando, J., "Proliferation, differentiation, and tube formation by endothelial progenitor cells in response to shear stress," *J. Appl. Physiol.*, Vol. 95, No. 5, 2003, pp. 2081–2088.

[102] Li, Y. J., Batra, N. N., You, L., Meier, S. C., Coe, I. A., Yellowley, C. E., and Jacobs, C. R., "Oscillatory fluid flow affects human marrow stromal cell proliferation and differentiation," *J. Ortho. Res.*, Vol. 22, No. 6, 2004, pp. 1283–1289.

[103] Yamamoto, K., Sokabe, T., Watabe, T., Miyazono, K., Yamashita, J. K., Obi, S., Ohura, N., Matsushita, A., Kamiya, A., and Ando, J., "Fluid shear stress induces differentiation of Flk-1-positive embryonic stem cells into vascular endothelial cells in vitro," *Am. J. Physiol.: Heart and Circ. Physiol.*, Vol. 288, No. 4, 2005, pp. H1915–H1924.

[104] Tourovskaia, A., Figueroa-Masot, X., and Folch, A., "Differentiation-on-a-chip: A microfluidic platform for long-term cell culture studies," *Lab Chip*, Vol. 5, No. 1, 2005, pp. 14–19.

[105] Chung, B. G., Flanagan, L. A., Rhee, S. W., Schwartz, P. H., Lee, A. P., Monuki, E. S., and Jeon, N. L., "Human neural stem cell growth and differentiation in a gradient-generating microfluidic device," *Lab Chip*, Vol. 5, No. 4, 2005, pp. 401–406.

[106] Thompson, D. M., King, K. R., Wieder, K. J., Toner, M., Yarmush, M. L., and Jayaraman, A., "Dynamic gene expression profiling using a microfabricated living cell array," *Anal. Chem.*, Vol. 76, No. 14, 2004, pp. 4098–4103.

[107] Prokop, A., Prokop, Z., Schaffer, D., Kozlov, E., Wikswo, J., Cliffel, D., and Baudenbacher, F., "NanoLiterBioReactor: Long-term mammalian cell culture at nanofabricated scale," *Biomed. Microdevices*, Vol. 6, No. 4, 2004, pp. 325–339.

[108] Pihl, J., Sinclair, J., Sahlin, E., Karlsson, M., Petterson, F., Olofsson, J., and Orwar, O., "Microfluidic gradient-generating device for pharmacological profiling," *Anal. Chem.*, Vol. 77, No. 13, 2005, pp. 3897–3903.

[109] Goncalves, M. B., Boyle, J., Webber, D. J., Hall, S., Minger, S. L., and Corcoran, J. P., "Timing of the retinoid-signalling pathway determines the expression of neuronal markers in neural progenitor cells," *Dev. Biol.*, Vol. 278, No. 1, 2005, pp. 60–70.

[110] Rohwedel, J., Guan, K., and Wobus, A. M., "Induction of cellular differentiation by retinoic acid in vitro," *Cells Tissues Organs*, Vol. 165, Nos. 3–4, 1999, pp. 190–202.

[111] Aulthouse, A. L., Beck, M., Griffey, E., Sanford, J., Arden, K., Machado, M. A., and Horton, W. A., "Expression of the human chondrocyte phenotype in vitro," *In Vitro Cell. Dev. Biol.*, Vol. 25, No. 7, 1989, pp. 659–668.

[112] Bissell, D. M., Arenson, D. M., Maher, J. J., and Roll, F. J., "Support of cultured hepatocytes by a laminin-rich gel: Evidence for a functionally significant subendothelial matrix in normal rat liver," *J. Clin. Invest.*, Vol. 79, No. 3, 1987, pp. 801–812.

[113] Dang, S. M., Gerecht-Nir, S., Chen, J., Itskovitz-Eldor, J., and Zandstra, P. W., "Controlled, scalable embryonic stem cell differentiation culture," *Stem Cells*, Vol. 22, No. 3, 2004, pp. 275–282.

[114] Engler, A. J., Sen, S., Sweeney, H. L., and Discher, D. E., "Matrix elasticity directs stem cell lineage specification," *Cell*, Vol. 126, No. 4, 2006, pp. 677–689.

[115] Hwang, N. S., Varghese, S., Zhang, Z., and Elisseeff, J., "Chondrogenic differentiation of human embryonic stem cell-derived cells in arginine-glycine-aspartate-modified hydrogels," *Tissue Eng.*, Vol. 12, No. 9, 2006, pp. 2695–2706.

[116] Li, Y. J., Chung, E. H., Rodriguez, R. T., Firpo, M. T., and Healy, K. E., "Hydrogels as artificial matrices for human embryonic stem cell self-renewal," *J. Biomed. Mater. Res. A*, Vol. 79, No. 1, 2006, pp. 1–5.

[117] Magyar, J. P., Nemir, M., Ehler, E., Suter, N., Perriard, J. C., and Eppenberger, H. M., "Mass production of embryoid bodies in microbeads," *Annals NY Acad. Sci.*, Vol. 944, 2001, pp. 135–143.

[118] Maguire, T., Novik, E., Schloss, R., and Yarmush, M., "Alginate-PLL microencapsulation: Effect on the differentiation of embryonic stem cells into hepatocytes," *Biotechnol. Bioeng.*, Vol. 93, No. 3, 2006, pp. 581–591.

[119] Wang, X., Wang, W., Ma, J., Guo, X., Yu, X., and Ma, X., "Proliferation and differentiation of mouse embryonic stem cells in APA microcapsule: A model for studying the interaction between stem cells and their niche," *Biotech. Prog.*, Vol. 22, No. 3, 2006, pp. 791–800.

[120] Mann, B. K., Schmedlen, R. H., and West, J. L., "Tethered-TGF-beta increases extracellular matrix production of vascular smooth muscle cells," *Biomaterials*, Vol. 22, No. 5, 2001, pp. 439–444.

[121] Yim, E. K., and Leong, K. W., "Proliferation and differentiation of human embryonic germ cell derivatives in bioactive polymeric fibrous scaffold," *J. Biomat. Sci.: Polymer Ed.*, Vol. 16, No. 10, 2005, pp. 1193–1217.

[122] Shive, M. S., and Anderson, J. M., "Biodegradation and biocompatibility of PLA and PLGA microspheres," *Adv. Drug Deliv. Rev.*, Vol. 28, No. 1, 1997, pp. 5–24.

[123] Jain, R. A., "The manufacturing techniques of various drug-loaded biodegradable poly(lactide-co-glycolide) (PLGA) devices," *Biomaterials*, Vol. 21, No. 23, 2000, pp. 2475–2490.

[124] Varde, N. K., and Pack, D. W., "Microspheres for controlled release drug delivery," *Expert Opin. Biol. Ther.*, Vol. 4, No. 1, 2004, pp. 35–51.

[125] Heffernan, M. J., and Murthy, N., "Polyketal nanoparticles: A new pH-sensitive biodegradable drug delivery vehicle," *Bioconjug. Chem.*, Vol. 16, No. 6, 2005, pp. 1340–1342.

[126] Kim, B. Y., Kang, H. S., and Kim, J. D., "Thermo-sensitive microparticles of PNIPAM-grafted ethylcellulose by spray-drying method," *J. Microencapsulation*, Vol. 19, No. 5, 2002, pp. 661–669.

[127] Meers, P., "Enzyme-activated targeting of liposomes," *Adv. Drug Deliv. Rev.*, Vol. 53, No. 3, 2001, pp. 265–272.

[128] Berkland, C., Kim, K., and Pack, D. W., "Fabrication of PLG microspheres with precisely controlled and monodisperse size distributions," *J. Control. Release*, Vol. 73, No. 1, 2001, pp. 59–74.

[129] Berkland, C., King, M., Cox, A., Kim, K., and Pack, D. W., "Precise control of PLG microsphere size provides enhanced control of drug release rate," *J. Control. Release*, Vol. 82, No. 1, 2002, pp. 137–147.

[130] Mahoney, M. J., and Saltzman, W. M., "Transplantation of brain cells assembled around a programmable synthetic microenvironment," *Nat. Biotechnol.*, Vol. 19, No. 10, 2001, pp. 934–939.

CHAPTER 13

Neuroscience on a Chip: Microfabrication for In Vitro Neurobiology

Nianzhen Li, Anna Tourovskaia, and Albert Folch

13.1 Introduction

The nervous system owes its information-processing capabilities to a bewilderingly complex three-dimensional spatial arrangement and interconnected function of a large variety of neuronal cell types and glial cells surrounded by richly heterogeneous, dynamic biochemical and biophysical environments. Traditional neural cell cultures, by contrast, consist of a random, planar arrangement of neurons and/or glial cells surrounded by a foreign medium and a biochemically simple substrate. This oversimplification of the cellular microenvironment makes the testing of neuroscience hypotheses in vitro particularly intractable.

Micro- and nanotechnology, comprising any technique that manufactures miniature components and devices at micrometer to nanometer resolution, have the intrinsic ability (or potential) to precisely control and separate the various biochemical and biophysical interactions that regulate cell behavior. Microfabrication has been extensively used to monitor and modulate the behavior of many cell types [1–8]; however, its application to neuroscience is in its infancy despite some early reports [9].

In this chapter, we survey efforts undertaken to address neurobiology questions in vitro using micro- and nanofabrication approaches, with focus on the two areas that have been studied most intensively: (1) microengineered neurite growth and neuronal polarity, and (2) microengineered cell-cell interactions.

13.2 Microengineered Neurite Growth and Neuronal Polarity

Neurons have an elaborate and highly organized morphology, typically consisting of a number of short dendrites conveying information toward the cell body and a single long axon conveying information away from the cell body. This molecular, structural, and functional polarized organization forms the basis for directional rapid signaling and for bidirectional trophic signaling [10]. During embryonic development, a neuron must first establish the polarity (i.e., specify one neurite to become the axon, a process termed "axogenesis"), then project the axon along specific paths to make precise connections with their targets. The navigation of axon tips ("growth cones") is guided by insoluble factors in the surrounding extracellular matrix (ECM) or cells and by concentration gradients of diffusing factors. Several families of guidance molecules have been identified and purified [11–13]. However, the elucidation of path-finding decisions is complicated by the simultaneous presence of different guidance factors in a dynamic environment and by changes in cellular responses with time (whether artifactual or developmental). With traditional methods, it is very difficult or impossible to design and modulate the extracellular environment with micron-scale accuracy. In addition, existing cell-based in vitro assays have very low throughputs.

The first step in neurite growth studies is to be able to spatiotemporally control, monitor, and engineer the growth of neuronal processes. In this section, we review the use of micro-/nanofabrication techniques to control the adhesion and growth of neurons in vitro.

13.2.1 Guidance by Insoluble Factors

13.2.1.1 Micropatterning Biomolecules on Surfaces

Neuronal adhesion and axon outgrowth are guided in part by contact factors (ECM proteins and cell-adhesion molecules) in vivo as well as in vitro. Surface micropatterns of adhesive molecules have been used to probe the permissiveness and growth-promoting characteristics of various substrates. In 1975, Letourneau made the first neuronal micropattern by depositing Pd through standard metal grids, commercially available for electron microscopy (EM), onto polyornithine- or collagen-coated surfaces [14]. Embryonic day eight (E8) dorsal root ganglion (DRG) neurons from chickens preferred polyornithine or collagen to Pd. Hammarback, Letourneau, et al. later created micropatterns of "bioactive" and inactive laminin by ultraviolet light inactivation of laminin-coated coverslips masked with EM grids; inactive laminin areas could be further adsorbed with agarose-albumin [15, 16]. Neurons only adhered to the laminin-coated regions, but some neurites extended across intervening agarose-albumin regions. Gundersen patterned nerve growth factor (NGF), laminin, or fibronectin on coverslips with micropipettes and found that chick DRG neurites were guided by adsorbed NGF and laminin but not by fibronectin [17, 18].

Kleinfeld, Kahler, and Hockberger pioneered the use of micropatterns of self-assembled monolayers (SAMs) to guide axonal growth [9]. High-resolution micropatterns of alternating aminosilane and alkylsilane SAM regions were produced by photolithography on glass substrates. Dissociated neurons (embryonic mouse spinal cells or perinatal rat cerebellar cells) were confined to square regions on the scale of 50 μm or on lines with widths less than 10 μm in serum-containing (5 to 10 percent) medium. The patterned growth of cerebellar cells was maintained up to twelve days in vitro; granule cells and Purkinje neurons developed normal electrical excitability.

Since then, many studies on neuronal patterning have been conducted using a variety of surface-modification methods with synthetic and biological molecules on various substrates (see [6, 19, 20]). Examples of neuronal patterns are shown in Figure 13.1(a, b). Shear et al. developed a clever "in situ microfabrication"

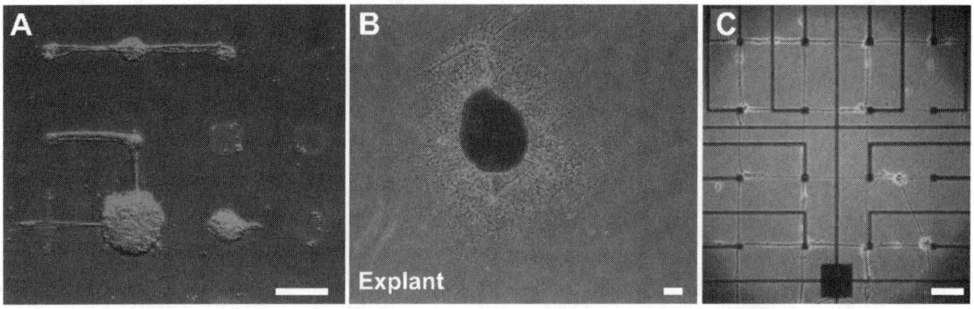

Figure 13.1 Examples of micropatterned neuronal cultures: (a) Dissociated mice embryonic cortical neurons (E12, 7 DIV) confined to various poly-D-lysine (PDL) patterns surrounded by a cell-repellent poly(ethylene glycol) thin-gel background. The image is an overlay of a DIC image of the cells and a fluorescence image of the PDL patterns. PDL was fluorescently labeled with Alexa Fluor-546 (lighter patterns in background). (*Source:* unpublished image from Li and Folch.) (b) An embryonic mice cortical explant (E12, 7 DIV) grew on PDL tracks (lighter line patterns in background) surrounded by bare glass. (*Source:* unpublished image from Li and Folch.) (c) Rat E18 hippocampal neurons (14 DIV, initial seeding density of 200 cells/mm^2) grew as networks on patterned poly-L-lysine (PLL) microelectrode arrays; the PLL patterns were fabricated by Jun et al. using microcontact printing with polydimethylsiloxane (PDMS) stamps. (*Source:* [22], reprinted with permission from Elsevier.) All scale bars: 100 μm.

technique for dynamic guidance of axon growth [21]. Multiphoton excitation was used to focally excite noncytotoxic photosensitizers, such as flavin adenine dinucleotide, that promote protein crosslinking into matrices with feature sizes greater than or equal to ≥250 nm. Barriers, growth lanes, and pinning structures comprising crosslinked proteins could be fabricated during cell growth without compromising the viability of neurons.

These studies have yielded insights on the adhesion and guidance of neuronal growth by insoluble factors and may also be used for selective immobilization of neurons in cell-based sensors [see, for example, Figure 13.1(c)] [22].

13.2.1.2 Microscale Instructional Roles of Substrates

Most micropatterning techniques rely on patterning substrates with alternating adhesive and nonadhesive regions. Adhesion alone may only discriminate the permissiveness for neuronal survival and growth but does not instruct the growth path or growth rate of axons. Micropatterns of several permissive substrates can be used to probe the relative growth-promoting or instructing role of these substrates. Lemmon et al. cultured chick retinal explant on three different natural substrates: L1, N-cadherin, and laminin [23]; the neurite growth rates, the degree of neurite fasciculation, the relative growth cone adhesiveness, and the choices neurites made between any two substrates were examined. Although L1, N-cadherin, and laminin have very different adhesivenesses, axon growth rate or the degree of fasciculation was not correlated with the substrate adhesiveness; furthermore, neurites showed little selectivity to the alternative lanes of these three substrates. Li and Folch studied neuronal growth on planar micropatterns of two growth-permissive biochemical substrates, poly-D-lysine (PDL) and diluted Matrigel, both adsorbed onto glass, and found that most mouse cortical neurons preferred to adhere and grow neurites on PDL regions as compared to the adjacent regions of Matrigel (a commercially available ECM extract); occasionally, some somas adhered on Matrigel regions, and some neurons extended axons from PDL to Matrigel regions [24]. On micropatterns of PDL and laminin, however, neurons showed little selectivity between the two substrates.

Micropatterned gradients have recently helped unveil an interesting mechanism of axon-substrate interaction for ephrin, a well-known family of membrane-bound ligands involved in axon guidance. In vivo, a nasal-to-temporal expression gradient of EphA receptors in the retina and a complementary anterior-to-posterior expression gradient of repellent ephrinA ligands in the optical tectum are critical in the formation of retinotectal projections [25–27]. To understand how growth cones "read" gradients of membrane-bound guidance molecules, Von Philipsborn et al. designed gradients consisting of microcontact-printed submicron-sized ephrinA spots or lines of various sizes and spacings [28]. Silicon masters were fabricated by low-voltage electron-beam lithography, followed with a lift-off process and reactive ion etching. Patterns were then transferred to the glass coverslips using ephrin-coated PDMS replicas of the masters. After the printing process, the coverslips were then coated with laminin. Growth cones of chick temporal (but not nasal) retinal axons were shown to stop at a distinct zone in the gradient while still undergoing filopodial activity (Figure 13.2). The position of the stop zone depended on both the steepness of the gradient and on the amount of substrate-bound ephrin per unit surface area. Quantitative analysis of axon outgrowth suggested that the "stop reac-

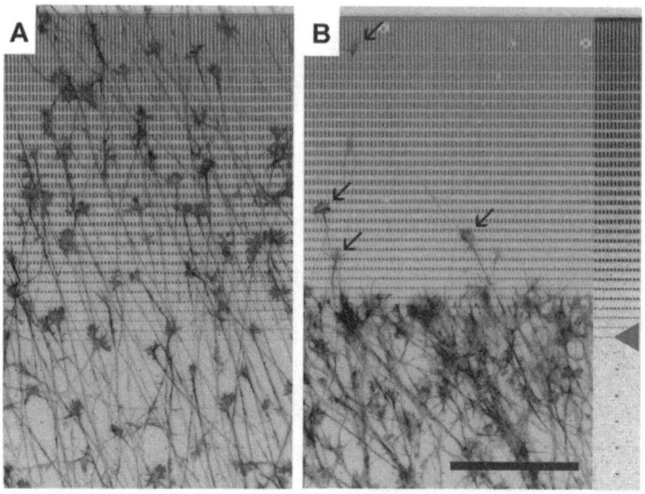

Figure 13.2 Surface-bound ephrin gradients to study axon guidance in retinal ganglion cell (RGC) explants. The discontinuous gradients were fabricated by microcontact printing of ephrin dots. Ephrin gradient (dot patterns in images) is visulaized with antibody staining, and axioms are visualized with phalloidin-stained actin. (a) For RGCs from the nasal retinal half, growth cones are not sensitive to the ephrin gradient. (b) By contrast, for RGCs from the temporal retinal half, growth cones stop at a distinct point in the ephrin gradient. As new axons continuously leave the explant, most growth cones are observed in the area between the explant and the stop zone. A few growth cones do not respond to the gradient (black arrows). The arrowhead to the right marks the beginning of the ephrin gradient. Scale bar: 100 μm. (*Source:* [28], reproduced with permission from the Company of Biologists.)

tion" was controlled by a combination of the local ephrin concentration and the total amount of ephrin encountered by the axon.

Cell-adhesiveness constraints presented by certain micropattern designs have been observed to induce modification of neuronal morphology and electrophysiological functions. Clark, Britland, and Connolly found that chick DRG neurons were multipolar with highly branched neurites on planar substrates but were typically bipolar with reduced/absent neurite branching on 25 μm–wide laminin tracks [29]. Using micropatterns of poly-L-lysine (PLL) and collagen IV, Romanova et al. found that cultured *Aplysia* neurons also showed a reduction in primary neurite number and branching [30]. The electrical activity of the neurons but not the neuropeptide content was influenced by the micropatterns. Chang, Brewer, and Wheeler reported that micropatterned neuronal cultures (grown on microelectrode arrays to monitor electrical activity) developed three- to fivefold-larger average firing rates, suggesting that cell micropatterning could be used to increase networked activity in neuron-based biosensors [31].

By studying the growth of embryonic hippocampal neurons on different shapes of polylysine microgrids, Withers et al. found that the micropattern geometry itself can modulate growth cone motility and branching [32]. When a growth cone extended from a narrow restrictive line (2 or 5 μm wide) to a more spacious node (15 μm–diameter circle), the growth cone often paused for extended periods of time, and microtubules appeared to defasciculate; once beyond the node, filopodia and lamellipodia persisted at that site, sometimes forming a collateral branch. However, growth from a narrow line onto an intersection of the same width induced branching with minimal hesitation. The report also revealed an intrinsic preference for branches to form at angles less than 90°, similar to the straightness preference that was observed by Li and Folch [24] (see discussion in Section 13.2.2).

13.2.1.3 Polarity Establishment Studied with Surface Micropatterning

A unique and critical characteristic of a neuron is its highly polarized morphology. How neuronal polarity is established or initiated (i.e., how the neuron decides that one of the processes will be the axon) is not well understood, despite the existence of an extensive catalog of structural, molecular, and functional differences between axons and dendrites. The "polarization program," activated during the early stages of neuronal differentiation, must have an intrinsic/stochastic component since cultured embryonic neurons and neuroblasts form distinct axons and dendrites even in seemingly homogenous environments [33]. However, little is known about how the polarization program is initiated or whether it can be modified by environmental signals (including biochemical gradients, injury, etc.).

Microfabrication technology has aided researchers in axon specification studies. Stenger et al. suggested that surface geometry could control the polarity of embryonic hippocampal neurons by using micropatterns of growth-permissive aminosilane SAM surrounded by a nonpermissive fluoroalkylsilane SAM background [34]. The micropattern contained arrays of 25 μm–diameter islands at the center of a cross of 5 μm–wide paths; three arms of the cross were discontinuous (10 μm–long segments) and one arm was continuous. Interestingly, most neurons grew their axons along the continuous path and dendrites along the other three paths, as identified by immunostaining of process-specific markers. In contrast, Vogt et al. found that the same micropattern geometry from Stenger et al.'s study failed to induce polarity for embryonic cortical neurons [35]. The micropattern was created by microstamping of ECM gel/poly-lysine mixture on a background of polystyrene. Neuronal polarity was determined by double patch-clamp measurements of synaptic communication. Two major experimental differences could have contributed to the conflicting results from the two groups: (1) only the micropattern geometry, not the surface composition or the cell type, was the same in both reports; and (2) while the edges of the two contrasting surface areas in substrates created by microcontact printing (Vogt et al.) are very abrupt (as shown by atomic force microscopy imaging), laser ablation (Stenger et al.) appears to create a surface gradient (of permissive molecules) at the edges of the pattern.

Esch, Lemmon, and Banker demonstrated that local extracellular signals can direct axon specification [36]. On substrates patterned with stripes of PLL and either laminin (LN) or the neuron-glia cell-adhesion molecule (NgCAM), hippocampal neurons initiated undifferentiated neurites equally on both substrates; however, neurites that contacted LN or NgCAM (but not on PLL) preferentially became axons [Figure 13.3(a–c)]. Axon specification was correlated with the speed of neurite growth (neurites grew faster on LN or NgCAM) [Figure 13.3(d–g)]. In addition, on alternating stripes of LN and NgCAM, cells with their somata on LN usually formed axons on NgCAM, whereas those with somata on NgCAM preferentially formed axons on LN, suggesting that the crossing from one axon-promoting substrate to another also provides a signal to specify the axon.

Dertinger et al. showed that gradients of substrate-bound LN had orienting effects for axons of rat hippocampal neurons [37]. Linear LN gradients were depos-

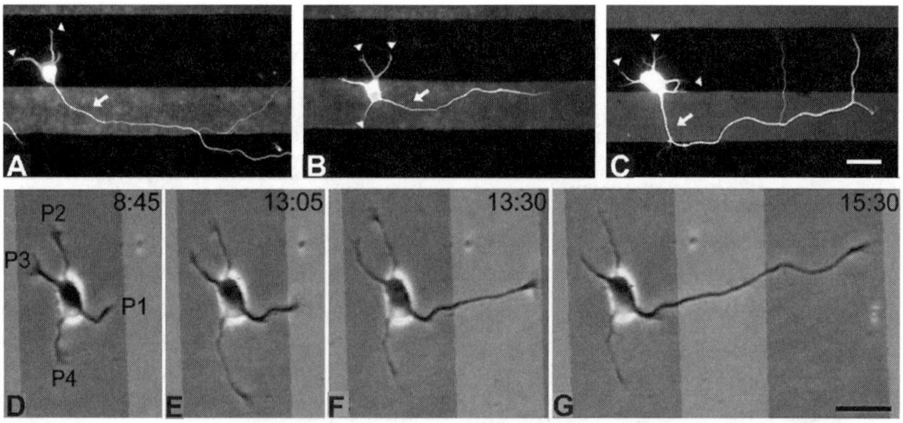

Figure 13.3 Studies of neuronal polarity using micropatterned substrates. Rat hippocampal neurons were cultured on alternating stripes of PLL and LN. (a–c) After twenty-four hours in culture, axons almost always formed on LN (arrows), although minor processes grew on both substrates (arrowheads). NgCAM had an effect similar to that of LN (not shown). Shown are superimposed fluorescent images of neurons and the patterns. Neurons were revealed by tubulin immunostaining, and the substrate patterns were revealed by immunostaining for LN (i.e., PLL appears dark, and LN appears light in the image). (d–g) Time-lapse images showing that contact with LN (e) lead a neurite to grow more rapidly than others and eventually become an axon (f, g). Shown are superimposed phase-contrast micrographs of neurons and corresponding fluorescence images of LN patterns (lighter stripes). All scale bars: 25 μm. (*Source:* [36]. © 2007 Society for Neuroscience.)

ited on PLL-coated glass substrates using a PDMS-based microfluidic gradient generator [38]. Compared to axons growing on a homogeneous laminin layer, axons growing on surface-bound LN gradients showed a preference to orient in the direction of increasing surface density of LN, but only when the slope was sufficiently large. This axon-orienting effect of LN gradients may result from LN-induced axon specification and/or axon guidance. In contrast, Buettner and Pittman reported that neurite outgrowth of dissociated neonatal rat sympathetic neurons were relatively insensitive to concentrations of LN ranging from 0.01 to 1.0 μg/cm^2 [39]. The contradicting results in these two reports may reflect the responsivity of the growth cone to a narrow range of LN gradient or may be due to the different neuronal types used (or both). Future investigation of the actual concentration range and gradient of specific ECM proteins in vivo is needed.

13.2.1.4 Intracellular Mechanisms of Contact Guidance

Although it is well known that axons are guided by ECM proteins and cell-adhesion molecules in vivo and in vitro, an understanding of the intracellular signaling pathway linking the contact cues to the eventual axon growth/turning is still lacking. Culturing mouse embryonic peripheral nerve explants on micropatterned poly-l-ornithine/laminin (PLO/LN), Turney and Bridgman unveiled the role of myosin II in mediating laminin-induced axon guidance [40]. Neurites of control explants were observed to turn or branch at the border of PLO/LN to PLO and rarely grew into PLO regions. However, when treated with the myosin II–inhibitor blebbstatin, outgrowth continued across the borders. Moreover, neurites from myosin II knockout mice ignored PLO/LN borders, and transient expression of myosin II in mutant neurons recovered the ability of neurites to turn at the border. These results collectively indicate that growth cones integrate myosin II–dependent contraction for rapid, coordinated turning.

13.2.2 Guidance by Microtopography

In vivo neurons are guided by the three-dimensional topography of their environment [41, 42], so there has been a lot of interest in using micro-/nanofabrication for precisely controlling the effect of surface topography on axon growth for various cell types.

Like other cell types, neurons often respond to microtopographical steps by turning and aligning to the edges. Clark et al. found that chick embryo cerebral neurons on Perspex plastic substrate grew processes in alignment with a single step of 1 to 5 μm in depth or repeated 2 μm–deep, 8 μm–wide grooves separated by 20 μm–wide ridges [43, 44]. Increasing step heights reduced the percentage of axon crossing.

However, two groups reported that besides alignment to microtopographical cues, sometimes neurons oriented neurites orthogonally to very narrow and shallow (micron or submicron) microgrooves. Nagata, Kawana, and Nakatsuji plated dissociated neuroblasts on PLL/LN-coated quartz substrates containing micromachined grooves [45]. Various types of central nervous system (CNS) neuroblasts, but not PNS neurons, oriented their processes and migrated both perpendicularly and in parallel to the axis of the microgrooves. Perpendicular orientation was frequently observed when the microgrooves had depths between 0.3 and 0.8 μm and a width of 1 μm, which may mimic the process by which axons align tightly in bundles. Rajnicek, Britland, and McCaig [46] further reported that embryonic *Xenopus* spinal cord neurons grew parallel to fused quartz microgrooves as shallow as 14 nm and as narrow as 1 µm; in contrast, rat hippocampal neurites grew parallel to deep, wide grooves (130 to 1,100 nm deep, 4 µm wide) but perpendicular to shallow, narrow ones (14 to 520 nm deep, 1 μm wide). The perpendicular orientation of neurites to microtopographical features is consistent with some specific cell organizations in vivo. For example, thin dendritic espalier trees of Purkinje cells orient perpendicularly with respect to parallel fibers (axonal bundles of granule cells) in the cerebellar cortex. Nagata et al. [47] demonstrated that on micropatterned substrates (either PLL patterned or topographically patterned), dissociated granule cells aligned axon bundles along the patterned microtracks, and dendrites of Purkinje cells extended at right angles to the neurites of granule cells. In three-dimensional microexplant cultures, a similar perpendicular orientation was also observed. Purkinje cells are known to have elaborate dendritic trees, and a single Purkinje cell has one of the largest numbers of afferent synapses among the vertebrate CNS neurons. The perpendicular orientation to parallel fibers may serve as an architectural base to produce networks capable of immense processing power.

The responses of neurons to microtopography may depend on both the geometry of the topographical features and the neuronal cell types. Li and Folch found that topographical guidance of mouse embryonic cortical neurons to PDL-coated micromolded PDMS grooves (rectangular cross section) was dependent on the groove depth (Figure 13.4) [24]. In response to steps of 22 to 69 μm in height, the vast majority of axons appeared to be guided by surface topography by turning and remaining inside the grooves or staying on the top surface (the "plateaus"). On the other hand, almost all neurons on shallow grooves (1 to 5 μm in height) disregarded the topographical steps and could extend axons freely into and out of the grooves [Figure 13.4(a)]. This marked disregard for topography in shallow grooves is in clear contrast with the aforementioned Clark reports [43, 44] and the Rajnicek, Britland, and McCaig report [46]; nevertheless, the reduction in the percentage of axon crossing as the step height increases was also observed by Clark et al. The differences between the three findings may be attributable to: (1) differences in neuronal cell type (murine cortical E11–14 versus rat hippocampal E16 or Xenopus spinal cord versus chick cerebral E8); (2) polylysine coverage/affinity for the substrate (resulting in adhesiveness differences); (3) microscale edge sharpness; (4) substrate compliance (elastomeric PDMS versus stiff quartz or Perspex plastic); and/or (5) the width of the gaps (Rajnicek, Britland, and McCaig report ridge widths of 1 to 4 μm). Clearly, a more thorough study of the involved variables is needed before conclusions can be extracted from those differences.

A few reports suggest that neurons become more responsive to shallow groove depths as the gap width decreases [48, 49]. Dowell-Mesfin et al. [49] reported that rat embryonic hippocampal axons followed the orthogonal patterns of polylysine-coated silicon 2.0 μm–wide pillars with 1.5 μm gaps, but not the patterns of pillars with 4.5 μm gaps. Mohoney et al. found that PC12 cells grown in glass-bottomed microchannels with polyimide walls (11 μm in height and 20 to 60

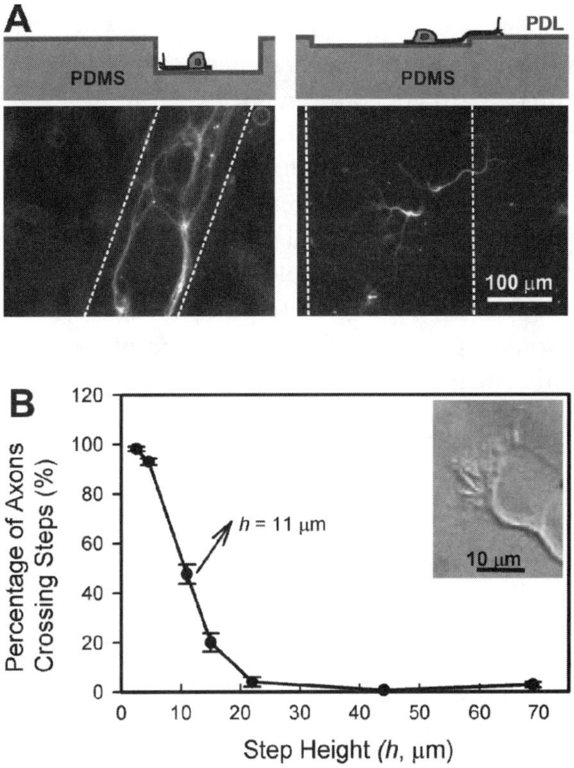

Figure 13.4 Substrate microtopography guides neurite growth: (a) Axons turn at the edges of deep microfabricated steps (left, depth $h = 69\ \mu$m) but disregard shallow steps (right, $h = 4.6\ \mu$m). Cross-sectional schematic representation (top row) and α-tubulin immunofluorescence micrographs (bottom row) illustrating the typical axon turning behavior at the edges of PDL-coated PDMS grooves. Edges are drawn in dashed lines. (b) The percentage of axons that cross a step increases as the groove depth h decreases, with a threshold at $h = 11\ \mu$m, similar to the size of a typical murine cortical neuron growth cone (shown in the inset DIC image, E13, 3 DIV). (*Source:* modified from [24], reprinted with permission from Elsevier.)

μm in width) showed a reduced complexity of neuronal branching, and neurites were oriented parallel to the channel walls; the magnitude of the effect on neurite growth depended on microchannel width, with the strongest effects at the smallest width (20 μm). In a different range of groove widths (much wider at 50 to 350 μm), a different cell type, and a different substrate (PDMS), however, the width did not influence turning behavior [24]. Again, these studies are not necessarily comparable given that the cells and the substrates were different.

Overall, it is not clear yet whether responses to artificial microtopographies have an obvious relevance to the growth mechanisms prevailing in vivo. Nevertheless, the geometry-dependence of topographical guidance may teach us some general principles of axon navigation. For example, in contrast with the strongly opposite responses to deep (>15 μm) and shallow (<10 μm) PDMS steps (PDL coated), the axons (of embryonic cortical) neurons showed a range of responses to steps of intermediate heights (11 and 15 μm): some axons (whether inside the grooves or on the plateaus) were guided by the surface topography and turned along the wall, but others disregarded the topography and were able to overcome the steps, advancing to the upper or lower surfaces [24]. The threshold in step height when a turning percentage of 50 percent was observed is ~10 to 11 μm, similar to the typical width of a growth cone [Figure 13.4(b)], suggesting that the height at which the vertical wall is no longer an obstacle for axons is determined by the three-dimensional exploratory range of the growth cone. The same report indicates that axons have an intrinsic preference for the "straightest" pathway possible. It was observed that when turning to tall (>22 μm) steps, axons almost always decided on the minimal angle of the possible turning choices (for example, an axon that approached the edge at a 25° angle with the edge line would turn 25° on the same plane—it would not turn 155° on the same plane or turn 90° down or up the wall—resulting in growth along the edge of the vertical wall). When approaching an edge 11 μm in height, the incidence of axon turning along the edge appeared to be inversely related to the angle of approach (α), suggesting that as α increases, it becomes increasingly more difficult for axons to turn along the edge than it is to bridge to the next plane. This "straightness preference" is likely a manifestation of a fundamental process in the underlying cytoskeletal dynamics.

Rajnicek and McCaig studied the cellular signaling pathways in the "orthogonal guidance" of hippocampal neurons by narrow quartz grooves (see above) [50]. Higher concentrations of two different protein kinase C inhibitors and three different calcium channel blockers all inhibited axon growth perpendicular to the grooves. However, perpendicular alignment persisted in the presence of a stretch-activated calcium channel blocker and inhibitors of G proteins, of protein tyrosine kinases, and of protein kinase A and G. Importantly, the frequency of perpendicular alignment of hippocampal neurites depended on the age of the embryos from which neurons were isolated, suggesting that contact guidance might be developmentally regulated.

13.2.3 Guidance by Gradients of Soluble Factors

The navigation of axons is guided in part by concentration gradients of diffusing factors [11, 51]. Several families of guidance molecules have been identified and purified in the last two decades. Traditional methods to expose cultured cells to diffusible gradients include the use of biological gels (e.g., collagen [52], fibrin [53], or agarose [54]), glass micropipettes [55], and a variety of static-chamber devices (e.g., the chambers of Boyden [56], Zigmond [57], or Zicha, Dunn, and Brown [58]). However, all of these methods generate gradients that are not quantitatively defined or reproducible.

Microfabricated devices are enabling researchers to create more precise and reproducible gradients. Dertinger et al. [38] have developed a microfluidic gradient generator that delivers multiple laminar-flow streams of fluid, each carrying different concentrations of a given solute. Since solutions in the microfluidic network are continuously renewed, the gradients are spatiotemporally constant and straightforward to quantify by fluorescence microscopy, which makes the devices particularly useful for cell-migration studies [59]. However, the use of this device for cellular studies can be impaired by the presence of flow-induced shear and drag forces, which could alter intracellular signaling through changes in cell shape [60] or, more directly, could confound the results by biasing cell motility [61]. Additionally, cells are enclosed in a microfluidic channel, which becomes a challenge for long-term cell viability (e.g., neurons that have to be allowed to recover/differentiate for days prior to application of the gradient) and hinders physical access to single cells (e.g., patch-clamp recording, intracellular injection, atomic force microscopy) during the course of the experiment. Last but not least, in continuous flow the data can only be averaged over proximal cells (since the gradients become smoother downstream), and only gradients perpendicular to flow are possible. Other microfluidic gradient generators may help address some of these concerns. A gradient generator based on "microjets" allows for creating tunable gradients on open surfaces and constant over arbitrarily long (but not wide) areas [62]. Microvalves have been used to create evolving radial gradients for neutrophil studies [63]. Very complex spatiotemporal gradients (including turning them on and off and inverting them) are possible with the use of tunable chaotic micromixers [64]. These gradient generators have not been tested yet for neuronal studies.

Rosoff et al. created gradients of NGF in three-dimensional gel by "printing" drops of NGF solution in a series of ten lines (1 mm apart) with increasing droplet density onto the surface of a thin collagen gel [65]. The gel is used both as the growth-supporting surface and as a barrier for diffusion; the NGF distribution steps evolve in time and are smoothened by diffusion, but the rate of change slows down. For rat DRG explants, a greater percentage of axons extended toward high concentrations for steeper gradients. Significant guidance occurred even for a 0.1 percent difference across 10 μm (typical size of a growth cone in a three-dimensional gel) but only within a narrow range of ligand concentrations (0.1 to 10 nM). In contrast, previous work reported a requirement of 1 to 2 percent minimum concentration difference in order for growth/motility to occur [66, 67] (as low as 0.5 percent for neutrophils [68, 69]).

Gradients can also be used to stimulate distinct cell membrane domains with subcellular resolution. Taylor et al. implemented in PDMS microchannels the old Campenot chamber that featured the capability to fluidically isolate axons from neuron cell bodies [70, 71]. The device consists of two chambers interconnected with a set of parallel PDMS microchannels 10 μm in width, 3 μm in height, and of various lengths. Neurons were initially seeded only in one chamber, then grew neurites into the other chamber through the microchannels. Due to the high fluid-flow resistance of the microchannels, very slow flow could be controlled to counteract diffusion, thus fluidically isolating the axons from the somas. The axons or somas could be harvested separately. Using this platform, it was found that presynaptic (Synaptophysin) but not postsynaptic (CaMK 2a) mRNA is localized to developing rat cortical and hippocampal axons. The platform could also be used to model CNS axonal injury and regeneration.

13.2.4 Integration of Multiple Guidance Cues

In vivo, particularly during development, neurons are exposed to a dynamic three-dimensional environment with multiple insoluble and soluble factors. Until recently, most studies have focused on neuronal responses to a single guidance cue (either a surface biochemical cue or topographical variation or a single soluble factor); thus, the responses of growth cones to multiple cues on three-dimensional substrata are virtually unknown.

Using a micropatterned LN substrate with an orthogonal dc electric field, Britland and McCaig [72] demonstrated that embryonic *Xenopus* spinal cord neurons were able to detect and integrate simultaneous electrical and adhesive guidance cues. Application of a dc field alone induced filopodial asymmetry and caused 70 percent of growth cones to turn and orient toward the cathode; however, on micropatterned LN tracks, after dc field application, 62 percent of cells remained aligned with the LN tracks, while only 35 percent of neurites oriented cathodally. Interestingly, some neurites appeared to bifurcate, with one branch growing along the laminin track and the other growing toward the cathode.

Li and Folch studied the integration of surface biochemical cues and microtopographical guidance cues in three dimensions by axons of mouse embryonic cortical neurons [24]. Competing growth options were provided as: (1) PDL coatings on microfabricated PDMS steps, and (2) complementary features of Matrigel. When only one guidance cue was present, axons displayed a preference for PDL over Matrigel, and axons also responded to a PDMS step in the fashion that preferred the straightest path within the exploratory range of the growth cone (Figure 13.4). When the two preferences conflicted with each other (at the edge of the plateaus), axons chose to grow straight into Matrigel; when the straight path was not permissive, the axon turned in the direction that minimized the turning angle (Figure 13.5). These results show that growth cones are able to integrate permissiveness and topographical cues.

13.3 Microengineered Cell-Cell Signaling

Intercellular signaling is critical for the normal development and physiology of the CNS. In addition to precisely controlling the surface and fluidic environment presented to cells, micro-/nanofabrication enables the engineering of the interaction of a cell with other cells, providing an invaluable tool for probing cell-cell signaling mechanisms.

In the following section, we discuss examples of applications of micro-/nanofabrication in neural cell-cell signaling, categorized by neuron-target signaling (synapse formation), glia-glia interactions, and neuron-glia coculture.

13.3.1 Synaptogenesis on a Chip

The specialized intercellular junction between an axon terminal and its target (dendrites of another neuron, muscle, or gland cells) is called a synapse. After axons reach their appropriate targets, synaptogenesis begins, resulting in formation of pre- and postsynaptic specializations. Presynaptic specializations include synaptic vesicles that contain neurotransmitters, active zones (containing exocytotic and endocytotic machinery), and calcium channels, while postsynaptic specializations include aggregates of neurotransmitter receptors and signal transduction molecules. The formation and maintenance of synapses is critical for the proper functioning of

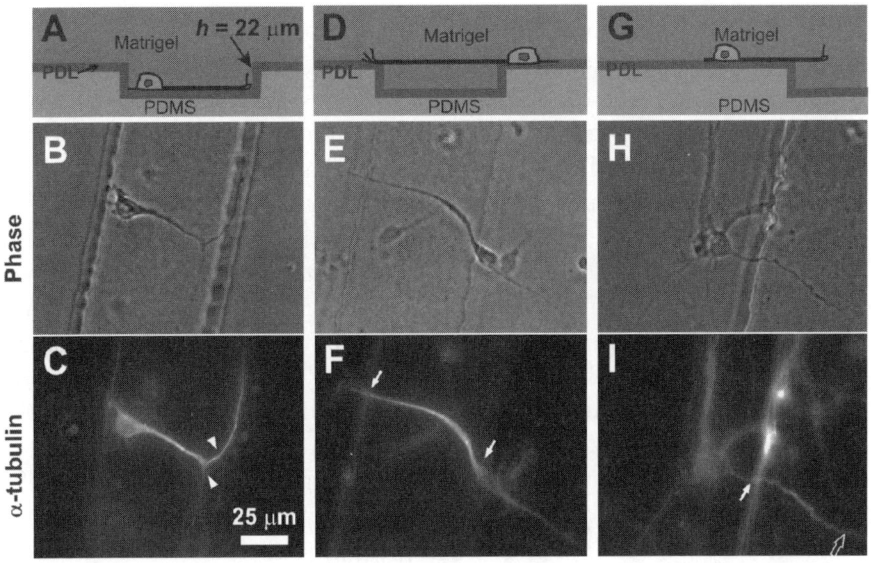

Figure 13.5 Axon growth on microstructured PDMS substrates with competing topographical and biochemical cues. Cross-sectional schematic representation (top row), phase-contrast (middle row), and corresponding α-tubulin immunofluorescence micrographs (bottom row) illustrating that neurons integrate both topographical and biochemical cues on microgrooved PDMS substrates covered with the growth-supporting gel Matrigel. In the absence of topographical cues, axons [both inside the grooves (a–c) and on the top plateau surface (g–l)] preferred PDL to Matrigel and kept growing attached to the PDL-coated PDMS surface. When axons reached the groove walls with competing topographical and biochemical cues, in-groove axons turned [arrowheads in (c)] along the wall (a–c), while on-plateau axons extended into Matrigel. (d–f) Crossing the gel and landing onto the other side of grooves. (g–l) Extending out of focus deeper into the gel. Solid white arrows: axons crossing the edges; hollow white arrows: axons growing into the gel, out of focus. (*Source:* modified from [24], reprinted with permission from Elsevier.)

the nervous system; yet, the mechanisms underlying the formation, stability, and plasticity of synapses are not fully understood.

During synaptogenesis, information is exchanged locally between pre- and postsynaptic cells, resulting in exact apposition of pre- and postsynaptic elements to ensure proper signal transmission. Micro-/nanofabrication technologies enable precise control of the location at which a neural or effector cell is stimulated and have thus been employed in studies of intercellular signaling underlying synapse formation.

13.3.1.1 Neuromuscular Synaptogenesis

At the neuromuscular junction, an essential step of synaptogenesis is the clustering of initially diffuse acetylcholine (ACh) receptors (AChRs) to a small domain on the postsynaptic membrane that directly apposes the nerve terminal upon innervation. The metabolic stability of AChRs also changes developmentally: AChRs turn over much more slowly (~days) at mature adult synapses compared to AChRs in uninnervated myotubes, at newly formed synapses, and in denervated muscle fibers (~hours) [73–75]. While the classical textbook view is that the clustering of AChRs is initiated by the nerve's release of the proteoglycan agrin [10, 76–80], recently

AChR clusters have been found in the embryo in the absence of innervation [81–84], as well as in pure myotube cultures [85–88], which indicates that the initiation of AChR cluster formation does not require agrin signaling. Other synaptogenic nerve-derived molecules, such as neuregulin and Ach, were also shown to affect AChR synthesis and aggregation, as well as stability of AChR clusters [89–91]; however, the exact functions of these molecules are not clear.

In order to evoke AChR clustering response in vitro, most studies have relied on bathing the entire surface of the myotube with a solution containing agrin or other synaptogenic molecules [92]. However, there is the concern that global stimulation may induce artifactual signaling. It has been demonstrated that global exposure to agrin in vitro induces the formation of AChR clusters at multiple random locations on myotubes. AChR clusters, on the other hand, form preferentially to agrin-stimulated small sites in studies where agrin is released by nerve cells [93, 94], expressed by neighboring CHO cells, or is immobilized onto beads [95]. Although these methods initially proved to be valuable means for studying synaptogenic molecules, there were serious shortcomings: (1) the position of the agrin stimulus on the cell could not be manipulated or relocated; (2) the onset and duration of the agrin stimulus was not well defined; and (3) the concentration of agrin was unknown. The results were thus limited to a largely qualitative characterization.

1. *Agrin micropatterns:* In a recent study, agrin molecules were microcontact-printed in a micron-scale grid onto a tissue-culture polystyrene substrate prior to seeding muscle cells [96, 97]. Over time, AChR clusters formed at sites where myotubes contacted the agrin pattern. In this high-throughput approach, immobilized agrin is presented on a physiologically relevant size scale; however, this method has important shortcomings: (1) in order to be printed, agrin must be dried; (2) the stimulus is fixed in space and time and cannot be removed, relocated, or reintroduced; (3) the concentration of agrin presented to cells is not known (agrin was visualized on micropatterns by mixing it with fluorescent bovine serum albumin); and, finally, (4) there is no control over the onset or duration of the stimulation (the cells are exposed to agrin since their first contact with the cell-culture surface), while in vivo neural agrin is released by the motor terminal at a precise developmental stage shortly after formation of the myotubes.
2. *Microfluidic devices:* Tourovskaia et al. developed a microfluidic approach to focally stimulate micropatterned myotubes with synapse-organizing molecules and quantitatively studied the effects of agrin on the clustering and the stability of AChRs (Figure 13.6) [98, 99]. Muscle cells were matured inside the microfluidic device for more than one week and directed to fuse into an array of parallel isolated myotubes by microtracks of Matrigel on a protein-repellent background. The precise geometrical arrangement of the myotubes was more physiological than random cultures (in vivo skeletal muscle consists of highly aligned muscle fibers) and allowed for easy identification and high-throughput monitoring of single myotubes over time [99].

   After a relatively short (one-hour) focal agrin stimulus, the vast majority of AChR clusters were found to localize to the area where the agrin stream

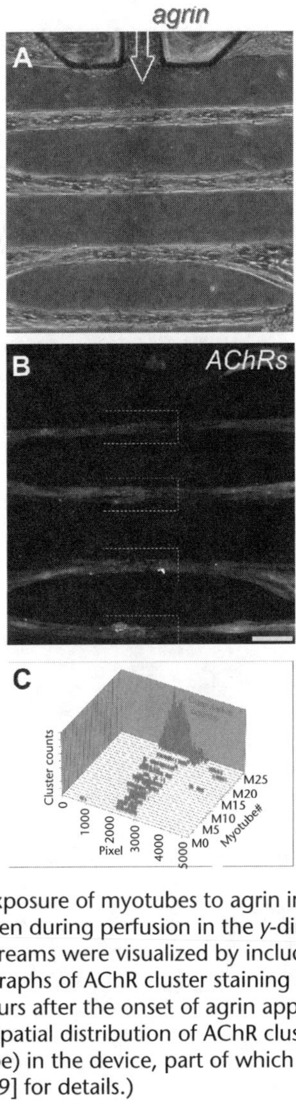

Figure 13.6 Microfluidic focal exposure of myotubes to agrin induces localized AChR clustering: (a) Phase-contrast micrograph taken during perfusion in the *y*-direction (arrows denote the flow direction); the agrin-containing streams were visualized by including a red food-coloring dye in the medium. (b) Fluorescence micrographs of AChR cluster staining (fluorescently conjugated α-bungarotoxin, or BTX) at six hours after the onset of agrin application (10 nM for one hour) in the same myotubes as in (a). (c) Spatial distribution of AChR cluster counts on myotubes (each row represents a different myotube) in the device, part of which is shown in (b). Scale bar: 50 μm. (*Source:* Dr. A. Tourovskaia; see [99] for details.)

contacted the myotubes [Figure 13.6(b, c)] [99]. A gradual increase of the AChR clustering area toward downstream was observed, most likely due to the broadening of the diffusion profile of the agrin stream. Although fluidic flow used in this method is nonphysiological, the method offers great control over the magnitude and spatiotemporal characteristics of the stimulus: it can be started at, and applied for, a specified time, easily removed, and, in principle, relocated and/or applied at separate different locations simultaneously (allowing for studies of competitive innervation).

3. *Planar nanoholes devices:* Planar nanoholes devices, microchip devices featuring micron- and submicron-scale apertures, present an attractive alternative to the aforementioned flow-based microfluidic devices due to: (1) delivery of the stimulus at potentially higher spatial resolution, provided the cells obstruct or form a seal with the aperture; (2) the absence of potentially

detrimental or confounding effects typically associated with exposure to flow (induced shear stress, and dilution of secreted growth factors, etc.); and (3) the simplicity of fluidic connections (absence of flow-control systems) [100]. Recently, Kosar et al. demonstrated the use of micro-/nanofabricated planar apertures to focally stimulate a small area of the membrane of a myotube with agrin [101]. The device consisted of a biocompatible silicon nitride surface featuring a microarray of micron-scale apertures that were fluidically addressed from both sides (Figure 13.7). Since the apertures were organized into microarrays, multiple experiments could be run in parallel on the same device. After the silicon nitride surface ("front side" of the chip) was coated with ECM proteins, C2C12 mouse myoblast cells were seeded on the surface. Agrin was applied to a selective area of the myotubes only

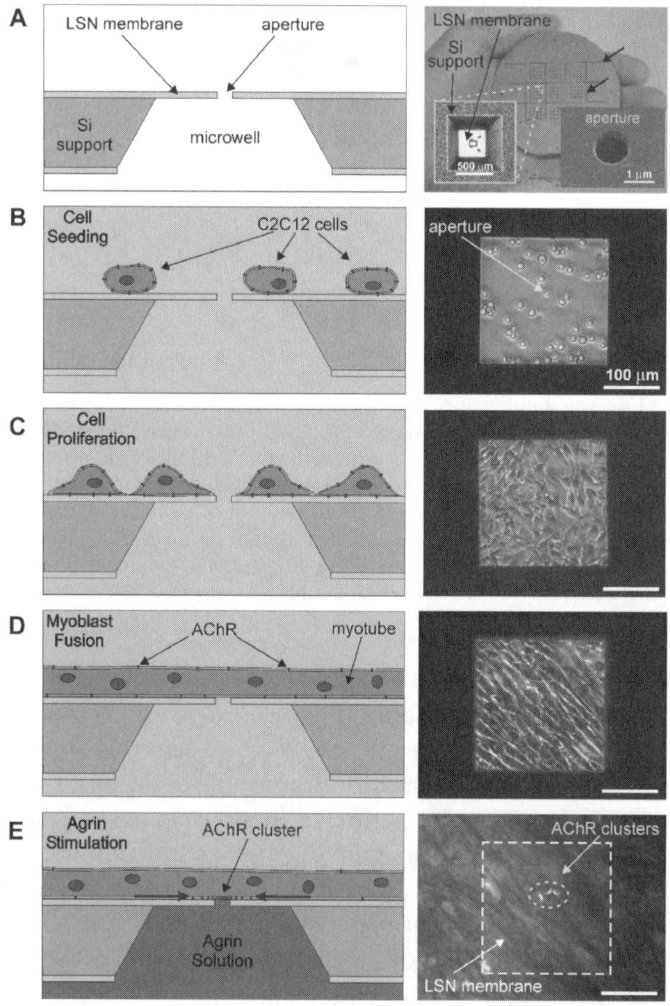

Figure 13.7 Focal delivery of agrin to C2C12 myotubes through nanoholes. Images on the right are micrographs of C2C12 cells on a suspended LSN membrane of a nanohole device taken at each step (depicted schematically on the left): (a) Seeding of C2C12 cells on the device, (b) spreading and proliferation of the cells, (c) differentiation of the cells into myotubes, and (d) focal delivery of agrin and aggregation of AChRs. (*Source:* Dr. F. Kosar; see [101] for details.)

through the apertures from underneath the cell-culture surface. AChR clusters formed only around the aperture [Figure 13.7(d)]. This device offers potential for great control over various experimental parameters, such as the concentration of the stimulus, time of onset and duration of the stimulus, and the surface area of the membrane region where the stimulus is applied. While the device cannot relocate the stimulus to an arbitrary location with respect to the cells, the system could prove invaluable for application of a sequence of signals in precise registration with each other, as occurs at the developing and mature synapses.

A similar, but implantable, "artificial synapse" device has been developed to deliver neurotransmitter pulses to retinal neurons, with the goal of providing treatment for primary diseases of the retina like retinitis pigmentosa and age-related macular degeneration [102, 103]. Neurotransmitter pulses are routed through microchannels by electro-osmosis and then ejected onto the neuronal tissue [100].

13.3.2 Microengineered Glial Cell Interactions and Glia-Neuron Cocultures

Besides neurons, glial cells, which include astrocytes, oligodendrocytes, Schwann cells, and microglia, serve many important functions of the nervous system. Glial cells are closely associated with neurons and play key roles in the development, maintenance, and regeneration of neurons. Microfabrication techniques, with the ability to control cell placement and orientation, have facilitated investigations of interactions between glial cells and between glial cells and neurons.

In vivo, astrocytes, a type of glial cell, closely wrap around neuronal synapses, provide neuronal migration cues, act as "housekeepers" of the extracellular environment, and actively modulate neuronal activity and synaptic transmission. To modulate the activity of adjacent astrocytes and neurons, an important signaling event in astrocytes is the intercellular calcium wave that propagates from cell to cell via two different pathways: gap junctions and extracellular diffusible messengers. Takano et al. presented an elegant approach to dissect the two signaling pathways using micropatterned cultures (Figure 13.8) [104]. Primary cortical astrocytes on PLL-coated coverslips were cultured with microfluidically deposited agarose lines. Due to the nonadhesive nature of agarose, confluent domains of astrocytes grew only on the PLL regions. When an astrocyte was mechanically stimulated, a calcium wave was propagated to the cells within the same domain and across spatially disconnected neighboring astrocyte domains [Figure 13.8(a)], confirming the involvement of some extracellular messenger, which was found to be ATP. In the presence of ATP receptor antagonist, the calcium wave was reduced in the same lane and completely blocked in adjacent lanes [Figure 13.8(b)], indicating that both the long-range extracellular ATP pathway and the short-range gap-junction pathway are involved in mediating normal calcium waves.

Schwann cells, the astrocytelike cells in the peripheral nervous system, enhance axonal regeneration following nerve injury in vivo and provide a favorable substrate for neurite outgrowth in vitro. A variety of regeneration-promoting cell-adhesion molecules are expressed on the surface of Schwann cells, including heparin-sulfate proteoglycan, neural cell-adhesion molecule, and L1. Thompson

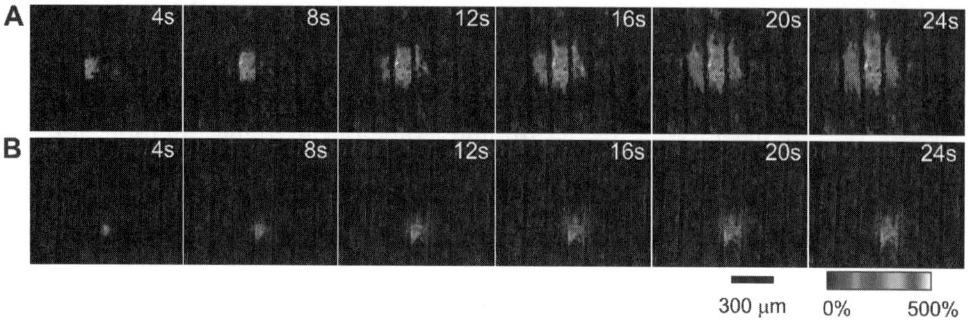

Figure 13.8 Probing glial intercellular communication pathways with micropatterned substrates. The figure shows two sets of fluorescence (dF/F_0) image sequences for confluent lanes of astrocytes (loaded with calcium indicator Fluo-3 AM); the astrocytes attach selectively to 110 μm–wide lanes of adsorbed PLL that were separated by 40 μm–wide stripes of an agarose film. (a) In control conditions, after mechanical stimulation, a calcium wave was initiated and then spread to astrocytes in the same lane and in adjacent lanes. (b) After astrocytes were loaded with 50 μM PPADS, an ATP receptor antagonist, a mechanical-stimulation-induced calcium wave propagated only to a few cells in the same lane but not to adjacent other lanes, confirming that extracellular diffusible ATP is involved in mediating the intercellular calcium waves. (*Source:* Dr. H. Takano; see [104] for details.)

and Buettner found that cultured Schwann cells spread in alignment with laminin tracks [105], and dissociated rat spinal neurons grew neurites along the orientation of the underlying Schwann cells monolayer [106] [Figure 13.9(a, b)]. Since neurons were plated after Schwann cells grew confluently on the laminin stripes, the orientation of neurites was caused by contact with Schwann cells, possibly through both topographical and molecular mechanisms. Miller, Jeftinija, and Mallapragada also observed that Schwann cells enhance the growth and alignment of the neurites of rat DRG neurons on laminin-coated microgrooves made of the biodegradable polymer poly(D, L-lactic acid) [107].

The effects of glia on the differentiation and growth of adult rat hippocampal neural progenitor cells has also been investigated with micropatterned polymer polystyrene surfaces. Recknor, Sakaguchi, and Mallapragada found that astrocytes enhanced the growth of the neural progenitor cells in cocultures; astrocytes aligned to micropatterns and led to the same neurite orientation of the progenitor cells [108].

Chang, Brewer, and Wheeler studied the effect that electrophysiological activity of neuronal networks on multiple electrode arrays was enhanced in micropatterned cultures with respect to random cultures [109]. It was found that even in serum-free media, patterned cultures had up to five times more astrocytes and three times more neurons than random cultures, suggesting that the enhanced activity of patterned neurons might be due to glia-neuron interaction.

13.4 Conclusions and Future Directions

Conceptually speaking, microfabrication techniques offer the potential to modulate, on a cellular and subcellular level, the biochemical composition and the topography of the substrate, the medium surrounding each cell, as well as the type of cells neighboring each cell. We have reviewed examples of how microfabrication technology

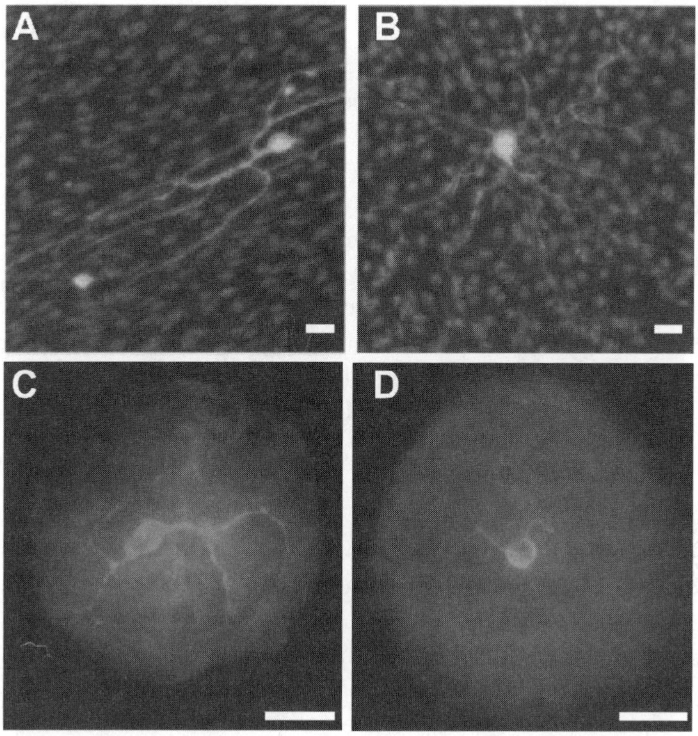

Figure 13.9 Microengineered glia-neuron cocultures: (a, b) Neurite outgrowth on (a) micropatterned Schwann cell monolayer and (b) on unaligned Schwann cell monolayer. Note that in (a) nuclei of Schwann cell are aligned in the same direction, and neurites showed an orientation similar to that of the Schwann cells. In (b), both Schwann cell nuclei and neurites extend randomly. Neitites are visualized with neurofilament antibody staining, and nuclei from all cells are stained with DAPI; scale bars: 50 μm. (*Source:* [106], reprinted with the permission of the *Annals of Biomedical Engineering.*) (c, d) Astrocyte-neuron cocultures on microfabricated microislands. Neuronal dendrites are visualized by MAP-2 staining, and astocyte microislands are visualized by actin staining; scale bars: 25 μm. (*Source:* Drs. X. Figueroa-Masot and H. Watari.)

can provide a unique platform to ask neurobiology questions that cannot be addressed by traditional cell-culture techniques. Surface biochemical and contact guidance of axon growth and specification has been studied with microfabricated substrates. Various microfluidic devices have been developed to study the effects of soluble factors on neuronal growth and synaptic development. In addition, intercellular signaling pathways have been probed using micropatterned cultures.

Similar to traditional cell-culture techniques, most current microengineered cultures employ two-dimensional substrates that cannot reproduce the complex three-dimensional networked organization found in vivo. Recent work in three-dimensional culture systems has exposed significant limitations with studying cells on two-dimensional surfaces. Therefore, the development of microengineered three-dimensional cultures could improve the physiological relevance of in vitro neuroscience.

We believe that, since micro-/nanofabrication is a relatively new technology for most neurobiologists, there is still a plethora of questions that would benefit from exploring a microfabrication approach. For example, a single neuron in a small adhesive island surrounded by nonadhesive substrate [Figure 13.9(c, d)] can be forced to form synapses on itself ("autapses"), a phenomenon that has been used for studying basic neurotransmission mechanisms. Also, neuron micropatterns in combination with microfluidic devices may allow for the arrangement of cells on surfaces of sensor devices used in environmental control and drug screening. Well-defined neuronal networks (i.e., where the relative position of cells and their axons/dendrites are determined by artificial design) could yield valuable insights into the effect of cell-cell communication on basic processes such as synapse development, synaptic transmission and potentiation, certain cell pathologies, and drug efficacy. Cocultures of different cell types with controlled placement of each cell type could allow for better understanding of the role of cell-cell interactions in brain function and more efficient neural-regeneration strategies after traumatic injury.

We anticipate a growing number of neuroscience studies using nanofabrication and microfluidics as the expertise gap between biologists and engineers narrows. The majority of the work reviewed here has focused on the "small-scale benefit" of using micro-/nanotechnology for precisely controlling the local environment of cells. However, micro-/nanotechnologies also provide an inherent "high-throughput" benefit (of operation as well as of fabrication) because microdevices: (1) can be batch-fabricated at low cost per device; (2) consume smaller amounts of reagents; (3) lighten the need for large cell numbers, thus require fewer animal sacrifices (an increasing concern in our society); and (4) produce statistically richer data. Therefore, we envision that future "neuroscience-on-a-chip" applications will take full advantage of the micro-/nanotechnologies by utilizing hundreds or thousands of cells cultured, controlled, and monitored in parallel on a small area and/or volume. These applications will be widely used in basic neurobiological investigations and in high-throughput cell-based biomedical assays.

References

[1] Bhatia, S. N., Balis, U. J., Yarmush, M. L., and Toner, M., "Effect of cell-cell interactions in preservation of cellular phenotype: Co-cultivation of hepatocytes and non-parenchymal cells," *Faseb J.*, Vol. 13, No. 14, 1998, pp. 1883–1890.

[2] Voldman, J., Gray, M. L., and Schmidt, M. A., "Microfabrication in biology and medicine," *Ann. Rev. Biomed. Eng.*, Vol. 1, 1999, p. 401.

[3] Folch, A., and Toner, M., "Microengineering of cellular interactions," *Ann. Rev. Biomed. Eng.*, Vol. 2, 2000, pp. 227–256.

[4] Whitesides, G. M., Ostuni, E., Takayama, S., Jiang, X. Y., and Ingber, D. E., "Soft lithography in biology and biochemistry," *Ann. Rev. Biomed. Eng.*, Vol. 3, 2001, pp. 335–373.

[5] Bhadriraju, K., and Chen, C. S., "Engineering cellular microenvironments to improve cell-based drug testing," *Drug Discov. Today*, Vol. 7, No. 11, 2002, pp. 612–620.

[6] Li, N., Tourovskaia, A., and Folch, A., "Biology on a chip: Microfabrication for studying the behavior of cultured cells," *Crit. Rev. Biomed. Eng.*, Vol. 31, Nos. 5–6, 2003, pp. 423–488.

[7] Park, T. H., and Shuler, M. L., "Integration of cell culture and microfabrication technology," *Biotechnol. Prog.*, Vol. 19, No. 2, 2003, pp. 243–253.

[8] Andersson, H., and van den Berg, A., "Microfabrication and microfluidics for tissue engineering: State of the art and future opportunities," *Lab Chip*, Vol. 4, No. 2, 2004, pp. 98–103.

[9] Kleinfeld, D., Kahler, K. H., and Hockberger, P. E., "Controlled outgrowth of dissociated neurons on patterned substrates," *J. Neurosci.*, Vol. 8, No. 11, 1988, pp. 4098–4120.

[10] Kandel, E. R., Schwartz, J. H., and Jessell, T. M., *Principles of Neural Science*, 4th ed, New York: McGraw-Hill, 2000, pp. 1087–1114.

[11] Tessier-Lavigne, M., and Goodman, C. S., "The molecular biology of axon guidance," *Science,* Vol. 274, No. 5290, 1996, p. 1123.

[12] Yu, T. W., and Bargmann, C. I., "Dynamic regulation of axon guidance," *Nat. Neurosci.*, Vol. 4, 2001, pp. 1169–1176.

[13] Song, H., and Poo, M., "The cell biology of neuronal navigation," *Nat. Cell. Biol.*, Vol. 3, No. 3, 2001, pp. E81–E88.

[14] Letourneau, P. C., "Cell-to-substratum adhesion and guidance of axonal elongation," *Dev. Biol.*, Vol. 44, 1975, pp. 92–101.

[15] Hammarback, J. A., and Letourneau, P. C., "Neurite extension across regions of low cell-substratum adhesivity: Implications for the guidepost hypothesis of axonal pathfinding," *Dev. Biol.*, Vol. 117, No. 2, 1986, pp. 655–662.

[16] Hammarback, J. A., McCarthy, J. B., Palm, S. L., Furcht, L. T., and Letourneau, P. C., "Growth cone guidance by substrate-bound laminin pathways is correlated with neuron-to-pathway adhesivity," *Dev. Biol.*, Vol. 126, No. 1, 1988, pp. 29–39.

[17] Gundersen, R. W., "Sensory neurite growth cone guidance by substrate adsorbed nerve growth factor," *J. Neurosci. Res.,* Vol. 13, Nos. 1–2, 1985, pp. 199–212.

[18] Gundersen, R. W., "Response of sensory neurites and growth cones to patterned substrata of laminin and fibronectin in vitro," *Dev. Biol.*, Vol. 121, No. 2, 1987, pp. 423–431.

[19] Jung, D. R., Kapur, R., Adams, T., Giuliano, K. A., Mrksich, M., Craighead, H. G., and Taylor, D. L., "Topographical and physicochemical modification of material surface to enable patterning of living cells," *Crit. Rev. Biotechnol.*, Vol. 21, No. 2, 2001, pp. 111–154.

[20] Corey, J. M., and Feldman, E. L., "Substrate patterning: An emerging technology for the study of neuronal behavior," *Exp. Neurol.*, Vol. 184, Suppl. 1, 2003, pp. S89–S96.

[21] Kaehr, B., Allen, R., Javier, D. J., Currie, J., and Shear, J. B., "Guiding neuronal development with in situ microfabrication," *Proc. Natl. Acad. Sci. USA*, Vol. 101, No. 46, 2004, pp. 16104–16108.

[22] Jun, S. B., Hynd, M. R., Dowell-Mesfin, N., Smith, K. L., Turner, J. N., Shain, W., and Kim, S. J., "Low-density neuronal networks cultured using patterned poly-l-lysine on microelectrode arrays," *J. Neurosci. Methods.*, Vol. 160, No. 2, 2007, pp. 317–326.

[23] Lemmon, V., Burden, S. M., Payne, H. R., Elmslie, G. J., and Hlavin, M. L., "Neurite growth on different substrates—permissive versus instructive influences and the role of adhesive strength," *J. Neurosci.*, Vol. 12, No. 3, 1992, pp. 818–826.

[24] Li, N., and Folch, A., "Integration of topographical and biochemical cues by axons during growth on microfabricated 3-D substrates," *Exp. Cell Res.,* Vol. 311, No. 2, 2005, pp. 307–316.

[25] Drescher, U., Kremoser, C., Handwerker, C., Loschinger, J., Noda, M., and Bonhoeffer, F., "In vitro guidance of retinal ganglion cell axons by RAGS, a 25 kDa tectal protein related to ligands for Eph receptor tyrosine kinases," *Cell,* Vol. 82, No. 3, 1995, pp. 359–370.

[26] Cheng, H. J., Nakamoto, M., Bergemann, A. D., and Flanagan, J. G., "Complementary gradients in expression and binding of ELF-1 and Mek4 in development of the topographic retinotectal projection map," *Cell,* Vol. 82, No. 3, 1995, pp. 371–381.

[27] Connor, R. J., Menzel, P., and Pasquale, E. B., "Expression and tyrosine phosphorylation of Eph receptors suggest multiple mechanisms in patterning of the visual system," *Dev. Biol.*, Vol. 193, No. 1, 1998, pp. 21–35.

[28] von Philipsborn, A. C., Lang, S., Loeschinger, J., Bernard, A., David, C., Lehnert, D., Bonhoeffer, F., and Bastmeyer, M., "Growth cone navigation in substrate-bound ephrin gradients," *Development*, Vol. 133, No. 13, 2006, pp. 2487–2495.

[29] Clark, P., Britland, S., and Connolly, P., "Growth cone guidance and neuron morphology on micropatterned laminin surfaces," *J. Cell. Sci.*, Vol. 105, Pt. 1, 1993, pp. 203–212.

[30] Romanova, E. V., Fosser, K. A., Rubakhin, S. S., Nuzzo, R. G., and Sweedler, J. V., "Engineering the morphology and electrophysiological parameters of cultured neurons by microfluidic surface patterning," *Faseb. J.*, Vol. 18, No. 11, 2004, pp. 1267–1269.

[31] Chang, J. C., Brewer, G. J., and Wheeler, B. C., "Modulation of neural network activity by patterning," *Biosens. Bioelectron.*, Vol. 16, Nos. 7–8, 2001, pp. 527–533.

[32] Withers, G. S., James, C. D., Kingman, C. E., Craighead, H. G., and Banker, G. A., "Effects of substrate geometry on growth cone behavior and axon branching," *J. Neurobiol.*, Vol. 66, No. 11, 2006, pp. 1183–1194.

[33] Dotti, C. G., Sullivan, C. A., and Banker, G. A., "The establishment of polarity by hippocampal neurons in culture," *J. Neurosci.*, Vol. 8, No. 4, 1988, pp. 1454–1468.

[34] Stenger, D. A., Hickman, J. J., Bateman, K. E., Ravenscroft, M. S., Ma, W., Pancrazio, J. J., Shaffer, K., Schaffner, A. E., Cribbs, D. H., and Cotman, C. W., "Microlithographic determination of axonal/dendritic polarity in cultured hippocampal neurons," *J. Neurosci. Methods*, Vol. 82, No. 2, 1998, p. 167.

[35] Vogt, A. K., Stefani, F. D., Best, A., Nelles, G., Yasuda, A., Knoll, W., and Offenhausser, A., "Impact of micropatterned surfaces on neuronal polarity," *J. Neurosci. Methods*, Vol. 134, No. 2, 2004, pp. 191–198.

[36] Esch, T., Lemmon, V., and Banker, G., "Local presentation of substrate molecules directs axon specification by cultured hippocampal neurons," *J. Neurosci.*, Vol. 19, No. 15, 1999, pp. 6417–6426.

[37] Dertinger, S. K., Jiang, X., Li, Z., Murthy, V. N., and Whitesides, G. M., "Gradients of substrate-bound laminin orient axonal specification of neurons," *Proc. Natl. Acad. Sci. USA*, Vol. 99, No. 20, 2002, pp. 12542–12547.

[38] Dertinger, S. K. W., Chiu, D. T., Jeon, N. L., and Whitesides, G. M., "Generation of gradients having complex shapes using microfluidic networks," *Anal. Chem.*, Vol. 73, No. 6, 2001, pp. 1240–1246.

[39] Buettner, H. M., and Pittman, R. N., "Quantitative effects of laminin concentration on neurite outgrowth in vitro," *Dev. Biol.*, Vol. 145, No. 2, 1991, pp. 266–276.

[40] Turney, S. G., and Bridgman, P. C., "Laminin stimulates and guides axonal outgrowth via growth cone myosin II activity," *Nat. Neurosci.*, Vol. 8, No. 6, 2005, pp. 717–719.

[41] Silver, J., and Sidman, R. L., "A mechanism for the guidance and topographic patterning of retinal ganglion cell axons," *J. Comp. Neurol.*, Vol. 189, No. 1, 1980, pp. 101–111.

[42] Krayanek, S., and Goldberg, S., "Oriented extracellular channels and axonal guidance in the embryonic chick retina," *Dev. Biol.*, Vol. 84, No. 1, 1981, pp. 41–50.

[43] Clark, P., Connolly, P., Curtis, A. S., Dow, J. A., and Wilkinson, C. D., "Topographical control of cell behaviour: I. Simple step cues," *Development*, Vol. 99, No. 3, 1987, pp. 439–448.

[44] Clark, P., Connolly, P., Curtis, A. S., Dow, J. A., and Wilkinson, C. D., "Topographical control of cell behaviour: II. Multiple grooved substrata," *Development*, Vol. 108, No. 4, 1990, pp. 635–644.

[45] Nagata, I., Kawana, A., and Nakatsuji, N., "Perpendicular contact guidance of CNS neuroblasts on artificial microstructures," *Development*, Vol. 117, No. 1, 1993, pp. 401–408.

[46] Rajnicek, A., Britland, S., and McCaig, C., "Contact guidance of CNS neurites on grooved quartz: Influence of groove dimensions, neuronal age and cell type," *J. Cell. Sci.*, Vol. 110, Pt. 23, 1997, pp. 2905–2913.

[47] Nagata, I., Ono, K., Kawana, A., and Kimura-Kuroda, J., "Aligned neurite bundles of granule cells regulate orientation of Purkinje cell dendrites by perpendicular contact guidance in two-dimensional and three-dimensional mouse cerebellar cultures," *J. Comp. Neurol.*, Vol. 499, No. 2, 2006, pp. 274–289.

[48] Curtis, A., and Wilkinson, C., "Topographical control of cells," *Biomaterials*, Vol. 18, No. 24, 1998, pp. 1573–1583.

[49] Dowell-Mesfin, N. M., Abdul-Karim, M. A., Turner, A. M., Schanz, S., Craighead, H. G., Roysam, B., Turner, J. N., and Shain, W., "Topographically modified surfaces affect orientation and growth of hippocampal neurons," *J. Neural Eng.*, Vol. 1, No. 2, 2004, pp. 78–90.

[50] Rajnicek, A., and McCaig, C., "Guidance of CNS growth cones by substratum grooves and ridges: Effects of inhibitors of the cytoskeleton, calcium channels and signal transduction pathways," *J. Cell. Sci.*, Vol. 110, Pt. 23, 1997, pp. 2915–2924.

[51] Song, H. J., and Poo, M. M., "Signal transduction underlying growth cone guidance by diffusible factors," *Curr. Opin. Neurobiol.*, Vol. 9, No. 3, 1999, p. 355.

[52] Brown, A. F., "Neutrophil granulocytes: Adhesion and locomotion on collagen substrata and in collagen matrices," *J. Cell. Sci.*, Vol. 58, 1982, pp. 455–467.

[53] Wilkinson, P. C., and Lackie, J. M., "The influence of contact guidance on chemotaxis of human neutrophil leukocytes," *Exp. Cell Res.*, Vol. 145, No. 2, 1983, pp. 255–264.

[54] Nelson, R. D., Quie, P. G., and Simmons, R. L., "Chemotaxis under agarose: A new and simple method for measuring chemotaxis and spontaneous migration of human polymorphonuclear leukocytes and monocytes," *J. Immunol.*, Vol. 115, No. 6, 1975, pp. 1650–1656.

[55] Gundersen, R. W., and Barrett, J. N., "Characterization of the turning response of dorsal root neurites toward nerve growth factor," *J. Cell. Biol.*, Vol. 87, No. 3, Pt. 1, 1980, pp. 546–554.

[56] Boyden, S., "The chemotactic effect of mixtures of antibody and antigen on polymorphonuclear leucocytes," *J. Exp. Med.*, Vol. 115, 1962, pp. 453–466.

[57] Zigmond, S. H., "Ability of polymorphonuclear leukocytes to orient in gradients of chemotactic factors," *J. Cell. Biol.*, Vol. 75, No. 2, Pt. 1, 1977, pp. 606–616.

[58] Zicha, D., Dunn, G. A., and Brown, A. F., "A new direct-viewing chemotaxis chamber," *J. Cell. Sci.*, Vol. 99, Pt. 4, 1991, pp. 769–775.

[59] Li Jeon, N., Baskaran, H., Dertinger, S. K., Whitesides, G. M., Van de Water, L., and Toner, M., "Neutrophil chemotaxis in linear and complex gradients of interleukin-8 formed in a microfabricated device," *Nat. Biotechnol.*, Vol. 20, No. 8, 2002, pp. 826–830.

[60] Chen, C. S., Mrksich, M., Huang, S., Whitesides, G. M., and Ingber, D. E., "Geometric control of cell life and death," *Science,* Vol. 276, No. 5317, 1997, pp. 1425–1428.

[61] Walker, G. M., Sai, J., Richmond, A., Stremler, M., Chung, C. Y., and Wikswo, J. P., "Effects of flow and diffusion on chemotaxis studies in a microfabricated gradient generator," *Lab Chip*, Vol. 5, No. 6, 2005, pp. 611–618.

[62] Keenan, T. M., Hsu, C. H., and Folch, A., "Microfluidic 'jets' for generating steady-state gradients of soluble molecules on open surfaces," *Appl. Phys. Lett.*, Vol. 89, No. 11, 2006.

[63] Frevert, C. W., Boggy, G., Keenan, T. M., and Folch, A., "Measurement of cell migration in response to an evolving radial chemokine gradient triggered by a microvalve," *Lab Chip*, Vol. 6, No. 7, 2006, pp. 849–856.

[64] Hsu, C. H., and Folch, A., "Spatio-temporally-complex concentration profiles using a tunable chaotic micromixer," *Appl. Phys. Lett.*, Vol. 89, No. 14, 2006.

[65] Rosoff, W. J., Urbach, J. S., Esrick, M. A., McAllister, R. G., Richards, L. J., and Goodhill, G. J., "A new chemotaxis assay shows the extreme sensitivity of axons to molecular gradients," *Nat. Neurosci.*, Vol. 7, No. 6, 2004, pp. 678–682.

[66] Parent, C. A., and Devreotes, P. N., "A cell's sense of direction," *Science,* Vol. 284, No. 5415, 1999, pp. 765–770.

[67] Firtel, R. A., and Chung, C. Y., "The molecular genetics of chemotaxis: Sensing and responding to chemoattractant gradients," *Bioessays*, Vol. 22, No. 7, 2000, pp. 603–615.

[68] Moghe, P. V., Nelson, R. D., and Tranquillo, R. T., "Cytokine-stimulated chemotaxis of human neutrophils in a 3-D conjoined fibrin gel assay," *J. Immunol. Methods*, Vol. 180, No. 2, 1995, pp. 193–211.

[69] Foxman, E. F., Campbell, J. J., and Butcher, E. C., "Multistep navigation and the combinatorial control of leukocyte chemotaxis," *J. Cell. Biol.*, Vol. 139, No. 5, 1997, pp. 1349–1360.

[70] Taylor, A. M., Blurton-Jones, M., Rhee, S. W., Cribbs, D. H., Cotman, C. W., and Jeon, N. L., "A microfluidic culture platform for CNS axonal injury, regeneration and transport," *Nat. Methods*, Vol. 2, No. 8, 2005, pp. 599–605.

[71] Taylor, A. M., Rhee, S. W., Tu, C. H., Cribbs, D. H., Cotman, C. W., and Jeon, N. L., "Microfluidic multicompartment device for neuroscience research," *Langmuir*, Vol. 19, No. 5, 2003, pp. 1551–1556.

[72] Britland, S., and McCaig, C., "Embryonic Xenopus neurites integrate and respond to simultaneous electrical and adhesive guidance cues," *Exp. Cell Res.*, Vol. 226, No. 1, 1996, pp. 31–38.

[73] Fambrough, D. M., "Control of acetylcholine receptors in skeletal muscle," *Physiol. Rev.*, Vol. 59, No. 1, 1979, pp. 165–227.

[74] Steinbach, J. H., Merlie, J., Heinemann, S., and Bloch, R., "Degradation of junctional and extrajunctional acetylcholine receptors by developing rat skeletal-muscle," *Proc. Natl. Acad. Sci. USA*, Vol. 76, No. 7, 1979, pp. 3547–3551.

[75] Berg, D. K., and Hall, Z. W., "Fate of alpha-bungarotoxin bound to acetylcholine receptors of normal and denervated muscle," *Science,* Vol. 184, No. 4135, 1974, pp. 473–475.

[76] McMahan, U. J., "The agrin hypothesis," *Cold Spring Harb. Symp. Quant. Biol.*, Vol. 55, 1990, pp. 407–418.

[77] McMahan, U. J., Horton, S. E., Werle, M. J., Honig, L. S., Kroger, S., Ruegg, M. A., and Escher, G., "Agrin isoforms and their role in synaptogenesis," *Curr. Opin. Cell. Biol.*, Vol. 4, No. 5, 1992, pp. 869–874.

[78] Sanes, J. R., and Lichtman, J. W., "Development of the vertebrate neuromuscular junction," *Ann. Rev. Neurosci.*, Vol. 22, 1999, pp. 389–442.

[79] Bezakova, G., Rabben, I., Sefland, I., Fumagalli, G., and Lomo, T., "Neural agrin controls acetylcholine receptor stability in skeletal muscle fibers," *Proc. Natl. Acad. Sci. USA*, Vol. 98, No. 17, 2001, pp. 9924–9929.

[80] Kummer, T. T., Misgeld, T., and Sanes, J. R., "Assembly of the postsynaptic membrane at the neuromuscular junction: Paradigm lost," *Curr. Opin. Neurobiol.*, Vol. 16, No. 1, 2006, pp. 74–82.

[81] Peng, H. B., and Nakajima, Y., "Membrane particle aggregates in innervated and noninnervated cultures of Xenopus embryonic muscle cells," *Proc. Natl. Acad. Sci. USA*, Vol. 75, No. 1, 1978, pp. 500–504.

[82] Harris, A. J., "Embryonic growth and innervation of rat skeletal muscles: III. Neural regulation of junctional and extra-junctional acetylcholine receptor clusters," *Philos. Trans. R. Soc. Lond. B Biol. Sci.*, Vol. 293, No. 1065, 1981, pp. 287–314.

[83] Lin, W., Burgess, R. W., Dominguez, B., Pfaff, S. L., Sanes, J. R., and Lee, K. F., "Distinct roles of nerve and muscle in postsynaptic differentiation of the neuromuscular synapse," *Nature*, Vol. 410, No. 6832, 2001, pp. 1057–1064.

[84] Yang, X., Arber, S., William, C., Li, L., Tanabe, Y., Jessell, T. M., Birchmeier, C., and Burden, S. J., "Patterning of muscle acetylcholine receptor gene expression in the absence of motor innervation," *Neuron*, Vol. 30, No. 2, 2001, pp. 399–410.

[85] Fischbach, G. D., and Cohen, S. A., "The distribution of acetylcholine sensitivity over uninnervated and innervated muscle fibers grown in cell culture," *Dev. Biol.*, Vol. 31, No. 1, 1973, pp. 147–162.

[86] Sytkowski, A. J., Vogel, Z., and Nirenberg, M. W., "Development of acetylcholine receptor clusters on cultured muscle cells," *Proc. Natl. Acad. Sci. USA*, Vol. 70, No. 1, 1973, pp. 270–274.

[87] Bekoff, A., and Betz, W. J., "Acetylcholine hot spots: Development on myotubes cultured from aneural limb buds," *Science,* Vol. 193, No. 4256, 1976, pp. 915–917.

[88] Kummer, T. T., Misgeld, T., Lichtman, J. W., and Sanes, J. R., "Nerve-independent formation of a topologically complex postsynaptic apparatus," *J. Cell. Biol.*, Vol. 164, No. 7, 2004, pp. 1077–1087.

[89] Misgeld, T., Kummer, T. T., Lichtman, J. W., and Sanes, J. R., "Agrin promotes synaptic differentiation by counteracting an inhibitory effect of neurotransmitter," *Proc. Natl. Acad. Sci. USA*, Vol. 102, No. 31, 2005, pp. 11088–11093.

[90] Ngo, S. T., Balke, C., Phillips, W. D., and Noakes, P. G., "Neuregulin potentiates agrin-induced acetylcholine receptor clustering in myotubes," *Neuroreport*, Vol. 15, No. 16, 2004, pp. 2501–2505.

[91] Trinidad, J. C., and Cohen, J. B., "Neuregulin inhibits acetylcholine receptor aggregation in myotubes," *J. Biol. Chem.*, Vol. 279, No. 30, 2004, pp. 31622–31628.

[92] Meier, T., and Wallace, B. G., "Formation of the neuromuscular junction: Molecules and mechanisms," *Bioessays*, Vol. 20, No. 10, 1998, pp. 819–829.

[93] Frank, E., and Fischbach, G. D., "Early events in neuromuscular junction formation in vitro: Induction of acetylcholine receptor clusters in the postsynaptic membrane and morphology of newly formed synapses," *J. Cell. Biol.*, Vol. 83, No. 1, 1979, pp. 143–158.

[94] Anderson, M. J., and Cohen, M. W., "Nerve-induced and spontaneous redistribution of acetylcholine receptors on cultured muscle cells," *J. Physiol.,* Vol. 268, No. 3, 1977, pp. 757–773.

[95] Bromann, P. A., Zhou, H., and Sanes, J. R., "Kinase- and rapsyn-independent activities of the muscle-specific kinase (MuSK)," *Neurosci. Lett.*, Vol. 125, No. 2, 2004, pp. 417–426.

[96] Jones, G., Herczeg, A., Ruegg, M. A., Lichtsteiner, M., Kroger, S., and Brenner, H. R., "Substrate-bound agrin induces expression of acetylcholine receptor epsilon-subunit gene in cultured mammalian muscle cells," *Proc. Natl. Acad. Sci. USA*, Vol. 93, No. 12, 1996, pp. 5985–5990.

[97] Cornish, T., Branch, D. W., Wheeler, B. C., and Campanelli, J. T., "Microcontact printing: A versatile technique for the study of synaptogenic molecules," *Mol. Cell. Neurosci.*, Vol. 20, No. 1, 2002, pp. 140–153.

[98] Tourovskaia, A., Figueroa-Masot, X., and Folch, A., "Differentiation-on-a-chip: A microfluidic platform for long-term cell culture studies," *Lab Chip*, Vol. 5, No. 1, 2005, pp. 14–19.

[99] Tourovskaia, A., Kosar, T. F., and Folch, A., "Local induction of acetylcholine receptor clustering in myotube cultures using microfluidic application of agrin," *Biophys. J.*, Vol. 90, No. 6, 2006, pp. 2192–2198.

[100] Peterman, M. C., Noolandi, J., Blumenkranz, M. S., and Fishman, H. A., "Localized chemical release from an artificial synapse chip," *Proc. Natl. Acad. Sci. USA*, Vol. 101, No. 27, 2004, pp. 9951–9954.

[101] Kosar, T. F., Tourovskaia, A., Figueroa-Masot, X., Adams, M. E., and Folch, A., "A nanofabricated planar aperture as a mimic of the nerve-muscle contact during synaptogenesis," *Lab Chip*, Vol. 6, No. 5, 2006, pp. 632–638.

[102] Peterman, M. C., Mehenti, N. Z., Bilbao, K. V., Lee, C. J., Leng, T., Noolandi, J., Bent, S. F., Blumenkranz, M. S., and Fishman, H. A., "The artificial synapse chip: A flexible retinal interface based on directed retinal cell growth and neurotransmitter stimulation," *Artif. Org.*, Vol. 27, No. 11, 2003, pp. 975–985.

[103] Peterman, M. C., Bloom, D. M., Lee, C., Bent, S. F., Marmor, M. F., Blumenkranz, M. S., and Fishman, H. A., "Localized neurotransmitter release for use in a prototype retinal interface," *Invest. Ophthalmol. Vis. Sci.*, Vol. 44, No. 7, 2003, pp. 3144–3149.

[104] Takano, H., Sul, J. Y., Mazzanti, M. L., Doyle, R. T., Haydon, P. G., and Porter, M. D., "Micropatterned substrates: Approach to probing intercellular communication pathways," *Anal. Chem.*, Vol. 74, No. 18, 2002, pp. 4640–4646.

[105] Thompson, D. M., and Buettner, H. M., "Oriented Schwann cell monolayers for directed neurite outgrowth," *Ann. Biomed. Eng.*, Vol. 32, No. 8, 2004, pp. 1120–1130.

[106] Thompson, D. M., and Buettner, H. M., "Neurite outgrowth is directed by schwann cell alignment in the absence of other guidance cues," *Ann. Biomed. Eng.*, Vol. 34, No. 1, 2006, pp. 161–168.

[107] Miller, C., Jeftinija, S., and Mallapragada, S., "Micropatterned Schwann cell-seeded biodegradable polymer substrates significantly enhance neurite alignment and outgrowth," *Tissue Eng.*, Vol. 7, No. 6, 2001, pp. 705–715.

[108] Recknor, J. B., Sakaguchi, D. S., and Mallapragada, S. K., "Growth and differentiation of astrocytes and neural progenitor cells on micropatterned polymer films," *Annals NY Acad. Sci.*, Vol. 1049, 2005, pp. 24–27.

[109] Chang, J. C., Brewer, G. J., and Wheeler, B. C., "Neuronal network structuring induces greater neuronal activity through enhanced astroglial development," *J. Neural Eng.*, Vol. 3, No. 3, 2006, pp. 217–226.

CHAPTER 14

Self-Assembly of Nanomaterials for Engineering Cell Microenvironment

Hossein Hosseinkhani, Mohsen Hosseinkhani, Karthikeyan Subramani, and Shuguang Zhang

14.1 Overview

This chapter reviews the applications of self-assembling proteins and peptides toward designing novel biomimetic nanomaterials and their potential applications in nanobiotechnology and biomedicine. Numerous proteins and peptides have been emerging as nanobiomaterials due to their ability to self-assemble into nanoscale structures like nanotubes, nanovesicles, helical ribbons, and fibrous scaffolds. Self-assembly can be defined as the ability of certain multimeric biological structures to assemble from their component parts through random movements of the molecules and formation of weak chemical bonds between surfaces with complementary shapes. In other words, self-assembly can be defined as the spontaneous organization of individual components into an ordered structure without human intervention [1, 2]. A few examples of self-assembled structures that can be found at the microscopic level are phospholipid bilayer of human cell membranes [3], RNA [4], and DNA complexes [5]. The detergent molecules exhibit self-assembly phenomena due to their amphiphilic properties. The molecular building mechanisms underlying the formation of bacteriophage and viral particles are based on self-assembly. The other common examples of self-assembly phenomena tailored by nature are lipid molecules forming oil droplets in water, four haemoglobin polypeptides forming a functional haemoglobin protein, and the combination of RNA and ribosomal proteins to form a functional ribosome.

Molecular self-assembly is a powerful phenomenon borrowed from nature by scientists for fabricating novel supramolecular architectures. Molecular self-assembly is mainly governed by weak noncovalent bonds like electrostatic interactions (ionic bonds), hydrogen bonds, hydrophobic interactions, water-mediated hydrogen bonds, and van der Waals interactions [2–4]. Although these forces are weak, their collective interactions can produce structurally and chemically stable structures. Self-assembly of biological molecules like proteins, peptides, nucleic acids, lipids, and other cellular components governs the biological structure and function of a living cell. Cellular events like amyloid fibril formation, antigen-antibody recognition, chromatin assembly, and phospholipid membrane self-assembly are excellent examples of molecular self-assembly.

14.2 Proteins and Peptides

Proteins are fundamental components of all living cells. They can be classified as a group of complex organic macromolecules containing carbon, hydrogen, nitrogen, oxygen, and sulphur. They are composed of one or more chains of amino acids. Amino acids contain an amino (-NH2) and a carboxyl (-COOH) group. Two or more amino acids linked by a peptide bond form a peptide molecule. Large numbers of peptide molecules arrange themselves in different fashions to make up different kinds of proteins. Enzymes, hormones, and antibodies are a few examples of biological substances that are made up of proteins and required for the proper functioning of a living organism. Structural analysis of protein molecules has revealed that they take up various shapes to form a stable macroscopic structure. Nature has used proteins to build a vast array of structures like keratin, collagen, coral, pearl, shell, and the like. Molecular self-assembly is exhibited by proteins and peptides. In the past few decades, numerous researches have been done to understand the structural characteristics that influence the self-assembly of protein and peptide molecules. The potential applications of self-assembling proteins and peptides have been effectively utilized to design novel biomimetic nanomaterials that have tremendous applications in the fields of nanobiotechnology and biomedicine.

14.3 Self-Assembly of Proteins and Peptides

Various authors have discussed in detail several self-assembling peptide and protein systems that self-assemble to form various nanostructures like nanotubes, vesicles, helical ribbons, and fibrous scaffolds. These structures are analysed to design and fabricate new materials that will have potential applications in biomedical nanotechnology. The findings discussed in this chapter have been grouped under different headings: "Self-assembly of Proteins and Peptides," "Findings About Amphiphilic and Surfactantlike Peptides," "Findings About Three-Dimensional Peptide Matrix Scaffolds," and "Use of Peptide Hydogels in Regenerative Biology and Three-Dimensional Cell Culture." A group of scientists has designed artificial proteins that self-assemble to form hydrogels (Figure 14.1). The artificial proteins designed by the researchers were made up of an ionic self-complementary peptide group that had an alternating polar and nonpolar fashion of arrangement of peptide molecules. These peptides formed stable β-strand and β-sheet structures, which self-assembled to form nanofibres. These nanofibres formed interwoven matrices that further formed a scaffold hydrogel with high water content. Hydrogel has water as its dispersion medium and responds to changes in pH and other environmental factors. These protein hydrogels can be used for advanced wound closure and tissue repair in regenerative medicine and tissue engineering. Biodegradable protein hydrogels can act as drug-delivery systems delivering pharmaceutical protein complexes in the treatment of diseases like cancer.

Another group of scientists has designed peptide nanotubes using surfactantlike peptides (Figure 14.2). These peptide nanotubes can be used as templates for growing metal nanocrystals; thus, nanowires can be fabricated. Peptide nanotubes can

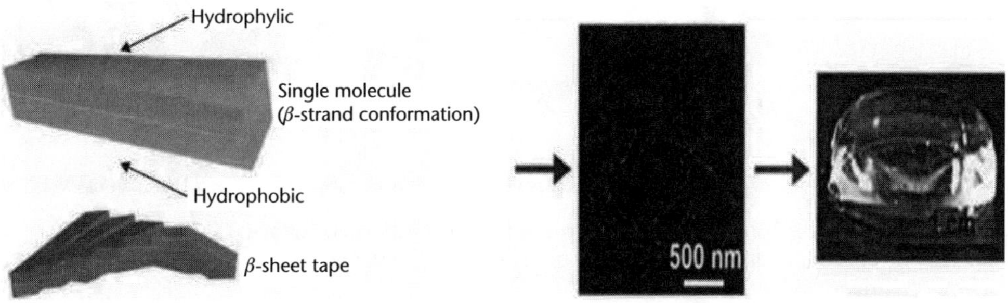

Figure 14.1 Amphiphilic peptides in β-strand and β-sheet conformation self-assemble into interwoven matrices that further form a scaffold hydrogel. (*Source:* www.nature.com/nbt/journal/v21/n10/fig_tab/nbt874_F1.html.)

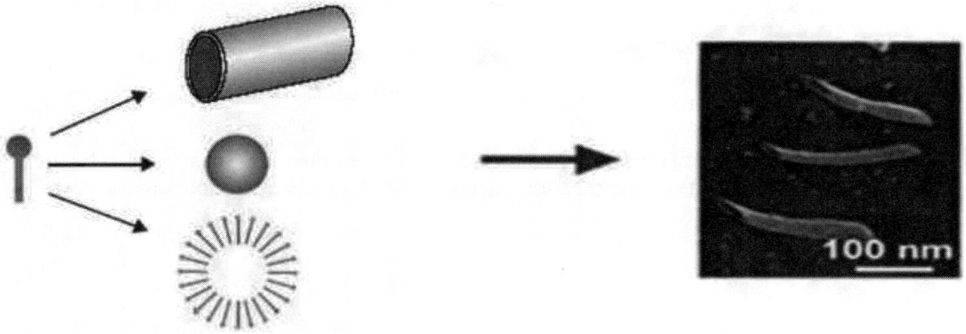

Figure 14.2 Surfactantlike peptides self-assemble to form nanotubes. (*Source:* www.nature.com/nbt/journal/v21/n10/fig_tab/nbt874_F1.html.)

also serve as ion channels when incorporated into the phopholipid bilayer of the cell membrane.

Recent results also refer to the researches done into biomimetic protein structures and peptide systems that can form complexes with metals and semiconducting elements. Surface-binding peptides can bind covalently with metal surfaces like gold (Figure 14.3). These peptides can also form complexes with DNA, which in turn can be bound to a metal surface. This property can be exploited to design and fabricate nanobiosensors.

Certain peptides with strong dipoles undergo drastic conformational changes between the α-helical structure and the β-sheet. These are called molecular switch peptides. Gold nanoparticles can be attached to these dipolar peptides to fabricate tiny molecular switches (Figure 14.4).

14.4 Findings About Amphiphilic and Surfactantlike Peptides

Amphiphilic molecules have a hydrophilic (polar) and a hydrophobic (nonpolar) component. In the presence of water, they self-assemble into distinct structures whose shape is largely determined by the size and shape of the hydrophilic polar head. The paper describes in detail the authors' research into amphiphilic and

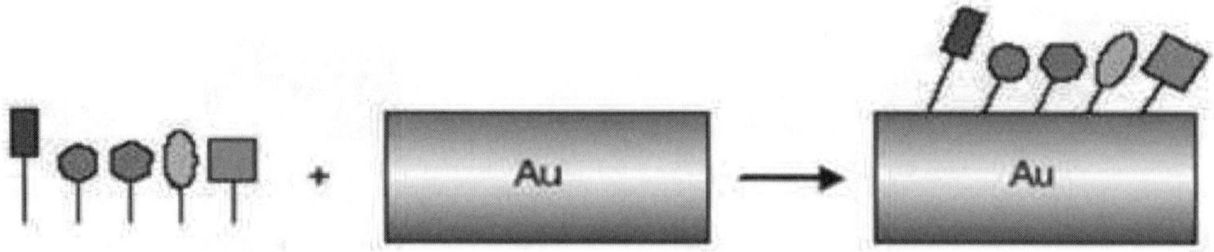

Figure 14.3 Surface-binding peptides binding with gold surface. (*Source:* http://cba.media.mit.edu/publications/articles/02.00.zhang.pdf.)

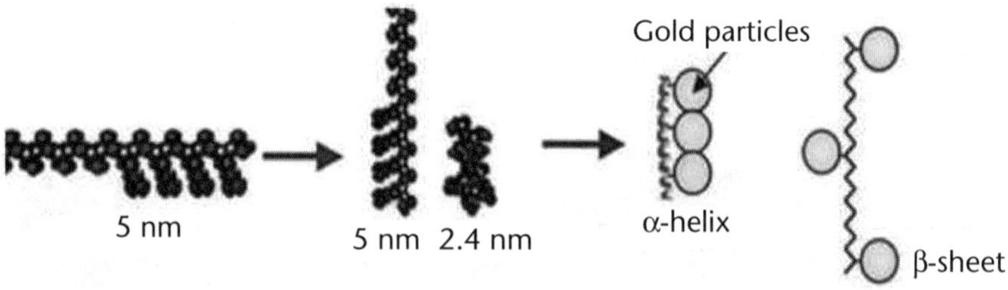

Figure 14.4 Helical dipolar peptides undergoing conformational changes can be used as tiny molecular switches with gold nanoparticles attached. (*Source:* www.puramatrix.com/NatureBiotechny_2003.pdf.)

surfactantlike peptides conducted in their laboratory. These molecules have one or two amino acids in the polar region and four or more consecutive hydrophobic amino acids at their nonpolar end. One such example discussed is the V6D amino acid complex. The V6D amino acid sequence (VVVVVVD) has six valine (V) residues that are hydrophobic and an aspartic acid (D) residue that is negatively charged (Figures 14.5 and 14.6).

Valine (α-aminoisovaleric acid) is an essential amino acid, and aspartic acid is a nonessential amino acid. This V6D peptide complex formed various nanostructures in aqueous solution like nanotubes and nanovesicles. The samples of the aqueous solution were frozen in liquid propane (–180°C) and surface-coated with a thin layer of platinum and carbon to preserve the structures formed, and when they investi-

Figure 14.5 Molecular model of surfactantlike peptide V6D (VVVVVVD). (*Source:* www.templeton.org/biochem-finetuning/papers/zhang_paper.doc.)

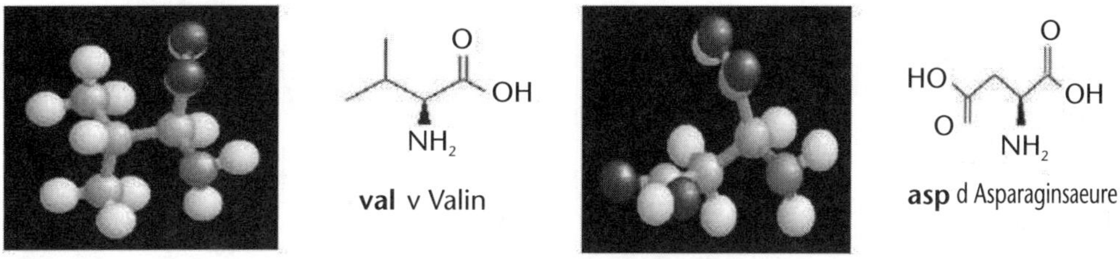

Figure 14.6 Three-dimensional structure and structural formula of valine and aspartic acid. (*Source:* http://ntri.tamuk.edu/cell/chapter3/val3d.html,http://ntri.tamuk.edu/cell/chapter3/asp3d.html.)

gated them using transmission electron microscopy (TEM), the researchers observed nanotubes and nanovesicles. The nanotubes measured about 30 to 50 nm (Figure 14.7). It has been indicated that the self-assembled structure can be modified by changing the sequence of the amino acids in the peptide chain and the environmental factors. These peptide nanotubes can be incorporated into self-assembled membranes for use in bionanosensor devices (Figure 14.8).

It is also suggested that these surfactant peptides can be engineered for improved functionality by using techniques like biotinylation. Biotinylation is a process of incorporation of biotinyl groups into molecules to visualise specific substrates by incubating them with biotin-labeled probes and avidin or streptavidin. It is a rapid method of detecting nucleic acids for use in the Western blot technique. When these surfactant peptide nanostructures are made to undergo the process of biotinylation, they can be bound to streptavidin-coated an inorganic metal surface. Histidine-tagged peptides and proteins can be bound to nickel surfaces. Thus, by utilizing the standard techniques in peptide chemistry, these nanostructures can be attached to metallic surfaces.

14.5 Findings About Three-Dimensional Peptide Matrix Scaffolds

A wider variety of self-assembling proteins and peptides have inspired researchers to fabricate nanoscale fibres and fibre network scaffolds. The main factor influenc-

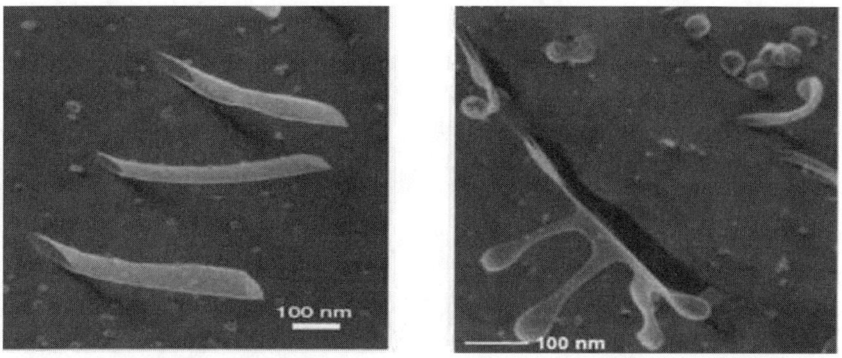

Figure 14.7 Quick-freeze/deep-etch transmission electron micrographs of V6D peptide complex: (left) Nanotubes with a diameter of 30 to 50 nm, and (right) nanovesicles and nanotubes. (*Source:* http://web.mit.edu/lms/www/PDFpapers/Zhang_MaterialsToday3BEBE.pdf.)

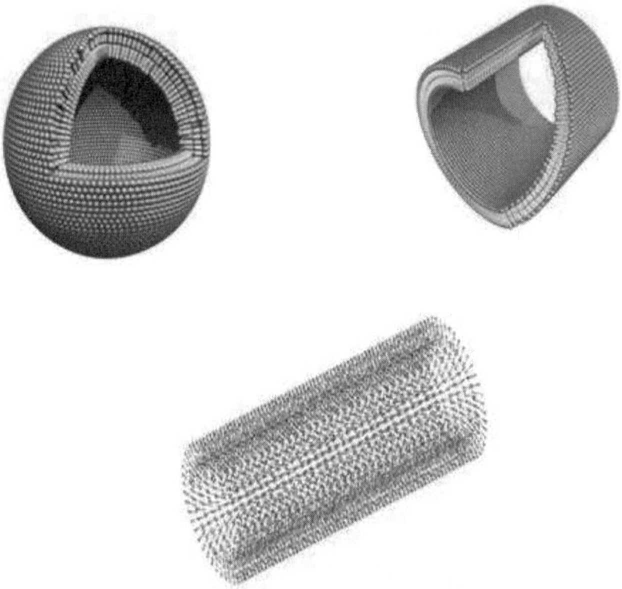

Figure 14.8 Molecular models of V6D peptide nanovesicle and nanotube. (*Source:* www.templeton.org/biochem-finetuning/papers/zhang_paper.doc.)

ing the entire process was the chirality of the individual building components of the peptide complex. Two molecules are said to be chiral if their mirror images are not superimposed. Researchers have studied the KFE8 peptide complex, which is an eight-residue peptide complex. The KFE8 has the peptide chain sequence of FKFEFKFE (Figure 14.9).

This structure represents a group of self-assembling peptides that spontaneously self-assemble under certain physiological conditions. This peptide self-assembles in aqueous solution into left-handed helical ribbons when the peptide backbone is twisted in the opposite direction (Figure 14.10). It has been reported that when certain amino acids in the hydrophobic side chains were replaced with another amino acid sequence, little change was observed. When a positively charged lysine (Lys) was replaced with a positively charged arginine (Arg), or when a negatively charged glutamate (Glu) was replaced with a negatively charged aspartate (Asp), very little

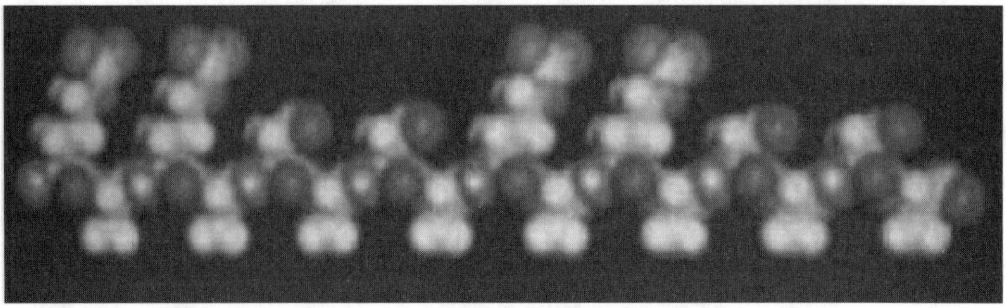

Figure 14.9 Molecular model of KFE8 peptide. (*Source:* http://web.mit.edu/lms/www/PDFpapers/Zhang_MaterialsToday3BEBE.pdf.)

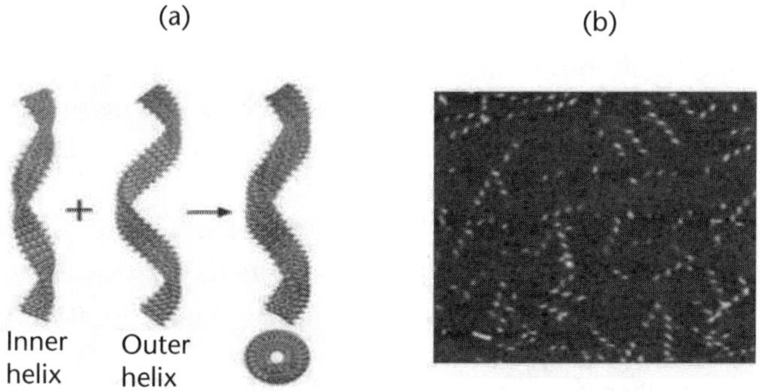

Figure 14.10 (a) The inner and outer β-helices of KFE8 peptide form a double-sheet helix with hydrophobic side chains sandwiched between the two layers. (b) Atomic force microscopy image (500×500 nm) of peptide solution deposited over mica. (*Source:* http://cba.media.mit.edu/publications/articles/02.00.zhang.pdf.)

change was observed in the nanofibres that were formed. But when the positively charged residue was replaced with a negatively charged residue or vice versa, the peptides did not self-assemble. When the alanines were replaced with hydrophobic residues, there was a greater tendency to self-assemble and form peptide matrices with enhanced strength.

This has led researchers to concentrate their attention on understanding the basis of protein-conformational diseases. Protein-conformational diseases are a group of disorders characterized by the accumulation of malformed protein structures in cells. Proteins must fold into a proper three-dimensional structure to carry out their normal functions. When they do not fold properly, they form malfolded protein structures that accumulate in cells, leading to pathological conditions. Alzheimer's disease, prion diseases, and Parkinson's disease are a few examples of protein-conformational diseases. Thus, by understanding the mechanism of formation of peptide nanofibres and the factors controlling their self-assembly, researchers aim to formulate a remedy for protein-conformational diseases.

The authors have divided the self-assembled peptide fibres into three theoretical models. The first model is the molecular model, where the β-sheet peptides self-assemble into helical ribbons. The second model is the semicontinuum model, where the peptides self-assemble to form elastic tapelike structures composed of bricklike building blocks. The third model is the fully continuum model, where the peptides self-assemble to form tubules. These approaches have helped researchers to learn more about the mechanisms underlying the formation of different structures when peptides and proteins self-assemble and will ultimately guide them to design efficient peptide-based and protein-based biomaterials.

14.6 Use of Peptide Hydogels in Regenerative Biology and Three-Dimensional Cell Culture

The design of materials that can regulate cell behavior such as proliferation and differentiation is a key component for the fabrication of tissue engineering scaffolds. From the viewpoint of immune system response of the body, the implanted biomaterials should mimic the structure as reported by Ma et al. [5]. Therefore, in order to mimic the biological function of ECM proteins, the scaffold materials used in tissue engineering need to be chemically functionalized to promote tissure regeneration as ECM does.

The initial report showed that nanoscaled features influenced cell behaviors [6]. Nanoscaled surface topography has been found to promote osteoblast adhesions [7]. It has been demonstrated that osteoblast adhesion, proliferation, alkaline phosphatase activity, and ECM secretion on carbon nanofibers increased with decreasing fiber diameter in the range of 60 to 200 nm, whereas the adhesion of other kinds of cells, such as chondorocytes, fibroblasts, and smooth muscle cells, was not influenced [8, 9]. It has been supposed that the nanoscaled surface affects the conformation of adsorbed adhesion proteins such as vitronectin, thus affecting cell behavior [10]. In addition, the nanoscaled dimensions of cell membrane receptors such as integrins should also be considered.

It has been reported that collagen and elastin as ECM proteins are made from fibers in dimension smaller than micrometers [5]. It seems that artificial nanoscaled fibers have great potential application in the field of biomaterials and tissue engineering. It has been reported that there are three different approaches toward the formation of nanofibrous materials; phase separation, electrospinning, and self-assembly [11]. Phase separation and self assembling of biomolecules can generate smaller diameter nanofibers in the same range of natural ECM, while electrospinning generate large diamter nanofibers on the upper end of the range of natural ECM [11]. One of the common approaches to produce fibers similarly to ECM proteins such as collagen is self-assembly. It has been shown that peptide amphiphile that contain a carbon alkyl tail and several other fucntional peptide formed nanofibers through self-assembly by mixing cell suspensions in media with dilute aqueous solutions of the peptide [12]. These self-assembled nanofibers have been used recently to study selective differentiation of progenitor cells [13]. Another type of peptide containing 16 alternative hydophobic and hydrophilic amino acids was fabricated to self-assemble into nanofibers under appropriate pH values [14]. Nanoscaled fibers produced by self assembly of peptide amphiphile may be a promising approach in designing the next generation of biomaterials for drug delivery and tissue engineering.

For successful tissue regeneration, the cells constituting the tissue to be regenerated, such as matured, progenitor, and precursor, are necessary. Considering the proliferation activity and differentiation potential of cells, stem cells are practically promising. Among them, mesenchymal stem cells (MSCs) have been widely investigated for use by themselves or in combination with the scaffolds necessary for the promotion of cell proliferation and differentiation. It was found that MSCs have an inherent nature to differentiate into not only osteogenic linage cells but also chondrogenic, myogenic, adipogenic, and neurogenic lineages [15–19]. It has been recognized that induction of tissue regeneration based on tissue engineering can be achieved by the following three key steps: the proliferation of cells, the seeding of cells and proliferation in a suitable scaffold, and the maintenance of the differentiation phenotype of the engineered tissues [20]. The property of scaffold material for cell attachment is one of the major factors contributing their morphology, proliferation, and functioning, as well as the subsequent tissue organization [21]. At first, cells attach to the material surface of scaffold, then spread and proliferate. A three-dimensional scaffold can provide larger surface area for cell attachment and spreading than a two-dimensional scaffold (i.e., tissue-culture plate). Xie, Yang, and Kniss [22] have reported that the initial rate of cells growth was higher for the two-dimensional culture, but once the cells reached confluence, their proliferation stopped. However, the cells' growth in the three-dimensional scaffold was continued for longer periods than in two-dimensional scaffolds. Other reports have demonstrated that cell proliferation was superior in the three-dimensional scaffold than the two-dimensional one [23–27]. It would be beneficial for biomedical applications if scaffold materials could promote the adhesion and growth of cells on their surfaces. The sequence of arginine-glycine-aspartic acid (RGD) has been discovered to be a cell-attachment sequence in various adhesive proteins present in the ECM and found in many proteins, such as fibronectin, collagen type I, vitronectin, fibrin, and the Von Willebrand factor [28]. It is well recognized that the sequence of RGD interacts with various types of integrin receptors of mammalian cells. Ever since the RGD sequence was discovered to be a cell-attachment sequence in adhesive proteins of the ECM, there have been several efforts to synthesize bioactive peptides incorporating RGD for therapeutic purposes [29].

Micro- and nanopatterned scaffolds have been investigated less well with regard to stem cells, although two recent studies highlight their attractiveness [30]. In one study, Silva et al. included a five amino acid, laminin-specific, cell-binding domain (which binds to specific integrins on the cell surface) at the hydrophilic head of their amphiphiles and showed that neural stem cells could be induced to differentiate into neurons when cultured within peptide gel [13]. In contrast, cells grown in control scaffolds without the laminin-specific domain or on two-dimensional tissue culture plastic-coated with laminin solution differentiated much less. This was hypothesized to be largely a result of the density of the cell-binding ligands to which the cells were exposed, indicating clearly the importance of ECM in influencing cell function. Our recent studies have indicated that when the laminin-specific domain in the amphiphilic molecule was replaced with the amino acid sequence, RGD, a common cell-binding domain in many ECM proteins, especially collagen, differentiation of MSCs to osteoblasts was significantly enhanced compared with amphililic nanofibers without this sequence or to two-dimensional controls (Figure 14.11) [31]. This is because the interaction of MSC integrin receptors with RGD of the peptide enhanced cell attachment on peptide nanofibers. The proliferation of cells in the three-dimensional scaffold needs an oxygen and nutrition supply. In this circumstance, the three-dimensional scaffold materials should provide such an environment for cells. The artificial scaffolds formed by self-assembling molecules not only provide suitable support for cell proliferation but also serve as a medium through which diffusion of soluble factors and migration of cells can occur. The result of the cell attachment and proliferation revealed that diffusion of nutrients, bioactive factors, and oxygen through these highly hydrated networks is sufficient for the survival of large numbers of cells for extended periods of time.

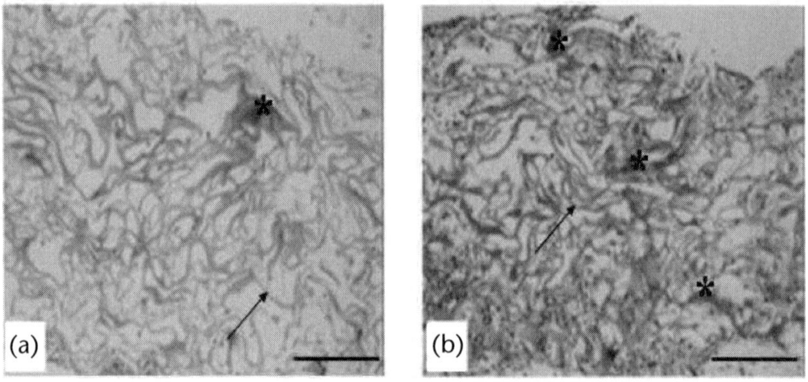

Figure 14.11 Histological cross sections of MSC attached to the self-assembled peptide nanofibers (a) without RGD and (b) with RGD incorporation four weeks after the culture in the osteogenic medium. The scale bar measures 100 μm in higher magnification views of center in gel. Arrows indicate the residual gel of the self-assembled peptide nanofibers. Asterisks indicate the newly formed bone. (*Source:* www.sciencedirect.com/science/journal/01429612/2006.pdf.)

As understood from the findings, proteins and peptides can self-assemble into various structures like nanotubes, nanovesicles, and three-dimensional peptide matrices with interwoven nanofibres. Macroscopic three-dimensional peptide matrices can be engineered to form various shapes by changing the peptide sequence. Self-assembled peptide materials encouraged cell proliferation and differentiation. These peptide materials were also able to support various types of cell attachments. The ability of the peptides to support attachment of mouse neuronal cells has been fully studied. The primary mouse neuron cells formed active connections with the peptide scaffolds that formed a valuable area of research for studying about neuron regeneration. In regenerative medicine, these peptide matrices were used to cultivate chondrocyte ECM that can be used to repair cartilage tissue. Thus, cartilage-tissue engineering has been done by placing the primary chondrocytes and MSCs into these self-assembled peptide hydrogels to produce collagen and glycosaminoglycans. These peptide matrices can also be used in the regeneration of bone by incorporating a phophorylated serine, which can attract and organize calcium ions to form hydroxyapatite crystals, and functionalizing them with a cell-adhesion motif like arginine-glycine-aspartic acid complex. Research studies have not been limited only to natural amphiphilic peptides. Many research trails have indicated on the synthesizing of complex amphiphilic peptides by joining hydrophilic peptides into long alkyl chains. The peptide end of the molecule was designed to function and regulate biomineralization. Bone is produced as a result of the deposition of calcium and phosphate ions to form hydroxyapatite crystals. This process is known as mineralization. Serine is a nonessential amino acid. When a phophorylated serine was incorporated with the synthetic amphiphilic peptide complex, it served to attract and organize calcium and phosphate ions to form hydroxyapaptite crystals. Furthermore, the peptides the synthetic amphiphilic peptide have been functionalized by adding a cell-adhesion motif. It was the RGD that was attached to the C-terminus of the peptide. This can be used to study the ability of the bone cells to differentiate, proliferate, and adhere to a biomaterial surface like titanium. Titanium is the most widely used biomaterial surface to produce orthopaedic implants, dental implants, and hip replacements. Inspite of its excellent biocompatibility, titanium implants still fail. Most orthopaedic implants have a lifetime of fifteen years at the maximum. In order to produce a newer version of titanium implants that can stay in the body for a longer period of time, its surface has to be modified with nano-size surface patterns so that bone cells (osteoblasts) differentiate and migrate into these patterns for better bone-implant adhesion. For such a purpose, these synthetic amphiphiles can be used to regulate and control the osteoblasts.

Another use of self-assembled peptides is in tissue engineering. Tissue engineering is designed to regenerate natural tissues or to create biological substitutes for defective or lost organs by making use of cells. Considering the use of cells in the body, there is no doubt that a sufficient supply of nutrients and oxygen to the transplanted cells is vital for their survival and functional maintenance [32]. Without a sufficient supply, only a small number of cells preseeded in the scaffold or migrated into the scaffold from the surrounding tissue would survive. Rapid formation of a vascular network at the transplanted site of cells must be a promising way to provide cells with the vital supply. This process of generating new microvasculature, termed neovascularization, is a process observed physiologically in development and wound healing [33]. It is recognized that basic fibroblast growth factor (bFGF) functions to promote such an angiogenesis process [33, 34]. The growth factors stimulate the appropriate cells (e.g., endothelial cells) already present in the body to migrate from the surrounding tissue, proliferate, and finally differentiate into blood vessels [33]. However, one cannot always expect sustained angiogenesis activity when these proteins are only injected in solution form, probably because of their rapid diffusional excretion from the injected site. One possible way to enhance the in vivo efficacy is to achieve its controlled release over an extended period by incorporating the growth factor into a polymer carrier. If this carrier is biodegraded and harmonized with tissue growth, it will work as a scaffold for tissue regeneration in addition to serving as a carrier matrix for growth factor release. Some studies have demonstrated that bFGF promoted angiogenesis when used in combination with delivery matrices and scaffold [35–40]. Our recent study has indicated that a three-dimensional network of self-assembled nanofibers was formed by mixing bFGF suspension with an aqueous solution of peptide amphiphile as an injectable carrier for controlled release of growth factors and used it for feasibility of prevascularization by the bFGF release from the three-dimensional networks of nanofibers in improving the efficiency of tissue regeneration [41]. Previous work has encapsulated bFGF within alginate, gelatin, agarose/heparin, collagen, and poly(etyhylene-co-vinyl acetate) carriers [40, 42, 43]. According to the results of these studies, it is conceivable to incorporate the angiogenic factor to a sustained releasing system and to use it prior to the implantation. The bFGF incorporated these releasing system requires surgery for implantation, which is not welcome. On the contrary, the bFGF incorporated in self-assembled peptides could be delivered to living tissues simply by injecting a liquid (i.e., peptide amphiphile solutions) and bFGF solution. The injected solutions would form a solid scaffold at the injected site of the tissue, and the releases bFGF would induce significant angiogenesis around the injected site, in marked contrast to bFGF injection alone or PA injection alone [41].

14.7 Applications of Synthetic Amphiphilic Peptides in Other Fields of Nanotechnology

Various authors have referred to the potential use of self-assembling amphiphilic peptides in various other areas of nanotechnology. The amphiphilic peptide molecules can be used as scaffolds to fabricate nanowires. A nanowire is a solid, metallic, cylinderlike structure with a diameter ranging from ten to several hundred nanometers. Nanowires can potentially be applied in electronics, chemistry, and biomedical engineering and therefore have greater applications in nanotechnology. Amphiphilic peptide nanotubes can be used as templates for metallization. One such example of the peptide sequence used was the histidine-rich peptide nanotube. This structure was metallized with gold nanocrystals, and the organic peptide scaffold was removed to make a conducting gold nanowire (Figure 14.12). Specific peptide sequences recognize specific metal ions. Thus, by varying the peptide sequence, much more efficient metal coatings can be coated on peptide nanotubes. Examples of other efficient metals that can be coated on peptide nanotubes include silver, platinum, copper, and nickel.

Various other methods to attach the peptides and the gold nanocrystals have recently been developed. A group of researchers fabricated nanotubes from bolaamphiphile peptides. They used crystalline glycylglycine bolaamphiphile tubules and studied the pH-sensitive structural transformation of the peptide tubules. These bolaamphiphiles undergo self-assembly in solution and form tubular structures with diameters ranging from 20 nm to 1 μm. The peptide nanotube was coated with a metalloporphyrin compound. Metalloporphyrins are a group of chemical compounds that have a metallic group attached to a porphyrin ring. Porphyrins are a group of closely related tetrapyrollic pigments occurring widely in nature and play a vital role in various biological processes. The heme component of the haemoglobin is a good example of this group of compounds. Researchers utilized the electron-transferring property of the metallophophyrins to bind to the pep-

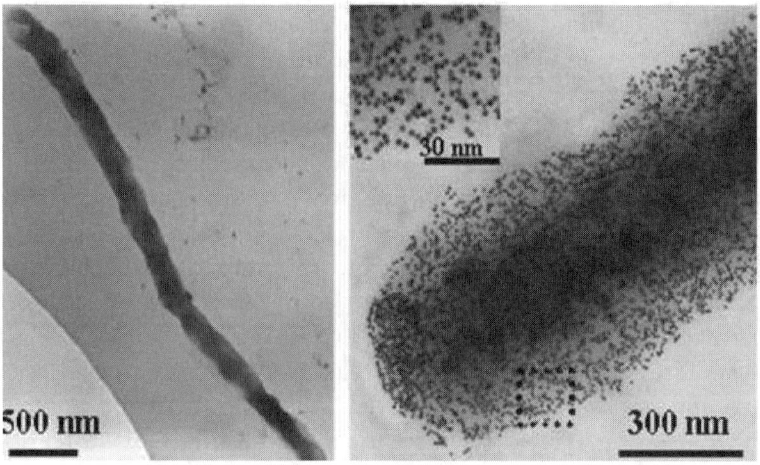

Figure 14.12 The TEM image of gold nanowire (left) made from histidine-rich peptide sequence as template. The image on the right shows gold nanocrystal particles coated over peptide nanotube. (*Source:* http://oasys2.confex.com/acs/225nm/techprogram/P611826.HTM.)

tide nanotubes. They were able to grow and immobilize the peptide nanotubes on a gold substrate. They also demonstrated that the peptide nanotubes can be coated with avidin, which enabled them to bind to gold surfaces treated with biotinylated self-assembled monolayers (SAMs) (Figures 14.13 and 14.14).

Thus, a wider range of applications and the design flexibility of self-assembling protein and peptide complexes, together with the researchers' knowledge of cell biology, biochemistry, and molecular biology, have enabled them to identify potential new applications in the field of material science and nanotechnology.

14.8 Conclusion

The increasing interest in bionanotechnology has stimulated researchers to scrutinize biological elements and learn from nature. Self-assembly of biological molecules forms the basic principle in the formation of complex biological structures. The topic discussed above on nanostructured biological materials through self-assembly of peptides and proteins gives us wide knowledge of the basic principle underlying the molecular self-assembly of proteins and peptides. It gives a

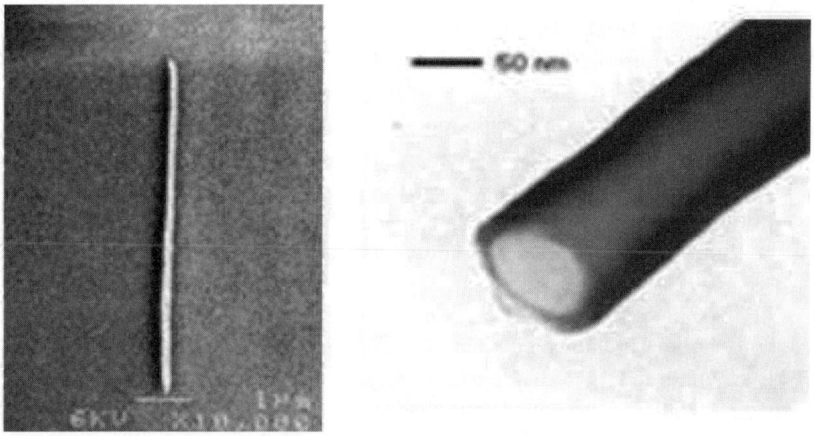

Figure 14.13 Scanning electron micrograph (SEM) and TEM image of a peptide nanotube. (*Source:* http://patsy.hunter.cuny.edu/matsui.html.)

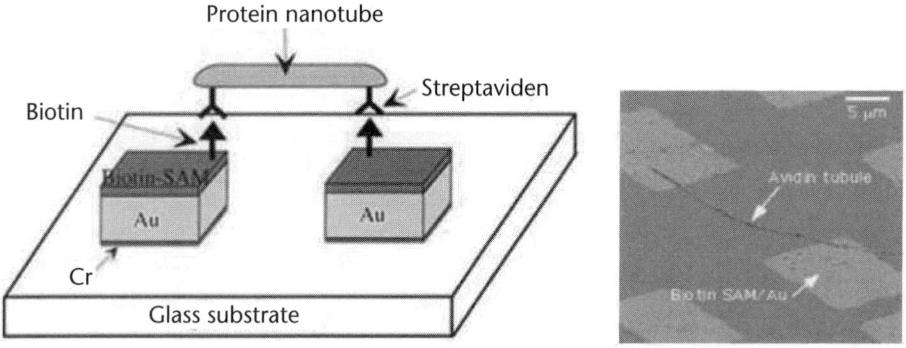

Figure 14.14 Schematic set-up of the process and SEM image of protein nanotube immobilized on biotin-SAM/Au surface. (*Source:* http://patsy.hunter.cuny.edu/matsui.html.)

general overview of different kinds of self-assembling protein and peptide systems. The research discussed in this chapter clearly guides the reader as to the potential applications of these self-assembled structures for fabricating a wider range of novel biomaterials for use in bionanotechnology. With appropriate references and examples, it opens the reader's mind to incorporating a wider range of knowledge about self-assembling proteins and peptide systems. These macroscopic structures have inspired researchers to use them in various areas of science like electronics, biotechnology, nanotechnology, and medicine.

References

[1] Kirby, S. B., and Umit, Y., "Protein-based hydrogels as advanced wound closure systems," www.asc2002.com/summaries/d/DP-15.pdf.

[2] Jeffrey, D. H., Juan, R. G., Ronald, A. M., Ghadiri, M. R., "Self-assembling peptide nanotubes," at www.ruf.rice.edu/~jdh/pdf/JACS-SAPN.pdf.

[3] Shuguang, Z., "Thoughts on prebiotic molecular self-organizations in water," at www.templeton.org/biochem-finetuning/papers/zhang_paper.doc.

[4] Steve, S. S., Sylvian, V., and Shuguang, Z., "Structures, functions and applications of amphiphilic peptides," at http://cba.media.mit.edu/publications/articles/02.12.santoso.pdf.

[5] Ma, Z., Kotaki, M., Inai, R., and Ramakrishna, S., "Potential of nanofiber matrix as tissue-engineering scaffolds," *Tissue Eng.*, Vol. 11, Nos. 1–2, 2005, pp. 101–109.

[6] Rosenberg, M. D., "Cell guidance by alterations in monomolecular films," *Science,* Vol. 139, No. 3553, 1963, pp. 411–412.

[7] Webster, T. J., Siegel, R. W., and Bizios, R., "Osteoblast adhesion on nanophase ceramics," *Biomaterials*, Vol. 20, No. 13, 1999, pp. 1221–1227.

[8] Pricea, R. L., Waidb, M. C., Haberstroha, K. M., and Webstera, T. J., "Selective bone cell adhesion on formulations containing carbon nanofibers," *Biomaterials, Vol. 24, No. 11, 2003, pp. 1877–1887.*

[9] Elias, K. L., Price, R. L., and Webster, T. J., "Enhanced functions of osteoblasts on nanometer diameter carbon fibers," *Biomaterials*, Vol. 23, No. 15, 2002, pp. 3279–3287.

[10] Webster, T. J., Schalder, L. S., Siegel, R. W., and Bizios, R., "Mechanisms of enhanced osteoblast adhesion on nanophase alumina involve vitronectin," *Tissue Eng.*, Vol. 7, No. 3, 2001, pp. 291–301.

[11] Smith, L. A., and Ma, P. X., "Nano-fibrous scaffolds for tissue engineering," *Colloids and Surfaces B: Biointerfaces*, Vol. 39, No. 3, 2004, pp. 125–131.

[12] Hartgerink, J. D., Beniash, E., and Stupp, S. I., "Self-assembly and mineralization of peptide-amphiphile nanofibers," *Science,* Vol. 294, No. 5547, 2001, pp. 1684–1688.

[13] Silva, G. A., et al., "Selective differentiation of neural progenitor cells by high-epitope density nanofibers," *Science,* Vol. 303, No. 5662, 2004, pp. 1352–1355.

[14] Hong, Y., Legge, R. L., Zhang, S., and Chen, P., "Effect of amino acid sequence and pH on nanofiber formation of self-assembling peptides EAK16-II and EAK16-IV," *Biomacromolecules,* Vol. 4, No. 5, 2003, pp. 1433–1442.

[15] Vandenburgh, H., et al., "Tissue-engineered skeletal muscle organoids for reversible gene therapy," *Hum. Gene Ther.*, Vol. 7, No. 17, 1996, pp. 2195–2200.

[16] Petite, H., et al., "Tissue-engineered bone regeneration," *Nat. Biotechnol.*, Vol. 18, 2000, pp. 959–963.

[17] Peretti, G. M., et al., "Cell-based tissue-engineered allogeneic implant for cartilage repair," *Tissue Eng.*, Vol. 6, No. 5, 2000, pp. 567–576.

[18] Woerly, S., Plant, G. W., and Harvey, A. R., "Neural tissue engineering: From polymer to biohybrid organs," *Biomaterials*, Vol. 17, 1996, pp. 301–310.

[19] Ishaug, S. L., et al., "Bone formation by three-dimensional stromal osteoblast culture in biodegradable polymer scaffolds," *J. Biomed. Mater. Res.*, Vol. 36, 1997, pp. 17–28.

[20] Minuth, W. W., Sittinger, M., and Kloth, S., "Tissue engineering: Generation of differentiated artificial tissues for biomedical applications," *Cell Tissue Res.*, Vol. 291, 1998, pp. 1–11.

[21] Ingber, D. E., et al., "Cell shape, cytoskeletal mechanics, and cell cycle control in angiogenesis," *J. Biomechan.*, Vol. 28, 1995, pp. 1471–1484.

[22] Xie, Y., Yang, S. T., and Kniss, D. A., "Three-dimensional cell-scaffold constructs promote efficient gene transfection: Implications for cell-based gene therapy," *Tissue Eng.*, Vol. 7, 2001, pp. 585–598.

[23] Yoshikawa, T., et al., "Analysis of gene expression in osteogenic cultured marrow/hydroxyapatite construct implanted at ectopic sites: A comparison with the osteogenic ability of cancellous bone," *J. Biomed. Mater. Res.*, Vol. 41, 1998, pp. 568–573.

[24] Noshi, T., et al., "Enhancement of the in vivo osteogenic potential of marrow/hydroxyapatite composites by bovine bone morphogenetic protein," *J. Biomed. Mater. Res.*, Vol. 52, 2000, pp. 621–630.

[25] Goldstein, A. S., et al., "Effect of convection on osteoblastic cell growth and function in biodegradable polymer foam scaffolds," *Biomaterials*, Vol. 22, 2001, pp. 1279–1288.

[26] Ma, T., et al., "Effects of pore size in 3-D fibrous matrix on human trophoblast tissue development," *Biotechnol. Bioeng.*, Vol. 70, 2000, pp. 606–618.

[27] Mueller-Klieser, W., "Three-dimensional cell cultures: From molecular mechanisms to clinical applications," *Am. J. Physiol.*, Vol. 273, 1997, pp. 1109–1123.

[28] Ruoslahti, E., and Pierschbacher, M. D., "New perspectives in cell adhesion: RGD and integrins," *Science,* Vol. 238, 1987, pp. 491–497.

[29] Williams, J. A., "Disintegrins: RGD-containing proteins which inhibit cell/matrix interactions (adhesion) and cell/cell interactions (aggregation) via the integrin receptors," *Pathol. Biol.*, Vol. 40, 1992, pp. 813–821.

[30] Evans, N. D., Gentleman, E., and Polak, J. M., "Scaffolds for stem cells," *Mater. Today*, Vol. 12, 2006, pp. 26–33.

[31] Hosseinkhani, H., et al., "Osteogenic differentiation of mesenchymal stem cells in self-assembled peptide amphiphile nanofibers," *Biomaterials*, Vol. 27, 2006, pp. 4079–4086.

[32] Colton, C. K., "Implantable biohybrid artificial organs," *Cell Transplant.*, Vol. 4, 1995, pp. 415–436.

[33] Polverini, P. J., "The pathophysiology of angiogenesis," *Crit. Rev. Oral Biol. Med.*, Vol. 6, 1995, pp. 230–247.

[34] Ware, J. A., and Simons, M., "Angiogenesis in ischemic heart disease," *Nat. Med.*, Vol. 3, 1997, pp. 158–164.

[35] Thompson, J. A., et al., "Site-directed neovessel formation in vivo," *Science,* Vol. 241, 1988, pp. 1349–1352.

[36] Tabata, Y., Nagano, A., and Ikada, Y., "Biodegradation of hydrogel carrier incorporating fibroblast growth factor," *Tissue Eng.*, Vol. 5, 1999, pp. 127–138.

[37] Tabata, Y., and Ikada, Y., "Vascularization effect of basic fibroblast growth factor released from gelatin hydrogels with different biodegradabilities," *Biomaterials*, Vol. 20, 1999, pp. 2169–2175.

[38] Cai, S., et al., "Injectable glycosaminoglycan hydrogels for controlled release of human basic fibroblast growth factor," *Biomaterials*, Vol. 26, 2005, pp. 6054–6067.

[39] Dogan, A. K., Gumusderelioglu, M., Aksoz, E., "Controlled release of EGF and bFGF from dextran hydrogels in vitro and in vivo," *J. Biomed. Mater. Res. B: Appl. Biomat.*, Vol. 74, 2005, pp. 504–510.

[40] Edelman, E. R., et al., "Controlled and modulated release of basic fibroblast growth factor," *Biomaterials*, Vol. 12, 1991, pp. 619–626.

[41] Hosseinkhani, H., et al., "Enhanced angiogenesis through controlled release of basic fibroblast growth factor from peptide amphiphile for tissue regeneration," *Biomaterials*, Vol. 27, 2006, pp. 5836–5844.

[42] Downs, E. C., et al., "Calcium alginate beads as a slow release system for delivering angiogenic molecules in vivo and in vitro," *J. Cell Physiol.*, Vol. 152, 1992, pp. 422–429.

[43] Iwakura, A., et al., "Gelatin sheets incorporating basic fibroblast growth factor enhances healing of devascularized sternum in diabetic rats," *Circulation*, Vol. 104, 2001, pp. 1325–1329.

CHAPTER 15

Microvascular Engineering: Design, Modeling, and Microfabrication

Jeffrey T. Borenstein, Eli J. Weinberg, Joseph P. Vacanti, and Mohammad R. Kaazempur-Mofrad

15.1 Introduction

Progress in the field of tissue engineering has been limited by the inability to produce thick, complex vascularized tissues in vitro, a central requirement for the generation of replacement tissues and organs in the laboratory. The sheer complexity, minimum dimension, and wide range of size scales of blood vessel networks represent towering challenges for the field of tissue engineering. Individual blood vessels with diameters ranging from 1 mm to several centimeters have been produced in vitro with varying degrees of success, but networks of vessels with dimensions ranging down to 10ì in diameter present challenges related to the formation of complex interconnected structures across a vast range of size scales. As will be described in this chapter, numerous approaches have been pursued to address this challenge, including the release of multiple growth factors from amorphous porous matrices. These methods have shown extremely promising results, including the sprouting of capillary beds within both synthetic and natural matrices. However, the long-term viability and lack of interconnectedness of the microvessels remain issues that limit these systems. More highly engineered approaches, in which scaffolds are constructed with embedded microchannel networks, have recently emerged. In these systems, scaffolds are formed with embedded networks of three-dimensionally coordinated channels, which can be seeded with endothelial and smooth muscle cells (SMCs) using microfluidic techniques. Such systems have been constructed using both nondegradable and biodegradable substrates and have been seeded and cultured for periods of weeks in vitro. The principal drawback of these techniques is related to the fundamentally two-dimensional nature of most scaffold-fabrication processes. The following describes these approaches and others toward to the generation of microvascular networks in engineered tissues. The chapter begins with an overview of the requirement for vascularization in engineered tissues, along with a brief discussion of the elements of engineered vascular networks, such as growth factors, matrices, and scaffolds. Next, the design of engineered networks is described in the context of tissue requirements for the transport of oxygen, nutrients, and flow and the development of both two- and three-dimensional models. Modeling of flow in engineered networks is then described, accounting for the complex nature of blood rheology and considerations such as shear stress. Finally, techniques for microfabricating vascular network structures are reviewed in a range of scaffolding materials and through the use of a spectrum of processing techniques.

15.1.1 Requirement for Vascularization in Complex Engineered Tissues and Organs

One of the principal challenges for engineering thick, complex tissues such as vital organs has been the need for a vascular bed capable of supplying the developing tissue with sufficient oxygen and nutrients [1]. In order to replicate the essential properties of the vasculature of highly metabolic tissues and organs such as heart, lung, and liver, this vascular supply must extend down to a vast microcirculation reaching to within 100μ of each region of the growing tissue. For this reason, most successful examples of tissue-engineered constructs involve skin, cartilage, and other tissues that are either avascular or that can be vascularized upon implantation. For complex, highly vascularized organs, numerous approaches have been developed in tissue engineering, including the release of angiogenic growth factors [2], the use of specialized matrices capable of triggering vascularization, and microengineered scaffolds comprising three-dimensional templates for vascular structures [3–5]. This chapter focuses on the latter of these three approaches, in which scaffolds are designed, constructed, and seeded so as to enable the directed formation of microvascular networks to supply oxygen and nutrients to the growing tissue.

15.1.2 Growth Factors, Matrices, and Engineered Scaffolds

Until the emergence of microfabrication-based tissue engineering, most approaches in the field utilized scaffolds with macroscale organization but lacking microscale structure beyond the basic material properties, pore size distribution, and pore density needed to support cell adhesion, proliferation, and function. Both inert and bioactive scaffolds have been used successfully in numerous applications, including skin, bone, cartilage, cornea, and tubular structures of the cardiopulmonary circulation, digestive tract, and urinary systems [6, 7]. Much of the work has involved biodegradable scaffolds prepared as foams, gels, fiber meshes, and other amorphous structures. Difficulties in nourishing tissues grown within these scaffolds have been associated with hypoxic conditions, principally because cell-culture models established within these constructs have not supported the formation and maintenance of viable microcirculatory networks. Therefore, numerous efforts over the past decade have focused on methods to establish a microvasculature within growing engineered tissues.

One of the principal approaches toward the establishment of microvascular networks within engineered tissues has entailed the introduction of growth factors from the scaffolding material itself. In 2001, Richardson et al. produced a seminal publication describing a method for introducing multiple growth factors from a scaffold in a stepwise fashion [2]. Figure 15.1 illustrates the approach toward microvascularization. This technique supports the complex interplay of processes necessary to establish a robust, functional vasculature, including the triggering of specific endothelial-cell (EC) behaviors, migration, and proliferation within the nascent channel architecture. Vascular endothelial growth factor (VEGF) is a

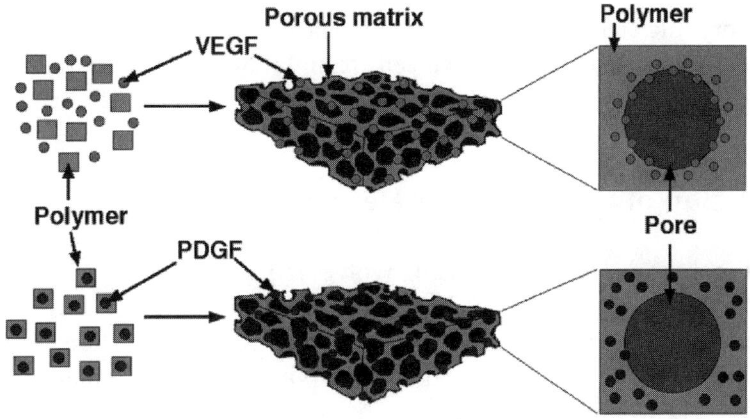

Figure 15.1 Dual growth factor–delivery strategy for microvasculature formation using VEGF and PDGF introduced in a stepwise fashion, as outlined in [2].

heavily used stimulant for angiogenesis, but the steps that follow, in which SMCs are recruited to the vessel wall and the EC/SMC coculture system is established and initiates matrix deposition, require additional stimulating factors. In the Richardson study, platelet derived growth factor (PDGF) is combined with VEGF within a poly(lactic-co-glycolic acid) (PLGA) scaffold, and microvascular structures are established and maintained. Both the spatial and temporal aspects of growth factor introduction are critical and must be orchestrated effectively in order to produce sustainable vessels. Another study investigated the introduction of fibroblast growth factor (FGF) in a hepatocyte culture model. In vivo studies of sustained VEGF delivery during hepatocyte culture also provide evidence for angiogenic action.

In addition to growth factor–based approaches, other groups have identified the significance of coculture systems. Lavik et al. have demonstrated that a coculture of brain endothelial cells and neural progenitor cells leads to stable perfused networks for periods of several weeks [8], when cultured in an open, macroporous, hydrogel-based matrix. Human embryonic stem cells have been differentiated and cultured to form tubelike structures; these structures have been implanted in mice with promising early results. The role of the matrix is critical in the development of functional engineered tissues, and the case of endothelialized artificial blood vessels is no different. Chrobak, Potter, and Tien have observed the establishment of barrier function and resistance of leukocyte adhesion in endothelialized channels formed in micromolded collagen gels [9]. Artificial vessels of SMCs have been formed on nylon strands and cultured to form concentric layers over periods of weeks. Additional cues, such as the application of pulsatile flow and pulsatile radial stress, have also been demonstrated to be significant elements for vascular-tissue engineering. Modularized approaches are also being explored, in which collagen gel rods are seeded with endothelial cells to form a microvascular bed [10]. These modules are then assembled into a larger tubular structure to comprise a microvascularized blood vessel. Cell sourcing represents another critical element of microvascular tissue engineering. Levenberg et al. have demonstrated the formation of functional vasculature using endothelial cells derived from human embryonic stem cells (HESC) [11]. The Scadden group has recently established a means to produce endothelial cells from HESCs using a scalable method that obviates the need to form an intermediate embryoid body [12].

15.2 Design of Microvascular Networks

Microvascular-network design involves the integration of a large number of considerations and parameters, including fluid transport, mechanical properties, and chemical considerations, such as surface chemistry. These elements are crucial because they control the viability, stability, and functionality of cell-seeded constructs [13], and they also govern the ability of the resulting microvessel networks to effectively distribute oxygen, nutrients, and waste products within a three-dimensional system. The cell microenvironment as represented by the scaffolding and surrounding matrix present mechanical, spatial and topographic, and chemical cues that control the behavior of cells [14] and, ultimately, the microvascular networks formed from these systems. Integration of these cues into a robust framework is essential for the success of techniques aimed at producing microvessels for either in vitro systems [15] or implantable constructs.

15.2.1 Scaffold Design Considerations: Mechanics, Transport, and Chemistry

Numerous considerations are involved in the design of microvascular networks for engineered tissues [16]. The predominant considerations involve adequate distribution of oxygen and nutrients to the entire volume of growing tissue; without vascularization, engineered tissues become necrotic at thicknesses approaching 1 mm. Beyond basic considerations of distribution, the fluid mechanics of the vasculature are also critical. Maintenance of physiologic levels of blood pressure and velocity within the channel networks is also critical, as is a uniform distribution of hematocrit. Wall shear stress is another important parameter as it is associated with atherogenic or atheroprotective behavior, depending upon the magnitude and stability of shear, especially at channel bifurcations [17, 18]. Arterial, venous, and capillary flows differ in many important ways, as do the phenotypes of the endothelial cells, pericytes, and SMCs along the channel walls. All of these considerations determine the specific properties of the microvasculature for engineered tissues. Structure-related mechanical properties of microvascular networks are equally important considerations for engineered tissues. Venous distensibility is a critical feature associated with regulation of blood volume, pressure, and electrolyte concentration. The mechanical strength of arterial vessels, required to sustain the high pressure and flow, must be very high. The surface presented to the cell must emulate the mechanical stimuli applied to vessel-lining cells in physiologic systems [19]. Therefore, scaffolds and matrices must be selected based on considerations such as the extent to which they are elastomeric.

Bulk transport within the scaffold represents another important consideration for the design of a microvascular construct. The simplest approaches utilize nonporous polymer scaffolding, relying entirely upon transport of oxygen and nutrients within the microchannel networks to feed and sustain the cultured cells. In some cases, oxygen transport may occur through the bulk of the scaffold, such as in the hepatocyte culture systems reported by LeClerc, Sakai, and Fujii [20]. The oxygen permeability of polydimethylsiloxane (PDMS) is high enough that cells within a closed microchannel network may receive sufficient oxygen through the scaffold walls. Recently reported microfluidic scaffolds utilizing calcium alginate hydrogels [21] possess high transport rates not only for oxygen but also small molecule nutrients; these approaches present exciting possibilities for combining microchannel and bulk transport for growing tissues. However, most scaffolding materials do not possess high oxygen-transport rates; therefore, oxygen and nutrients must be provided by the fluid flow or other means. Typically, matrices utilize porous structures such as a woven mesh, gel, or foam. Scaffolds may also incorporate porous membranes as part of the microfluidic structure, thereby allowing molecular transport passage across compartments or channel networks.

Chemical signals presented to cells within microvascular networks represent another critical element of the scaffold design. The use of PDMS scaffolding, for instance, is problematic for endothelial-cell culture because of the fundamentally hydrophobic nature of the PDMS surface and the difficulty in attaching matrices, such as collagen and fibronectin, to the channel walls. For in vitro culture systems, a significant body of work has been aimed at establishing functional PDMS surfaces [22–24]. These approaches take various forms, including the use of oxygen plasma functionalization, chemical vapor deposition of various species, grafting of functional groups, and light-assisted modification of surfaces with photocurable monomers to create patterned functionalized structures. Such methods can be used to modify the surface to reduce the prevalence of fouling or to attach groups that are more amenable to endothelial-cell culture. Polyethylene oxide (PEO) may be attached directly to the PDMS surface to reduce fouling or clotting. The Sheardown group, for instance, has cultured human umbilical vein endothelial cells (HUVECs) on PDMS surfaces using a PEO linker, followed by attachment of functional groups such as cell-adhesion peptides [22].

15.2.2 Morphometry of the Microcirculation

A large body of work has centered on the characterization of the morphometry of the microcirculation in highly vascularized tissues, such as heart muscle and the lung [25–28]. These studies have used techniques such as optical sectioning to trace actual capillary, arteriole, and venule patterns within the branching networks and to use this data to build models for the average length of the vessels, length between intersections, width, and branching angles. Kassab and Fung have used catalyzed plastic to infuse the coronary capillaries in a porcine model, removing the tissue once the plastic has hardened into a cast that can be visualized to produce the parameters of interest [27]. Huang et al. conducted a similar study on human pulmonary vasculature using silicone casting; an image of a cast is given in Figure 15.2 [26]. This investigation utilized an order-based system in which each level of the arterial and venous tree was characterized with an order number; in total, fifteen orders of arterial vessels and venous vessels were defined. These classifications of the arterial and venous branching networks were used to generate a fractal-based model for the microvasculature, based on scaling laws described in Section 15.2.3.

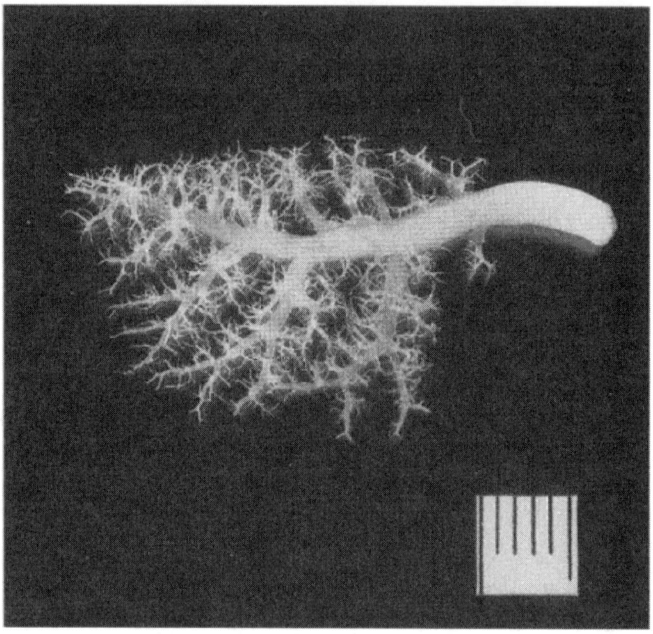

Figure 15.2 Polymer cast of microvasculature of a rat lung, showing the branching networks and geometries found in physiologic systems, used as the basis for morphometric models. (*Source:* [32].)

15.2.3 Scaling Laws for Microvascular Networks

Numerous models have been developed to account for the scaling of blood-vessel length, diameter, and frequency of bifurcations. These models are based upon first-principles considerations, as well as on morphometric analysis as described in Section 15.2.2. One of the guiding principles for vascular morphometry is known as Murray's law [29, 30], which states that the distribution of vessel diameters throughout the network is governed by minimum energy considerations. This principle minimizes the total work involved in sustaining blood flow and blood volume; increasing blood vessel radii require less work to maintain blood flow but more work to sustain blood volume. This law translates to the fact that the cube of the diameter of the parent vessel at a bifurcation equals the sum of the cubes of the diameters of the two daughter vessels and that the wall shear stress is constant throughout the network.

As is described in the microfabrication discussion (see Section 15.4), the ability to mimic the geometric features of physiologic microvessels and networks may be limited by processing capabilities. One such limitation involves the cross section of engineered vessels; physiologic blood vessels have rounded cross sections, while microfabricated systems are often rectangular or trapezoidal. Emerson et al. have generated an analytical model for Murray's law within microvessels that diverge from the classical cylindrical geometry [31]. Such systems may be limited to a constant depth for varying-width vessels and may further be constructed with roughly rectangular cross sections with sharp corners and abrupt transitions. The work of Emerson et al. addresses these limitations and provides a biomimetic set of design rules capable of describing fluid dynamics within microvascular networks of arbitrary geometry.

Several scaling laws for the nature of the branching networks within the microvasculature have been developed and applied to mammalian vascularized tissues. These systems include the Weibel, Horsfield, and Strahler models [32]. Jiang et al. have described the various advantages and shortcomings of these model systems and have generated a connectivity matrix that accounts for much of the observed branching behavior in the microcirculation [32]. The simplest models treat all vessel segments in the same manner; namely, each element is in parallel, and successive elements differ from one another by small stepwise diametric changes. However, actual microvasculature does not follow these simple rules, and the observed hemodynamics differ sharply from these predictive models. Therefore, advanced models such as the diameter-defined Strahler system break the branching networks into vessel segments and vessel elements, where a segment represents the length between bifurcations and an element comprises a unit joining segments in a serial fashion. Further, physiologic vascular networks commonly exhibit large changes in dimensions between successive elements, referred to as small twigs on a large trunk. These observations, which are critical to the replication of actual microvascular networks, lead to the need for more complex tools to design and model physiologic systems.

15.2.4 Two-Dimensional Designs

Two-dimensional designs for microvascular networks have been developed using a range of basic assumptions and design rules. These models have been governed by considerations such as uniform distribution of fluid flow among all of the smallest-order vessels, uniform distribution of hematocrit within the entire vascular bed, and adherence to Murray's law for minimum work considerations for microvessel formation, as well as approaches that seek to limit coagulation or hemolysis. Early designs described a hierarchy of several orders of vessel dimensions ranging from 1 mm-wide inlets down to 15μ- to 35μ-wide microchannels, with simple bifurcations and rectangular geometries, as illustrated in Figure 15.3 [3]. Control of wall

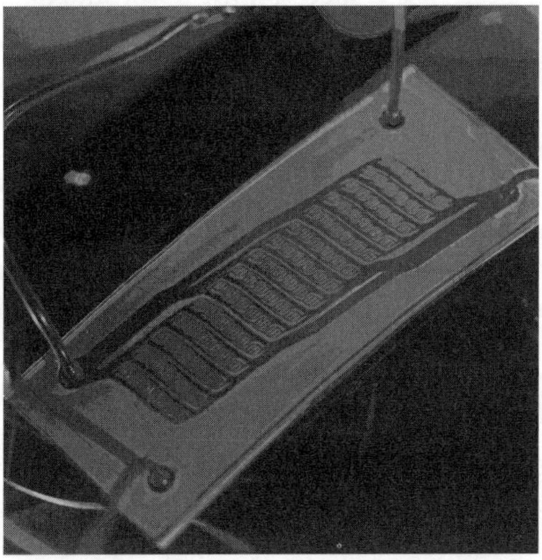

Figure 15.3 Image of microvascular network designed to provide uniform flow across multiple generations of vessel diameters, constructed in a PDMS scaffold and infused with red dye.

shear stress has led to a new generation of designs that eliminate a roughly hundred-fold variation in fluid shear across preexisting networks [33]. Models accounting for mass transport and nutrient requirements for growing tissues have also been proposed [34]. Here the microvascular requirements for a tissue-engineered skin product have been evaluated and addressed using assumptions based on rectangular or square duct geometries. In another study, Wang and Hsu addressed rectangular and circular duct geometries, evaluating velocity profiles in each case across the range of different branch segments and size scales [35].

One of the important features of most engineered microvascular scaffolds is the influence of hydrostatic pressure on the geometry of the channels themselves. Vessel distensibility, a critical element of the regulation of blood pressure, is reflected in microvascular scaffolds constructed using flexible, elastomeric materials such as PDMS [36]. Thin films of PDMS result in relatively large changes in dimension as physiologic pressures are applied to the fluid networks, thereby reducing the resistance to flow. From a mechanotransduction standpoint, the elastomeric nature of the surface is important because of the role of mechanical interaction on cell signaling [19]. From a microfluidic standpoint, vessel distensibility provides another level of control over the fluid dynamics of cell seeding and blood flow in microchannels. The vessel thickness and distensibility can be modulated to provide the desired degree of hoop stress to the seeded construct as part of the engineered design.

15.2.5 Three-Dimensional Designs

The fundamentally two-dimensional nature of most microfabrication techniques presents a severe limitation to the ability to produce scalable three-dimensional microvascular network systems. Numerous approaches are being utilized to address this limitation, many of which are beyond the scope of this discussion. The simplest method for obtaining a three-dimensional system from two-dimensional microvessel networks is to produce large numbers of two-dimensional layers of microvessels and to integrate them via vertical pipes in the third dimension. This approach has been used successfully to produce three-dimensional systems for endothelial-cell culture [37], hepatocyte culture [20], and renal filtration systems [38]. Kartalov et al. have recently provided a powerful set of tools capable of generating three-dimensional microfluidic networks with vertical vias connecting stacked layers with crossovers using simple, monolithic fabrication techniques [39]. Crossovers such as underpasses and overpasses are described, as is a construct known as a microfluidic septum, which may be viewed as an intentionally defective via. These tools overcome some of the limitations associated with large-scale three-dimensional integration of microchannel networks for microvasculature and other applications.

In each of the above cases, the vertical interconnects are formed with right-angle geometries with respect to the horizontal microvessel networks. One of the principal challenges associated with three-dimensional systems involves the density of small-diameter vessels representative of the capillaries. Limitations imposed by the fabrication technologies involve the use of a small number of relative large-bore vertical connections between horizontal microvessel layers; therefore, the space between the two-dimensional layers is devoid of small vessels. This limitation results in a density of small vessels that is well below that seen in physiological systems, where there is no such restriction on small vessels in two-dimensional layers. Design approaches have been developed that address this limitation, but successful fabrication of such designs has yet to be demonstrated.

15.3 Computational Models for Microvascular Networks

Physiologic microvascular networks are very complex systems that have proven to be very difficult to replicate using tissue-engineering techniques. Methods relying on growth factors result in networks with large numbers of small-diameter vessels on the order of capillaries, but these vessels do not possess the geometric order or hydrodynamic behavior representative of anatomic vasculature. Much like the vascular networks present within cancerous tumors, engineered vascular networks generated through the introduction of VEGF or other growth factors are tangled structures lacking in long-range order. It is the goal of computational modeling of microvascular networks to generate designs with long-range order and fluid dynamic behavior similar to that seen in physiologic networks. Models for microvascular networks must, however, be designed in ways that are consistent with existing fabrication techniques. As was discussed in Section 15.2.5, truly three-dimensional vascular network structures are difficult to produce due to limitations in existing microfabrication techniques.

15.3.1 Fluid Mechanics of Microvasculature

The rheology of blood is complex for numerous reasons, including its mixed-phase nature, the distensibility of vessel walls and compressibility of erythrocytes, and small-scale phenomena such as the Fahraeus-Lindquist effect [40]. The rich and complex array of phenomena associated with the behavior of blood in small channels presents a challenge for the computational fluid modeling of these systems. Furthermore, interactions between the endothelial cells lining the channel walls and the fluid mechanical forces exerted by the flowing blood present additional challenges for engineering vascular networks. Low, oscillatory shear stresses have been connected to endothelial-cell activation and the formation of atherosclerotic plaques; therefore, control of shear stress has become another design consideration for engineering microvascular networks [33]. Clotting and inflammatory response represent additional challenges for engineering microvasculature; in particular, engineered constructs made using artificial materials are known to suffer from clotting and require the introduction of anticoagulants in order to operate effectively for long periods of time. The principle known as Virchow's triad [41] suggests that three principal mechanisms are involved in clotting: (1) patient-related effects; (2) interactions between blood and contacting surfaces; and (3) fluid mechanical effects. The latter two effects must be considered carefully when engineering microvascular structures or networks because the fluid mechanics of artificial networks and the surfaces present in these constructs must be engineered in order to minimize clotting.

15.3.2 Complex Fluid Mechanical Phenomena

Modeling of blood flow in microvascular networks can be approached using simple fluid mechanical techniques in which vessels are approximated as series and parallel connections of fluidic resistances. Each of these resistances may represent simple pipe flow of a Newtonian fluid or may account for more complex behavior, as discussed in Section 15.3.1. The non-Newtonian nature of blood flow suggests the application of an apparent viscosity highly dependent on the local percentage of erythrocytes (hematocrit). Kaazempur-Mofrad et al. [42] have generated a computational model capable of accounting for the non-Newtonian rheology of blood as well as the solid fraction of erythrocytes [42].

Other considerations involve the circular geometry of physiologic vessels and the noncircular shape of most microfabricated vascular networks. Typical microfabrication techniques result in rectangular vessels, perhaps with curved corners to avoid high stresses. These rectangular geometries are simple to model using either analytic solutions or numerical approximations modifying the standard Hagen-Pouseuille law. However, these fabrication processes often result in rectangular vessels for which the depth of the vessel does not scale with the width due to limitations in the processing technique. Such deviations from physiologic behavior result in higher fluidic resistances in the engineered networks than in natural systems, as well as unusual fluidic behavior that can lead to increases in clotting or phase separation within the channels.

In fluid mechanical behavior in network structures, the presence and nature of channel intersections, both in physiologic systems and in artificial engineered microchannels, often predominate. Morphometric analysis discussed in Section 15.3.1 has been used to investigate the specific geometries observed in microvessel intersections, but the properties of the fabrication techniques used to build microchannel network scaffolds have typically limited the ability to replicate the detailed architecture of vessel bifurcations. The most important consideration for vessel bifurcations relates to the effect these structures have on clotting behavior; therefore, sharp corners and large angles between vessels present specific problems for systems used to flow whole blood.

15.3.3 The Role of Shear Stress

One of the most important parameters controlling endothelial-cell behavior in blood vessels and engineered systems is the level of fluid shear stress. The importance of biomechanical stimuli on the behavior of endothelial cells in vivo and in vitro has been well established and includes components such as hydrostatic pressure, cyclic stretch (hoop stress), and wall shear stress. Flow-induced alignment represents one of the most obvious behaviors of endothelium within vascular networks, and it has also been established that endothelial-cell phenotype is determined by forces such as laminar shear stress [43]. Cell signaling [44] is strongly affected by alterations in shear stress levels, as is gene expression, and remodeling of the cell wall has also been observed in response to long-term changes in flow. In addition to these developmental phenomena, shear stress has been strongly correlated with disease states such as clotting and the formation of atherosclerotic plaques [45]. For instance, atherosclerosis has been observed to be a geometrically focal disease, with a concentration of plaque formation at the outer walls of vessel bifucations, where flow is often disturbed, and the shear stress may oscillate and even change sign [17]. These observations clearly demonstrate that control of wall shear stress within engineered microvascular networks represents an important consideration for the formation of systems that exhibit robust, sustainable physiologic behavior.

Initial approaches toward engineered microvascular networks focused on the establishment of uniform flow within the constructs. Simple pipe flow models with parallel and series resistor networks have been used to generate microvascular networks for which the flow within all of the vessels of the smallest dimension is the same [3]. However, analysis of the wall shear stress in these systems indicates that a range of shear stress greater than a hundredfold exists within the various capillary microchannel structures. Therefore, an iterative shear-based algorithm has been generated that is capable of driving the microvessel networks toward a uniform-flow, uniform-shear condition (Figure 15.4) [40]. Designs have been produced that are capable of maintaining physiologic shear levels (15 to 30 dynes/cm^2) throughout 95 percent of a microvascular network containing roughly five thousand microchannels.

15.4 Microfabrication Technology for Vascular Network Formation

Microfabrication technology has emerged over the past two decades, with applications ranging from automotives, aerospace, and defense to biomedical microsystems, such as implantable sensors and drug-delivery devices, lab-on-a-chip microfluidic devices, and scaffolds for tissue engineering [46]. In early biomedical

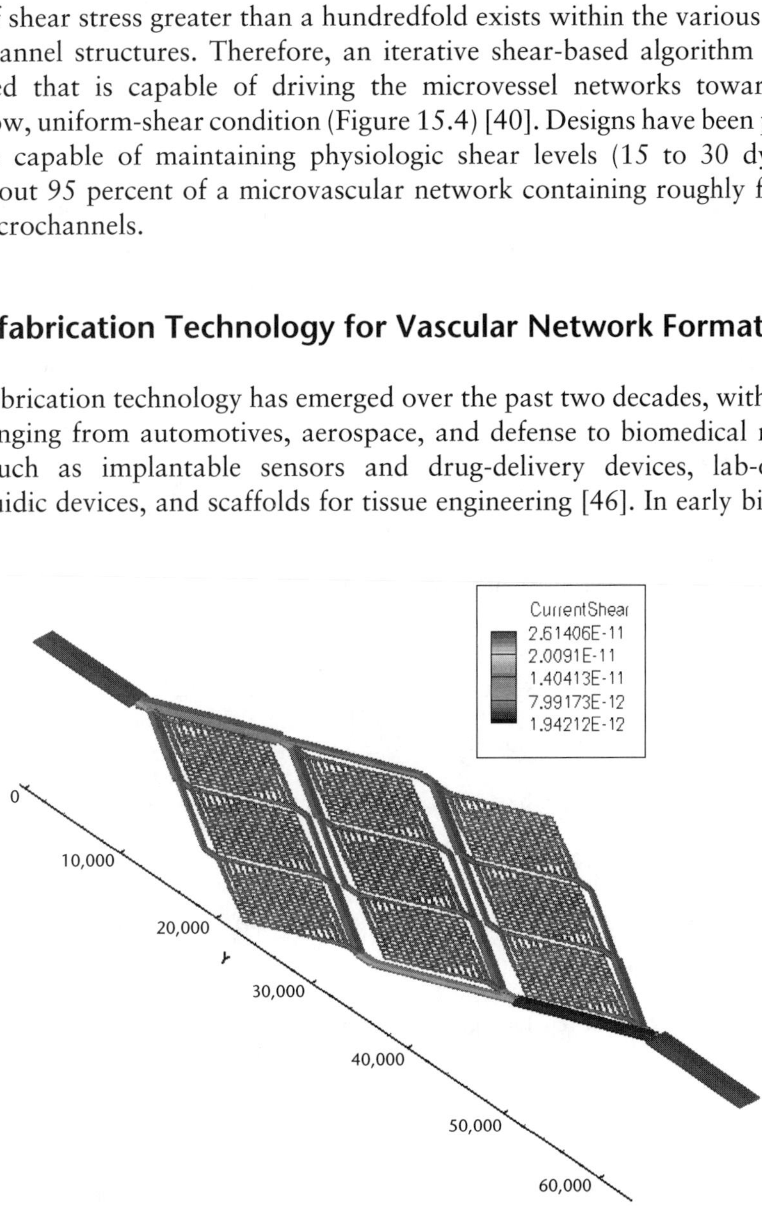

Figure 15.4 False color map indicating distribution of wall shear stress within a microvascular network designed to provide constant shear across several generations of vessel dimensions. (*Source:* [40].)

microsystems, silicon micromachining predominated due to the availability of standard processes originally developed for microelectronics manufacturing. More recently, the field of soft lithography and polymer-based biomicroelectromechanical systems has emerged [47–52]. These capabilities have enabled the fabrication of precision structures and devices in polymeric materials suitable for cell culture and scaffolds for tissue engineering. These polymeric materials range from the commonly used biocompatible silicone rubber substrate PDMS to biodegradable systems such as polycaprolactone (PCL) [53], PLGA [4, 54], polyglycerol sebacate (PGS) [5, 37, 55], and silk fibroin [56]. Two-dimensional tissue-engineering scaffolds have been extended into the third dimension through stacking or layering techniques, as well as methods in which cells and biological materials are introduced in ways that are interspersed with the mechanical structuring processes. Table 15.1 summarizes the range of microvascular network-formation microfabrication techniques described herein, along with relative strengths and limitations.

15.4.1 Polymer Micromolding of Scaffolding Layers for Vascular Networks

The field of replica molding has emerged over the past decade, pioneered by Quake and Scherer [57], Whitesides et al. [58] and Jo and Beebe [59]. Replica molding is commonly used in a technique known as soft lithography, in which the precision-molded structures are used to print or deposit biologic materials, such as surface adhesion molecules on substrates to organize cells into engineered structures. Seminal work in the field of soft lithography includes the discovery by Chen et al. that geometric control of cell shape governs apoptosis [60], as well as the establish-

Table 15.1 Summary of Microfabrication Techniques Used to Build Microvascular Networks Along with Relative Strengths and Limitations As a Guide for Future Directions for Research in This Area

Technique/Scaffolding Material	*Strengths*	*Limitations*	*Principal Reference*
Microfluidic network replica molded in PDMS	Flexible, transparent, scalable	Nondegradable, limited biocompatibility, difficulty in scaling aspect ratio	[3]
Microfluidic networks in PLGA	Biodegradable, Food and Drug Adminstration (FDA) approved	Rigid scaffold, inflammatory response	[4]
Microfluidic networks in PGS	Biodegradable, elastomeric	Rapid degradation rate	[36]
Microfluidic networks in PCL	Biodegradable	Slow degradation, complex assembly process	[53]
Microfluidic calcium alginate hydrogel	Bulk nutrient-transport capacity	Mechanical stability	[70]
Solvent bonding of poly(lactic acid), PLGA layers	Biodegradable, FDA approved	Rigid scaffold, inflammatory response	[67]
Electroplated master molding of networks	Scalable aspect ratio for channel networks	Process control and cost	[64]
Laser-microfabricated artificial vasculatures	Scalable aspect ratio for channel networks	Unclear applicability to range of biomaterials	[30]
Negative photoresist master for PDMS/polymethylmethacrylate	Simple, scalable	Control of geometry	[62]
Computer-controlled system for shearing	Precise control	Complex approach for tissue engineering	[73]

ment of spatially modulated coculture of hepatocytes and fibroblasts by Bhatia et al. [61]. More recently, polymer micromolding has been used to fashion scaffolds replicating the geometry of physical structures such as capillary beds, liver sinusoids and renal proximal tubules.

Microvascular networks in polymer scaffolds were first reported by Borenstein, King, et al. [3, 4]. These scaffolds were built using PDMS, and therefore many of the cell-seeding challenges associated with PDMS substrates were factors. The naturally hydrophobic nature of PDMS can be temporarily reversed to yield highly hydrophilic surfaces using oxygen plasma treatment, but surfaces quickly return to their natural hydrophobic nature after periods of several hours. Surface coatings such as fibronectin, collagen, and Matrigel can be used to improve cell adhesion and stabilization of function on PDMS scaffold surfaces. Therefore, King, et al. were able to demonstrate endothelial-cell culture on PDMS microengineered scaffolds for periods of up to four weeks. The principal advantages of PDMS scaffolds include the ease of fabrication and the mechanical properties; the elastomeric nature of PDMS is mechanically highly suitable for cell attachment and mechanotransduction based on its similarity to basement membrane. Scaffolds for microvascular networks have also been reported in polycarbonate, as well as in polymethylmethacrylate by Wang et al. [62].

Techniques for producing the master molds for polymeric microvascular scaffolds have also been developed by several groups. The original work was done using etched silicon wafers in order to replicate the cylindrical geometries of blood vessels. Isotropic silicon etching using reactive ion etching equipment produces U-shaped profiles, but replica molding leads to the inverse geometry rather than cylindrical channels in the polymer. Inductively coupled plasma etching produces rectangular geometries [63], translating to sharp corners in the microvessels that present difficulties for confluent endothelial-cell seeding. The most common master-mold fabrication technique, SU-8 epoxy resin photolithography, is preferred due to the ability to produce very deep channels (up to 1 mm in depth), as well as multiple channel depths through the use of repeated exposure and development steps. In addition, the SU-8 lithography process does not require expensive etching or other clean room equipment beyond a mask aligner.

Techniques aimed at addressing the geometric limitations of master molds formed by SU-8 lithography and etching technologies have also been pursued. LaVan, George, and Langer [64] have reported a method for forming circular channel geometries and varying the channel depth in a smooth manner using electroplating. Seed metal patterns are formed on the surface of a silicon wafer with intentional, precisely defined gaps between sections of the pattern. Once the electroplating has expanded in both height and width to the point that these gaps in the seed layer are bridged, these regions now contacted by the plating begin to plate themselves, thereby enabling a multitude of channel diameters to be produced using a single continuous process. Another novel approach toward constructing microengineered scaffolds for artificial vasculatures has been reported by Lim et al. [30]. Here a direct-write technique was employed using a Nd:YAG laser to produce novel microchannel architectures that more closely follow Murray's law than do typical molded geometries. Figure 15.5 shows an image of an eight-generation microvascular network fabricated using a direct-write laser technique. Other

Figure 15.5 Artificial microvasculature formed using the Nd:YAG direct-write laser technique, showing eight generations of microvessels designed to obey Murray's law. (*Source:* [30].)

approaches capable of high-volume production, including microstereolithography, high-precision CNC machining, injection molding, and electric discharge machining are all potential avenues for generating molds and scaffolds for microvascular networks [65].

15.4.2 Biodegradable Systems

The nondegradable polymer PDMS has been the material of choice for most studies of cell culture on microfabricated substrates. Capillarylike structures have been reported in PDMS, with endothelialization of the walls of the channels approaching near confluence for periods of days to weeks. However, these PDMS systems are principally used for in vitro cell-culture studies, rather than implantable systems, because of the large volume of polymeric material that will remain a permanent part of the structure. This large volume of polymer scaffolding limits the ultimate size these systems can achieve, as well as structurally interfering with the establishment of basement membrane and cell-cell contacts that will be required to establish the mechanical integrity and robustness of the engineered tissue construct. In light of these constraints, numerous groups have investigated the use of microfabricated scaffolds consisting of fully biodegradable materials.

An early study of a biodegradable microsystem for tissue engineering was reported by Armani and Liu [53], in which PCL layers containing microchannels were joined together to form a closed three-dimensional system. However, this system was not fully biodegradable due to the use of a gold film as an adhesive film between layers. The first fully biodegradable microfabricated scaffold was reported by King et al. [54], in which a three-dimensional PLGA structure was fabricated, then seeded with endothelial cells. The field of biodegradable microfluidics was soon launched with other groups reporting novel techniques for PLGA scaffold fabrication [66] and polydiaxanone [67]. Typical mixing ratios of lactic to glycolic acid of 85:15 result in a degradation time of several months, reasonable for many tissue-engineering applications. The PCL scaffolds reported by Armani and Liu have degradation times exceeding one year but may have reduced inflammatory response, especially considering the large volume of scaffold material present in the system.

A large body of research has emerged demonstrating the critical role played by the mechanical properties of the substrate. Mechanotransduction, the means by which mechanical forces are transmitted to overlying cells in both directions, has been established as an important mechanism by which cells interact with their surrounding microenvironment. Therefore, it is desirable to investigate scaffolding materials possessing properties similar to physiologic basement membrane; one such property is the flexible nature of basement membrane. A biodegradable elastomer, PGS, has been reported [68]; this substrate produces reduced inflammatory response relative to other biomaterials and possesses a high degree of distensibility. Hepatocyte culture in PGS has been reported by Bettinger et al. in microfluidic structures [5], as well as in nanoscale contact guidance structures [55]. Endothelialized capillary networks have been reported by Fidkowski et al. [37]. Figure 15.6 shows an image of an endothelialized microchannel network in PGS from [29]; note that the endothelial cells have formed a confluent monolayer around the inner lumen of the channels. Other biodegradable materials have also been used for vascular tissue-engineering applications; Wagner et al. [69] have reported progress in forming vascular structures using poly(ester urethane urea) scaffolding.

15.4.3 Applications of Microvascular Network Systems

Numerous early applications of microvascular network systems have been reported, including aforementioned systems aimed at providing a vascular bed for engineered tissues. Other applications involve clinical applications in the field of wound healing, as well as several innovative approaches in which precise control of

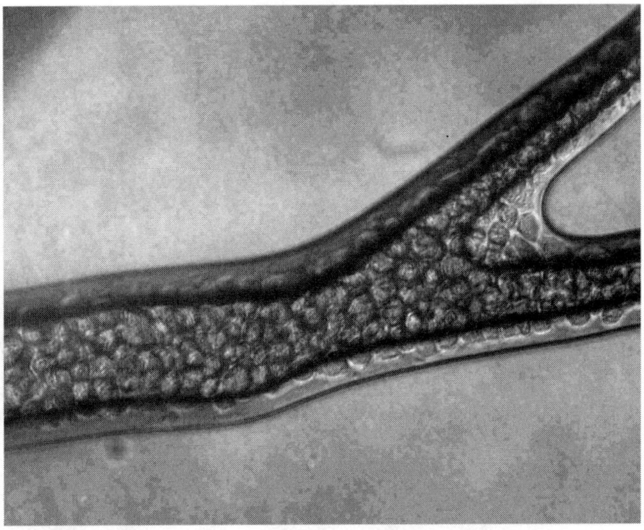

Figure 15.6 Endothelialized capillary microchannel in PGS scaffold. Endothelial cells fully line the inner lumen of the microchannel, which is of rectangular geometry approximately 50μ in width and depth. (*Source:* [36].)

fluid flow has been utilized to study a range of cell-related phenomena in microchannels. The latter approaches have principally utilized PDMS microfluidic systems to apply controlled levels of fluid shear to cells cultured on the surfaces of the constructs, while clinical applications tend to favor polymer scaffolds with superior biocompatibility and perhaps resorbable systems.

Microfluidic biomaterials research has been extended to hydrogels such as calcium alginate by Cabodi et al. [21]; an image of a microchannel network formed in this scaffold is shown in Figure 15.7. These structures have applications in microvascular-network formation but differ from conventional polymer scaffolding in several important ways. The hydrogel scaffold has the advantage of bulk diffusion of oxygen and nutrients due to the open structure within the material but suffers from a relative lack of mechanical integrity. One of the most exciting potential applications for these systems is in the field of wound healing [70], where the importance of micromechanical forces and fluid flow has recently been identified [71]. Microfluidic biomaterials such as these calcium alginate hydrogels could be formed into a wound dressing to deliver nutrients and remove waste from a wound bed, utilizing both transport and flow through the channel networks, as well as bulk transport across the hydrogel to maximize the efficacy of the treatment.

Devices capable of precisely controlling shear have proven to be very useful systems for analyzing cell behavior in microfluidic environments. Lu et al. have studied the behavior of fibroblasts in PDMS microfluidic channels under controlled shear conditions in order to study cell adhesion as a function of numerous experimental parameters [72]. These parameters included channel geometry, surface chemistry, and the presence of various growth factors. Clearly, such systems serve a very useful purpose as testbeds for probing and optimizing cell behavior in microchannels and could be a powerful tool for identifying and establishing the appropriate conditions for endothelial and SMC seeding in microvascular network devices. Another approach involving the use of controlled shear microfluidic devices has been reported by Song et al. [73]. In this work, alignment and elongation of endothelial

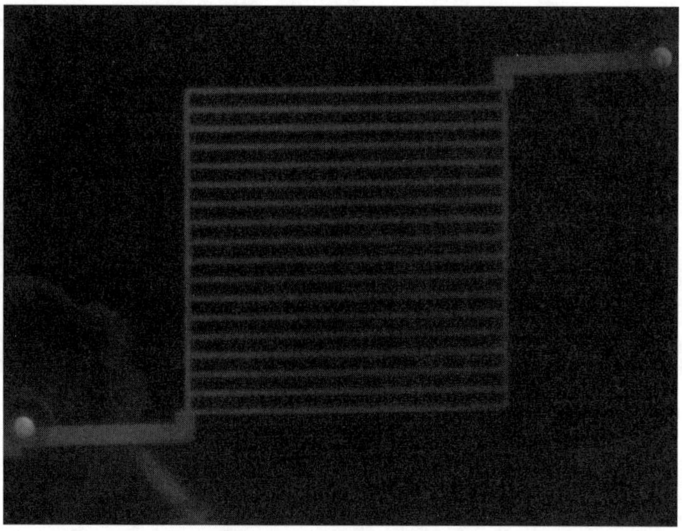

Figure 15.7 Rhodamine dye perfusion through microchannel network formed in micromolded calcium alginate hydrogel as described [68].

cells in culture were studied as a function of applied shear stress, using a system modified from a Braille display. The Braille display is a computer-controlled system of piezoelectric pins that are raised and lowered in alignment with PDMS structures so as to form an array of microfluidic pumps and valves. Endothelial-cell morphology can therefore be investigated as a function of applied shear within a pulsatile flow system. Vollmer et al. have reported the development of artificial capillary beds for sensing and delivery of oxygen for artificial-lung applications [74]. These structures address issues such as the choice of vessel dimension for optimal oxygen transfer efficiency. Ismagilov et al. [75] have developed a system for testing coagulation in PDMS microfluidic networks as a function of clotting activators and fluid mechanical parameters such as shear rate. This system does not contain endothelial cells but instead addresses clotting phenomena as a function of fluid dynamics and surface activation, as well as the distribution of clotting components within flowing blood.

15.5 Conclusion

Microvascular networks represent an important element of engineered cell-based structures, both for direct tissue-engineering applications and indirect in vitro diagnostic and discovery platforms. Numerous approaches have been undertaken to form microvascular network structures. Some of these approaches rely more heavily on signaling from the scaffold, matrix, neighboring cells, and circulating biomolecules to produce a "self-assembled" network capable of being anastomosed to larger vascular structures. Other approaches are more dependent upon engineering tools to form a framework for a "top-down" microvascular network construct in which the template for the microchannels is provided within a scaffold prior to the introduction of cells. All of these approaches present challenges associated with the integration of three-dimensional networks of robust vessel constructs possessing the mechanical properties and transport phenomena required for such systems. Future developments may rely on combinations of aspects of both of these approaches, along with advances in the cell sourcing challenges to produce long-term, viable vascularized constructs capable of supporting large-scale engineered tissues.

Acknowledgments

We gratefully acknowledge the support of the Center for Integration of Medicine and Innovative Technology, Department of the Army, Cooperative Agreement DAMD-17-02-2-0006, and Draper Laboratory.

References

[1] Kaihara, S., Borenstein, J. T., and Koka, R., "Silicon micromachining to tissue-engineer branched vascular channels for liver fabrication," *Tissue Eng.*, Vol. 6, No. 2, 2000, pp. 105–117.

[2] Richardson, T. P., Peters, M. C., Ennett, A. B., and Mooney, D. J., "Polymeric system for dual growth factor delivery," *Nat. Biotech.*, Vol. 19, 2001, p. 1029.

[3] Borenstein, J. T., Terai, H., King, K. R., Weinberg, E. J., Kaazempur-Mofrad, M. R., and Vacanti, J. P., "Microfabrication technology for vascularized tissue engineering," *Biomed. Microdevices*, Vol. 4, 2002, pp. 167–175.

[4] King, K. R., Wang, C. J., Kaazempur-Mofrad, M. R., Vacanti, J. P., and Borenstein, J. T., "Biodegradable Microfluidics," *Adv. Mater.*, Vol. 16, 2004, pp. 2007–2012.

[5] Bettinger, C. J., Weinberg, E. J., Wang, Y., Kulig, K., Vacanti, J. P., Borenstein, J. T., and Langer, R., "Three-dimensional microfluidic tissue engineering scaffolds using a flexible biodegradable polymer," *Adv. Mater.*, Vol. 16, 2006, pp. 165–169.

[6] Langer, R., and Vacanti, J., "Tissue engineering," *Science,* Vol. 260, 1993, p. 920.

[7] Vacanti, J. P., and Langer, R., "Tissue engineering: The design and fabrication of living replacement devices for surgical reconstruction and transplantation," *Lancet,* Vol. 354, 1S, July 1999, 23SI–34SI.

[8] Ford, M. C., Bertram, J. P., Hynes, S. R., Michaud, M., Li, Q., Young, M., Segal, S. S., et al., "A macroporous hydrogel for the coculture of neural progenitor and endothelial cells to form functional vascular networks in vivo," *Proc. Nat. Acad. Sci.*, Vol. 103, 2006, pp. 2512–2517.

[9] Chrobak, K. M., Potter, D. R., and Tien, J., "Formation of perfused, functional microvascular tubes in vitro," *Microvasc. Res.,* Vol. 71, 2006, pp. 185–196.

[10] McGuigan, A. P., and Sefton, M. V., "Vascularized organoid engineered by modular assembly enables blood perfusion," *Proc. Nat. Acad. Sci.*, Vol. 103, 2006, pp. 11461–11466.

[11] Levenberg, S., Golub, J. S., Amit, M., Itskovitz-Eldor, J., and Langer, R., "Endothelial cells derived from human embryonic stem cells," *Proc. Nat. Acad. Sci.*, Vol. 99, 2002, pp. 4391–4396.

[12] Wang, Z. Z., Au, P., Chen, T., Shao, Y., Daheron, L. M., Bai, H., Arzigian, M., et al., "Endothelial cells derived from human embryonic stem cells form durable blood vessels in vivo," *Nat. Biotech.*, Vol. 25, 2007, pp. 317–318.

[13] Khademhosseini, A., Langer, R., Borenstein, J. T., and Vacanti, J. P., "Microscale technologies for tissue engineering and biology," *Proc. Nat. Acad. Sci.*, Vol. 103, 2006, pp. 2480–2487.

[14] Desai, T. A., "Micro- and nanoscale structures for tissue engineering constructs," *Med. Eng. Phys.*, Vol. 22, 2000, pp. 595–606.

[15] Khademhosseini, A., Yeh, J., Eng, G., Kazi, H., Borenstein, J. T., Karp, J., Farokhzad, O., et al., "Cell docking inside microwells within reversibly sealed microfluidic channels for multiphenotype cellular arrays," *Lab Chip*, Vol. 5, 2005, pp. 1380–1386.

[16] Kaazempur-Mofrad, M. R., Weinberg, E. J., Borenstein, J. T., and Vacanti, J. P., "Tissue engineering: Multiscaled representation of tissue architecture and function," in *Complex Syst. Sci. Biomed.*, Springer, 2003, pp. 737–761.

[17] Malek, A. M., and Izumo, S., "Mechanism of endothelial cell shape change and cytoskeletal remodeling in response to fluid shear stress," *J. Cell Sci.*, Vol. 109, 1996, pp. 713–726.

[18] Gimbrone, M. A., "Endothelial dysfunction, hemodynamic forces, and atherosclerosis," *Thromb. Haemost.*, Vol. 82, No. 2, 1999, pp. 722–726.

[19] Discher, D. E., Janmey, P., and Wang, Y.-L., "Tissue cells feel and respond to the stiffness of their substrate," *Science,* Vol. 310, 2005, pp. 1139–1143.

[20] LeClerc, E., Sakai, Y., and Fujii, T., "Cell culture in 3-dimensional microfluidic structure of PDMS," *Biomed. Microdevices*, Vol. 5, 2003, pp. 109–114.

[21] Cabodi, M., Choi, N. W., Gleghorn, J. P., Lee, D., C. S., Bonassar, L. J., and Stroock, A. D., "A microfluidic biomaterial," *J. Amer. Chem. Soc.*, Vol. 127, 2005, pp. 13788–13789.

[22] Chen, H., Brook, M. A., and Sheardown, H., "Cell interactions with PDMS surfaces modified with cell adhesion peptides using a generic method," *Mater. Sci. Forum*, Vol. 539, 2006, pp. 705–709.

[23] Hu, S., Ren, X., Bachman, M., Sims, C. E., Li, G. P., and Allbritton, N. L., "Surface-directed, graft polymerization within microfluidic channels," *Anal. Chem.*, Vol. 76, 2004, pp. 1865–1870.

[24] Sui, G., Wang, J., Lee, C.-C., Lu, W., Lee, S. P., Leyton, J. V., Wu, A. M., et al., "Solution-phase surface modification in intact Poly(dimethylsiloxane) microfluidic channels," *Anal. Chem.*, Vol. 78, 2006, pp. 5543–5551.

[25] Escaned, J., Segovia, J., Flores, A., Aragoncillo, P., Salas, C., Alfonso, F., Lopez, M., et al., "Assessment of coronary microcirculation in cardiac allografts: A comparison of intracoronary physiology, intravascular ultrasound and histological morphometry," *J. Heart Lung Transplant.*, Vol. 20, 2000, p. 204.

[26] Huang, W., Yen, R. T., McLaurine, M., and Bledsoe, G., "Morphometry of the human pulmonary vasculature," *J. Appl. Physiol.*, Vol. 81, 1996, p. 2123.

[27] Kassab, G. S., Rider, C. A., Tang, N. J., and Fung, Y. C., "Morphometry of pig coronary arterial trees," *Am. J. Physiol.*, Vol. 265, 1993, pp. 350–365.

[28] Minnch, B., and Lametschwandtner, A., "Lengths measurements in microvascular corrosion castings: Two-dimensional versus three-dimensional morphometry," *Scanning Vol.*, Vol. 22, 2000, pp. 173–177.

[29] Murray, C. D., "The physiological principle of minimum work: I. The vascular system and the cost of blood volume," *Proc. Nat. Acad. Sci.*, Vol. 12, 1926, pp. 207–214.

[30] Lim, D., Kamotani, Y., Cho, B., Mazumder, J., and Takayama, S., "Fabrication of microfluidic mixers and artificial vasculatures using a high-brightness diode-pumped Nd:YAG laser direct write method," *Lab Chip*, Vol. 3, 2003, pp. 318–323.

[31] Emerson, D. R., Cieslicki, K., Gu, X., and Barber, R. W., "Biomimetic design of microfluidic manifolds based on a generalised Murray's law," *Lab Chip*, Vol. 6, 2006, pp. 447–454.

[32] Jiang, Z. L., Kassab, G. S., and Fung, Y. C., "Diameter-defined Strahler system and connectivity matrix of the pulmonary arterial tree," *J. Appl. Physiol.*, Vol. 76, 1994, pp. 882–892.

[33] Weinberg, E. J., Borenstein, J. T., Kaazempur-Mofrad, M. R., Orrick, B., and Vacanti, J. P., "Design and fabrication of a constant shear microfluidic network for tissue engineering," *MRS Symp. Proc.*, Vol. 820, 2004, pp. 121–126.

[34] Janakiraman, V., Mathur, K., and Baskaran, H., "Optimal planar flow network designs for tissue-engineered constructs with built-in vasculature," *Ann. Biomed. Eng*, Vol. 35, 2006, pp. 337–347.

[35] Wang, G.-J., and Hsu, Y.-F., "Structure optimization of microvascular scaffolds," *Biomed. Microdevices*, Vol. 10, 2005, pp. 51–58.

[36] Weinberg, E. J., Kaazempur-Mofrad, M. R., and Borenstein, J. T., "Numerical model of flow in distensible microfluidic network," in *Computational Fluid and Solid Mechanics*, Vol. 2, Oxford: Elsevier Science, 2003, pp. 1569–1572.

[37] Fidkowski, C., Kaazempur-Mofrad, M. R., Borenstein, J. T., Vacanti, J. P., Langer, R., and Wang, Y., "Endothelialized microvasculature based on a biodegradable elastomer," *Tissue Eng.*, Vol. 11, 2005, pp. 302–309.

[38] Kaazempur-Mofrad, M. R., Krebs, N. J., Weinberg, E. J., Vacanti, J. P., and M.-A., B. R. R. S. J. T. Borenstein, "A MEMS-based renal replacement system," *Proc. 2004 Sens. Actuators Conf.*, 2004, pp. 67–70.

[39] Kartalov, E. P., Walker, C., Taylor, C. R., Anderson, W. F., and Scherer, A., "Microfluidic vias enable nested bioarrays and autoregulatory devices in Newtonian fluids," *Proc. Nat. Acad. Sci.*, Vol. 103, 2006, pp. 12280–12284.

[40] Kaazempur-Mofrad, M. R., Vacanti, J. P., and Kamm, R. D., "Computational modeling of blood flow and rheology in fractal microvascular networks,"in *Computational Fluid and Solid Mechanics*, Vol. 2, Oxford: Elsevier Science, 2001, pp. 864–867.

[41] Wagner, W. P., and Gage, K. L., "Thrombosis and thromboembolism in stent grafts: Evaluating Virchow's triad," in J. Vossoughi, N. Kipshidze, J. W. Moses, and J. W. Karanian (eds.), *Stent Graft Update*, Medical and Engineering Publishers, 2000, Chapter 9.

[42] Kaazempur-Mofrad, M. R., Weinberg, E. J., Borenstein, J. T., and Vacanti, J. P., "Tissue engineering: Multiscaled representation of tissue architecture and function," *Complex Syst. Biomed.*, 2003, p. 25.

[43] Garcia-Cardena, G., Comander, J., Anderson, K. R., Blackman, B. R., and Gimbrone, M. A., "Biomechanical activation of vascular endothelium as a determinant of its functional phenotype," *Proc. Nat. Acad. Sci.*, 2001, pp. 98, 2001, pp. 4478–4485.

[44] Helmlinger, G., Berk, B. C., and Nerem, R. M., "Calcium responses of endothelial cell monolayers subjected to pulsatile and laminar flow differ," *Am. J. Physiol.*, Vol. 269, 1995, pp. 367–375.

[45] Girard, P. R., and Nerem, R. M., "Shear stress modulates endothelial cell morphology and F-actin organization through the regulation of focal adhesion-associated proteins," *J. Cell. Physiol.*, Vol. 163, 1995, pp. 179–193.

[46] Gabriel, K. J., "Microelectromechanical systems," *Proc. IEEE*, Vol. 86, 1998, p. 1534.

[47] Whitesides, G. M., and Stroock, A. D., "Flexible methods for microfluidics," *Physics Today*, Vol. 54, No. 6, 2001, pp. 42–48.

[48] Bhatia, S. N., and Chen, C. S., "Tissue engineering at the micro-scale," *Biomed. Microdevices*, Vol. 2, No. 2, 1999, pp. 131–144.

[49] Griffith, L. G., and Swartz, M. A., "Capturing complex 3-D tissue physiology in vitro," *Nat. Rev.: Mol. Cell Biol.*, Vol. 7, 2006, pp. 211–223.

[50] Nguyen, N. T., and Wereley, S., *Fundamentals and Applications of Microfluidics*, Norwood, MA: Artech House, 2002.

[51] Raghavan, S., and Chen, C. S., "Micropatterned environments in cell biology," *Adv. Mater.*, Vol. 16, 2004, pp. 1303–1313.

[52] Tsang, V. L., and Bhatia, S. N., "Three-dimensional tissue fabrication," *Adv. Drug Del. Rev.*, Vol. 56, 2004, pp. 1635–1647.

[53] Armani, D. K., and Liu, C., "Microfabrication technology for polycaprolactone, a biodegradable polymer," *J. Microelectromech. Syst.*, Vol. 10, No. 1, 2000, pp. 80–84.

[54] King, K. R., Wang, C. C., Shin, M., Vacanti, J. P., and Borenstein, J. T., "Biodegradable polymer microfluidics for tissue engineering microvasculature," Vol. 729, 2002.

[55] Bettinger, C. J., Orrick, B. K., Langer, R., and Borenstein, J. T., "Tunable contact guidance using feature geometry in biodegradable substrates," *Biomaterials*, Vol. 27, 2006, pp. 2558–2565.

[56] Bettinger, C. J., Cyr, K. M., Matsumoto, A., Langer, R., Borenstein, J. T., and Kaplan, D. L., "Silk fibroin microfluidic devices," *Adv. Mater.* In press.

[57] Quake, S. R., and Scherer, A., "From micro- to nanofabrication with soft materials," *Science,* Vol. 290, 2000, p. 1536.

[58] Kane, R. S., Takayama, S., Ostuni, E., Ingber, D. E., and Whitesides, G. M., "Patterning proteins and cells using soft lithography," *Biomaterials*, Vol. 20, 1999, p. 2363.

[59] Jo, B.-H., and Beebe, D. J., "Fabrication of three-dimensional microfluidic systems by stacking molded PDMS layers," *SPIE*, Vol. 3877, 1999, p. 222.

[60] Chen, C. S., Mrksich, M., Huang, S., Whitesides, G. M., and Ingber, D. E., "Geometric control of cell life and death," *Science,* Vol. 276, 1997, p. 1425.

[61] Bhatia, S. N., Balis, U. J., et al., "Effect of cell-cell interactions in preservation of cellular phenotype," *Faseb J.*, Vol. 13, 1999, pp. 1883–1900.

[62] Wang, G. J., Chen, C. L., Shu, S. H., and Chaing, Y. L., "Bio-MEMS-fabricated artificial capillaries for tissue engineering," *Microsyst. Technol.*, Vol. 12, 2005, pp. 120–127.

[63] Ayon, A. A., Chen, K.-S., Lohner, K. A., Spearing, S. M., Sawin, H. H., and Schmidt, M. A., "Deep reactive ion etching of silicon," *Mat. Res. Soc. Symp. Proc.*, Vol. 546, 1999, p. 51.

[64] LaVan, D. A., George, P. M., and Langer, R., "Simple, three-dimensional microfabrication of electrodeposited structures," *J. Gesellschaft Deutscher Chemiker*, Vol. 42, 2003, pp. 1262–1265.

[65] Heckele, M., and Schomburg, W. K., "Review of micro molding of thermoplastic polymers," *J. Micromech. Microeng.*, Vol. 14, 2007, pp. R1–R14.

[66] Ryu, W., Min, S., Fasching, R. J., and Prinz, F. B., "3-D multi-layered micro-fabricated tissue scaffolds of biodegradable polymers," *Soc. Biomater. Conf.*, Pittsburgh, PA, 2006.

[67] Ryu, W., Min, S. W., Hammerick, K. E., Vyakarnam, M., Greco, R. S., Prinz, F. B., and Fasching, R. J., "The construction of three-dimensional micro-fluidic scaffolds of biodegradable polymers by solvent vapor based bonding of micro-molded layers," *Biomaterials*, Vol. 28, 2007, pp. 1174–1184.

[68] Wang, Y., Guillermo, A., Ameer, A., Sheppard, J., and Langer, R., "A tough biodegradable elastomer," *Nat. Biotechnol.*, Vol. 20, 2002, pp. 602–606.

[69] Stankus, J. J., Soletti, L., Fujimoto, K., Hong, Y., Vorp, D. A., and Wagner, W. R., "Fabrication of cell microintegrated blood vessel constructs through electrohydrodynamic atomization," *Biomaterials,* Vol. 28, 2007, pp. 2738–2746.

[70] Cabodi, M., Havenstrite, K. L., Curtis, V., Suzanne, S., and Stroock, A. D., "A microfluidic wound dressing and wound analysis tool," *ASME Summer Bioeng. Conf.*, Vail, CO, 2005.

[71] Saxena, V., Hwang, C. W., Huang, S., Eichbaum, Q., Ingber, D. E., and Orgill, D. P., *Plast. Reconstruct. Surg.*, Vol. 114, 2004, p. 1086.

[72] Lu, H., Koo, L. Y., Wang, W. M., Lauffenburger, D. A., Griffith, L. G., and Jensen, K. F., "Microfluidic shear devices for quantitative analysis of cell adhesion," *Anal. Chem.*, Vol. 76 2004, pp. 5257–5264.

[73] Song, J. W., Gu, W., Futai, N., Warner, K. A., Nor, J. E., and Takayama, S., "Computer-controlled microcirculatory support system for endothelial cell culture and shearing," *Anal. Chem.*, Vol. 77, 2005, pp. 3993–3999.

[74] Vollmer, A. P., Probstein, R. F., Gilbert, R., and Thorsen, T., "Development of an integrated microfluidic platform for dynamic oxygen sensing and delivery in a flowing medium," *Lab Chip*, Vol. 5, 2005, pp. 1059–1066.

[75] Kastrup, C. J., Runyon, M. K., Shen, S., and Ismagilov, R. F., "Modular chemical mechanism predicts spatiotemporal dynamics of initiation in the complex network of hemostasis," *Proc. Nat. Acad. Sci.*, Vol. 103, 2006, pp. 15747–15752.

CHAPTER 16

Nanotechnology for Inducing Angiogenesis

Sangamesh G. Kumbar, Roshan James, Lakshmi S. Nair, and Cato T. Laurencin

16.1 Introduction

The growing demand for transplants far exceeds the availability, leading to an acute shortage of organs and tissues available to those in need. In the United States, 89,286 patients are wait-listed for transplantation as of August 2005, and nearly 4,000 new patients are added each month (www.OPTN.org). Only 27,037 patients received organ transplants in 2004, and many passed away waiting for that "gift of life." There is a great need for the development of functional organs and tissues to meet these shortages. Tissue engineering has emerged with a hope to replace a wide range of dysfunctional tissues as alternatives to harvested tissues and prostheses. Tissue engineering has been defined by Laurencin et al. as the application of biological, chemical, and engineering principles toward the repair, restoration, or regeneration of tissues using cells, scaffolds, and growth factors alone or in combination [1]. Since isolated cells cannot reassemble into functional tissues, biodegradable biomaterial scaffolds are used to organize cells into a three-dimensional form to create functional tissue. The scaffolds serve to mimic the structure and functions of native extracellular matrix (ECM) and guide the proper organization of cells into complex tissue. Studies so far have demonstrated the feasibility of culturing different cell types on biocompatible scaffolds in vitro, and these have been implanted in vivo in an attempt to repair small-scale injuries [2–4]. However, it remains a challenge to translate these early successes into regeneration of complex tissues and organs due to the inability of the engineered scaffold-cell construct to functionally integrate into surrounding tissues. A major hurdle in developing engineered tissue is the inability to vascularize these artificial constructs. A vascular system is highly essential to enable transport of nutrients and oxygen to cells within the porous structure of the scaffold and remove metabolic wastes in a timely fashion.

Angiogenesis and vasculogenesis in combination are involved in establishing a new vasculature. Angiogenesis is the formation of new blood vessels from preexisting vessels, while vasculogenesis is the formation of capillary beds in the absence of a preexisting vascular network. Angiogenesis involves the migration and proliferation of free mature endothelial cells (EC) from the basement membrane to form sprouts from preexisting vessels. In vasculogenesis, circulating endothelial progenitor cells (EPC) are recruited to neovascularization sites and differentiate *in situ* into mature ECs. A schematic presentation of the neovascularization process is presented in Figure 16.1. A tissue of a few cubic millimeters in size cannot depend on diffusion alone for the supply of nutrients and oxygen. Cells are found to be metabolically inactive or necrotic if they are more than 200 μm away from the vascular network [5–8]. Folkman et al. reported the inability of tumors to grow beyond a few cubic millimeters in volume by directing the ingrowth of capillaries from adjacent blood vessels unless they grow their own blood vessels [5–8]. These findings resulted in a better understanding of the processes involved during the formation of new capillaries [8–11]. Angiogenesis plays a significant role in tumor growth and metastasis, rheumatoid arthritis, psoriasis, scleroderma, diabetic retinopathy, retrolental fibroplasia, and neovascular glaucoma [12–18]. The restriction or control of vascular-network establishment is an active research area for the treatment of such disorders. Angiogenesis also contributes to the establishment and maintenance of the body defense mechanism by creating collateral vessels. For instance, new vessels sprouting from preexisting capillaries penetrate the necrotic area of myocardial infarcts and protect the heart from ischemic damage [14]. Angiogenesis plays an active role in body defensive mechanisms during the female reproduction cycle,

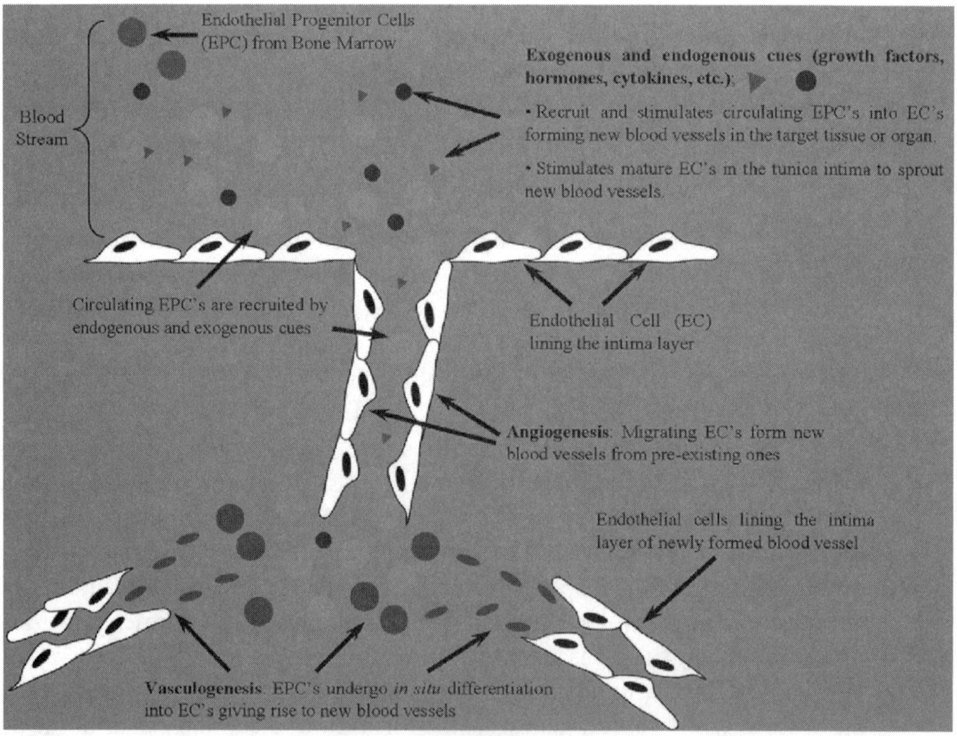

Figure 16.1 Angiogenesis: In response to stimulants such as growth factors, hormones and cytokines, ECs from the tunica intima will be stimulated to migrate and sprout new blood vessels from the parental or preexisting vascular network. Vasculogenesis: Bone marrow-derived EPCs circulating in the bloodstream will be recruited and directed to neovascularization sites by the exogenous and endogenous factors. The recruited cells will undergo differentiation in situ into the endothelial phenotype These endothelial-like cells at neovascularization sites mature to become ECs that form new blood vessels. Growth factors, cytokines, or hormones are released endogenously in response to tissue ischemia or are administered exogenously for therapeutic neovascularization.

placental tissue formation, and embryo implantation. Angiogenesis is an actively pursued research area in engineering organs and tissues.

Current strategies to vascularize tissue-engineered constructs utilize growth factors such as vascular endothelial growth factor (VEGF) and fibroblast growth factor (FGF) to induce angiogenesis following implantation [19–21]. Several efforts were made to induce angiogenesis by preseeding suitable scaffolds with endothelial cells alone or in combination with growth factors. A critical challenge in designing tissue-engineering scaffolds is to provide suitable surface properties that will promote selective cell adhesion and direct a complex organization and reorganization of cells into the desired functional organ or tissue. Cells attach to scaffold surfaces via cytoskeletal receptors located in the cell membrane, and attachment is governed by the ability of scaffold surfaces to absorb specific cell-adhesion proteins from the surrounding plasma and other body fluids. Cell adhesion stimulates multiple biochemical signaling pathways within the cell, and exploitation of this behavior provides an opportunity to design scaffolds that can elicit desired tissue growth. Cell-scaffold interactions can be engineered by modulating scaffold architecture and surfaces at the basic level to direct controlled and highly precise interactions with proteins and cells. Such scaffolds may contain specific chemical or structural cues that can potentially control and modulate cell-cell and cell-scaffold interactions to favor tissue formation.

In native tissue, the extracellular matrix is composed of several functional groups, specific proteins, growth factors, and many tropic agents that together influence various cellular functions, namely, adhesion, proliferation, migration, differentiation, and cell shape [22–32]. In addition, the ECM possesses a complex topography at the nanodimension comprising hierarchically arranged collagen, laminin, and other fibrils. The basement membrane is an integrated component of the ECM and measures approximately 200 nm in thickness. Various intertwining nanotopographic features, such as fibers, ridges, and pores, constitute the ECM. These topographical features measured 178 ± 57 nm, 127 ± 54 nm, 52 ± 28 nm, and 82 ± 49 nm for basement membrane height, interpore distance, and fiber and pore diameters, respectively [33, 34]. Scanning electron micrographs (SEMs) of mouse ECM obtained after enzymatic separation of the mouse epidermis show dense collagen nanofiber meshwork as evident in Figure 16.2(a) [35]. Figure 16.2(b, c) shows the dermis comprising a fibroblast attached to the collagen nanofiber mesh [Figure 16.2(b)] and migration of dendritic cells through the collagen mesh after forty-eight hours of culture [Figure 16.2(c)]. These micrographs represent the actual nanotopographic features experienced by cells in the living tissue. These ECM-like nanotopographical features have been manufactured using various fabrication techniques. The fabricated nanoarchitecture has significantly modulated the cellular adhesion [22], proliferation [23], morphology [24, 25], motility [26], endocytotic activity [27], and gene regulations [28] of several cell types, including fibroblasts [23, 25, 28, 29], osteoblasts [22, 30], osteoclasts [31], ECs [36], smooth muscle cells (SMCs) [37], epithelial cells [32], and epitenon cells [38].

This chapter discusses the various techniques used to fabricate ECM-like nanodimensional features. Techniques such as electrospinning [39–51], chemical etching [52–57], self-assembly [58, 59], and lithography [24–26] have been used to fabricate nanostructures that will modulate ECs, EC progenitors, and SMC behav-

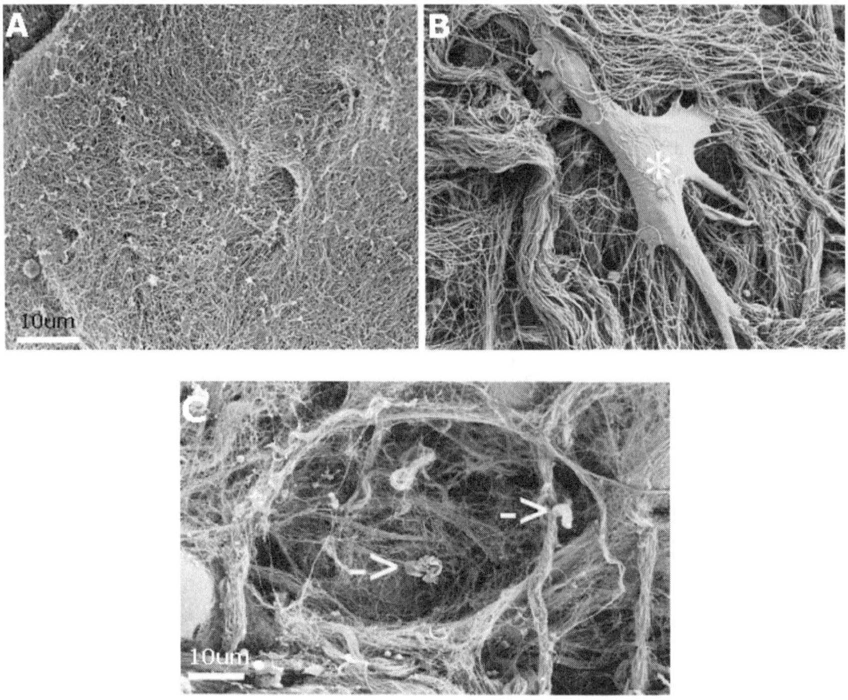

Figure 16.2 SEM micrographs of mouse epidermis after enzymatic separation: (a) collagen meshwork of basement membrane, (b) dermis with attached fibroblast (asterisk), and (c) migration of dendritic cells (arrows) through dermal collagen meshwork after skin explant culture for forty-eight hours. It is evident from these pictures that microdimension cells are exposed to nanotopographic features in vivo. (*Source:* [35], reprinted with permission from the Society for Investigative Dermatology.)

ior to induce vascularization [36, 37, 39–66]. Techniques such as grafting and coating were employed to immobilize various proteins, peptides, and other biomolecules. Immobilized proteoglycans created a cell-adhesive multifunctional surface that mimicked the binding domains of the ECM environment. The oligopeptide arginine-glycine-aspartic acid (RGD) sequence is a part of adhesive proteins such as fibronectin and vitronectin and can support EC adhesion and spreading [67]. Surface functionalization with RGD and other bioactive molecules can promote the spatial and temporal organization of various cells by selective adhesion to generate a functional tissue. Such surface modifications accompany changes in surface chemical composition, surface energy, surface roughness, and nanotopography that might potentially facilitate neovascularization [68–80]. Table 16.1 summarizes the advantages and disadvantages of the techniques used to fabricate nano-size architectures and how they modulate EC behavior.

16.2 Nanostructured Scaffolds and Angiogenesis

16.2.1 Electrospinning

Electrospinning is a popular and versatile technique to fabricate polymeric fiber scaffolds with controlled architecture and properties [81, 82]. Due to the mild fabri-

Table 16.1 Summary of Different Nanotopographic Features Used to Induce Angiogenesis.

Technique	*Topographic Features*	*Merits*	*Shortcomings*	*Observed Change in Cellular Behaviors*	*References*
Electrospinning	Possible to fabricate aligned and random nanofibers, microfibers, and nanomicrofibers	Simple, facile, wide range of polymers can be readily electrospun; possible to incorporate bioactive molecules	Can create only fibrous structures	Higher proliferation rate, strong alignment, migration, and orientation	[39–51]
Self-assembly	Possible to create fibers of both nano- and microdimension depending on the start-up material	Simple, facile, and can create complex functional structures with self-healing ability	No direct process control and difficult to fabricate complex structures	Higher rate of adhesion, proliferation, differentiation, and phenotype expression	[58, 59]
Chemical etching	Nature and time of exposure etchants largely decides the feature's dimension	Simple, facile, and inexpensive	No direct process control and difficult to fabricate complex structures	Higher rate of adhesion, migration, proliferation, and phenotype expression	[52–57]
Lithographic techniques	Can create precise geometries and patterns without any mask	Expensive equipment, more time, lower resolution, hard to pattern large surface area	Expensive and complicated instrumentation and nanodimensions at the higher side	Higher proliferation rate, strong alignment, migration, and orientation	[60–62]
Polymer demixing	Pits, islands, or ribbons of a wide range of heights or depths can be created	Possible to pattern large surface area with low cost and high efficiency	Only pits, islands, or ribbons can be created	Higher rate of adhesion, migration, proliferation, and phenotype expression	[63–66]
Functionalization	Different cell-binding domains of the ECM can be immobilized on the surface	Simple and possible to modify surface	Suitable functionalization technique needs to be adopted to maintain bioactivity	Higher rate of adhesion, migration, proliferation, and phenotype expression	[68–90]

cation process, electrospinning has been used to develop polymeric scaffolds loaded with therapeutic agents as sustained drug-/protein-delivery vehicles. The versatility of the process also enables the fabrication of scaffolds containing random or aligned fibers of both nano and micron sizes alone or in combination, using the same experimental setup. Such electrospun nanofiber scaffolds fabricated from various polymers were used to study EC and SMC behavior in an effort to induce angiogenesis to develop tissue-engineered vascular grafts [39–51].

Electrospun poly(L-lactic acid) (PLLA) nano- and microfibers having nano- to micron-scale surface roughness, along with smooth surface solvent cast films, were used to investigate EC-scaffold interaction [39]. Endothelial cells exhibited enhanced proliferation and a typical EC morphology showing a cobblestone appearance with higher confluence levels on flat surfaces than on surfaces with nano- or microroughness. Less and sporadic distribution of ECs was observed on electrospun nanofibers presumably due to inferior cell attachment to electrospun surfaces as opposed to smooth surfaces, as evident from Figure 16.3. As seen from Figure 16.3(c), ECs organized into capillarylike microtubes on smooth surfaces rep-

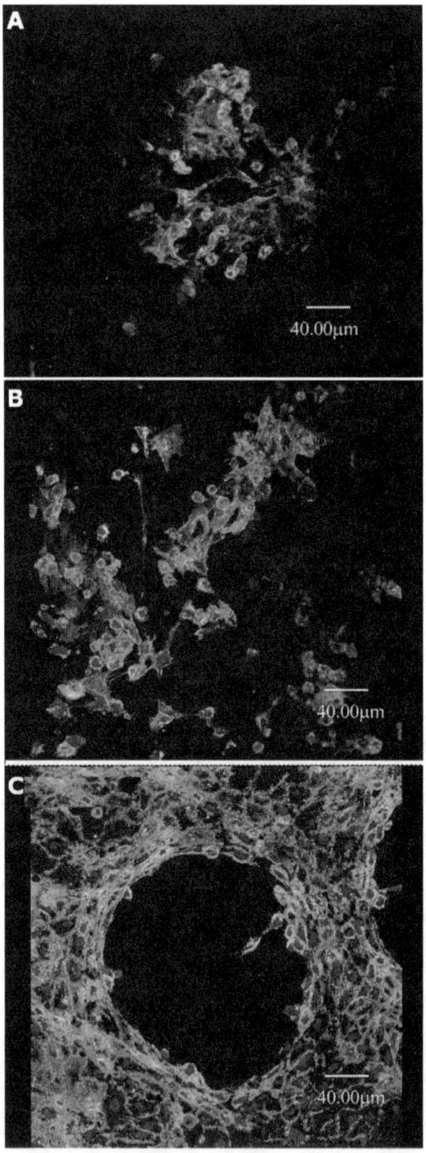

Figure 16.3 Confocal micrographs of ECs immunostained with cell-surface adhesion protein CD31 after three days of culture on PLLA surfaces of varying surface roughness. EC on PLLA (a) nanofibers, (b) microfibers, and (c) smooth film. EC showed a well-spread morphology and enhanced proliferation on smooth surfaces as compared to electrospun nano- and microfibers. (*Source:* © 2004 Wiley Periodicals, reprinted with permission.)

resenting functional endothelialization. Several factors can contribute to this varied response of ECs toward biomaterial scaffolds; therefore, it is important to understand the mechanism of EC interactions with materials while designing tissue-engineering scaffolds to repair or regenerate small-diameter blood vessels. To improve the EC-adhesive properties of electrospun PLLA nanofibers, various methodologies were investigated either by developing PLLA-blend nanofibers or coating PLLA nanofibers with other biopolymers, such as collagen, elastin, and gelatin [40–46]. Nanofibers of PLLA-collagen blends supported the adhesion, spreading, and enhanced viability of human coronary artery endothelial cells better than PLLA nanofibers alone. The ECs maintained normal phenotype, and the improved cell viabilities on blend nanofibers may be due to the presence of collagen on the surface [40]. Coating these PLLA-collagen blend nanofibers with a layer of collagen enhanced endothelialization as observed by the spread cell morphology is shown in Figure 16.4(a), and increased cell viability is shown in Figure 16.4(b, c) [41]. The improved endothelialization is presumably due to the bioactivity of the coated collagen layer. Though electrospinning is a proven mild fabrication technique to encapsulate bioactive agents [81, 82], the lower bioactivity of the PLLA-collagen blend scaffolds may be presumably due to the denaturation of the collagen in organic solvent used for electrospinning. Another study investigated the effect of collagen-coated aligned and random nanofibers of poly(L-lactic acid)-co-poly(ε-caprolactone) (PLLA-PCL) on human coronary artery endothelial-cell behavior [36]. Aligned nanofibers showed greater tensile modulus and strength than random fibers, and ECs oriented parallel to the aligned nanofibers with elongated morphology. ECs preserved their phenotype expression and functions on both the nanofiber matrices. In a similar study, electrospun nanofiber scaffolds fabricated from blends of poly(lactide-co-glycolide) (PLAGA) with collagen I and elastin showed compliance characteristics similar to native arteries [37]. These blended nanofiber scaffolds supported EC and SMC adhesion and showed enhanced proliferation as evidenced by SEM micrographs and hematoxylin and eosin staining (Figure 16.5). Aligned and random PCL nanofiber surfaces modified by gelatin grafting showed enhanced EC proliferation and spreading compared to unmodified PCL nanofiber matrices. ECs oriented along the aligned nanofiber matrices in the direction of the fiber alignment [42]. On the gelatin modified PCL nanofiber matrices, ECs maintained the expression of three characteristic markers, namely platelet–endothelial-cell adhesion molecule–1 (PECAM-1), intercellular adhesion molecule–1 (ICAM-1), and vascular cell adhesion molecule–1 (VCAM-1) [42]. These studies demonstrated the effect of surface chemistries on nano-/microfabricated scaffolds on EC and SMC adhesion, proliferation, and phenotypic expression.

In addition to the surface chemistry, the fiber diameter has also shown to have significant influences on cell behavior. Nano- and microfibers fabricated by electrospinning of poly(L-lactide-co-epsilon-caprolactone) copolymers with different compositions showed that fiber size directed the spreading and proliferation of human umbilical vein endothelial cells (HUVECs) [43]. ECs adhered with well-spread morphology on fiber matrices having diameters in the range of 300 to 1,200 nm, while reduced adhesion and no signs of proliferation were observed on microfibers of 7 μm diameter (Figure 16.6) [44]. The study indicates the favorable

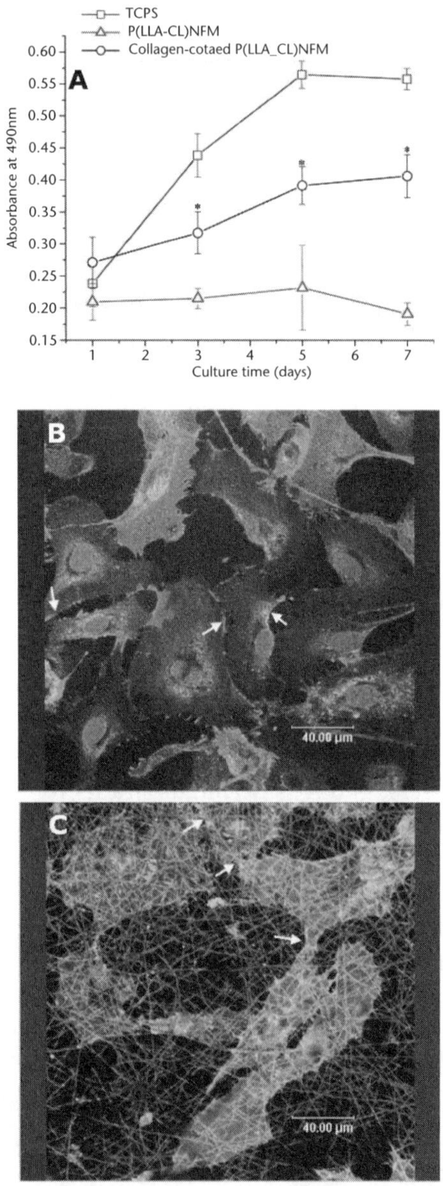

Figure 16.4 (a) Viability of HCAEC on nanofiber structures at the seven-day time point. Immunofluorescent staining of PECAM-1(CD31) on (a) HCAEC and (b) TCPS (control). (c) RBITC-labeled collagen-coated nanofibers appear red in color. Significantly higher cell viability was observed on collagen-coated nanofibers, and arrows indicate strong PECAM-1 expression at the cell-cell interface. (*Source:* [41], reprinted with permission from Elsevier.)

influences nanoscale features have on HUVEC adhesion and proliferation compared to micron-scale features.

In addition to the biomaterial properties, the cell type can also have a profound effect on vasculogenesis. Blends of α-hydroxyesters, in particular PLLA, poly(L-glutamic acid) (PGA), and PLAGA with PCL or polyurethane (PU), were electrospun into nanofibers and evaluated for their EC performance [44–50]. It has been found that ECs derived from human EPCs seeded onto PGA-PLLA nonwoven

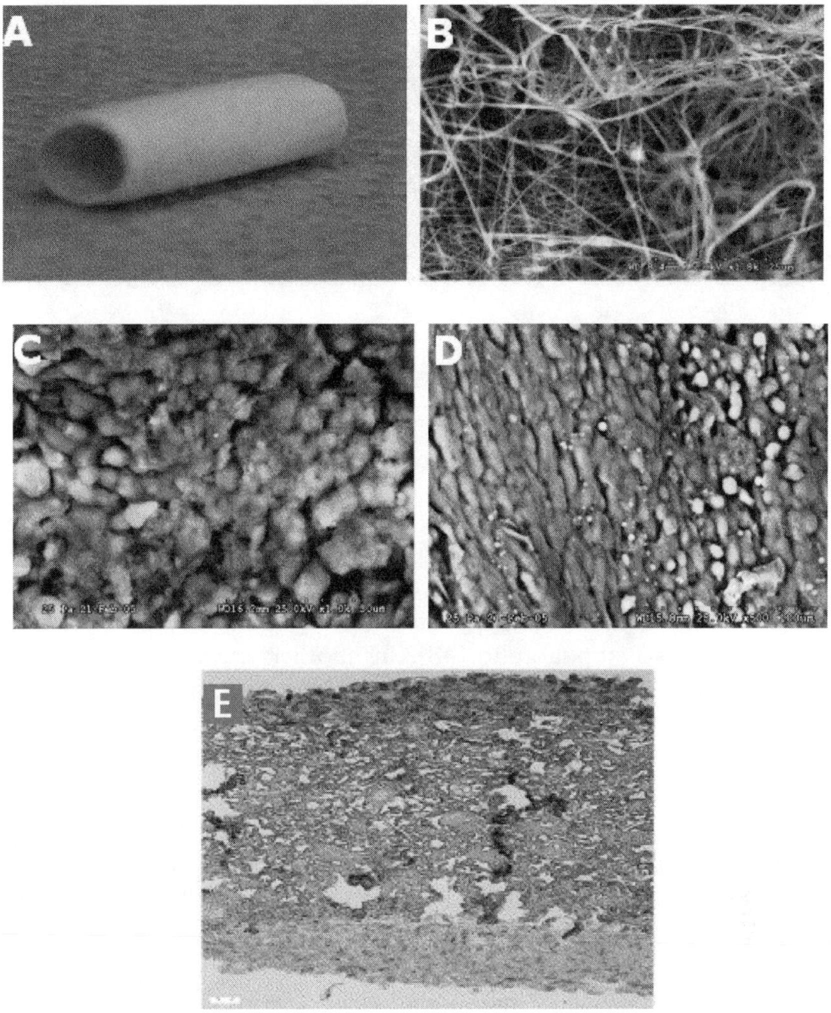

Figure 16.5 (a) Collagen-elastin-PLAGA blend nanofiber scaffolds. SEM micrographs of (b) blend nanofiber scaffold, (c) bovine endothelial cell–seeded scaffold, (d) bovine SMC-seeded scaffold, and (e) Hematoxylin and Eosin staining of the SMCs (upper surface) and ECs (lower surface). The blended nanofiber scaffold supported EC and SMC adhesion and has compliance characteristics similar to native arteries. (*Source:* [37], reprinted with permission from Elsevier.)

fiber matrices did not form microvessels [44]. However, when seeded with human SMCs, capillarylike structures were formed throughout the scaffold measuring about 76.5 ± 35 microvessels/mm^2. The study indicates that EPCs along with SMCs seeded on nanoscale dimensions may be well suited for vasculogenesis in tissue-engineering scaffolds [44]. Electrospun nanofibers of PLLA-PCL copolymer with an l-lactide to ε-caprolactone ratio of 75:25 supported the adhesion and proliferation of both ECs and SMCs to form an integrated, three-dimensional cellular network with no phenotypic change [45, 46]. Wet-spun PCL fibers coated with polyurethane nanofibers by electrospinning to form composite fiber matrices were evaluated for EC behavior [47]. Composite nanofiber matrices supported strong EC attachment, and cells proliferated to form a monolayer with characteristic cobblestone morphology. Furthermore, EC characteristics were confirmed by release of von Willebrand

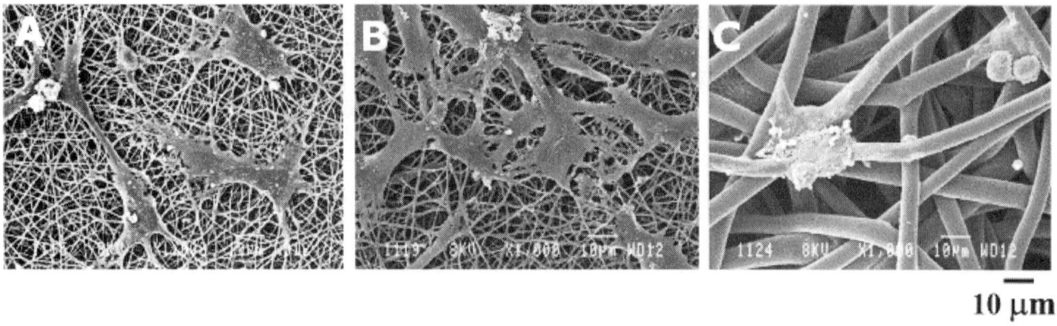

Figure 16.6 SEM images of HUVECs on poly(L-lactide-co-epsilon-caprolactone) 50/50 fiber matrices after seven days of culture: (a) 300 nm, (b) 1.2 μm, and (c) 7 μm. ECs showed a well-spread morphology on fiber diameters in the range of 300 to 1,200 nm, whereas a rounded morphology is evident on microfibers. (*Source:* [43], reprinted with permission from Elsevier.)

factor, nitric oxide, and ICAM-1 under physiological stimuli [47]. Similarly, nanofibers of poly(ester urethane)urea and collagen type I at various ratios resulted in enhanced adhesion and proliferation of SMCs over the control, which had a smooth surface topography [48]. In another study, poly(ethylene terephthalate) (PET) nanofibers were surface-modified by grafting the surface with either poly(methacrylic acid) or gelatin to improve EC performance [49]. Gelatin grafting onto the nanofiber surfaces improved EC adhesion, proliferation, and phenotype expression over control PET nanofibers [50]. Gelatin and poly(methacrylic acid) grafting onto nanofiber surfaces might increase the protein binding onto the nanofiber scaffold, thereby promoting cell adhesion. Such surface modifications can help to overcome the incomplete EC covering of small-diameter vascular grafts and improve endothelialization.

In addition to organic polymers, certain inorganic polymers such as polyphosphazenes with amino acid ester side groups were also evaluated for EC interaction. Electrospun tubular nanofiber scaffolds of poly[(ethyl phenylalanato)$_{1.4}$ (ethyl glycinato)$_{0.6}$ phosphazene] supported EC adhesion, and at day four, higher EC proliferation was observed on these surfaces as compared with the control (i.e., tissue-culture polystyrene) [51]. Polyphosphazene nanofiber scaffolds grafted with amino acid side chains better mimicked the ECM and resulted in improved EC response. Tubular nanofiber scaffolds attained confluence after day sixteen, forming an EC monolayer as evidenced by microscopic observations. However, ECs failed to migrate through the tubular scaffold wall, and the inner part of the scaffold remained acellular [51].

16.2.2 Self-Assembly

Molecular self-assembly is a simple, facile, and scalable bottom-up approach to fabricating nanofiber scaffolds using bioactive molecules. Several nanofiber scaffolds fabricated using this approach were evaluated for EC and SMC behavior to aid the process of angiogenesis [58, 59]. Self-assembled RAD16-I (AcN-RADARADARADARADA-CONH2) peptide nanofiber surfaces were functionalized with various motifs of peptide sequences present in protein components of basement membranes to study EC function [58]. In particular, the peptide sequence motifs laminin 1 (YIGSR, RYVVLPR) and collagen IV (TAGSCLRKFSTM) enhanced the rate of formation of confluent EC monolayers over the control. On these modified surfaces, ECs maintained their phenotype with enhanced nitric oxide release and expression of laminin 1 and collagen IV [58]. In another study, self-assembled peptide amphiphile (PA) nanofibers in combination with heparin and growth factors VEGF/FGF-2 were tested for their efficacy to induce angiogenesis in rat cornea [59]. In brief, PA-heparin gel with and without growth factors (GFs) was injected into a surgically created pocket in the rat cornea to produce self-assembled nanofibers as shown in Figure 16.7(a). Heparin-binding to nanofibers help to conjugate FGF-2, and the conjugates released FGF-2 in a

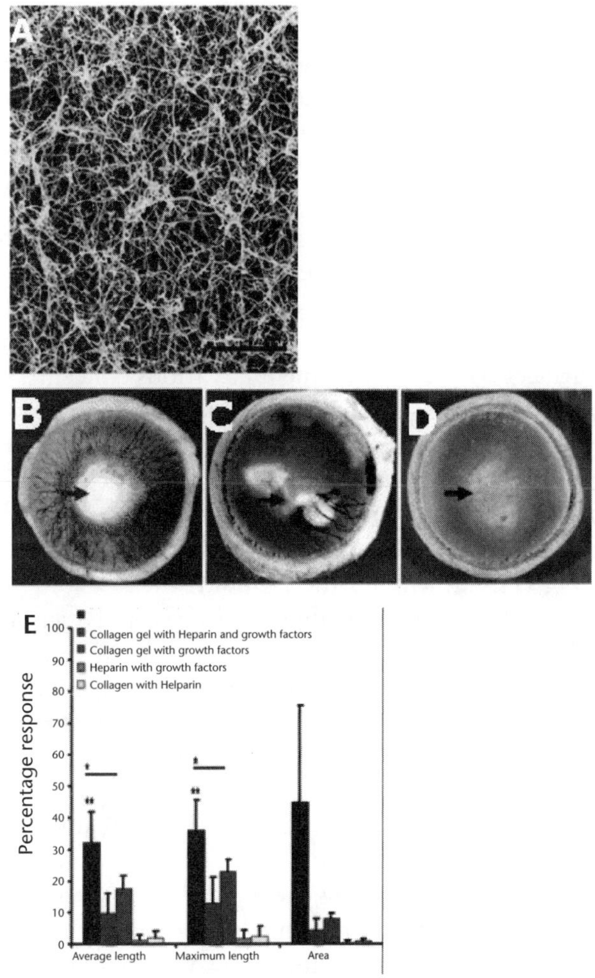

Figure 16.7 SEM micrograph of (a) heparin-nucleated PA nanofibers, (b) PA-heparin nanofiber with FGF-2 showing extensive neovascularization in the rat cornea at day ten, (c) collagen-heparin nanofiber scaffold with growth factor, and (d) collagen with growth factor alone; (c) showed less neovascularization than (b). (e) Graph showing neovascularization response in terms of the average and maximum length of new blood vessels and the area of corneal neovascularization. PA-heparin scaffold with growth factor showed significantly higher neovascularization over all experimental groups. (*Source:* [59], reprinted with permission from the American Chemical Society.)

controlled manner for longer duration. Such a delayed FGF-2 release could presumably be due to the high surface area available for growth factor binding and resulted in longer effective bioactivity. Neovascularization response at day ten is shown in Figure 16.7(b–e), and it is clear that the heparin-PA nanofibers combined with GF promote significant neovascularization over all experimental groups. The study also demonstrated that the physical encapsulation of GF in a matrix could significantly enhance these nanofibers' bioavailability [59]. PA-heparin nanofibers without GF resulted in undetectable levels of neovascularization, thus confirming no inflammatory response to these nanofibers [59]. Designing nanostructures with the right chemistry and the capability to incorporate bioactive agents without altering their efficacy is highly crucial for eliciting the desired responses.

16.2.3 Chemical Etching

Chemical etching is a simple and inexpensive surface-modification process to create nanoscale roughness on surfaces. Etchants like sodium hydroxide (NaOH), nitric acid (HNO_3), or hydrofluoric acid (HF) are typically used to etch away the surfaces. Varying dimensions of surface roughness comprising pits and projections are created by varying the exposure time to the etchants. The EC and SMC interactions with these nanofeatures have been evaluated in an effort to elucidate the effect of nanofeatures on inducing angiogenesis [52–57]. Webster et al. in a series of studies evaluated the effect of chemically etched polymer surfaces having nanoscale roughness toward EC behavior. Micron- to nanometer-sized surface roughness created on PLAGA surfaces by NaOH treatment at various concentrations and exposure times was evaluated for SMCs (endothelial and aortic smooth muscle cells), bladder smooth muscle [52], and EC responses [52, 53]. Enhanced adhesion and proliferation was observed for SMCs while ECs showed decreased adhesion and proliferation as compared to untreated PLAGA smooth surfaces. These observed changes, however, can be due to a combined effect of surface roughness and chemistry as a result of NaOH treatment. Further, to study the topographical effects on cell behavior, surface roughness was created by using a polymer/elastomer casting method without affecting the surface chemistry [52, 53]. Both SMCs and ECs showed increased adhesion and proliferation compared to untreated PLAGA. This study demonstrated the effect of biomaterial surface chemistry and topography on vascular cell behavior, and such an understanding is crucial in creating tissue-engineered vascular grafts.

In a related study, nanometer roughness was created on PLAGA and PU surfaces via chemical etching. Enhanced SMC adhesion was observed on surfaces with nanosized roughness created by chemical etching over their conventional counterparts [54]. Such observed changes might be due to the changes in surface chemistry and the nanosized roughness created; however, roughness was found to be the major contributor. In an effort to elucidate the possible mechanisms underlying improved EC and SMC behavior on surfaces with nanoscale roughness, protein absorption studies were carried out on nanostructured PLAGA surfaces [55]. Significantly more quantity of vitronectin and fibronectin from serum was adsorbed on nanostructured PLAGA surfaces as compared to control PLAGA sheets (untreated). These cell-adhesive proteins presumably resulted in better performance of ECs and SMCs on the nanostructured surfaces compared to the smooth surface. This was further corroborated by blocking cell-binding epitopes of fibronectin and vitronectin, which resulted in decreased vascular cell adhesion on nanostructures compared to the untreated PLAGA control [55].

Similarly, nanoscale roughness created by chemical etching of PLAGA, PU, and PCL surfaces resulted in enhanced adhesion and proliferation of SMCs when compared with untreated smooth surfaces [56]. PCL films were treated with NaOH in an effort to increase the surface hydrophilicity and roughness in nanodimensions. Such surface modified PCL was evaluated for both EC and SMC behavior as a candidate for developing vascular grafts. Nanoroughness with improved hydrophilicity favored enhanced EC and SMC adhesion and proliferation, and these cells maintained their characteristic morphology compared to untreated PCL [57]. Both cell types maintained their mitochondrial function and its redox activity as demonstrated by the reduction of MTT reagent (Figure 16.8). The studies demonstrated that the ability of nanostructures to modulate protein adsorption could potentially alter the bioactivity of the biomaterial surfaces and elicit desired endothelialization for vascular tissue-engineering applications.

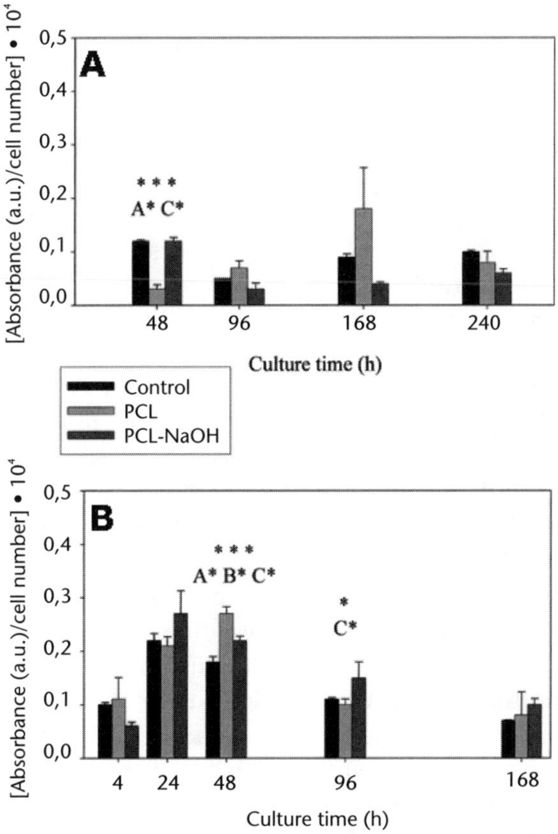

Figure 16.8 Mitochondrial redox activity as evaluated by reduction of MTT reagent in (a) EC and (b) SMC cultures on PCL films. ECs and SMCs at short culture times of twenty-four to forty-eight hours showed higher mitochondrial activity; however, for longer culture times, such observed differences were absent. (*Source:* [57], reprinted with permission from WILEY-VCH Verlag GmbH and Co. KGaA, Weinheim.)

16.2.4 Lithographic Techniques

Various lithographic techniques are used to create precise nanotopographic features containing grooves, ridges, and round nodes from nano to several microns in size on polymer surfaces. Poly(glycerol-sebacate), a biodegradable and biocompatible flexible polymer, was fabricated into surfaces having 500 nm rounded features using photolithography and plasma etching to study the effect of cell orientation and morphology [60]. Governed by topographical cues, bovine aortic endothelial cells were observed to be aligned and elongated on the surface, a phenomenon referred to as contact guidance. In another study, the plasma lithography technique was used to create square- and rectangular-shaped island patterns on a polymer surface to investigate the effect of geometrical influences on SMC behavior in the presence and absence of serum [61]. SMCs spanned more than one island and were confined to areas of the adhesive pattern. A well-ordered actin cytoskeleton was formed by spreading-restricted cells, and the pattern geometry influenced the shape of the nucleus [62].

Efforts were also made to create an endothelialized network with a vascular geometry on a biocompatible poly(dimethyl siloxane) (PDMS). High-resolution PDMS templates were produced by replica-molding from micromachined silicon wafers. ECs under dynamic culture conditions proliferated well in confined geometries, expressed markers for CD31 and von Willebrand factor (vWF), and attained confluency in four days [62].

16.2.5 Polymer Demixing

Polymer demixing can create various nanotopographic features, such as pits, islands, and ribbons of varying heights and depths on material surfaces. Such nanotopography created by polymer demixing of polystyrene and poly(4-bromostyrene) was evaluated for EC behavior [63]. On three islands of heights 13, 35, and 95 nm, cells showed better spread morphology than on flat surfaces of similar chemistry as seen from Figure 16.9(a, b). Among the different heights investigated, 13 nm islands elicited better cell response characterized by well-spread morphologies, proliferation, and well-defined cytoskeleton, as seen in Figure 16.9(c–f). The observed EC morphology and cell characteristics decreased with increasing island height. The cytoskeleton is modulated by signal transduction, and a well-defined cytoskeleton supports the observed EC spreading and proliferation on 13 nm islands. It is evident from this study that nanoislands can elicit specific morphological and cytoskeleton response to enhance endothelialization. Nanostructures created using various techniques, namely, polymer demixing, embossing, and photolithography, were evaluated for human endothelial cell (HGTFN) and mouse fibroblast (3T3) behavior [64]. Nanoislands of polystyrene and poly(4-bromostyrene) of varying heights, namely, 18, 40, and 100 nm, showed height-dependent attachment behavior of both the cell types [Figure 16.10(a)]. Nanoislands of 18 nm height showed better EC response (i.e., significantly higher EC attachment), and cells showed filopodia extension as seen from Figure 16.10(b) [64]. The observed filopodia extensions by cells indicate cellular movement on scaffold surfaces, suggesting increased interaction and scaffold compatibility.

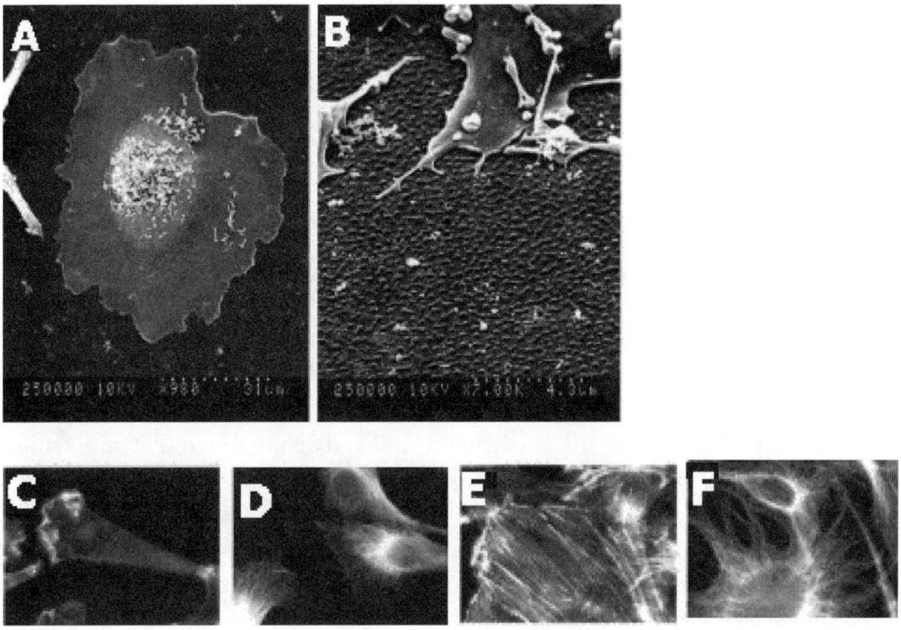

Figure 16.9 (a) SEM micrographs of EC on plane control surface with well-spread normal morphology, and (b) on 13 nm islands, cells show filopodia interactions with nanofeatures. Fluorescent images of (c) actin cytoskeleton after forty-eight hours on flat control, (d) tubulin cytoskeleton after seventy-two hours on flat control, (e) actin cytoskeleton after forty-eight hours on 13 nm islands, and (f) tubulin cytoskeleton after seventy-two hours on 13 nm islands (bar = 50 μm). A well-organized cytoskeleton was observed on cells aligning around the arcuate features. (*Source:* [63], reprinted with permission from Elsevier Science Ltd.)

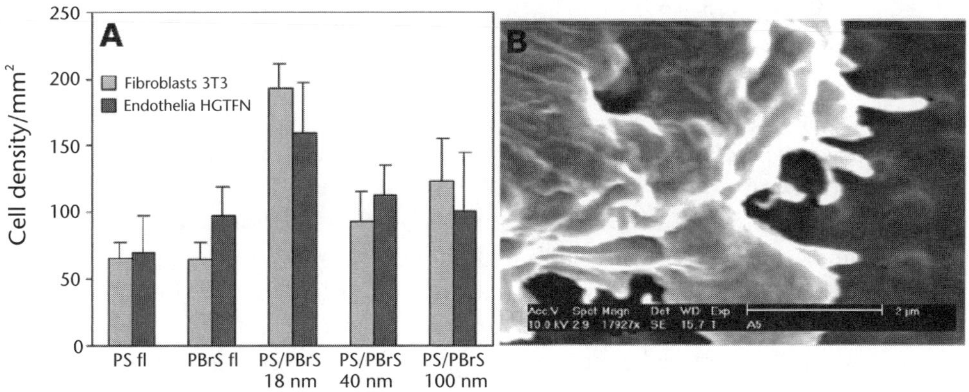

Figure 16.10 (a) Cell count expressed as density per millimeter squared on nanoislands of different heights after forty-eight hours of culture. Nanoisland of 18 nm height showed better EC response. (b) SEM micrograph showing well-spread ECs with filopodia adhered to the top of the islands. (*Source:* [64], reprinted with permission from Kluwer Academic Publishers.)

Poly(carbonate-urethanes) copolymers with different compositions resulted in various degrees of phase separation at the nanoscale based on time and film casting temperatures. A greater degree of nanophase separation resulted in better EC attachment and proliferation [65]. In another study, porous three-dimensional PLAGA scaffolds created by salt leaching were chemically etched to produce nanoscale roughness and evaluated for SMC response to these modified surfaces. Nanoscale surface roughness resulted in enhanced cell adhesion and growth with increased production of extracellular proteins such as collagen and elastin by the cells [66]. These studies indicate the feasibility of designing scaffolds containing specific geometries to bring about desirable cell-cell and cell-biomaterial interactions and, thereby, promoting accelerated tissue formation.

16.3 Functionalized Smooth Surfaces and Angiogenesis

Smooth biomaterial surfaces functionalized using bioactive molecules have been shown to create nanoscale surface roughness in addition to increased biofunctionality. Several chemical immobilization and physical-coating techniques of bioactive agents have been investigated to develop bioactive surfaces with nanoscale roughness. This part of the chapter summarizes various such efforts made to enhance the endothelialization of biomaterials [68–80].

Efforts were made to mimic the bonding domains of the ECM by modifying the scaffold surfaces with various cell-adhesive peptides and other bioactive agents [68–73] to enhance endothelialization. Polyurethane membranes having varying degrees of surface roughness (nanoscale) were created by grafting polyethylene glycol (PEG) chains of different lengths and cell-adhesive Gly-Arg-Gly-Asp (GRGD) peptides. The modified surfaces showed nanoscale roughness that varied as follows: PU < PU-PEG(2000)-GRGD < PU-PEG(mix) [68], where PU-PEG(mix) is obtained by grafting PEG having different molecular weights of 1,100, 2,000 and 5,000 with a molar ratio of 1:2:1 on PU surfaces. HUVEC cells showed increased adhesion and growth with increasing surface roughness that was significantly higher than with smooth PU surfaces. Among these surfaces, PU-PEG(mix)-GRGD registered the highest cell density, while PU-PEG(2000) registered the sparsest. Cell-adhesive peptide sequence GRGD combined with nanofeatures may be better suited to induce endothelialization.

In another study, RGD-peptide-sequence-functionalized poly(lactic acid-co-lysine) (PLAL) surfaces supported adhesion and proliferation of bovine aortic endothelial cells. It has been found that EC behavior can effectively be controlled based on the extent of functionalization. For instance, surface modification of PLAL with RGD increased effective cell coverage area over four hours from $77 \pm 2\ \mu m^2$ to $405 \pm 29\ \mu m^2$. The effective cell coverage area of the polymer surfaces were as follows: PLLA $116 \pm 11\ \mu m^2$, PLAL $87 \pm 4\ \mu m^2$, and RDG-grafted PLAL $105 \pm 4\ \mu m^2$ after four hours of culturing [69, 70]. It is evident from this study that nanoscale roughness created on the scaffold surface by surface modification essentially caused increased cell adhesion and coverage over the control. In another study, polymers composed of poly(vinyl amine) backbone with hexanal branches and varying ratios of cell-binding peptide (RGD) to carbohydrate (maltose) were tested for EC migration and proliferation behavior [71]. Surface peptide density influenced EC growth and migration. The highest peptide density resulted in increased EC proliferation, while EC migration decreased. Such modified functional surfaces with nanoscale roughness could favor endothelialization of tissue-engineering constructs [71]. In a similar study, PU surfaces modified with RGD sequence supported EC adhesion and proliferation superior to TCPS [72]. Sequential coating of PU small-diameter sponge with epoxide crosslinked thin layer of gelatin, followed by cellulose binding domain containing RGD sequence (CBD-RGD), improved the functions of HUVECs on the surface [73]. The surface modification also resulted in reduction of platelet activity toward the surface and helped to maintain EC differentiated morphology [73]. The RGD and gelatin functionalization of the surface induced nanoscale surface roughness, thereby enhancing the observed EC activity.

To overcome adverse tissue responses associated with vascular tissue-engineering implants, efforts were made to incorporate various bioactive agents to improve EC lining (endothelialization). Curcumin, an anti-inflammatory drug also known for its anti-SMC proliferative activity, was incorporated into PLLA thin sheets, creating nanoscale surface roughness. These PLLA matrices were further modified by adsorptive coating of adhesive proteins, including fibronectin, collagen I, vitronectin, laminin, and Matrigel, to improve EC adhesion and proliferation. Surfaces coated with these adhesive proteins having nanoscale surface roughness resulted in higher EC adhesion and proliferation [74]. Scaffolds with anti-SMC proliferative drug showed improved endothelialization that might potentially help to overcome many of the associated problems with implants, including thrombosis, inflammation, and restenosis. In another study, PLLA surface was subjected to modification using various techniques, such as aminolysis; collagen immobilization with glutaraldehyde; chondroitin sulfate and collagen layer-by-layer assembly; photoinduced grafting copolymerization of hydrophilic methacrylic acid; and further immobilization of collagen with 1-ethyl-3-(3-dimethylamino propyl) carbodiimide hydrochloride [75]. All these surface modifications potentially created nanoscale surface roughness with multifunctional bioactive surfaces based on the type of modification. HUVECs showed improved attachment, spreading morphology, and secretion of phenotypic markers vWF and 6-keto-PGF(1 alpha) on modified surfaces over the PLLA smooth surface. Among these modifications, collagen immobilization showed the most positive cell response than the other modifications, presumably due to the collagen bioactivity.

In another attempt, nanoscale surface roughness was created on aminated PLLA by the layer-by-layer polyelectrolyte deposition of chitosan and poly(styrene sulfonate, sodium salt) (PSS) [76]. PLLA fuctionalized with chitosan improved cytocompatibility toward ECs, and cells showed elongated morphology and maintained EC function characterized by vWF secretion. Three or five bilayers of PSS/chitosan with EC cultured on chitosan showed better attachment, proliferation, and EC activity compared to one bilayer of PSS/chitosan or the PLLA control [76].

Another surface-coating method was also explored in an attempt to enhance endothelial-cell adhesion and proliferation to design surfaces for vascular repair. Polyelectrolytes such as PSS, poly(allylamine hydrochloride) (PAH), PGA, and poly(D-lysine) were used as coating material alone or in combination, and HUVECs were seeded on these surfaces [77]. The polyelectrolyte multilayered films showed increased initial cell attachment over the polyelectrolyte monolayer, and cells maintained their normal morphology and phenotype on these surfaces. The polyelectrolyte multilayer composition PSS/PAH showed the highest attachment and EC proliferation [77]. Such polyelectrolyte coating can create potential surface roughness in the nanoscale and promote cell colonization. Nanoporous epoxy crosslinked gelatin coating on PU scaffolds fabricated by traditional salt-leaching methods with a polymer/salt ratio of 2:1 showed compliance close to natural vessels [78]. Nanoporous coating resulted in enhanced attachment and spreading of EC with minimum platelet activation required for any potential scaffold material for vascular grafts. Nanoporous structures and gelatin might have provided the required physicochemical properties needed for the vascular grafts. Poly(ethersulfone) (PES) hollow fibers and their surfaces coated with gelatin or fibronectin supported EC adhesion and proliferation and expressed various EC markers, including E-selectin and ICAM-1. Under the simulated conditions of angiogenesis, ECs monolayered on PES fibers migrated to form microvessel-like structures [79].

Plasma process has also been investigated to develop functionalized surfaces with nanofeatures. PET surfaces were treated by plasma process in the presence of various gas mixtures, and the functionalized surfaces were characterized for EC behavior [80]. Such surface modifications improved PET-EC compatibility characterized by enhanced adhesion, growth, and observed well-spread flattened morphology compared to unmodified PET [80]. The favorable responses toward modified surfaces can be attributed to the increase in nanoscale surface roughness, as well as to the increased adsorption of cell-adhesive proteins.

16.4 Conclusion

The acute shortage of donor organs and tissues available for transplantation raises the need to develop engineered organs and tissues. The success of tissue-engineering approachs to regenerate or grow the desired organs or tissues depends on the extent of vascularization of the cell-biomaterial construct. Traditionally, angiogenic growth factors such as VEGF, basic FGF, and a cocktail of several growth factors in combination with scaffolds are explored for this purpose. Efforts were also made to induce angiogenesis by preseeding scaffolds with endothelial cells and angiogenic factors. However, endothelialization has always been hampered since scaffold surfaces are not optimized for endothelial-cell attachment, or often cells detach from the scaffold surface upon exposure to blood circulation. Recent studies have provided significant insights into the influence of topographic features in regulating cell behavior. Topographic features provide essential chemical and physical cues that cells can recognize and elicit desired cellular functions, including preferential adhesion, migration, proliferation rate, and expression of specific cell phenotypes to bring the desired effect. Endothelial cells showed encouraging cellular behavior to the nanotopographic features, and there is great need to combine nanotopographical features with cell-bonding proteins or angiogenic factors to induce angiogenesis. Studies thus far reported are single-cell type, and attempts need to be made to reconstruct specialized tissue using more than one cell type. Such efforts that utilize these multifunctional nanotopographic features along with bioactive factors and use of more than one cell type with EC may be an alternative to aid the development of engineered organs.

References

[1] Laurencin, C. T., Ambrosio, A. M., Borden, M. D., and Cooper, J. A., Jr., "Tissue engineering: Orthopedic applications," *Ann. Rev. Biomed. Eng.*, Vol. 1, 1999, pp. 19–46.

[2] Laurencin, C. T., Khan, Y., Kofron, M., El-Amin, S., Botchwey, E., Yu, X., and Cooper, J. A., Jr., "The ABJS Nicolas Andry Award: Tissue engineering of bone and ligament: A 15-year perspective," *Clin. Orthop. Relat. Res.*, Vol. 447, 2006, pp. 221–236.

[3] Gilbert, J. C., Takada, T., Stein, J. E., Langer, R., and Vacanti, J. P., "Cell Transplantation of Genetically Altered Cells on Biodegradable Polymer Scaffolds in Syngeneic Rats," *Transplantation*, Vol. 56, No. 2, 1993, pp. 423–427.

[4] Mauney, J. R., Jaquiery, C., Volloch, V., Heberer, M., Martin, I., and Kaplan, D. L., "In vitro and in vivo evaluation of differentially demineralized cancellous bone scaffolds combined with human bone marrow stromal cells for tissue engineering," *Biomaterials*, Vol. 26, No. 16, 2005, pp. 3173–3185.

[5] Folkman, J., "Tumor angiogenesis: Therapeutic implications," *New Eng. J. Med.*, Vol. 285, No. 21, 1971, pp. 1182–1186.

[6] Folkman, J., "Tumor angiogenesis," *Advances Cancer Res.*, Vol. 19, 1974, pp. 331–358.

[7] Folkman, J., "The role of angiogenesis in tumor growth," *Semin. Cancer Biol.*, Vol. 3, No. 2, 1992, pp. 65–71.

[8] Folkman, J., "Role of angiogenesis in tumor growth and metastasis," *Semin. Oncol.*, Vol. 26, No. 6, 2002, pp. 15–18.

[9] Hanahan, D., and Folkman, J., "Patterns and emerging mechanisms of the angiogenic switch during tumorigenesis," *Cell*, Vol. 86, No. 3, 1996, pp. 353–364.

[10] Risau, W., "Mechanisms of angiogenesis," *Nature*, Vol. 386, No. 6626, 1997, pp. 671–674.

[11] Girling, J. E., and Rogers, P. A., "Recent advances in endometrial angiogenesis research," *Angiogenesis*, Vol. 8, No. 2, 2005, pp. 89–99.

[12] Li, W. W., Talcott, K. E., Zhai, A. W., Kruger, E. A., and Li, V. W., "The role of therapeutic angiogenesis in tissue repair and regeneration," *Advances in Skin and Wound Care*, Vol. 18, No. 9, 2005, pp. 491–500.

[13] Frank, S., Hubner, G., Breier, G., Longaker, M. T., Greenhalgh, D. G., and Werner, S., "Regulation of vascular endothelial growth factor expression in cultured keratinocytes: Implications for normal and impaired wound healing," *J. Biolog. Chem.*, Vol. 270, No. 21, 1995, pp. 12607–12613.

[14] Rivard, A., and Isner, J. M., "Angiogenesis and vasculogenesis in treatment of cardiovascular disease," *Mol. Med.*, Vol. 4, No. 7, 1998, pp. 429–440.

[15] Reynolds, L. P., Killilea, S. D., and Redmer, D. A., "Angiogenesis in the female reproductive system," *Fed. Am. Soc. Exp. Biol.*, Vol. 6, No. 3, 1992, pp. 886–892.

[16] Reynolds, L. P., and Redmera, D. A., "Angiogenesis in the placenta," *Biol. Reprod.*, Vol. 64, 2001, pp. 1033–1040.

[17] Colville-Nash, P. R., and Scott, D. L., "Angiogenesis and rheumatoid arthritis: Pathogenic and therapeutic implications," *Annals Rheum. Dis.*, Vol. 51, No. 7, 1992, pp. 919–925.

[18] Funatsu, H., Yamashita, H., Noma, H., Shimizu, E., Yamashita, T., and Hori, S., "Stimulation and inhibition of angiogenesis in diabetic retinopathy," *Jap. J. Opthalmol.*, Vol. 45, No. 6, 2001, pp. 577–584.

[19] Laschke, M. W., Harder, Y., Amon, M., Martin, I., Farhadi, J., Ring, A., Torio-Padron, N., Schramm, R., Ruecker, M., Junker, D., Haeufel, J. M., Carvalho, C., Heberer, M., Germann, G., Vollmar, B., and Menger, M. D., "Angiogenesis in tissue engineering: Breathing life into constructed tissue substitutes," *Tissue Eng.*, Vol. 12, No. 8, 2006, pp. 2093–2104.

[20] Ennett, A. B., and Mooney, D. J., "Tissue engineering strategies for in vivo neovascularisation," *Expert Opin. Biol. Ther.*, Vol. 2, No. 8, 2002, pp. 805–818.

[21] Sahota, P. S., Burn, J. L., Brown, N. J., and MacNeil, S., "Approaches to improve angiogenesis in tissue-engineered skin," *Wound Rep. Regen.*, Vol. 12, No. 6, 2004, pp. 635–642.

[22] Dalby, M. J., McCloy, D., Robertson, M., Wilkinson, C. D., and Oreffo, R. O., "Osteoprogenitor response to defined topographies with nanoscale depths," *Biomaterials*, Vol. 27, No. 8, 2006, pp. 1306–1315.

[23] Dalby, M. J., Riehle, M. O., Johnstone, H. J., Affrossman, S., and Curtis, A. S., "Polymer-demixed nanotopography: Control of fibroblast spreading and proliferation," *Tissue Eng.*, Vol. 8, No. 6, 2002, pp. 1099–1108.

[24] Andersson, A. S., Backhed, F., von Euler, A., Richter-Dahlfors, A., Sutherland, D., and Kasemo, B., "Nanoscale features influence epithelial cell morphology and cytokine production," *Biomaterials*, Vol. 24, No. 20, 2003, pp. 3427–3436.

[25] Dalby, M. J., Riehle, M. O., Sutherland, D. S., Agheli, H., and Curtis, A. S., "Morphological and microarray analysis of human fibroblasts cultured on nanocolumns produced by colloidal lithography," *Eur. Cells Mater.*, Vol. 13, No. 9, 2005, pp. 1–8.

[26] Chen, C. S., Mrksich, M., Huang, S., Whitesides, G. M., and Ingber, D. E., "Geometric control of cell life and death," *Science*, Vol. 276, No. 5317, 1997, pp. 1425–1428.

[27] Berry, C. C., Dalby, M. J., Oreffo, R. O., McCloy, D., and Affrosman, S., "The interaction of human bone marrow cells with nanotopographical features in three-dimensional constructs," *J. Biomed. Mater. Res. A.*, Vol. 79, No. 2, 2006, pp. 431–439.

[28] Dalby, M. J., Yarwood, S. J., Johnstone, H. J., Affrossman, S., and Riehle, M. O., "Fibroblast signaling events in response to nanotopography: A gene array study," *IEEE Trans. Nanobiosci.*, Vol. 1, No. 1, 2002, pp. 12–17.

[29] Dalby, M. J., Yarwood, S. J., Riehle, M. O., Johnstone, H. J., Affrossman, S., and Curtis, A. S., "Increasing fibroblast response to materials using nanotopography: Morphological and genetic measurements of cell response to 13-nm-high polymer demixed islands," *Exp. Cell Res.*, Vol. 276, No. 1, 2002, pp. 1–9.

[30] Price, R. L., Haberstroh, K. M., and Webster, T. J., "Enhanced functions of osteoblasts on nanostructured surfaces of carbon and alumina," *Med. Biolog. Eng. Comput.*, Vol. 41, No. 3, 2003, pp. 372–375.

[31] Webster, T. J., Ergun, C., Doremus, R. H., Siegel, R. W., and Bizios, R., "Enhanced osteoclast-like cell functions on nanophase ceramics," *Biomaterials*, Vol. 22, No. 11, 2001, pp. 1327–1333.

[32] Andersson, A. S., Backhed, F., von Euler, A., Richter-Dahlfors, A., Sutherland, D., and Kasemo, B., "Nanoscale features influence epithelial cell morphology and cytokine production," *Biomaterials*, Vol. 24, No. 20, 2003, pp. 3427–3436.

[33] Abrams, G. A., Murphy, C. J., Wang, Z. Y., Nealey, P. F., and Bjorling, D. E., "Ultrastructural basement membrane topography of the bladder epithelium," *Urol. Res.*, Vol. 31, No. 5, 2003, pp. 341–346.

[34] Abrams, G. A., Goodman, S. L., Nealey, P. F., Franco, M., and Murphy, C. J., "Nanoscale topography of the basement membrane underlying the corneal epithelium of the rhesus macaque," *Cell Tissue Res.*, Vol. 299, No. 1, 2000, pp. 39–46.

[35] Stoitzner, P., Pfaller, K., Stössel, H., and Romani, N., "A close-up view of migrating langerhans cells in the skin," *J. Invest. Dermatol.*, Vol. 118, No. 1, 2002, pp. 117–125.

[36] He, W., Yong, T., Ma, Z. W., Inai, R., Teo, W. E., and Ramakrishna, S., "Biodegradable polymer nanofiber mesh to maintain functions of endothelial cells," *Tissue Eng.*, Vol. 12, No. 9, 2006, pp. 2457–2466.

[37] Stitzel, J., Liu, J., Lee, S. J., Komura, M., Berry, J., Soker, S., Lim, G., Van Dyke, M., Czerw, R., Yoo, J. J., and Atala, A., "Controlled fabrication of a biological vascular substitute," *Biomaterials*, Vol. 27, No. 7, 2006, pp. 1088–1094.

[38] Gallagher, J. O., McGhee, K. F., Wilkinson, C. D. W., and Riehle, M. O., "Interaction of animal cells with ordered nanotopography," *IEEE Trans. Nanobiosci.*, Vol. 1, No. 1, 2002, pp. 24–28.

[39] Xu, C., Yang, F., Wang, S., and Ramakrishna, S., "In vitro study of human vascular endothelial cell function on materials with various surface roughness," *J. Biomed. Mater. Res. A.*, Vol. 17, No. 1, 2004, pp. 154–161.

[40] He, W., Yong, T., Teo, W. E., Ma, Z., and Ramakrishna, S., "Fabrication and endothelialization of collagen-blended biodegradable polymer nanofibers: Potential vascular graft for blood vessel tissue engineering," *Tissue Eng.*, Vol. 11, Nos. 9–10, 2005, pp. 1574–1588.

[41] He, W., Ma, Z., Yong, T., Teo, W. E., and Ramakrishna, S., "Fabrication of collagen-coated biodegradable polymer nanofiber mesh and its potential for endothelial cells growth," *Biomaterials*, Vol. 26, No. 36, 2005, pp. 7606–7615.

[42] Ma, Z., He, W., Yong, T., and Ramakrishna, S., "Grafting of gelatin on electrospun poly(caprolactone) nanofibers to improve endothelial cell spreading and proliferation and to control cell orientation," *Tissue Eng.*, Vol. 11, Nos. 7–8, 2005, pp. 1149–1158.

[43] Kwon, I. K., Kidoaki, S., and Matsuda, T., "Electrospun nano- to microfiber fabrics made of biodegradable copolyesters: Structural characteristics, mechanical properties and cell adhesion potential," *Biomaterials*, Vol. 26, No. 18, 2005, pp. 3929–3939.

[44] Wu, X., Rabkin-Aikawa, E., Guleserian, K. J., Perry, T. E., Masuda, Y., Sutherland, F. W., Schoen, F. J., Mayer, J. E., Jr, and Bischoff, J., "Tissue-engineered microvessels on three-dimensional biodegradable scaffolds using human endothelial progenitor cells," *Am. J. Physiol.: Heart Circ. Physiol.*, Vol. 287, No. 2, 2004, pp. H480–H487.

[45] Mo, X. M., Xu, C. Y., Kotaki, M., and Ramakrishna, S., "Electrospun P(LLA-CL) nanofiber: A biomimetic extracellular matrix for smooth muscle cell and endothelial cell proliferation" *Biomaterials*, Vol. 25, No. 10, 2004, pp. 1883–1890.

[46] Xu, C., Inai, R., Kotaki, and M., Ramakrishna, S., "Electrospun nanofiber fabrication as synthetic extracellular matrix and its potential for vascular tissue engineering," *Tissue Eng.*, Vol. 10, Nos. 7–8, 2004, pp. 1160–1168.

[47] Williamson, M. R., Black, R., and Kielty, C., "PCL-PU composite vascular scaffold production for vascular tissue engineering: Attachment, proliferation and bioactivity of human vascular endothelial cells," *Biomaterials*, Vol. 27, No. 19, 2006, pp. 3608–3616.

[48] Stankus, J. J., Guan, J., and Wagner, W. R., "Fabrication of biodegradable elastomeric scaffolds with sub-micron morphologies," *J. Biomed. Mater. Res. A*, Vol. 70, No. 4, 2004, pp. 603–614.

[49] Ma, Z., Kotaki, M., Yong, T., He, W., and Ramakrishna, S., "Surface engineering of electrospun polyethylene terephthalate (PET) nanofibers towards development of a new material for blood vessel engineering," *Biomaterials*, Vol. 26, No. 15, 2005, pp. 2527–2536.

[50] Ma, Z., Kotaki, M., Inai, R., Ramakrishna, S., "Potential of nanofiber matrix as tissue-engineering scaffolds," *Tissue Eng.*, Vol. 11, Nos. 1–2, 2005, pp. 101–109.

[51] Carampin, P., Conconi, M. T., Lora, S., Menti, A. M., Baiguera, S., Bellini, S., Grandi, C., and Parnigotto, P. P., "Electrospun polyphosphazene nanofibers for in vitro rat endothelial cells proliferation," *J. Biomed. Mater. Res. A*, Vol. 80, No. 3, 2007, pp. 661–668.

[52] Miller, D. C., Thapa, A., Haberstroh, K. M., and Webster, T. J., "Endothelial and vascular smooth muscle cell function on poly(lactic-co-glycolic acid) with nano-structured surface features," *Biomaterials*, Vol. 25, No. 1, 2004, pp. 53–61.

[53] Miller, D. C., Thapa, A., Haberstroh, K. M., and Webster, T. J., "Enhanced functions of vascular and bladder cells on poly-lactic-co-glycolic acid polymers with nanostructured surfaces," *IEEE Trans. Nanobiosci.*, Vol. 1, No. 2, 2002, pp. 61–66.

[54] Thapa, A., Miller, D. C., Webster, T. J., and Haberstroh, K. M., "Nano-structured polymers enhance bladder smooth muscle cell function," *Biomaterials*, Vol. 24, No. 17, 2003, pp. 2915–2926.

[55] Miller, D. C., Haberstroh, K. M., and Webster, T. J., "Mechanism(s) of increased vascular cell adhesion on nanostructured poly(lactic-co-glycolic acid) films," *J. Biomed. Mater. Res. A*, Vol. 73, No. 4, 2005, pp. 476–484.

[56] Thapa, A., Webster, T. J., and Haberstroh, K. M., "Polymers with nano-dimensional surface features enhance bladder smooth muscle cell adhesion," *J. Biomed. Mater. Res. A*, Vol. 67, No. 4, 2003, pp. 1374–1383.

[57] Serrano, M. C., Portoles, M. T., Vallet-Regi, M., Izquierdo, I., Galletti, L., Comas, J. V., and Pagani, R., "Vascular endothelial and smooth muscle cell culture on NaOH-treated poly(epsilon-caprolactone) films: A preliminary study for vascular graft development," *Macromol. Biosci.*, Vol. 5, No. 5, 2005, pp. 415–423.

[58] Genove, E., Shen, C., Zhang, S., and Semino, C. E., "The effect of functionalized self-assembling peptide scaffolds on human aortic endothelial cell function," *Biomaterials*, Vol. 26, No. 16, 2005, pp. 3341–3351.

[59] Rajangam, K., Behanna, H. A., Hui, M. J., Han, X., Hulvat, J. F., Lomasney, J. W., and Stupp, S. I., "Heparin binding nanostructures to promote growth of blood vessels," *Nano Lett.*, Vol. 6, No. 9, 2006, pp. 2086–2090.

[60] Bettinger, C. J., Orrick, B., Misra, A., Langer, R., and Borenstein, J. T., "Microfabrication of poly(glycerol-sebacate) for contact guidance applications," *Biomaterials*, Vol. 27, No. 12, 2006, pp. 2558–2565.

[61] Goessl, A., Bowen-Pope, D. F., and Hoffman, A. S., "Control of shape and size of vascular smooth muscle cells in vitro by plasma lithography," *J. Biomed. Mater. Res.*, Vol. 57, No. 1, 2001, pp. 15–24.

[62] Shin, M., Matsuda, K., Ishii, O., Terai, H., Kaazempur-Mofrad, M., Borenstein, J., Detmar, M., and Vacanti, J. P., "Endothelialized networks with a vascular geometry in microfabricated poly(dimethyl siloxane)," *Biomed. Microdevices*, Vol. 6, No. 4, pp. 269–278.

[63] Dalby, M. J., Riehle, M. O., Johnstone, H., Affrossman, S., and Curtis, A. S., "In vitro reaction of endothelial cells to polymer demixed nanotopography," *Biomaterials*, Vol. 23, No. 14, 2002, pp. 2945–2954.

[64] Barbucci, R., Pasqui, D., Wirsen, A., Affrossman, S., Curtis, A., and Tetta, C., "Micro and nano-structured surfaces." *J. Mater. Sci.: Mater. Med.*, Vol. 14, No. 8, 2003, pp. 721–725.

[65] Hsu, S. H., and Kao, Y. C., "Cell attachment and proliferation on poly(carbonate urethanes) with various degrees of nanophase separation," *Macromol. Biosci.*, Vol. 4, No. 9, 2004, pp. 891–900.

[66] Pattison, M. A., Wurster, S., Webster, T. J., and Haberstroh, K. M., "Three-dimensional, nano-structured plga scaffolds for bladder tissue replacement applications," *Biomaterials*, Vol. 26, No. 15, 2005, pp. 2491–2500.

[67] Isahara, K., and Yamamoto, M., "The interaction of vascular endothelial cells and dorsal root ganglion neurites is mediated by vitronectin and heparan sulfate proteoglycans," *Brain Res.: Dev. Brain Res.*, Vol. 84, No. 4, 1995, pp. 164–178.

[68] Chung, T. W., Liu, D. Z., Wang, S. Y., Wang, S. S., "Enhancement of the growth of human endothelial cells by surface roughness at nanometer scale," *Biomaterials*, Vol. 24, No. 25, 2003, pp. 4655–4661.

[69] Quirk, R. A., Chan, W. C., Davies, M. C., Tendler, S. J., and Shakesheff, K. M., "Poly(L-lysine)-GRGDS as a biomimetic surface modifier for poly(lactic acid)," *Biomaterials*, Vol. 22, No. 8, 2001, pp. 865–872.

[70] Cook, A. D., Hrkach, J. S., Gao, N. N., Johnson, I. M., Pajvani, U. B., Cannizzaro, S. M., and Langer, R., "Characterization and development of RGD-peptide-modified poly(lactic acid-co-lysine) as an interactive, resorbable biomaterial," *J. Biomed. Mater. Res.*, Vol. 35, No. 4, 1997, pp. 513–523.

[71] Sagnella, S. M., Kligman, F., Anderson, E. H., King, J. E., Murugesan, G., Marchant, R. E., and Kottke-Marchant, K., "Human microvascular endothelial cell growth and migration on biomimetic surfactant polymers," *Biomaterials*, Vol. 25, Nos. 7–8, 2004, pp. 1249–1259.

[72] Wang, D. A., Feng, L. X., Ji, J., Sun, Y. H., Zheng, X. X., and Elisseeff, J. H., "Novel human endothelial cell-engineered polyurethane biomaterials for cardiovascular biomedical applications," *J. Biomed. Mater. Res. A*, Vol. 65, No. 4, 2003, pp. 498–510.

[73] Hsu, S. H., Sun, S. H., and Chen, D. C., "Improved retention of endothelial cells seeded on polyurethane small-diameter vascular grafts modified by a recombinant RGD-containing protein," *Artif. Org.*, Vol. 27, No. 12, 2003, pp. 1068–1078.

[74] Patel, H. J., Su, S. H., Patterson, C., and Nguyen, K. T., "A combined strategy to reduce restenosis for vascular tissue engineering applications," *Biotech. Prog.*, Vol. 22, No. 1, 2006, pp. 38–44.

[75] Zhu, Y., Gao, C., Liu, Y., and Shen, J., "Endothelial cell functions in vitro cultured on poly(L-lactic acid) membranes modified with different methods," *J. Biomed. Mater. Res. A*, Vol. 69, No. 3, 2004, pp. 436–443.

[76] Zhu, Y., Gao, C., He, T., Liu, X., and Shen, J., "Layer-by-layer assembly to modify poly(l-lactic acid) surface toward improving its cytocompatibility to human endothelial cells," *Biomacromolecules*, Vol. 4, No. 2, 2003, pp. 446–452.

[77] Boura, C., Menu, P., Payan, E., Picart, C., Voegel, J. C., Muller, S., and Stoltz, J. F., "Endothelial cells grown on thin polyelectrolyte mutlilayered films: An evaluation of a new versatile surface modification," *Biomaterials*, Vol. 24, No. 20, 2003, pp. 3521–3530.

[78] Hsu, S., Tseng, H., and Wu, M., "Comparative in vitro evaluation of two different preparations of small diameter polyurethane vascular grafts," *Artific. Org.*, Vol. 24, No. 2, 2000, pp. 119–128.

[79] Unger, R. E., Peters, K., Huang, Q., Funk, A., Paul, D., and Kirkpatrick, C. J., "Vascularization and gene regulation of human endothelial cells growing on porous polyethersulfone (PES) hollow fiber membranes," *Biomaterials*, Vol. 26, No. 17, 2005, pp. 3461–3469.

[80] Ramires, P. A., Mirenghi, L., Romano, A. R., Palumbo, F., and Nicolardi, G., "Plasma-treated PET surfaces improve the biocompatibility of human endothelial cells," *J. Biomed. Mater. Res. A*, Vol. 51, No. 3, 2000, pp. 535–539.

[81] Li, W. J., Laurencin, C. T., Caterson, E. J., Tuan, R. S., and Ko, F. K., "Electrospun nanofibrous structure: A novel scaffold for tissue engineering," *J. Biomed. Mater. Res.*, Vol. 60, No. 4, 2002, pp. 613–621.

[82] Kumbar, S. G., Nair, L. S., Bhattacharyya, S., and Laurencin, C. T., "Polymeric nanofibers as novel carriers for the delivery of therapeutic molecules," *J. Nanosci. Nanotechnol.*, Vol. 6, Nos. 9–10, 2006 pp. 2591–2607.

CHAPTER 17

Micropatterning Approaches for Cardiac Biology

Nicholas A. Geisse, Adam W. Feinberg, Po-ling Kuo, Sean P. Sheehy, Mark-Anthony Bray, and Kevin Kit Parker

17.1 Introduction

Our understanding of the relationship between cell boundary conditions and cell function has been greatly accelerated with the implementation of soft lithography in order to control the cellular microenvironment. Using these techniques, it has been shown that imposed boundary conditions can influence biological processes as diverse as gene-expression profiles and cell fates [1], subcellular localization of chemical signaling pathways and organization of cytoskeletal architecture [2], and the direction and magnitude of cell-tractional forces [3]. Beyond insights at the single-cell level, soft lithography has been applied to the study of cell-cell interactions and to the generation of engineered tissue constructs [4, 5]. Indeed, engineering the cell microenvironment provides investigators unprecedented abilities to investigate the relationship between cell structure and function. These structure-function relationships are no less important in the myocardium, where the relationship between cell shape and boundary conditions is crucial for the proper functioning of both the individual cardiac myocyte and the tissue syncytium. For example, mammalian ventricular cardiac myocytes in vivo are generally cylindrical, with approximate dimensions of 100 to 120 μm in length, 15 to 20 μm in width, and 10 µm in thickness. Cardiac myogenesis occurs by developing electrical and mechanical connections between adjacent myocytes while maintaining uniaxial alignment with respect to one another. This uniformity of structure is supported by experimental data and computer-modeling studies, which hypothesize that the shape and connectivity of individual myocytes greatly influences electrical-impulse propagation throughout the myocardium [6–8]. Although cellular dimensions vary across species and individuals, healthy myocytes generally have a length:width (or aspect) ratio of ~7:1. Various cardiomyopathies are accompanied by deviations from this ratio [9], and myocyte shape changes are concomitant with declining contractile performance. These maladaptive responses may occur due to a variety of stimuli [10], but at present the causal relationship between this structural change and dysfunction at the myocyte, tissue, and organ level has yet to be directly established.

Microscale surface-engineering techniques have laid the foundation for investigating the role that cellular boundary conditions play in potentiating various aspects of myocyte physiology and pathophysiology and have had a long evolution due to the difficulty of creating micro- and nanoscale features for cell attachment. Early studies focused on creating mechanically microtextured cell-culture substrates for patterned cell growth. One of the earliest techniques, developed by Lieberman et al., focused on creating long, cylindrical fibers of a cardiac myocyte preparation in order to perform electrical measurements, and these were created by culturing myocytes on manually cut trenches in soft agar culture substrates [11, 12]. Using this technique, Lieberman et al. were able to create strands of electrically conductive cardiac tissue that could be investigated with electrophysiology techniques. Other methods of manual alteration of the cell microenvironement include microabrasion (a method reproduced in Section 17.3.1 of this chapter), where pliant culture substrates such as polystyrene or polyvinylchloride are mechanically scratched, providing three-dimensional surface topography for directing cell growth and spatially orienting cell-cell communication [13].

Despite these methods for altering the cell-culture microenvironment to spatially direct and organize cardiac myocyte growth, it was not until the development of photolithography and soft lithography techniques [14] and their subsequent applications to biological samples [15] that the investigator was given the ability to design surfaces with micron to nanometer precision. Here, surfaces composed of a polymer photoresist are etched by exposure to ultraviolet light shone through a photomask. The design placed on the photomask is transferred to the substrate by either stabilizing or destabilizing the photoresist, depending on the type used. After processing, the surface provides a near perfect relief or reverse-relief of the photomask, which can be designed with micron-sized features with little loss of pattern fidelity. Adaptation of this technique utilized the precise spatial control afforded by photolithography, and implementation technology for controlling the cardiac myocyte microenvironment was pioneered by Kleber et al. [16–18], who created patterned surfaces in biocompatible photoresist in order to study the relationship between electrical conduction and tissue structure. With these techniques, it is possible to control the size of an engineered myocyte strand to the resolution of a cell width, while recapitulating the anisotropic cellular architecture found in the in vivo myocardium. This is important since it is thought that the structure and cytoskeletal architecture of the cardiac myocyte greatly influence, if not determine, myocyte contractility and intercellular electrical coupling [19–21]. A wide variety of stimuli can evoke physiological responses in the heart, and, in particular, it has been shown that mechanical forces play a central role in cardiac myocyte regulation and function [10]. However, the relationship between cardiac myocyte shape, cytoskeletal architecture, and physiology remains unclear due to the difficulty of culturing cardiac myocytes and controlling the cellular microenvironment in vitro.

In the following discussion, we describe several techniques adapted for use in our laboratory to study structure-function relationships in single cardiac myocytes and two-dimensional cardiac tissue constructs. We include a discussion of myocyte isolation and culture maintenance and the special considerations needed for cardiac myocytes. Central to our structure-function studies is the use of the microcontact printing (μCP) technique in order to control myocyte shape by controlling the availability and spatial distribution of extracellular matrix (ECM). The μCP technique allows for fabrication of shape-controlled myocytes and two-dimensional tissues, and it is shown that the boundary conditions determined by μCP influence intercellular organization. Coupling the μCP technique with fluorescence microscopy on deformable substrates and with novel image-processing techniques, the tractional forces exerted by the myocyte onto its substrate can be observed. We close this discussion with a look at the prospective techniques in cell and tissue engineering.

17.2 Isolation and Culture of Cardiac Myocytes

In vitro cell-culture models are a powerful tool for studying the activities and characteristics of cellular processes and how they respond to various controlled conditions. Isolated adult rat cardiac myocytes have mature myofibrillar networks that maintain their structure and function when enzymatically dissociated from the whole heart. As such, adult cardiac myocytes settle on to surfaces but do not respond to chemical or mechanical microenviromental cues either by forming focal contact adhesions with the substrate or by conforming to the shape of the substrate. This inability to form attachments with the substrate limits adult myocyte viability to approximately one day in vitro, after which myocyte viability decreases while the rate of dedifferentiation increases. In contrast, freshly isolated neonatal rat ventricular myocytes (NNRVMs) have a more malleable cytoskeletal structure, allowing them to conform to the chemical and mechanical characteristics of the cell-culture substrate. This flexibility, coupled with techniques to engineer the cell microenvironment, has resulted in the use of NNRVMs for the study of various myocardial structural and functional properties, such as action potential duration [22], impulse propagation [17], calcium-induced calcium release [23], signaling pathways, and gene expression [24]. Additionally, cultured NNRVMs have been used to develop models for the study of pathological conditions in the heart, like ischemia-reperfusion injury [25], hypertrophy [26], and arrythmogenesis [13].

Cultured NNRVMs are an excellent cell type for in vitro cardiac research because they retain a limited measure of developmental plasticity that their adult counterparts do not and because they can survive in culture for longer periods. Good sterile technique and swift completion of the culture process are important to ensure maximum yield and longevity of healthy, viable myocytes. The procedure that follows is briefly outlined in Figure 17.1. All procedures should be conducted within a biosafety cabinet using supplies and equipment that have been autoclave-sterilized to prevent contamination. Cardiac myocytes are postmitotic cells and have a limited lifetime in vitro dependent upon the conditions under which they are maintained, and this is further complicated by the temporal duration of attachment compared to other adherent cells, such as fibroblasts (days compared to hours, respectively). All procedures described in this text are in compliance with the Harvard University Care and Use Committee, and experiments presented here were performed within the appropriate approved guidelines.

17.2.1 Harvesting and Isolating NNRVMs

Prior to the myocyte isolation and culture procedure, the appropriate culture medium must be prepared in order to preserve the cells during isolation. An excel-

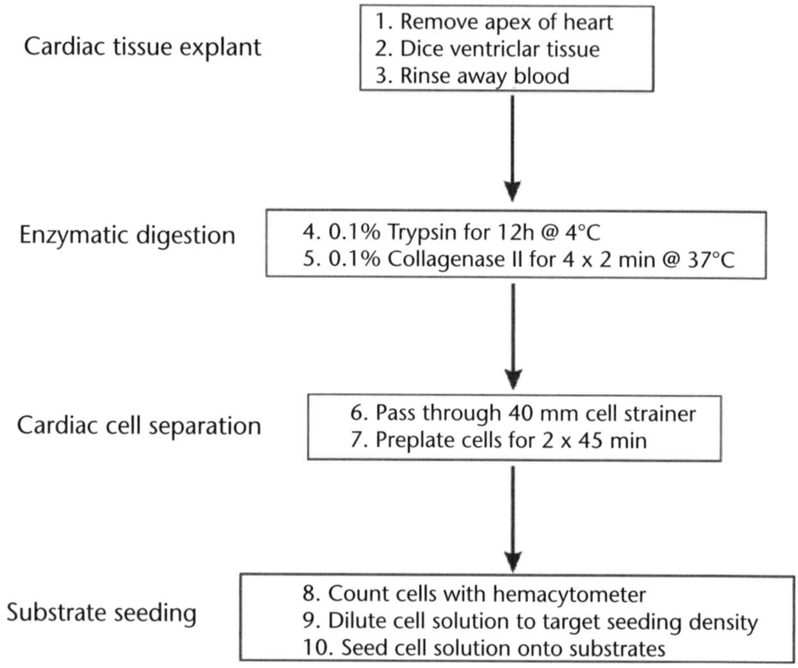

Figure 17.1 *Flowchart for isolation of NNRVMs.* Full isolation and culture of NNRVMs involves several steps performed over the course of approximately eighteen hours. The procedure involves manual tissue dissection, followed by enzymatic degradation to separate cells from the ECM and each other. Further steps are taken to remove nonmyocyte adherent cells. See text for details.

lent preservative solution is made by supplementing medium 199 with 10 mM HEPES buffer solution, 10mM MEM nonessential amino acids, 2 mM L-glutamine solution, 20 mM glucose, 1.5 μM vitamin B12, 50 units/mL penicillin, and 10 percent fetal bovine serum. Isolation of cells begins with extraction from the organism, which must be sterilized with ethanol to minimize the transfer of contaminants from the surface of its skin onto the surgical equipment. In general, the organism is sacrificed via decapitation, and the heart is removed with a mid-sternal ventral incision. Once the heart is extracted, ventricular tissue is excised and rinsed in Hank's Balanced Salt Solution (HBSS). Homogenization of the tissue is done manually with scissors or scalpels, and the tissue is then diced and kept submerged in HBSS. Subsequently, the tissue is digested in a solution of 0.1 percent trypsin in HBSS and incubated with agitation for approximately twelve hours at 4°C. After digestion, the tissue is serially dissociated into individual cells by multistep incubation in a 0.1 percent collagenase type II solution for two minutes each at 37°C.

Cells are isolated and resuspended by centrifugation and strained in order to remove large, undigested sections of tissue. Perhaps the greatest difficulty faced when making NNRVM preparations is the presence of contaminating cells, such as fibroblasts, vascular smooth muscle, and capillary epithelial cells. In order to minimize these contaminations, the cell solution can be preplated twice for forty-five minutes each time in 75 mm^2 tissue-culture flasks. Once the preplating steps are complete, the myocyte solution is ready to be diluted and plated onto a variety of microengineered cell-culture substrates.

17.3 Engineering the Cellular Microenvironment In Vitro

17.3.1 Microabrasion of the Cardiac Myocyte Culture Substrate

A simple method for controlling tissue anisotropy is via mechanical etching of the culture substrate, which provides physical cues for myocyte attachment and assembly [13]. Unlike methods that utilize large trenches either cut into a substrate [11, 12] or fabricated via photolithography [16–18], microabrasions that can guide myocyte orientation can be etched into the surface using commonly available materials. Polyvinyl chloride (PVC) transparent coverslips are abraded using fine, one hundred grit sandpaper. Atomic force microscopy shows that a one-pass abrasion creates individual trenches distributed across the surface that have a depth and peak of ~0.5 μm from the surface [Figure 17.2(a)]. Multipass abrasions generate trenches along the entire surface that are ~3.0 to 4.0 μm in height [Figure 17.2(b)]. Both substrates can be treated with a solution of fibronectin (FN) and used for cell plating. Cardiac myocytes that are seeded onto these substrates show an alignment with the

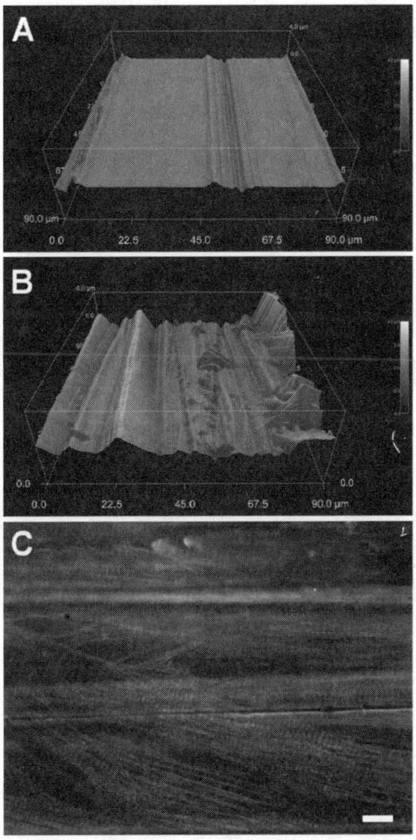

Figure 17.2 *Microabrasion of the cardiac myocyte culture substrate.* Atomic force micrographs of microabraded PVC coverslips show geometrical cues in the direction of the abrasion. (a) Single-pass microabrasion creates small, individual trenches across the surface that are approximately 0.5 μm deep. (b) Multiple-pass microabrasion results in a completely rough culture substrate with trenches in the direction of the abrasion. Height variation across this substrate is 3.0 to 4.0 μm. c) Cardiac myocytes cultured on these substrates and stained for actin filaments, sarcomeric α-actinin, and the nucleus show myofibrillar alignment along the direction of the microabrasions, visualized as the dark horizontal line. Scale bar = 10 μm.

microabraded lines, as shown in Figure 17.2(c). Microabraded trenches are easily identified in both the transmitted light (not shown) and the fluorescence images, as shown by the dark horizontal line in Figure 17.2(c). Myofibrillar staining of actin (green) and sarcomeric α-actinin (red) show that these structures align in parallel with and perpendicularly to the microabraded lines, respectively. A similar technique has been successfully used to generate constructs for the measurement of action potential propagation in two-dimensional ansisotropic tissues [13]. Although this technique is suitable for creating anisotropic tissues, microabrasion does not allow the investigator control over cell geometry and tissue anisotropy on the scale of single-cell widths.

Our experience has led us to the conclusion that μCP is generally superior both to the microabrasion technique and to other soft lithography techniques, such as microfluidics, for building ECM templates of myocyte shape and two-dimensional tissue architecture. In μCP, a stamp is used to "ink" a surface with a pattern of ECM proteins that the dissociated myocytes settle on and remodel in response to the geometric cues in the ECM and their neighboring cells. A description of the principles that must be kept in mind before designing a μCP pattern, as well as a discussion of the chemical properties of the elastomer, is useful prior to explanation of this technique.

17.3.2 Microcontact Printing and Soft Lithography

One of the primary advantages of microcontact printing is that it allows for cardiac myocytes to be grown on layers of a silicone polymer. In many cases, polydimethylsiloxane (PDMS) is the silicone of choice. The mechanical properties of PDMS ($E \sim 1$ MPa) allow for cardiac myocytes to be cultured in vitro for longer periods than on glass ($E \sim 70$ GPa) or tissue-culture-grade polystyrene ($E \sim 3$ GPa). Myocytes rebuild their myofibrillar networks upon being seeded onto surfaces, forming focal contact adhesions and adopting a cell shape dependent on the two-dimensional boundary conditions of the ECM. Myocyte contractions result in cyclic loading of the focal contact adhesions between the myocytes and the surface, generating large stresses between the cell and ECM, between the ECM and substrate, and within the substrate. The delamination is believed to be caused by enzymatic degradation of the ECM by matrix metalloproteinases produced by the myocytes and/or fibroblasts and likely accelerated by the high stress conditions. In contrast, the lower modulus PDMS surface absorbs some of the strain, delaying myocyte delamination until approximately nine days in vitro, thereby extending the experimental window.

One of the critical advantages of μCP is that it allows for patterned cells to be grown on traditional culture substrates, facilitating analysis by a variety of techniques, including high numerical aperture optical and fluorescence microscopy. Previous techniques utilized alkane-thiol-based μCP printing strategies to control cell shape [2]. These techniques are less than ideal for many live-cell-imaging methods due to the opacity of the culture substrate. By exploiting the control of the wettability of PDMS [27], μCP can be performed using either glass or PDMS as a cell-culture substrate, both of which are transparent and permit live-cell fluorescence imaging. For our studies, we perform μCP on glass microscope coverslips that are coated with a thin (~17 μm) layer of PDMS.

Standard soft lithography procedures are used to fabricate the PDMS stamps used for microcontact printing [14]. A graphical overview of the process from stamp production to cell seeding is presented in Figure 17.3(a). Silicon wafers (3″ diameter) are spin-coated with a 2 to 3 μm thick layer of SU-8 photoresist (MicroChem Corp., Newton, MA), followed by a prebake at 65°C for two minutes

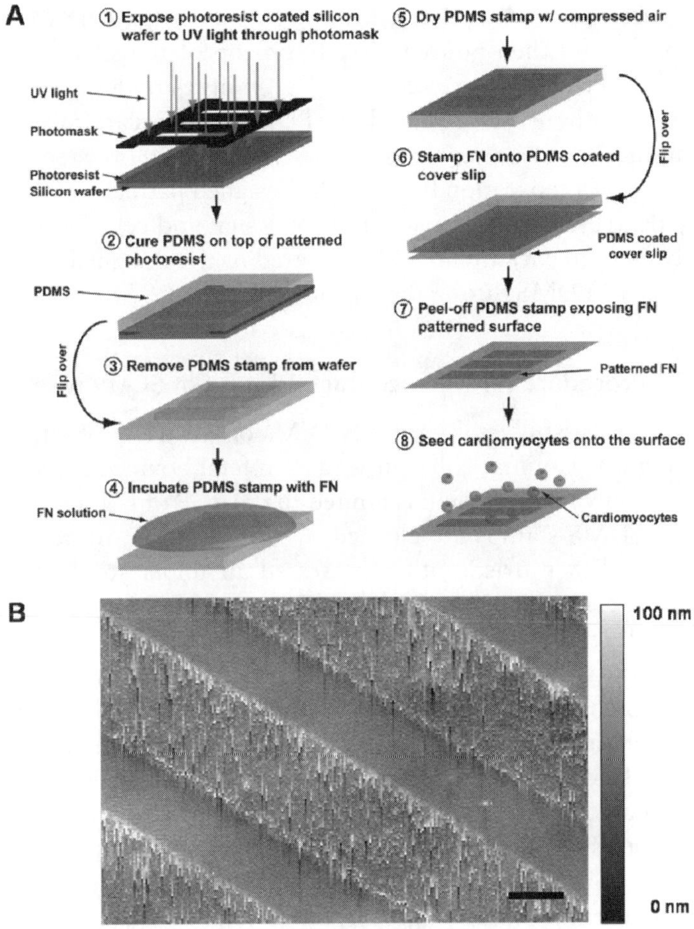

Figure 17.3 *Schematic of the stamp fabrication and microcontact printing process for ECM proteins on to PDMS coated coverslips*: (a) Overview of the process: (1) A silicon wafer is coated in photoresist and photolithographically patterned by exposing the photoresist with ultraviolet light passed through a photomask. (2) The photoresist is developed, thus removing the ultraviolet-exposed regions, and the remaining topography is copied by casting PDMS prepolymer on top. (3) The cured PDMS is peeled off of the silicon wafer, creating a stamp with microtopography. (4) The PDMS stamp is inked by incubation with an ECM protein solution for one hour. (5) The ECM protein solution is rinsed off in ddH_2O and then dried under compressed air. (6) The inked PDMS stamp is placed pattern-side down on the PDMS-coated coverslip, transferring the ECM protein to the substrate in defined patterns in a process adapted from [27]. (8) Once the ECM patterned coverslip is washed in buffer, myocytes are seeded. (b) Atomic force deflection micrograph of 20 μm–wide micropatterned lines of FN reveal a thin, ~3 nm layer of protein on the PDMS surface.

and soft-bake at 95°C for four minutes. Photoresist-coated silicon wafers are transferred to a mask aligner, where they are patterned via exposure to ultraviolet light through a photomask (Karl Suss MicroTec, Munich, Germany) using contact photolithography [Figure 17.3(a), step 1]. After exposure, wafers are developed in propylene glycol monomethyl ether acetate, then washed for one minute in isopropyl alcohol, resulting in a micropatterned layer of SU-8 photoresist on top of the silicon wafer. The surface of the wafer is then passivated by silanazation to prevent the PDMS from permanently binding to the wafer during stamp preparation. PDMS stamps for microcontact printing are prepared by curing PDMS on top of the micropatterned wafers [Figure 17.3(a), step 2]. Sylgard 184 PDMS (Dow Corning) is prepared and then poured directly on top of the silanized and micropatterned wafer. The PDMS is degassed for thirty minutes under vacuum and then cured at 65°C for four hours. Once cured, the PDMS layer is carefully peeled off of the wafer and cut into stamps of appropriate size [Figure 17.3(a), step 3]. Although there are four degrees of separation between the designed pattern and the final stamp, there is a high degree of fidelity between mask design and cell shape. Final cell size is only slightly (~6 percent) attenuated compared to the designed shape, mostly due to the shrinkage of PDMS during the curing process.

17.3.3 Procedure for Microcontact Printing of ECM Proteins

We have successfully cultured NNRVMs on a variety of patterned ECM proteins, including FN, laminin, collagen, and even fetal bovine serum. For purposes of clarity, the following discussion is limited to the μCP of FN. Before incubation with the protein, PDMS stamps are cleaned by sonication in 50 percent ethanol for thirty minutes and then dried with compressed air under sterile conditions. The PDMS stamps are "inked" with a droplet of 50 μg/ml of FN in DI water and incubated for one hour [Figure 17.3(a), step 4]. Excess protein is removed by washing twice in DI water, and stamps are then dried using filtered compressed air [Figure 17.3(a), step 5]. PDMS-coated coverslips (prepared as described above) are ultraviolet-ozone treated for eight minutes to sterilize the surface immediately prior to microcontact printing. Under sterile conditions, the stamp is brought into contact with the PDMS-coated coverslip for two minutes and then removed prior to cell seeding [Figure 17.3(a), steps 6–8]. Proteins are transferred to the surface of the PDMS coated coverslips, forming an approximately 3 nm–thick layer of the protein in the pattern designated by the stamp [Figure 17.3(b)]. The PDMS-coated coverslip is now patterned with protein and is further functionalized as per the specific experimental protocol. To restrict cell growth to the protein-patterned area, the PDMS-coated coverslip is incubated in a blocking solution of 1 percent (w/v) Pluronics F-127 (BASF Corp., Mount Olive, NJ) in DI water solution for fifteen minutes and then washed three times with phosphate buffered saline (PBS). The stamped, PDMS-coated coverslips can be stored covered in PBS for up to forty-eight hours prior to seeding with cardiac myocytes. The seeding density for myocytes on individual islands with liberal spacing in between the islands on a 25 mm coverslip may be on the order of 10^5 myocytes, whereas for a two-dimensional anisotropic tissue, 10^6 myocytes may be required. Once prepared, myocytes can be plated at an appropriate density determined by experimental needs.

Prior to μCP, studies requiring contiguous cardiac tissue assembled from NNRVMs were generally isotropic [Figure 17.4(a–c), left panels] cultured on an isotropic pattern of ECM. Now, with a μCP template, a two-dimensional anisotropic tissue that more accurately depicts the laminar structure of the ventricles is possible [Figure 17.4(a–c), right panels]. Fabrication of anisotropic two-dimensional tissues is possible by following μCP of FN lines with a step designed to provide a background layer of FN. This background layer allows myocytes to adhere between the patterned lines. After the PDMS-coated coverslip is stamped, the coverslip is coated with a layer of low-density FN (2.5 μg/ml of FN in DI water) for fifteen minutes and rinsed three times with PBS. The Pluronics blocking step is omitted for this method of substrate preparation. The printed lines provide geometric cues necessary to guide the orientation of the myocytes, while the background FN ensures that myocytes will align on both sides of the stamped line. When the myocytes are seeded, they will remodel their shape and spontaneously align with respect to the FN lines and their neighboring myocytes [Figure 17.4(a), right panel]. Upon contact with the cell membranes [Figure 17.4(b), right panel],

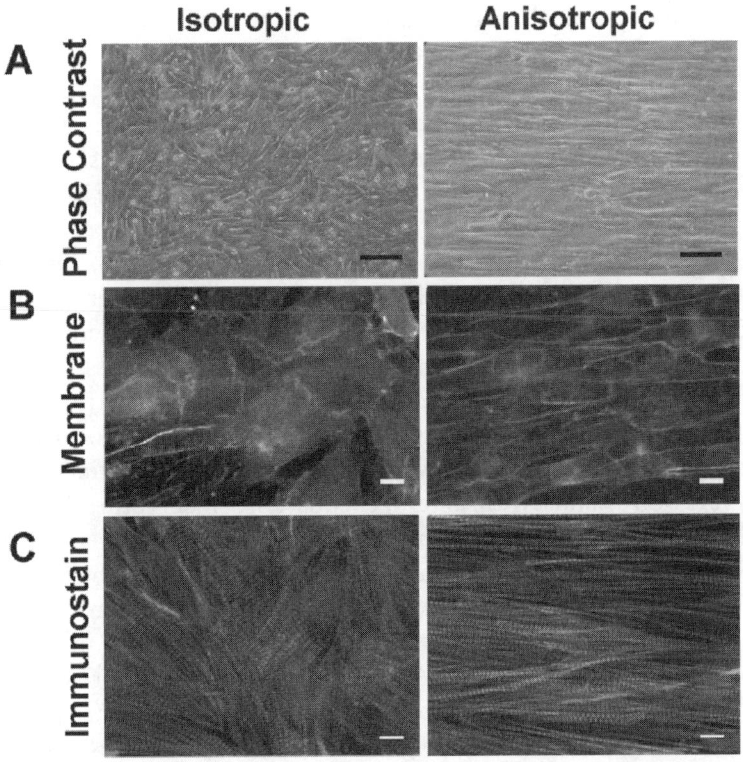

Figure 17.4 *Myofibrillar architecture of two-dimensional tissue constructs*: (a, left) NNRVMs cultured according to the above protocol on uniform layers of FN exhibit isotropic cellular orientation when compared to (a, right) NNRVMs cultured on lines of FN that result in an anisotropic orientation, as shown by phase-contrast microscopy. (b) Staining the myocytes with a fluorescent membrane dye (di-8-ANNEPS, Molecular Probes) shows that the myocyte edges also respond to geometrical cues in the FN. (c) Fixed and stained isotropic (left) and anisotropic (right) reveal that myofibrillar architecture is also oriented based on these cues. Immunostain labels for actin filaments, sarcomeric α-actinin, and the nucleus were used to visualize myofibrils and the nuclei. Scale bars for (a) are 100 μm; scale bars for (b) and (c) are 10 μm.

myocytes will form gap-junction connections with neighboring myocytes. As this electrical continuity is achieved, myocytes begin to beat spontaneously, inducing sarcomeregenesis and the reorganizing of existing sarcomeres. When this process is complete, the sarcomere Z-lines, as indicated by sarcomeric α-actinin immunostaining, will be observed to register both intra- and intercellularly with an orientation perpendicular to the longitudinal axis of the myocytes and the tissue that they form [Figure 17.4(c), right panel]. On the rigid PDMS coverslip, their contraction will be isometric but can be detected by the displacement of the myocyte nucleus.

Previous cardiac biology on cultured myocytes relied on either freshly harvested myocytes or pleomorphic myocytes [Figure 17.5(a)]. With the implementation of μCP for engineering the cardiac myocyte microenvironment, a variety of myocyte shapes are now possible [Figure 17.5(b–f)]. When patterning single myocytes, it is important that the mask design take into account the requirement of large numbers of cardiac myocytes to establish the paracrine signaling networks required for myocyte survival in vitro. This can be accomplished either by seeding a high density of myocyte islands per unit area or by the inclusion of a "conditioning layer" in the design, as noted by Rohr, Scholly, and Kleber [16]. This layer is a region of the mask that, when stamped, leads to an area of the coverslip surface being occupied by a small, confluent region of isotropic myocytes.

When the myocytes make contact with the FN islands, they will spread slowly in comparison to other cell types. This is because sarcomeres will need to disassemble at the leading edge where the myocyte is reaching for the island border. Unlike other cell types, myocytes will not extend lamellipodia from the island periphery. As mentioned previously, myocytes may require coaxing by the addition of epinephrine to induce beating and hypertrophy in order to achieve the island shape. Whereas myocyte shape is well controlled in the plane of the coverslip, in the vertical direction the nucleus will appear as a prominent bump, generally in the center of the myocyte [Figure 17.5(b)], as observed in other cell types [2]. Myocytes will then reorganize their cytoskeleton with respect to the geometric cues in the FN island. After the actin network is stabilized and secured by focal adhesion complexes, myofibrillogenesis will follow the actin template for distinct and repeatable architectures on islands with heterogeneous border curvature [Figure 17.5(c, d)]. On islands with homogeneous curvature, myocytes, like other cell types, will lack the spatial cues required to stabilize their cytoskeletal architecture [Figure 17.5(e)], resulting in a widely varying myofibrillar pattern from myocyte to myocyte. As in the two-dimensional tissue, contraction will be isometric and is detectable by monitoring nuclear displacement. Arrays of myocytes that approximate the broad ranges of aspect ratios observed in the ventricular myocardium are possible and will allow high-throughput screening of pharmacological candidates and/or toxins, as well as facilitate statistical studies of myocyte behavior [Figure 17.5(f)]. In this regard, the isometric contraction of the myocytes on the rigid substrate is advantageous during high-speed fluorescent video microscopy with Ca^{2+} indicators or voltage-sensitive dyes due to lack of motion artifact. For contractility studies, we have adapted the traction force microscopy technique [28] for the unique requirements of the rapidly contracting myocyte.

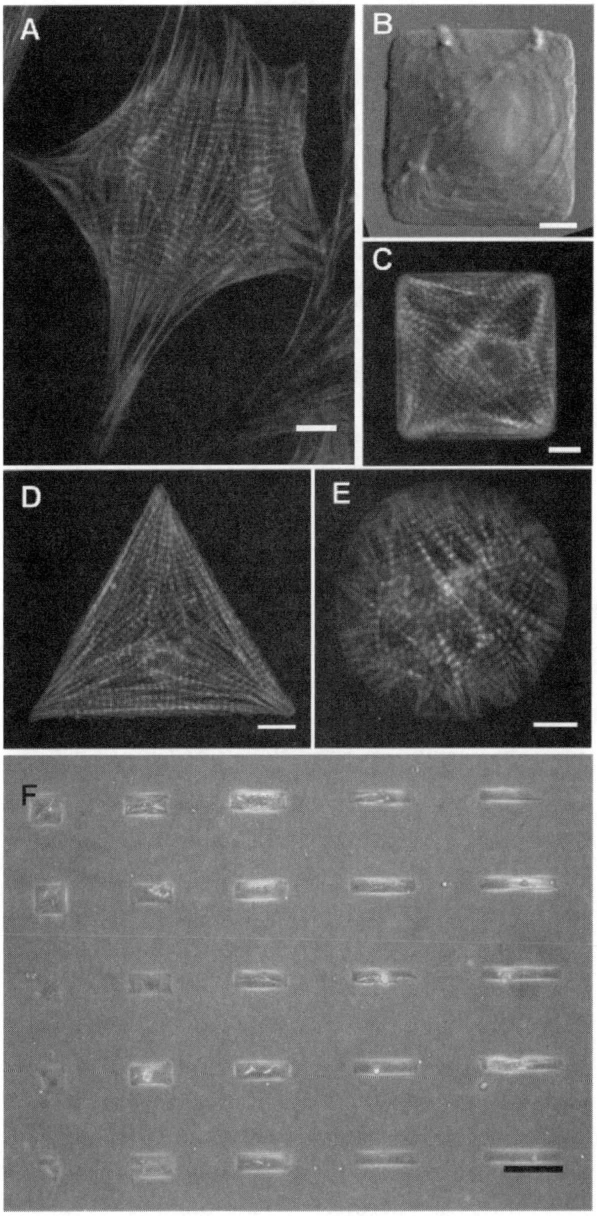

Figure 17.5 *Morphology of individual NNRVMs*: (a) NNRVMs plated at low density on a uniform monolayer of FN display a random shape and alignment of myofibrils. (b) Atomic force microscopy of living NNRVMs plated onto 50 × 50 μm square FN islands shows the myocyte maintains a relatively flat (~0.5 to 1 μm) structure along the edges, with a raised section (2.9 μm) corresponding to the location of the nucleus. (c) NNRVMs prepared identically to (b) take the shape of the island, and myofibrils align to the long axis and the edges of the pattern. Using this technique for micropatterning single myocytes, (d) triangular and (e) circular myocytes can be generated, each having different degrees of structural orientation based on imposed boundary conditions. Stains for (a–e) are phalloidin; sarcomeric α-actinin; DNA (DAPI). (f) Patterned myocytes (here, with varying aspect ratios) can be oriented closely to each other in culture, increasing throughput for experimental techniques. atomic force microscopy image in (b) were taken under physiological conditions with an MFP-3-D-IO (Asylum Research, USA). Scale bars for (a–e) are 10 μm; scale bar for (f) is 100 μm.

17.4 Traction Force Microscopy for Cardiac Myocytes

Cardiac myocytes are designed to contract in order to pump blood. As described previously, it has been hypothesized that the mechanical properties of individual myocytes may be related to their shape [29]. In particular, the tractional forces that myocytes exert on both the ECM and each other may play an important role in overall tissue and organ contractility. This relationship is important when considering the design of myocardial tissue constructs, as well as in understanding the regulation of a variety of pathophysiological processes such as hypertrophy, dilated cardiomyopathy, and heart failure.

Quantitative measurement of the tractional forces exerted by mammalian cells on the cell-substrate interface have been measured by the use of either a continuous surface [28] or a discrete array of vertical microneedles [30]. In the continuous-surface approach, the cell-culture substrate is embedded with fluorescent beads that can be tracked with fluorescence optical microscopy. The displacements of the beads are measured and then correlated to the deformations generated by cell contractions. From these data, a continuous force field is reconstructed by using standard numerical techniques for solving ill-posed problems. Balaban et al. [3] developed a similar approach by micropatterning the surface of the substrate to measure the contractile force exerted on one focal adhesion by a single cardiac myocyte. Recently, Wang et al. [31] combined the continuous-surface approach with the micropatterning method developed by Ostuni et al. [32] to measure the traction force of shape-controlled cells. Another method of measuring cellular tractional forces is the discrete-array approach, where cells are cultured on an array of elastomeric posts [30]. In this technique, as the cells contract, the displacement of the posts is measured using optical microscopy. Based on the predetermined material composition and structure of the posts, a simple spring equation is applied to relate the measured deflection the corresponding traction force. Zhao and Zhang [33] used this approach to measure the contractile force of a single cardiac myocyte. They validated the inotropic effect of β-adrenergic stimulators by showing that a greater subcellular force generation was exerted when myocytes were perfused with isoproterenol. The drawback of the discrete-array approach is that the density of the microneedles under the cell is low due to fabrication limitations. This limitation significantly affects cell adhesion and locomotion compared to a flat and continuous surface [34]. While the microneedle technique is the most computationally efficient from the perspective of the investigator, the spacing between the microneedles represents a formidable obstacle for use with cardiac myocytes as they lack the motile capabilities of other cells that can easily extend lamellipodia and reach across free space to adhere to the adjacent needle head. We have found traction force microscopy on a continuous substrate to be superior for examining the structure-function relationships in shape-controlled NNRVMs. Not only do these techniques represent an excellent methodology for understanding cell function, but combining them with methods to control and engineer the cell microenvironment allows the investigator to systematically study structure-function relationships in cell biology.

Here, we describe our method of quantitatively measuring the contractile force of micropatterned myocytes cultured on a continuous surface. In this approach, fluorescent beads are attached to the surface of a flat elastomer that is suitable for cardiac myocyte cell culture and verified by atomic force microscopy [Figure 17.6(a, b)] as opposed to being embedded throughout the volume of the substrate. Using this substrate, the shape of cardiac myocytes can be controlled by μCP. High-speed fluorescence video microscopy imaging of substrate deformation (measured by bead movement) due to myocyte contraction is then analyzed by an algorithm modified from Schwarz et al. to estimate the contractile force field of the micropatterned myocytes [Figure 17.6(c, d)] [35].

17.4.1 Substrate Preparation for Traction Force Microscopy

In this technique, PDMS is prepared at a higher base:curing agent ratio in order to lower its stiffness. As before, the polymer is coated onto a glass coverslip for analysis with optical microscopy. A similarly prepared block of PDMS was measured with a Bohlin CVOR Rheometer (Malvern Instruments, Malvern, UK) and was found to have a bulk modulus of ~12 to 15 kPa; this modulus is assumed to be the same for the coated coverslips. Fluorescence latex beads that are 0.2 μm in diameter with an excitation-emission spectrum of λ_{ex} = 540 to 580 nm and λ_{em} = 608 to 683 nm (Molecular Probes, Eugene, OR) are diluted 1:5,000 into a (PBS) solution, pH 7.40. The PDMS-coated coverslips are ultraviolet-ozone treated (Model 342,

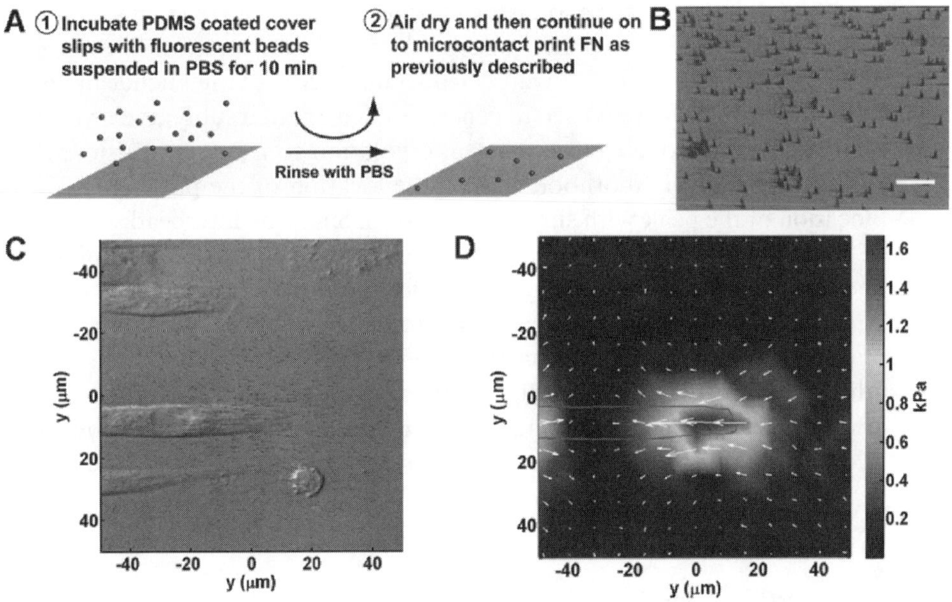

Figure 17.6 *Calculation of traction forces of micropatterned cardiac myocytes*: (a) Overview of the procedure for fabricating bead-coated PDMS substrates. (b) Atomic force micrograph of these substrates demonstrating a flat surface with 0.2 μm–diameter beads distributed across the surface. Scale bar = 5 ìm. (c) A DIC image of micropatterned myocytes superimposed with the displacements of the fluorescence beads. The myocytes were cultured on 10 μm micropatterned FN lines spaced 10 μm apart. The length and direction of the vectors represent the magnitude and the direction of bead displacements, respectively. The displacement vectors show that the substrate is pulled toward the longitudinal end of the myocyte but curled away at the middle of the cell. (d) The traction force map of the substrate. Relatively higher tractions are found to be concentrated at the long end of the myocyte.

Jelight Company, Irvine, CA) for two minutes and then incubated with the bead solution for ten minutes to allow bead attachment onto the surface of the PDMS. The bead solution is rinsed from the coverslip with PBS, air dried, and then micropatterned with FN using μCP as described in the section 17.3.3.

17.4.2 Identification and Tracking of Fluorescent Beads

Several image-processing procedures are involved in extracting the bead locations from the fluorescence image. A top-hat filter is applied to remove variations in the background intensity across an image. This also allows foreground objects (i.e., beads) that are smaller than a given size to be enhanced, while the intensity of background objects in the image is diminished. A thresholding approach [36] automatically determines an upper boundary (T_{UB}) grayscale value to define a bead in the image. The image is binarized with increasing threshold values to generate a curve composed of the average area of the thresholded regions as a function of threshold value. The lower boundary (T_{LB}) grayscale value is defined as that value where the average object area is maximum. The optimal threshold is finally selected as the maximum value in the range of $[T_{LB}, T_{UB}]$. Next, a generalized second-derivative test is used to detect relative intensity maxima within the image.

The grayscale image is converted into a phase map permitting the localization of singular points without regard to the local intensity [37]. These singular points correspond to intensity peaks (i.e., the bead locations) in a given image. Singular points that: (1) lie above the optimal threshold; and (2) are contained in the relative maxima are then determined. A watershed segmentation of the Euclidean distance transform of the point map is used to generate a demarcated subregion around each bead. For each subregion, a three-point interpolation to a two-dimensional Gaussian is used to fit the neighborhood around the location of the peak intensity, giving the location of the peak with subpixel precision. Since the interbead separation (usually 5 to 10 pixels) is bigger than the displacement of a single bead (~1 pixel), individual beads are tracked over time by looking for the closest detected bead in the successive image. Incorrect tracking is handled automatically via Delaunay triangulation of the bead points and looking for changes in the grid connectivity. By assuming that the displacements of neighboring beads are on the same spatial scale if the beads are sufficiently close together, exaggerated displacements produced by errant beads detected in one frame, but missed in the next, can be identified and eliminated.

17.4.3 Calculation of Traction Forces

The algorithm developed by Schwarz et al. is adapted to quantitatively measure the peak systolic forces of the patterned myocytes [35]. The substrate is considered to be an elastic half-space under tangential traction on the surface since the typical displacements of the beads are negligible (a few hundred nanometers) as compared to the substrate thickness (~50 μm). Because the substrates are assumed to be isotropic and incompressible, the Poisson's ratio (ν) is close to 0.5, and no out-of-plane deformations occur during tangential traction. Consequently, the whole elastic formulation becomes linear as well as two-dimensional and can therefore be described by a Fredholm integral equation of the first kind:

$$u_i(r) = \int dr' G_{ij}(r - r') F_j(r') \tag{17.1}$$

where $F(r')$ and $U(r')$ denote the stress and the displacement field, respectively; the summation is applied with $1 \le i, J \le 2$ since the problem is two-dimensional. In our case, is the Green function of the elastic isotropic half-space and can be derived from the Boussinesq solution:

$$G_{ij}(r) = \frac{3}{4\pi E r}\left(\delta_{ij} + \frac{x_i x_j}{r^2}\right) \tag{17.2}$$

where x_i represents the Cartesian coordinate of the displacement field referenced against a concentrated force exerted at the origin, $r \equiv \sqrt{x_i x_i}$, E is the Young modulus, and δ_{ij} is the Kronecker delta. The experimentally measured bead displacements are used to inversely solve the traction field from (17.1). A lattice composed of $8 \times 8\ \mu m^2$ squares is applied to the whole image to approximate the discretized localization of traction forces. Since the typical bead displacement is less than 1 μm and the Young's modulus of the material is ~10 to 15 kPa, the approximated force points are separated at least by 4 μm, as suggested by Schwarz et al. [35]. The χ^2 statistics were employed to choose a solved force pattern such that its resultant displacement field is the best fit to the measured one. Since the inverse problem is ill posed, zero-order Tikhonov regularization is applied to stabilize the solution by minimizing χ^2 under the additional constraint that the forces should not become exceedingly large, according to the following equation:

$$\min_F \left\{ |GF - u|^2 + \lambda^2 |F|^2 \right\} \tag{17.3}$$

where λ is the regularization parameter and serves to filter out small displacement data to stabilize the solution. L-curve criterion is used to determine λ [38] and choose the value of λ at which the residual norm $R = |GF - u|^2$ starts to increase significantly as a function of λ. The chosen value of λ thus corresponds to the optimal balance between data agreement and regularization.

Using this, bead-displacement data from epifluorescence microscopy can be converted into tractional forces, as seen in Figure 17.6(c, d). DIC optical microscope images of a single contracting myocyte can be overlaid with bead-displacement maps [Figure 17.6(c)], relating the maximal displacement with cellular structure. Further, these displacements indicate that maximal tractional forces occur just below the terminal end of the micropatterened myocytes and that there is significant force transduction from the cell to the substratum. The advantage of this technique versus traditional methods that require the gluing of freshly harvested myocytes to the ends of suspended pipettes mounted in force transducers is that the previous method allowed only for the measurement of contraction along one axis, whereas traction force microscopy allows for examination of the functional consequences of myocyte shape, as well as a broad array of vector mathematical measurement of the traction field. Temporal data, such as the rate of contraction and relaxation, is easily handled by the image-analysis technique described, allowing examination of the behavior of subcompartments within the myocyte.

17.5 Conclusions and Future Perspectives

Understanding structure-function relationships is critical in cell biology in general, but it is of particular importance in the myocardium. Pathological examination of failing hearts reveals various changes in the cardiac tissue microenvironment that may provide clues in the search for new drug targets to curb the progression of cardiomyopathies and suppress arrhythmias. Alterations in the expression of ECM proteins, cell shape and alignment, and gap-junction distribution indicate that stress and strain patterns within the myocardium have contributed to both structural and electrical remodeling. To date, the geometric complexity of the heart and the unique dynamics of the cardiac tissue microenvironment have made it difficult to elucidate the mechanochemical and mechanoelectrical signaling pathways that underlie these events. Thus, new techniques developed from the materials sciences have made possible experiments involving the control of cellular structure to investigate the resulting effects on function and cell-ECM or cell-cell interactions [39]. In the heart, several pathologies are correlated to structural changes seen in myocytes [9, 40], suggesting that maladaptive remodeling on the single myocyte level produces regulatory or contractile dysfunction of the myocardial syncytium. Until the implementation of techniques to control cellular architecture and geometry by engineering the cell microenvironment, investigators did not possess the tools to control cellular structure to such a precise degree.

Microcontact printing represents a simple technique for controlling the cell microenvironment. For myocytes, control of myofibrillar architecture can be achieved at the single-cell and two-dimensional tissue level. Cardiac myocytes respond to ECM geometry by orienting their myofibrillar architecture based on micropatterned spatial cues, and this can be exploited to reconstitute structural phenotypes found in vivo in either healthy or failing hearts. Beyond the flexibility of experimental design imparted by these techniques, fabrication of two-dimensional and three-dimensional tissue constructs for regenerative purposes is also a focus of cell and tissue engineering [5].

The use of three-dimensional scaffolds [41, 42] and substrates [43–45] represents promising technologies for future understanding of cell dynamics in a three-dimensional environment. For example, it has been demonstrated not only that two-dimensional separation between cells is important for expression of connexin-43 and N-cadherin, but the vertical (i.e., depth) dimension of the substrate plays an important role in the proper expression and localization of these genes [44]. Further, expression of cell-attachment proteins and cytoskeletal structure seems to be altered when cells are cultured on a three-dimensional substratum rather than the traditional two-dimensional microenvironment [45]. These results suggest intriguing trends, and future work may result in a methodology to create three-dimensionally oriented tissues in vitro.

Within the past five years, innovations in the field of thermosensitive polymers has opened new prospects for tissue engineering by controlling the cell microenvironment. Shimizu et al. [46, 47] have developed a methodology for culturing cardiac myocytes on layers of poly(N-isopropylacrylimide) (PIPAAm), a thermosensitive, biocompatible polymer. Using this technique, cells are grown in vitro on a PIPAAm coated substrate and can be released by simply lowering the temperatrure of the culture conditions rather than by traditional enzymatic degradation. Using this technique, pulsatile grafts of cardiac tissue are liberated from the culture substrate and can be used as needed. Combining these techniques with others that control the architecture and morphology of engineered tissue represents a powerful methodology for generating oriented, substrate, and scaffold-free muscle grafts.

Further advances in tissue engineering for myocardial tissues are manifest [48–54]. It is clear that the field has advanced greatly since the introduction of microfabrication and soft lithography techniques borrowed mainly from the materials sciences. Both two-dimensional and three-dimensional manipulation of the cell-culture microenvironment has greatly enhanced our understanding of structure-function relationships in the cell, particularly the cardiac myocyte. Not only do these techniques serve for making simplified models of individual cardiac myocytes, but extensions of these techniques can result in the fabrication of simple two-dimensional tissues and more complex pseudo-three-dimensional tissues free of the culture substrate/scaffold. Implementing a variety of these techniques will be a powerful means of understanding the function of the myocardium, as well as serve as a prelude to clinical and therapeutic applications of cardiac-tissue engineering.

Acknowledgments

The work presented here has been supported by the Harvard University Nanoscale Science and Engineering Center and the Harvard University Material Research Science and Engineering Centers of the National Science Foundation (NSF) under NSF award numbers PHY-0117795 and DMR-0213805, and by the Defense Advanced Research Projects Agency Biomolecular Motors Program, project number FA9550-05-1-0015 for engineered muscle activators cells and tissues.

References

[1] Chen, C. S., Mrksich, M., Huang, S., Whitesides, G. M., and Ingber, D. E., "Geometric control of cell life and death," *Science,* Vol. 276, 1997, pp. 1425–1428.

[2] Parker, K. K., Brock, A. L., Brangwynne, C., Mannix, R. J., Wang, N., Ostuni, E., Geisse, N. A., Adams, J. C., Whitesides, G. M., and Ingber, D. E., "Directional control of lamellipodia extension by constraining cell shape and orienting cell tractional forces," *Faseb J.,* Vol. 16, 2002, pp. 1195–1204.

[3] Balaban, N. Q., Schwarz, U. S., Riveline, D., Goichberg, P., Tzur, G., Sabanay, I., Mahalu, D., Safran, S., Bershadsky, A., Addadi, L., and Geiger, B., "Force and focal adhesion assembly: A close relationship studied using elastic micropatterned substrates," *Nat. Cell. Biol.,* Vol. 3, 2001, pp. 466–472.

[4] Bhatia, S. N., Balis, U. J., Yarmush, M. L., and Toner, M., "Effect of cell-cell interactions in preservation of cellular phenotype: Cocultivation of hepatocytes and nonparenchymal cells," *Faseb. J.,* Vol. 13, 1999, pp. 1883–1900.

[5] Khademhosseini, A., Langer, R., Borenstein, J., and Vacanti, J. P., "Microscale technologies for tissue engineering and biology," *Proc. Natl. Acad. Sci. USA,* Vol. 103, 2006, pp. 2480–2487.

[6] Spach, M. S., Heidlage, J. F., Dolber, P. C., and Barr, R. C., "Electrophysiological effects of remodeling cardiac gap junctions and cell size: Experimental and model studies of normal cardiac growth," *Circ. Res.,* Vol. 86, 2000, pp. 302–311.

[7] Kleber, A. G., and Rudy, Y., "Basic mechanisms of cardiac impulse propagation and associated arrhythmias," *Physiol. Rev.*, Vol. 84, 2004, pp. 431–488.

[8] Spach, M. S., Miller, W. T., III, Geselowitz, D. B., Barr, R. C., Kootsey, J. M., and Johnson, E. A., "The discontinuous nature of propagation in normal canine cardiac muscle: Evidence for recurrent discontinuities of intracellular resistance that affect the membrane currents," *Circ. Res.,* Vol. 48, 1981, pp. 39–54.

[9] Gerdes, A. M., and Capasso, J. M., "Structural remodeling and mechanical dysfunction of cardiac myocytes in heart failure," *J. Mol. Cell. Cardiol.*, Vol. 27, 1995, pp. 849–856.

[10] Lammerding, J., Kamm, R. D., and Lee, R. T., "Mechanotransduction in cardiac myocytes," *Annals NY Acad. Sci.*, Vol. 1015, 2004, pp. 53–70.

[11] Lieberman, M., Roggeveen, A. E., Purdy, J. E., and Johnson, E. A., "Synthetic strands of cardiac muscle: Growth and physiological implication," *Science,* Vol. 175, 1972, pp. 909–911.

[12] Lieberman, M., Sawanobori, T., Kootsey, J. M., and Johnson, E. A., "A synthetic strand of cardiac muscle: Its passive electrical properties," *J. Gen. Physiol.,* Vol. 65, 1975, pp. 527–550.

[13] Bursac, N., Parker, K. K., Iravanian, S., and Tung, L., "Cardiomyocyte cultures with controlled macroscopic anisotropy: A model for functional electrophysiological studies of cardiac muscle," *Circ. Res.,* Vol. 91, 2002, pp. e45–e54.

[14] Xia, Y., and Whitesides, G. M., "Soft lithography," *Angewandte Chemie—Inter. Ed.*, Vol. 37, 1998, pp. 550–575.

[15] Whitesides, G. M., Ostuni, E., Takayama, S., Jiang, X., and Ingber, D. E., "Soft lithography in biology and biochemistry," *Ann. Rev. Biomed. Eng.*, Vol. 3, 2001, pp. 335–373.

[16] Rohr, S., Scholly, D. M., and Kleber, A. G., "Patterned growth of neonatal rat heart cells in culture: Morphological and electrophysiological characterization," *Circ. Res.,* Vol. 68, 1991, pp. 114–130.

[17] Rohr, S., Kleber, A. G., and Kucera, J. P., "Optical recording of impulse propagation in designer cultures: Cardiac tissue architectures inducing ultra-slow conduction," *Trends Cardiovasc. Med.*, Vol. 9, 1999, pp. 173–179.

[18] Gillis, A. M., Fast, V. G., Rohr, S., and Kleber, A. G., "Mechanism of ventricular defibrillation: The role of tissue geometry in the changes in transmembrane potential in patterned myocyte cultures," *Circulation*, Vol. 101, 2000, pp. 2438–2445.

[19] Kaplan, S. R., Gard, J. J., Protonotarios, N., Tsatsopoulou, A., Spiliopoulou, C., Anastasakis, A., Squarcioni, C. P., McKenna, W. J., Thiene, G., Basso, C., Brousse, N., Fontaine, G., and Saffitz, J. E., "Remodeling of myocyte gap junctions in arrhythmogenic right ventricular cardiomyopathy due to a deletion in plakoglobin (Naxos disease)." *Heart Rhythm*, Vol. 1, 2004, pp. 3–11.

[20] Kaplan, S. R., Gard, J. J., Carvajal-Huerta, L., Ruiz-Cabezas, J. C., Thiene, G., and Saffitz, J. E., "Structural and molecular pathology of the heart in Carvajal syndrome," *Cardiovasc. Pathol.*, Vol. 13, 2004, pp. 26–32.

[21] Kanno, S., and Saffitz, J. E., "The role of myocardial gap junctions in electrical conduction and arrhythmogenesis," *Cardiovasc. Pathol.*, Vol. 10, 2001, pp. 169–177.

[22] Sathaye, A., Bursac, N., Sheehy, S., and Tung, L., "Electrical pacing counteracts intrinsic shortening of action potential duration of neonatal rat ventricular cells in culture," *J. Mol. Cell. Cardiol.*, Vol. 41, 2006, pp. 633–641.

[23] Lipp, P., and Niggli, E., "Modulation of Ca

release in cultured neonatal rat cardiac myocytes: Insight from subcellular release patterns revealed by confocal microscopy," *Circ. Res.,* Vol. 74, 1994, pp. 979–990.

[24] Kennedy, R. A., Kemp, T. J., Sugden, P. H., and Clerk, A., "Using U0126 to dissect the role of the extracellular signal-regulated kinase 1/2 (ERK1/2) cascade in the regulation of gene expression by endothelin-1 in cardiac myocytes," *J. Mol. Cell. Cardiol.*, Vol. 41, 2006, pp. 236–247.

[25] Arutunyan, A., Swift, L. M., and Sarvazyan, N., "Initiation and propagation of ectopic waves: Insights from an in vitro model of ischemia-reperfusion injury," *Am. J. Physiol.: Heart Circ. Physiol.*, Vol. 283, 2002, pp. H741–H749.

[26] Simpson, P., "Norepinephrine-stimulated hypertrophy of cultured rat myocardial cells is an alpha 1 adrenergic response," *J. Clin. Invest.*, Vol. 72, 1983, pp. 732–738.

[27] Tan, J. L., Liu, W., Nelson, C. M., Raghavan, S., and Chen, C. S., "Simple approach to micropattern cells on common culture substrates by tuning substrate wettability," *Tissue Eng.*, Vol. 10, 2004, pp. 865–872.

[28] Dembo, M., and Wang, Y. L., "Stresses at the cell-to-substrate interface during locomotion of fibroblasts," *Biophys. J.*, Vol. 76, 1999, pp. 2307–2316.

[29] Russell, B., Motlagh, D., and Ashley, W. W., "Form follows function: How muscle shape is regulated by work," *J. Appl. Physiol.*, Vol. 88, 2000, pp. 1127–1132.

[30] Tan, J. L., Tien, J., Pirone, D. M., Gray, D. S., Bhadriraju, K., and Chen, C. S., "Cells lying on a bed of microneedles: An approach to isolate mechanical force," *Proc. Natl. Acad. Sci. USA*, Vol. 100, 2003, pp. 1484–1489.

[31] Wang, N., Ostuni, E., Whitesides, G. M., and Ingber, D. E., "Micropatterning tractional forces in living cells," *Cell Motil. Cytoskeleton*, Vol. 52, 2002, pp. 97–106.

[32] Ostuni, E., Kane, R., Chen, C. S., Ingber, D. E., and Whitesides, G. M., "Patterning mammalian cells using elastomeric membranes," *Langmuir*, Vol. 16, 2000, pp. 7811–7819.

[33] Zhao, Y., and Zhang, X., "Cellular mechanics study in cardiac myocytes using PDMS pillars array," *Sens. Actuators A: Physical*, Vol. 125, 2006, pp. 398–404.

[34] du Roure, O., Saez, A., Buguin, A., Austin, R. H., Chavrier, P., Silberzan, P., and Ladoux, B., "Force mapping in epithelial cell migration," *Proc. Natl. Acad. Sci. USA*, Vol. 102, 2005, pp. 2390–2395.

[35] Schwarz, U. S., Balaban, N. Q., Riveline, D., Bershadsky, A., Geiger, B., and Safran, S. A., "Calculation of forces at focal adhesions from elastic substrate data: The effect of localized force and the need for regularization," *Biophys. J.*, Vol. 83, 2002, pp. 1380–1394.

[36] Otsu, N., "A thresholding selection method from gray-level histogram," *IEEE Trans. Syst. Man Cybernet.*, Vol. 9, 1979, pp. 62–66.

[37] Bray, M. A., Lin, S. F., Aliev, R. R., Roth, B. J., and Wikswo, J. P., Jr., "Experimental and theoretical analysis of phase singularity dynamics in cardiac tissue," *J. Cardiovasc. Electrophysiol.*, Vol. 12, 2001, pp. 716–722.

[38] Hansen, P. C., "Rank-deficient and discrete ill-posed problems: Numerical aspects of linear inversion," *Society for Industrial and Applied Mathematics*, Philadelphia, PA, 1997.

[39] Raghavan, S., and Chen, C. S., "Micropatterned environments in cell biology," *Adv. Mater.*, Vol. 16, 2004, pp. 1303–1313.

[40] Gerdes, A. M., "Cardiac myocyte remodeling in hypertrophy and progression to failure," *J. Card. Fail.*, Vol. 8, 2002, pp. S264–S268.

[41] Christman, K. L., and Lee, R. J., "Biomaterials for the treatment of myocardial infarction," *J. Am. Coll. Cardiol.*, Vol. 48, 2006, pp. 907–913.

[42] Park, H., Radisic, M., Lim, J. O., Chang, B. H., and Vunjak-Novakovic, G., "A novel composite scaffold for cardiac tissue engineering," *In Vitro Cell. Dev. Biol.: Anim.*, Vol. 41, 2005, pp. 188–196.

[43] Deutsch, J., Motlagh, D., Russell, B., and Desai, T. A., "Fabrication of microtextured membranes for cardiac myocyte attachment and orientation," *J. Biomed. Mater. Res.*, Vol. 53, 2000, pp. 267–275.

[44] Motlagh, D., Hartman, T. J., Desai, T. A., and Russell, B., "Microfabricated grooves recapitulate neonatal myocyte connexin43 and N-cadherin expression and localization," *J. Biomed. Mater. Res. A*, Vol. 67, 2003, pp. 148–157.

[45] Motlagh, D., Senyo, S. E., Desai, T. A., and Russell, B., "Microtextured substrata alter gene expression, protein localization and the shape of cardiac myocytes," *Biomaterials*, Vol. 24, 2003, pp. 2463–2476.

[46] Shimizu, T., Yamato, M., Isoi, Y., Akutsu, T., Setomaru, T., Abe, K., Kikuchi, A., Umezu, M., and Okano, T., "Fabrication of pulsatile cardiac tissue grafts using a novel 3-dimensional cell sheet manipulation technique and temperature-responsive cell culture surfaces," *Circ. Res.*, Vol. 90, 2002, p. e40.

[47] Shimizu, T., Sekine, H., Isoi, Y., Yamato, M., Kikuchi, A., and Okano, T., "Long-term survival and growth of pulsatile myocardial tissue grafts engineered by the layering of cardiomyocyte sheets," *Tissue Eng.*, Vol. 12, 2006, pp. 499–507.

[48] Radisic, M., Park, H., Shing, H., Consi, T., Schoen, F. J., Langer, R., Freed, L. E., and Vunjak-Novakovic, G., "Functional assembly of engineered myocardium by electrical stimulation of cardiac myocytes cultured on scaffolds," *Proc. Natl. Acad. Sci. USA*, Vol. 101, 2004, pp. 18129–18134.

[49] Zimmermann, W. H., Schneiderbanger, K., Schubert, P., Didie, M., Munzel, F., Heubach, J. F., Kostin, S., Neuhuber, W. L., and Eschenhagen, T., "Tissue engineering of a differentiated cardiac muscle construct," *Circ. Res.*, Vol. 90, 2002, pp. 223–230.

[50] Leor, J., Aboulafia-Etzion, S., Dar, A., Shapiro, L., Barbash, I. M., Battler, A., Granot, Y., and Cohen, S., "Bioengineered cardiac grafts: A new approach to repair the infarcted myocardium?" *Circulation*, Vol. 102, 2000, pp. III56–III61.

[51] Carrier, R. L., Papadaki, M., Rupnick, M., Schoen, F. J., Bursac, N., Langer, R., Freed, L. E., and Vunjak-Novakovic, G., "Cardiac tissue engineering: Cell seeding, cultivation parameters, and tissue construct characterization," *Biotechnol. Bioeng.*, Vol. 64, 1999, pp. 580–589.

[52] Papadaki, M., Bursac, N., Langer, R., Merok, J., Vunjak-Novakovic, G., and Freed, L. E., "Tissue engineering of functional cardiac muscle: Molecular, structural, and electrophysiological studies," *Am. J. Physiol.: Heart Circ. Physiol.*, Vol. 280, 2001, pp. H168–H178.

[53] Eschenhagen, T., Fink, C., Remmers, U., Scholz, H., Wattchow, J., Weil, J., Zimmermann, W., Dohmen, H. H., Schafer, H., Bishopric, N., Wakatsuki, T., and Elson, E. L., "Three-dimensional reconstitution of embryonic cardiomyocytes in a collagen matrix: A new heart muscle model system," *Faseb. J.*, Vol. 11, 1997, pp. 683–694.

[54] Khademhosseini, A., Eng, G., Yeh, J., Kucharczyk, P. A., Langer, R., Vunjak-Novakovic, G., and Radisic, M., "Microfluidic patterning for fabrication of contractile cardiac organoids," *Biomed. Microdevices*, Vol. 9, 2007, pp. 149–157.

CHAPTER 18

Microreactors for Cardiac Tissue Engineering

Rohin K. Iyer, Brian Plouffe, Shashi K. Murthy, and Milica Radisic

18.1 Introduction

The inability of myocardium to regenerate after injury necessitates the need for development of alternative treatment options. Engineered three-dimensional cardiac tissues or cardiac myofiber arrays can thus serve as models to study normal and pathological tissue function in vitro or to test the effect of drugs on tissue structure and function.

The myocardium is a highly differentiated tissue, 1 cm thick in humans, composed of cardiac myocytes and fibroblasts with a dense supporting vasculature and collagen-based extracellular matrix (ECM). Cardiac myocytes form a three-dimensional syncytium that enables propagation of electrical signals across specialized intracellular junctions to produce coordinated mechanical contractions that pump blood through the systemic circulation [1]. Only 20 to 40 percent of the cells in the heart are cardiac myocytes, but they occupy 80 to 90 percent of the heart volume. The average cell density in the native rat myocardium is on the order of 5×10^8 cells/cm^3. Morphologically, intact cardiac myocytes have an elongated, rod-shaped appearance. The contractile apparatus of cardiac myocytes consists of sarcomeres arranged in parallel myofibrils. The high metabolic activity of cardiac myocytes is supported by the high density of mitochondria. Electrical-signal propagation is provided by specialized intercellular connections known as gap junctions [1].

The control of heart contractions is almost entirely self-contained. Groups of specialized cardiac myocytes (pacemakers), the fastest of which are located in the sinoatrial node, drive periodic contractions of the heart. The majority of cells in the myocardium are nonpacemaker cells, and they respond to the electrical stimuli generated by pacemaker cells. Excitation of each cardiac myocyte is followed by an increase in the amount of cytoplasmic calcium that triggers mechanical contraction. The propagation of the electrical excitation through the tissue by ion currents in the extracellular and the intercellular space results in synchronous contractions that enable the pumping of blood from the heart.

Cell-patterning approaches thus attempt to reproduce the appropriate structure of cardiac myofibers at several different length scales so that the in vivo–like function of cells in culture can be maintained. At the micrometer scale, cells in engineered cardiac tissue must be coupled by functional gap junctions and capable of propagating electrical impulses in order to prevent arrhythmia upon implantation. Finally, at the subcellular level, the excitation-contraction machinery of individual cardiomyocytes must be functional.

In this chapter, we focus on the engineering of cardiomyocyte arrays in both two- and three-dimensional culture, microreactors that enable application of controlled physical and biochemical stimuli, and microfluidic methods for separation of heterogeneous cell populations in the heart.

18.2 Patterned Cardiomyocyte Cultures in Two Dimensions

A starting point for studying the growth, organization, and alignment of cardiac tissue in vitro is a microstructured two-dimensional culture of cardiomyocytes. Such two-dimensional cultures can be quickly and inexpensively patterned on conventional transparent surfaces used in cell culture, such as glass or tissue-culture polystyrene (TCP), allowing them to be easily assayed or imaged with conventional microscopy tools. The microstructured features of these two-dimensional patterns promote directional alignment of cardiomyocytes in an architecture that resembles that of aligned cardiac myofibres in the native heart.

Cell and protein arrays have been generated on various substrates using soft lithographic approaches such as self-assembled monolayers (SAMs) [2–8], microstamped proteins [9], biological [10–13] and comb polymers [14], microfluidic channels [15], and elastomeric membranes [16, 17]. Since native myocardium consists of elongated cardiomyocytes arranged into aligned myofibers, significant efforts have been focused on reproducing the elongated phenotype in two-dimensional substrates. Attachment and alignment has been modulated using grooved and pegged surfaces [18]. Micropatterning substrates and microdevices have the potential to enhance the function of cardiomyocytes by controlling topographical features and spatial presentation of surface molecules. For instance, fibroblast overgrowth in cultures of primary cardiomyocytes was prevented using topographical cues [19]. Elongated phenotype was achieved by patterning photoresist lanes on glass substrate and growing the myocytes onto the exposed glass [20, 21]. Myocyte cultivation on patterned substrates generated important insight into the role of fibroblasts [22] in electrical-signal propagation and the effect of electrical-field shocks on changes in transmembrane potential [23].

Radisic et al. developed a simple system for creating patterned two-dimensional cell-culture substrates based on photocrosslinkable chitosan [24]. A mask was designed in a commercial vector-graphics package (Adobe Illustrator, Adobe Systems Inc.) and printed onto an acetate transparency sheet with a high-resolution printer. The resulting mask was used to create patterned areas on glass or TCP by selectively exposing photocrosslinkable chitosan to ultraviolet (UV) light. After UV exposure, uncrosslinked chitosan was washed away, revealing patterned chitosan islands that were cell-repellent. The empty lanes flanked by the chitosan micropatterns were conducive to cell attachment and could be seeded with cells.

As shown in Figure 18.1, various shapes (lanes, squares, triangles, circles) could be easily patterned on these two commonly used culture surfaces. The lane widths were simply altered, at a micrometer resolution, simply by varying the exposure

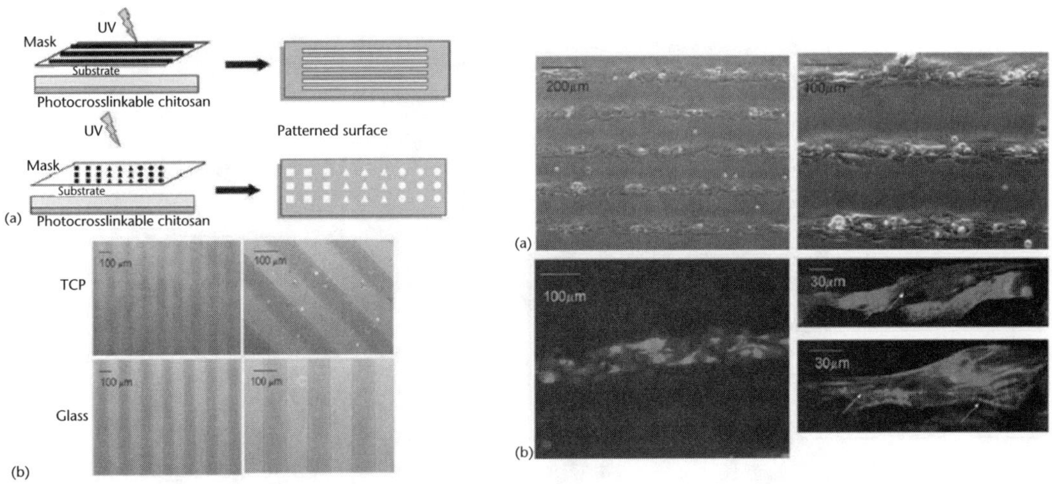

Figure 18.1 (Left) Experimental set-up: (a) Photocrosslinkable chitosan was patterned on glass and TCP by selective exposure to UV light, resulting in lanes with micrometer resolution (TCP). (b) Crosslinked chitosan patterns were created on two surfaces most commonly used in tissue culture: TCP and glass. For visualization, chitosan lanes were stained with eosin. (Right) Patterning of cardiomyocytes: (a) Neonatal rat cardiomyocytes were seeded on chitosan-patterned glass surfaces at eight days of culture. The cells adhered to glass and formed confluent cell lanes that exhibited spontaneous contractions. (b) Patterned cardiomyocytes express cardiac troponin I and exhibit a developed contractile apparatus (arrows) [24].

time to UV light for the same mask. The patterns were stable for up to eighteen days in culture. Cardiomyocytes, patterned in lanes 68 to 99 μm wide exhibited expression of cardiac troponin I and well-developed contractile apparatus and contracted synchronously in response to electrical-field stimulation.

McDevitt et al. used microcontact printing of laminin lanes to direct the growth of spatially organized cardiomyocyte cultures, which were created on biodegradable, elastomeric polyurethane films [25]. Photolithographic techniques were used to etch lane patterns of various widths into photoresist on silicon wafers, which were used as negative templates for the creation of poly(dimethyl siloxane) (PDMS) stamps. The PDMS stamps were coated with laminin-1 and placed face down onto polystyrene or poly(lactic-co-glycolic acid) (PLGA) surfaces that had been coated in a cell-repellent coating, bovine serum albumin (BSA). The resulting laminin lanes, which ranged from 5 to 50 μm in width, directed the elongated growth of two-dimensional cardiac cultures resembling native myofibres, as shown in Figure 18.2.

The authors showed that the organoids formed bipolar contacts and intercalated discs containing N-cadherin and Connexin-43 and found that the 5 to 15 μm lane widths resulted in organoids with optimal aspect ratio and a three-dimensional appearance. The aspect ratio (length:width) of the patterned cardiomyocytes decreased with increasing lane width, and the extent of bridging between lanes was dependent on the interlane spacing. By forty-eight hours, entire lanes were spontaneously contracting and were characterized by a high degree of myofibril alignment. Similarly, work by Motlagh et al. has also shown that myocytes grown on microgrooved substrates of similar dimension ($10 \times 5 \times 5$ μm) showed N-cadherin and Connexin-43 expression resembling that found in the neonatal heart [26].

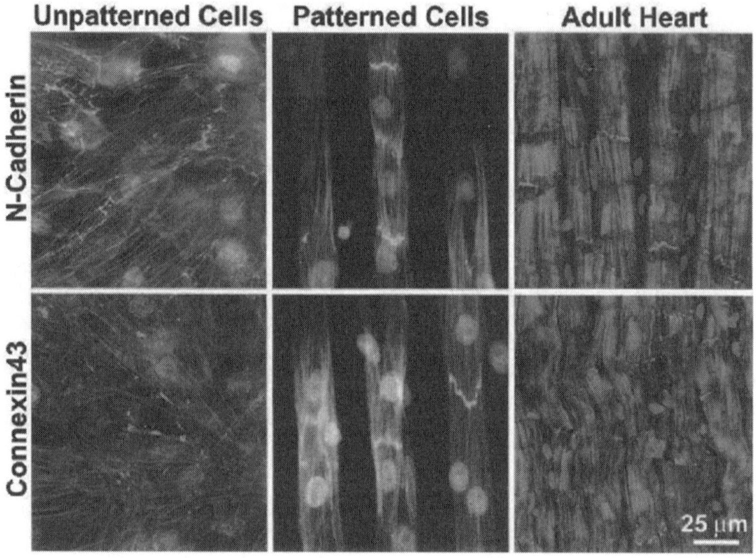

Figure 18.2 Immunofluorescent staining shows differences between patterned and unpatterned cardiomyocytes but striking similarities between patterned cultures and the architecture of the native heart. N-cadherin (top row), Connexin43 (bottom row), and actin (phalloidin); nuclei are stained with DAPI [25].

18.3 Patterned Cardiomyocyte Cultures in Three Dimensions

Patterning of cardiomyocytes in three dimensions may more accurately emulate the behavior of the cells in native heart tissue, which is three-dimensional in nature. Some microscale techniques for three-dimensional patterning and growth of cardiomyocyte cultures are discussed in this section.

18.3.1 Cardiac Organoids

Patterning can be used to engineer cylindrical cardiac organoids tens to hundreds of microns in diameter and several millimeters long that more closely resemble three-dimensional myofibres. Since they have physiologically relevant dimensions, such three-dimensional organoids are highly amenable to rapid-screening studies and can serve as in vitro models for pharmacological and pathophysiological discovery. They can also act as important tools for screening studies in cardiac-tissue engineering.

Microfluidic patterning was used to obtain three-dimensional structures with many different cell types. Microfluidic patterning through two-dimensional and three-dimensional networks has been used previously to obtain complex structures of *E. coli*, erythrocytes, bovine capillary endothelial cells, and human bladder cancer cells on planar surfaces [27, 28]. Geometric parameters essential for formation of functional neuronal networks were investigated via microfluidic patterning of poly-L-lysine and collagen 4 [29]. Three-dimensional structures were achieved using photoreaction injection molding in microfluidic channels of cells (e.g., fibroblasts) encapsulated in arginine-glycine-aspartic acid (RGD)–modified poly(ethylene glycol) [30]. In addition, microfluidic patterning allows for formation of three-dimensional structures consisting of multiple cell types (e.g., fibroblasts, smooth muscle cells, and endothelial cells) [31, 32]. In this approach, a desired cell type resuspended in an appropriate ECM is applied into a microfluidic network. Following the contraction of the biopolymer matrix by cells another layer with a different cell type is applied to the microfluidic network to build a tissue with multiple cell types arranged in the *z*-direction.

Radisic et al. have engineered three-dimensional contractile organoids by way of a microfluidic patterning approach [33]. Microchannels were used as inductive templates for organoid formation. Hyalorunan (HA) was used as a cell-repulsive protein lane, while fibronectin (FN) was used as a cell-adhesive protein layer in the spaces between these lanes. As shown in Figure 18.3, PDMS moulds were patterned from silicon masters, resulting in three-dimensional lanes 100 μm in width and 60 μm in height. The PDMS templates were cleaned with soap and water, sterilized with ethanol, plasma-treated to confer hydrophilicity upon them, and reversibly bonded to glass substrates. Subsequently, a few drops of HA solution were placed at the edge of the PDMS lanes, allowing the solution to infiltrate the PDMS channels by capillary action. The HA solution was allowed to dry for several hours, after which the PDMS templates were peeled away, leaving behind cell-repulsive HA lanes. The resulting microchannels between these HA lanes were treated with FN solution to promote cell adhesion.

The repellent properties of HA were characterized by assessing attachment of BSA, FN, and NIH 3T3 fibroblasts to HA, showing that the HA lanes did not promote protein adsorption or cell attachment. Cardiomyocytes seeded on the glass substrates preferentially attached to the FN-coated lanes, as well as at the interface between the FN- and HA-coated regions, but elongated significantly more at the interface between the two regions forming elongated organoids. These interfacially formed organoids displayed increasing amplitude of contraction with time in culture. All organoids had completely detached from the glass substrates by day three to four, by which time individual cells had elongated to lengths on the order of ~100 μm, and the organoids themselves had achieved lengths on the order of several millimeters. Organoids were fixed in paraformaldehyde and stained for cardiac

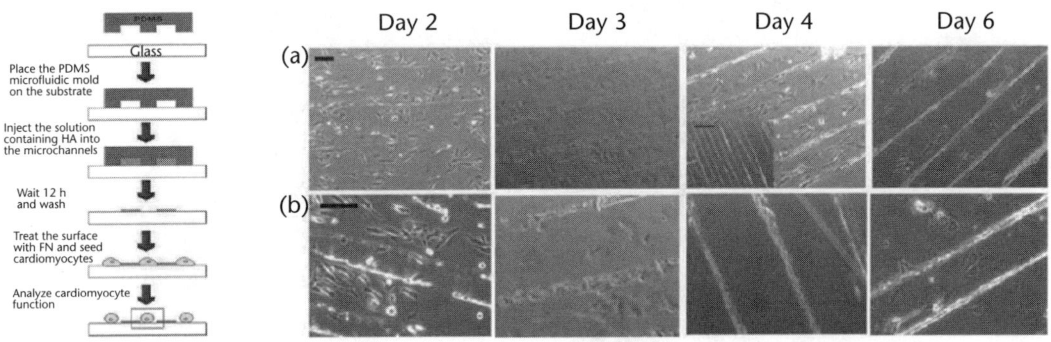

Figure 18.3 (Left) Method for microfluidic patterning of cardiac organoids. (Right) Progression of cardiac organoid formation on HA patterned surfaces: (a) Images taken at 100×. Day four inset image taken at 40× illustrates several-millimeter-long cardiac organoids. (b) Images taken at 200×. Scale bars for (a, b) = 100 μm. Inset scale bar = 1 mm [33].

troponin I, showing cross striations. This microfluidic patterning approach may aid in guiding the formation of parallel arrays of contractile organoids, which may be useful for drug discovery studies.

18.3.2 Cardiac Tissue Engineering

In a novel, scaffold-free approach, Baar et al. have used delamination and self-organization of cardiomyocyte monolayers to engineer three-dimensional cardiac organoids [34]. PDMS was poured into 35 mm plates and allowed to cure. The PDMS layer was coated with the ECM protein laminin to promote cell attachment. Two laminin-coated silk suture threads were pinned at either end of the dish to act as anchors for the delaminated cell layer. Cardiomyocytes isolated from neonatal rat ventricles were seeded on the laminin-coated PDMS layer at a density of 0.4×10^6 cells/cm^2. A confluent, beating monolayer was formed and began to detach from the hydrophobic PDMS surface after seven days in culture, as shown in Figure 18.4.

The monolayer had completely wrapped itself around the two anchors by ten days in culture, organizing itself into a cylindrical cardiac organoid. The resulting structure was 24 mm long and 100 μm in diameter and resembled papillary muscle. The ability of the cardiac organoids to generate force and the response to both electrical and β-adrenergic stimulation was characterized. The authors also showed that the organoids expressed an early embryonic phenotype, characterized by expression of both α- and β-tropomyosin, but only low levels of both SERCA2a and the mature isoform of cardiac troponin T.

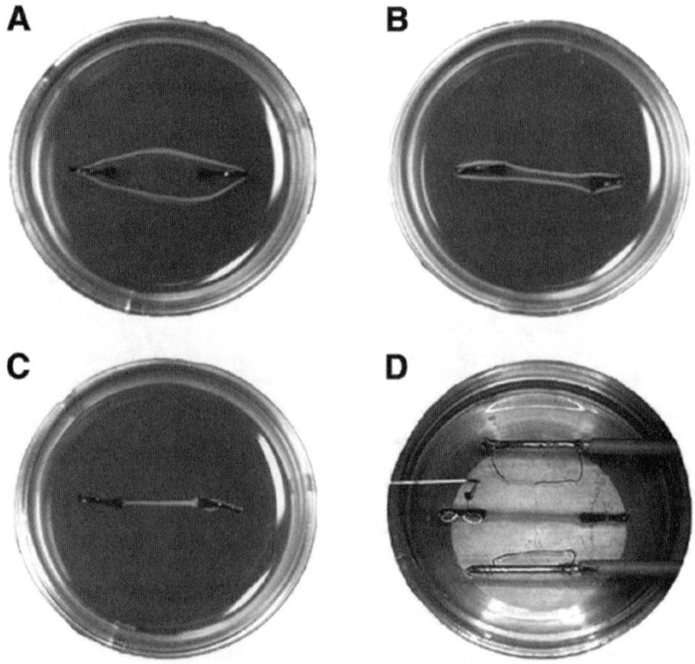

Figure 18.4 (a–c) Delamination of cardiomyocyte monolayers with time results in formation of a cardiac organoid. (d) Functional testing using platinum electrodes [34].

While the majority of three-dimensional patterning work in cardiac-tissue engineering was aimed at producing ordered arrays of organoids, a number of researchers applied the principles of microfabrication to produce micro- and nanostructured fibrous scaffolds for use in cardiac-tissue engineering. Such scaffolds have a porous geometry and exhibit a microstructured topology with a large surface area, making them highly conducive to cell attachment and infiltration. In addition, their close resemblance to the ECM makes these scaffolds ideally suited for cardiac tissue-engineering studies.

As an example, Shin et al. have developed engineered cardiac tissues that they term cardiac nanofibrous meshes, or CNMs [35]. Their metholodology involves culturing neonatal cardiomyocytes on nanofibrous poly(caprolactone) (PCL) meshes. The nanofibrous meshes were made by electrostatic fibre spinning, more commonly referred to as electrospinning. The set-up consisted of a metal needle suspended above a grounded collector. A polymer solution was fed to the needle through a constant flow syringe, and a large electric field was established between the needle and the collector, causing the solvent to evaporate and the polymer solution to be deposited on the collector as a series of thin, unordered, nanostructured fibers. The PCL mesh was stretched around a wire ring, which acted as a passive mechanical load to the myocytes. Cardiomyocytes began contracting three days after seeding and could be cultured in vitro for fourteen days. They were able to penetrate into the porous scaffolds and stained positively for cardiac troponin I, actin, tropomyosin, and gap junctions, demonstrating that a primarily cardiac phenotype was maintained and that fibroblast overgrowth did not occur.

In order to control myocyte alignment, Zong et al. [36] fabricated several variants of biodegradable PLGA-based electrospun scaffolds for cardiac-tissue engineering. The micro- and nanostructured architecture of these scaffolds was controlled to facilitate both isotropic and anisotropic growth of cardiomyocytes along the fibres. The electrospun scaffolds were either unoriented or oriented. The oriented scaffolds were created simply by applying uniaxial stretch to the unoriented scaffolds as a postprocessing step. Oriented scaffolds were shown to have a parallel arrangement of fibres compared to the random arrangement of fibres in their corresponding unoriented scaffolds. The oriented scaffolds were further shown to promote significant aligmnment of cardiomyocytes along the direction of fibres (Figure 18.5).

Varying the chemical composition and hydrophobicity of the scaffolds was found to influence cardiomyocyte attachment and alignment, as well as contractile function. For example, cardiomyocytes had a preference for a hydrophobic surface, and poly(L-lactic acid) (PLLA) scaffolds were found to promote the formation of more mature sarcomeres with better contractile properties than PLGA+PEG-PLA or PLA10GLA90+PLLA (75 percent + 25 percent) scaffolds.

In the preceding sections, we have discussed several micropatterned systems in which only one cell type was used, namely, cardiomyocytes. There is now evidence to suggest that myocyte-nonmyocyte interactions may be crucial for cardiac-tissue engineering. As discussed in Section 18.4, a microsystems approach can facilitate the study of these important interactions in a rapid and cost-effective manner in both two and three dimensions.

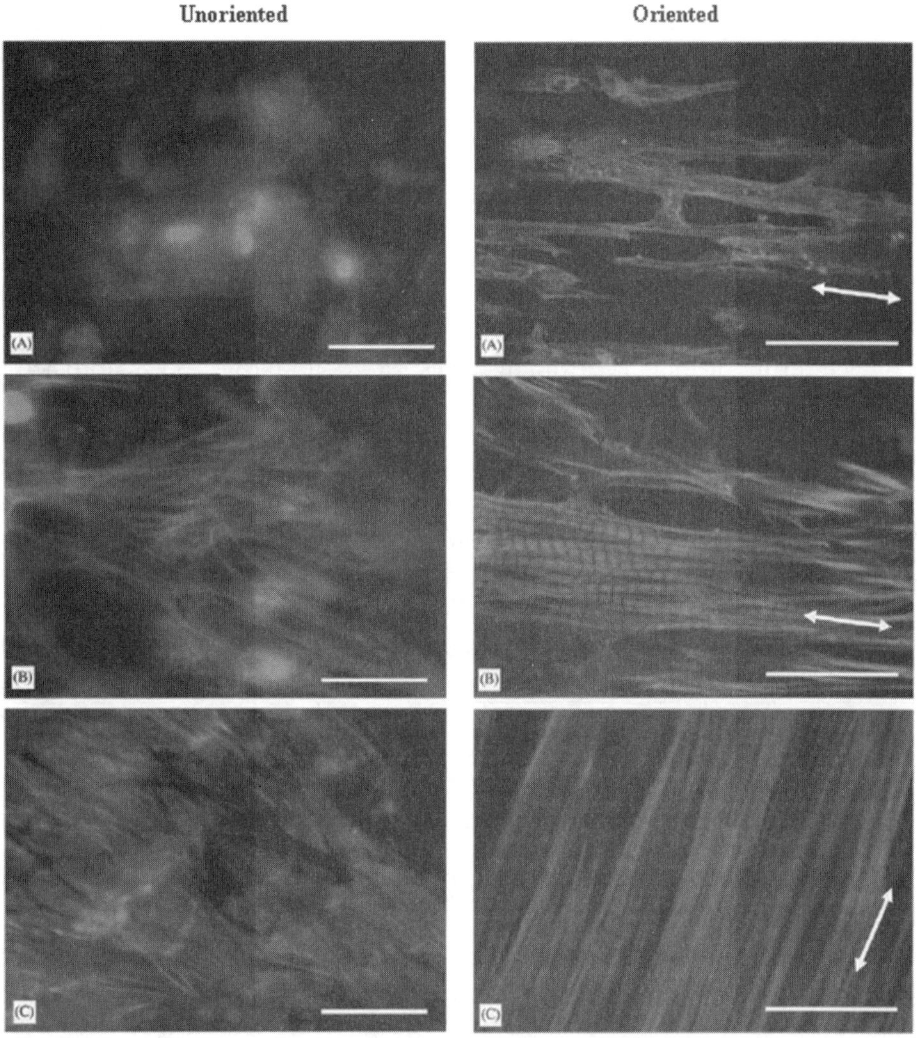

Figure 18.5 Isotropic and anistropic growth of cardiomyocytes on nanofibrous scaffolds: (Left) Unoriented scaffolds did not promote cell alignment, leading to isotropic cultures of myocytes. (Right) Oriented scaffolds (arrows indicate fibre orientation) guide the anistropic growth of cardiomyocytes. Cardiomyocytes cultured on (a) PLGA+PEG-PLA and (b) PLA10GLA90+PLLA were less aligned than those cultured on (c) PLLA scaffolds, which also exhibited better contractile function. Scale bars are 20 μm. [36].

18.4 Microsystems for Co- and Tricultures in Two and Three Dimensions

Microscale technologies are ideally suited for rapid screening of numerous culture variables, including the simultaneous growth of multiple cell types. Since native myocardium is made up of fibroblasts (FBs), endothelial cells (ECs), and cardiomyocytes (CMs), co- and triculture of the three main cardiac subpopulations in vitro may result in engineered cardiac tissues with an architecture more similar to native myocardium. There is evidence to suggest tissue-engineered muscle based on multiple cell populations promotes better structural organization, vascularization, and cell survival [37–41]. The use of microsystems can facilitate such studies by guiding the formation of microstructured, aligned two-dimensional and three-dimensional co- and tricultures of the three main cardiac cell types. Furthermore, these microstructured co- and tricultures can be cost-effectively screened for a large number of conditions while using minimal reagent volumes and cell numbers.

Camelliti, McCulloch, and Kohl used a microfluidic approach to engineer two-dimensional cardiac tissues containing both cardiac myocytes and fibroblasts [42]. Microstructured PDMS molds were pressed against the surface of culture dishes and flushed with a solution of collagen type I, resulting in 30 μm–wide collagen tracks. To create intersecting collagen tracks at prescribed angles, the process was repeated with a second layer of collagen tracks. The dishes were sterilized and immersed in Pluronic F-108 solution to prevent nonspecific cell and protein adhesion in the separating areas between the tracks. Preplating was carried out at 37°C for either fifteen minutes or two hours. This way, the nonadherent cells collected after preplating would be either a coculture consisting of both myocytes and fibroblasts (fifteen-minute preplating) or a predominantly a single culture enriched for myocytes (two-hour preplating). The cells were seeded on microstuctured tracks and on nonpatterened dishes as a control. The authors found that the fibroblasts and myocytes displayed a more adultlike phenotype, with evidence of cross-striation patterns. In addition, end-to-end gap-junctional coupling could be seen between myocytes, and the spatial arrangement of cells in the microstructured cocultures closely resembled the architecture of native myocardium (Figure 18.6).

A major limitation of cardiac cell coculture is the overgrowth of the nonmyocytes. In an attempt to limit the use of pharmacological agents, microtopographical cues were used to inhibit the overgrowth of fibroblasts in myocyte-nonmyocyte cocultures (Figure 18.7) [19]. Microstructured silicone surfaces were patterned with 10 μm–high vertical "pegs" by photolithographic techniques and were coated with laminin to promote cell attachment. Neonatal rat ventricular cells were preplated to enrich for myocytes. The resulting

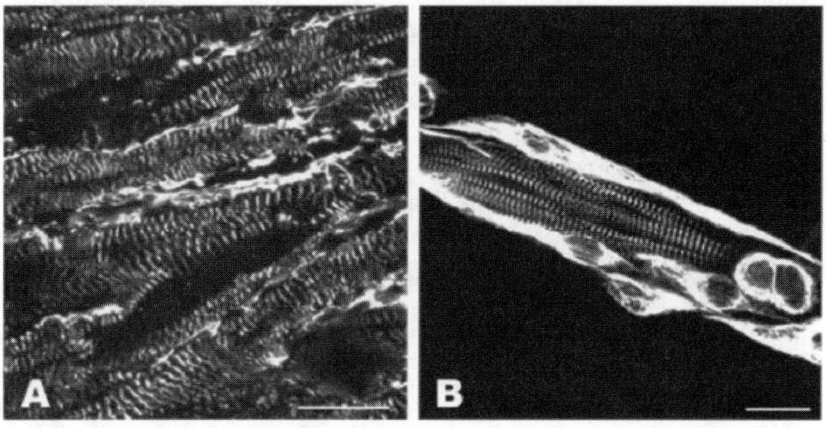

Figure 18.6 Immunostaining of fibroblasts (bright, elongated cells, vimentin) interacting with myocytes (striated cells, myomesin) in (a) rabbit ventricular tissue and (b) microstructured cocultures. In both cases, fibroblasts are seen to interact with striated myocytes as parallel bundles, displaying an in vivo–like morphology [42].

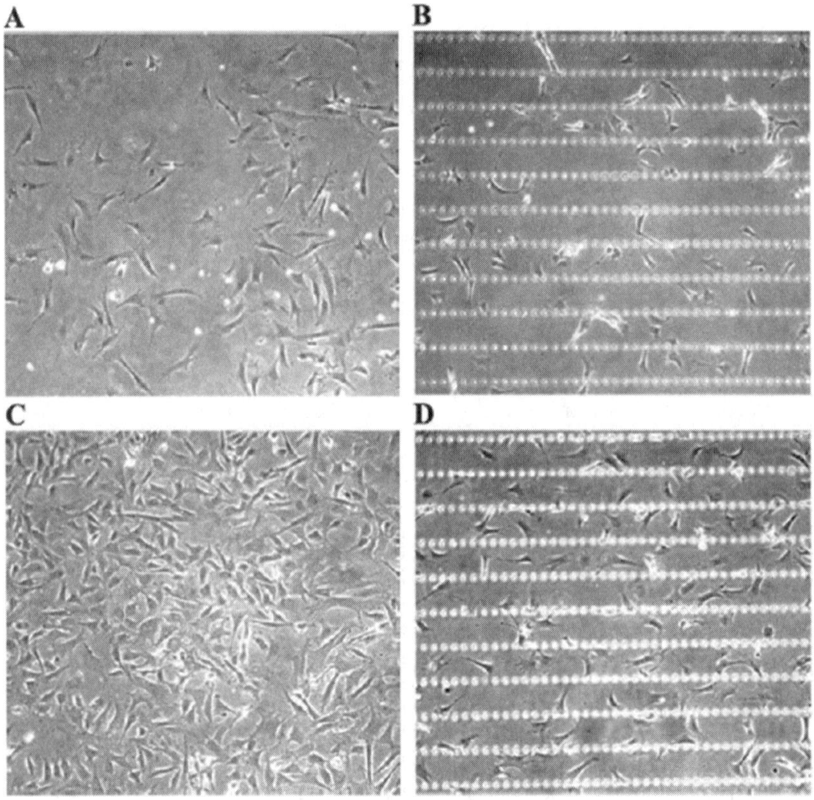

Figure 18.7 Silicone patterned with microtopographical features ("pegs") prevents fibroblast overgrowth in myocyte-fibroblast coculture: (a) day two cocultures on flat silicone membranes, (b) day two cocultures on 10 μm micropegged silicone membranes, (c) day five cocultures on flat silicone membranes, and (d) day five cocultures on 10 μm micropegged silicone membranes [19].

myocyte-enriched cell suspension (containing both fibroblasts and myocytes) was subsequently seeded on the pegged surfaces and compared to flat surfaces as a control. After four days of culture, total DNA was measured to quantify cell proliferation, showing 37 percent less proliferation of fibroblasts on micropegged surfaces compared to flat surfaces. This percentage was further reduced by addition of the DNA replication inhibitor AraC. The ratio of aminopropeptide of collagen-expressing cells (fibroblasts) to myosin-heavy chain-expressing cells (myocytes) was also decreased by 50 percent on pegged surfaces, and a 40 percent increase in Connexin-43 expression was also confirmed by Western blotting, suggesting more myocyte-myocyte interaction. The authors suggest that that the observed inhibition may be due to contact inhibition mediated via the Rho signal transduction pathway.

Microtextured substrates are evidently important to eliciting a host of in vivo–like cell responses. For example, cardiomyocyte shape, protein localization and gene expression have also been shown to be influenced by such three-dimensional microtopographical cues [43]. These cues may become important factors in the design of improved scaffolds for cardiac-tissue engineering.

Iyer et al. developed a simple microscale triculture system to allow for engineering of cardiac organoids with dimensions resembling native myofibres (~1 mm in length and ~100 μm in diameter) (Figure 18.8) [44]. In this approach, cells were cul-

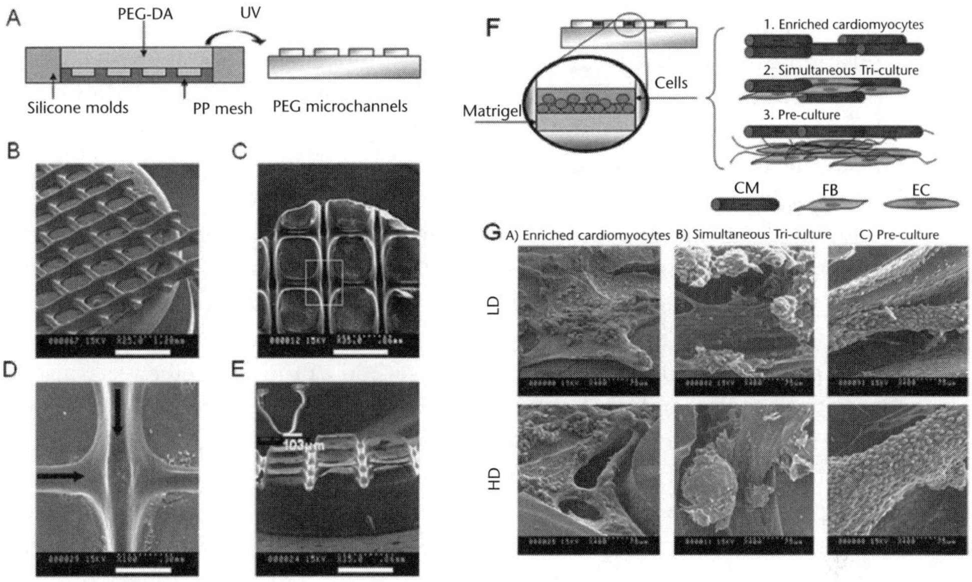

Figure 18.8 PEG microchannels act as co- and triculture templates for engineering cardiac organoids [44].

tivated in three-dimensional microchannels fabricated in photocrosslinkable poly(ethylene glycol) (PEG). We hypothesized that seeding cardiomyocytes on preformed networks of nonmyocytes (FBs and ECs) would improve structural and functional properties of engineered cardiac organoids compared to simultaneously seeding the three cell types (FB, EC, and CM) or seeding enriched cardiomyocytes alone.

Microchannels were patterned by photopolymerization of poly(ethylene glycol) diacrylate (PEG, 700 Da) using a commercially available polypropylene grid as a master [Figure 18.8(a–b)]. The resulting PEG microchannels were 100 to 200 μm in diameter and 4 mm in length [Figure 18.8(c–e)]. The microchannels were coated with 5 μL of Matrigel and allowed to gel at 37°C prior to the addition of cells. NIH 3T3 cells or cardiac FBs and D4T endothelial cells were used. For preculture, nonmyocytes (FBs and ECs) were cultivated in the microchannels for two days prior to the addition of CM. Preculture was compared to organoids based on simultaneous seeding of FB, EC, and CM (simultaneous triculture) or seeded with enriched CM alone (enriched cardiomyocytes) [Figure 18.8(f)]. As shown in Figure 18.8(g), preculture resulted in cylindrical, compact cardiac organoids that contained elongated cardiomyocytes. These organoids expressed Connexin-43 and had contractile properties similar to organoids formed from enriched cardiomyocytes. In contrast, simultaneous triculture resulted in organoids with clusters containing cardiomyocytes growing separately from elongated FBs and ECs [Figure 18.8(g)] and exhibited a complete lack of contractile activity and absence of Connexin-43.

In another novel study, a modified hanging-drop culture technique was employed to engineer spherical cardiac microtissues from myocyte-enriched and mixed myocyte-nonmyocyte cocultures [45]. In this approach, isolated neonatal rat ventricular cells were plated either as a heterogeneous coculture containing 75 percent nonmyocytes and 25 percent cardiomyocytes (corresponding to the native composition of the rodent heart) or a nearly homogeneous population of myocytes (95 percent purity), enriched by density gradient separation. Neonatal mouse ventricular myocytes were also studied, but the separation was achieved by preplating, resulting in only 50 percent purity for the enriched population. Gravity-enabled assembly of spheroids was achieved by cultivating the cells in inverted dishes, analogous to the hanging-drop method. The resulting spheroids began spontaneously contracting after forty-eight hours of cultivation. The size of the spheroids after four days ranged from 100 μm in diameter to over 300 μm in diameter and could be directly varied by increasing the cell density from twelve hundred to ten thousand cells per hanging drop. Aggregates above 200 μm in diameter were also shown to secrete vascular endothelial growth factor due to hypoxia occurring within the spheroid, suggesting the potential for induction of angiogenesis by soluble factor secretion. While the spheroid size was comparable between mixed-cell aggregates and pure myocyte-derived aggregates, the mixed-cell aggregates displayed higher collagen type I expression (due to the presence of fibroblasts in the mixture) and a characteristic spatial preference of myocytes at the outer periphery of the spheroids (as evidenced by sarcomeric α-actinin staining).

18.5 Microbioreactors for Culture of Cardiac Organoids

The native heart experiences a host of physiological cues in the form of mechanical stretching (from the rhythmic contractions of the heart), electrical stimulation (generated by the pacemaker cells and transmitted from cell to cell as action potentials), and biochemical stimulation (through soluble and matrix-mediated factors). Microbioreactor systems that supply these cues in vitro may therefore aid in engineering heart tissues with improved functionality and ultrastructural organization. Several microbioreactors employing mechanical, electrical, and biochemical stimuli are discussed in this section. Quantifying how cardiac cells and tissues respond to these cues is also essential to improving cardiac tissue-engineering methodologies.

18.5.1 Mechanical Stimuli

Mechanical stretch plays an extremely important role in the growing neonatal heart. Following birth, the increased demand on the heart to pump blood through the circulation results in significant cardiomyocyte hypertrophy [46]. Two-dimensional cultures of cardiomyocytes have also been shown to undergo hypertrophy, binucleation, and parallel organization in vitro with the application of unidirectional mechanical stretch [47]. Microsystems may enable a better understanding of how such mechanical stimuli enhance the growth and organization of cardiac cells when applied in vitro. These systems are already being designed specifically as culture platforms for ordered arrays of cardiomyocytes for studies in mechanobiology [18]. Due to their small size, microsystems also allow for the measurement and delivery of these mechanical cues at the level of individual cells. They can also be designed to harness the forces generated by cardiomyocytes to create novel micropumps operating on biochemical sources of energy. We discuss these and other interesting applications of mechanical stimuli in the context of microsystems.

The application of static uniaxial strain appears to be significant to the structural reorganization of cardiomyocytes. In a study conducted by Yu and Russell, microstructured three-dimensional wells were used to study alignment and addition of sarcomeres in neonatal cardiomyocytes under the application of uniaxial, static strain [48]. Microgrooved parylene masters with features 5 μm deep and 10 μm apart were used to create three-dimensional microwells out of PDMS. Since PDMS is a deformable elastomer, the microgrooves could be easily stretched using a mechanical stretcher. The microchannels were sterilized and incubated in a solution of fibronectin to promote cell adhesion along the grooves. Cardiomyocytes were cultivated in the grooves and were subject to 10 percent uniaxial, static strain that was maintained for between one and four hours. Cell morphology was characterized at various time points by immunostaining for α-actinin, N-cadherin, and F-actin. The authors found that the application of static strain resulted in disruptions in the periodic structure of sarcomeres compared to unstretched controls. Over time, they observed that these disruptions were repaired and remodeled, resulting in bright staining for F-actin in the disrupted areas three to four hours after initiation of strain. End-to-end myocyte coupling at intercalated discs was also disrupted, showing a tortuous pattern of N-cadherin expression compared to straight junctions in unstretched cultures. The authors suggest that these morphological changes accommodate the formation of new sarcomeres and may aid in the process of lengthening of myocytes in response to mechanical stretch.

The orientation of cardiomyocytes relative to the principal axis of stretch also has a large impact on the response of these cells to external mechanical cues. Gopalan et al. applied anisotropic biaxial strain to microstructured cultures of myocytes to assess the effect of mechanical strain on myocytes when applied in parallel with or transverse to the orientation of the myofibrils [49]. PDMS microchannel molds were fabricated from etched silicon wafers and sealed against a deformable silicone membrane, leaving the PDMS microchannels open at both ends. Collagen solution was placed at one end of the microchannels and was evenly distributed throughout the channels by applying a vacuum at the other end. The silicone membrane was connected to an elliptically shaped cell stretcher device with a 2:1 aspect ratio, enabling it to deliver a 10 percent:5 percent maximum:minimum ratio of anisotropic static strain either parallel with or transverse to the direction of the microgrooves, as shown in Figure 18.9.

The authors found that the application of maximal strain in the transverse direction produced more marked changes in morphology and protein expression compared to the longitudinal direction or to unstretched controls. In the transverse case, F-actin staining became continuous and showed loss of striation patterns. N-cadherin and Connexin-43 staining were also upregulated, as was staining for atrial natriuretic factor (ANF), a marker of hypertrophy. These results were further confirmed by Western blotting, which revealed higher protein expression levels of F-actin, N-cadherin, Connexin-43, and ANF.

Microstructured surfaces have been used to measure and harness the force-generating properties of individual beating cardiac cells [50, 51]. As an example, Tanaka et al. generated ordered arrays of micropillars in PDMS and modeled

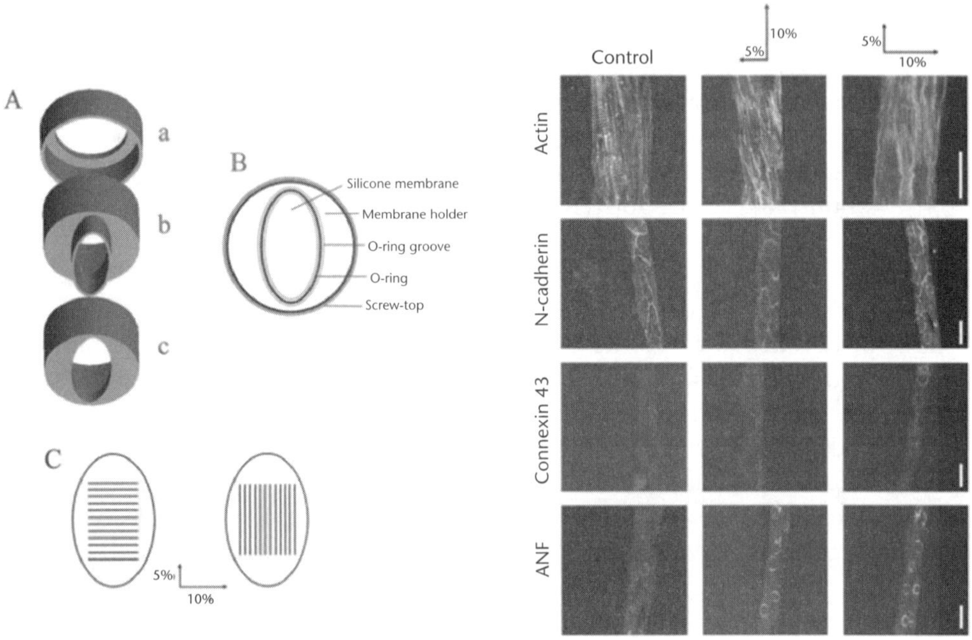

Figure 18.9 (Left) (a) Experimental setup showing the three components of the cell-stretch device: (a) screw top, (b) indenter ring, and (c) membrane holder. (b) An elastic membrane is fixed to the membrane holder section with an O-ring. (c) Schematic showing the two possible directions of maximal strain. (Right) Immunostaining of patterned cultures of cardiomyocytes subject to maximal strain either parallel with (middle column) or tranverse to (right column) the orientation of the myofibrils unstretched controls (left column). Loss of striations, homogeneous actin staining, and overall brighter staining for N-cadherin, Connexin-43, and atrial natriuretic factor (ANF) are seen mainly under transverse strain conditions [49].

them as springs to estimate the capacity of individual cardiomyocytes to drive biomicroactuators [50]. The spring constants of the pillars were determined by relating their dimensions to their Young's modulus and their moment of inertia. From Hooke's law, the spring constant of the pillars was multiplied by the displacements of the pillars caused by beating cardiomyocytes as visualized by microscopy. The authors estimated that the cardiomyocytes could generate displacements as large as 2.8 μm and forces exceeding 3.5 μN for up to one week. The advantage of such a biomicroactuator system is that it is driven by cardiomyocytes, which converts biochemical energy, stored in the form of glucose, into mechanical energy.

Work by the same group is now focused on harnessing the contractile forces of cardiomyocytes to power on-chip pumps for microfluidics. In one study, a bioactuated pump was created by coupling a contractile sheet of cardiomyocytes to a push bar connected via a flexible diaphragm to a microfluidic channel [52]. As shown in Figure 18.10, the cultured cardiomyocytes exerted contractile forces upon a push bar, which deformed a thin diaphragm. A directional fluid flow of around 2 nL/min. was achieved using this system in combination with cantilever-type microcheck valves. The authors found that contraction frequency increased with an increase in temperature from 30°C to 40°C, but this was accompanied by a concomitant decrease in contraction amplitude.

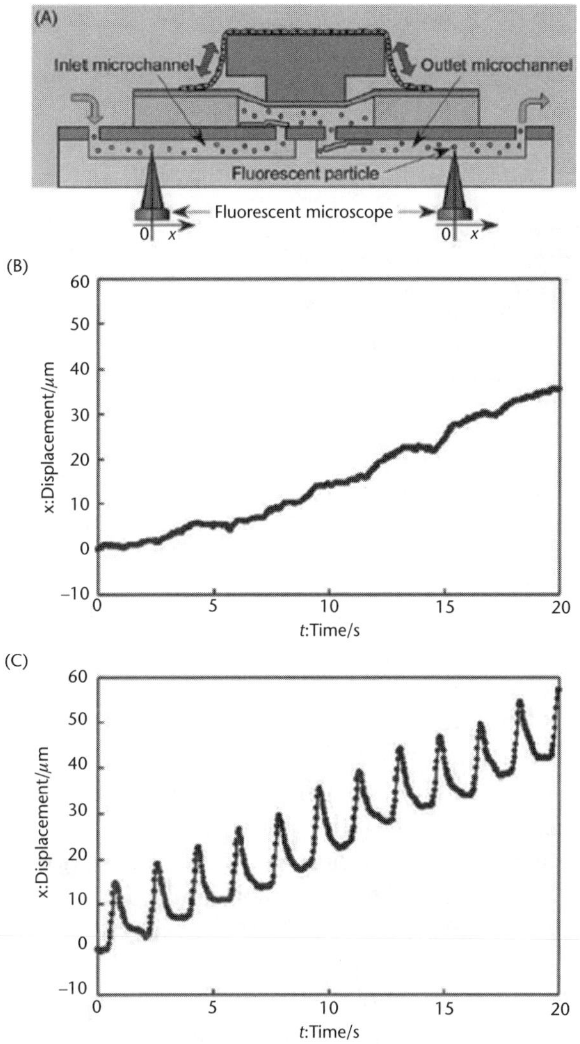

Figure 18.10 Directional fluid pumping achieved in a cardiomyocyte sheet-powered pump: (a) schematic view of microchannel geometry showing direction of fluid motion and polystyrene tracking particles, (b) displacement over time of one particle near the inlet of the microchannel, and (c) displacement over time of one particle near the outlet of the microchannel [52].

More recently, the same authors created a bio/artificial hybrid microspherical pump by wrapping a cardiomyocyte sheet around a deformable hollow PDMS sphere [53]. A Teflon capillary was threaded through a dissolvable sugar ball to create an inlet and outlet hole for fluid flow. PDMS was cured around the ball, and the sugar ball was subsequently dissolved by flushing the inside of the sphere with water for several hours. The resulting hollow PDMS microsphere served as a deformable chamber that could be pumped by a cardiomyocyte sheet, as shown in Figure 18.11.

The PDMS sphere was sterilized in ethanol and UV light and incubated in fibronectin to promote cell attachment. A detachable cardiomyocyte monolayer was produced by culturing cardiomyocytes on the thermoresponsive polymer poly(N-isopropylacrylamide) (PIPAAm). Upon lowering the temperature of the cell

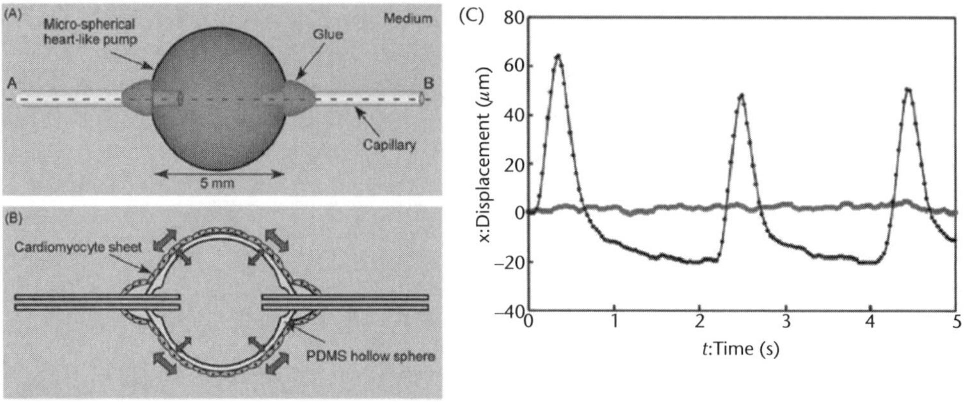

Figure 18.11 A bioactuated spherical pump: (Left) Schematic representation of the pump: (a) Two capillaries were used as an inlet and outlet for flow and were glued to the PDMS to prevent leakage. (b) A beating cardiomyocyte sheet was wrapped around the PDMS chamber to create a pulsatile pump. (Right) (c) Displacement of a fluorescent polystyrene microsphere in the absence of valve is pulsatile (dark line) compared to its displacement prior to cell-sheet transplantation (light grey) [53].

monolayer to below 32°C, the monolayer could be nonenzymatically detached and wrapped around the spherical PDMS chamber as a single sheet. The contractile sheet attached to the sphere after just one hour of incubation at 37°C, after which spontaneous contractions of the sphere immediately began generating an oscillating fluid flow. However, this fluid flow was not directional since the pump was not equipped with check valves of any kind. The fluid flow was visualized by videomicroscopy and by tracking of polystyrene microspheres in the chamber. While the beating frequency and displacement of microspheres fluctuated greatly from day to day and did not display any regular pattern, the authors provided estimates of expected flow rates that would be achieved if valves were incorporated into the design. These flow rates ranged from 0.01 to 0.1 μL/min.

18.5.2 Electrical Stimuli

The electrical activity of the heart drives the synchronous beating of the entire muscle. The spontaneous contractions observed in individual cardiomyocytes in vitro can be synchronized through the application of external electrical fields. We have shown that electrical-field stimulation enhances functional and ultrastructural organization of engineered cardiac tissues [54]. Microelectrodes are now being incorporated into microsystems and used in a number of interesting applications. These electrodes can be used to measure the intrinsic electrical activity of cardiomyocytes or to stimulate cardiomyocytes and measure their contractile response. By confining cells to small volumes between the electrodes, it is possible to study these electrical phenomena at the level of single cells without any confounding external influences such as cell-cell coupling. Electrical fields have also been used to control the spatial deposition of individual cardiomyocytes at the microscale. These topics are further discussed in this section.

The use of commercially available substrate-integrated microelectrode arrays (MEAs) is now gaining popularity as an easy way to measure cellular electrical activity. Banach et al. used such a system to measure the electrical activity of embryonic stem (ES) cells as they differentiated into cardiomyocytes over time [55]. Differentiating embryoid bodies (EBs) were plated on the MEAs, and their electrophysiological properties were measured. The authors showed an increase in spontaneous beating frequency from 1 to 5 Hz during the first ten days of culture, accompanied by a decrease in action-potential duration and rise time and an increased conduction velocity due to developmental expression of Connexin-43.

Werdich et al. fabricated a hybrid microfluidic/electrode device that could confine a single cardiomyocyte to a volume of 100 pL between a series of recording electrodes for measuring extracellular potentials [56]. Microelectrodes were fabricated by photolithography and coupled to a series of PDMS microfluidic channels designed to control the localization of individual cells between the electrodes. The incorporation of microfluidics in this design was a novel feature since it allowed for perfusion of the cell for the duration of the experiment, thereby preventing initiation of apoptosis. This perfusion system could also be used to enable targeted delivery of drugs to individual cells trapped within defined chambers of the device. The microfluidic channels also allowed efficient maneuvering of single cells into the space between the electrodes without requiring pipetting of individual cells. Adult murine ventricular cardiomyocytes were isolated, and the cell suspension was pipetted into a 4 mm–diameter reservoir on the chip. Individual cells could then be maneuvered into a cell trap flanked by several electrodes (as shown in Figure 18.12) by creating pressure differentials through the manual operation of external syringes.

Extracellular potential readings were obtained by measuring the potentials across the cell using three recording electrodes and a reference electrode. The potential readings correlated well with visually observed contractions during contraction waves (Figure 18.12). The authors also showed that the dye fluorescein could be

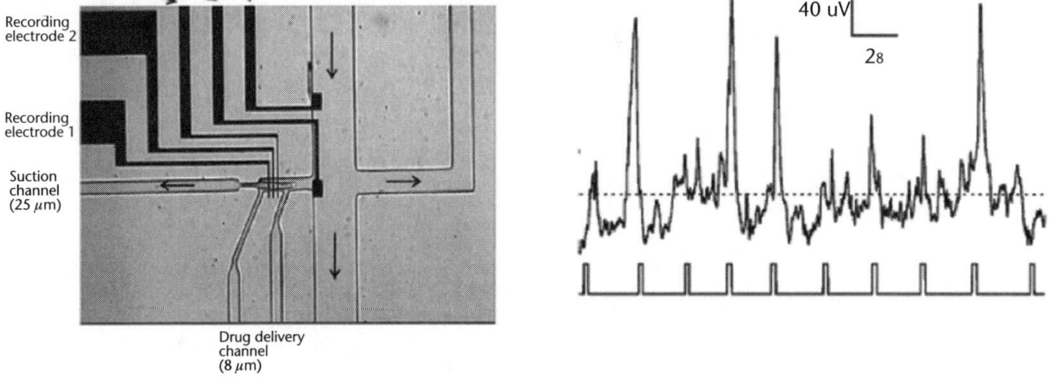

Figure 18.12 (Left) A single cardiomyocyte can be trapped in an electrode chamber by manual application of pressure gradients in the directions indicated by the arrows. The cell trap is 125 mm long and 25 mm wide, corresponding to a volume of about 100 pL. (Right) Extracellular potential readings taken with platinum electrodes showing conduction waves. The square wave plot under the trace corresponds to visually observed contractions [56].

delivered in a targeted manner to an individual cell, verifying that drug delivery could, in principle, be achieved at the single-cell level and that the response of the cell to the drug could be measured using the electrode set-up.

Klauke, Smith, and Cooper used a planar array of electrodes to stimulate individual myocytes [57]. Their design comprised PDMS microchannels 40 μm wide, 10 mm long, and 10 μm deep that were embedded with planar electrodes. The electrodes were 40 μm wide, 20 μm long, and 100 nm thick and could be individually addressed. Electrical-field stimulation was provided relative to a single, centrally located reference electrode. Individual myocytes could be confined to the space between the electrodes, which had a volume ranging from 100 pL to 5 nL. The geometry and spacing of the electrodes (Figure 18.13) was chosen so that even a small voltage of 0.5V could be applied across a 200 µm gap, allowing field strengths of up to 25 V/cm to be to produced while still not exceeding the thresholds for electrolysis of the cells (>1V). Action potentials were quantified using the voltage-sensitive dye Di-8-ANNEPPS, by measuring Ca^{2+} transients, or by measuring cell shortening. The authors applied symmetric biphasic pulses to individual cells and found that the average electric-field strength required to induce synchronous contractions (without harming the cells through electrolysis or electroporation) was 27 V/cm ± 10 V/cm.

An interesting approach adopted by Yang et al. [93] was to use microelectrode arrays to anistropically direct myocyte spatial organization using a combination of electro-orientation and dielectrophoresis. Electro-orientation theory specifies that a nonspherical object (such as a rod-shaped myocyte) experiences minimal torque when its major axis is aligned along an electric field line. Dielectrophoresis (DEP) is the migration of uncharged particles toward (positive DEP) or away (negative DEP) from an electrode due to polarization of the particle by a nonuniform electric field [58]. A series of thin-film, interdigitated electrodes was fabricated by standard photolithography techniques. Rod-shaped ventricular myocytes were isolated from adult Wistar rats and delivered to the chamber containing the electrodes by way of a syringe, as shown in Figure 18.14.

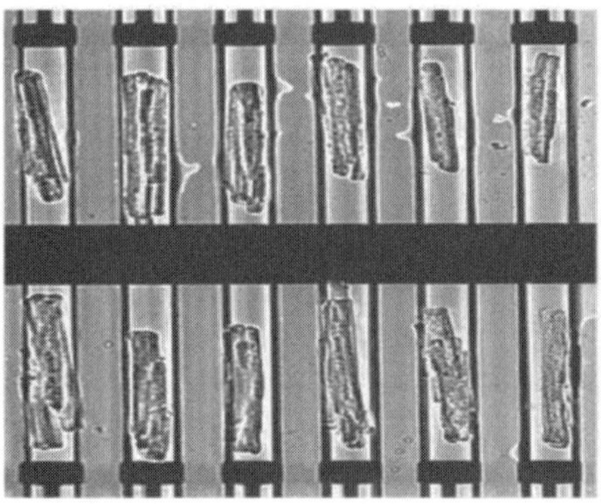

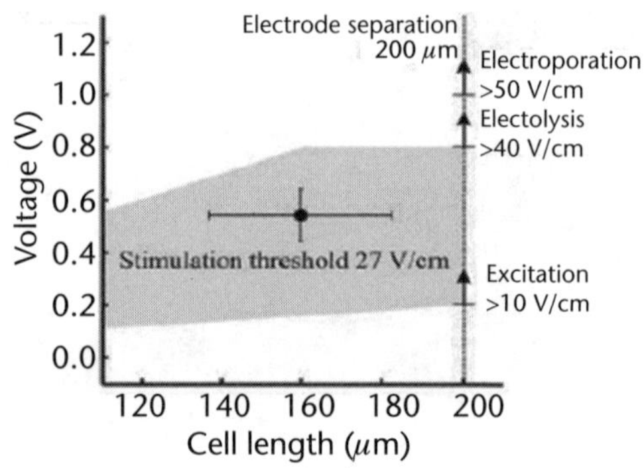

Figure 18.13 Electrode geometry and contractile response of individual myocytes to electrical stimulation: (Left) Phase-contrast image showing individual myocytes confined to microchannels and flanked by the electrodes. (Right) An analysis of contraction length as a function of applied voltage across the electrode separation distance of 200 μm. The mean stimulation threshold was measured to be 27 V/cm [57].

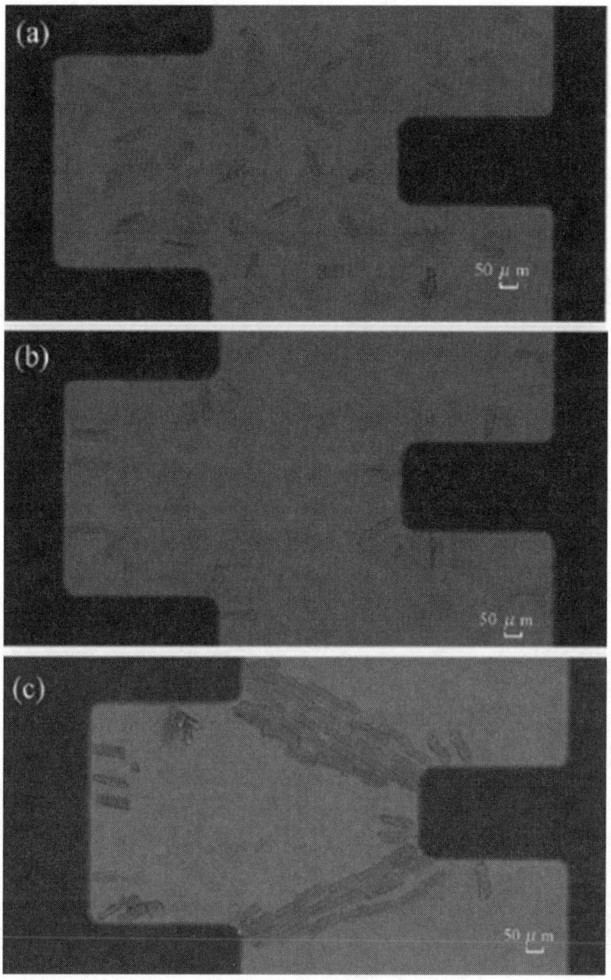

Figure 18.14 Electro-orientiation and DEP were used to direct the organization of cardiomyocytes in an interdigitated-castellated electrode setup: (a) Prior to the application of an Ac electric field, cells appear unoriented. (b, c) With the application of a sinusoidal waveform at a frequency of 2 MHz, cells become oriented and organized: (b) $2V_{peak\ to\ peak}$ amplitude, 2.1×10^3 cells/mL, and (c) $4V_{peak\ to\ peak}$ amplitude, 5.4×10^4 cells/mL. The higher seeding density and signal amplitude in (c) resulted in a more highly aligned assembly of rod-shaped myocytes.

The cells appeared randomly orientated prior to the application of an ac voltage. At a cell density of 2.1×10^3 cells/mL, a 2 MHz sinusoidal waveform of $2V_{peak\ to\ peak}$ amplitude caused cells to orient themselves along the edges of the electrodes. By increasing the cell density to 5.4×10^4 cells/mL and the amplitude of the waveform to $4V_{peak\ to\ peak}$, the cells bridged the gap between the electrodes and adopted a tissue-like morphology.

18.5.3 Biochemical Stimuli

In the native heart, cardiac cells experience a host of biochemical stimuli, ranging from paracrine signaling through cytokines to cell-cell and cell-matrix interactions [59–61]. In the context of microsystems for cardiac-tissue engineering, biochemical stimuli can be defined as synthetic or biologically derived molecules that elicit cellular responses in cardiac cells. The advantage with using microsystems to deliver these biochemical stimuli is the high degree of control over targeted spatiotemporal delivery of these molecules. This section highlights two studies in which biochemical stimuli could be delivered to muscle cells as soluble factors.

Kaji, Nishizawa, and Matsue demonstrated a technique for the delivery of localized chemical stimuli to cultures of cardiomyocytes via a microfluidic approach [62]. The device consisted of three carefully aligned layers: a glass slide at the bottom, an acrylic layer at the top containing three inlets and an outlet, and a silicone layer sandwiched in between, bearing a stencil of the microfluidic channels. Microcontact printing was used to pattern FN lanes on the glass substrate. The lanes were seeded with myocytes derived from chick embryo hearts and loaded with 10 μM fluo-3 to enable intracellular Ca^{2+} imaging. The resulting two-dimensional cultures preferentially adhered to the FN-coated areas.

The authors used this system to study the time course of cytosolic Ca^{2+} fluorescence prior to and during stimulation with 100 μM 1-octanol, a gap-junction blocker. They demonstrated localized delivery of 1-octanol, which inactivated only the part of the micropatterned cardiomyocyte culture where it was delivered (as evidenced by a decrease in intracellular Ca^{2+} fluorescence). The delivery was so well localized that it did not affect the cells immediately adjacent (Figure 18.15), which retained their synchronous pulsatile activity, demonstrating remarkable spatial control. Thus, the described microfluidic approach could be used to achieve targeted and localized delivery of drugs, cytokines, or other biochemical stimuli to predefined areas in a two-dimensional patterned culture to facilitate in vitro screening studies.

Tourovskaia, Figueroa-Masot, and Folch employed a microfluidic perfusion bioreactor to direct myoblast differentiation [63]. PDMS channels were used to etch microchannels onto two-inch borosilicate glass discs by selectively exposing an interpenetrating network (IPN) of P(AAm-co-EG) to an oxygen plasma. The resulting microchannels were 50 μm deep and 15 to 40 μm wide. The same PDMS channels were used to deliver poly-D-lysine and a 1:6 mixture of Matrigel:culture medium to the etched regions between the IPN patterns to preserve protein biological activity and to promote cell adhesion, respectively. Following this, the PDMS channel mask was removed, and a separate microfluidic template was placed atop the patterned glass substrate. C2C12 myoblasts were seeded through the inlet of the microfluidic channels, and the medium was switched to differentiating medium after the myoblasts reached confluence (after one day of culture). Using this system, hydrodynamic focusing could be employed to "squeeze" the central fluid flow (containing a soluble factor or a cell-labeling molecule), thereby enabling an easy method of directing heterogeneous flows along the center microfluidic channel versus the side channels.

In both of the above-mentioned studies, a microfluidic device enabled the delivery of biochemical stimuli to patterned cardiomyocytes cultures. Microfluidic systems are also ideally suited for separating heterogeneous mixtures of cardiac cell suspensions (as in the case of native heart cell isolates), and a number of novel separation process have been employed for cardiac-tissue engineering. The use of microsystems for cell separation will be discussed Section 18.6.

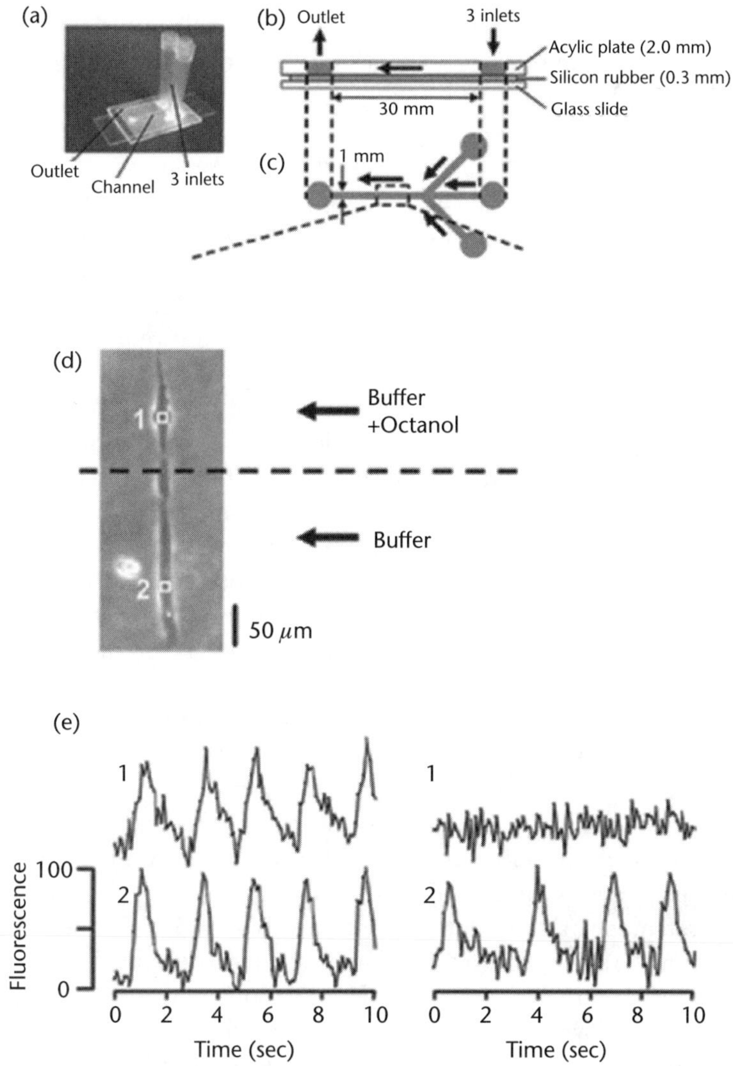

Figure 18.15 (a) Macroscopic view of the microfluidic device. The three layers of the device are seen from (b) the side view and (c) the top view. (d) Phase contrast image showing the effect of the gap-junction blocker 1-octanol on cell morphology. (e) Fluorescence intensity of intracellular Ca^{2+}. (e, left) Prior to localized delivery of 1-octanol and (e, right) two minutes following delivery of 1-octanol.

18.6 Microfluidic Devices for Cardiac Cell Separation

An important consideration in culturing cardiomyocytes on scaffolds in vitro is the purity of cardiomyocytes in the heterogeneous suspension obtained from the digestion of a donor cardiac tissue sample. The molecular, structural, and electrophysiological properties of engineered cardiac tissue have been shown to improve by increasing the concentration of cardiomyocytes in the cell suspension seeded onto scaffolds [64]. The ability to effectively enrich or isolate cardiomyocytes becomes all the more significant if only limited quantities of donor tissue are available. Cardiomyocytes are the largest cells in the myocardium, with diameters ranging from roughly 13 to 19 μm. The nonmyocytes collectively range in size from 3 to 12 μm.

The most commonly used separation technique in the area of cardiac-tissue engineering is preplating. This technique is used to remove fibroblasts and smooth muscle cells from the heterogeneous myocardial cell suspension by differential adhesion following digestion of the donor tissue with collagenase solutions [65, 66]. In preplating, the cell suspension is kept in a tissue-culture plate for a period of fifteen to seventy-five minutes to allow removal of fibroblasts and smooth muscle cells by fast and preferential attachment to the tissue-culture plastic surface. In addition to these two cell types, the tissue-culture plate surface can also contain islet-1+ cardiac progenitor cells, as reported recently by Laugwitz et al. [65]. This technique suffers from several shortcomings: (1) while the cell suspension is certainly enriched in cardiomyocytes and endothelial cells, some of these cells can become adhered to the surface along with the other cell types (i.e., the myocyte enrichment process is not efficient); (2) this technique is not systematic in that there are no parameters that can be manipulated to control the enrichment process; and (3) this process does not provide a facile way to further separate the nonmyocyte populations that are adhered onto the tissue-culture plate surface. Indeed, as a result of (1) and (3), if cultures with high nonmyocyte content are desired, three to seven days of proliferation are required to allow the nonmyocytes to overgrow the cardiomyocytes present on the plate surface. Aside from the disadvantage of having to wait during the additional three-to-seven-day time period, there is a risk of gene-expression changes in the nonmyocyte cells. All of the above shortcomings can be particularly acute if only a small amount of donor cardiac tissue is available. Another commonly used separation technique is the use of cell strainers [67–70]. This technique is even more nonsystematic and prone to limitations (2) and (3) described above.

The above constraints and limitations motivate the development of a new separation technique that can: (1) isolate cardiomyocytes with high efficiency from the myocardial suspension; (2) provide pure populations of the nonmyocyte cell subpopulations, including cardiac progenitors, and (3) accomplish separation without compromising the function (including cell phenotype and gene expression) and viability of the cardiomyocytes and progenitor cells. Furthermore, this separation technique should be easy to use in both laboratory and clinical settings.

Microfluidic devices have recently emerged as effective tools for cell separation. These devices are fabricated using lithographic techniques originally developed in the semiconductor industry. The advent of soft lithography resulted in a process whereby an elastomeric polymer (typically PDMS) can be poured onto a lithographically patterned mold, cured, and then peeled off. The patterned polymer can then be bonded to glass slides to create functional fluidic devices. The ability to create replicas in this manner is termed "rapid prototyping" and is widely recognized for its cost-effective and facile nature [71–73]. Microfluidic devices fabricated by this process can range from simple parallel-plate flow chambers to highly complex systems with features such as sieves, valves, and electrical and optical interfaces [74–82]. Unlike macroscale size-based separation approaches, such as the use of cell strainers, the microscale geometry of the flow channels in these devices ensures that fluid flow is laminar. This in turn results in predictable and reproducible cell movement.

Among the numerous microfluidic cell-separation technologies developed in recent years are approaches based on size and adhesion [83]. Sethu, Sin, and Toner developed a sieve-based separation device for the depletion of leukocytes from whole blood [84]. This device consists of a central flow channel connected to channels on either side by rectangular sieves. When blood is flowed into this middle channel, these sieves allow the disk-shaped erythrocytes to pass from the middle to the side channels, while retaining the spherical leukocytes in the middle channel. The main advantages of a size-based approach are that it does not require the presence of cell-specific markers to achieve separation and that it is the least invasive. This approach can be particularly useful in the isolation of cardiomyocytes, which are larger than the other cell types present in the myocardium.

Murthy et al. recently demonstrated how antibody-coated microfluidic devices can be used to separate leukocyte subpopulations (T and B lymphocytes) to high purity (>97 percent) [85]. In this work, the differences in antigen expression between the two cell types (which are of the same size) were exploited to separate the lymphocyte subpopulations by attaching antibodies to the surfaces of flow channels. The role of flow rate and shear stress in the antibody-mediated capture of target cells was also quantified. This approach can be useful in the separation of nonmyocytes.

The systematic nature of microfluidic cell separation is illustrated by the availability of operational parameters that can be manipulated to influence the separation process in terms of purity and/or throughput. These parameters include flow rate and sieve dimensions in the case of the size-based approach and shear stress and surface composition in the case of adhesion-based separation.

18.6.1 Size-Based Separation

Murthy et al. have shown recently that a sieve-based device can be utilized to enrich cardiomyocytes from a heterogeneous myocardial suspension obtained from neonatal rat hearts [86]. As shown in Figure 18.16, this device consists of a central flow channel separated from two adjacent channels by rectangular sieves that are 40 μm long and 5 μm in height. The design of this device is such that the large cardiomyocytes are expected to remain in the main (middle) channel, whereas the smaller nonmyocytes are expected to flow through the sieves into the adjacent (side) channels.

Figure 18.17 shows histograms of cell-size distributions in the suspensions recovered from the middle channel and the side channels, as well as the original suspension. The original suspension [Figure 18.17(a)] shows a bimodal distribution with peaks in the 7 to 9 μm region and the 15 to 17 μm region. The larger cells are mostly cardiomyocytes, whereas the smaller cells consist of a number of different cell types, as determined by flow cytometry and expression of troponin I (a cardiomyocyte marker). The data in Figure 18.17(c) shows that the concentration of the small cells in the side channel output is certainly enriched relative to the original suspension as indicated by the single peak in the 7 to 9 μm range. The suspension recovered from the middle channel, however, presents a different picture [Figure 18.17(b)]. Several small cells remain, and there are fewer large cells than would be expected. This result is very likely a consequence of sieves becoming

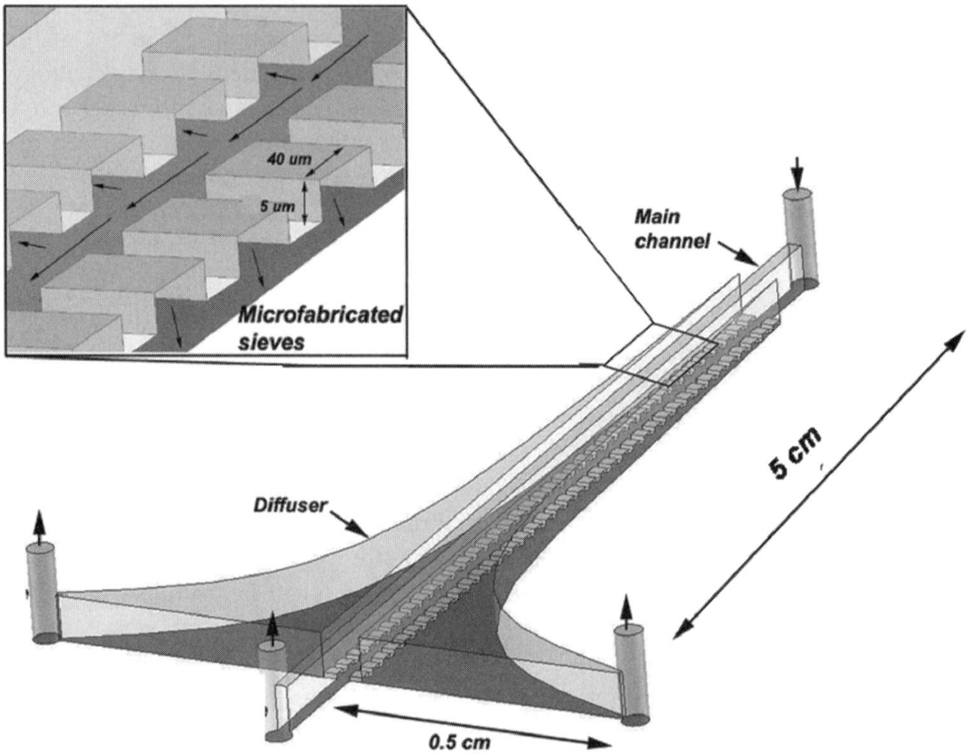

Figure 18.16 Schematic diagram of the sieve-based microfluidic device utilized by Murthy et al. for cardiomyocyte enrichment [86]. This device separates cells based on size by allowing small cells to pass through sieves between the main (middle) and side flow channels. Larger cells do not pass through, and these are recovered from the main channel.

clogged with the large cardiomyocytes, thereby preventing the passage of the smaller cells. The insufficient cardiomyocyte enrichment and the problem of clogging motivate the development of a new design for microfluidic separation. Nevertheless, this design represents a first-pass attempt and provides many insights into the behavior in flow of cardiac cells and their separation characteristics.

Flow through the microfluidic devices did not affect cell viability for all cell types. Furthermore, cell culture indicated that the recovered cells from both the middle and side channels retained their attachment ability and function after microfluidic fractionation. Figure 18.18 shows fluorescence micrographs of the cells recovered from the middle and side channels compared to the original suspension. To identify cell subpopulations, the cultures were stained for the cardiomyocyte marker troponin I and the nonmyocyte marker vimentin. The original suspension and the middle channel output both contain a mixture of cardiomyocytes and nonmyocytes. As highlighted by the arrows in Figure 18.18(b), cardiomyocytes are large and contain well-developed contractile apparatus. The side channel output mostly comprises nonmyocytes, which tend to spread on the culture dish surface [Figure 18.18(c)]. The myocytes obtained from the middle channel showed spontaneous contractile activity after forty-eight hours in culture.

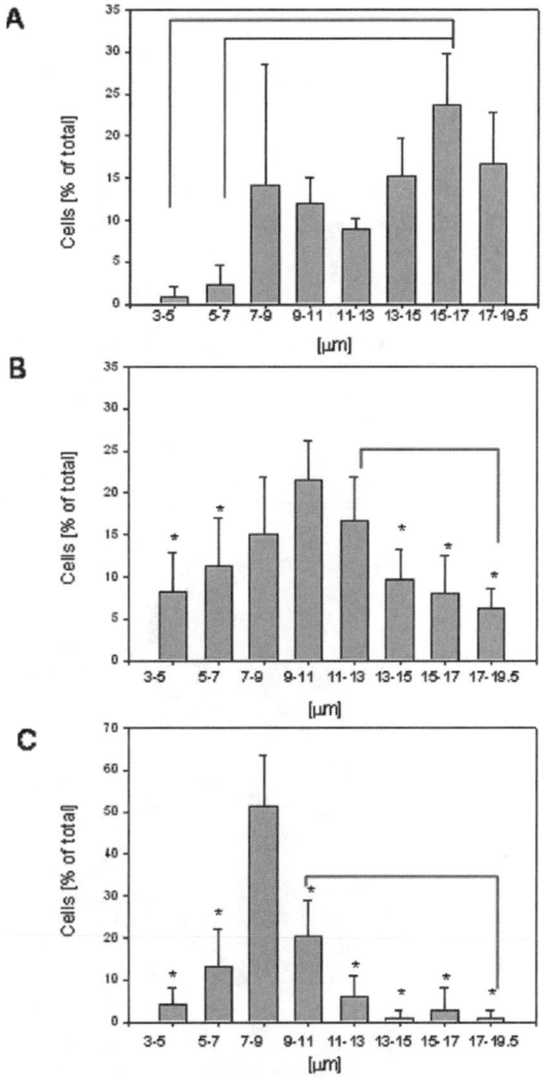

Figure 18.17 Size distributions for the cardiac cells in (a) the original suspension, (b) the middle channel output, and (c) the side channel output [86]. In (b), "*" indicates significantly less than 9 to 11 μm, and in (c), "*" indicates significantly less than 7 to 9 μm, as determined by Tukey's test with one-way ANOVA with $p < 0.05$ being considered significant.

18.6.2 Adhesion-Based Separation

Size-based separation techniques cannot be used to separate the different subpopulations of nonmyocytes present in the myocardial suspension since they are of similar size. However, these cells, which include smooth muscle cells, endothelial cells, and fibroblasts, can be separated on the basis of selective adhesion to peptide-coated surfaces. Smooth muscle and endothelial cells, for example, are known to preferentially adhere to surfaces coated with the peptides Val-Ala-Pro-Gly and Arg-Glu-Asp-Val, respectively [87, 88]. Surfaces coated with the peptide Arg-Gly-Dsp-Ser (RGDS) are known to bind a number of cell types, including

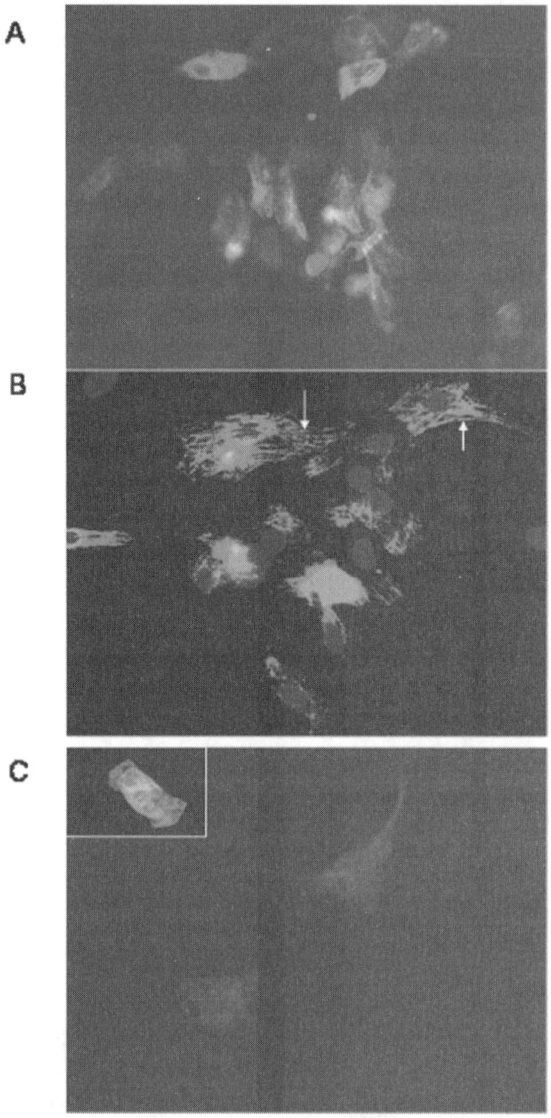

Figure 18.18 Fluorescence micrographs of cultured neonatal rat cardiac cells from (a) the original suspension, (b) the middle channel, and (c) the side channel [86]. Cells are stained for troponin I and vimentin, along with a nuclear stain (DAPI).

fibroblasts [89, 90]. For adhesion-based separation in a flow system, the effect of fluid shear forces on cell adhesion must also be considered [91]. Figure 18.19 shows the geometry of a flow chamber designed such that the shear stress along the longitudinal axis of the device decreases linearly as a function of distance. This type of device is a useful tool in assessing the combined effects of surface composition and fluid shear stress on cell adhesion.

The potential to exploit differential adhesion of nonmyocyte cell populations using microfluidic devices coated with peptides is illustrated in Figure 18.20. This data was obtained using immortalized cell lines, namely the A7r5 smooth muscle cells derived from the mouse aorta, the H5V mouse cardiac endothelial cells, and the

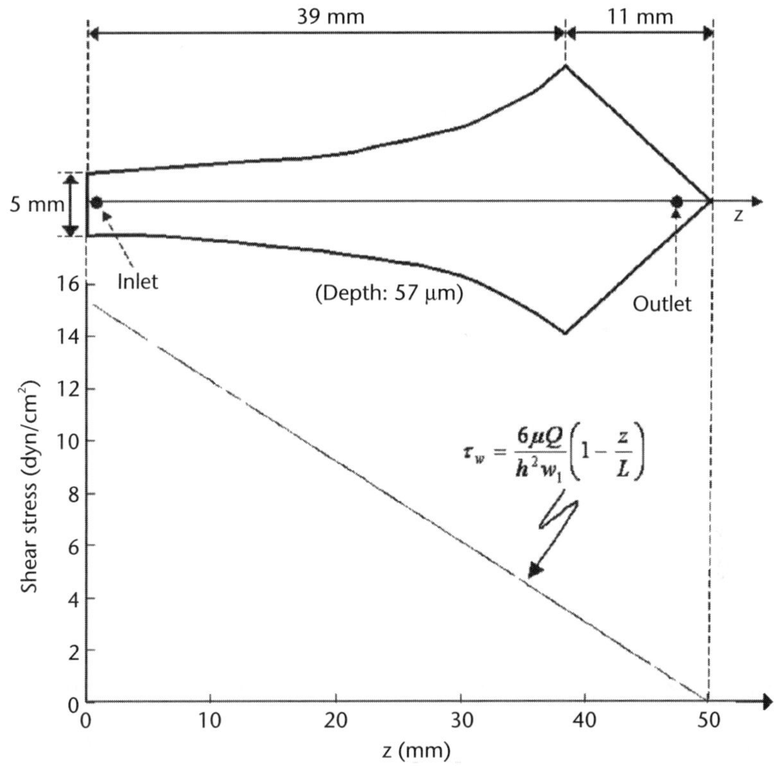

Figure 18.19 Geometry of a flow chamber capable of generating a linear shear stress gradient. In the equation given for shear stress, Q is the volumetric flow rate, μ the viscosity of water, h the chamber depth, w_1 the inlet width, and L the chamber length. This geometry is based on equations derived by Usami et al. [92].

3T3-J2 mouse fibroblast cell line. The microfluidic device used for this work had the geometry shown in Figure 18.19, and the data for each cell type was collected in a separate experiment (i.e., the three different cell types were not mixed). The peptide RGDS was immobilized onto the surface of the microfluidic device using a three-step process described by Murthy et al. [85]. All three cell types show decreasing adhesion as shear stress increases. Fibroblasts have the highest degree of adhesion relative to endothelial cells and smooth muscle cells for shear stresses below about 1.6 dyn/cm^2. While this experiment does not actually involve a separation of different cell types, it suggests the feasibility of using peptide-coated microfluidic devices. This experiment, specifically, suggests that a suspension containing three nonmyocyte cell types can be depleted of fibroblasts and endothelial cells using an RGDS-coated microfluidic device, thereby enriching the smooth muscle cell content.

18.7 Looking Forward

Considering that the application of soft lithography, microcontact printing, and micropatterning to cardiac-tissue engineering has occurred relatively recently, it is

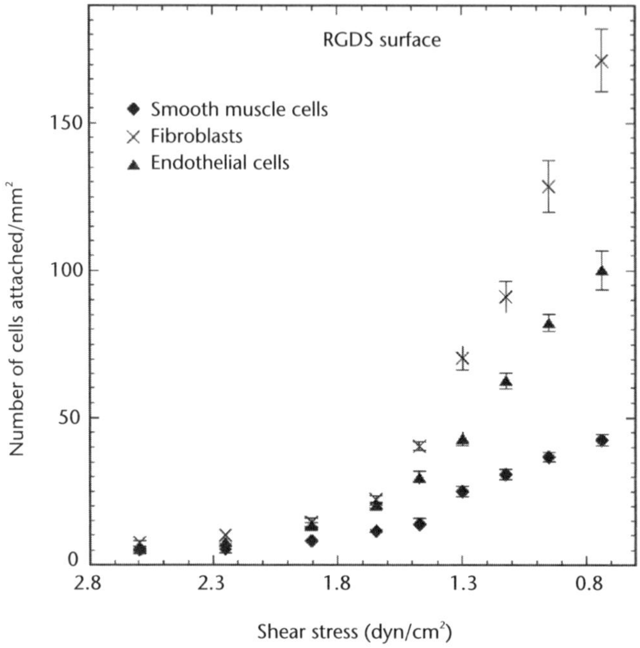

Figure 18.20 Cell adhesion as a function of shear stress in linear shear stress gradient microfluidic devices coated with RGDS. Experiments were performed with one cell type at a time; no mixed suspensions were used. For each cell type, adhesion values represent averages over ten independent experiments, and error bars denote the standard error (standard deviation/$\sqrt{n}$, where $n = 10$).

both surprising and encouraging to see the diversity of applications that have already come out of the amalgation of these disciplines. Despite the knowledge already acquired as to the importance of two-dimensional and three-dimensional microtopographies to cardiac cell behaviour, a better elucidation of the precise mechanisms at work is needed. The ultimate goal of this work would be to controllably reproduce these cues in three-dimensional scaffolds to exert the same influences at a macroscopic level to facilitate tissue engineering of a beating cardiac patch. Given their compact size, minimal need for reagent volumes, and cell numbers, the microbioreactors possess an immense potential for aiding further drug discovery, as well as studies in stem cell differentiation and pathophysiology. We anticipate that these studies will make use of advanced microdevices that combine multiple stimuli (mechanical, electrical, and biochemical) in a single system.

18.8 Conclusion

Microfabricated systems can guide the formation of structured two-dimensional and three-dimensional cultures of cardiomyocytes, as well as co- and tricultures of cardiomyocytes with nonmyocytes found in the native heart. Such microstructured cultures can serve as ideal in vitro models for pathophysiological studies and for screening studies of novel drugs or therapeutics. Microbioreactors supplying physiological cues in the form of mechanical, electrical, or biochemical stimuli, or a combination thereof, can serve as excellent prototypic models for the design of macroscale bioreactors. Such bioreactors may ultimately aid in engineering a clinically thick (~1 cm), beating cardiac patch through the application of biomimetic conditioning. Microsystems have also been applied to a number of separation processes to separate multiple cardiac cell populations from heterogeneous mixtures of cells. Continued improvements in microfabrication technology and a better understanding of the precise biological mechanisms triggered in cardiac cells by these microscale stimuli will undoubtedly foster many more new and exciting discoveries in the area of cardiac-tissue engineering.

References

[1] Severs, N. J., "The cardiac muscle cell," *Bioessays*, Vol. 22, No. 2, 2000, pp. 188–199.

[2] Singhvi, R., Kumar, A., Lopez, G. P., Stephanopoulos, G. N., Wang, D. I., Whitesides, G. M., and Ingber, D. E., "Engineering cell shape and function," *Science,* Vol. 264, No. 5159, 1994, pp. 696–698.

[3] Mrksich, M., and Whitesides, G. M., "Using self-assembled monolayers to understand the interactions of man-made surfaces with proteins and cells," *Ann. Rev. Biophys. Biomol. Struct.*, Vol. 25, 1996, pp. 55–78.

[4] Mrksich, M., Dike, L. E., Tien, J., Ingber, D. E., and Whitesides, G. M., "Using microcontact printing to pattern the attachment of mammalian cells to self-assembled monolayers of alkanethiolates on transparent films of gold and silver," *Exp. Cell Res.*, Vol. 235, No. 2, 1997, pp. 305–313.

[5] Whitesides, G. M., Ostuni, E., Takayama, S., Jiang, X., and Ingber, D. E., "Soft lithography in biology and biochemistry," *Ann. Rev. Biomed. Eng.*, Vol. 3, 2001, pp. 335–373.

[6] Prime, K. L., and Whitesides, G. M., "Self-assembled organic monolayers: Model systems for studying adsorption of proteins at surfaces," *Science,* Vol. 252, No. 5010, 1991, pp. 1164–1167.

[7] Mrksich, M., Chen, C. S., Xia, Y., Dike, L. E., Ingber, D. E., and Whitesides, G. M., "Controlling cell attachment on contoured surfaces with self-assembled monolayers of alkanethiolates on gold," *Proc. Natl. Acad. Sci. USA*, Vol. 93, No. 20, 1996, pp. 10775–10778.

[8] Chen, C. S., Mrksich, M., Huang, S., Whitesides, G. M., and Ingber, D. E., "Geometric control of cell life and death," *Science,* Vol. 276, No. 5317, 1997, pp. 1425–1428.

[9] Hyun, J., Zhu, Y. J., Liebmann-Vinson, A., Beebe, T. P., and Chilkoti, A., "Microstamping on an activated polymer surface: Patterning biotin and streptavidin onto common polymeric biomaterials," *Langmuir*, Vol. 17, No. 20, 2001, pp. 6358–6367.

[10] Bhatia, S. N., Balis, U. J., Yarmush, M. L., and Toner, M., "Microfabrication of hepatocyte/fibroblast co-cultures: Role of homotypic cell interactions," *Biotechnol. Prog.*, Vol. 14, No. 3, 1998, pp. 378–387.

[11] Bhatia, S. N., Balis, U. J., Yarmush, M. L., and Toner, M., "Probing heterotypic cell interactions: Hepatocyte function in microfabricated co-cultures," *J. Biomater. Sci. Polym. Ed.*, Vol. 9, No. 11, 1998, pp. 1137–1160.

[12] Bhatia, S. N., Yarmush, M. L., and Toner, M., "Controlling cell interactions by micropatterning in co-cultures: Hepatocytes and 3T3 fibroblasts," *J. Biomed. Mater. Res.*, Vol. 34, No. 2, 1997, pp. 189–199.

[13] Bhatia, S. N., Balis, U. J., Yarmush, M. L., and Toner, M., "Effect of cell-cell interactions in preservation of cellular phenotype: Cocultivation of hepatocytes and nonparenchymal cells," *Faseb J.*, Vol. 13, No. 14, 1999, pp. 1883–1900.

[14] Hyun, J., Ma, H., Zhang, Z., Beebe, T. P., and Chilkoti, A., "Universal route to cell micropatterning using an amphiphilic comb polymer," *Adv. Mat.*, Vol. 15, Nos. 7–8, 2003, pp. 576–579.

[15] Takayama, S., McDonald, J. C., Ostuni, E., Liang, M. N., Kenis, P. J. A., Ismagilov, R. F., and Whitesides, G. M., "Patterning cells and their environments using multiple laminar fluid flows in capillary networks," *Proc. Natl. Acad. Sci. USA*, Vol. 96, No. 10, 1999, pp. 5545–5548.

[16] Ostuni, E., Kane, R., Chen, C. S., Ingber, D. E., and Whitesides, G. M., "Patterning mammalian cells using elastomeric membranes," *Langmuir*, Vol. 16, No. 20, 2000, pp. 7811–7819.

[17] Folch, A., Jo, B. H., Hurtado, O., Beebe, D. J., and Toner, M., "Microfabricated elastomeric stencils for micropatterning cell cultures," *J. Biomed. Mater. Res.*, Vol. 52, No. , 2000, pp. 346–353.

[18] Deutsch, J., Motlagh, D., Russell, B., and Desai, T. A., "Fabrication of microtextured membranes for cardiac myocyte attachment and orientation," *J. Biomed. Mater. Res.*, Vol. 53, No. 3, 2000, pp. 267–275.

[19] Boateng, S. Y., Hartman, T. J., Ahluwalia, N., Vidula, H., Desai, T. A., and Russell, B., "Inhibition of fibroblast proliferation in cardiac myocyte cultures by surface microtopography," *Am. J. Physiol.: Cell. Physiol.*, Vol. 285, No. 1, 2003, pp. C171–C182.

[20] Rohr, S., Scholly, D. M., and Kleber, A. G., "Patterned growth of neonatal rat heart cells in culture: Morphological and electrophysiological characterization," *Circ. Res.*, Vol. 68, No. 1, 1991, pp. 114–130.

[21] Thomas, S. P., Bircher-Lehmann, L., Thomas, S. A., Zhuang, J., Saffitz, J. E., and Kleber, A. G., "Synthetic strands of neonatal mouse cardiac myocytes: Structural and electrophysiological properties," *Circ. Res.*, Vol. 87, No. 6, 2000, pp. 467–473.

[22] Gaudesius, G., Miragoli, M., Thomas, S. P., and Rohr, S., "Coupling of cardiac electrical activity over extended distances by fibroblasts of cardiac origin," *Circ. Res.*, Vol. 93, No. 5, 2003, pp. 421–428.

[23] Fast, V. G., Rohr, S., and Ideker, R. E., "Nonlinear changes of transmembrane potential caused by defibrillation shocks in strands of cultured myocytes," *Am. J. Physiol.: Heart Circ. Physiol.*, Vol. 278, No. 3, 2000, pp. H688–H697.

[24] Karp, J. M., Yeo, Y., Geng, W., Cannizarro, C., Jan, K., Kohane, D. S., Vunjak-Novakovic, G., Langer, R., and Radisic, M., "A photolitographic method to create cellular micropatterns," *Biomaterials*, in press.

[25] McDevitt, T. C., Angello, J. C., Whitney, M. L., Reinecke, H., Hauschka, S. D., Murry, C. E., and Stayton, P. S., "In vitro generation of differentiated cardiac myofibers on micropatterned laminin surfaces," *J. Biomed. Mater. Res.*, Vol. 60, No. 3, 2002, pp. 472–479.

[26] Motlagh, D., Hartman, T. J., Desai, T. A., and Russell, B., "Microfabricated grooves recapitulate neonatal myocyte connexin43 and N-cadherin expression and localization," *J. Biomed. Mater. Res. A*, Vol. 67, No. 1, 2003, pp. 148–157.

[27] Takayama, S., McDonald, J. C., Ostuni, E., Liang, M. N., Kenis, P. J., Ismagilov, R. F., and Whitesides, G. M., "Patterning cells and their environments using multiple laminar fluid flows in capillary networks," *Proc. Natl. Acad. Sci. USA*, Vol. 96, No. 10, 1999, pp. 5545–5548.

[28] Chiu, D. T., Jeon, N. L., Huang, S., Kane, R. S., Wargo, C. J., Choi, I. S., Ingber, D. E., and Whitesides, G. M., "Patterned deposition of cells and proteins onto surfaces by using three-dimensional microfluidic systems," *Proc. Natl. Acad. Sci. USA*, Vol. 97, No. 6, 2000, pp. 2408–2413.

[29] Romanova, E. V., Fosser, K. A., Rubakhin, S. S., Nuzzo, R. G., and Sweedler, J. V., "Engineering the morphology and electrophysiological parameters of cultured neurons by microfluidic surface patterning," *Faseb J.*, Vol. 18, No. 11, 2004, pp. 1267–1269.

[30] Koh, W. G., Itle, L. J., and Pishko, M. V., "Molding of hydrogel multiphenotype cell microstructures to create microarrays," *Anal. Chem.*, Vol. 75, No. 21, 2003, pp. 5783–5789.

[31] Tan, W., and Desai, T. A., "Layer-by-layer microfluidics for biomimetic three-dimensional structures," *Biomaterials*, Vol. 25, Nos. 7–8, 2004, pp. 1355–1364.

[32] Tan, W., and Desai, T. A., "Microfluidic patterning of cells in extracellular matrix biopolymers: Effects of channel size, cell type, and matrix composition on pattern integrity," *Tissue Eng.*, Vol. 9, No. 2, 2003, pp. 255–267.

[33] Khademhosseini, A., Eng, G., Yeh, J., Kucharczyk, P. A., Langer, R., Vunjak-Novakovic, G., and Radisic, M., "Microfluidic patterning for fabrication of contractile cardiac organoids," *Biomed. Microdevices*, Vol. 9, No.2, 2007, pp.149–157.

[34] Baar, K., Birla, R., Boluyt, M. O., Borschel, G. H., Arruda, E. M., and Dennis, R. G., "Self-organization of rat cardiac cells into contractile 3-D cardiac tissue," *Faseb J.*, Vol. 19, No. 2, 2005, pp. 275–277.

[35] Shin, M., Ishii, O., Sueda, T., and Vacanti, J. P., "Contractile cardiac grafts using a novel nanofibrous mesh," *Biomaterials*, Vol. 25, No. 17, 2004, pp. 3717–3723.

[36] Zong, X., Bien, H., Chung, C. Y., Yin, L., Fang, D., Hsiao, B. S., Chu, B., and Entcheva, E., "Electrospun fine-textured scaffolds for heart tissue constructs," *Biomaterials*, Vol. 26, No. 26, 2005, pp. 5330–5338.

[37] Zimmermann, W. H., Schneiderbanger, K., Schubert, P., Didie, M., Munzel, F., Heubach, J. F., Kostin, S., Nehuber, W. L., and Eschenhagen, T., "Tissue engineering of a differentiated cardiac muscle construct," *Circ. Res.*, Vol. 90, No. 2, 2002, pp. 223–230.

[38] Naito, H., Melnychenko, I., Didie, M., Schneiderbanger, K., Schubert, P., Rosenkranz, S., Eschenhagen, T., and Zimmermann, W. H., "Optimizing engineered heart tissue for therapeutic applications as surrogate heart muscle," *Circulation*, Vol. 114, No. 1, 2006, pp. I72–I78.

[39] Narmoneva, D. A., Vukmirovic, R., Davis, M. E., Kamm, R. D., and Lee, R. T., "Endothelial cells promote cardiac myocyte survival and spatial reorganization: Implications for cardiac regeneration," *Circulation*, Vol. 110, No. 8, 2004, pp. 962–968.

[40] Levenberg, S., Rouwkema, J., Macdonald, M., Garfein, E. S., Kohane, D. S., Darland, D. C., Marini, R., van Blitterswijk, C. A., Mulligan, R. C., D'Amore, P. A., and Langer, R., "Engineering vascularized skeletal muscle tissue," *Nat. Biotechnol.*, Vol. 23, No. 7, 2005, pp. 879–884.

[41] Caspi, O., Lesman, A., Basevitch, Y., Gepstein, A., Arbel, G., Habib, I. H., Gepstein, L., and Levenberg, S., "Tissue engineering of vascularized cardiac muscle from human embryonic stem cells," *Circ. Res.*, Vol. 100, No. 2, 2007, pp. 263–272.

[42] Camelliti, P., McCulloch, A. D., and Kohl, P., "Microstructured cocultures of cardiac myocytes and fibroblasts: A two-dimensional in vitro model of cardiac tissue," *Microsc. Microanal.*, Vol. 11, No. 3, 2005, pp. 249–259.

[43] Motlagh, D., Senyo, S. E., Desai, T. A., and Russell, B., "Microtextured substrata alter gene expression, protein localization and the shape of cardiac myocytes," *Biomaterials*, Vol. 24, No. 14, 2003, pp. 2463–2476.

[44] Iyer, R. K., Chiu Loraine, L. Y., and Radisic, M., "Microfabricated poly(ethylene glycol) templates enable rapid screening of tri-culture conditions for cardiac tissue engineering," *J. Biomater. Res. A*, 2008, (in press).

[45] Kelm, J. M., Ehler, E., Nielsen, L. K., Schlatter, S., Perriard, J. C., and Fussenegger, M., "Design of artificial myocardial microtissues," *Tissue Eng.*, Vol. 10, Nos. 1–2, 2004, pp. 201–214.

[46] Anversa, P., Olivetti, G., and Loud, A. V., "Morphometric study of early postnatal development in the left and right ventricular myocardium of the rat: I. Hypertrophy, hyperplasia, and binucleation of myocytes," *Circ. Res.*, Vol. 46, No. 4, 1980, pp. 495–502.

[47] Vandenburgh, H., Solerssi, R., Shansky, J., Adams, J., and Henderson, S., "Mechanical stimulation of organogenic cardiomyocyte growth in vitro," *Am. J. Physiol.*, Vol. 270, 1996, pp. C1284–C1292.

[48] Yu, J. G., and Russell, B., "Cardiomyocyte remodeling and sarcomere addition after uniaxial static strain in vitro," *J. Histochem. Cytochem.*, Vol. 53, No. 7, 2005, pp. 839–844.

[49] Gopalan, S. M., Flaim, C., Bhatia, S. N., Hoshijima, M., Knoell, R., Chien, K. R., Omens, J. H., and McCulloch, A. D., "Anisotropic stretch-induced hypertrophy in neonatal ven-

tricular myocytes micropatterned on deformable elastomers," *Biotechnol. Bioeng.*, Vol. 81, No. 5, 2003, pp. 578–587.

[50] Tanaka, Y., Morishima, K., Shimizu, T., Kikuchi, A., Yamato, M., Okano, T., and Kitamori, T., "Demonstration of a PDMS-based bio-microactuator using cultured cardiomyocytes to drive polymer micropillars," *Lab Chip*, Vol. 6, No. 2, 2006, pp. 230–235.

[51] Zhao, Y., and Zhang, X., "Cellular mechanics study in cardiac myocytes using PDMS pillars array," *Sens. Actuators A: Physical*, Vol. 125, No. 2, 2006, pp. 398–404.

[52] Tanaka, Y., Morishima, K., Shimizu, T., Kikuchi, A., Yamato, M., Okano, T., and Kitamori, T., "An actuated pump on-chip powered by cultured cardiomyocytes," *Lab Chip*, Vol. 6, No. 3, 2006, pp. 362–368.

[53] Tanaka, Y., Sato, K., Shimizu, T., Yamato, M., Okano, T., and Kitamori, T., "A micro-spherical heart pump powered by cultured cardiomyocytes," *Lab Chip*, Vol. 7, No. 2, 2007, pp. 207–212.

[54] Radisic, M., Park, H., Shing, H., Consi, T., Schoen, F. J., Langer, R., Freed, L. E., and Vunjak-Novakovic, G., "Functional assembly of engineered myocardium by electrical stimulation of cardiac myocytes cultured on scaffolds," *Proc. Natl. Acad. Sci. USA*, Vol. 101, No. 52, 2004, pp. 18129–18134.

[55] Banach, K., Halbach, M. D., Hu, P., Hescheler, J., and Egert, U., "Development of electrical activity in cardiac myocyte aggregates derived from mouse embryonic stem cells," *Am. J. Physiol.: Heart Circ. Physiol.*, Vol. 284, No. 6, 2003, pp. H2114–H2123.

[56] Werdich, A. A., Lima, E. A., Ivanov, B., Ges, I., Anderson, M. E., Wikswo, J. P., and Baudenbacher, F. J., "A microfluidic device to confine a single cardiac myocyte in a sub-nanoliter volume on planar microelectrodes for extracellular potential recordings," *Lab Chip*, Vol. 4, No. 4, 2004, pp. 357–362.

[57] Klauke, N., Smith, G. L., and Cooper, J., "Stimulation of single isolated adult ventricular myocytes within a low volume using a planar microelectrode array," *Biophys. J.*, Vol. 85, No. 3, 2003, pp. 1766–1774.

[58] Radisic, M., Iyer, R. K., and Murthy, S. K., "Micro- and nanotechnology in cell separtion," *Int. J. Nanomed.*, Vol. 1, No. 1, 2006, pp. 3–14.

[59] Banerjee, I., Yekkala, K., Borg, T. K., and Baudino, T. A., "Dynamic interactions between myocytes, fibroblasts, and extracellular matrix," *Annals NY Acad. Sci.*, Vol. 1080, 2006, pp. 76–84.

[60] Kuzuya, M., and Kinsella, J. L., "Induction of endothelial cell differentiation in vitro by fibroblast-derived soluble factors," *Exp. Cell Res.*, Vol. 215, No. 2, 1994, pp. 310–318.

[61] Hsieh, P. C., Davis, M. E., Lisowski, L. K., and Lee, R. T., "Endothelial-cardiomyocyte interactions in cardiac development and repair," *Ann. Rev. Physiol.*, Vol. 68, 2006, pp. 51–66.

[62] Kaji, H., Nishizawa, M., and Matsue, T., "Localized chemical stimulation to micropatterned cells using multiple laminar fluid flows," *Lab Chip*, Vol. 3, No. 3, 2003, pp. 208–211.

[63] Tourovskaia, A., Figueroa-Masot, X., and Folch, A., "Differentiation-on-a-chip: A microfluidic platform for long-term cell culture studies," *Lab Chip*, Vol. 5, No. 1, 2005, pp. 14–19.

[64] Papadaki, M., Bursac, N., Langer, R., Merok, J., Vunjak-Novakovic, G., and Freed, L. E., "Tissue engineering of functional cardiac muscle: Molecular, structural, and electrophysiological studies," *Am. J. Physiol.: Heart Circ. Physiol.*, Vol. 280, No. 1, 2001, pp. H168–H178.

[65] Laugwitz, K. L., Moretti, A., Lam, J., Gruber, P., Chen, Y. H., Woodard, S., Lin, L. Z., Cai, C. L., Lu, M. M., Reth, M., Platoshyn, O., Yuan, J. X. J., Evans, S., and Chien, K. R., "Postnatal isl1+cardioblasts enter fully differentiated cardiomyocyte lineages," *Nature*, Vol. 433, No. 7026, 2005, pp. 647–653.

[66] Wang, J. X., Fan, J., Cheung, F., Laschinger, C., Seth, A., and McCulloch, C., "Alpha-smooth muscle actin (SMA) is required for mechanical force-induced p38 activation," *Faseb J.*, Vol. 18, No. 8, 2004, pp. C39–C39.

[67] Beltrami, A. P., Barlucchi, L., Torella, D., Baker, M., Limana, F., Chimenti, S., Kasahara, H., Rota, M., Musso, E., Urbanek, K., Leri, A., Kajstura, J., Nadal-Ginard, B., and Anversa, P., "Adult cardiac stem cells are multipotent and support myocardial regeneration," *Cell*, Vol. 114, No. 6, 2003, pp. 763–776.

[68] Linke, A., Muller, P., Nurzynska, D., Casarsa, C., Torella, D., Nascimbene, A., Castaldo, C., Cascapera, S., Bohm, M., Quaini, F., Urbanek, K., Leri, A., Hintze, T. H., Kajstura, J., and Anversa, P., "Stem cells in the dog heart are self-renewing, clonogenic, and multipotent and regenerate infarcted myocardium, improving cardiac function," *Proc. Nat. Acad. Sci. USA*, Vol. 102, No. 25, 2005, pp. 8966–8971.

[69] Oh, H., Bradfute, S. B., Gallardo, T. D., Nakamura, T., Gaussin, V., Mishina, Y., Pocius, J., Michael, L. H., Behringer, R. R., Garry, D. J., Entman, M. L., and Schneider, M. D., "Cardiac progenitor cells from adult myocardium: Homing, differentiation, and fusion after infarction," *Proc. Nat. Acad. Sci. USA*, Vol. 100, No. 21, 2003, pp. 12313–12318.

[70] Orlic, D., Kajstura, J., Chimenti, S., Jakoniuk, I., Anderson, S. M., Li, B. S., Pickel, J., McKay, R., Nadal-Ginard, B., Bodine, D. M., Leri, A., and Anversa, P., "Bone marrow cells regenerate infarcted myocardium," *Nature*, Vol. 410, No. 6829, 2001, pp. 701–705.

[71] Sia, S. K., and Whitesides, G. M., "Microfluidic devices fabricated in poly(dimethylsiloxane) for biological studies," *Electrophoresis*, Vol. 24, No. 21, 2003, pp. 563–3576.

[72] McDonald, J. C., and Whitesides, G. M., "Poly(dimethylsiloxane) as a material for fabricating microfluidic devices," *Acct. Chem. Res.*, Vol. 35, No. 7, 2002, pp. 491–499.

[73] Whitesides, G. M., Ostuni, E., Takayama, S., Jiang, X. Y., and Ingber, D. E., "Soft lithography in biology and biochemistry," *Ann. Rev. Biomed. Eng.*, Vol. 3, 2001, pp. 335–373.

[74] Hashimoto, M., Hupert, M. L., Murphy, M. C., Soper, S. A., Cheng, Y. W., and Barany, F., "Ligase detection reaction/hybridization assays using three-dimensional microfluidic networks for the detection of low-abundant DNA point mutations," *Anal. Chem.*, Vol. 77, No. 10, 2005, pp. 3243–3255.

[75] Hansen, C. L., Sommer, M. O. A., and Quake, S. R., "Systematic investigation of protein phase behavior with a microfluidic formulator," *Proc. Nat. Acad. Sci. USA*, Vol. 101, No. 40, 2004, pp. 14431–14436.

[76] Beebe, D. J., Mensing, G. A., and Walker, G. M., "Physics and applications of microfluidics in biology," *Ann. Rev. Biomed. Eng.*, Vol. 4, 2002, pp. 261–286.

[77] Lagally, E. T., and Mathies, R. A., "Integrated genetic analysis microsystems," *J. Phys. D Appl. Phys.*, Vol. 37, No. 23, 2004, pp. R245–R261.

[78] Fiorini, G. S., and Chiu, D. T., "Disposable microfluidic devices: Fabrication, function, and application," *Biotechniques*, Vol. 38, No. 3, 2005, pp. 429–446.

[79] Tegenfeldt, J. O., Prinz, C., Cao, H., Huang, R. L., Austin, R. H., Chou, S. Y., Cox, E. C., and Sturm, J. C., "Micro- and nanofluidics for DNA analysis," *Anal. Bioanal. Chem.*, Vol. 378, No. 7, 2004, pp. 1678–1692.

[80] Metzker, M. L., "Emerging technologies in DNA sequencing," *Genome Res.*, Vol. 15, No. 12, 2005, pp. 1767–1776.

[81] Pihl, J., Karlsson, M., and Chiu, D. T., "Microfluidic technologies in drug discovery," *Drug Discov. Today*, Vol. 10, No. 20, 2005, pp. 1377–1383.

[82] Toner, M., and Irimia, D., "Blood-on-a-chip," *Ann. Rev. Biomed. Eng.*, Vol. 7, 2005, pp. 77–103.

[83] Radisic, M., Iyer, R. K., and Murthy, S. K., "Micro- and nanotechnology in cell separation," *Int. J. Nanomed.*, Vol. 1, No. 1, 2006, pp. 3–14.

[84] Sethu, P., Sin, A., and Toner, M., "Microfluidic diffusive filter for apheresis (leukapheresis)," *Lab Chip*, Vol. 6, No. 1, 2006, pp. 83–89.

[85] Murthy, S. K., Sin, A., Tompkins, R. G., and Toner, M., "Effect of flow and surface conditions on human lymphocyte isolation using microfluidic chambers," *Langmuir*, Vol. 20, No. 26, 2004, pp. 11649–11655.

[86] Murthy, S. K., Sethu, P., Vunjak-Novakovic, G., Toner, M., and Radisic, M., "Size-based microfluidic enrichment of neonatal rat cardiac cell populations," *Biomed. Microdevices*, Vol. 8, 2006, pp. 231–237.

[87] Gobin, A. S., and West, J. L., "Val-ala-pro-gly, an elastin-derived non-integrin ligand: Smooth muscle cell adhesion and specificity," *J. Biomed. Mater. Res. A*, Vol. 67A, No. 1, 2003, pp. 255–259.

[88] Hubbell, J. A., Massia, S. P., Desai, N. P., and Drumheller, P. D., "Endothelial cell-selective materials for tissue engineering in the vascular graft via a new receptor," *Biotechnology*, Vol. 9, No. 6, 1991, pp. 568–572.

[89] Sakiyama-Elbert, S. E., and Hubbell, J. A., "Functional biomaterials: Design of novel biomaterials," *Ann. Rev. Mater. Res.*, Vol. 31, 2001, pp. 183–201.

[90] Kam, L., Shain, W., Turner, J. N., and Bizios, R., "Selective adhesion of astrocytes to surfaces modified with immobilized peptides," *Biomaterials*, Vol. 23, No. 2, 2002, pp. 511–515.

[91] Murthy, S. K., Olsen, B. D., and Gleason, K. K., "Peptide attachment to vapor deposited polymeric thin films," *Langmuir*, Vol. 20, No. 11, 2004, pp. 4774–4776.

[92] Usami, S., Chen, H. H., Zhao, Y. H., Chien, S., and Skalak, R., "Design and construction of a linear shear-stress flow chamber," *Annals Biomed. Eng.*, Vol. 21, No. 1, 1993, pp. 77–83.

[93] Yang, M., Zhang, X., "Electrical Assisted Patterning of Cardiac Myocytes with Controlled Microscopic Anisotropy Using a Microfluidic Dielectrophoresis Chip," *Sensor and Actuators*, Vol. 135, Issue 1, 2007, pp. 73–79.

CHAPTER 19

Nanoengineered Hydrogels for Stem Cell Cartilage Tissue Engineering

H. Janice Lee, Shyni Varghese, Nathaniel Hwang, and Jennifer Elisseeff

19.1 Hydrogel Microenvironments

19.1.1 Natural and Synthetic Hydrogel Scaffolds

Hydrogels are three-dimensional insoluble networks of hydrophilic polymer chains that swell in aqueous solutions and, therefore, have the ability to retain a large quantity of water and biological fluids. The elastic hydrogel network holds the solvent inside the matrix by osmotic forces, while the liquid prevents the polymer network from collapsing into a compact mass. The combination of these two parameters, namely, the osmotic forces and the elastic retractivity, defines the properties of the gels. Depending upon their chemical composition, crosslinking density, and hydrophobicity, hydrogels can vary in consistency from viscous fluids to fairly rigid solids. Basically, hydrogels are wet, soft, and capable of undergoing large deformation. Hydrogels have very similar macromolecular structure to a native tissue, which is a network of various extracellular matrix (ECM) components. The hydrogel-like nature of tissues also helps the organisms to transport oxygen, nutrients, and other bioactive moieties easily and effectively while retaining their solid nature. Moreover, hydrogels can be engineered to have a nanoscale pore size to accommodate cell encapsulation, as the natural ECM does.

Hydrogels can be broadly divided into two categories, natural and synthetic, based on their sources. Natural hydrogels are hydrophilic networks made out of naturally derived polymer as shown in Figure 19.1. For example, hydrogels made from ECM components, such as collagen [1], hyaluronic acid (HA) [2], and chondroitin sulfate [3], are used as scaffolds for tissue-engineering applications, including cartilage and skin regeneration. In particular, Matrigel matrix (commercially available from BD Biosciences) is a soluble form of basement membrane extracted from mouse tumors that contains several components of ECM proteins. The major components of Matrigel include laminin, collagen 1V, heparan sulfate, proteoglycans, and entactin. At room temperature, Matrigel polymerizes to produce biologically active hydrogel resembling the mammalian cellular basement membrane. Cells are known to behave as they do in vivo when they are cultured in Matrigel matrix; hence, Matrigel has been used as a model system to study cell behavior in three-dimensional environments [4]. Matrigel indeed provides a physiologically relevant environment for studying cell morphology, biochemical function, cell migration and invasion, and gene expression [5, 6]. In the case of rhesus mon-

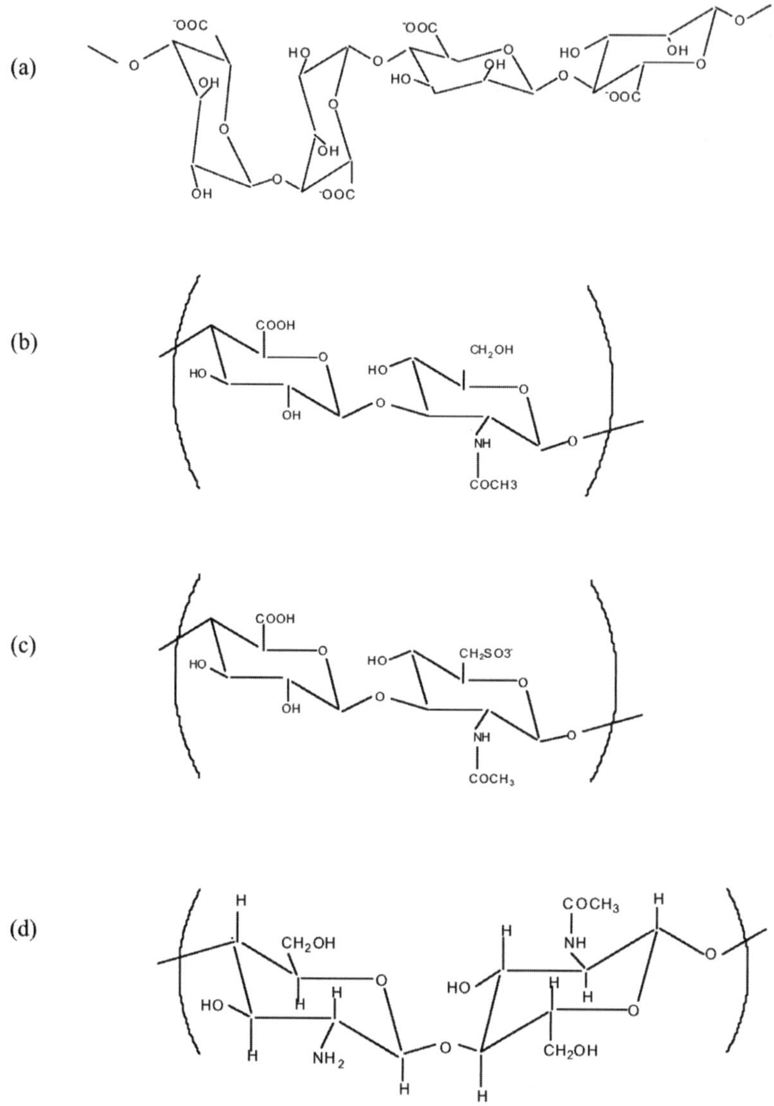

Figure 19.1 Chemical structure of natural polymers that have been used to create hydrogels: (a) alginate, (b) hyaluronic acid, (c) chondroitin sulfate, and (d) chitosan [9].

key embryonic stem (ES) cells, the use of Matrigel has been shown to induce cell growth and differentiation [7]. In another study, Xu et al., showed that the presence of Matrigel along with mouse-fibroblast-conditioned medium maintains human ES cells in an undifferentiated state in a feeder-free culture system [8].

Other natural sources of materials used to synthesize hydrogels include alginate from seaweed, chitosan from chitin, a polysaccharide found in the exoskeletons of shellfish, and silk from silkworms (*Bombyx mori*). Alginate beads have been used to deliver chondrocytes [10, 11], hepatocytes [12], and islets of Langerhans [13] into the body for cell therapy. Alginate has also been previously used for wound dressing [14] and as a scaffold for musculoskeletal-tissue engineering [15–18]. From the ongoing discussion, it is clear that using natural polymers as tissue-engineering scaffolds offers a wide range of advantages, such as biological signaling, cell adhesion, cell-responsive degradation, and remodeling of tissue. These materials, however, generally lack adequate mechanical properties, compromising their utilization as unique scaffold materials. Another major concern with using natural materials is the possibility of immunorejection and the transfer of various viruses, though proper screening and purification can overcome these limitations.

On the other hand, the most appealing factor about synthetic hydrogels is that their properties, such as mechanical strength, porosity, degradation profile, and even biologically active sites, can be molecularly tailored. Poly(ethylene glycol diacrylate) (PEGDA), poly(vinyl alcohol), poly(lactic acid) (PLA), poly(lactic-co-glycolic acid) (PLGA), poly(hydroxyl ethyl methacrylate), and poly(anhydride), as shown in Figure 19.2, are just a few examples of synthetic polymers used to create hydrogels and have found applications as tissue-engineering scaffolds.

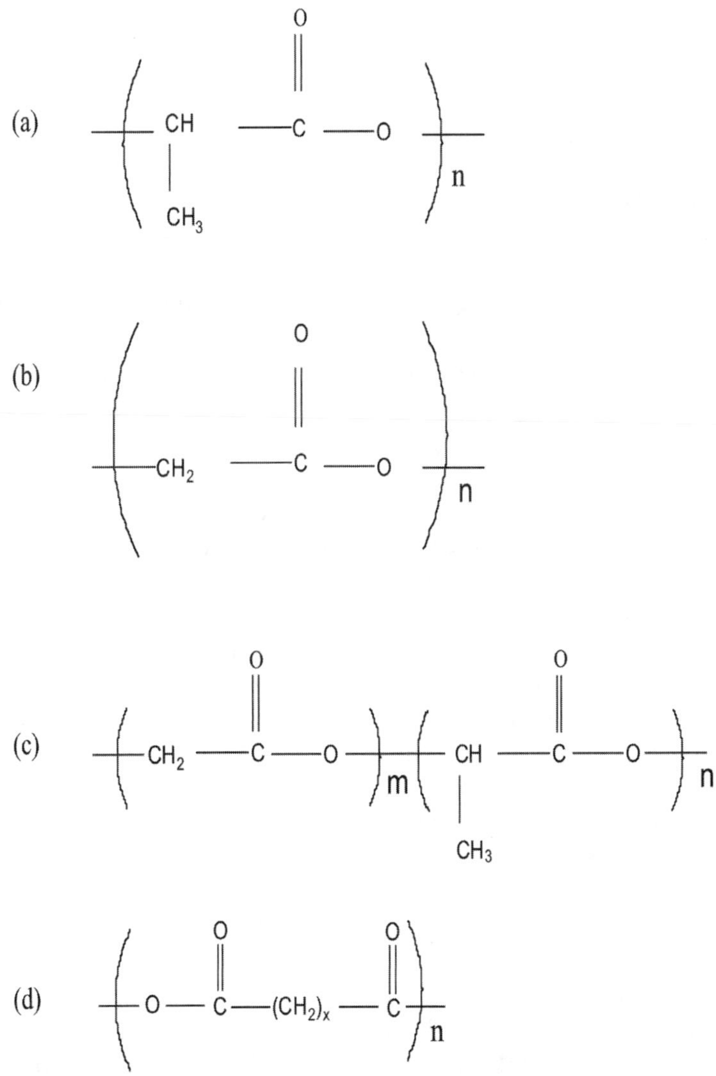

Figure 19.2 Chemical structure of (a) PLA, (b) poly(glycolic acid), (c) PLGA, and (d) poly(anhydride) [9].

Polyethylene glycol (PEG) hydrogel, for example, is widely used in various biomedical applications because of its biocompatibility, hydrophilicity, and resistance to protein adhesion and cell adhesion [19, 20]. It has been reported that PEG-modified proteins exhibit decreased immunogenicity and antigenicity at an increased circulation time in the body [21]. PEGDA with a molecular weight less than 20,000 Da can be dissolved in body fluids and eliminated from the body via excretion through the kidneys [22]. The resistance of PEG to protein adhesions has been widely used to create nonfouling surfaces, where PEG chains are immobilized onto the surface by covalent bonding or adsorption [23, 24]. The biocompatibility and hydrophilicity of PEG has been explored to impart these properties to other materials by copolymerizing with PEG [25].

Recent years have witnessed a surge of interest in using synthetic biological hybrid hydrogels as tissue-engineering scaffolds, integrating principles from cell and molecular biology. A simple mixture of natural ECM and other materials has been studied, but a physical network of ECM and a synthetic material without chemical crosslinking results in a heterogeneous distribution of ECM components throughout the scaffold. On the other hand, the advantages of a synthetic biological hybrid scaffold provide a more uniform and controllable environment for cell encapsulation [26]. This new category of hybrid hydrogels has been developed with molecular cues mimicking certain aspects of the structure and function of natural microenvironments for cell encapsulation, such as incorporating peptides into the network. Natural proteins are subject to denaturation and degradation, and their chemical reactivity is unpredictable, but peptides, in general, are considered more stable and easier to conjugate to synthetic hydrogels while preserving protein functions. For example, the integrin-binding peptide arginine-glycine-aspartic acid (RGD) has been extensively investigated to compensate for PEGDA's bioinert properties that prevent its interaction with proteins and cells. Hern and Hubbell seeded fibroblasts on RGD-modified PEGDA surface and illustrated significantly more adherence and spreading [27]. Studies by Yang et al. and Burdick and Anseth presented increased bone formation of encapsulated osteoblasts and mesenchymal stem cells (MSCs), respectively, in three-dimensional PEGDA scaffold conjugated with RGD [28, 29]. Other biological domains attached to modify inert hydrogels include collagen mimetic peptides [30, 31] and laminin-derived peptides (YIGSR and IKVAV) [32, 33]. Various studies have also accelerated the degradation of synthetic hydrogels through the introduction of degradable groups onto the polymer, although some degradation is observed only when the material is in contact with cells or implanted [34, 35]. Ulbrich, Strohalm, and Kopecek built enzyme-sensitive synthetic hydrogels by stepwise copolymerization of hydrophilic polymers and proteolytically sensitive peptide or protein building blocks, which responded to encapsulated cell activity [36].

19.1.2 Methods to Form Hydrogels

Hydrogel networks are formed through either chemical or physical crosslinking. Methods for chemical crosslinking include ionic or covalent bonds, while those for physical crosslinking include entanglements, crystallites, hydrogen bonds, and hydrophobic interactions. Alginate hydrogels, for example, are chemically crosslinked by ionic bonds (Figure 19.3) and have pore sizes in the range of 1 to 30 nm, depending on the concentration [37]. Alginate undergoes gelation under physiological conditions in the presence of a small concentration of divalent cations, such as Ca^{2+}, Ba^{2+}, and Sr^{2+}, as the ionic bonds are formed between the carboxylic group located on the polymer backbone and the cation. The metal-induced gelation of alginate is attributed to the ability of glucuronic acid chains to form an "egg box"–shaped structure in the presence of divalent cations [38]. Such ionically crosslinked alginate hydrogels, however, mainly undergo degradation by uncontrolled dissolution of polymer chains. Some of these resulting chains have a high molecular weight and, therefore, cannot be readily eliminated by the body. In order to achieve bioresorbable dissolution products, high-molecular-weight alginate chains are broken into smaller segments by treating them with γ-irradiation prior to their gelation [39]. Irradiated alginate hydrogels were found to support bone formation in vivo while allowing the excretion of the polymer chains by kidneys [15]. In another approach, partial oxidation of alginate polymer chains has been used to

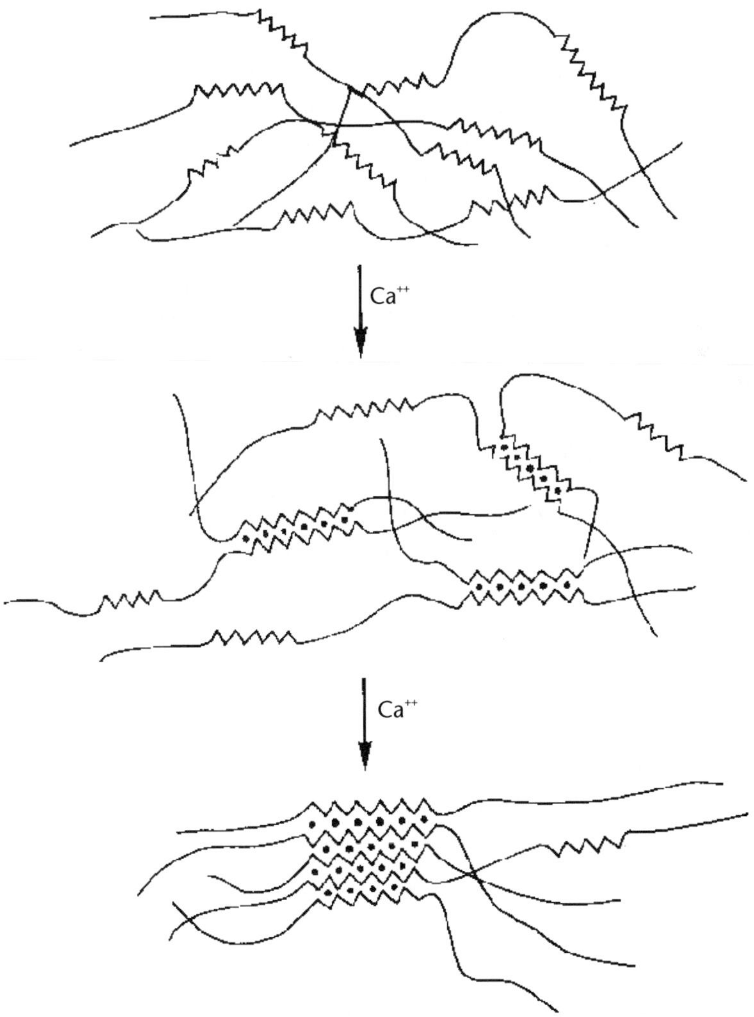

Figure 19.3 Gel fomation via ionic bonds: "egg-box" model of alginate hydrogel [38].

render them hydrolytically degradable [40]. In addition to metal-induced gelation, alginate hydrogels can also be prepared by crosslinking alginate with other monomers [41, 42]. Moreover, alginate gels have been patterned to have nanochannels with the surface coated by adhereing peptides; the primary rat dorsal root ganglia cells were seeded and exhibited oriented axonal growth in the patterned alginate [43].

Hydrogels have also been created by a reaction between different functional groups present in the reacting oligomers [44, 45]. Sperinde and Griffith have reported the formulation of a novel PEG-based hydrogel in which the network is formed by enzymatic crosslinking that uses transglutaminase, a natural enzyme, as a catalyst [46, 47]. Tranglutaminase is a calcium-dependent enzyme, and it is ubiquitous in body fluid. Other chemically crosslinked hydrogels via covalent bonds are usually created by radical polymerization from a vinyl moiety in the starting monomer and oligomer mixtures. One such approach widely used to create chemically crosslinked hydrogels involves photopolymerization using ultraviolet (UV) light [48], visible light [49], and γ-radiation [50]. PEG hydrogels are created by UV photopolymerization when the PEG precursor contains acrylate termini such as PEGDA, where the acrylate group functions as the crosslinkable moiety, as illustrated in Figure 19.4. The concentration and molecular weight of a monomer govern the pore size of a PEGDA hydrogel, which ranges from 2 to 10 nm [51].

Other approaches involving physical entanglements, hydrophilic-hydrophobic interactions [52–54], crystallization [55, 56], hydrogen bonding [57, 58], molecular recognition [59], and self-assembly [60–62] have been widely explored to create physically crosslinked hydrogels. Collagen monomers form triple helices and align to build collagen fibers via electrostatic interactions. Therefore, at room temperature, these collagen fibers are soluble only in acidic conditions. Only when the solution is neutralized do the collagen fibers precipitate to form physical entanglements and rebuild ionic interactions. Collagen gels, however, have weak mechanical prop-

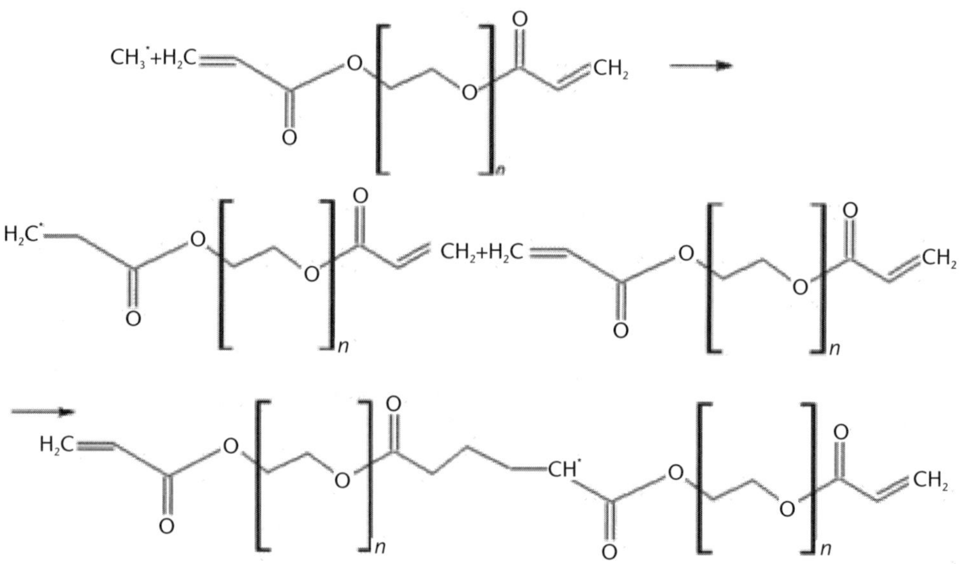

Figure 19.4 Photopolymerization of PEGDA. A free radical is formed from the photoinitiator, which then attacks the double bond in PEGDA and initiates a chain reaction [48].

erties like most physically crosslinked hydrogels because of intrinsically weaker interactions compared to chemical crosslinking [63]. Therefore, several methods to modify collagen gels have been suggested: chemical glycation or heat treatment to add chemical crosslinkages and application of magnetic field to permanently align collagen fibers. These modifications, however, compromise an important advantage of physical crosslinking: reversibility. Recently, considerable work has been devoted to the development of novel biomaterials through molecular self-assembly. Zhang et al. have fabricated self-assembling peptide hydrogels with 50 to 200 nm pore size [64, 65]. The hydrogels use ionic self-complementary peptides, which form β-sheet structures in aqueous solution with two distinct surfaces of hydrophilic or hydrophobic properties [65]. The hydrophobic residues shield themselves from water and self-assemble in water in a manner similar to that seen in the case of tertiary structural protein folding in vivo. The unique structural feature of these peptides is that they form organized complementary electrostatic interactions with regular repeats on the hydrophilic surface, which, while retaining ionic water, enhances the mechanical strength of the peptide gel. A variety of cells, such as chondrocytes, neurons, osteoblasts, and stem cells, have been successfully cultured in this self-assembling peptide gel [60, 64, 66–68].

19.1.3 Hydrogels for Cartilage Tissue Engineering

Cartilage lacks the ability to self-repair or regenerate. As a result, reconstructive orthopedic and plastic surgeons are continually faced with the challenge of replacing lost cartilage due to trauma, disease, or congenital abnormalities. Hyaline cartilage is found on diarthrodial or synovial joints, for example, the knee and temporomandibular joint, where the viscoelastic properties of cartilage provide a frictionless brearing that allows joint motion. Cartilage also provides structural support in craniofacial tissues, including the nose and ear. Failure of articular cartilage to provide its bearing function can have debilitating effects, such as the progression of osteoarthritis. Approximately fifty million Americans suffer from osteoarthritis. Over one million surgical procedures involve cartilage replacement and 176 million artificial implants are placed each year to restore cartilage function, with a significant fraction involving facial and dental implants. Current therapies, such as tissue grafting and joint prostheses, have been effective in alleviating much of the disability associated with loss of cartilage, however, these therapies suffer from serious limitations, such as lack of donor tissue, as well as rejection and susceptibility to infection in the case of allogenic and alloplastic grafts. Cells in the musculoskeletal system continually remodel tissue and respond to changes in the physical and mechanical environment of the body, which purely nonbiological prosthetics are incapable of fulfilling. Tissue engineering provides a solution to the problems faced with nonbiological cartilage replacements. Furthermore, photoencapsulation of living cells in hydrogels is a promising strategy for the administration of a biological cartilage replacement in a minimally invasive manner to decrease patient morbidity, rehabilitation, and numerous problems associated with extensive surgery. Furthermore, the photopolymerization process offers spatial and temporal control over the formation of the implant in situ, enabling the creation of more complex structures to mimic native tissue. For applications in plastic surgery, where subcutaneous tissue augmentation is desired, the cell-polymer solution may be injected and photopolymerized transdermally [69].

The increased interest in hydrogels for tissue-engineering applications is attributed to their water content, viscoelasticity akin to the native tissues, biocompatibility, and ability to permit diffusion of nutrients and bioactive molecules. In addition to hydrogels' structural stability, their mechanical and biological properties can also easily be tailored molecularly to incorporate suitable biological signals that can stimulate cell proliferation and differentiation as discussed previously. Moreover, the soft viscoelastic nature of hydrogels minimizes irritation to the surrounding tissues and prevents stress shielding [70–72]. Especially, anchorage-independent cells like chondrocytes exhibit good cell viability within hydrophilic scaffolds like hydrogels [73].

Articular cartilage is a highly specialized connective tissue containing chondrocytes embedded in an ECM that is highly enriched with large proteoglycans and type II collagen and has an additional unique function in resisting compression. Chondrocytes are known to dedifferentiate into fibroblastlike cells during their in vitro monolayer expansion, and various biomimicking synthetic hydrogels have been used to effectively redifferentiate the dedifferentiated chondrocytes, where the hydrogels provide a three-dimensional environment to the chondrocytes [10, 74]. The chondrocytes that are encapsulated within the hydrogels retain their spherical cell morphology and produced the cartilage-specific markers: collagen type II and aggrecan. Our laboratory uses PEGDA scaffolds for cartilage engineering where photopolymerization of PEGDA is used to encapsulate chondrocytes, MSCs, and ES cells [75–77]. Our results indicate that a PEGDA hydrogel serves as a three-dimensional support and permits the chondrogenic differentiation of encapsulated cells.

19.2 Stem Cell Encapsulation in Hydrogels

19.2.1 Stem Cells

Stem cells have generated a great deal of excitement and promise as a potential source of cells for cell-based therapeutic strategies, primarily owing to their intrinsic ability to self-renew and differentiate into functional cell types that constitute the tissue in which they exist. Compared to mature differentiated somatic cells, stem cells are more likely to guarantee adequate regeneration and cell turnover at the transplantation site for an extended period of time, possibly a lifetime. Adult and embryonic stem cells are categorized depending on their sources: ES cells are derived from the inner cell mass of an embryo's blastocysts, and adult stem cells derive from various adult tissues, such as bone marrow, muscle, and brain.

Scientists discovered ways to obtain or derive stem cells from early mouse embryos more than twenty years ago. Many years of detailed study of the biology of mouse stem cells led to the discovery in 1998 of methods to isolate human ES (hES) cells from human embryos and grow the cells in the laboratory [78]. The embryos used in these studies were created for infertility purpose through in vitro fertilization procedures, and when they were no longer needed for that purpose, they were donated for research with the informed consent of the donors. ES cells, derived from the inner cell mass of a blastocyst, are versatile cells capable of differentiating into cells of all three germ-layer lineages [78, 79]. These cells are capable of unlimited symmetrical self-renewal, thus providing an unlimited cell source for tissue-engineering applications. However, one of the challenges in stem cell regenerative therapy is the control of the differentiation process. Differentiation of ES cells through embryoid bodies (EBs) parallels embryonic development, with EBs recapitulating the early embryonic development phase. Recently, biological signals in the form of growth factors, as well as ECM, were shown to induce EB differentiation toward the chondrogenic lineage. These attributes suggest that ES cells may represent a useful cell source for cartilage-tissue engineering [80].

In some adult tissues, such as bone marrow, muscle, and brain, discrete populations of adult stem cells generate replacements for cells that are lost through normal wear and tear, injury, or disease. In particular, one particular type of adult stem cells, MSCs, contributes to the regeneration of mesenchymal tissues, including bone, cartilage, muscle, marrow stroma, tendon, and adipose, as shown in Figure 19.5 [81, 82]. MSCs were originally isolated from bone marrow but are known to reside in a large number of adult tissues, including trabecular bone, muscle, adipose tissue, periosteum, synovial membrane, articular cartilage, and deciduous teeth. While embryonic stem cells are immortal and could potentially provide an unlimited supply of differentiated chondrocytes and chondroprogenitor cells for transplantation, the self-renewing and proliferating capacities of MSCs are limited [83]. MSCs, however, are considered a readily accepted source of stem cells because such cells have already demonstrated efficacy in multiple types of cellular therapeutic strategies, including applications in treating children with osteogenesis imperfecta [84], hematopoietic recovery [85], and bone-tissue regeneration strategies [28]. More importantly, unlike embryonic stem cells, these cells may be directly

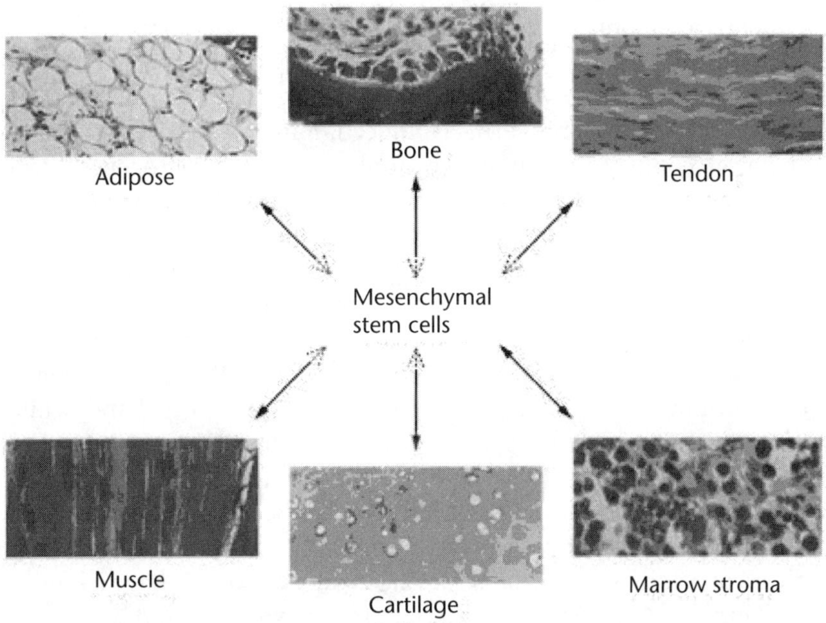

Figure 19.5 Multilineage differentiation potential of adult human MSCs. The arrows are presented as bidirectional, indicating dedifferentiation or transdifferentiation [82].

obtained from individual patients, thereby eliminating the complications associated with immune rejection of allogenic tissue [86]. On the other hand, although still a topic of debate, the cell number, proliferation, and differentiation potential of MSCs appear to decrease with advancing age [87]. It has also been shown that MSCs derived from the bone marrow of patients with advanced osteoarthritis show significantly reduced proliferating capacity and chondrogenic and adipogenic activity compared with those from healthy donors [88]. In this case, however, chondrogenic potential did not seem to change with age in healthy donors. The potential decrease of MSC quantity and quality with age and disease might limit the use of autologous MSCs in clinical settings, and this possibility warrants further studies.

19.2.2 Controlling Stem Cell Differentiation in Hydrogel Microenvironments

When the tissue defect sites are large and the cell migration from the residual surrounding tissues is minimal or impeded, an implantation of both cells and a functional scaffold is necessary for repair. In this approach, the scaffold provides the initial structural support to the encapsulated cells and guides their proliferation and differentiation into the desired tissue or organ. A cell population that has the ability to proliferate and produce the required matrix is placed onto the defect site in combination with a biomaterial. A number of cell sources exist for this purpose, and these include fully differentiated cells isolated from tissue (e.g., chondrocytes for cartilage repair), adult and embryonic stem cells described earlier. The polymer scaffold acts as an artificial ECM and provides a favorable niche or microenvironment for the cells to grow toward the desired lineage. Additionally, it also serves as a carrier for the transport of cells into the defect site and also confines the cells to the defect site. Cells can either be seeded onto a solid fibrous or porous scaffold or encapsulated within a gelatinous scaffold. In both cases, the cells are suspended or attached to the scaffold; then, they proliferate, migrate, and secrete ECM.

Differentiation of adult and embryonic stem cells is generally controlled by various cues from the microenvironment. In addition to cell-matrix interaction, an addition of exogenous growth factors or signaling molecules directs the process of stem cell chondrogenesis in hydrogels. Indeed, numerous growth factors, cytokines, and chemical compounds have been implicated in chondrogenesis. Moreover, many growth factors are also known to induce osteogenesis as the process of chondrogenesis is so closely intertwined with osteogenesis. Therefore, the optimal combination and temporal administration of growth factors will be required for the stable differentiation of stem cells into chondrocytes. For example, some bone morphogenic proteins (BMPs) have been shown to induce the chondrogenic phenotype in cultured bone marrow–derived human MSC pellets when combined with transforming growth factor (TGF)–β3 [89]. On the other hand, while TGF-β3 alone was sufficient for chondrogenic differentiation of human MSCs derived from bone marrow, synovium-derived human MSCs required BMP2 and dexamethasone for optimal MSC chondrogenesis [90]. Dexamethasone, a synthetic glucocorticoid, is an example of a nonproteinaceous chemical compound promoting chondrogenic differentiation of stem cells [91]. Chemical compounds have a longer active half-life in solution compared to growth factors or cytokines and can be manufactured by more controlled chemical reactions than growth factors or cytokines, which have to be synthesized in living cells. Other examples of chemical compounds for chondrogenesis include steroid derivatives (thyroid hormones) [92, 93], vitamin C (ascorbic acid) [94], vitamin D (1,25-dihydroxy vitamin D_3) [95], and a protein kinase C inhibitor (staurosporine) [96, 97].

In addition to bioactive factors, various forms of physical stimuli have been shown to affect stem cell differentiation. For example, cyclic hydrostatic loading has been shown to enhance MSC chondrogenic differentiation [98]. Another approach to control stem cell differentiation is to lower the oxygen tension. Articular cartilage chondrocytes normally exist in much lower oxygen tension than standard in vitro culture conditions. Mimicking the low oxygen tension of the natural cartilage environment has been shown to promote cartilage-specific matrix production by MSCs in vitro [99]. One approach to providing the necessary environment of biochemical and biomechanical stimulation is the use of bioreactors. For example, Chen et al. successfully utilized a rotary bioreactor to rapidly expand human MSCs [100]. By using combinations of growth factors and mechanical stimulation, cartilage-tissue engineering has generated tissue constructs with mechanical strength and proteoglycan and collagen contents approaching, although not quite as high as, those of native articular cartilage.

19.2.3 Unique Requirements and Responses of Stem Cells in Hydrogels

Hydrogels composed of various synthetic and natural polymers, such as PEGDA, silk, collagen, and the like, have been explored for the differentiation of MSCs and ES cells into chondrocytes [76, 101, 102]. In our laboratory, we investigated the ability of MSCs to undergo chondrogenesis both in vivo and in vitro [76]. Our studies and others demonstrate the potential of MSCs as an alternative cell source for cartilage-tissue engineering, given the limitations of chondrocytes. In the case of in vivo studies, Sharma et al. used high molecular weight hyaluronan (HA) along with PEGDA oligomer solution to create a semi-interpenetrating network for the scaffold. In addition to increasing the viscosity of the starting solution for easy handling, the presence of HA also facilitates the differentiation of MSCs to chondrocytes. Moreover, the presence of HA in the hydrogel downregulates collagen type I expression, indicating that HA plays a crucial role in modulating chondrocyte phenotype. These observations may be attributed to known biological functions of HA such as ligand-specific interactions with the cells. Hegewald et al. also observed beneficial effect of HA on the chondrogenic differentiation of equine MSCs [103].

Recently it has been reported that the encapsulation of ES cells within a scaffold system could increase its differentiation efficiency and allow formation of three-dimensional tissues [5, 104, 105]. Guiding the differentiation of ES cells using biomaterial scaffolds is a relatively unexplored area that has great potential in tissue engineering. Hwang et al. compared the chondrogenic potential of ES cells in three-dimensional environments, a PEGDA hydrogel scaffold, with their monolayer culture [104]. They observed enhanced chondrogenesis in the three-dimensional system as compared to the conventional monolayer system. These findings further support the argument that the differentiation of ES cells can be modulated by three-dimensional culture. Langer et al. also observed a similar trend in three-dimensional scaffolds leading to the formation of neural, hepatic, and cartilage tissues by ES cells [5]. They further investigated the role of mechanical properties of the scaffold on hES cell differentiation by using matrices such as Matrigel and a three-dimensional PLGA-co PLA scaffold coated with Matrigel. Although Matrigel provides a three-dimensional environment to the encapsulated cells, it failed to support three-dimensional tissue formation compared to the Matrigel coated PLGA-PLA scaffold. These findings indicate the importance of the mechanical stiffness of the scaffold, as well as the presence of biochemical cues in the scaffold, for hES cell differentiation. Philip et al. studied the effect of various ECM components on the differentiation of rhesus ES cells [7]. Their in vivo results show that the presence of Cartrigel, an extract of cartilage matrix components, resulted in enhanced musculoskeletal differentiation of the ES cells. These results indicate that the differentiation of ES cells can be directed to a particular lineage by using scaffolds containing ECM components derived from similar tissues. Xu et al. developed a feeder-free culture system for the proliferation of hES cells while retaining their undifferentiated state by exploiting the biological functions of Matrigel, which is rich in ECM components [8].

The ongoing discussion clearly indicates the need to design biomaterial scaffolds with desired biological cues and physical properties for the controlled differentiation of ES cells. To identify interactions between stem cells with a large array of biomaterials and growth factors, Anderson, Levenberg, and Langer applied high-throughput microarray technology to screening biomaterials [106]. The advantage of this approach is that it enables the screening of large libraries of biomaterials in small quantities. Their findings indicate that certain biomaterials significantly induce the differentiation of hES cells into epithelial-like cells, whereas other biomaterials support differentiation only in the presence or absence of specific growth factors.

19.3 Coculture Microenvironments for Directing Stem Cell Differentiation and Tissue Development

19.3.1 Novel Systems for Three-Dimensional Cell Coculture

The advantages of hydrogels as a scaffold for cartilage-tissue engineering have been discussed in the previous sections. In this section, we discuss how the versatility of hydrogels could be harnessed to create multilayer cell-laden scaffolds that control the spatial organization of different cell types within a three-dimensional system. Such multilayer hydrogels can be used as a tool to investigate various fundamental biological issues, such as cell-cell interactions in morphogenesis and pathology. Additionally, multilayer hydrogels can provide a structural model of stratified tissues. For instance, articular cartilage has a zonal architecture where chondrocytes in each distinct zone vary in their morphology, metabolic activity, and function. Therefore, single-layer chondrocyte-seeded scaffolds may not yield cartilage tissue with mechanical and functional properties commensurate with native tissue because of the limitation in recreating the zonal architecture of normal articular cartilage. To this end, our lab has utilized multilayer hydrogel systems to regenerate cartilage tissues akin to native tissue by encapsulating cells from different layers of cartilage within different hydrogel strips. To create the multilayer cell-laden hydrogels, polymer solutions containing chondrocytes from different zones were photopolymerized independently in the manner shown in Figure 19.6 [107]. Here, the ability of polymer chains to diffuse in, and mechanically entangle with, the "solidified" hydrogel layer onto which the cell-polymer solution was added was used to integrate the layers.

Another challenge in cartilage-tissue engineering is the integration of the engineered articular cartilage with the surrounding tissue. Cartilage-to-cartilage interfaces do not integrate completely due to the dense matrix and paucity of cells at the interface. On the other hand, implantation of osteochondral tissues has been considered to be beneficial as the bony region can serve as an anchor for the implant. Multilayer hydrogels are also used as a potential scaffold for creating osteochondral tissue with integrated cartilage and bone tissue in a controlled manner [75]. Here, bilayered cell-laden hydrogels containing respective layers of chondrocytes and MSCs were created and then cultured in osteogenic-inducing conditions where chondrocytes retained their phenotype while MSCs underwent osteogenic differentiation. The multilayer hydrogel technology could be extended to engineer other stratified tissues, such as blood vessels and corneal tissues, as well.

Dreier et al. applied a hydrogel coculture system to understand the paracrine interaction between human articular chondrocytes and macrophages during cartilage degradation [108]. Strictly speaking, instead of using an integrated bilayer hydrogel system, the authors placed a hydrogel containing chondrocytes on top of a hydrogel containing macrophages. Employing this coculture system, the authors deciphered that both chondrocytes and monocytes mutually induce alteration in their biosynthetic balance, which further plays a crucial role in cartilage degradation.

Hydrogel coculture systems have also been used to regulate the lineage-specific differentiation of stem cells. Recently, our lab demonstrated that the morphogenetic

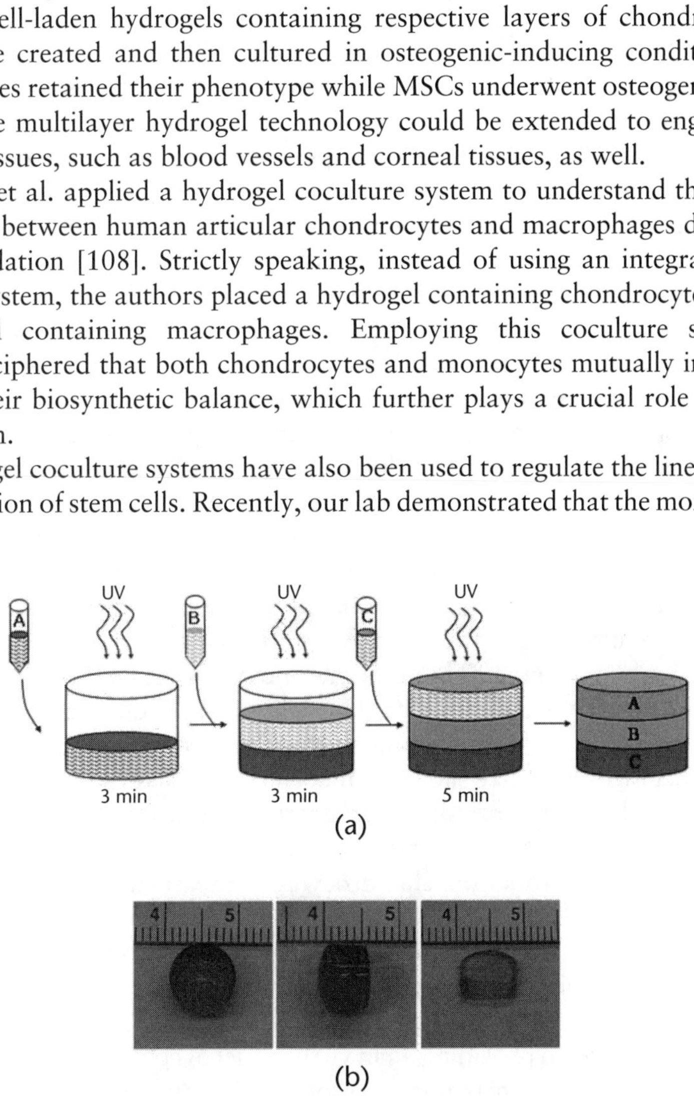

Figure 19.6 Schematic drawings showing the technique to make multilayered constructs: (a) encapsulating processes, and (b) the produced constructs [107].

factors from chondrocytes were able to induce chondrogenesis of embryonic germ-derived stem cells in a controlled manner, even in the absence of any exogenous growth factors. In most cases, paracrine signaling is effective within a short range of distances; therefore, it is advantageous to have the cells in close proximity as the morphogenic signals break down rapidly over larger distances. Such a close proximity between the cells may not be easily achieved in conventional coculture systems, such as a transwell system. Additionally, coculture systems, such as bilayered hydrogels, provide three-dimensional environments to the cells, which could be important because their native stage cells reside in a three-dimensional milieu. Some of our preliminary results also suggest that the effect of morphogenetic factors could not be completely mimicked by conditioned medium or growth factors.

19.3.2 Coculture Systems to Direct Embryonic Stem Cell Differentiation

In recent years, adult stem cells isolated from postnatal tissues and embryonic stem cells isolated from the blastocyst stage of an embryo have spurred tremendous interest due to their potential clinical applications in regenerative medicine. However, the lack of lineage-specific surface markers and gene expression limits the direct transplantation or use of these undifferentiated or uncommitted stem cells for regenerative application. Lack of expression for organ-specific cell markers, such as gap-junction connexin proteins that mediate cell-cell communications and surface ligands for interaction with ECMs, could pose limitations for the integration and engraftment of these cells within the recipient organs or tissue.

Chondogenesis is a complex process that is controlled by cellular interactions with the surrounding matrix, growth and morphogenic factors, and other environmental cues that modulate cellular signaling pathways and transcription of specific genes in a temporal-spatial manner. Chondrogenic differentiations of stem cells, especially ES cells, require microenvironments composed of multifacet direction for differentiation. A number of studies indicate direct chondrogenic differentiation of stem cells by coculture with a terminally differentiated chondrocyte cell population. Direct coculture may influence the surface receptors of cocultured cells to come into direct physical contact. The autocrine and paracrine factors secreted by different cell types may readily interact with stem cells to lead a direct and efficient transduction of molecular signals and result in expression of tissue-specific markers. Indeed, a number of studies have shown that coculture and cotransplantation of different cell types could be a promising therapeutic approach in regenerative medicine. For example, a study by Vats et al. demonstrated that microenvironments created by morphogenic factors from chondrocytes were sufficient to induce chondrogenic differentiation of hES cells [109]. Coculture with limb bud progenitor cells resulted in enhanced chondrogenic differentiation of ES cells. On the other hand, coculture with synovial-lining macrophages and embryonic calvarial cells was reported to stimulate chondrogenic differentiation of MSCs [110].

In addition to providing an in vitro model of coculture-mediated differentiation, recent studies indicate that in vivo cotransplantation results in modulation of differentiation potential, as well as vascularization of transplanted tissue constructs. Levenberg et al. achieved significant vascularization of engineered tissues with enhanced cell viability by cotransplantation of human ES cell derivatives with endothelial cells and mouse embryonic feeder cells [4, 111]. Kaigler et al. showed that endothelial cotransplantation with MSCs resulted in osteogenic differentiation of MSCs, along with the formation of microvasculature that delivered nutrient to developing bone [112, 113].

As discussed previously, it has been demonstrated that morphogenic signals from chondrocytes can regulate MSC differentiation, and the study showed that these factors promote the osteochondrogenic potential of MSCs [114, 115]. Moreover, chondrocyte-conditioned-medium-expanded MSCs maintain chondrocytic cell morphology and long-term survival in hydrogels when implanted in vivo, indicating that morphogenetic factors have long-lasting programming or instructive effects on MSCs. Given the inefficient differentiation protocols, coculture-mediated differentiation shows a promising methodology to direct chondrogenic differentiation and tissue formation from stem cells.

In summary, various methods to modify nanostructures of hydrogels have been developed to accommodate cell seeding. Proper cell attachment and viability have been the topic at issue, and a microenvironment specific to encapsulated cells has been created. Hydrogels for stem cell encapsulation have been extensively investigated, including coculture microenvironments. Finally, the feasibility of guiding the differentiation of ES cells using biomaterial scaffolds has been demonstrated, and further studies are envisioned to tailor hydrogels to accommodate the specific needs of encapsulated cells.

References

[1] Wallace, D. G., and Rosenblatt, J., "Collagen gel systems for sustained delivery and tissue engineering," *Adv. Drug Deliv. Rev.*, Vol. 55, No. 12, 2003, pp. 1631–1649.

[2] Burdick, J. A., Chung, C., Jia, X., Randolph, M. A., and Langer, R., "Controlled degradation and mechanical behavior of photopolymerized hyaluronic acid networks," *Biomacromolecules*, Vol. 6, No. 1, 2005, pp. 386–391.

[3] Li, Q., Williams, C. G., Sun, D. D., Wang, J., Leong, K., and Elisseeff, J. H., "Photocrosslinkable polysaccharides based on chondroitin sulfate," *J. Biomed. Mater. Res. A*, Vol. 68, No. 1, 2004, pp. 28–33.

[4] Levenberg, S., and Langer, R., "Advances in tissue engineering," *Curr. Top. Dev. Biol.*, Vol. 61, 2004, pp. 113–134.

[5] Levenberg, S., Huang, N. F., Lavik, E., Rogers, A. B., Itskovitz-Eldor, J., and Langer, R., "Differentiation of human embryonic stem cells on three-dimensional polymer scaffolds," *Proc. Natl. Acad. Sci. USA*, Vol. 100, No. 22, 2003, pp. 12741–12746.

[6] Lavik, E., and Langer, R., "Tissue engineering: Current state and perspectives," *Appl. Microbiol. Biotechnol.*, Vol. 65, No. 1, 2004, pp. 1–8.

[7] Philp, D., Chen, S. S., Fitzgerald, W., Orenstein, J., Margolis, L., and Kleinman, H. K., "Complex extracellular matrices promote tissue-specific stem cell differentiation," *Stem Cells*, Vol. 23, No. 2, 2005, pp. 288–296.

[8] Xu, C., Inokuma, M. S., Denham, J., Golds, K., Kundu, P., Gold, J. D., and Carpenter, M. K., "Feeder-free growth of undifferentiated human embryonic stem cells," *Nat. Biotechnol.*, Vol. 19, No. 10, 2001, pp. 971–974.

[9] Varghese, S., and Elisseeff, J. H., "Hydrogels for musculuskeletal tissue engineering," *Adv. Polym. Sci.*, Vol. 203, 2006, pp. 95–144.

[10] Grunder, T., Gaissmaier, C., Fritz, J., Stoop, R., Hortschansky, P., Mollenhauer, J., and Aicher, W. K., "Bone morphogenetic protein (BMP)-2 enhances the expression of type II collagen and aggrecan in chondrocytes embedded in alginate beads," *Osteoarth. Cartil.*, Vol. 12, No. 7, 2004, pp. 559–567.

[11] Marijnissen, W. J., van Osch, G. J., Aigner, J., van der Veen, S. W., Hollander, A. P., Verwoerd-Verhoef, H. L., and Verhaar, J. A., "Alginate as a chondrocyte-delivery substance in combination with a non-woven scaffold for cartilage tissue engineering," *Biomaterials*, Vol. 23, No. 6, 2002, pp. 1511–1517.

[12] Selden, C., and Hodgson, H., "Cellular therapies for liver replacement," *Transplant. Immunol.*, Vol. 12, Nos. 3–4, 2004, pp. 273–288.

[13] Lanza, R. P., and Chick, W. L., "Transplantation of encapsulated cells and tissues," *Surgery*, Vol. 121, No. 1, 1997, pp. 1–9.

[14] Doyle, J. W., Roth, T. P., Smith, R. M., Li, Y. Q., and Dunn, R. M., "Effects of calcium alginate on cellular wound healing processes modeled in vitro," *J. Biomed. Mater. Res.*, Vol. 32, No. 4, 1996, pp. 561–568.

[15] Alsberg, E., Kong, H. J., Hirano, Y., Smith, M. K., Albeiruti, A., and Mooney, D. J., "Regulating bone formation via controlled scaffold degradation," *J. Dent. Res.*, Vol. 82, No. 11, 2003, pp. 903–908.

[16] Awad, H. A., Wickham, M. Q., Leddy, H. A., Gimble, J. M., and Guilak, F., "Chondrogenic differentiation of adipose-derived adult stem cells in agarose, alginate, and gelatin scaffolds," *Biomaterials*, Vol. 25, No. 16, 2004, pp. 3211–3222.

[17] Rowley, J. A., Madlambayan, G., and Mooney, D. J., "Alginate hydrogels as synthetic extracellular matrix materials," *Biomaterials*, Vol. 20, No. 1, 1999, pp. 45–53.

[18] Diduch, D. R., Jordan, L. C., Mierisch, C. M., and Balian, G., "Marrow stromal cells embedded in alginate for repair of osteochondral defects," *Arthroscopy*, Vol. 16, No. 6, 2000, pp. 571–577.

[19] Bryant, S. J., and Anseth, K. S., "Hydrogel properties influence ECM production by chondrocytes photoencapsulated in poly(ethylene glycol) hydrogels," *J. Biomed. Mater. Res.*, Vol. 59, No. 1, 2002, pp. 63–72.

[20] Peppas, N. A., Huang, Y., Torres-Lugo, M., Ward, J. H., and Zhang, J., "Physicochemical foundations and structural design of hydrogels in medicine and biology," *Ann. Rev. Biomed. Eng.*, Vol. 2, 2000, pp. 9–29.

[21] Reddy, K. R., "Controlled-release, pegylation, liposomal formulations: New mechanisms in the delivery of injectable drugs," *Ann. Pharmacother.*, Vol. 34, Nos. 7–8, 2000, pp. 915–923.

[22] Bailey, F. E., Koelske, J. V., *Poly(ethylene oxide)*, Acad. Press, Vol. 6, 1976, p. 103.

[23] Dalsin, J. L., Hu, B. H., Lee, B. P., and Messersmith, P. B., "Mussel adhesive protein mimetic polymers for the preparation of nonfouling surfaces," *J. Am. Chem. Soc.*, Vol. 125, No. 14, 2003, pp. 4253–4258.

[24] Ratner, B. D., and Bryant, S. J., "Biomaterials: Where we have been and where we are going," *Ann. Rev. Biomed. Eng.*, Vol. 6, 2004, pp. 41–75.

[25] Kim, B. S., Hrkach, J. S., and Langer, R., "Biodegradable photo-crosslinked poly(ether-ester) networks for lubricious coatings," *Biomaterials*, Vol. 21, No. 3, 2000, pp. 259–265.

[26] Spector, S. N. e. a. I. Y., "Matrix collagen type and pore size influence behavior of seeded canine chondrocytes (collagen-GAG copolymer)," *Biomaterials*, Vol. 18, 1997, p. 769.

[27] Hern, D. L., and Hubbell, J. A., "Incorporation of adhesion peptides into nonadhesive hydrogels useful for tissue resurfacing," *J. Biomed. Mater. Res.*, Vol. 39, No. 2, 1998, pp. 266–276.

[28] Yang, F., Williams, C. G., Wang, D. A., Lee, H., Manson, P. N., and Elisseeff, J., "The effect of incorporating RGD adhesive peptide in polyethylene glycol diacrylate hydrogel on osteogenesis of bone marrow stromal cells," *Biomaterials*, Vol. 26, No. 30, 2005, pp. 5991–5998.

[29] Burdick, J. A., and Anseth, K. S., "Photoencapsulation of osteoblasts in injectable RGD-modified PEG hydrogels for bone tissue engineering," *Biomaterials*, Vol. 23, No. 22, 2002, pp. 4315–4323.

[30] Lutolf, M. P., Weber, F. E., Schmoekel, H. G., Schense, J. C., Kohler, T., Muller, R., and Hubbell, J. A., "Repair of bone defects using synthetic mimetics of collagenous extracellular matrices," *Nat. Biotechnol.*, Vol. 21, No. 5, 2003, pp. 513–518.

[31] Lee, H. J., Lee, J. S., Chansakul, T., Yu, C., Elisseeff, J. H., and Yu, S. M., "Collagen mimetic peptide-conjugated photopolymerizable PEG hydrogel," *Biomaterials*, Vol. 27, No. 30, 2006, pp. 5268–5276.

[32] Jun, H. W., and West, J. L., "Modification of polyurethaneurea with PEG and YIGSR peptide to enhance endothelialization without platelet adhesion," *J. Biomed. Mater. Res. B Appl. Biomater.*, Vol. 72, No. 1, 2005, pp. 131–139.

[33] Gunn, J. W., Turner, S. D., and Mann, B. K., "Adhesive and mechanical properties of hydrogels influence neurite extension," *J. Biomed. Mater. Res. A*, Vol. 72, No. 1, 2005, pp. 91–97.

[34] Wang, D. A., Williams, C. G., Li, Q., Sharma, B., and Elisseeff, J. H., "Synthesis and characterization of a novel degradable phosphate-containing hydrogel," *Biomaterials*, Vol. 24, No. 22, 2003, pp. 3969–3980.

[35] Metters, A. T., Anseth, K. S., and Bowman, C. N., "Fundamental studies of biodegradable hydrogels as cartilage replacement materials," *Biomed. Sci. Instrum.*, Vol. 35, 1999, pp. 33–38.

[36] Ulbrich, K., Strohalm, J., and Kopecek, J., "Polymers containing enzymatically degradable bonds: VI. Hydrophilic gels cleavable by chymotrypsin," *Biomaterials*, Vol. 3, No. 3, 1982, pp. 150–154.

[37] Klein, L., Stock, J., and Vorlop, K. G., "Pore size and properties of spherical Ca-alginate biocatalysts," *Eur. J. Appl. Microbiol. Biotechnol.*, Vol. 18, 1983, pp. 86–91.

[38] Grant, G., Morris, E., Rees, D., Smith, P., and Thom, D., "Biological interactions between polysaccharides and divalent cations—egg-box model," *Febs. Lett.*, Vol. 32, No. 1, 1973, pp. 195–198.

[39] Kong, H. J., Kaigler, D., Kim, K., and Mooney, D. J., "Controlling rigidity and degradation of alginate hydrogels via molecular weight distribution," *Biomacromolecules*, Vol. 5, No. 5, 2004, pp. 1720–1727.

[40] Bouhadir, K. H., Lee, K. Y., Alsberg, E., Damm, K. L., Anderson, K. W., and Mooney, D. J., "Degradation of partially oxidized alginate and its potential application for tissue engineering," *Biotechnol. Prog.*, Vol. 17, No. 5, 2001, pp. 945–950.

[41] Lee, K. Y., Bouhadir, K. H., and Mooney, D. J., "Evaluation of chain stiffness of partially oxidized polyguluronate," *Biomacromolecules*, Vol. 3, No. 6, 2002, pp. 1129–1134.

[42] Eiselt, P., Yeh, J., Latvala, R. K., Shea, L. D., and Mooney, D. J., "Porous carriers for biomedical applications based on alginate hydrogels," *Biomaterials*, Vol. 21, No. 19, 2000, pp. 1921–1927.

[43] Luo, Y., and Shoichet, M. S., "A photolabile hydrogel for guided three-dimensional cell growth and migration," *Nat. Mater*, Vol. 3, No. 4, 2004, pp. 249–253.

[44] Yamamoto, M., Tabata, Y., Hong, L., Miyamoto, S., Hashimoto, N., and Ikada, Y., "Bone regeneration by transforming growth factor beta1 released from a biodegradable hydrogel," *J. Control. Release*, Vol. 64, Nos. 1–3, 2000, pp. 133–142.

[45] Reyes, J. M., Herretes, S., Pirouzmanesh, A., Wang, D. A., Elisseeff, J. H., Jun, A., McDonnell, P. J., Chuck, R. S., and Behrens, A., "A modified chondroitin sulfate aldehyde adhesive for sealing corneal incisions," *Invest. Ophthalmol. Vis. Sci.*, Vol. 46, No. 4, 2005, pp. 1247–1250.

[46] Sperinde, J., and Griffith, L., "Control and prediction of gelation kinetics in enzymatically cross-linked poly(ethylene glycol) hydrogels," *Macromolecules*, Vol. 33, No. 15, 2000, pp. 5476–5480.

[47] Sperinde, J., and Griffith, L., "Synthesis and characterization of enzymatically cross-linked poly(ethylene glycol) hydrogels," *Macromolecules*, Vol. 30, No. 18, 1997, pp. 5255–5264.

[48] Liu, V., and Bhatia, S., "Three-dimensional photopatterning of hydrogels containing living cells," *Biomed. Microdevices*, Vol. 4, No. 4, 2002, pp. 257–266.

[49] Sawhney, A. S., Pathak, C. P., and Hubbell, J. A., "Interfacial photopolymerization of poly(ethylene glycol)-based hydrogels upon alginate-poly(l-lysine) microcapsules for enhanced biocompatibility," *Biomaterials*, Vol. 14, No. 13, 1993, pp. 1008–1016.

[50] Bhalerao, V., Varghese, S., Lele, A., and Badiger, M., "Thermoreversible hydrogel based on radiation induced copolymerisation of poly(N-isopropyl acrylamide) and poly(ethylene oxide)," *Polymer*, Vol. 39, No. 11, 1998, pp. 2255–2260.

[51] Cruise, G. M., Scharp, D. S., and Hubbell, J. A., "Characterization of permeability and network structure of interfacially photopolymerized poly(ethylene glycol) diacrylate hydrogels," *Biomaterials*, Vol. 19, No. 14, 1998, pp. 1287–1294.

[52] Varghese, S., Lele, A., and Mashelkar, R., "Designing new thermoreversible gels by molecular tailoring of hydrophilic-hydrophobic interactions," *J. Chem. Phys.*, Vol. 112, No. 6, 2000, pp. 3063–3070.

[53] Fisher, J., Jo, S., Mikos, A., and Reddi, A., "Thermoreversible hydrogel scaffolds for articular cartilage engineering," *J. Biomed. Mater. Res. A*, Vol. 71A, No. 2, 2004, pp. 268–274.

[54] Mata, J., Joshi, T., Varade, D., Ghosh, G., and Bahadur, P., "Aggregation behavior of a PEO-PPO-PEO block copolymer plus ionic surfactants mixed systems in water and aqueous salt solutions," *Colloids and Surfaces A: Physicochem. Eng. Aspects*, Vol. 247, Nos. 1–3, 2004, pp. 1–7.

[55] Cascone, M., Lazzeri, L., Sparvoli, E., Scatena, M., Serino, L., and Danti, S., "Morphological evaluation of bioartificial hydrogels as potential tissue engineering scaffolds," *J. Mater. Sci: Mater. Med.*, Vol. 15, No. 12, 2004, pp. 1309–1313.

[56] Ricciardi, R., Auriemma, F., Gaillet, C., De Rosa, C., and Laupretre, F., "Investigation of the crystallinity of freeze/thaw poly(vinyl alcohol) hydrogels by different techniques," *Macromolecules*, Vol. 37, No. 25, 2004, pp. 9510–9516.

[57] Bell, C., and Peppas, N., "Modulation of drug permeation through interpolymer complexed hydrogels for drug delivery applications," *J. Control. Release*, Vol. 39, Nos. 2–3, 1996, pp. 201–207.

[58] Mathur, A. M., Hammonds, K. F., Klier, J., and Scranton, A. B., "Equilibrium swelling of poly(methacrylic acid-g-ethylene glycol) hydrogels: Effect of swelling medium and synthesis conditions," *J. Control. Release*, Vol. 54, No. 2, 1998, pp. 177–184.

[59] Miyata, T., Asami, N., and Uragami, T., "A reversibly antigen-responsive hydrogel," *Nature*, Vol. 399, No. 6738, 1999, pp. 766–769.

[60] Holmes, T. C., de Lacalle, S., Su, X., Liu, G., Rich, A., and Zhang, S., "Extensive neurite outgrowth and active synapse formation on self-assembling peptide scaffolds," *Proc. Natl. Acad. Sci. USA*, Vol. 97, No. 12, 2000, pp. 6728–6733.

[61] Xu, C., Breedveld, V., and Kopecek, J., "Reversible hydrogels from self-assembling genetically engineered protein block copolymers," *Biomacromolecules*, Vol. 6, No. 3, 2005, pp. 1739–1749.

[62] Hartgerink, J. D., Beniash, E., and Stupp, S. I., "Self-assembly and mineralization of peptide-amphiphile nanofibers," *Science*, Vol. 294, No. 5547, 2001, pp. 1684–1688.

[63] Aigner, T., and Stove, J., "Collagens—major component of the physiological cartilage matrix, major target of cartilage degeneration, major tool in cartilage repair," *Adv. Drug Deliv. Rev.*, Vol. 55, No. 12, 2003, pp. 1569–1593.

[64] Bokhari, M. A., Akay, G., Zhang, S., and Birch, M. A., "The enhancement of osteoblast growth and differentiation in vitro on a peptide hydrogel-polyHIPE polymer hybrid material," *Biomaterials*, Vol. 26, No. 25, 2005, pp. 5198–5208.

[65] Zhang, S., "Fabrication of novel biomaterials through molecular self-assembly," *Nat. Biotechnol.*, Vol. 21, No. 10, 2003, pp. 1171–1178.

[66] Kisiday, J., Jin, M., Kurz, B., Hung, H., Semino, C., Zhang, S., and Grodzinsky, A. J., "Self-assembling peptide hydrogel fosters chondrocyte extracellular matrix production and cell division: Implications for cartilage tissue repair," *Proc. Natl. Acad. Sci. USA*, Vol. 99, No. 15, 2002, pp. 9996–10001.

[67] Gelain, F., Bottai, D., Vescovi, A., and Zhang, S., "Designer self-assembling peptide nanofiber scaffolds for adult mouse neural stem cell 3-dimensional cultures," *PLoS One*, Vol. 1, 2006, p. e119.

[68] Semino, C. E., Merok, J. R., Crane, G. G., Panagiotakos, G., and Zhang, S., "Functional differentiation of hepatocyte-like spheroid structures from putative liver progenitor cells in three-dimensional peptide scaffolds," *Differentiation*, Vol. 71, Nos. 4–5, 2003, pp. 262–270.

[69] Elisseeff, J., Anseth, K., Sims, D., McIntosh, W., Randolph, M., and Langer, R., "Transdermal photopolymerization for minimally invasive implantation," *Proc. Natl. Acad. Sci. USA*, Vol. 96, No. 6, 1999, pp. 3104–3107.

[70] Hoffman, A. S., "Hydrogels for biomedical applications," *Adv. Drug Deliv. Rev.*, Vol. 54, No. 1, 2002, pp. 3–12.

[71] Hennink, W. E., and van Nostrum, C. F., "Novel crosslinking methods to design hydrogels," *Adv. Drug Deliv. Rev.*, Vol. 54, No. 1, 2002, pp. 13–36.

[72] Smetana, K., Jr., "Cell biology of hydrogels," *Biomaterials*, Vol. 14, No. 14, 1993, pp. 1046–1050.

[73] Bryant, S. J., Anseth, K. S., Lee, D. A., and Bader, D. L., "Crosslinking density influences the morphology of chondrocytes photoencapsulated in PEG hydrogels during the application of compressive strain," *J. Orthop. Res.*, Vol. 22, No. 5, 2004, pp. 1143–1149.

[74] Miot, S., Woodfield, T., Daniels, A. U., Suetterlin, R., Peterschmitt, I., Heberer, M., van Blitterswijk, C. A., Riesle, J., and Martin, I., "Effects of scaffold composition and architecture on human nasal chondrocyte redifferentiation and cartilaginous matrix deposition," *Biomaterials*, Vol. 26, No. 15, 2005, pp. 2479–2489.

[75] Sharma, B., and Elisseeff, J. H., "Engineering structurally organized cartilage and bone tissues," *Ann. Biomed. Eng.*, Vol. 32, No. 1, 2004, pp. 148–159.

[76] Williams, C. G., Kim, T. K., Taboas, A., Malik, A., Manson, P., and Elisseeff, J., "In vitro chondrogenesis of bone marrow–derived mesenchymal stem cells in a photopolymerizing hydrogel," *Tissue Eng.*, Vol. 9, No. 4, 2003, pp. 679–688.

[77] Hwang, N. S., Varghese, S., Theprungsirikul, P., Canver, A., and Elisseeff, J., "Enhanced chondrogenic differentiation of murine embryonic stem cells in hydrogels with glucosamine," *Biomaterials*, Vol. 27, No. 36, 2006, pp. 6015–6023.

[78] Thomson, J. A., Itskovitz-Eldor, J., Shapiro, S. S., Waknitz, M. A., Swiergiel, J. J., Marshall, V. S., and Jones, J. M., "Embryonic stem cell lines derived from human blastocysts," *Science*, Vol. 282, No. 5391, 1998, pp. 1145–1147.

[79] Hoffman, L. M., and Carpenter, M. K., "Characterization and culture of human embryonic stem cells," *Nat. Biotechnol.*, Vol. 23, No. 6, 2005, pp. 699–708.

[80] Elisseeff, J. H., "Embryonic stem cells: Potential for more impact," *Trends Biotechnol.*, Vol. 22, No. 4, 2004, pp. 155–156.

[81] Song, L., and Tuan, R. S., "Transdifferentiation potential of human mesenchymal stem cells derived from bone marrow," *Faseb J.*, Vol. 18, No. 9, 2004, pp. 980–982.

[82] Chen, F. H., Rousche, K. T., and Tuan, R. S., "Technology insight: Adult stem cells in cartilage regeneration and tissue engineering," *Nat. Clin. Pract. Rheumatol.*, Vol. 2, No. 7, 2006, pp. 373–382.

[83] Minguell, J. J., Erices, A., and Conget, P., "Mesenchymal stem cells," *Exp. Biol. Med.* (Maywood), Vol. 226, No. 6, 2001, pp. 507–520.

[84] Chamberlain, J. R., Schwarze, U., Wang, P. R., Hirata, R. K., Hankenson, K. D., Pace, J. M., Underwood, R. A., Song, K. M., Sussman, M., Byers, P. H., and Russell, D. W., "Gene targeting in stem cells from individuals with osteogenesis imperfecta," *Science*, Vol. 303, No. 5661, 2004, pp. 1198–1201.

[85] Koc, O. N., Gerson, S. L., Cooper, B. W., Dyhouse, S. M., Haynesworth, S. E., Caplan, A. I., and Lazarus, H. M., "Rapid hematopoietic recovery after coinfusion of autologous-blood stem cells and culture-expanded marrow mesenchymal stem cells in advanced breast cancer patients receiving high-dose chemotherapy," *J. Clin. Oncol.*, Vol. 18, No. 2, 2000, pp. 307–316.

[86] Pittenger, M. F., Mackay, A. M., Beck, S. C., Jaiswal, R. K., Douglas, R., Mosca, J. D., Moorman, M. A., Simonetti, D. W., Craig, S., and Marshak, D. R., "Multilineage potential of adult human mesenchymal stem cells," *Science*, Vol. 284, No. 5411, 1999, pp. 143–147.

[87] Sethe, S., Scutt, A., and Stolzing, A., "Aging of mesenchymal stem cells," *Ageing Res. Rev.*, Vol. 5, No. 1, 2006, pp. 91–116.

[88] Murphy, J. M., Dixon, K., Beck, S., Fabian, D., Feldman, A., and Barry, F., "Reduced chondrogenic and adipogenic activity of mesenchymal stem cells from patients with advanced osteoarthritis," *Arthritis Rheum.*, Vol. 46, No. 3, 2002, pp. 704–713.

[89] Sekiya, I., Larson, B. L., Vuoristo, J. T., Reger, R. L., and Prockop, D. J., "Comparison of effect of BMP-2, -4, and -6 on in vitro cartilage formation of human adult stem cells from bone marrow stroma," *Cell Tissue Res.*, Vol. 320, No. 2, 2005, pp. 269–276.

[90] Shirasawa, S., Sekiya, I., Sakaguchi, Y., Yagishita, K., Ichinose, S., and Muneta, T., "In vitro chondrogenesis of human synovium-derived mesenchymal stem cells: Optimal condition and comparison with bone marrow–derived cells," *J. Cell. Biochem.*, Vol. 97, No. 1, 2006, pp. 84–97.

[91] Johnstone, B., Hering, T. M., Caplan, A. I., Goldberg, V. M., and Yoo, J. U., "In vitro chondrogenesis of bone marrow–derived mesenchymal progenitor cells," *Exp. Cell Res.*, Vol. 238, No. 1, 1998, pp. 265–272.

[92] Siebler, T., Robson, H., Shalet, S. M., and Williams, G. R., "Dexamethasone inhibits and thyroid hormone promotes differentiation of mouse chondrogenic ATDC5 cells," *Bone*, Vol. 31, No. 4, 2002, pp. 457–464.

[93] Robson, H., Siebler, T., Stevens, D. A., Shalet, S. M., and Williams, G. R., "Thyroid hormone acts directly on growth plate chondrocytes to promote hypertrophic differentiation and inhibit clonal expansion and cell proliferation," *Endocrinology*, Vol. 141, No. 10, 2000, pp. 3887–3897.

[94] Farquharson, C., Berry, J. L., Mawer, E. B., Seawright, E., and Whitehead, C. C., "Ascorbic acid-induced chondrocyte terminal differentiation: The role of the extracellular matrix and 1,25-dihydroxyvitamin D," *Eur. J. Cell. Biol.*, Vol. 76, No. 2, 1998, pp. 110–118.

[95] Tsonis, P. A., "1,25-Dihydroxyvitamin D3 stimulates chondrogenesis of the chick limb bud mesenchymal cells," *Dev. Biol.*, Vol. 143, No. 1, 1991, pp. 130–134.

[96] Kulyk, W. M., and Reichert, C., "Staurosporine, a protein kinase inhibitor, stimulates cartilage differentiation by embryonic facial mesenchyme," *J. Craniofac. Genet. Dev. Biol.*, Vol. 12, No. 2, 1992, pp. 90–97.

[97] Kulyk, W. M., "Promotion of embryonic limb cartilage differentiation in vitro by staurosporine, a protein kinase C inhibitor," *Dev. Biol.*, Vol. 146, No. 1, 1991, pp. 38–48.

[98] Angele, P., Yoo, J. U., Smith, C., Mansour, J., Jepsen, K. J., Nerlich, M., and Johnstone, B., "Cyclic hydrostatic pressure enhances the chondrogenic phenotype of human mesenchymal progenitor cells differentiated in vitro," *J. Orthop. Res.*, Vol. 21, No. 3, 2003, pp. 451–457.

[99] Lennon, D. P., Edmison, J. M., and Caplan, A. I., "Cultivation of rat marrow–derived mesenchymal stem cells in reduced oxygen tension: Effects on in vitro and in vivo osteochondrogenesis," *J. Cell. Physiol.*, Vol. 187, No. 3, 2001, pp. 345–355.

[100] Chen, X., Xu, H., Wan, C., McCaigue, M., and Li, G., "Bioreactor expansion of human adult bone marrow–derived mesenchymal stem cells," *Stem Cells*, Vol. 24, No. 9, 2006, pp. 2052–2059.

[101] Meinel, L., Karageorgiou, V., Hofmann, S., Fajardo, R., Snyder, B., Li, C., Zichner, L., Langer, R., Vunjak-Novakovic, G., and Kaplan, D. L., "Engineering bone-like tissue in

vitro using human bone marrow stem cells and silk scaffolds," *J. Biomed. Mater. Res. A*, Vol. 71, No. 1, 2004, pp. 25–34.

[102] Niemeyer, P., Krause, U., Punzel, M., Fellenberg, J., and Simank, H. G., "Mesenchymal stem cells for tissue engineering of bone: 3-D-cultivation and osteogenic differentiation on mineralized collagen," *Z Orthop. Ihre Grenzgeb*, Vol. 141, No. 6, 2003, pp. 712–717.

[103] Hegewald, A. A., Ringe, J., Bartel, J., Kruger, I., Notter, M., Barnewitz, D., Kaps, C., and Sittinger, M., "Hyaluronic acid and autologous synovial fluid induce chondrogenic differentiation of equine mesenchymal stem cells: A preliminary study," *Tissue Cell*, Vol. 36, No. 6, 2004, pp. 431–438.

[104] Hwang, N. S., Kim, M. S., Sampattavanich, S., Baek, J. H., Zhang, Z., and Elisseeff, J., "Effects of three-dimensional culture and growth factors on the chondrogenic differentiation of murine embryonic stem cells," *Stem Cells*, Vol. 24, No. 2, 2006, pp. 284–291.

[105] Liu, H., and Roy, K., "Biomimetic three-dimensional cultures significantly increase hematopoietic differentiation efficacy of embryonic stem cells," *Tissue Eng.*, Vol. 11, Nos. 1–2, 2005, pp. 319–330.

[106] Anderson, D. G., Levenberg, S., and Langer, R., "Nanoliter-scale synthesis of arrayed biomaterials and application to human embryonic stem cells," *Nat. Biotechnol.*, Vol. 22, No. 7, 2004, pp. 863–866.

[107] Kim, T. K., Sharma, B., Williams, C. G., Ruffner, M. A., Malik, A., McFarland, E. G., and Elisseeff, J. H., "Experimental model for cartilage tissue engineering to regenerate the zonal organization of articular cartilage," *Osteoarth. Cartil.*, Vol. 11, No. 9, 2003, pp. 653–664.

[108] Dreier, R., Wallace, S., Fuchs, S., Bruckner, P., and Grassel, S., "Paracrine interactions of chondrocytes and macrophages in cartilage degradation: Articular chondrocytes provide factors that activate macrophage-derived pro-gelatinase B (pro-MMP-9)," *J. Cell. Sci.*, Vol. 114, Pt. 21, 2001, pp. 3813–3822.

[109] Vats, A., Bielby, R. C., Tolley, N., Dickinson, S. C., Boccaccini, A. R., Hollander, A. P., Bishop, A. E., and Polak, J. M., "Chondrogenic differentiation of human embryonic stem cells: The effect of the micro-environment," *Tissue Eng.*, Vol. 12, No. 6, 2006, pp. 1687–1697.

[110] Heng, B. C., Cao, T., and Lee, E. H., "Directing stem cell differentiation into the chondrogenic lineage in vitro," *Stem Cells*, Vol. 22, No. 7, 2004, pp. 1152–1167.

[111] Levenberg, S., Golub, J. S., Amit, M., Itskovitz-Eldor, J., and Langer, R., "Endothelial cells derived from human embryonic stem cells," *Proc. Natl. Acad. Sci. USA*, Vol. 99, No. 7, 2002, pp. 4391–4396.

[112] Murphy, W. L., Simmons, C. A., Kaigler, D., and Mooney, D. J., "Bone regeneration via a mineral substrate and induced angiogenesis," *J. Dent. Res.*, Vol. 83, No. 3, 2004, pp. 204–210.

[113] Kaigler, D., Krebsbach, P. H., Wang, Z., West, E. R., Horger, K., and Mooney, D. J., "Transplanted endothelial cells enhance orthotopic bone regeneration," *J. Dent. Res.*, Vol. 85, No. 7, 2006, pp. 633–637.

[114] Gerstenfeld, L. C., Barnes, G. L., Shea, C. M., and Einhorn, T. A., "Osteogenic differentiation is selectively promoted by morphogenetic signals from chondrocytes and synergized by a nutrient-rich growth environment," *Connect. Tissue Res.*, Vol. 44, Suppl. 1, 2003, pp. 85–91.

[115] Gerstenfeld, L. C., Cruceta, J., Shea, C. M., Sampath, K., Barnes, G. L., and Einhorn, T. A., "Chondrocytes provide morphogenic signals that selectively induce osteogenic differentiation of mesenchymal stem cells," *J. Bone Miner. Res.*, Vol. 17, No. 2, 2002, pp. 221–230.

CHAPTER 20

Microscale Approaches for Bone Tissue Engineering

Jeffrey M. Karp, Alborz Mahdavi, Seungpyo Hong, Ali Khademhosseini, and Robert Langer

20.1 Introduction

Given the high incidence of bone trauma, cancer, and adult and congenital disease, which is associated with over five hundred thousand bone-graft procedures each year in North America, it is not surprising that bone is the most commonly transplanted tissue, second only to blood [1]. The challenge to addressing this great demand is compounded by limitations with current therapeutic treatments, including autologous graft procedures, where inadequate supplies exist, especially in elderly patients. Harvesting donor bone tissues often induces additional trauma, which has associated risks, such as donor site morbidity and infection [2].

Although allograft bone currently represents an essential alternative to autologous bone grafts, it is less effective in stimulating bone regeneration since it is typically treated to remove potential immunogenic and disease-related moieties and, thus, is thus devoid of the living cells and active growth factors that provide significant biological activity within autografts. In addition to autografts and allografts, artificial prostheses have been used extensively, but these are subject to wear, fatigue, and eventual failure. Given the enormity of these deficiencies and the importance of developing new therapies and bone-disease prevention measures, the World Health Organization declared 2002 to 2011 the Bone and Joint Decade [1].

The promising field of bone-tissue engineering holds promise for generating sufficient autologous tissues to graft into a defect site, especially those defects that would not heal if left empty (critical-sized defects), without the need to harvest donor tissue. Typical bone-tissue-engineering strategies involve the combination of biodegradable polymeric scaffolds containing growth factors and a source of osteogenic cells, which may be harvested from a bone marrow aspirate.

Given that cells in vitro have limited ability to grow in three-dimensional orientations, scaffolds are obligatory to achieve physiologically relevant three-dimensional architectures and the requisite cell-cell interactions to create functional tissues. The scaffolds, although temporary, can be engineered to support diverse cellular functions, including migration, proliferation, and differentiation of osteoprogenitor cells, and to aid in the organization of these cells in three dimensions. Once implanted, the scaffold must maintain its volume through resisting mechanical deformation caused by migrating cells or movement of surrounding tissues. Ideally, bone will form on the scaffold, which may act as a vehicle for cells and/or an "active" template for the developing bone, after which the scaffold material will degrade, and the patient will be left with functional bone tissue.

A major challenge in orthopedic and craniofacial medicine is to understand the biological mechanisms that can be utilized to achieve and/or accelerate bone regeneration. One of the guiding principles in the field of biomaterials and tissue engineering is that a successful clinical outcome can be achieved through engineering materials to manipulate the cellular microenvironment and ultimately control cell function. In certain cases, it may be desirable to control the placement of cells and factors in three dimensions for designing multicomponent engineered tissues such as osteochondral implants. This may be achieved through precision multiscale control of materials, architecture, and cells with biology as the guiding design principle [3].

Combining scaffolds with cells that have been expanded in vitro is an active area of research and has been extensively explored [4–11]. However, these typical strategies have varied success rates when applied to critical-sized defects. In addition, implantation of empty macroporous scaffolds that are devoid of cells typically does not improve the healing response [8, 12, 13]. The vast differences in the composition of scaffolds employed in the field of bone engineering makes it challenging to compare results between studies, and this limits what can be learned and applied by other researchers in the field. For example, typical scaffold-based strategies may differ with respect to materials, porosity, surface chemistry, morphology, degradation rate, pore sizes, and mechanical properties. To better understand how individual properties affect bone regeneration, more controlled studies are required, and increased understanding of the osteogenic microenvironment may help provide rational design criteria for scaffolds.

20.2 Importance of Cell-Cell Interactions for Regulating Osteogenesis

Our current understanding of cell-cell interactions within the osteogenic compartment has been primarily derived from gross mixtures of cell populations or isolation of cell populations via various means (e.g., Boyden chambers, chambered culture dishes, etc.) [14, 15]. To gain further understanding of these types of phenomena, techniques have been developed to spatially control the interaction of two or more cell types through coculture systems [16–18]. Specifically, using a hepatocyte model system, Bhatia et al. demonstrated the ability to control the level of homotypic interaction in cultures of a single cell type and the degree of heterotypic contact in cocultures. Similary numerous technologies have been devised to elucidate the effects of cell–extracellular matrix (ECM) interactions, cell-cell interactions, and the effects of soluble stimuli on cell function [18–21].

Recent studies have demonstrated that osteoblast interactions with other cell types, including endothelial cells, chrondrocytes, fibroblasts, or hematopoietic cells, may ultimately modulate the proliferation, migration, self-renewal, and differentiation of these cells [14, 15, 22–24]. Through use of coculture techniques, it has been shown that paracrine signaling between osteoblasts and chondrocytes allows for these two cell types to regulate each other's growth and differentiation [24]. The complexity of cell-cell interactions may be compounded by a multitude of factors, including the differentiation state of the cells, which can influence the biological outcome. For example, differentiated osteogenic cells that line the endosteal bone surface have been proposed as a regulator of the hematopoietic stem cell niche within bone marrow [22]. Interestingly, biodegradable scaffolds seeded with osteogenic cultures of bone marrow–derived cells and subcutaneously implanted into nude mice have been demonstrated to recapitulate host marrow formation within the pores of the scaffold and may represent an important consideration for bone-tissue-engineering approaches [25]. This system seems to serve as an in vivo model to interrogate interactions between osteogenic and hematapoietic compartments. However, interactions between individual cells or colonies of cells are difficult to control or probe in this and other similar systems.

Bone formation in vitro can result from sample sizes as small as a single cell that undergoes a proliferation-differentiation sequence. Although it has been suggested that heterotypic cell-cell interactions play an important role in this process [14, 15, 26], very little is known about the mechanisms that initiate bone formation in vitro. Current knowledge is partially derived from experiments with model systems comparing bone formation in vitro between monolayer and micromass cultures. These studies have shown that cells in micromass cultures have a more rapid decrease in the rates of proliferation and differentiation, which lead to increased levels of collagen assembly and mineralization [26]. This difference has been attributed to an increased level of cell-cell contacts in the micromass cultures. In addition, bone-nodule formation has been demonstrated to be dependent on the initial cell density, and the number of bone nodules produced has been shown to follow a linear relationship (with cell-seeding density) in limiting dilution analysis only at high cell densities [15]. This was attributed to more than one cell type being limiting for stromal osteoprogenitor differentiation and to a dependence on heterotypic cell-cell interactions. In an attempt to better understand these phenomena, cocultures of adherent and nonadherent bone marrow cells and cocultures of adherent bone marrow cells and fibroblasts have been performed [15, 23, 27]. However, these experiments have relied on using "randomly distributed" cultures, which have a number of limitations, including a lack of control over cell-cell interactions. For example, cells are free to migrate throughout the cultures, and differential proliferation rates typically exist that can vary cell-cell interactions in an uncontrolled manner. Also, to study the effects of cell number in these cultures, the cell-seeding density must be altered. In addition to cell contact, several studies have looked at the importance of autocrine and paracrine signaling in directing stem cell differentiation along osteogenic pathways [28–31]. For example, differential adhesion of osteoblasts to different microtopographies produces changes in transforming growth factor (TGF)-β production and other regulatory molecules, which ultimately effects osteogenesis [5]. To extend this knowledge further, new techniques are required to control the positioning of cells in culture, and methods of controlling local cell density are desirable.

Various approaches have been used to control cellular microenvironments and local cell density. The most established in vitro systems for interrogating cellular microenvironments has focused on the liver as described above. To better understand liver function, microfabrication techniques have been used to localize fibroblasts and hepatocytes into patterned configurations to examine cell-cell homotypic [32] and heterotypic [33] interactions. Through creating islands of cell-adhesive regions on a substrate, improved control over both homotypic and heterotypic cell-cell interactions has been achieved compared to the typical "randomly distributed" cultures. Microfabrication techniques allow the isolation of cell number as an independent variable and the prevention of cell migration, whereas in "randomly distributed" cultures, this is not possible.

Application of existing microfabrication techniques will undoubtedly help us better understand the role of homotypic and heterotypic cell-cell interactions during in vitro bone formation, and this may prove useful for therapeutic applications. For example, typical bone-tissue-engineering strategies involve an initial in vitro component where populations of cells, containing osteoprogenitors, are seeded onto three-dimensional biodegradable scaffolds. By understanding how cell-cell interactions control bone formation in vitro, one may be able to achieve ubiquitous colonization of the scaffold with mineralized bone tissue prior to in vivo implantation. Furthermore, it is clear that several cell types can influence osteogenesis through direct cell-cell interactions, and these may be further elucidated through microscale-based control of the local cell environment. Engineering of cell cultures at the microscale level allows for better control of cell output and the ability to study a wide range of factors that influence cell-culture output.

20.3 Use of Substrate Properties to Control Osteogenesis

In addition to defined cell-cell interactions, surface properties, including topography, matrix elasticity, surface chemistry, and presentation of specific ligands, can be utilized to potentially enhance the tissue-formation and reorganization phases of bone generation. These properties can be utilized for microscale control of bone formation and are briefly described below.

20.3.1 Micro- and Nanotopography

It is clear that surface micro- and nanotopography can influence osteogenic cell adhesion, migration, differentiation, matrix production, and autocrine and paracrine signaling events [34–39]. In addition to increasing one's control over osteogenic cell function, patterned substrates may help maintain cellular function during long-term in vitro experimentation or potentially after implantation [40]. Although it is debatable whether osteoblasts ever "see" the implant surface [41], the impact of such topographical features can influence the clinical outcome [39, 42]. There is growing evidence that the osteogenic response to an implant is a function of the initial blood interactions with the implant (i.e., platelet interactions and attachment properties of the blood clot) [43, 44], whereas the direct interaction of microtopography with osteoblasts may be more relevant to bone-tissue-engineering strategies that involve the placement of osteoblasts directly onto a substrate.

In addition to using topographical features for influencing cell function, patterned topography can influence cell orientation [38, 45–50]. These substrates can be used for the study of important factors in osteoblast response and to enhance the development of precisely defined implant surfaces. For example, rat bone marrow–derived osteoblasts seeded on smooth and microgrooved polystyrene substrates were shown to align within grooves above 5 μm but not below 2 μm [46]. Osteoblast response was also examined on polyimide substrates containing patterned microtopography and chemistries through hot-embossing imprint lithography followed by microcontact printing [51]. Recently, kinetic analysis of adhesion, spreading, and proliferation was performed with human bone-derived MG63 osteogenic cells to evaluate the effect of microscale and nanoscale topographies, separately and in combination, on model titanium surfaces [50]. Through using well-defined model surfaces, it was found that microscale and nanoscale roughness exerts a synergistic effect on cell proliferation where surface topologies smaller than 10 μm in feature size are recognizable by single cells and influence focal-adhesion contacts. In a related study using hierarchical synthesis of materials, micropatterned silicon surfaces were coated with nanostructured hydroxyapatite and shown to exhibit favorable responses, including alignment on surfaces containing parallel ridges (Figure 20.1) [38]. Similar results were also obtained for microtextured collagen substrates [45]. In this study, collagen was applied to microtextured surfaces using solvent casting, followed by stabilization via combinations of either dehydrothermal treatment, carbodiimide crosslinking, or gluteraldehyde crosslinking followed by treatment with calcium phosphate coatings. Additionally, rat-calvaria-derived osteogenic cells expressed higher levels of ALP activity and showed increased bone nodule formation (with grooves from 10 to 30 μm) on both titanium- and hydroxyapatite-coated microgrooved silicon wafers in comparison to smooth surfaces [47]. Similar enhancement of ALP activity and mineralization has been demonstrated with microgrooved poly(lactic acid) (PLA) surfaces in comparison to smooth PLA surfaces [48]. It is becoming clear that surface topography not only influences osteogenic cell function but can be controlled and, thus, represents as a powerful tool for engineering surfaces to enhance bone formation [52]. Another active area of research aims to uncover the effects of mechanics on the osteoblast function, and this may be achieved through fixing the substrate mechanical properties during preparation or through applying defined mechanical strain (stagnant or cyclical) [49, 53–57].

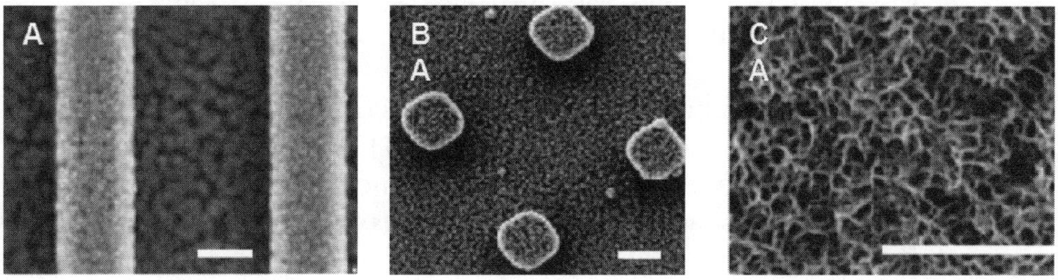

Figure 20.1 Micro- and nanoscale hierarchical structures, including (a) parallel ridges and (b) posts formed through use of photolithography and reactive ion etching on silicon wafers, followed by silanization and succinylation chemical modifications, then mineralization in a supersaturated calcium phosphate solution. (c) A high magnification image of the resultant nanotextured hydroxyapatite coating. Scale bars = 4 μm. (*Source:* [38].)

20.3.2 Substrate Mechanical Properties

Osteogenic cells are highly responsive to mechanical stimuli during the initial phases of bone regeneration. Responsiveness of the cells to the mechanical environment of the implant has been shown to guide cell function, including proliferation, gene and protein expression for ECM components, and the production of soluble factors [54–57]. Clinically, implants must resist compressive stress and tensile strain to maintain a permissive environment for bone formation [54, 58], where the implant surface serves as an interface for force transmission [59]. This presents a unique opportunity to impact bone regeneration through optimizing specific biophysical stimuli via implant design [60].

An active area of research is the field of mechanobiology, which aims to uncover how the mechanical signals regulate stem cell differentiation [54, 61]. For example, to understand the role of matrix elasticity on mesenchymal stem cell (MSC) behavior, Engler et al. have recently created cell-adhesive substrates with varying elasticities through polymerizing polyacrylamide gels with varying concentrations of bis-acrylamide as a crosslinker, followed by a thin coating of collagen type-1 to allow for cell attachment [62]. They demonstrated that the underlying elasticity of the matrix exerts strong effects on the lineage specification and commitment of sparse cultures of native MSCs. Specifically, matrices with elastic modulus of at least 25 to 40 kPa yielded polygonal cell morphologies with upregulation of Cbfa-1 transcriptional factor, which is an early marker of osteogenesis. Other methods for driving the differentiation of stem cells toward the osteogenic lineage, including physical and other means such as substrate composition, have been reviewed elsewhere [63].

20.3.3 Substrate Composition and Shape

It has been well established that interactions between adherent osteogenic cells and the chemistry of their substrate can regulate survival, proliferation, differentiation, and bone formation [64]. Osteogenic cells express several proteins on their surfaces, including integrins that are involved in signal transduction from the ECM to the cell. For example, the cells adhering to fibronectin utilize the cell-adhesive integrin $\alpha_5\beta_1$, which is critical for cell survival and proliferation [65]. Considerable work has focused on understanding cell-ECM interactions [66, 67] and designing surfaces with specific ligands that control osteogenic cell function, as reviewed previously [64, 68–70].

In addition to immobilized cell-adhesive ligands, growth factors have been patterned to influence cell function (Figure 20.2) [71]. Among others, an inkjet deposition system has been used to synthesize two-dimensional patterns of FGF-2 onto fibrin substrates using native binding affinities. MG-63 human preosteoblastic cells responded with enhanced proliferation rates within regions containing FGF-2. It may be possible to use a similar approach to pattern growth factors to stimulate differentiation toward multiple lineages on the same substrate. This can be useful for engineering multicomponent systems, such as osteochondral constructs that contain both osteoblasts and chrondrocytes.

Substrate shape also has a profound effect on cell function [72, 73]. Recently, a simple method using cell-repellant photocrosslinkable chitosan was used to create

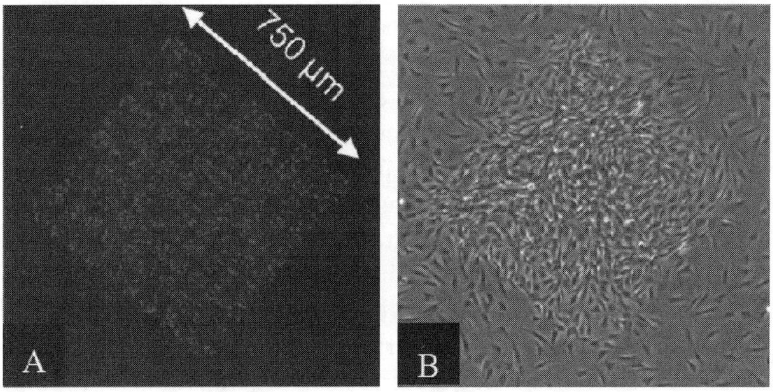

Figure 20.2 (a) Fluorescence image of FGF-2 array patterns on fibrin films, and (b) light micrograph showing enhanced osteogenic cell proliferation within patterned regions. (*Source:* [71].)

patterned surfaces of osteoblasts on glass or thermanox-treated glass (Figure 20.3) [74]. The potential for osteogenic cells to express hallmarks of bone formation in this system was examined by studying the responses of SAOS-2 osteoblasts in a series of square, circular, and trianglular geometries. This work demonstrates the potential to control bone-nodule shape via the presentation of specific patterns on the underlying surface, which may be an important design criteria for achieving bone formation on solid surfaces.

20.4 Techniques for Translating Two-Dimensional Systems to Three-Dimensional Scaffolds

Based on the recent technological advancements and knowledge gained using two-dimensional microscale systems, several methods for controlling three-dimensional cell-cell and cell-scaffold interactions have been introduced [75–78]. Current techniques can achieve three-dimensional control generally within

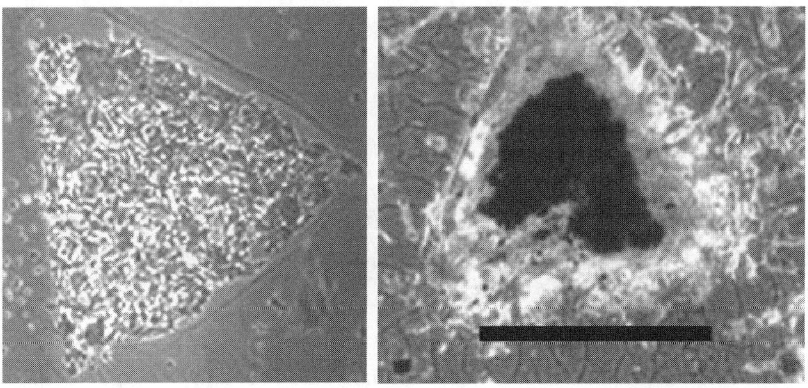

Figure 20.3 Shape control of osteogenic cells and produced mineralized ECM: (Left) Alkaline phosphatase stained SAOS-2 cells isolated with micropatterned triangular wells within a chitosan film on a TCPS substrate. (Right) von Kossa and alkaline phosphatase after fourteen days under osteogenic culture conditions. Scale bar = 350 μm. (*Source:* [74].)

the micrometer (most common) or nanometer range. For example, in one study, circular patterns (≥200 μm in diameter) of osteogenic cells were created on porous three-dimensional scaffolds with control over size, spacing, and pattern geometry through use of agarose stamps [Figure 20.4(a)] [79]. This approach may be useful to promote the organization of cells in three dimensions and to exploit cell-cell interactions explored in two dimensions. Other techniques for controlling the microarrangement and placement of cells in three dimensions include cell printing [80–82], with most strategies relying on the use of standard inkjet printing technologies. These techniques can easily be interfaced with computer-aided design (CAD) and computer-aided manufacturing (CAM) processes to produce well-defined two-dimensional and three-dimensional organization of multiple cell types with micrometer resolution. In addition to adapting inkjet technology to biological printing, a noncontact, orifice-free, laser-based printing technique called biological laser printing (BioLP) has been specifically developed for biological printing [83]. The BioLP technique can also be interfaced with CAD/CAM systems like the inkjet technology. Moreover, it exhibits high spatial accuracy better than 5 μm and may accommodate multiple cell types with high cell viability. Precise control of the placement of cells in three dimensions has also been achieved using a technique called "controlled cell assembly" [Figure 20.4(b)] [77]. This high-resolution (~10 μm) method utilizes a three-dimensional micropositioning system with a pressure-controlled syringe to deposit cell/biomaterial structures. This powerful approach enables placement of cells, scaffold materials, nutrients, therapeutic drugs, genes, and growth factors in a spatially and temporally relevant manner.

In addition to techniques for controlling the placement of cells in three dimensions, methods now exist for translating micro-/nanotopography to three-dimensional constructs through the use of a polymer demixing process [Figure 20.4(c)] [84]. Polymer demixing is a relatively fast and cheap chemical method for fabricating a large area of nanotopography, making this technique ideal for cell testing. Like two-dimensional systems, three-dimensional topography influ-

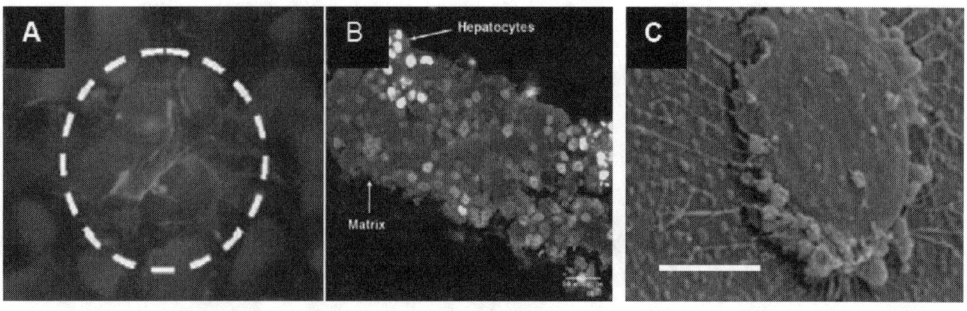

Figure 20.4 (a) High-magnification light micrograph of a hydroxyapatite scaffold patterned with 700 μm circular cell islands [79]. (b) Hepatocyte/gelatin/sodium alginate construct fabricated by the controlled cell-assembler technique. Note that the location of hepatocytes is precisely controlled on the matrix. (*Source:* [77].) (c) Electron micrograph of a bone marrow–derived cell cultured on a nanotextured three-dimensional substrate (scale bar =15 μm). (*Source:* [84].)

ences cell adhesion, spreading, and cytoskeletal organization. Cells preferentially interact with textured islands prepared by the polymer demixing as shown in Figure 20.4(c). The polymer demixing technology represents a simple technique for translating two-dimensional topography into three-dimensional systems.

A two-step technique involving fiber bonding and electrospinning represents another interesting approach to fabricating three-dimensional scaffolds with micro- and nanoscale topography [85]. By cominding starch/polycaprolactone nanofibers and microfibers, improved cell colonization, cell viability, and alkaline phosphatase activity of osteoblasts was observed compared to native microfiber scaffolds.

Through rapid-prototyping techniques, it is possible to design substrates of defined porosity and microarchitecture [86]. Rapid-prototyping techniques have been used for multiple materials, including ceramics [87] and thermoplastic polymers [88], and allow for construction of complex bone-tissue-engineering scaffolds that have a range of material properties within the same scaffold. By using this approach, one can combine the strength of hydroxyapatite materials with the flexibility and biodegradability of polymeric materials. In addition, the internal and external structures of the scaffold can be independently defined through rapid prototyping. Layer-by-layer approaches to fabricating three-dimensional scaffold microfeatures include solid free-form fabrication techniques such as stereolithography, selected laser sintering, single- and multiple-nozzle deposition, fused deposition modeling, and three-dimensional printing, as reviewed elsewhere [3]. A similar approach to these printing technologies is precise extrusion manufacturing, where the material, which is usually a thermoplastic polymer, is extruded from a micronozzle to allow for layer-by-layer construction of three-dimensional microenvironments and scaffolds [89].

Recent progress with three-dimensional fabrication technologies permits precise control of the microarchitecture of the scaffolds and the placement of cells and factors in three dimensions. With an integrated understanding of the optimal two-dimensional and three-dimensional microenvironmental cues that control osteogenic function, these three-dimensional techniques will prove useful for clinically relevant scaffold design and development.

20.5 Conclusion

Several questions relating to the regulation of osteogenesis through precise interactions of cells with each other or their environment at the micro- and nanoscale levels have remained unanswerable until recently. While our current understanding of the bone microenvironment has mainly focused on standard petri dish–based cell-culture and molecular-biology techniques, recent experiments have demonstrated that state-of-the-art micro- and nanoscale-based approaches can be used as tools to model the bone microenvironment and perhaps uncover new knowledge for advancing cell-construct-based therapeutics. It is clear that we are entering a new era where micro- and nanoscale technologies will become commonplace within bioengineering laboratories and enable the development of complex tissue constructs that mimic the desired biology.

References

[1] www.boneandjointdecade.org.

[2] Rose, F. R., and Oreffo, R. O., "Bone tissue engineering: Hope vs. hype," *Biochem. Biophys. Res. Commun.*, Vol. 292, No. 1, 2002, pp. 1–7.

[3] Sun, W., Yan, Y., Lin, F., and Spector, M., "Biomanufacturing: A US-China National Science Foundation–sponsored workshop," *Tissue Eng.*, Vol. 12, No. 5, 2006, pp. 1169–1181.

[4] Holy, C. E., Shoichet, M. S., and Davies, J. E., "Engineering three-dimensional bone tissue in vitro using biodegradable scaffolds: Investigating initial cell-seeding density and culture period," *J. Biomed. Mater. Res.*, Vol. 51, No. 3, 2000, pp. 376–382.

[5] Boyan, B. D., Lossdorfer, S., Wang, L., Zhao, G., Lohmann, C. H., Cochran, D. L., and Schwartz, Z., "Osteoblasts generate an osteogenic microenvironment when grown on surfaces with rough microtopographies," *Eur. Cell. Mater.*, Vol. 6, 2003, pp. 22–27.

[6] Hutmacher, D. W., Sittinger, M., and Risbud, M. V., "Scaffold-based tissue engineering: Rationale for computer-aided design and solid free-form fabrication systems," *Trends Biotechnol.*, Vol. 22, No. 7, 2004, pp. 354–362.

[7] Goshima, J., Goldberg, V. M., and Caplan, A. I., "The origin of bone formed in composite grafts of porous calcium phosphate ceramic loaded with marrow cells," *Clin. Orthop.*, No. 269, 1991, pp. 274–283.

[8] Petite, H., Viateau, V., Bensaid, W., Meunier, A., de Pollak, C., Bourguignon, M., Oudina, K., Sedel, L., and Guillemin, G., "Tissue-engineered bone regeneration," *Nat. Biotechnol.*, Vol. 18, No. 9, 2000, pp. 959–963.

[9] Viswanathan, S., Benatar, T., Rose-John, S., Lauffenburger, D. A., and Zandstra, P. W., "Ligand/receptor signaling threshold (LIST) model accounts for gp130-mediated embryonic stem cell self-renewal responses to LIF and HIL-6," *Stem Cells*, Vol. 20, No. 2, 2002, pp. 119–138.

[10] Meinel, L., Betz, O., Fajardo, R., Hofmann, S., Nazarian, A., Cory, E., Hilbe, M., McCool, J., Langer, R., Vunjak-Novakovic, G., Merkle, H. P., Rechenberg, B., Kaplan, D. L., and Kirker-Head, C., "Silk based biomaterials to heal critical sized femur defects," *Bone*, Vol. 39, No. 4, 2006, pp. 922–931.

[11] Karp, J. M., Shoichet, M. S., and Davies, J. E., "Bone formation on two-dimensional poly(DL-lactide-co-glycolide) (PLGA) films and three-dimensional PLGA tissue engineering scaffolds in vitro," *J. Biomed. Mater. Res. A*, Vol. 64, No. 2, 2003, pp. 388–396.

[12] Louisia, S., Stromboni, M., Meunier, A., Sedel, L., and Petite, H., "Coral grafting supplemented with bone marrow," *J. Bone Joint Surg. Br.*, Vol. 81, No. 4, 1999, pp. 719–724.

[13] Grande, D. A., Breitbart, A. S., Mason, J., Paulino, C., Laser, J., and Schwartz, R. E., "Cartilage tissue engineering: Current limitations and solutions," *Clin. Orthop.*, 1999, pp. S176–S185.

[14] Purpura, K. A., Aubin, J. E., and Zandstra, P. W., "Sustained in vitro expansion of bone progenitors is cell density dependent," *Stem Cells*, Vol. 22, No. 1, 2004, pp. 39–50.

[15] Aubin, J. E., "Osteoprogenitor cell frequency in rat bone marrow stromal populations: Role for heterotypic cell-cell interactions in osteoblast differentiation," *J. Cell. Biochem.*, Vol. 72, No. 3, 1999, pp. 396–410.

[16] Bhatia, S. N., Yarmush, M. L., and Toner, M., "Controlling cell interactions by micropatterning in co-cultures: Hepatocytes and 3T3 fibroblasts," *J. Biomed. Mater. Res.*, Vol. 34, No. 2, 1997, pp. 189–199.

[17] Bhatia, S. N., Balis, U. J., Yarmush, M. L., and Toner, M., "Microfabrication of hepatocyte/fibroblast co-cultures: Role of homotypic cell interactions," *Biotechnol. Prog.*, Vol. 14, No. 3, 1998, pp. 378–387.

[18] Bhatia, S. N., Balis, U. J., Yarmush, M. L., and Toner, M., "Effect of cell-cell interactions in preservation of cellular phenotype: Cocultivation of hepatocytes and nonparenchymal cells," *Faseb J.*, Vol. 13, No. 14, 1999, pp. 1883–1900.

[19] Khademhosseini, A., Langer, R., Borenstein, J., and Vacanti, J. P., "Microscale technologies for tissue engineering and biology," *Proc. Natl. Acad. Sci. USA*, Vol. 103, No. 8, 2006, pp. 2480–2487.

[20] Whitesides, G. M., Ostuni, E., Takayama, S., Jiang, X., and Ingber, D. E., "Soft lithography in biology and biochemistry," *Ann. Rev. Biomed. Eng.*, Vol. 3, 2001, pp. 335–373.

[21] Mrksich, M., "What can surface chemistry do for cell biology?" *Curr. Opin. Chem. Biol.*, Vol. 6, No. 6, 2002, pp. 794–797.

[22] Calvi, L. M., Adams, G. B., Weibrecht, K. W., Weber, J. M., Olson, D. P., Knight, M. C., Martin, R. P., Schipani, E., Divieti, P., Bringhurst, F. R., Milner, L. A., Kronenberg, H. M., and Scadden, D. T., "Osteoblastic cells regulate the haematopoietic stem cell niche," *Nature*, Vol. 425, No. 6960, 2003, pp. 841–846.

[23] Baksh, D., Davies, J. E., and Zandstra, P. W., "Soluble factor cross-talk between human bone marrow–derived hematopoietic and mesenchymal cells enhances in vitro CFU-F and CFU-O growth and reveals heterogeneity in the mesenchymal progenitor cell compartment," *Blood*, Vol. 106, No. 9, 2005, pp. 3012–3019.

[24] Nakaoka, R., Hsiong, S. X., and Mooney, D. J., "Regulation of chondrocyte differentiation level via co-culture with osteoblasts," *Tissue Eng.*, Vol. 12, No. 9, 2006, pp. 2425–2433.

[25] Gomi, K., Kanazashi, M., Lickorish, D., Arai, T., and Davies, J. E., "Bone marrow genesis after subcutaneous delivery of rat osteogenic cell-seeded biodegradable scaffolds into nude mice," *J. Biomed. Mater. Res. A*, Vol. 71, No. 4, 2004, pp. 602–607.

[26] Fritz, J., Baller, M. K., Lang, H. P., Rothuizen, H., Vettiger, P., Meyer, E., Guntherodt, H., Gerber, C., and Gimzewski, J. K., "Translating biomolecular recognition into nanomechanics," *Science*, Vol. 288, No. 5464, 2000, pp. 316–318.

[27] Kinto, N., Iwamoto, M., Enomoto-Iwamoto, M., Noji, S., Ohuchi, H., Yoshioka, H., Kataoka, H., Wada, Y., Yuhao, G., Takahashi, H. E., Yoshiki, S., and Yamaguchi, A., "Fibroblasts expressing sonic hedgehog induce osteoblast differentiation and ectopic bone formation," *Febs. Lett.*, Vol. 404, Nos. 2–3, 1997, pp. 319–323.

[28] Jaiswal, N., Haynesworth, S. E., Caplan, A. I., and Bruder, S. P., "Osteogenic differentiation of purified, culture-expanded human mesenchymal stem cells in vitro," *J. Cell. Biochem.*, Vol. 64, No. 2, 1997, pp. 295–312.

[29] Haynesworth, S. E., Baber, M. A., and Caplan, A. I., "Cytokine expression by human marrow–derived mesenchymal progenitor cells in vitro: Effects of dexamethasone and IL-1 alpha," *J. Cell. Physiol.*, Vol. 166, No. 3, 1996, pp. 585–592.

[30] Stains, J. P., and Civitelli, R., "Cell-cell interactions in regulating osteogenesis and osteoblast function," *Birth Defects Res. C: Embryo Today*, Vol. 75, No. 1, 2005, pp. 72–80.

[31] Nakaoka, R., Hsiong, S. X., and Mooney, D. J., "Regulation of chondrocyte differentiation level via co-culture with osteoblasts," *Tissue Eng.*, Vol. 2006.

[32] Bhatia, S. N., Balis, U. J., Yarmush, M. L., and Toner, M., "Microfabrication of hepatocyte/fibroblast co-cultures: Role of homotypic cell interactions," *Biotechnol. Prog.*, Vol. 14, No. 3, 1998, pp. 378–387.

[33] Bhatia, S. N., Balis, U. J., Yarmush, M. L., and Toner, M., "Probing heterotypic cell interactions: Hepatocyte function in microfabricated co-cultures," *J. Biomater. Sci. Polym. Ed.*, Vol. 9, No. 11, 1998, pp. 1137–1160.

[34] Hamilton, D. W., Wong, K. S., and Brunette, D. M., "Microfabricated discontinuous-edge surface topographies influence osteoblast adhesion, migration, cytoskeletal organization, and proliferation and enhance matrix and mineral deposition in vitro," *Calcif. Tissue Int.*, Vol. 78, No. 5, 2006, pp. 314–325.

[35] Sato, M., and Webster, T. J., "Nanobiotechnology: implications for the future of nanotechnology in orthopedic applications," *Expert Rev. Med. Devices*, Vol. 1, No. 1, 2004, pp. 105–114.

[36] Zhao, G., Zinger, O., Schwartz, Z., Wieland, M., Landolt, D., and Boyan, B. D., "Osteoblast-like cells are sensitive to submicron-scale surface structure," *Clin. Oral Implants Res.*, Vol. 17, No. 3, 2006, pp. 258–264.

[37] Dalby, M. J., McCloy, D., Robertson, M., Wilkinson, C. D., and Oreffo, R. O., "Osteoprogenitor response to defined topographies with nanoscale depths," *Biomaterials*, Vol. 27, No. 8, 2006, pp. 1306–1315.

[38] Tan, J., and Saltzman, W. M., "Biomaterials with hierarchically defined micro- and nanoscale structure," *Biomaterials*, Vol. 25, No. 17, 2004, pp. 3593–3601.

[39] Christenson, E. M., Anseth, K. S., van den Beucken, J. J., Chan, C. K., Ercan, B., Jansen, J. A., Laurencin, C. T., Li, W. J., Murugan, R., Nair, L. S., Ramakrishna, S., Tuan, R. S., Webster, T. J., and Mikos, A. G., "Nanobiomaterial applications in orthopedics," *J. Orthop. Res.*, Vol. 25, No. 1, 2007, pp. 11–22.

[40] Desai, T. A., "Micro- and nanoscale structures for tissue engineering constructs," *Med. Eng. Phys.*, Vol. 22, No. 9, 2000, pp. 595–606.

[41] Davies, J. E., "Understanding peri-implant endosseous healing," *J. Dent. Educ*, Vol. 67, No. 8, 2003, pp. 932–949.

[42] Schwartz, Z., Nasazky, E., and Boyan, B. D., "Surface microtopography regulates osteointegration: The role of implant surface microtopography in osteointegration," *Alpha Omegan*, Vol. 98, No. 2, 2005, pp. 9–19.

[43] Kikuchi, L., Park, J. Y., Victor, C., and Davies, J. E., "Platelet interactions with calcium-phosphate-coated surfaces," *Biomaterials*, Vol. 26, No. 26, 2005, pp. 5285–5295.

[44] Park, J. Y., Gemmell, C. H., and Davies, J. E., "Platelet interactions with titanium: Modulation of platelet activity by surface topography," *Biomaterials*, Vol. 22, No. 19, 2001, pp. 2671–2682.

[45] Ber, S., Torun Kose, G., and Hasirci, V., "Bone tissue engineering on patterned collagen films: An in vitro study," *Biomaterials*, Vol. 26, No. 14, 2005, pp. 1977–1986.

[46] Matsuzaka, K., Walboomers, X. F., Yoshinari, M., Inoue, T., and Jansen, J. A., "The attachment and growth behavior of osteoblast-like cells on microtextured surfaces," *Biomaterials*, Vol. 24, No. 16, 2003, pp. 2711–2719.

[47] Perizzolo, D., Lacefield, W. R., and Brunette, D. M., "Interaction between topography and coating in the formation of bone nodules in culture for hydroxyapatite- and titanium-coated micromachined surfaces," *J. Biomed. Mater. Res.*, Vol. 56, No. 4, 2001, pp. 494–503.

[48] Matsuzaka, K., Walboomers, F., de Ruijter, A., and Jansen, J. A., "Effect of microgrooved poly-l-lactic (PLA) surfaces on proliferation, cytoskeletal organization, and mineralized matrix formation of rat bone marrow cells," *Clin. Oral Implants Res.*, Vol. 11, No. 4, 2000, pp. 325–333.

[49] Wang, J. H., Grood, E. S., Florer, J., and Wenstrup, R., "Alignment and proliferation of MC3T3-E1 osteoblasts in microgrooved silicone substrata subjected to cyclic stretching," *J. Biomech.*, Vol. 33, No. 6, 2000, pp. 729–735.

[50] Zinger, O., Anselme, K., Denzer, A., Habersetzer, P., Wieland, M., Jeanfils, J., Hardouin, P., and Landolt, D., "Time-dependent morphology and adhesion of osteoblastic cells on titanium model surfaces featuring scale-resolved topography," *Biomaterials*, Vol. 25, No. 14, 2004, pp. 2695–2711.

[51] Charest, J. L., Eliason, M. T., Garcia, A. J., and King, W. P., "Combined microscale mechanical topography and chemical patterns on polymer cell culture substrates," *Biomaterials*, Vol. 27, No. 11, 2006, pp. 2487–2494.

[52] Kenar, H., Kose, G. T., and Hasirci, V., "Tissue engineering of bone on micropatterned biodegradable polyester films," *Biomaterials*, Vol. 27, No. 6, 2006, pp. 885–895.

[53] Yoshikawa, T., Peel, S. A., Gladstone, J. R., and Davies, J. E., "Biochemical analysis of the response in rat bone marrow cell cultures to mechanical stimulation," *Biomed. Mater. Eng.*, Vol. 7, No. 6, 1997, pp. 369–377.

[54] Carter, D. R., Beaupre, G. S., Giori, N. J., and Helms, J. A., "Mechanobiology of skeletal regeneration," *Clin. Orthop. Relat. Res.*, 1998, pp. S41–S55.

[55] Mullender, M., El Haj, A. J., Yang, Y., van Duin, M. A., Burger, E. H., and Klein-Nulend, J., "Mechanotransduction of bone cells in vitro: Mechanobiology of bone tissue," *Med. Biol. Eng. Comput.*, Vol. 42, No. 1, 2004, pp. 14–21.

[56] Jacobs, C. R., "The mechanobiology of cancellous bone structural adaptation," *J. Rehabil. Res. Dev.*, Vol. 37, No. 2, 2000, pp. 209–216.

[57] Wozniak, M., Fausto, A., Carron, C. P., Meyer, D. M., and Hruska, K. A., "Mechanically strained cells of the osteoblast lineage organize their extracellular matrix through unique sites of alphavbeta3-integrin expression," *J. Bone Miner. Res.*, Vol. 15, No. 9, 2000, pp. 1731–1745.

[58] Loboa, E. G., Beaupre, G. S., and Carter, D. R., "Mechanobiology of initial pseudarthrosis formation with oblique fractures," *J. Orthop. Res.*, Vol. 19, No. 6, 2001, pp. 1067–1072.

[59] Cehreli, M. C., and Akca, K., "Mechanobiology of bone and mechanocoupling of endosseous titanium oral implants," *J. Long Term Eff. Med. Implants*, Vol. 15, No. 2, 2005, pp. 139–152.

[60] Simmons, C. A., Valiquette, N., and Pilliar, R. M., "Osseointegration of sintered porous-surfaced and plasma spray-coated implants: An animal model study of early postimplantation healing response and mechanical stability," *J. Biomed. Mater. Res.*, Vol. 47, No. 2, 1999, pp. 127–138.

[61] Estes, B. T., Gimble, J. M., and Guilak, F., "Mechanical signals as regulators of stem cell fate," *Curr. Top. Dev. Biol.*, Vol. 60, 2004, pp. 91–126.

[62] Engler, A. J., Sen, S., Sweeney, H. L., and Discher, D. E., "Matrix elasticity directs stem cell lineage specification," *Cell*, Vol. 126, No. 4, 2006, pp. 677–689.

[63] Heng, B. C., Cao, T., Stanton, L. W., Robson, P., and Olsen, B., "Strategies for directing the differentiation of stem cells into the osteogenic lineage in vitro," *J. Bone Miner. Res.*, Vol. 19, No. 9, 2004, pp. 1379–1394.

[64] Garcia, A. J., and Keselowsky, B. G., "Biomimetic surfaces for control of cell adhesion to facilitate bone formation," *Crit. Rev. Eukaryot. Gene Expr.*, Vol. 12, No. 2, 2002, pp. 151–162.

[65] Garcia, A. J., and Reyes, C. D., "Bio-adhesive surfaces to promote osteoblast differentiation and bone formation," *J. Dent. Res.*, Vol. 84, No. 5, 2005, pp. 407–413.

[66] Bennett, J. H., Moffatt, S., and Horton, M., "Cell adhesion molecules in human osteoblasts: Structure and function," *Histol. Histopathol.*, Vol. 16, No. 2, 2001, pp. 603–611.

[67] Siebers, M. C., ter Brugge, P. J., Walboomers, X. F., and Jansen, J. A., "Integrins as linker proteins between osteoblasts and bone replacing materials: A critical review," *Biomaterials*, Vol. 26, No. 2, 2005, pp. 137–146.

[68] Healy, K. E., Rezania, A., and Stile, R. A., "Designing biomaterials to direct biological responses," *Annals NY Acad. Sci.*, Vol. 875, 1999, pp. 24–35.

[69] Curran, J. M., Chen, R., and Hunt, J. A., "The guidance of human mesenchymal stem cell differentiation in vitro by controlled modifications to the cell substrate," *Biomaterials*, Vol. 27, No. 27, 2006, pp. 4783–4793.

[70] Winkelmann, M., Gold, J., Hauert, R., Kasemo, B., Spencer, N. D., Brunette, D. M., and Textor, M., "Chemically patterned, metal oxide based surfaces produced by photolithographic techniques for studying protein- and cell-surface interactions: I. Microfabrication and surface characterization," *Biomaterials*, Vol. 24, No. 7, 2003, pp. 1133–1145.

[71] Campbell, P. G., Miller, E. D., Fisher, G. W., Walker, L. M., and Weiss, L. E., "Engineered spatial patterns of FGF-2 immobilized on fibrin direct cell organization," *Biomaterials*, Vol. 26, No. 33, 2005, pp. 6762–6770.

[72] Chen, C. S., Mrksich, M., Huang, S., Whitesides, G. M., and Ingber, D. E., "Geometric control of cell life and death," *Science*, Vol. 276, No. 5317, 1997, pp. 1425–1428.

[73] McBeath, R., Pirone, D. M., Nelson, C. M., Bhadriraju, K., and Chen, C. S., "Cell shape, cytoskeletal tension, and RhoA regulate stem cell lineage commitment," *Dev. Cell*, Vol. 6, No. 4, 2004, pp. 483–495.

[74] Karp, J. M., Yeo, Y., Geng, W., Cannizarro, C., Yan, K., Kohane, D. S., Vunjak-Novakovic, G., Langer, R. S., and Radisic, M., "A photolithographic method to create cellular micropatterns," *Biomaterials*, Vol. 27, No. 27, 2006, pp. 4755–4764.

[75] Tsang, V. L., and Bhatia, S. N., "Three-dimensional tissue fabrication," *Adv. Drug Deliv. Rev.*, Vol. 56, No. 11, 2004, pp. 1635–1647.

[76] Yoshimoto, H., Shin, Y. M., Terai, H., and Vacanti, J. P., "A biodegradable nanofiber scaffold by electrospinning and its potential for bone tissue engineering," *Biomaterials*, Vol. 24, No. 12, 2003, pp. 2077–2082.

[77] Yan, Y., Wang, X., Pan, Y., Liu, H., Cheng, J., Xiong, Z., Lin, F., Wu, R., Zhang, R., and Lu, Q., "Fabrication of viable tissue-engineered constructs with 3-D cell-assembly technique," *Biomaterials*, Vol. 26, No. 29, 2005, pp. 5864–5871.

[78] Vozzi, G., Flaim, C., Ahluwalia, A., and Bhatia, S., "Fabrication of PLGA scaffolds using soft lithography and microsyringe deposition," *Biomaterials*, Vol. 24, No. 14, 2003, pp.2533–2540.

[79] Stevens, M. M., Mayer, M., Anderson, D. G., Weibel, D. B., Whitesides, G. M., and Langer, R., "Direct patterning of mammalian cells onto porous tissue engineering substrates using agarose stamps," *Biomaterials*, Vol. 26, No. 36, 2005, pp. 7636–7641.

[80] Nakamura, M., Kobayashi, A., Takagi, F., Watanabe, A., Hiruma, Y., Ohuchi, K., Iwasaki, Y., Horie, M., Morita, I., and Takatani, S., "Biocompatible inkjet printing technique for designed seeding of individual living cells," *Tissue Eng.*, Vol. 11, Nos. 11–12, 2005, pp. 1658–1666.

[81] Xu, T., Jin, J., Gregory, C., Hickman, J. J., and Boland, T., "Inkjet printing of viable mammalian cells," *Biomaterials*, Vol. 26, No. 1, 2005, pp. 93–99.

[82] Mironov, V., "Printing technology to produce living tissue," *Expert Opin. Biol. Ther.*, Vol. 3, No. 5, 2003, pp. 701–704.

[83] Barron, J. A., Wu, P., Ladouceur, H. D., and Ringeisen, B. R., "Biological laser printing: A novel technique for creating heterogeneous 3-dimensional cell patterns," *Biomed. Microdevices*, Vol. 6, No. 2, 2004, pp. 139–147.

[84] Berry, C. C., Dalby, M. J., Oreffo, R. O., McCloy, D., and Affrosman, S., "The interaction of human bone marrow cells with nanotopographical features in three-dimensional constructs," *J. Biomed. Mater. Res. A*, Vol. 79, No. 2, 2006, pp. 431–439.

[85] Tuzlakoglu, K., Bolgen, N., Salgado, A. J., Gomes, M. E., Piskin, E., and Reis, R. L., "Nano- and micro-fiber combined scaffolds: A new architecture for bone tissue engineering," *J. Mater. Sci. Mater. Med.*, Vol. 16, No. 12, 2005, pp. 1099–1104.

[86] Zhang, H., Hutmacher, D. W., Chollet, F., Poo, A. N., and Burdet, E., "Microrobotics and MEMS-based fabrication techniques for scaffold-based tissue engineering," *Macromol. Biosci.*, Vol. 5, No. 6, 2005, pp. 477–489.

[87] Nicolau, D. V., Taguchi, T., Tanigawa, H., and Yoshikawa, S., "Control of the neuronal cell attachment by functionality manipulation of diazo-naphthoquinone/novolak photoresist surface," *Biosensors and Bioelectronics*, Vol. 11, No. 12, 1996, pp. 1237–1252.

[88] Petzold, R., Zeilhofer, H. F., and Kalender, W. A., "Rapid protyping technology in medicine—basics and applications," *Comput. Med. Imaging Graph*, Vol. 23, No. 5, 1999, pp. 277–284.

[89] Hoque, M. E., Hutmacher, D. W., Feng, W., Li, S., Huang, M. H., Vert, M., and Wong, Y. S., "Fabrication using a rapid prototyping system and in vitro characterization of PEG-PCL-PLA scaffolds for tissue engineering," *J. Biomater. Sci. Polym. Ed.*, Vol. 16, No. 12, 2005, pp. 1595–1610.

CHAPTER 21

Nanoengineering for Bone Tissue Engineering

Lijie Zhang, Sirinrath Sirivisoot, Ganesh Balasundaram, and Thomas J. Webster

21.1 Introduction

As an emerging interdisciplinary field, bone-tissue engineering has evoked increasing interest from scientists wishing to develop biological substitutes that either restore, maintain, or improve bone-cell functions. It is widely known that natural bone is a well-organized matrix that consists of a protein-based soft hydrogel template (i. e., collagen) and hard inorganic components (specifically, hydroxyapatite). The special structure of natural bone supplies unique physical properties (such as Young's modulus, elasticity, and strength) in order to support various mechanical loadings [1]. Novel biomaterials that biomimic the nanostructure of natural bone have thus been investigated and have shown promise as improved orthopedic implants.

This chapter gives an introduction into the complications that occur for current orthopedic implants. Then, the main part of this chapter focuses on the contemporary development of nanomaterials for orthopedic applications. Important future research directions concerning nanoengineering are presented in the last section.

21.1.1 Problems with the Current Orthopedic Implants

With the rapid development of orthopedic-implant technology, various bone-implant surgeries and procedures (such as for healing bone fractures, repairing defects, and inserting hip and knee replacements) are routinely performed. The first hip replacement was conducted in 1923. However, the most significant breakthrough happened in the 1960s with the introduction of the ultra-low-friction cemented arthroplasty by John Charnley [2, 3]. Since then, as one of the most important surgery advances of the last century, this orthopedic-implant technology has rescued a myriad of people suffering intense pain and limited mobility from arthritis. The American Academy of Orthopedic Surgeons reported that in the United States alone, in just a four-year period, there was a 19.7 percent increase in hip replacements performed (from nearly 274,000 procedures in 1999 [including 168,000 total hip replacements and 106,000 partial hip replacements] to 328,000 procedures in 2003 [including 220,000 total hip replacements and 108,000 partial hip replacements]) [4]. The total number of hip replacements is projected to almost double by 2030 to 272,000 procedures [4], a staggering number and trend unavoidably accompanied by increasing hospitalization costs.

Although orthopedic implants have become more successful and safer, it is important to note that current orthopedic-implant technology has not been perfected. While 90 percent of hip and knee replacements can last ten to fifteen years on average and implant joints for some patients have endured for more than twenty years with no problems, occasionally hip or knee replacement implants fail immediately after surgery [5]. Considering the endurance of current orthopedic implants, younger and more active patients (men under sixty years and women under fifty-five years of age) will inevitably need more than one revision surgery in their lives. For example, in the United States, nearly 11 percent of hip replacements (36,000 revisions for 328,000 replacements) and 8 percent of knee replacements (33,000 revisions for 418,000 replacements) were revision procedures of previously failed hip and knee replacements in 2003 [5]. A similar trend occurs in other industrialized countries as well. Due to the strikingly increasing number of patients who need various medical implants and the relatively high percentage of revision procedures performed around the world, it is urgent to develop a new generation of orthopedic-implant technology and a set of materials that can significantly lengthen the service lifetime of orthopedic implants and, thus, dramatically reduce patient pain and health-insurance costs.

21.1.2 Reasons for Implant Failures

Many reasons can lead to implant failures. For example, in the early stage of implant surgeries, as acute complications, severe host responses (such as infections and inflammations), prosthesis dislocations, and surgery failures (such as improper placement and cement extrusion) may cause implant failure [6]. After several years, various additional reasons, including implant loosening, osteolysis (softening of the bone tissue), wear of articular bearing surfaces, and fractures, become the main reasons necessitating revision surgery [6]. In other words, based on clinical data collected by the Canadian Joint Replacement Registry in 2004, the most common reasons for revisions of a total hip replacement were loosening of the implant (55 percent), followed by osteolysis (33 percent), implant wear and tear (30 percent), instability/dislocation (17 percent), bone and implant fracture (12 percent), and infection (10 percent) [7]. Since some patients have more than one reason necessitating orthopedic-implant revision surgery, the aforementioned percentages do not add up to 100 percent.

21.1.2.1 Loosening of the Implant

Generally, one crucial criterion for the long-term success of orthopedic implants is the forming of sufficient bonding of the implant to juxtaposed bone, a process known as osseointegration, which was initially described as a direct structural and functional bone-to-implant contact under load by Branemark [8]. Osseointegration plays an important role in minimizing the motion-induced damage to surrounding tissues. Insufficient bonding to juxtaposed bone may lead to a mismatch of mechanical properties between implant materials and surrounding bone tissues [9].

As a result, a variety of implant-loosening conditions originating from stress and strain imbalances may frequently take place. Several factors, such as surface roughness of the fixture, overloading, and infection [10, 11], can be related to loss of acquired osseointegration and even failure of osseointegration. Most commonly, surface properties of implant materials have been intimately related to the success of osseointegration between an endosseous implant and bone. Numerous studies elucidate that surface roughness of implant materials affects the rate of osseointegration and biomechanical fixation [12–17]. Since traditional orthopedic-implant materials have been chosen based on their mechanical properties and biological inertness, there has not been enough concern about their chemical structure and surface properties. Thus, loosening of implants (such as of the socket or femoral component) has been frequently observed when conventional materials are used in total hip replacements.

On the other hand, the formation of a fibrous soft tissue capsule that originates from excessive secretion of fibrous tissue from inflammatory cells is another vital factor that leads to insufficient osseointegration and eventually causes loosening of implants (Figure 21.1) [9, 18]. After implanting biomaterials in a surgical site, it is inevitable to damage the surrounding tissue and cause injury. Then, various host

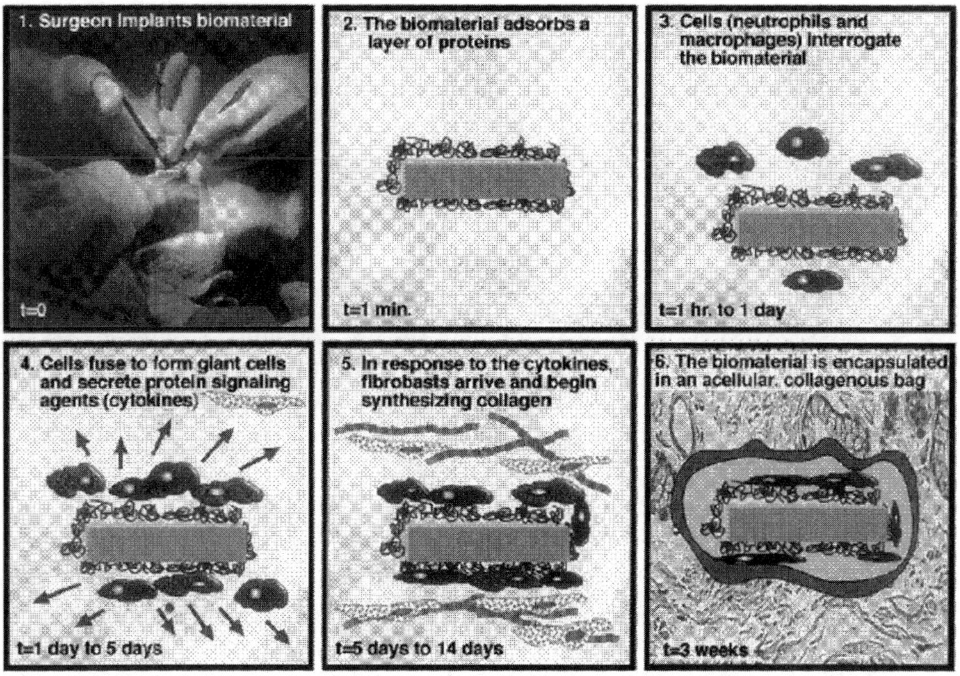

Figure 21.1 The foreign-body reaction toward implanted biomaterials: (1) A biomaterial is implanted and damage to surrounding tissue occurs. (2) In seconds to minutes, nonspecific proteins from body fluids are adsorbed on the implant surface. (3) Neutrophils and macrophages try to interrogate the material. (4) Failing to interrogate the material, cells fuse to form giant cells and concurrently secrete cytokines to signal other kinds of cells. (5, 6) Fibroblasts arrive and fabricate a collagenous bag to insulate the material from the surrounding tissue. (*Source:* [18].)

responses such as infections and inflammations are triggered (Figure 21.1). Initially, neutrophils and macrophages arrive at the injury site and begin to attempt to phagocytose the implant material; when they fail to engulf this large mass, they fuse together to form giant cells and interrogate again the implant to wall it off from the surrounding tissue. At the same time, giant cells send out chemical messengers (cytokines) to recruit other cells to "help" them. In response to cytokines, fibroblasts, which are in charge of secreting and synthesizing collagen into the extracellular matrix (ECM) comprising the soft tissue capsule [19], arrive and generate a collagen capsule, the first sign of a new tissue. In the final stage of the process, the implant is completely encapsulated in an acellular, avascular collagen bag about 50 to 200 μm thick [18]. The fibrous capsule prevents sufficient bonding between an implant surface and juxtaposed bone and frequently leads to clinical failure of orthopedic implants. Consequently, repelling fibroblasts that can cause an inflammatory reaction is necessary at the surface of a newly implanted orthopedic device [9]. Controlling protein adsorption by surface modification is an effective method to attract desirable cells onto the implant surface and minimize the functions of fibroblasts (such as their adhesion and secretion of fibrous tissue) [9, 20].

Therefore, numerous research efforts in bone-tissue engineering have focused on surface modifications (altering the implant surface topography and chemistry) of implant materials, including the development of novel nanophase materials (that is, materials with basic structural units, grains, particles, fibers, or other constituent components in the range of 1 to 100 nm [21]), such as nanophase metals, ceramics, polymers, and the like. These materials have been critical in the overall design and synthesis of bonelike composites that possess not only mechanical properties similar to those of physiological bone but also cytocompatible surface properties in order to more successfully promote cell adhesion and tissue integration [22–24].

21.1.2.2 Osteolysis

As a late-appearing complication, osteolysis induced by wear particles of implant materials has been regarded as one of the most common complications and among the leading reasons for revision surgeries after primary replacement [25–28]. It can cause severe bone loss, which eventually results in implant instabilities and failures. Metals, ceramics, and polymers, especially ultra-high-molecular-weight polyethylene, are widely accepted as a source of prosthetic wear particles and serve as a major contributor to bone resorption. These wear particles appear most commonly at the bearing surface but are also at host-implant or implant-implant interfaces. Figure 21.2 shows the osteolysis process caused by biomaterial wear debris [29, 30].

Macrophages are believed to play a key role during osteolysis (shown in Figure 21.1). But there exists a critical number of active macrophages necessary for osteolysis since only one disturbed macrophage and its neighboring osteoclast would not cause osteolysis. For instance, clinical experience indicates that osteolysis cannot happen if the wear rate of the polyethylene cup is below 0.05 mm per year. In contrast, osteolysis frequently occurs if the wear rate exceeds 0.2 mm per year [31]. Moreover, studies have also illustrated that there exists a critical size for bioactive particles, which can trigger cellular responses and finally result in osteolysis.

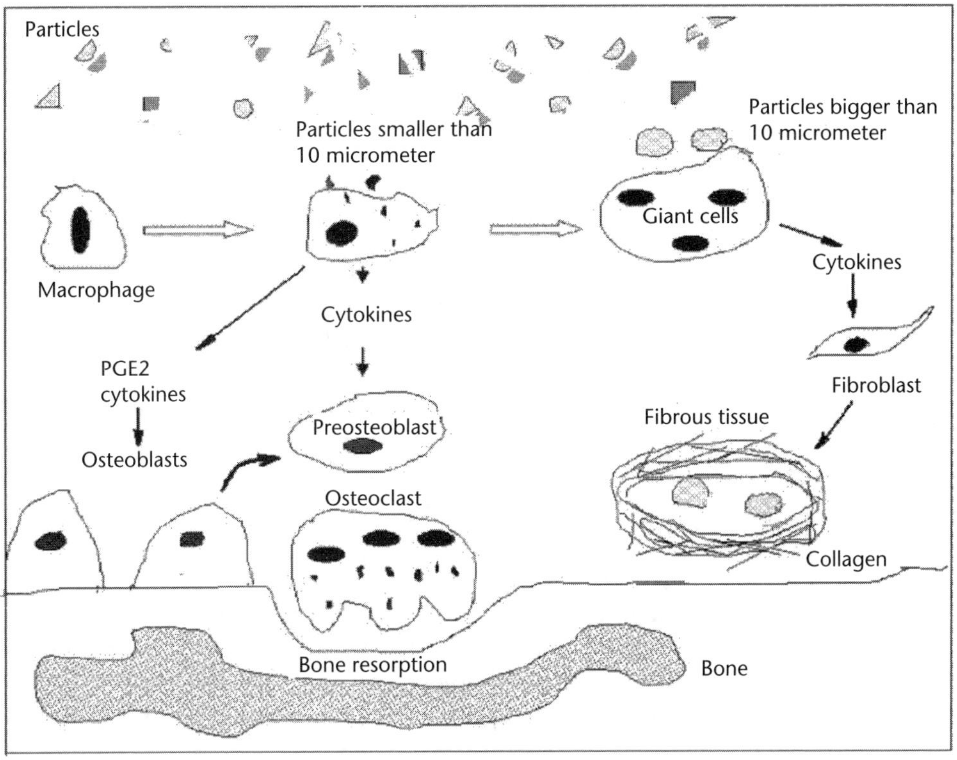

Figure 21.2 Schematic of biomaterial wear debris activating macrophages and inducing osteolysis. Particulate debris is generated from wear of arthroplasty components, metal corrosion, or polymer degradation. Particles less than 10 μm are engulfed directly by macrophages but cannot be digested easily. Consequently, macrophages begin to release interleukins (IL-1, IL-6), prostaglandin E_2 (PGE_2), and tumor necrosis factors (TNF), which then mediate bone resorption via osteoclasts [28]. On the other hand, macrophages form giant cells to phagocytose large particles (>10 mm) and then secrete cytokines to signal fibroblasts to synthesize collagen to result in a fibrous tissue capsule around the implant. (*Source:* [29, 30].)

Macrophages have evolved to detect and phagocytose bioactive particles in the size range of 0.5 to 10 μm. Particles larger than 10 μm or less than 0.5 μm are relatively less active for macrophages. Particles greater than 10 μm will cause giant cell responses, which lead to the formation of fibrous tissue (see details in Section 21.1.2.1) without osteolysis. In addition, particles with sharp edges are more active than spherical particles [20, 28].

It is apparent that minimizing the generation of wear particles in the interfacial tissues is a basic approach to preventing osteolysis. Several possible strategies may address the problem of osteolysis [25]: designing novel artificial joint materials with more wear-resistant features, preventing debris from accessing the bone-implant interface, and decreasing cellular reaction to debris. There are a few research papers that provide preliminary results of the effects of nanoparticles on osteoblasts and on other cells [32–34]; such studies show a less adverse effect on osteoblast viability for nano- compared to microparticle wear debris. Although the effects of nanoparticulate wear debris at the bone-implant interface are not totally understood, to date, the biologically inspired features of nanomaterials have motivated wide investigations into their applications in orthopedic implants. Further investigation of the influence of nanoparticulate wear debris on bone-cell health is necessary to effectively minimize osteolysis, as well as to fully realize the benefits of nanotechnology in orthopedic implants [35, 36].

21.1.2.3 Other Complications (Wear of Articular Bearing Surfaces, Fractures, etc.)

The relative motion of some orthopedic implants is impossible to avoid, and the resulting mechanical load is extremely large (especially at the articulating surface of hip or knee implants). The more active and younger the patients who receive orthopedic implants are, the higher the risk of articulating-bearing surface wear and periprosthetic fractures [37]. In fact, wear of articulating-bearing surfaces has commonly evoked our attention. From the basic mechanical viewpoint, adhesive wear, abrasive wear, third-body wear, wear fatigue, and corrosion are five major mechanisms causing wear of articulating-bearing surfaces [38]. Modification of polyethylene (widely used as a material for acetabular components in hip replacements) and substitution of metal-metal or ceramic-ceramic bearings for polyethylene may both help to improve the wear-resistant property of implant materials.

Several investigations have revealed that periprosthetic femoral fractures accompanied with total hip replacements are more common after revision surgeries than after primary replacements [39, 40]. So, with the increasing number of revision surgeries, it is expected that periprosthetic femoral fractures will become more common. Since mechanical properties (including strength, toughness, and ductility) of implant materials are closely related to the above complications and have influences on the long-term success of an orthopedic implant, synthesizing various orthopedic-implant materials that biomimic the mechanical properties of natural bone is rather desirable.

It should be realized that the previously mentioned complications for orthopedic implants are widely related to each other. That is, the high rate of implant fractures is usually accompanied by a loose implant [41, 42]; loosening of implants can also be caused by osteolysis or inflammation. Wear particle generation (see Section 21.1.2.2) and inflection [5] often coexist with wear of articulating surfaces. Therefore, reducing the influence of one complication may have positive effects on preventing other complications and finally improving the long-term success of orthopedic implants.

21.2 The Role of Nanomaterials in Orthopedic Implants

Nanomaterials are materials with basic structural units, grains, particles, fibers, or other constituent components in the range of 1 to 100 nm [21]. Compared to respective conventional micron-scale materials, they can exhibit enhanced mechanical properties, cytocompatibility, and electrical properties. To date, a wide range of nanostructured ceramics, metals, polymers, organic materials, and composites have shown promise in lowering the rate of implant failures for orthopedic implants by promoting osseointegration. Thus, due to potentially numerous applications of novel nanomaterials in medical implants, nanomaterials have been proposed as the next generation of orthopedic-implant materials.

21.2.1 Ceramics

Ceramics are nonmetallic inorganic materials that have excellent cytocompatibility and possibly biodegradability in the physiological environment, which makes them attractive for orthopedic applications. For these reasons, they have been widely adopted in orthopedic implants (known as bioceramics for several decades). According to the tissue response in an osseous environment, bioceramics can be classified into three groups: bioactive ceramics (such as hydroxyapatite (HA), tricalcium phosphate (TCP), bioglasses, HA/TCP biphase ceramics, and glass-ceramics), biopassive ceramics (such as alumina, titania, and zirconia), and biodegradable ceramics (such as TCP).

Although traditional bioceramics have long served as bone substitutes and filler materials, structural forms, and surface coatings in orthopedic applications [43], there often exists a variety of implant failures due to insufficient osseointegration, osteolysis, and implant wear (see Section 21.1.2). There are many reasons to use nanophase ceramics to overcome these traditional implant failures. As we know, 70 percent of the human bone matrix is composed of inorganic crystalline hydroxyapatite, which is typically 20 to 80 nm long and 2 to 5 nm thick [19]. Additionally, other components in the bone matrix such as collagen and noncollagenous proteins (laminin, fibronectin, and vitronectin) are nanometer scale in dimension [9]. Therefore, novel nanophase ceramics (grain sizes less than 100 nm in diameter) that biomimic the nanostructure of natural bone have become quite popular in orthopedics. Present researchers have shown that nanostructured alumina, titania, ZnO and HA can greatly enhance osteoblast adhesion and promote calcium/phosphate mineral deposition [14–16, 44–48]. It is believed that the special surface topography, increased wettability, and better mechanical properties of nanoceramics may contribute to enhanced osteoblast functions. Obviously, nanostructured ceramics may become a new generation of more promising and efficient orthopedic material. Section 21.2.1.1 discusses the advantages of various nanostructured ceramics compared to conventional ceramics.

21.2.1.1 Surface Properties

With decreased grain size and decreased pore diameter, nanophase ceramics have increased surface area, surface roughness (shown in Figure 21.3), and number of grain boundaries at the surface. For example, from the results of extensive characterization studies (Table 21.1), increased surface roughness and improved surface wettability (decreased contact angles) are evident in nanophase compared to conventional alumina, HA, and titania [16]. In the case of ZnO, atomic force microscopy (AFM) root-mean-square roughness values of nanophase and microphase ZnO were 32 and 10 nm, respectively [49]. Additionally, a 23 nm grain size alumina had approximately 50 percent more surface area for cell adhesion than a 177 nm grain size alumina; similarly, a 32 nm grain size titania had nearly 35 percent

(a) 2.12 μm grain size titania

(b) 32 nm grain size titania

(c) 177 nm grain size alumina

(d) 23 nm grain size alumina

Figure 21.3 AFM images of titania and alumina: (a) and (c) represent conventional titania and alumina, respectively, and (b) and (d) represent nanophase titania and alumina, respectively. (*Source:* [14].)

Table 21.1 Nanophase and Conventional Ceramic Structure and Surface Wettability [16]

Material	*Porosity (%)*	*Pore Diameter (nm)*	*Surface Roughness (nm)*	*Contact Angle (degrees)*
24 nm alumina	4.5	0.69	20	6.4±0.7
45 nm alumina	3.4	1.11	19	10.8±1.3
167 nm alumina	2.4	2.94	17	18.6±0.9
39 nm titania	4.1	0.98	32	2.2±0.1
97 nm titania	3.8	1.91	24	18.1±3.2
4,520 nm titania	3.2	23.3	16	26.8±2.8
67 nm HA	1.1	0.66	17	6.1±0.5
132 nm HA	1.1	1.98	11	9.2±0.4
179 nm HA	1.1	3.10	10	11.5±1.1

more surface area than a 2.12 μm grain size titania [14], while nanophase ZnO had 25 percent more surface area than microphase ZnO [49].

In order to illustrate the advantages of nanophase ceramics in terms of surface properties for orthopedic and dental applications, calcium phosphate derivatives are perfect examples (such as HA, tricalcium phosphate, calcium carbonate, and bioglass, which have extensive applications in orthopedics [50]). Since they share a similar crystal structure and chemical composition to natural bone, calcium phosphate ceramics have been considered the most popular coating materials on traditional implant metals, as well as in pure form for bone-graft substitutes. For example, HA and TCP (α- or β-crystalline) can serve as bone-tissue-engineering scaffolds to facilitate new bone formation and as a coating on femoral (metal) or socket (polymer) prostheses to inhibit complications related to osteolysis [51]. HA and TCP have a relatively high rate of degradation in vivo as well, making them promising in drug-eluting biodegradable bone-graft therapies.

Furthermore, HA has been shown to have good osteoconductive properties since their surfaces can undergo selective chemical reactivity with surrounding tissues, resulting in a tight bond between bone and the implant [52]. Thus, new bone formation will be promoted by such a bonded interface [53, 54]. Moreover, it has been demonstrated that osseoinductivity of calcium phosphate ceramics is strongly related to their surface properties. Higher amounts of surface porosity, increased surface roughness, and improved wettability of nanophase ceramics have positive effects on the osseointegration of calcium phosphate ceramics with juxtaposed bone. The nanometer-sized grains and high volume fraction of grain boundaries in nanostructured HA can increase osteoblast adhesion, proliferation, and mineralization [55]. For example, Webster, Siegel, and Bizios showed that there was significantly enhanced osteoblast adhesion and strikingly inhibited fibroblast adhesion on nanophase HA (67 nm) compared to conventional HA (179 nm HA) after four-hour cell-culture studies [16]. In addition, enhanced osteoblast-like cell functions (such as the synthesis of tartrate-resistant acid phosphatase and formation of resorption pits) on nanophase HA have also been observed compared to conventional HA [46]. Since excessive secretion of fibrous tissue from fibroblasts causes the formation of a detrimental fibrous tissue capsule that can contribute to loosening of implants, decreased fibroblast functions on nanophase HA may contribute to a better bone-implant interface. Recently, a bioresorbable nanocrystalline HA paste was used as a valuable addition to TCP-HA ceramic granules for acetabular bone grafting in animal models. That study showed that the nanocrystalline HA paste resulted in better acetabular cup stability than the pure allografts and at the same time didn't cause adverse biological reactivity [56]. Histology slides in human cancellous bone revealed that nanocrystalline HA paste appeared to be a suitable bone substitute for bone defects or cavities since it showed good tissue incorporation, high biocompatibility, and rapid osseointegration [57].

To demonstrate the versatility of nanoceramics, similar tendencies also appeared for nanophase alumina, zirconia, and titania. A 51 percent increase in osteoblast adhesion and a 235 percent decrease in fibroblast adhesion was observed on alumina as grain size decreased from 167 to 24 nm in a four-hour period [16]. Moreover, osteoblast adhesion increased by 146 percent and 200 percent on nanophase zirconia (23 nm) and titania (32 nm) compared to microphase zirconia (4.9 mm) and titania (4.1 mm), respectively, when normalized to projected surface area [49]. Furthermore, increased collagen synthesis, alkaline phosphatase activity, and calcium mineral deposition by osteoblasts were observed on nanophase zirconia, titania, and alumina (theta+delta crystalline phase) compared to conventional equivalents [48, 49].

To determine the role of nanoceramics in fighting implant infections, various bacteria functions were recently investigated on nanophase ceramics. For instance, *Staphylococcus epidermidis* (a common bacterium in human skin) forms a thick ECM and eventually a biofilm around bone implants [58], which leads to implant failure. Thus, decreased *Staphylococcus epidermidis* functions on implants are desirable. Nanophase zinc oxide and titania have indeed been observed to significantly inhibit *Staphylococcus epidermidis* functions [49].

It was elucidated that the special surface topography and wettability of nanophase ceramics may be the main reasons for the promoted bone cell and inhibited bacteria functions on nanoceramics. At the tissue-implant interface, the first and necessary step for success is the adsorption of specific proteins onto biomaterial surfaces. Many researchers have demonstrated that the protein-adsorption process depends on surface properties (such as hydrophilicity, charge density, roughness, and surface energy) [59, 60]. For instance, on hydrophilic and high nanorough surfaces, studies have shown maximum vitronectin, fibronectin, and laminin adsorption [16, 59, 60]. Specifically, the highest concentration of vitronectin (a protein known to mediate anchorage-dependent cell adhesion) adsorption on nanophase ceramics may explain the subsequent enhanced osteoblast adhesion on nanophase ceramics [16]. In addition, proteins will interact with specific cell membrane receptors (integrins) to aid cell adhesion on the substrate and at the same time inhibit other nonspecific cell adhesion. It is apparent that surface properties are closely related to anchorage-dependent cell (e.g., osteoblasts and osteoclasts) adhesion to a biomaterial surface. After that, cell proliferation, differentiation, and ECM deposition can happen and will lead to a successful integration with surrounding tissue. Many current attempts, therefore, have focused on nanosurface modifications of conventional orthopedic materials in order to promote anchorage-dependent osteoblast adhesion on implant surfaces. Importantly, promising results have been observed on nanoceramics without any modification (it is their raw surface that works).

21.2.1.2 Mechanical Properties

Mechanical properties (such as hardness, ductility, stiffness, bending, and compressive and tensile strength) of biomaterials play a crucial role on the long-term success of orthopedic implants. Many attempts and efforts have ensued to improve the mechanical properties of current metals, ceramics, and polymers in order to biomimic those of physiological bone. Unlike metals, ceramics (such as alumina and HA) have little ductility and have low fracture toughness (0.8 to 1.2 MPa · $m^{-1/2}$ for HA). They cannot tolerate damage and are vulnerable to crack initiation and propagation. They cannot therefore be applied in loading-bearing applications. Table 21.2 shows the mechanical properties of traditional ceramics compared to natural bone as well as other implant materials.

To date, it is well known that decreasing grain sizes to the nanoscale (diameter greater than 12 to 20 nm) increases strength, hardness, and plasticity for both metals and ceramics. Theoretically, the well-known empirical Hall-Petch equation, which relates yield stress (σ_y) to average grain size (d), can exhibit remarkably increased strength when reducing grain size from the micronmeter to the nanometer regime [62]:

$$\sigma_y = \sigma_0 + \frac{k}{\sqrt{d}} \tag{21.1}$$

Table 21.2 Comparison of Mechanical Properties of Ceramics and Bone and Other Orthopedic Materials

Material	*Modulus of Elasticity/ Stiffness (GPa)*	*Band Strength (MPa)*	*Ductiltity (%)*	*Hardness (Knoop Hardness Number)*
Cortical bone	10–16	140	0–2	
Ceramics				
Alumina	380	550	0	2,100
Zirconia	170	350	0	1,160
Hydroxyapatite	345	200	0	
Metals				
Implant alloy (Co-Cr_Mo)	240	825	10	430
Stainless steel annealed (316L)	190	485	40	235
Ti-6Al-4V (ELI)	110	900	12	325
Polymers				
Polymethyl methacrylate (PMMA)	3	55	0–2	
Polyethylene (UHMW)	1	44	400	

Source: [61].

Here σ_0 is a friction stress, and k is a constant. Hardness and grain size also have the above similar relationships. Many experimental results partly confirmed the above theoretical predictions. For example, as grain size is reduced to nanoscale dimensions, the hardness of nanocrystalline metals (such as copper, palladium, and silver) typically increases and can be a factor of two to five higher than that of conventional metals [63–65]. Karch and Birringer reported that an increased value of hardness and remarkably high values of fracture toughness (~14 MPa $\cdot$ $m^{-1/2}$) were observed for nanocrystalline titania [62]. Reducing the grain size of tetragonal zirconia polycrystals from commercially used 0.3 μm to 10 nm enhanced deformation rates by a factor of two and provided a material with superplasticity properties [62]. Webster, Siegel, and Bizios also reported the bending strength of 23 nm grain size alumina increased compared to that of 177 nm grain size alumina [15]. However, since flaws such as pores, dislocations, and other defects of nano- crystalline ceramics frequently occur, sometimes strength or hardness values of nanocrystalline ceramics deviate from Hall-Petch predictions. This phenomena is called the inverse Hall-Petch effect, namely, a decrease in strength or hardness as grain size is decreased below ~10 nm [55]. Controlling manufacturing parameters may avoid the inverse Hall-Petch effect.

Currently, the most widely used bearing couples in hip replacements are metal-on-polyethylene components. However, as we know, wear debris originating from polyethylene causes foreign-body reactions around the implant that frequently influence the longevity of orthopedic implants. Consequently, many researchers have focused on evaluating alternative materials for articulating surfaces and lowering the wear rate of articulating components in order to reduce wear-related complications to improve the life expectancy of bone implants [66]. Generally, low wear rates can be achieved by using materials with high hardness. Clearly, ceramics have rather high hardness compared to other implant materials (Table 21.2). For this reason, ceramics such as alumina are widely evaluated as wear-resistant articulating materials considering their high hardness, chemical inertness, respectable strength, and fracture toughness [67]. For instance, metal-on-metal implants with ceramics coatings (such as nanostructured diamond), ceramic-on-ceramic, and ceramic-on-polyethylene couples have been studied (Table 21.3). It has been shown that the wear rates of ceramic-on-polyethylene and further ceramic-on-ceramic systems [68] are noticeably lower than for metal-on-polyethylene systems of total hip prostheses. Moreover, since grain size and porosity can be closely related to the wear behavior of ceramics, many investigations of the wear resistance of alumina ceramics have focused on the role of nanophase ceramics. By reducing the grain size as well as decreasing porosity, the wear characteristics can be significantly improved [69–73]. For example, Al_2O_3-SiC nanocomposites have been shown to have much greater resistance to surface fracture than conventional ceramics [67]. Polycrystalline phased-stabilized zirconia from nanoparticles has improved ductility, fracture toughness, and wear-resistance compared to current-generation zirconia [74].

In this light, nanostructured diamond and diamondlike carbon coatings on cobalt-chrome and titanium (Ti) alloys can be considered for potential orthopedic applications [55] due to their excellent mechanical properties compared to conventional micron-scale implant materials. They are one of the hardest, strongest, and most wear-resistant materials and also have suitable biocompatibility and corrosion resistance. Through using chemical vapor deposition technology, a nanostructured diamond film with high hardness, enhanced toughness, and good adhesion to alloys can be obtained. Catledge et al. have demonstrated that nanostructured diamond films can be tailored on metallic surfaces with hardness values ranging from 10 to 100 GPa by changing the feed gas (N_2/CH_4) ratio [75]. The above excellent properties of nanostructured diamond make it promising as a coating for load-bearing articulating surfaces [55] in orthopedic implants.

In summary, nanoceramics can offer significantly improved mechanical properties, good wear properties, and excellent biocompatibility compared to conventional ceramics. Although further investigations of nanoceramics are needed to justify and apply them to real orthopedic applications, nanoceramics are clearly promising future orthopedic materials.

21.2.2 Metals

Artificial joints or prostheses have employed a variety of metallic components because of their durability, physical strength, and physiological inertness. For example, titanium, Ti alloys (such as Ti6Al4V), metal alloys (such as CoCrMo), and stainless steel are commonly used in orthopedics. Most metallic components, such as hip ball-and-sockets joints are made of stainless steels or titanium, which allows

Table 21.3 Coefficient of Friction for Different Bearing Couples Used in Total Hip Replacements

Bearing Couples	*Cartilage-Cartilage*	*CoCr-UHMWPE*	*Zirconia-UHMWPE*	*Alumina-UHMWPE*	*CoCr-CoCr*	*Alumina-Alumina*
Coefficient of friction	0.002	0.094	0.09–0.11	0.08–0.12	0.12	0.05–0.1

Source: [74–76].

bone ingrowth. Commercially, pure titanium and Ti6Al4V alloys are two of the most common titanium-based implant biomaterials [77].

Because of the fact that titanium is not ferromagnetic, titanium implants can be safely examined with magnetic resonance imaging, which makes them especially useful for long-term implants. Moreover, titanium develops an oxide layer in contact with water or air, resulting in high biocompatibility. Thus, titanium and titanium alloys are considered the best choice for manufacturing permanent nonbiodegradable implants [77, 78]. Importantly, the modulus of elasticity of titanium can be exploited to closely match the modulus of the bone [79]. However, the stiffness of titanium alloys is still more than twice that of bone.

The micro- or nanoscale surface properties of metals just like ceramics, such as surface composition and topography, can affect bone formation. Coatings and/or surface modifications are interrelated with topography and surface energy. For this reason, it is very difficult to determine how these, as independent factors, affect the final bone formation result [77]. The topographical features obtained on the metal implant surface can range from nanometers to millimeters, which are far below the size of osteoblasts but have been shown to promote bone formation [80].

Numerous treatment processes, including machining or micromachining, particle blasting, Ti plasma spraying, HA plasma spraying, chemical or electrochemical etching, and anodization are available to modify Ti surface topography. In addition, anodizing titanium under different conditions can change the nature of the oxide layer (thickness, porosity, and crystallinity) on the nanoscale to strongly affect in vivo bone formation [77].

The topography of an implant surface can be defined in terms of form, waviness, and roughness (Figure 21.4), with the waviness and roughness often presented together under the term "texture" [81]. Affected by the surface roughness, the composition of the initially adsorbed protein layers and the orientation of the molecules adsorbed on the implant surfaces have been shown to be an important factor for osseointegration.

For example, compared with conventional Ti (Figure 21.5), Ti6Al4V (Figure 21.6), and CoCrMo (Figure 21.7), nanophase surfaces have been shown to increase osteoblast function (Figure 21.8; Table 21.4). The dimensions of nanometer surface features gave rise to larger amounts of interparticulate voids (with fairly homogeneous distribution) on nanophase Ti and Ti6Al4V, unlike the corresponding conventional Ti and Ti6Al4V compacts; rather, these latter compacts revealed fewer interparticulate voids with a nonhomogeneous distribution [17]. Similar to ceramics, these nanoproperties are promising for bone formation.

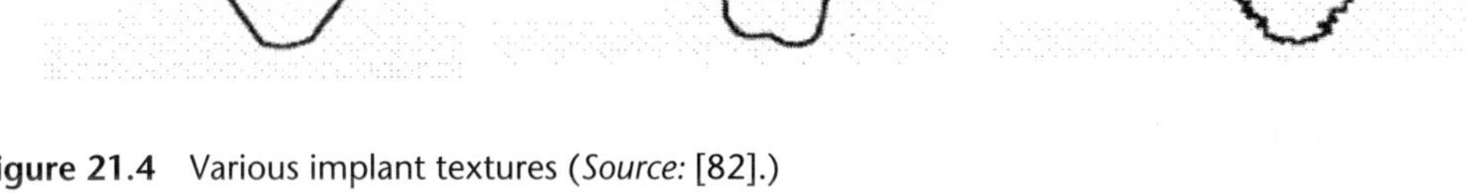

Figure 21.4 Various implant textures (*Source:* [82].)

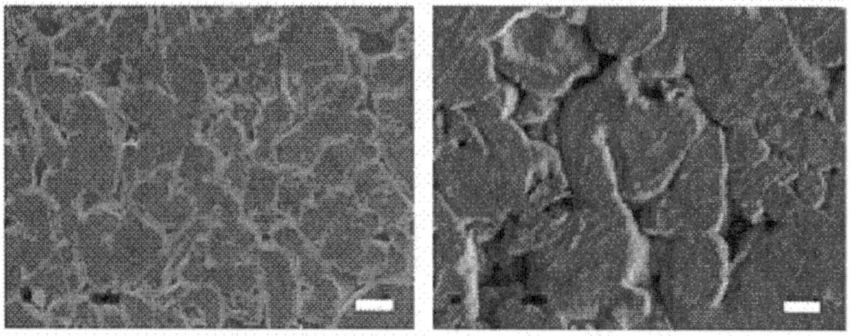

Nanophase Ti Conventional Ti

Figure 21.5 Scanning electron microscopy (SEM) images of Ti compacts. Increased nanostructured surface roughness was observed on nanophase compared to conventional Ti. Bar =1 μm for nanophase Ti and 10 μm for conventional Ti [17].

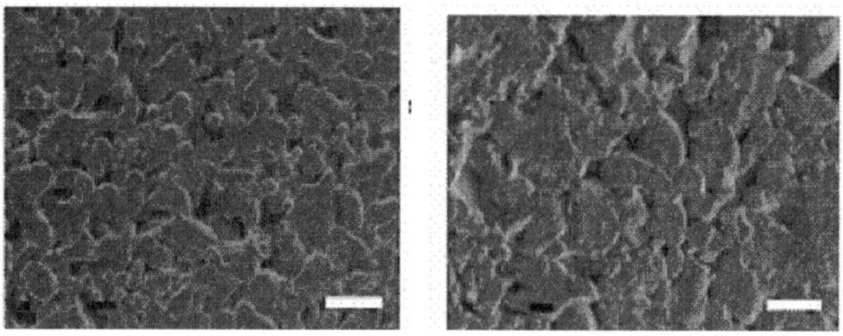

Nanophase Ti6Al4V Conventional Ti6Al4V

Figure 21.6 SEM images of Ti6Al4V compacts. Increased nanostructured surface roughness was observed on nanophase compared to conventional Ti6Al4V. Bar = 1 μm for nanophase Ti6Al4V and 10 μm for conventional Ti6Al4V [17].

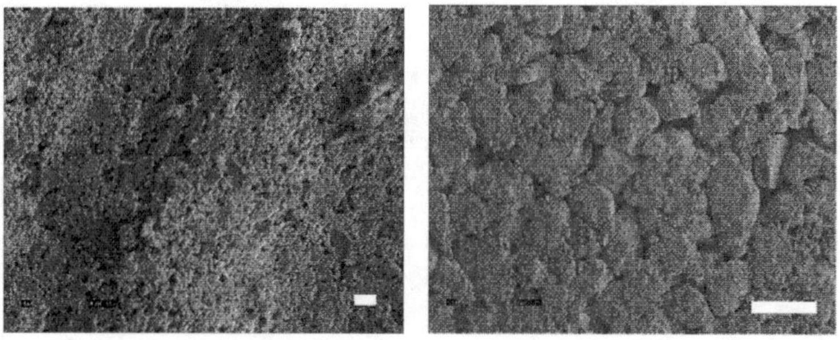

Nanophase CoCrMo Conventional CoCrMo

Figure 21.7 SEM images of CoCrMo compacts. Increased nanostructured surface roughness was observed on nanophase compared to conventional CoCrMo. Bar =1 μm for nanophase CoCrMo and 100 μm for conventional CoCrMo [17].

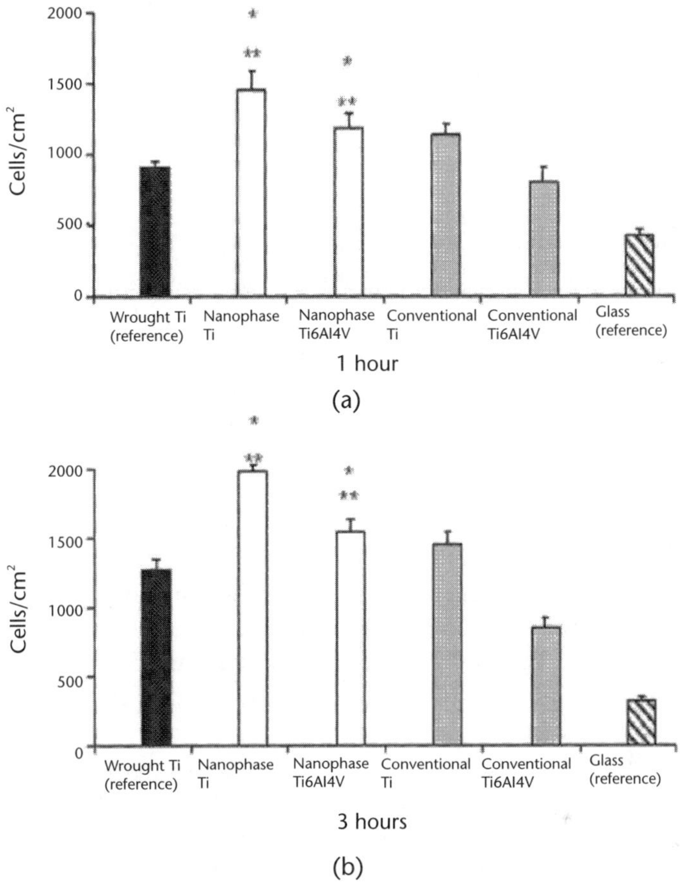

Figure 21.8 Increased osteoblast adhesion on nanophase Ti and Ti6Al4V compacts (a) 1 hour; (b) 3 hours. Data = mean +/– SEM; $n = 3$; $*p < 0 \cdot 01$ compared to respective conventional metal, and $**p < 0 \cdot 01$ compared to wrought Ti (reference) [17].

Table 21.4 Surface Roughness of Metal Compacts [17]

Substrate	*Surface Roughness (rms; nm)*
Ti (nano)	11.9
Ti (conv)	4.9
Ti6A14V (nano)	15.2
Ti6A14V (conv)	4.9
CoCrMo (nano)	356.7
CoCrMo (conv)	186.7

However, despite these promises, there may be some problems with certain titanium alloys. For example, Hong et al. reported that Ti surfaces are (excessively) highly thrombogenic [83], leading to more rapid collagenous encapsulation and faster remineralization. Some researchers have discussed the potential cytotoxicity associated with the vanadium element in Ti6Al4V [77]. When Ti6Al4V was compared with commercially pure Ti [84], gingival fibroblasts demonstrated a rounded cell shape and a reduced area of spreading on the alloy, presumably because of a minor toxicity to vanadium or aluminum. In this light, it is important to mention that fibroblast adhesion has been shown to be lower on nano- compared to conventional metals.

21.2.3 Polymers

Polymers are a class of synthetic materials characterized by their high versatility [85] and, thus, have been widely applied in orthopedics. They can mechanically secure components of hip replacements into bone (as in bone cements, such as injectable polymethylmethacrylate [PMMA]) and can line acetabular cups for smooth articulation with metallic femoral head components (such as in ultra-high-molecular-weight polyethylene [UHMWPE]). However, there also exist many disadvantages related to traditional polymers used in orthopedic implants. For example, PMMA suffers from fatigue-related cracking and impact-induced breakage due to its poor setting and fixation [86]. Furthermore, PMMA can also elicit a host response due to release of toxic monomers and necrosis of the surrounding tissue caused by exothermic polymerization in vivo [87]. Increased articulation-induced wear of UHMWPE frequently leads to osteolysis (see Section 21.1.2.2) [88].

Therefore, further development of various polymer alternatives that could improve osseointegration is necessary, not only for fracture fixation but also as bone-tissue-engineering scaffolds. For this reason, polymers such as polylactic acid (PLA), polyglycolic acid (PGA), poly(lactic-co-glycolic acid) (PLGA) and various hydrogels currently generate more interest from scientists to investigate their potentials as bone-tissue-engineering scaffolds to repair bone defects due to their excellent biocompatibility, suitable mechanical properties, biodegradability, and ease of modification for different applications.

An ideal bone-repair scaffold should be biocompatible to minimize local tissue response while maximizing osseointegration and biodegradability after new bone formation. As will be described, an important consideration in scaffold design is providing a polymer nanoscale framework for cellular interactions.

21.2.3.1 Nanostructured Biodegradable Polymers as Bone-Tissue-Engineering Scaffolds

One of the most common polymers used as a biodegradable biomaterial has been PLGA [85]. It has high biocompatibility, the ability to degrade into harmless monomer units, and a useful range of mechanical properties (depending on the copolymer ratio). Recently, this copolymer and its homopolymer derivatives, PLA and PGA, have received substantial attention for skeletal repair and regeneration. Various techniques (such as particulate leaching, textile technologies, and three-dimensional printing techniques) have successfully created three-dimensional porous matrices from PLGA and its derivatives [89–92]. For instance, with the simplicity of the polymer-casting and phase-separation processes, a robust nanofibrous poly(l-lactic acid) (PLLA) scaffold can be prepared [93]. These nanofibrous scaffolds have a collagenlike three-dimensional structure, high porosity, and a suitable surface structure for osteoblast attachment, proliferation, and differentiation.

In addition, several studies have shown that nanostructured PLGA (Figure 21.9) prepared by chemical etching procedures can promote vascular endothelial and smooth muscle cell adhesion compared to conventional PLGA [94, 95]. Further investigation has demonstrated that nanostructured PLGA can adsorb significantly more vitronectin and fibronectin from serum compared to conventional PLGA (just as the aforementioned nanoceramics and metals do) [93]. This increase in protein adsorption may help explain enhanced osteoblast adhesion observed on PLGA with nanosurface roughness. In addition, significantly increased chondrocyte functions (adhesion, proliferation, and matrix synthesis) have been observed on NaOH- treated nanostructured PLGA [97]. Furthermore, improved osteoblast functions have also been observed on nanophase titania/PLGA composites, which make them promising as a bone-tissue-engineering scaffold materials [98]. At the same time, it was found that fibroblast density decreased on nanophase PLGA, polyurethane, and polycaprolactone, which may inhibit the formation of a fibrous tissue capsule [99]. Obviously, because nanostructured PLGA has excellent biocompatibility properties (above that of traditional PLGA) and ease of degradability, it has been considered a new generation of tissue-engineering scaffold for numerous tissue-engineering composite applications, especially for bone and articular cartilage.

21.2.3.2 Nanostructured and Injectable Hydrogels as Bone-Tissue-Engineering Scaffolds

Polymeric hydrogels have excellent biocompatibility, which makes them useful in orthopedic applications (such as for injectable bone repair and cartilage reconstruction) [100]. Natural hydrogels (such as collagen and gelatin) are the main components of ECMs of mammalian tissues, including bone, cartilage and so on [101]. Through a technique called electrospinning, nanofibrous scaffolds [102] can be successfully created from various synthetic and natural polymers such as gelatin and collagen (Figure 21.10). Several studies have shown that protein and cell interactions are promoted by nanofibrous scaffolds that biomimic the ECM [103, 104].

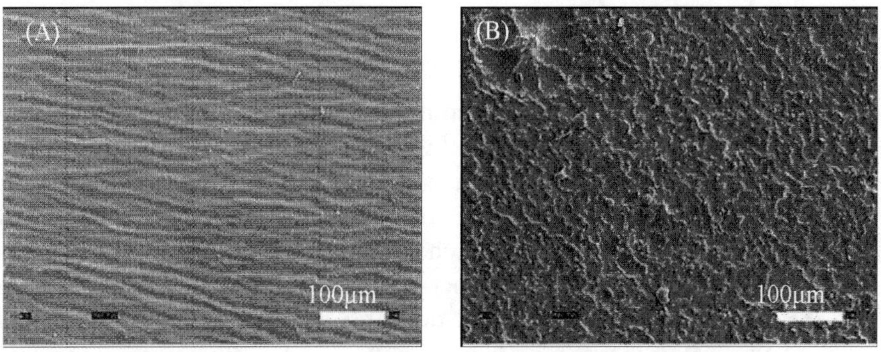

Figure 21.9 SEM images of chemically treated and conventional untreated PLGA surfaces: (a) conventional PLGA (feature dimensions ~10 to 15 μm) and (b) chemically treated nanostructured PLGA (feature dimensions ~50 to 100 nm). Scale bar = 100 μm. (*Source:* [95].)

Figure 21.10 SEM images of electrospun polyaniline-contained gelatin nanofibers: (a) polyaniline-gelatin-blend fibers with ratios of 15:85 and (b) polyaniline-gelatin-blend fibers with ratios of 60:40. This figure shows that the electrospun fibers are homogeneous, while 60:40 fibers were electrospun with beads [102]. Scale bars = 5 μm for (a) and 1 μm for (b).

On the other hand, studies have demonstrated that using self-assembling peptide KLD-12 hydrogels for encapsulating chondrocytes can support chondrocyte differentiation and promote the synthesis of a cartilage-like ECM for cartilage repair [105].

In addition, various synthetic hydrogels such as poly(2-hydroxyethyl methacrylate) (pHEMA) were modified to have nanofeatures for orthopedic applications. For example, some studies have created an injectable nanostructured pHEMA-hydrogel scaffold that incorporates novel self-assembled helical rosette nanotubes (HRNs) into hydrogels to fill bone fractures and repair bone defects [106]. The nanoscale, tubular architecture of HRNs on and in hydrogels can provide a topography that improves protein adsorption and an environment that resembles an ECM surrounding cells in vivo. Results revealed that HRNs coated on and embedded in hydrogels significantly improved osteoblast adhesion even at a low concentration of 0.001 mg/ml [106]. These HRN-incorporated hydrogels can also solidify at body temperatures, thus allowing for aqueous injection and in situ bone healing. It has thus been anticipated that such nanostructured hydrogels are promising for bone-tissue-engineering applications.

21.2.4 Composites

Bone is a natural nanostructured matrix composite made from calcium-phosphate apatite mineral that is reinforced with collagen fibers. Cells in the body are accustomed to interacting with composite materials that have nanostructured features; thus, one way to closely mimic the properties of natural bone tissues is by using nanocomposite biomaterials. Such composite biomaterials are engineered materials consisting of two or more distinct materials with significantly different physical or chemical properties that provide desirable overall mechanical, chemical, biological, and physical properties. Such composites may contain reinforcing phases embedded within a matrix phase. The reinforcing material can be either fibers or particles, for example, carbon nanofibers, carbon nanotubes, HRNs, or ceramic nanoparticles (such as hydroxyapatite or calcium phosphates). The matrix can be metal, ceramic, or polymer, but most biomedical composites have polymeric matrices. The majority of composites used in biomaterial applications are intended for the purposes of bone repair.

Polymer composite materials offer desired high-strength and bonelike elastic properties, potentially leading to a more favorable bone remodeling response. For example, polyethylene (PE) is a polymeric material which, in different grades (namely, high-density polyethylene (HDPE), ultra-high-molecular-weight polyethylene, and low-density polyethylene), can be used as an implant [107]. Unfortunately, osseointegration of this polymer is low; thus, normally its attachment to bone should be mechanical in nature. However, in many studies, bioactivated PE has been obtained by adding bioactive materials to PE, such as, hydroxyapatite-reinforced high-density polyethylene.

Minerals of calcium phosphates are frequently used as an additive to polymeric matrices. Calcium phosphates (CaPs) are a major constituent of natural bone. Therefore, they show great potential as a supplement in orthopedic applications, including bone replacements, bone cements, and scaffolds for tissue engineering [108, 109]. Thus, they are very popular as bioactive ceramic materials. For example, the chemical composition of HA makes its composite biomaterials more stable against resorption. On the other hand, TCP is partially resorbable and frequently used as a bone substitute [110]; in particular, β-TCP, is known to be biocompatible, osteoconductive, and osteoinductive [111].

β-TCP-reinforced high-density polyethylene (β-TCP/HDPE) is one of the new biomaterials that can replace bone tissues. Homaeigohar et al. concluded that inserting TCP particles into the structure of PE is beneficial for both structural compatibility and biological performance [112]. Also, the results from this study provided a basis for further studies in the in vivo field toward using this composite as a candidate as a bone-graft substitute in orthopedic surgery. Murugan and Ramakrishna demonstrated the possibilities of bone-paste production incorporating a composite of HA nanocrystals and chitosan, which is an organic polymer, using a wet chemical synthesis method at low temperature [113]. The rate of bioresorption of HA is improved by the addition of chitosan. It is anticipated that this kind of composite can act as a bioresorbable bone substitute with superior bioactivity and osteoconductivity in vivo. Moreover, Jabbari et al. grafted HA nanoparticles with hydrophilic unsaturated poly(ethylene glycol) oligomers to improve their suspension stability and interfacial bonding in an aqueous hydrogel solution. They hypothesized that hydrogel/apatite nanocomposites are the ideal biomaterial to mimic the physiochemical and biological properties of the bone matrix for bone regeneration [114]. Even doping nanostructured HA with ions such as carbonate, sodium, and magnesium to give a material closely resembling the composition of the surrounding hard tissue is possible.

Another highly studied ceramic material in composites is zirconium dioxide, or zirconia (ZrO_2). It enhances fracture toughness in polymers or other ceramics. Novel orthopedic bioceramic are zirconia-hydroxyapatite (ZrO_2-HA) composites, which have good biocompatibility, making them suitable for orthopedic biomaterials [115]. CaP and phosphate-based glass-composite coatings have been shown to strengthen ZrO_2 and to improve biocompatibility. The proliferation and alkaline phosphatase activities of osteoblasts on such composite coatings were improved by approximately 30 to 40 percent when compared to ZrO_2 substrates alone and were comparable to pure-HA coating. These findings suggest that the CaP and P-glass composites are potentially useful for coating traditional biomaterials used as hard-tissue implants due to their morphological and mechanical integrity, enhanced bioactivity, and the favorable responses of osteoblast-like cells [116]. Kong et al. studied HA-added zirconia-alumina (ZrO_2-Al_2O_3) nanocomposites in load-bearing orthopedic applications [117]. The HA-added zirconia-alumina nanocomposites contained biphasic calcium phosphates of HA/TCP and had higher flexural strength than conventionally mixed HA-added zirconia-alumina composites. The in vitro tests showed that the proliferation and differentiation of osteoblasts on this nanocomposite gradually increased as the amount of HA added increased.

Synthetic biodegradable polymers are widely utilized in the fabrication of scaffolds for bone-tissue engineering [118]. Because carbon nanotubes (CNTs) have a very high strength-to-weight ratio for greater mechanical properties, they can be used as a reinforcing material for polymer composites. For example, Webster et al. developed a carbon-nanofiber-reinforced polycarbonate urethane composite in an attempt to determine the possibility of using either carbon nanofibers or strips of CNTs as either neural or orthopedic prosthetic devices [119]. The results have shown that such composite biomaterials have the potential to promote osteoblast functions. Also, Shi et al. used single-walled carbon nanotubes as reinforcing materials for poly(propylene fumarate) (PPF) [120]. Such composites are used as an injectable, highly porous scaffold for load-bearing bone repair. Moreover, it is believed that electrical stimulation enhances osteoblast function on such CNT composites. Conductivity of PLA, which is an insulator, can be enhanced by adding CNTs. Studies have shown that osteoblast proliferation increases significantly on such composites after they are exposed to electrical stimulation [121]. Thus, great promise exists for many nanocomposites for orthopedic applications.

21.2.5 Role of Chemistry

In bone-tissue engineering, nanotechnology is also being used in conjunction with changes in implant surface chemistries. For example, for titanium, the surface oxide layer has many qualities regarded as important for optimal reactions with bone. The excellent biocompatibility of Ti-based materials can be attributed to the high corrosion resistance (low metal-ion release) and, moreover, to the good bone-binding ability of TiO_2, which covers implant surfaces. An electrochemical method known as anodization or anodic oxidation (Figure 21.11) is a well-established surface nanomodification technique for metals like titanium to produce protective layers that possibly enhance bone growth [122].

Essentially, electrochemical approaches were reported to form highly defined porous TiO_2 layers. Under specific electrochemical conditions, self-organized TiO_2 nanotube layers can be grown on Ti simply by the anodization of Ti in dilute HF electrolytes. With this approach, TiO_2 layers could be grown consisting of tubes with a diameter of ~100 nm and a length of ~500 nm. This oxide layer that forms on titanium implants can be manipulated chemically, and there has been speculation as to whether the biological properties of the oxide surface may then be changed and

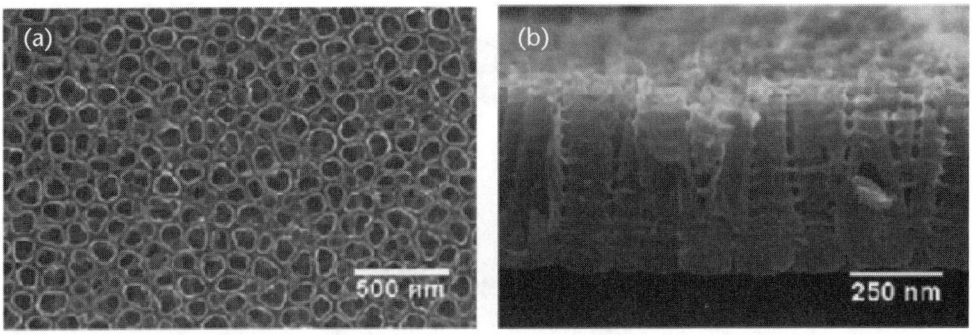

Figure 21.11 SEM image of the short-nanotube layer formed by anodization in 1M H_2SO_4+0.15 wt percent HF at 20V: (a) top-view, and (b) cross section. (*Source:* [122].)

even improved as a result. The significance of surface chemistry can be illustrated by the varied cellular responses reported on different titanium alloys, different grades of c. p. titanium, and different bulk metals. Chemical modification of titanium surfaces by their treatment in simulated body fluid, covalent attachment of biological molecules, changes in the surface ion content, and alkali treatment have all been reported to affect cellular responses to the implant [123–125]. It is noteworthy that this approach is successful not only for Ti but also for other value metals and for the formation of nanotube layers on biomedical Ti alloys such as Ti6Al4V and Ti6Al7Nb [126]. In this light, it is important to note that many studies to date highlighting the ability of nanomaterials to increase tissue growth have been performed on nano- compared to conventional materials of the same chemistry.

Another emerging area, still in the experimental stage, is the use of photolithography to produce micro- and nanofabricated surfaces [127]. Such surfaces, made of silicon and titanium, incorporate intentional surface chemical and topographical features in the nano- and micrometer scales and provide greater opportunities to control cell behavior.

The biological performance of biomedical implants strongly depends on the first interaction occurring when the implant surfaces come into contact with a biological environment. ECM proteins that contain the cell-binding domain arginine-glycine-aspartic acid (RGD) play a critical role in mediating cell behavior because they regulate gene expression by signal transduction set in motion by cell adhesion to proteins adsorbed on biomaterial [128]. Interactions between cell-membrane integrins and extracellular proteins are often facilitated by the RGD sequence. These interactions are important for the adhesion of many cells types. Integrin-mediated cell attachment influences and regulates cell migration, growth, differentiation, and apoptosis [128]. Various proteins (such as all types of collagen, fibronectin, and vitronectin) are known to be particularly important in mediating osteoblast adhesion; importantly, RGD is contained in all of these proteins and is recognized by cell membrane integrin receptors. Capitalizing on the use of such sequences may enhance cell targeting and adhesion to specific receptors, as well as also cell behavior via activation of specific signalling cascades, as shown in Figure 21.12.

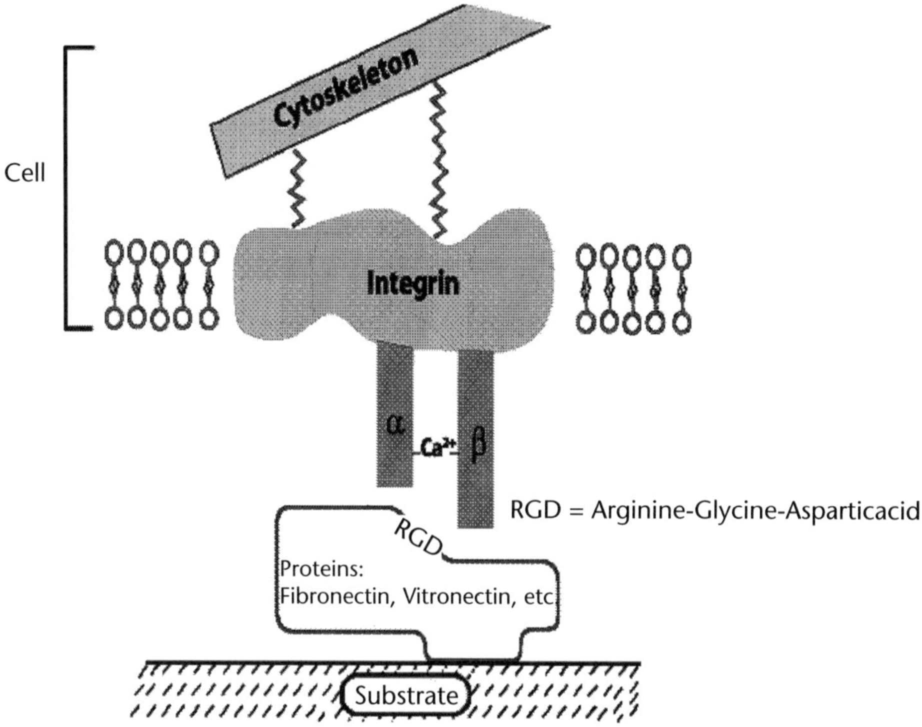

Figure 21.12 Surface properties affecting protein conformation and bioactivity. (*Source:* [128].)

The central hypothesis of biomimetic surface engineering is that peptides that mimic part of the ECM affect cell attachment to the material and that surfaces or three-dimensional matrices modified with these peptides can induce tissue formation according to the cell type seeded on the material. Therefore, extensive research over the last decade has been performed on the incorporation of adhesion-promoting oligopeptides onto the biomaterial surfaces. Further, the major advantage of using small peptides (such as RGD) with respect to larger peptides or proteins is their resistance to proteolysis and their ability to bind with high affinities to integrin receptors. Combining this approach (covalently linking peptides) with the use of nanomaterials has maximized cell responses.

21.3 Future Challenges

Nanotechnology applications are entering industrial production, mainly for diagnostics, drugs, and therapies. Particularly in orthopedic applications, nanometer-scale modifications of implant surfaces would improve durability and biocompatibility.

Although nanostructured implant materials may have many potential advantages in the context of bone-tissue engineering, it is important to remember that studies on nanophase materials have only just begun; there are still many other issues regarding health that must be answered. Most importantly, the influences of nanoparticulates on human health are not well understood, whether exposure occurs through the manufacturing or implantation of nanophase materials. Clearly, detailed studies in this context are required if nanoparticles are to be used in these systems [129].

Interactions of nanoparticles with biomolecules such as DNA, RNA, or proteins are also more likely with decreasing particle size. A number of reports on cellular uptake of micro- and nano-size particles have been published. Reports on particle uptake by endothelial cells, pulmonary epithelium, intestinal epithelium, alveolar macrophages, other macrophages, nerve cells, and other cells are available [129]. Further, it is expected that transport of nanoparticles across the blood-brain barrier is possible by either passive diffusion or by carrier-mediated endocytosis.

Nanoparticles may also become loose through the degradation of implanted polymeric materials through oxidation and/or hydrolysis, which accelerates exposure of materials [36]. Corrosion of metals once implanted can also cause release of nanoparticulates and, thus, contribute toxicity to biological environments. Oxidation accelerates the degradation of metals and the release of metal ions such as Al^{3+}, Ni^{2+}, Cr^{6+}, and Co^{2+}.

In addition, nanoparticles can be generated at artificial joints where friction between two surfaces is high. The outcome of micron-sized wear debris on bone health has been well studied for several decades. For instance, ultra-high-molecular-weight polyethylene often used as acetabular cups becomes brittle through oxidation and, thus, becomes susceptible to wear. Conventional-size wear debris (i.e., micron) triggers osteolysis as well as further wearing (third-body wear).

In contrast, the influence of nanoparticulate wear debris (or particles in general) on bone-cell health is only just beginning to be understood. It has been speculated that the effects of nanotubes on lungs are more toxic than quartz dust. One study showed that nanometer titania particles (50 nm) and carbon nanotubes (20 × 100 nm) induced morphological changes in neutrophils and decreased the overall cell survival rate [130].

On the contrary, other studies have demonstrated increased cell viability in the presence nanometer, compared with conventional, particles [36]. Specifically, nanophase materials did not stimulate cytotoxicity but even redeemed the toxic effects observed on osteoblast viability in the presence of conventionally sized particles. In this study, osteoblast proliferation was not negatively influenced by alumina and titania nanoparticles, whereas conventional particles of the same chemistry and crystalline phase increased cell death and slowed cell proliferation. The potential lack of nanoparticle toxicity was also demonstrated on coatings of pigment-grade titania particles as nanorods and nanodots of titania (<50 nm), which did not result in as much lung inflammation when compared with larger particle sizes (>300 nm) [36].

Obviously, further tests and studies concerning the toxicity of nanophase materials need to be conducted and expanded before the benefits of nanotechnology in orthopedic applications can be realized. However, it is important to emphasize that such studies need accurate comparisons when determining the health influences of nanometer compared with conventionally sized particles of the same size and crystallinity.

In summary, despite the challenges that lie ahead, significant evidence now exists elucidating that nanophase materials represent an important growing area of research that may improve bonding between an implant and surrounding bone. Even if nanophase materials do not provide the ultimate answer for increasing bone-cell responses, we have learned a tremendous amount of information concerning bone-cell recognition with nanostructured surfaces that will most certainly aid in improving orthopedic-implant efficacy.

References

[1] Joos, U., Wiesmann, H. P., Szuwart, T., and Meyer, U., "Mineralization at the interface of implants," *Int. J. Oral Maxillofac. Surg.*, Vol. 35, 2006, pp. 783–790.

[2] Charnley, J., "Anchorage of the fermoral head prosthesis to the shaft of the femur," *J. Bone Joint Surg.*, Vol. 42B, 1960, pp. 28–30.

[3] Charnley, J., "Surgery of the hip-joint: Present and future developments," *Br. Med. J.*, Vol. 5176, 1960, pp. 821–826.

[4] Frankowski, J. J., and Watkins-Castillo, S., "Primary total knee and hip arthroplasty projections for the U.S. population to the year 2030," *Am. Acad. Orthop. Surg.*, reviewed by T. P. Sculco, and S. M. Sporer, Vol. 10, August 23, 2002, p. 217.

[5] Charnlyy, J., "Surgery of the hip-joint: Present and future developments," *Br. Med. J.*, Vol. 19, 1999, pp. 821–826.

[6] Bozic, K. J., Rubash, H. E., Berry, J., Saleh, K. J., et al., "Modes of failure in revision hip and knee replacement," *Clin. Orthop. Relat. Res.*, Vol. 420, 2004, pp. 18–25.

[7] Rubash, H. E., and Berry, J., "Revisions of hip and knee replacements in Canada," *Can. Joint Repl. Reg. Anal. Bull.: Can. Inst. Health Inf.*, Vol. 603, 2004, pp. 1–20.

[8] Branemark, P. I., Adell, R., Albrektsson, T., Lekholm, U., et al., "Osseointegrated titanium fixtures in the treatment of edentulousness," *Biomaterials*, Vol. 4, 1983, pp. 25–28.

[9] Webster, T. J., *Nanophase ceramics: The future orthopedic and dental implant material: Advances in chemical engineering*, Acad. Press, Vol. 503, 2001, pp. 125–166.

[10] Esposito, M., Hirsh, J. M., Lekholm, U., and Thomsen, P. "Biological factors contributing to failures of osseointegrated oral implants: I. Success criteria and epidemiology," *Eur. J. Oral Sci.*, Vol. 106, 1998, pp. 527–551.

[11] Esposito, M., Hirsh, J. M., Lekholm, U., and Thomsen, P. "Biological factors contributing to failures of osseointegrated oral implants: II. Etiopathogenesis," *Eur. J. Oral Sci.*, Vol. 106, 1998, pp. 721–764.

[12] Cochran, D. L., Schenk, R. K., Lussi, A., Higginbottom, F. L., et al., "Bone response to unloaded and loaded titanium implants with a sandblasted and acid-etched surface: A histometric study in the canine mandible," *J. Biomed. Mater. Res.*, Vol. 40, 1998, pp. 1–11.

[13] Wennerberg, A., Hallgren, C., Johansson, C., and Danelli, S., et al., "Evaluation of screw-shaped implants each prepared with two surface roughnesses," *Clin. Oral Implants Res.*, Vol. 9, 1998, pp. 11–19.

[14] Webster, T. J., Siegel, R. W., and Bizios, R., "Osteoblast adhesion on nanophase ceramics," *Biomaterials*, Vol. 20, 1999, pp. 1221–1227.

[15] Webster, T. J., Siegel, R. W., and Bizios, R., "Design and evaluation of nanophase alumina for orthopaedic/dental applications," *Nanostruct. Mater.*, Vol. 12, 1999, pp. 983–986.

[16] Webster, T. J., Siegel, R. W., and Bizios, R., "Specific proteins mediate osteoblast adhesion on nanophase ceramics," *J. Biomed. Mater. Res.*, Vol. 51, 2000, pp. 475–483.

[17] Webster, T. J., and Ejiofor, J. U., "Increased osteoblast adhesion on nanophase metals: Ti, Ti6Al4V, and CoCrMo," *Biomaterials*, Vol. 25, 2004, pp. 4731–4739.

[18] Castner, D. G., and Ratner, B. D., "Biomedical surface science: Foundations to frontiers," *Surface Sci.*, Vol. 500, 2002, pp. 28–60.

[19] Kaplan, F. S., Hayes, W. C., Keaveny, T. M., Boskey, A., et al., "Form and function of bone: Orthopedic basic science," *Am. Acad. Orthop. Surg.*, Vol. 20, 1994, pp. 127–185.

[20] Chun, A. L., "Helical rosette nanotubes: An investigation towards its application in orthopaedics," Thesis, Purdue University, May 2006.

[21] Siegel, R. W., and Fougere, G. E., "Mechanical properties of nanophase metals," *Nanostruct. Mater.*, Vol. 6, 1995, pp. 205–216.

[22] Hayakawa, T., Yoshinari, M., Kiba, H., Yamamoto, H., et al., "Trabecular bone response to surface roughened and calcium phosphate (Ca-P) coated titanium implants," *Biomaterials*, Vol. 23, 2002, pp. 1025–1031.

[23] Meyer, U., Wiesmann, H. P., Fillies, T., and Joos, U., "Early tissue reaction at the interface of immediately loaded dental implants," *Int. J. Oral Maxillofac. Implants*, Vol. 18, 2003, pp. 489–499.

[24] Gu'ehennec, L. L., Soueidan, A., Layrolle, P., and Amouriq, Y., "Surface treatments of titanium dental implants for rapid osseointegration," *Dent. Mater.*,Vol. 10, 2006, pp. 12–16.

[25] Zhu, Y. H., Chiu, K. Y., and Tang, W. M., "Review article: Polyethylene wear and osteolysis in total hip arthroplasty," *J. Orthop. Surg.*, Vol. 9, 2001, pp. 91–99.

[26] Harris, W. H., "The problem is osteolysis," *Clin. Orthop.*, Vol. 311, 1995, pp. 46–53.

[27] David, V., Jarka, N., and Olga, K., "Periprosthetic osteolysis and its association with the molecule RANKL expression," *Physiolog. Res.*, Vol. 30, 2002, pp. 12–19.

[28] Bronson, J. G., "Breakthrough technology: Searching for artificial articular cartilage," *Orthoped Technol. Rev.*, Vol. 1, 1999, pp. 22–34.

[29] Buma, P., Donk, S., Slooff, T. J., and Schreurs, W., "Bone graft incorporation after reconstruction of bony defects with impacted morselized bone graft: Histology of animals and patients," *Ortop. Traumatol. Rehabil.*, Vol. 30, 2001, pp. 41–47.

[30] Hanson, S. R., Lalor, P. A., Niemi, S. M., Northup, S. J., et al., "Testing biomaterials. Biomaterials science," in B. D. Ratner, A. S. Hoffman, F. J. Schoen, and J. E. Lemons (eds.), *An Introduction to Materials and Medicine*, New York: Academic Press, 1994, p. 221.

[31] Anaheim and Calif. Rosemont, "Polyethylene: The past, present and future," *Orthopaed Proc., 66th Annual Meeting, Am. Acad. Orthop. Surg.*, Vol. 66, 2001, pp. 445–468.

[32] Gutwein, L. G., and Webster, T. J., "Increased viable osteoblast density in the presence of nanophase compared to conventional alumina and titania particles," *Biomaterials*, Vol. 25, 2004, pp. 4175–4183.

[33] Gutwein, L. G., and Webster, T. J., "Osteoblast and chrondrocyte proliferation in the presence of alumina and titania nanoparticles," *J. Nanopart. Res.*, Vol. 4, 2002, pp. 231–238.

[34] Tamura, K., Takashi, N., and Akasaka, T., et al., "Effects of micro-/nano particle size on cell function and morphology," *Key Engr. Mat.*, Vol. 20, 2004, pp. 254–256.

[35] Warheit, D. B., "Nanoparticles: Health impacts?" *Mater. Today*, February 2004, pp. 32–35.

[36] Sato, M., and Webster, T. J., "Nanobiotechnology: Implications for the future of nanotechnology in orthopedic applications," *Expert Rev. Med. Devices*, Vol. 1, 2004, pp. 105–114.

[37] Lindahl, H., Garellick, G., Regnér, H., Herberts, P., and Malchau, H., "Three hundred and twenty-one periprosthetic femoral fractures," *J. Bone Joint Surg. Am.*, Vol. 88, 2006, pp. 1215–1222.

[38] "What are the wear mechanisms and what control them?" in T. M. Wright and S. Goodman, *Implant Wear in Total Joint Replacement: Clinical and Biologic Issues, Material and*

Design Considerations, American Academy of Orthopedic Surgeons, Rosemont, IL, 2001, Chapter 22.

[39] Learmonth, I. D., "The management of periprosthetic fractures around the femoral stem," *J. Bone Joint Surg. Br.*, Vol. 86, 2004, pp. 13–19.

[40] Herberts, P., and Malchau, H., "Long-term registration has improved the quality of hip replacement: A review of the Swedish THR Register comparing 160,000 cases," *Acta Orthop. Scand.*, Vol. 71, 2000, pp. 111–121.

[41] Bethea, S. B., J. S., DeAndrade, J. R., Fleming, L. L., Lindenbaum, S., et al. "Proximal femoral fractures following total hip arthroplasty," *Clin. Orthop. Relat. Res.*, Vol. 170, 1982, pp. 95–106.

[42] Beals, R. K., and Tower, S. S., "Periprosthetic fractures of the femur: An analysis of 93 fractures," *Clin. Orthop. Relat. Res.*, Vol. 327, 1996, pp. 238–246.

[43] Lemons, J. E., "Bioceramics: Is there a difference?" *Clin. Orthop. Relat. Res.*, Vol. 261, 1990, pp: 153–158.

[44] Webster, T. J., Ergun, C., Doremus, R. H., Seigel, R. W., and Bizios, R., "Enhanced functions of osteoblasts on nanophase ceramics," *Biomaterials*, Vol. 21, 2000, pp. 1803–1810.

[45] Webster, T. J., Siegel, R. W., and Bizios, R., "Nanoceramic surface roughness enhances osteoblast and osteoclast functions for improved orthopedic/dental implant efficacy," *Scripta Mater.*, Vol. 44, 2001, pp. 1639–1642.

[46] Webster, T. J., Ergun, C., Doremus, R. H., Seigel, R. W., and Bizios, R., "Enhanced osteoclast-like functions on nanophase ceramics," *Biomaterials*, Vol. 22, 2001, pp. 1327–1333.

[47] Gutwein, L. G., and Webster, T. J., "Increased viable osteoblast density in the presence of nanophase compared to conventional alumina and titania particles," *Biomaterials*, Vol. 25, 2004, pp. 4175–4183.

[48] Webster, T. J., Hellenmeyer, E. L., Price, R. L., "Increased osteoblast functions on theta+delta nanofiber alumina," *Biomaterials*, Vol. 26, 2005, pp. 953–960.

[49] Colon, G., Ward, B. C., Webster, T. J., "Increased osteoblast and decreased Staphylococcus epidermidis functions on nanophase ZnO and TiO2," *J. Biomed. Mater. Res. A.*, Vol. 78, No. 3, 2006, pp. 595–604.

[50] Hamadouche, M., Sedel, L., "Ceramics in oethopaedics," *J. Bone Joint Surg. Br.*, Vol. 82B, 2000, pp. 1095–1099.

[51] Willert, H. G., Bertram, H., Buchhorn, G. H., "Osteolysis in alloarthroplasty of the hip: The role of bone cement fragmentation," *Clin. Orthop.*, Vol. 258, 1990, pp. 108–121.

[52] Hulbert, S. F., Hench, L. L., Forbers, D., and Bowman, L. S., "History of bioceramics," *Ceram. Int.*, Vol. 8, No. 4, 1982, pp. 131–140.

[53] Chang, C. K., Wu, J. S., Mao, D. L., and Ding, C. X., "Mechanical and histological evaluations of hydroxyapatite-coated and noncoated Ti-6Al-4V implants in tibia bone," *J. Biomed. Mater. Res.*, Vol. 56, No. 1, 2001, pp. 17–23.

[54] Cook, S. D., Thomas, K. A., Kay, J. F., and Jarcho, M., "Hydroxyapatite-coated porous titanium for use as an orthopedic biologic attachment system," *Clin. Orthop. Relat. Res.*, Vol. 230, 1988, pp. 303–312.

[55] Catledge, S. A., Fries, M. D., Vohra, Y. K., Lacefield, W. R., Lemons, J. E., Woodard, S., and Venugopalan, R., "Nanostructured ceramics for biomedical implants," *J. Nanosci. Nanotechnol.*, Vol. 2, 2002, pp. 293–312.

[56] Chris Arts, J. J., Verdonschot, N., Schreurs, B. W., Buma, P., "The use of a bioresorbable nano-crystalline hydroxyapatite paste in acetabular bone impaction grafting," *Biomaterials*, Vol. 27, 2006, pp: 1110–1118.

[57] Huber, F.-X., Belyaev, O., Hillmeier, J., Kock, H.-J., Huber, C., Meeder, P.-J., and Berger, I., "First histological observations on the incorporation of a novel nanocrystalline hydroxyapatite paste OSTIM in human cancellous bone," *BMC Musculoskel. Disord.*, Vol. 7, 2006, p. 50.

[58] Donlan, R. M., and Costerton, J. W., "Biofilms: Survival mechanisms to clinically relevant microorganisms," *Clin. Microbiol. Rev.,* Vol. 15, 2002, pp. 167–193.

[59] Luck, M., Paulke, B. R., Schröder, W., Blunk, T., and Müller, R. H., "Analysis of plasma protein adsorption on polymeric nanoparticles with the different surface characteristics," *J. Biomed. Mater. Res.,* Vol. 39, 1998, pp. 478–485.

[60] Lopes, M. A., "Hydrophobicity, surface tension, and zeta potential measurements of glass-reinforced hydroxyapatite composites," *J. Biomed. Mater. Res.,* Vol. 45, No. 4, 1999, pp. 370–375.

[61] Cooke, F. W., "Ceramics in orthopedic surgery," *Clin. Orthop. Relat. Res.,* Vol. 276, 1992, pp. 135–146.

[62] Weertman, J. R., and Averback, R. S., "Mechanical properties," in A. S. Edelstein and R. C. Cammarata (eds.), *Nanomaterials: Sythesis, Properties and Applications*, London: Institue of Physics Publishing, 1998, pp. 323.

[63] Nieman, G. W., "Processing and mechanical behavior of nanocrystalline Cu, Pd and Ag," Ph.D thesis, Northwestern University, Chicago, IL, 1991.

[64] Nieman, G. W., Weertman, J. R., and Siegel, R. W., "Microhardness of nanocrystalline palladium and copper produced by inert-gas condensation," *Scr. Metall.*, Vol. 23, 1989, pp. 2013–2018.

[65] Nieman, G. W., Weertman, J. R., and Siegel, R. W., "Mechanical behavior of nanocrystalline Cu and Pd," *J. Mater. Res.,* Vol. 6, 1991, pp. 1012–1027.

[66] Wallbridge, N., and Dowson, D., "The walking activity of patients with artificial hip joints," *Eng. Med.,* Vol. 11, 1982, pp. 95–96.

[67] Dogan, C. P., and Hawk, J. A., "Role of composition and microstructure in the abrasive wear of high-alumina ceramics," *Wear,* 225–229, 1999, pp. 1050–1058.

[68] Sedel, L., Kerboul, L., Christel, P., Meunier, A., and Witvoet, J., "Alumina-on-alumina hip replacement," *J. Bone Joint Surg. Br.*, Vol. 72B, 1990, pp. 658–663.

[69] Xu, H. H. K., Wei, L., and Jahanmir, S., "Influence of grain size on the grinding response of alumina," *J. Am. Ceram. Soc.,* Vol. 79, 1996, pp. 1307–1313.

[70] Xu, H. H. K., Jahanmir, S., and Wang, Y., "Effect of grain size on scratch interactions and material removal in alumina," *J. Am. Ceram. Soc.,* Vol. 78, 1995, pp. 881–891.

[71] Xiong, F., Manory, R. R., Ward, L., Terheci, M., and Lathabai, S., "Effect of grain size and test configuration on the wear behavior of alumina," *J. Am. Ceram. Soc.,* Vol. 80, 1997, pp. 1310–1312.

[72] Mukhopadhyay, A. K., and Mai, Y-W., "Grain size effect on abrasive wear mechanisms in alumina ceramics," *Wear,* 162–164, 1993, pp. 258–268.

[73] He, Y., Winnubst, L., Burggraaf, A. J., and Verweij, H., "Influence of porosity on friction and wear of tetragonal zirconia polycrystal," *J. Am. Ceram. Soc.,* Vol. 80, No. 2, 1997, pp. 377–380.

[74] Litsy, A. S., and Spector, M., "Form and function of cartilage," in S. P. Sinmon (ed.), *Orthopedic Basic Science*, American Academy of Orthopedic Surgeons, 1994, p. 482.

[75] Catledge, S. A., Borham, J., Vohra, Y. K., Lacefield, W. R., and Lemons, J. E., "Nanoindentation hardness and adhesion investigations of vapor deposited nanostructured diamond films," *J. Appl. Phys.,* Vol. 91, No. 8, 2002, pp. 5347–5352.

[76] Grim, M. J., "Orthopedic biomaterials," in *Standard Handbook of Biomedical Engineering and Design*, New York: McGraw-Hill Professional Publishing, 2002, pp. 15.1–15.22.

[77] Vanzillotta, P. S., Soares, G. A., Bastos, I. N., Simão, R. A., and Kuromoto, N. K., "Potentialities of some surface characterization techniques for the development of titanium biomedical alloys," *Mater. Res.,* Vol. 7, No. 3, 2004, pp. 437–444.

[78] Tengvall, P., Textor, M., and Brunette, D. M., *Titanium in Medicine*, Springer, 2001.

[79] Steinemann, S. G., "Titanium: The material of choice?" *Periodontol.,* Vol. 7, 2000, pp. 7–21.

[80] Rompen, E., Domken, O., Degidi, M., Farias Pontes, A. E., and Piattelli, A., "The effect of material characteristics, of surface topography and of implant components and connections on soft tissue integration: A literature review," *Clin. Oral Implants Res.*, Vol. 17, No. 2, 2006, pp. 55–67.

[81] Thomas, T., *Rough Surfaces*, 2nd ed., London: Imperial College Press, 1999.

[82] Wennerberg, A., and Albrektsson, T., "Suggested guidelines for the topographic evaluation of implants surfaces," *Int. J. Oral and Maxillofac. Implants*, Vol. 15, No. 3, 2000, pp. 331–344.

[83] Hong, J., Andersson, J., Ekdahl, K. N., Elgue, G., Axen, N., Larsson, R., and Nilsson, B., "Titanium is a highly thrombogenic biomaterial: Possible implications for osteogenesis," *Thromb. Haemost.*, Vol. 82, No. 1, 1999, pp. 58–64.

[84] Eisenbarth, E., Meyle, J., Nachtigall, W., and Breme, J., "Influence of the surface structure of titanium materials on the adhesion of fibroblasts," *Biomaterials*, Vol. 17, 1996, pp. 1399–1403.

[85] Laurencin, C. T., Ambrosio, A. M. A., Borden, M. D., and Cooper, J. A., Jr., "Tissue engineering: Orthopedic applications," *Ann. Rev. Biomed. Eng.*, 1999, pp. 19–46.

[86] Unwin, P. S., "The recent advantages in bone and joint implant technology," in M. Ward-Close (ed.), *Cost-Effective Titanium Component Technology for Leading-Edge Performance*, Bury St. Edmunds and London: Professional Engineers Publishing, 2000.

[87] Dunn, M. G., and Maxian, S. H., "Biomaterials used in orthopedic surgery," in R. S. Greco (ed.), *Implantation biology: The Host Response and Biomedical Devices*, Boca Raton: CRC Press, 1994, pp. 229–252.

[88] Davidson, J. A., Poggie, R. A., and Mishra, A. K., "Abrasive wear of ceramic, metal and UHMPE bearing surfaces from third-body bone, PMMA bone cement and titanium debris," *Biomed. Mater. Eng.*, Vol. 4, 1994, pp. 213–229.

[89] Coombes, A. D., and Heckman, J. D., "Gel casting of resorbable polymers," *Biomaterials*, Vol. 13, 1992, pp. 217–224.

[90] Devin, J. E., Attawia, M. A., and Laurencin, C. T., "Three-dimensional degradable porous polymer-ceramic matrices for use in bone repair," *J. Biomater. Sci. Polym. Ed.*, Vol. 7, 1996, pp. 661–669.

[91] Mikos, A. G., Thorsen, A. J., Czerwonka, L. A., Bao, Y., and Langer, R., "Preparation and characterization of poly(l-lactic acid) foams," *Polymer*, Vol. 35, 1994, pp. 1068–1077.

[92] Thompson, R. C., Yaszemski, M. J., Powers, J. M., and Mikos, A. G., "Fabrication of biodegradable polymer scaffolds to engineer trabecular bone," *J. Biomater. Sci. Polym. Ed.*, Vol. 7, 1995, pp. 23–38.

[93] Chen, V. J., and Ma, P. X., "Nano-fibrous poly(l-lactic acid) scaffolds with interconnected spherical macropores," *Biomaterials*, Vol. 25, 2004, pp. 2065–2073.

[94] Miller, D. C., Thapa, A., Haberstroh, K. M., and Webster, T. J., "Enhanced functions of vascular and bladder cells on poly-lactic-co-glycolic acid polymers with nanostructured surfaces," *IEEE Trans. Nanobiosci.*, Vol. 1, No. 2, 2002, pp. 61–66.

[95] Thapa, A., Miller, D. C., Webster, T. J., and Haberstroh, K. M., "Nano-structured polymers enhance bladder smooth muscle cell function," *Biomaterials*, Vol. 24, 2003, pp. 2915–2926.

[96] Miller, D. C., Haberstroh, K. M., and Webster, T. J., "Mechanism(s) of increased vascular cell adhesion on nanostructured poly(lactic-*co*-glycolic acid) films," *J. Biomed. Mater. Res. A.*, Vol. 73, No. 4, 2005, pp. 476–484.

[97] Park, G. E., Pattison, M. A., Park, K., and Webster, T. J., "Accelerated chondrocyte functions on NaOH-treated PLGA scaffolds," *Biomaterials*, Vol. 26, No. 16, 2005, pp. 3075–3082.

[98] Liu, H., Slamovich, E. B., and Webster, T. J., "Increased osteoblast functions among nanophase titania/poly(lactide-*co*-glycolide) composites of the highest nanometer surface roughness," *J. Biomed. Mater. Res. A*, Vol. 78, No. 4, 2006, pp. 798–807.

[99] Vance, R. J., Miller, D. C., Thapa. A., Haberstroh, K. M., and Webster, T. J., "Decreased fibroblast cell density on chemically degraded poly(lactic-co-glycolic acid), polyurethane, and polycaprolactone," *Biomaterials*, Vol. 25, 2004, pp. 2095–2103.

[100] Li, J., "Polymeric hydrogels," in T. S. Hin (ed.), *Engineering Materials for Biomedical Applications*, World Scientific, 2004, pp. 7.1–7.18.

[101] Lee, K. Y., and Mooney, D. J., "Hydrogels for tissue engineering," *Chem. Rev.*, Vol. 101, No. 7, 2001, pp. 1869–1879.

[102] Lia, M., Guob, Y., Weib, Y., MacDiarmidc, A. G., and Lelkes, P. I., "Electrospinning polyaniline-contained gelatin nanofibers for tissue engineering applications," *Biomaterials*, Vol. 27, 2006, pp. 2705–2715.

[103] Stevens, M. M., and George, J. H., "Exploring and engineering the cell surface interface," *Science*, Vol. 310, 2005, pp. 1135–1138.

[104] Bonzani, I. C., George, J. H., and Stevens, M. M., "Novel materials for bone and cartilage regeneration," *Curr. Opin. Chem. Biol.*, Vol. 10, 2006, pp. 1–8.

[105] Kisiday, J., Jin, M., Kurz, B., Hung, H., Semino, C., and Zhang, S., et al., "Self-assembling peptide hydrogel fosters chondrocyte extracellular matrix production and cell division: Implications for cartilage tissue repair," *Proc. Natl. Acad. Sci.*, Vol. 99, 2002, pp. 9996–10001.

[106] Zhang, L., Ramsaywack, S., Fenniri, H., and Webster, T. J., "Helical rosette nanotubes as a biomimetic tissue engineering scaffold material," American Institute of Chemical Engineers Annual Meeting, San Francisco, CA, November 12–17, 2006.

[107] Yamada, S., Heymann, D., Bouler, J. N., and Daculsi, G. D., "Osteoclastic resorption of calcium phosphate ceramics with different hydroxyapatite/tricalcium phosphate ratios," *Biomaterials*, Vol. 18, 1997, pp. 1037–1041.

[108] Ramachandra, R. R., Roopa, H. N., and Kannan, T. S., "Solid state synthesis and thermal stability of HAP and HAP-β-TCP composite ceramic powders," *J. Mater. Sci. Mater. Med.*, Vol. 8, 1997, pp. 511–518.

[109] Tampieri, A., Celloti, G., Szontagh, F., and Landi, E., "Sintering and characterization of HA and TCP bioceramics with control of their strength and phase purity," *J. Mat. Sci. Mat. Med.*, Vol. 8, 1997, pp. 29–37.

[110] Cuneyt, A. T., Korkusuz, F., Timucin, M., and Akkas, N., "An investigation of the chemical synthesis and high-temperature sintering behavior of calcium hydroxyapatite (HA) and tricalcium phosphate (TCP) bioceramic," *J. Mat. Sci. Mat. Med.*, Vol. 8, 1997, pp. 91–96.

[111] Hench, L. L., "Bioceramics," *J. Am. Ceram. Soc.*, Vol. 81, 1998, pp. 705–1728.

[112] Homaeigohar, S. S. H., Shokrgozar, M. A., Sadi, A. Y., Khavandi, A., Javadpour, J., and Hosseinalipour, M., "In vitro evaluation of biocompatibility of beta-tricalcium phosphate-reinforced high-density polyethylene; an orthopedic composite," *J. Biomed. Mater. Res. A*, Vol. 75, No. 1, 2005, pp. 14–22.

[113] Murugan, R., and Ramakrishna, S., "Bioresorbable composite bone paste using polysaccharide based nano hydroxyapatite," *Biomaterials*, Vol. 25, No. 17, 2004, pp. 3829–3835.

[114] Esmaiel, J., "Aqueous based hydrogel/apatite nanocomposite scaffolds for guided bone regeneration, topical H #256d (TH014)—biomedical applications of nanotechnology (bionanotechnology)," *American Institute of Chemical Engineers Annual Meeting*, Cincinnati, OH, October 30–November 4, 2005.

[115] Quan, R., Yang, D., and Miao, X., "Biocompatibility of graded zirconia-hydroxyapatite composite," *Chin. J. Reparat. Reconstr. Surg.*, Vol. 20, No. 5, 2006, pp. 569–573.

[116] Kim, H. W., Georgiou, G., Knowles, J. C., Koh, Y. H., and Kim, H. E., "Calcium phosphates and glass composite coatings on zirconia for enhanced biocompatibility," *Biomaterials*, Vol. 25, No. 18, 2004, pp. 4203–4213.

[117] Kong, Y. M., Bae, C. J., Lee, S. H., Kim, H. W., and Kim, H. E., "Improvement in biocompatibility of ZrO_2-Al_2O_3 nano-composite by addition of HA," *Biomaterials*, Vol. 26, No. 5, 2005, pp. 509–517.

[118] Mistry, A. S., and Mikos, A. G., "Tissue engineering strategies for bone regeneration," *Advances Biochem. Eng/Biotechnol.*, Vol. 94, 2005, pp. 1–22.

[119] Webster, T. J., Waid, M. C., McKenzie, J. L., Price, R. L., and Ejiofor, J. U., "Nano-biotechnology: Carbon nanofibres as improved neural and orthopaedic implants," *Nanotechnology*, Vol. 15, 2004, pp. 48–54.

[120] Shi, X., Hudson, J. L., Spicer, P. P., Tour, J. M., Krishnamoorti, R., and Mikos, A. G., "Injectable nanocomposites of single-walled carbon nanotubes and biodegradable polymers for bone tissue engineering," *Biomacromolecules*, Vol. 7, No. 7, 2006, pp. 2237–2234.

[121] Supronowicz, P. R., Ajayan, P. M., Ullmann, K. R., Arulanandam, B. P., Metzger, D. W., and Bizios, R., "Novel current-conducting composite substrates for exposing osteoblasts to alternating current stimulation," *J. Biomed. Mater. Res.*, Vol. 59, Vol. 3, 2002, pp. 499–506.

[122] Tsuchiya, H., Macak, J. M., Muller, L., Kunze, J., Muller, F., Greil, P., Virtanen, S., Schmuki, P., "Hydroxyapatite growth on anodic TiO2 nanotubes," *J. Biomed. Mater. Res. A*, Vol. 77, 2006, pp. 534–541.

[123] Serro, A. P., and Saramago, B., "Influence of sterilization on the mineralization of titanium implants induced by incubation in various biological model fluids," *Biomaterials*, Vol. 24, 2003, pp. 4749–4760.

[124] Barber, T. A., Golledge, S. L., Castner, D. G., and Healy, K. E., "Peptide modified p (AAm-co-EG/AAc) IPNs grafted to bulk titanium modulate osteoblast behavior in vitro," *J. Biomed. Mater. Res.*, Vol. 64, 2003, pp. 38–47.

[125] Lu, X., Leng, Y., Zhang, X., Xu, J., Qin, L., and Chan, C. W., "Comparative study of osteoconduction on micromachined and alkalitreated titanium alloy surfaces in vitro and in vivo," *Biomaterials*, Vol. 26, 2005, pp. 1793–1801.

[126] Macak, J. M., Tsuchiya, H., Taveira, L., Ghicov, A., and Schmuki, P., "Self-organized nanotubular oxide layers on Ti-6Al-7Nb and Ti-6Al-4V formed by anodization in NH4F solutions," *J. Biomed. Mater. Res.*, 2005, Vol. 75, pp. 928–933.

[127] Winkelmann, M., Gold, J., and Hauert, R., "Chemically patterned, metal oxide based surfaces produced by photolithographic techniques for studying protein- and cell-surface interactions: I. Microfabrication and surface characterization," *Biomaterials*, Vol. 24, 2003, pp. 1133–1145.

[128] Balasundaram, G., and Webster, T. J., "A perspective on nanophase materials for orthopedic implant applications," *J. Mater. Chem.*, Vol. 16, 2006, pp. 3737–3745.

[129] Hoet, P. H., Bruske-Hohlfeld, I., and Salata, O. V., "Nanoparticles—known and unknown health risks," *J. Nanobiotechnol.*, Vol. 2, 2004, pp. 2–12.

[130] Tamura, K., Takashi, N., and Akasaka, T., "Effects of micro-/nano particle size on cell function and morphology," *Key Eng. Mat.*, Vol. 254, 2004, pp. 919–922.

CHAPTER 22

Bioinspired Engineered Nanocomposites for Bone Tissue Engineering

Esmaiel Jabbari and Alireza S. Sarvestani

22.1 Introduction

The clinical needs for bone regeneration are diverse and include applications arising from resection of primary and metastatic tumors, bone loss after skeletal trauma, primary and revision total-joint arthroplasty with bone deficiency, spinal arthrodesis, and trabecular voids following osteoporotic insufficiency fractures [1]. Roughly one million patients with skeletal defects require bone-graft procedures to achieve healing each year in the United States. Current clinical methods of treating skeletal defects involve bone transplantation or the use of synthetic materials to restore continuity. Autograft tissue is in limited supply, and morbidity at the harvest site is a concern. Allografts present risk of disease transfer and immunogenic response from the host tissue. Stress shielding and particulate wear are potential concerns with the use of nondegradable polymers such as poly(methyl methacrylate) (PMMA) for filling defects in orthopedics. Synthetic, degradable, bioinspired composites are an attractive option because of minimum risk of disease transfer and the potential to design scaffolds with mechanical and biological properties independent of each application. In addition, the degradation of the composite can be designed to coincide with the tissue regeneration rate for a given application.

22.2 Bone Structure

Bone is a composite material consisting of a collagenous and a mineral phase [2]. The inorganic component, made up of apatite crystals with surface-active carbonate and phosphate groups, contributes approximately 65 percent of the wet weight of the bone, and the collagenous phase contributes about 20 percent [3]. Noncollagenous water-soluble glycoproteins and proteoglycans control the water content of the bone matrix to approximately 15 percent [4]. The collagenous phase gives bone its form and contributes to its ability to resist bending, while the mineral component resists compression [5]. Bone that has been demineralized is flexible and resistant to fracture, while bone with its collagenous matrix removed is brittle, and a slight deformation factures it [6, 7].

Bone exhibits many orders of hierarchical structures from macroscopic to microscopic, submicroscopic, nanoscale, and subnanostructure length scales [8]. These levels of organization include the cancellous and cortical macrostructures, the osteons and single trabeculae microstructures (10 to 500 μm), the lamellae submicrostructures (1–10 μm), the fibrillar collagen and embedded mineral nanostructures (100 to 1,000 nm), and the molecular structure of the constituent elements like mineral, collagenous, and noncollagenous proteins [9]. Despite these hierarchical structures, the properties of the bone on the macroscale are determined at the nanoscale with nanoapatite calcium phosphate crystals embedded in the soft collagen fibrils, as shown schematically in Figure 22.1.

The plate-shaped crystals of carbonate apatite in bone are 2 to 3 nm thick and tens of nanometers long and wide. The aligned collagen fibers in the soft matrix have different levels of organization from 1 to 4 nm for fibrils to 50 to 70 nm for fibers and 150 to 250 nm for bundles. The collagenous phase of the bone plays a central role in regulation of collagen fibril mineralization, modulation and control of cell division, cell migration, differentiation and maturation, maintenance of matrix integrity, growth factor modulation, and the extent of mineral-collagen interactions [10, 11]. Noncollagenous proteins in the extracellular matrix (ECM) serve complex functions during bone formation and remodeling [12]. The collagenous phase provides bone with a medium for a multitude of domains on proteins in the ECM to maintain their bioactivity and to facilitate communication with the cellular environment [13, 14]. For example, osteonectin, a major

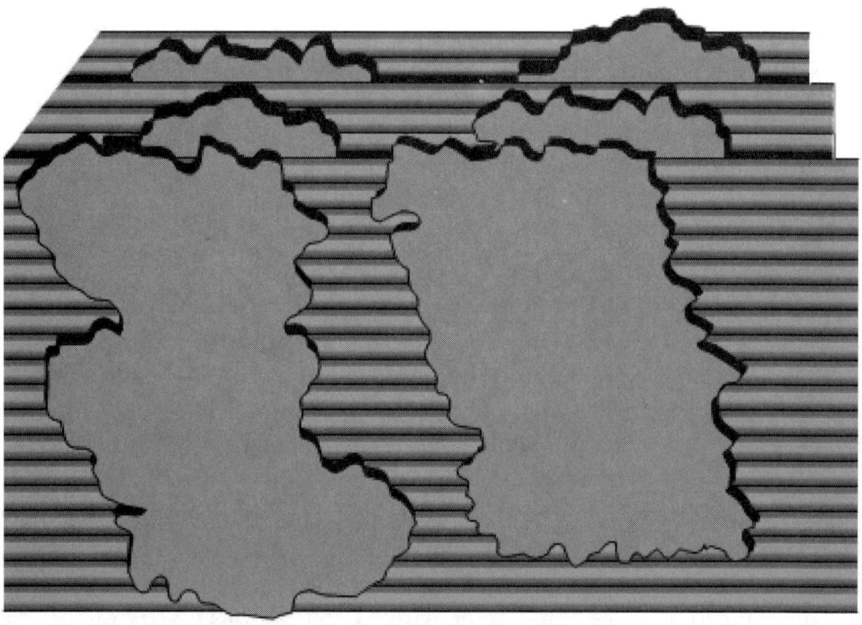

Figure 22.1 A schematic diagram of the collagen-mineral structure in calcified bone tissue. The bone tissue contains extended sheets of mineral positioned in length along the long axis of the collagen molecules and width along the periodic grooves perpendicular to the collagen axis and parallel to one another in the neighboring collagen layers. The sheets in the adjacent collagen layers fuse to form a thicker and broader collagen-mineral composite that can provide structural stability to the load-bearing bone tissue. The thickness of the mineral crystals is a few nanometers.

noncollagenous bone ECM protein, is involved in several functions, including linking of the mineral and collagenous phases, regulation of mineralization, cell motility, and matrix degradation. As another example, matrix metalloproteinases (MMPs) secreted by the cellular component are released to degrade the collagen network in bone remodeling [15, 16]. In fracture repair, progenitor stromal cells from the bone marrow or periosteum migrate to the fracture site as a result of degradation of the collagen network by MMPs and the release of migration factors [17]. An engineered material as a scaffold for bone regeneration should mimic the complexity of the bone's ECM by providing multiple bioactive groups to control ECM interactions and organized nanoscale structures with optimized arrangement and orientation of the components.

22.3 Degradable Polymers as Scaffolds for Bone Regeneration

The flexibility in the design of biodegradable polymers allows the synthesis of a wide range of polymers with varying mechanical, biologic, degradation, and rheologic properties. For instance, their mechanical and degradation properties can be manipulated by changing the molecular weight during synthesis and can thus be tailored to fit a particular application. Degradable polymers based on polyhydroxyalkanoates (PHAs) have been used as scaffolds for guided regeneration of femoral neck osteonecrosis, mandibula reconstruction, and fusion of spinal processes [18]. PHAs used in orthopedics include homopolymers of poly(L-lactic acid) (PLLA), poly(glycolic acid) (PGA), polycaprolactone (PCL), poly(trimethylene carbonate), poly(butylene terephthalate), poly(hydroxybutyrate), poly- (hydroxyvalerate), poly(dioxanone) (PDS), and their copolymers [19–24]. The development of drawn PLLA, self-reinforced PGA, self-reinforced PLLA, and PDS expanded the use of PHAs in load-bearing applications [25, 26]. An unsaturated lactide-co-glycolide macromer has been developed for fabrication of in situ crosslinkable scaffolds with degradation profile that can be tailored to a particular application [27]. A hybrid of poly(lactic-co-glycolic acid) (PLGA) and collagen has been used as membranes for guided bone regeneration to treat periodontesis [28]. A group of photopolymerizable poly(anhydrides) consisting of polymers of sebacic acid (SA) alone or copolymers of SA with 1,6-bis(p-carboxyphenoxy) hexane (CPH) have been developed [29, 30]. The combinations of different amounts of SA with CPH resulted in polymers with degradation properties designed for a specific application. Because these polymers degrade by surface erosion, bulk mechanical properties are maintained while undergoing degradation.

To develop materials for minimally invasive applications, in situ crosslinkable poly(propylene fumarate) (PPF), an unsaturated polyester that can be crosslinked through its fumarate bonds, has been developed. The maximal temperature during setting is 48°C, which is significantly less than that of PMMA bone cement [31–33]. A new macromer, poly(ε-caprolactone fumarate) (PCLF), is developed by replacing propylene glycol in PPF with poly(ε-caprolactone) [34]. Since each caprolactone unit provides four free rotating carbon-carbon bonds, the PCLF chain is more flexible than the PPF chain, which makes the fumarate and ester bonds more accessible for crosslinking and hydrolysis, respectively. The increased mobility of the fumarate bonds allows PCLF to self-crosslink without the addition of crosslinker. A star poly(dioxanone-co-glycolide) degradable copolymer endcapped with lysine diisocyanate crosslinker has been developed that conforms to the geometry of the bone defect by expansion of the implant [35]. After injection, the isocyanate groups at the end of each arm of the macromer react with water from the surrounding tissue to form a polyurea/urethane network and the liberation of carbon dioxide. Poly(ethylene glycol) hydrogels modified with arginine-glycine-aspartic acid (RGD) adhesive peptide sequences, coupled with photopolymerization, have been used as encapsulation matrices for osteoblasts [36]. Increased cell cytoskeleton reorganization was observed with the highest RGD concentration. A hydrogel composed of poly(aldehyde guluronate) as the macromer and adipic acid dihydrazide as the crosslinker has been developed as a carrier for osteoblasts [37]. Alginate hydrogels grafted with adhesion ligands have also been used for encapsulation of calvarial osteoblasts [38]. A novel in situ crosslinkable, enzymatically degradable poly(lactide-co-glyoclide ethylene oxide) macromer has been developed and used for encapsulation of bone marrow stromal cells [39]. These studies demonstrate the importance of coupling adhesion ligands to the hydrogel carrier for cell spreading and focal-point adhesion.

Immiscibility between two liquid phases has been used to develop interbody spine fusion elements that harden in situ after injection in the intervertebral lumen [40]. The first hardening phase was PMMA, while the second resorbable phase was a viscoelastic solution of polyvinyl alcohol or dextran. This study demonstrated that phase separation can be used to fabricate in situ hardening scaffolds with interconnected pore morphology, but persistence of the PMMA within the disc space is a major limitation.

22.4 Degradable Composite Scaffolds for Bone Regeneration

Bioactive ceramics have been used as scaffolds for filling bone defects [41]. Calcium phosphates such as hydroxyapatite (HA) and tricalcium phosphate have been shown to promote bone ingrowth and are biocompatible and osteoconductive. Their major drawback is that their initial strength is less than that of cancellous bone, which leads to difficulty in maintaining the composite within the defect. To overcome this issue, bioactive ceramics that harden in vivo–like dicalcium phosphate dihydrate, calcium magnesium phosphate, octacalcium phosphate, and calcium-deficient hydroxyapatite (CDHA) have been developed [42, 43]. Of these, CDHA promotes bone growth, has compressive strength higher than that of cancellous bone, and has been clinically tested [44, 45]. Potential drawbacks of calcium phosphates are fatigue fracture and their very low strength in shear and tension. This limits their use in bone regeneration to fractures that are subject to relatively uniform compressive loading [46].

Bioactive ceramics in particulate form or as mineralized bone fibers are often combined with biodegradable polymers. Natural polymers reinforced with calcium phosphates include collagen [47, 48], chitosan [49], and silk [50]; synthetic polymers include PLLA [51–53], PLGA [54], PPF [55], poly(ε-caprolactone) [56], and blends of PLGA with collagen [57]. These composite materials exhibit compressive strengths in the range of 2 to 30 MPa, suitable for replacement of human trabecular bone. Addition of HA to drawn PLLA not only improved biodegradability profile and strength, but it also increased compatibility with the bone tissue and reduced inflammation caused by the transient production of lactic and glycolic acid. Recent in vitro studies have demonstrated that the reinforcement of degradable polymers with calcium phosphates improves attachment, proliferation, and differentiated osteoblastic function of marrow stromal cells.

Cell-based approaches using osteoblasts [58], bone marrow–derived osteoprogenitor cells [59], and stem cells [60] are utilized for bone reconstruction. Cell-based approaches include implantation of the cell-scaffold hybrid or implantation after ex vivo culture for proliferation and differentiation of the osteoprogenitor cells [61–63]. The rate of ECM deposition and mineralization in cell-based systems depended on the presence of soft tissue for blood supply and the degradation rate, porosity, pore size, and extent of pore interconnectivity of the scaffold.

Although remarkable progress has been made in developing biodegradable composite matrices as scaffolds for bone regeneration [64, 65], these substrates do not provide signals to the migrating bone marrow stromal cells to differentiate into multiple mature phenotypes that are different from their tissue of origin. Furthermore, these composites cannot be used for reconstruction of load-bearing bone (i.e., femoral or tibial epiphyseal fractures) because they fail to mimic the bone's natural micro- and nanostructure. Due to the lack of vacularization, the formed mineralized tissue is prominently in the outer regions of the implanted scaffold.

22.5 Collagen nanostructure and its effect on differentiation of bone marrow stromal cells

Bone is a dynamic, highly vascularized tissue with the unique capacity to heal and adaptively remodel. Bone tissue has an elaborate vascular system. Even in the case of dense cortical bone tissue, the organization of the vascular canals ensures that every bone cell lies within 300 μm from a blood vessel. Disruption of the blood supply to bone due to disease or injury can cause necrosis and impairs repair [66]. For example, Diaphyses of long bones are supplied with blood from three sources, namely the nutrient arteries, arteries that penetrate the epiphysis and metaphysis, and the periosteal arteries. Thus, adequate vascularization is a prerequisite for generation of high quality bone. Avascular engineered cellular constructs have to rely on molecular diffusion and blood perfusion for nutrient supply which is a major limitation for transfer of tissue engineered constructs from in vitro to in vivo environment [67]. Since diffusion can provide nutrient supply within the range of 300 μm, cell viability in the center of constructs filling large bone defects is limited by lack of or suboptimal vascularization [68,69]. For example, a significant loss of osteoblasts, seeded in bovine cancellous bone matrix, was observed one week after implantation in a calvarial defect [70]. Therefore, induction of vascularization in tissue engineered constructs is critical to successful regeneration of large bone defects.

Engineering of bone tissue requires the interaction of osteo-progenitor cells seeded in an osteoconductive scaffold which is embedded with differentiation and growth factors to guide the process of regeneration. In the bone tissue, a variety of different cell types including osteoblasts, chondroblasts, marrow cells, vascular cells, osteoctytes, and osteoclasts are involved in the process of bone formation, maintenance, and remodeling [71]. While osteoclasts, derived from the hematopoietic component of the bone marrow, play a key role in bone remodeling [72], other cell types, derived from the stromal cell network, are involved in the bone formation process. For example, in bone repair, periosteal cells from the periosteum [73], stromal cells from the bone marrow [74,75], osteo-progenitor and chondroblast cells from the surrounding bone tissues [76], migrate to the regenerating site as a result of gradients of differentiation factors formed. These progenitor cells differentiate and pass various lineage states while proliferating rapidly to become preosteoblasts and finally osteoblasts to produce a mineralized extracellular matrix [77]. Adult stem cells, isolated from bone marrow, adipose tissue, muscle tissue, umbilical core blood, and even peripheral blood, and seeded in engineered scaffolds, have been used for bone regeneration [78]. In particular, pluripotent bone marrow stromal cells, easily isolated and expanded by variety of tissue culture techniques, are very attractive as a cell source for bone tissue regeneration. Despite the low density of BMS cells, isolated from the patient's bone marrow, and the decrease in their proliferative capacity with donor age, recent developments demonstrate that their proliferative capacity and osteogenic differentiation potential can be enhanced by cultivation in the presence of fibroblast growth factor-2 (FGF-2), by transfection with the telomerase enzyme, or by cultivation on extracellular matrices like denatured collagen type I [79]. Furthermore, the adherent stromal cells of the bone marrow contain vascular and neural progenitor cells as well as mesenchymal progenitor cells, in which the latter can be differentiated into skeletal cells like osteoblasts [80]. Previous work has demonstrated that these vascular progenitor cells and other bone marrow derived cells contribute to the growth of new vessels [81,82]. Therefore, stromal cells, isolated from the bone marrow cavity, remain the best choice as a cell source for bone tissue engineering with adult stem cells.

Differentiation of BMS cells towards osteogenic lineage in vitro is facilitated in the presence of dexamethasone, β-glycerol phosphate and ascorbic acid [83,84]. These differentiated cells show increased tissue-specific alkaline phosphatase activity and formation of mineralized bone-like nodules containing matricellular deposits [85]. Differentiation of stem cells not only depends on the growth media components but also on the type of substrate on which the cells are cultured [86,87]. It has been demonstrated that differentiation and formation of mineralized tissue is facilitated when BMS cells are seeded on 3-D type I collagen sponge and cultured in osteogenic media [88, 89]. Grafting a flexible type I collagen foam (70% porosity and 30–60 μm pore size) onto an injured myocardium induced angiogenesis and neo-vessel formation [90]. It has been shown that rats BMS cells, seeded on collagen (type I)/ glycosaminoglycan scaffold characterized by a narrow pore size distribution and equiaxed pore shape, supported osteogenesis and matrix mineralization in osteogenic media while the exposure of the BMS seeded scaffolds to chondrogenic factors supported chondrogenic differentiation and formation of collagen type II-rich extracellular matrix [91].

Rat BMS cells were harvested from the bone marrow of young adult male Wister rats [92]. After euthanasia, the femurs and tibias were aseptically excised from the hind limbs, cleaned of soft tissue, and washed in Dulbecco's Modified Eagle Medium (DMEM; 4.5 g/l glucose with L-glutamin and without sodium pyruvate) containing 5 times higher than for cell culture concentration of gentamicin sulfate (GS; 100 μg/ml) to avoid contamination during the harvest. Plugs of marrow were extracted by cutting the distal ends of tibias and proximal ends of femurs to expose the marrow cavity. Small holes were bored in the other end of each bone with an 18 gauge needle. The marrow was flushed out with 5 ml of cell basal (primary) media which consisted of DMEM supplemented with 10% fetal bovine serum (FBS), 20 μg/ml fungizone, and 20 μg/ml GS. The heat-inactivated FBS was screened for compatibility with rat bone marrow stromal cells. Plugs were dispersed by shear and resulting suspensions were passed through 22 gauge needle to obtain fine single cells suspension. Cell clumps were broken up by repeatedly pipetting and the cell suspensions were combined and centrifuged at 200×g for 5 min. The resulting supernatant was aspirated and cell pellets re-suspended in 12 ml basal media and aliquoted into T-75 flasks. The flasks were subsequently maintained in a humidified 5% CO2 incubator at 37°C. Cultures were washed with PBS and replaced with fresh media at 3 and 7 days to remove haematopoetic and other unattached cells from the flasks. The cells were rinsed with PBS, enzymatically lifted with 25 µl/cm2 of 0.05% trypsin/ 0.53 mM Ethylenediaminetetraacetic acid disodium salt (EDTA), and centrifuged at 400×g for 5 min. After two passages, attached marrow stromal cells were completely devoid of any non-adhering haematopoietic cells.

3-D collagen type I tubular scaffolds, with fiber angle of 18° relative to the tube long axis and pore size of 10 μM (Figure 22.2), were used to investigate differentiation of BMS cells cultured in osteogenic media [93,94]. BMS cells were seeded into the collagen tubes at a density of 2x106 cells and cultured in osteogenic media (10 mM sodium β-glycerol phosphate, 50 μg/ml L-ascorbic acid, and 10-8 M dexamethasone). At time intervals of 3, 6, and 9 days, the cell cultures were terminated and processed for RT-PCR, immunohistochemical and cytochemical analyses. Lineage specific proteins were localized by immunofluorescence using confocal laser scanning microscopy and mRNA transcript analysis was performed by Real-Time quantitative PCR (RT-qPCR).

The staining patterns for BMS cells seeded in collagen type 1 tubes after 9 days are shown in Figure 22.3 [94]. The early and late markers for bone related proteins

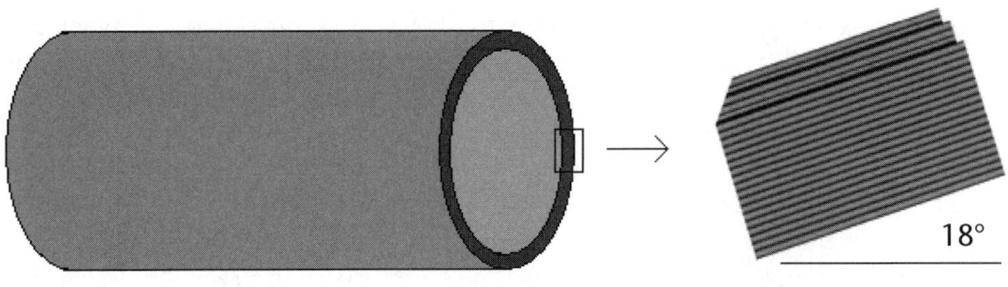

Figure 22.2 2-D collagen type I tubular scaffolds with fiber angle of 18° relative to the tube long axis.

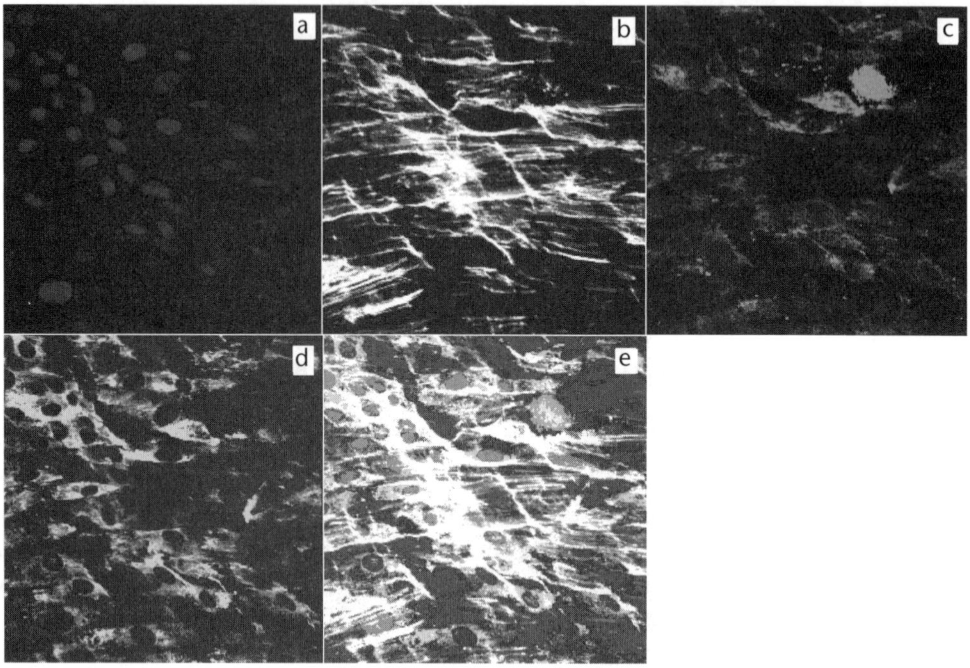

Figure 22.3 Expression pattern of osteogenic markers: (c) Osteonectin, (d) Osteopontin, and (e) merged image [94]. This figure shows sheets of differentiated marrow stromal cells seeded into the collagen tubes at a density of 2 x 106 cells and cultured in osteogenic media (10 mM sodium β-glycerol phosphate, 50 μg/ml L-ascorbic acid, and 10-8 M dexamethasone). Cells were stained for nuclei (blue, DAPI) and actin (green, Alexa 488 phalloidin).

included osteonectin (c) and osteopontin (d). BMS cells displayed the distinctive phenotype (cuboidal to polygonal mononuclear cells) and biosynthetic activity of terminally differentiated osteoblasts. These uniform sheets of cells were alkaline phosphatase positive and formed multiple microscopic foci of multilayered cellular organization as a result of cellular aggregation (images b and e) [94].

The gene expression profile of the BMS cells seeded on the collagen scaffolds was compared with those seeded on Petri dishes, as a reference substrate. The mRNA transcript levels of osteogenic markers for BMS cells seeded in collagen tube as a function of culture time are shown in Figures 22.4 [94]. The expression pattern of key osteogenic markers significantly differed in response to basal and osteogenic media. Tubes containing BMS cells when cultured in osteogenic media demonstrated an initial upregulation of osteopontin which returned to baseline level following 6 days. In contrast, tubes with BMS cells cultured in ostegenic media showed a sustained upregulation of osteocalcin beyond day 3. The Alkaline phosphatase activity and calcium content (alizarin red staining) significantly increased over the observed period of time in osteogenic medium.

Scaffolds cultured in osteogenic media showed areas of nodule formation by Alizarin red staining after 9 days. Surprisingly, sheets of α-smooth muscle positive cells were observed in the tubes seeded and cultured with BMS cells (Figure 22.5) [94]. Moreover, nascent capillary-like vessels were also seen amidst the osteoblasts in osteogenic cultures (Figure 22.5).

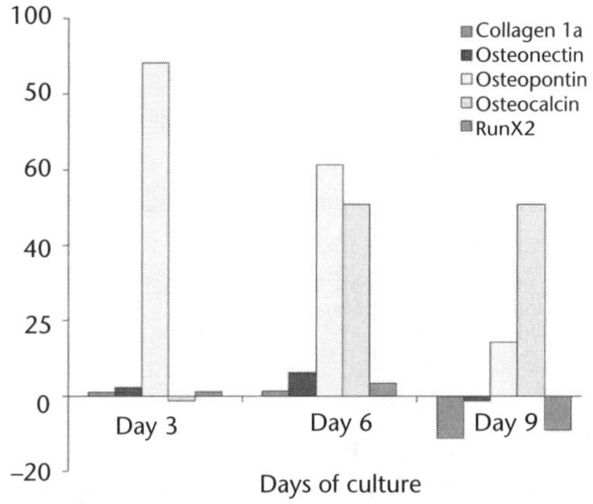

Figure 22.4 Quantitative real-time polymerase chain reaction (RT-qPCR) fold difference of various osteogenic markers, collagen 1α (Col1a1), osteonectin (Sparc), osteopontin (Spp1), osteocalcin (Bglap2) and runx2 (Runx2) as a function of time cultured in collagen type I tubular scaffold [94]. Tubes with BMS cells cultured in ostegenic media showed a sustained upregulation of osteocalcin beyond day 3. The Alkaline phosphatase activity and calcium content (alizarin red staining) significantly increased over the observed period of time in osteogenic medium.

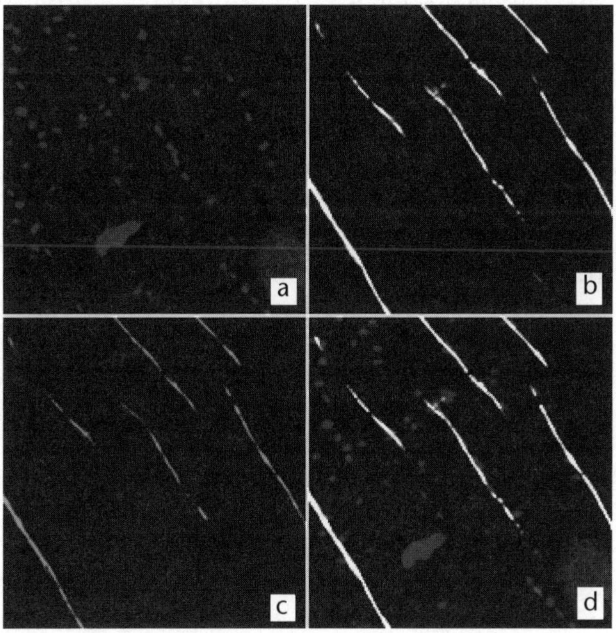

Figure 22.5 Expression pattern of vasculogenic markers α-SMA and tomato lectin (TL) [94]. In the tubes, nuclei appeared aligned (image a, DAPI). Localization of vascular progenitor cells cultured on the collagen tube scaffold by confocal microscopy is seen in composite image d. Sections of a day 9 tube demonstrated capillary-like structures positive for α-SMA (b) and tomato lectin (c). These elongated cord-like structures showed strong colocalization of α-SMA and actin (merged, d).

Bone marrow stromal cells (BMS) are fundamentally a heterogeneous population of cells with multilineage differentiation potential. When BMS cells are seeded in a three-dimensional tubular scaffold fabricated from aligned type I collagen strands, and cultured in osteogenic media, they underwent simultaneous maturation and differentiation into osteoblastic and vascular cell lineages [94]. In addition these cells produced mineralized matricellular deposits. The expression pattern of osteogenic markers in the collagen scaffolds demonstrated an initial upregulation of osteopontin which returned to baseline level following 6 days. In contrast, osteocalcin showed a sustained upregulation beyond day 3. Transcripts for type I collagen and osteonectin that are associated with the formation of the extracellular matrix were actively expressed until day 6 and gradually down regulated with their transcript being maintained at a low basal level during subsequent stages of osteoblastic differentiation. The alkaline phosphatase activity and calcium content significantly increased over the observed period of time in osteogenic medium [94]. BMS cells seeded in the collagen scaffold underwent simultaneous maturation and differentiation into osteogenic and vascular cell lineages and produced mineralized matricellular deposits. Sheets of abundant Pecam (CD31), Flk-1 (VEGFR-2), tomato lectin (TL/LEL) and α-smooth muscle actin (α-SMA) positive cells were observed in the collagen scaffolds. In our in-vitro study, concurrent differentiation of BMS cells to osteogenic and vascular lineages demonstrates that the substrate has a profound effect on guiding the differentiation pathway of BMS cells. It appears that the subnano-, nano, and microstructural organization of the 3-D collagen scaffold and the supplements of osteogenic media not only enhanced the osteoblastic differentiation but also supported vascularization. These findings demonstrate that the structural organization of the scaffold at the nano- and micro- scale affects the maturation and differentiation of BMS cells to multiple pathways including osteogenesis and vasculogenesis when cultured in osteogenic media.

22.6 Biomimetic Hydrogel Nanocomposites for Bone Regeneration

The scaffold for bone regeneration should provide structural support to withstand in vivo stresses. Hydrogels reinforced with calcium phosphate apatite fillers are very attractive for bone-tissue regeneration [95, 96]. They can exhibit the typical mechanical properties of the bone and replicate bone microstructure while possessing excellent biocompatibility [97]. Previous experimental studies imply that the matrix structure and molecular weight, filler size and volume fraction, and interfacial strength between the polymer and filler are the major factors affecting the viscoelastic properties of the polymer/ceramic biocomposites [98–100]. In general, composite systems with nanoparticles (<100 nm) exhibit a higher stiffness compared with those with microparticles at the same volume fraction (Figure 22.6) [98]. It has been shown that the mechanical performance of the composites is significantly improved by increasing the interfacial adhesion between the polymer and dispersed particles [99, 100]. Moreover, experiments show that while filled systems with large particles (>1 μm) exhibit no nonlinearity in a wide range of deformations, in nanofilled composites, nonlinearity can emerge even at small strain amplitudes [101, 102]. Molecular models predict that the viscoelastic response of hydrogels reinforced with nonpercolated nanoparticles is significantly affected by the filler size and matrix-particle energetic interactions [103].

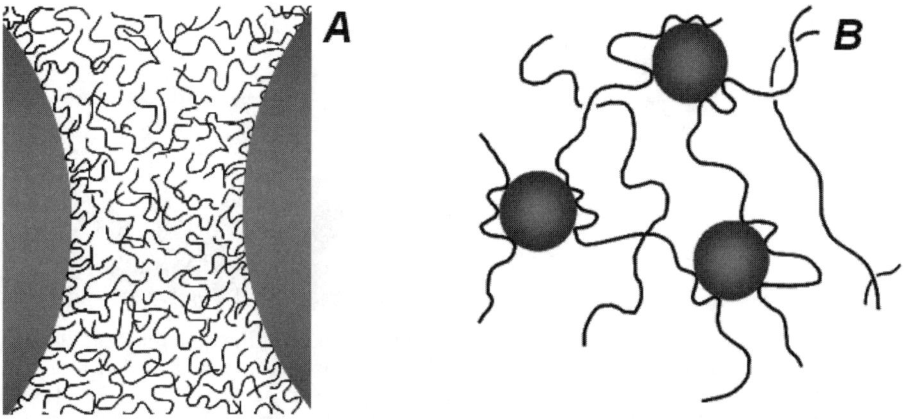

Figure 22.6 Schematic diagrams for comparing matrix-mineral scaling relations in (a) micro- and (b) nanoparticulate composites. (a) In microcomposites, the radius of gyration of the chains in the matrix is orders of magnitude smaller than the size of the apatite particles, and the presence of filler particles does not affect the relaxation times of the chains. (b) In nanocomposites, the radius of gyration of the chains is of the same order of magnitude as the size of the particles, which increases the relaxation time of the chains interacting with multiple filler particles.

Bone is a composite of collagen, a protein-based gelatinous network, and inorganic apatite crystals. The combination of a hard inorganic nanoparticulate phase and an elastic hydrogel network provides bone with unique mechanical properties while facilitating communication with the cellular environment [14, 104, 105]. The gelatinous phase gives bone its form and contributes to its ability to resist bending, while the mineral component resists compression [5]. Unique factors that contribute to the toughness of bone are the presence of nano-size apatite crystals, a dense network of collagen fibers on which the mineral crystals are deposited, and the acidic proteins with the ability to link the calcium-binding apatite crystals to collagen fibers to form a completely crosslinked network [106, 107]. One of the noncollagenous proteins in the gelatinous matrix with bone-specific functions is osteonectin, or bone connector, because it has a strong affinity for both collagen and hydroxyapatite, and it is believed to be a bone-specific nucleator of mineralization [108]. In the process of cartilage-to-bone transition during fracture repair, one of the key initial events is the switch from osteonectin to osteopontin mRNA in chondrocyte maturation, followed by the expression of mRNAs for osteonectin, and osteocalcin in the ossification front [109, 110].

It is believed that the first seventeen NH_2-terminal amino acids of osteonectin are responsible for binding to the bone collagen network [111], while a glutamic acid–rich sequence binds to HA nanoparticles (see Figure 22.7) [11]. Among all noncollagenous proteins, osteonectin has the highest affinity for calcium ions [106]. In the presence of osteonectin, a greater supersaturation of inorganic phosphates is required for precipitation of calcium phosphate and formation of hydroxyapatite under physiological conditions [108]. Furthermore, the apatite formed in the presence of osteonectin has the smallest crystal size [112]. These observations suggest that the switch from osteopontin to osteonectin mRNA in osteogenic cells of the ossification front is to limit the size of the apatite crystals and to produce a crosslinked composite by linking the collagen network to the mineral phase.

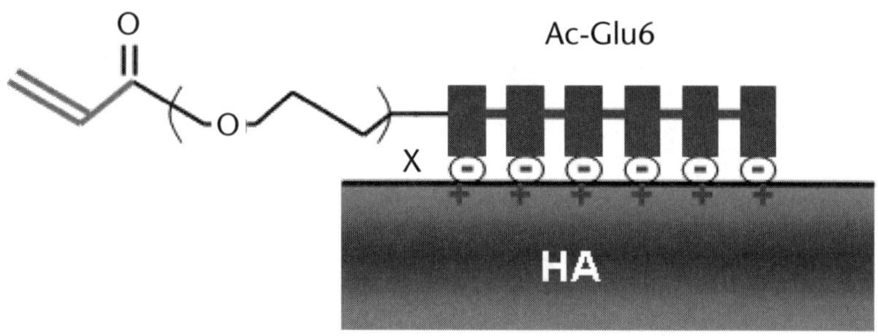

Figure 22.7 Schematic diagram to illustrate the binding of the glutamic acid–rich sequence of osteonectin protein in the bone ECM to HA nanoparticles. A glutamic acid–rich peptide (a sequence of six glutamic acids), functionalized with an acrylate group (Ac-Glu6) for covalent attachment to the matrix, was synthesized by solid-phase-Fmoc chemistry.

To test the effect of ionic interactions between the apatite crystals and gelatinous matrix, a glutamic acid–rich peptide (a sequence of six glutamic acids) derived from osteonectin, functionalized with an acrylate group (Ac-Glu6) for covalent attachment to the matrix, was synthesized by solid-phase Fmoc chemistry [113]. A similar procedure was used to synthesize the neutral Gly-Gly-Gly-Gly-Gly-Gly (Gly6) peptide and positively charged Lys-Lys-Lys-Lys-Lys-Lys (Lys6) peptide sequences with an acrylate group at one chain end (Ac-Gly6 and Ac-Lys6, respectively). The Ac-Glu6 sequence was attached to the surface of hydroxyapatite microparticles (mean diameter 50 μm) or nanoparticles (mean diameter 50 nm) by electrostatic interactions, as shown in Figure 22.7. Next, a novel in situ crosslinkable terpolymer poly(lactide ethylene oxide fumarate) (PLEOF) consisting of ultra-low-molecular-weight poly(L-lactide) and poly(ethylene glycol) blocks linked by fumaric acid units was added to the suspension [107, 114]. The mixture was injected between the parallel plates of a rheometer, and the gelation kinetics was measured.

Figure 22.8 shows the dependence of the shear modulus of the composites on the size of the dispersed HA particles. Nanoapatite composites (treated and untreated) displayed far larger stiffness compared with microcomposites at the same volume fraction. The modulus of the composites with micron-size particles did not appreciably change with the addition of Ac-Glu6. The contribution of hydrodynamic effect to the modulus of the composite can be predicted by the Guth-Smallwood equation [115], $G_0'(\Phi) = G_0'(0)(1+2.5\Phi)$, where $G_0'(0)$ and $G_0'(\Phi)$ are the storage modulus of the gel and that of the composite, respectively, and Φ is the HA volume fraction. The storage modulus of the composites prepared with micron-size particles can be reasonably predicted by the Guth-Smallwood equation. However, the large difference between the experimental and predicted results for composites prepared with nano-size HA implies that the reinforcement cannot be explained solely by the hydrodynamic effect in nanoparticulate systems.

To provide evidence for specific energetic affinity between the Ac-Glu6 peptide and HA and its effect on the modulus, similar measurements were performed on the composites treated with equal molar concentrations of Ac-Gly6 and Ac-Lys6 in place of Ac-Glu6. Contrary to the Glu6 peptide, Gly6 and Lys6 are neutral and positively charged, respectively, and they do not have ionic interactions with HA nanoparticles. HA/Ac-Gly6 and HA/Ac-Lys6 nanocomposites with 9 vol% apatite

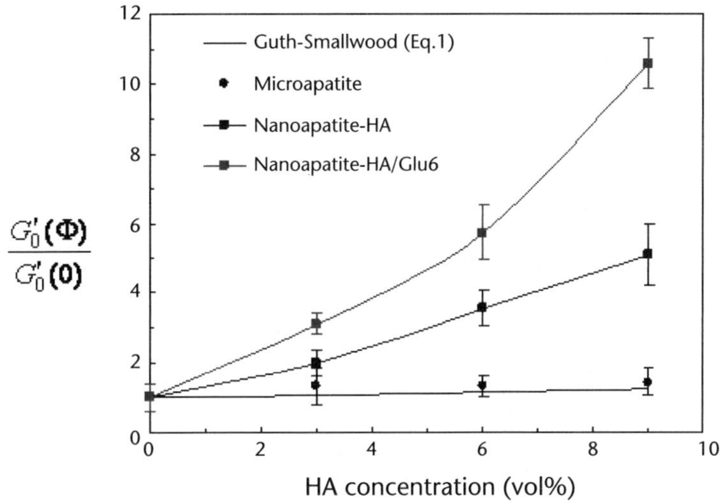

Figure 22.8 The dependence of the shear modulus of the PLEOF/Ac-Glu6/apatite composites on the size of the dispersed HA particles. Nanoapatite composites (treated and untreated) displayed far larger stiffness compared with microcomposites at the same volume fraction. The modulus of the composites with micron-size particles did not appreciably change with the addition of Ac-Glu6.

did not show a significant change in storage shear modulus compared to that without HA surface treatment, as shown in Figure 22.9. These results demonstrate that the increase in overall viscoelastic response of the nanocomposite is specific to the Ac-Glu6 peptide, and the reinforcement is amplified as the size of the nanoparticles is reduced from 5 μm to 50 nm.

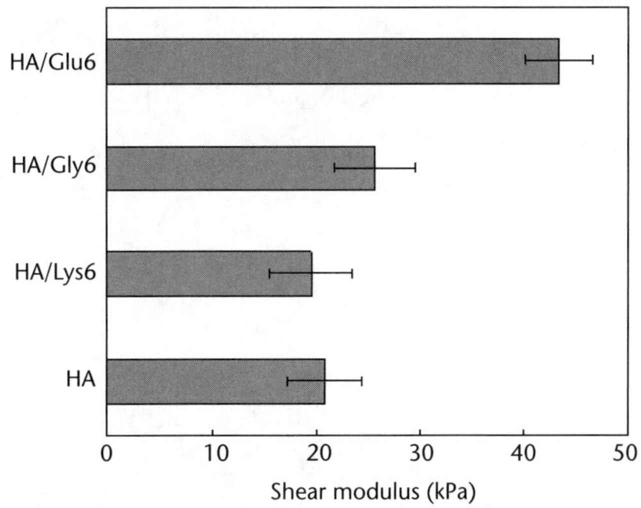

Figure 22.9 Comparison of the shear storage modulus of the nanocomposites treated with the Ac-Glu6 peptide to that of those treated with equal molar concentrations of Ac-Gly6 and Ac-Lys6. HA/Ac-Gly6 and HA/Ac-Lys6 nanocomposites with 9 vol% apatite did not show a significant change in storage shear modulus compared to that without HA surface treatment.

The ionic bonds formed by the interaction of Ac-Glu6 with HA have a relatively higher adsorption energy (compared to polar interactions in the absence of Ac-Glu6) and longer residence time on the HA particles, as shown in Figure 22.10. Thus, more matrix-filler crosslinks are expected to exist in Ac-Glu6-treated samples at any instant in time compared to untreated samples. This gives rise to a higher concentration of (physical) crosslinks in the vicinity of the apatite particles, leading to the higher modulus of the HA/Ac-Glu6 composites.

The transmission electron microscopy micrographs of HA dispersions in untreated and Ac-Glu6-treated samples support these conclusions. These micrographs demonstrate that treatment of HA nanoparticles with the Ac-Glu6 peptide gives rise to a more homogeneous microstructure of small aggregates or individual HA nanoparticles. These results demonstrate that one of the factors contributing to the toughness of bone is the crosslinking of the bone's gelatinous phase (simulated with the PLEOF hydrogel) and mineral phase (simulated by HA nanoparticles) by a specific peptide sequence derived from osteonectin (simulated by Ac-Glu6 with an acrylate group for attachment to the PLEOF network and the Glu6 peptide for bonding to HA particles). These findings demonstrate that the reinforcement effect of the Glu6 peptide, a sequence in the terminal region of osteonectin, is modulated by the size of the apatite crystals. These findings can be used to develop biomimetic composites for bone-tissue regeneration in load-bearing applications.

22.7 Conclusion

Bone tissue is composed of a heterogeneous mixture of cell types (osteoid and endothelial lineages) embedded within a mineralized ECM, which serves to maintain its

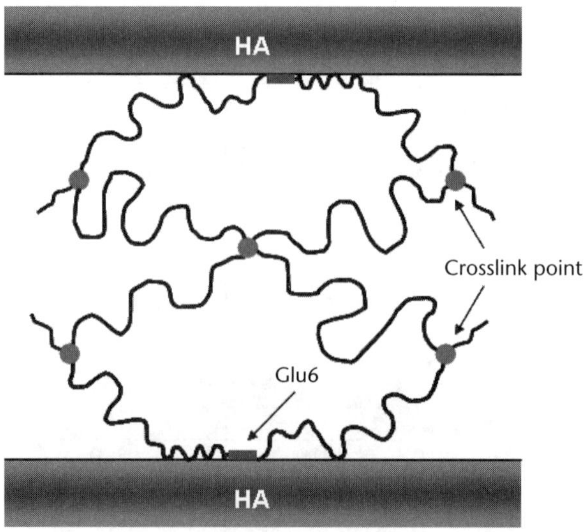

Figure 22.10 A schematic diagram showing the ionic bonds formed by the interaction of Ac-Glu6 with HA, with a relatively higher adsorption energy (compared to polar interactions in the absence of Ac-Glu6) and longer residence time on the HA particles. These ionic interactions give rise to a higher concentration of (physical) crosslinks in the vicinity of the apatite particles, leading to a higher modulus of the HA/Ac-Glu6 composites.

structural integrity. Unique factors that contribute to the toughness of bone are the presence of nano-size apatite crystals, a dense network of collagen fibers on which the mineral crystals are deposited, and the acidic proteins with the ability to link the calcium-binding apatite crystals to collagen fibers to form a completely crosslinked network. Although remarkable progress has been made in developing biodegradable composite matrices as scaffolds for bone regeneration, these substrates do not provide signals to the migrating bone marrow stromal cells to differentiate into multiple mature phenotypes that are different from their tissue of origin. Furthermore, these composites cannot be used for reconstruction of load-bearing bone (i.e., femoral or tibial epiphyseal fractures) because they do not mimic the bone's natural micro- and nanostructure. In this chapter, we demonstrate that the structural organization of the scaffold at the nano- and microscale, like the alignment of the collagen type I fibrils, affects the maturation and differentiation of BMS cells to multiple differentiation pathways, including osteogenesis and vasculogenesis, when cultured in osteogenic media. We further demonstrate that the reinforcement effect of a specific peptide sequence derived from osteonectin, a noncollagenous protein in the bone ECM with a strong affinity for both collagen and hydroxyapatite, is modulated by the size of the apatite crystals. These findings can be used to develop biomimetic composites by controlling biological (cell differentiation) and mechanical (structural stability) functions for bone-tissue engineering.

Acknowledgments

Preparaton of this chapter was supported by grants from the AO (Arbeitgemeinschaft Fur Osteosynthesefragen) Foundation and the Aircast Foundation.

References

[1] Yaszemski, M. J., Oldham, J. B., Lu, L., and Currier, B. L., "Clinical needs for bone tissue engineering technology," in J. E. Davis (ed.), *Bone Engineering*, Toronto: em squared, 2000, pp. 541–547.

[2] Glimcher, M. J., "The nature of the mineral component of bone and the mechanisms of calcification," in F. L. Coe and M. J. Favus (eds.), *Disorders of Bone and Mineral Metabolism*, New York: Raven Press, 1992, pp. 265–286.

[3] Herring, G. M., "Methods for the study of glycoproteins and proteoglycans of bone using bacterial collagenase: Determination of bone sialoprotein and chondroitin sulfate," *Calcif. Tissue Res.*, Vol. 24, 1977, pp. 29–36.

[4] Fernández-Seara, M. A., Wehrli, S. L., Takahashi, M., and Wehrli, F. W., "Water content measured by proton-deuteron exchange NMR predicts bone mineral density and mechanical properties," *J. Bone Miner. Res.*, Vol. 19, 2004, pp. 289–296.

[5] Bailey, A. J., and Knott, L., "Molecular changes in bone collagen in osteoporosis and osteoarthritis in the elderly," *Exp. Gerontol.*, Vol. 34, 1999, pp. 337–351.

[6] Summitt, M. C., and Reisinger, K. D., "Characterization of the mechanical properties of demineralized bone," *J. Biomed. Mater. Res.*, Vol. 67A, 2003, pp. 742–750.

[7] Kotha, S. P., Walsh, W. R., Pan, Y., and Guzelsu, N., "Varying the mechanical properties of bone tissue by changing the amount of its structurally effective bone mineral content," *Biomed. Mater. Eng.*, Vol. 8, 1998, pp. 321–334.

[8] Landis, W. J., "The strength of a calcified tissue depends in part on the moelcular structure and organization of its constituent mineral crystals in their organic matrix," *Bone*, Vol. 16, No. 5, 1995, pp. 533–544.

[9] Rho, J.-Y., Kuhn-Spearing, L., and Zioupos, P., "Mechanical properties and the hierarchical structure of bone," *Med. Eng. Phys.*, Vol. 20, 1998, pp. 92–102.

[10] Buckwalter, J. A., and Cooper, R. R., "Bone biology: II. Formation, form, modeling, remodeling, and regulation of cell function," *J. Bone Joint Surg.*, Vol. 77A, 1995, pp. 1276–1289.

[11] Fujisawa, R., Wada, Y., Nodasaka, Y., and Kuboki, Y., "Acidic amino acid–rich sequences as binding sites of osteonectin to hydroxyapatite crystals," *Biochim. Biophys. Acta.*, Vol. 1292, No. 1, 1996, pp. 53–60.

[12] Sikavitsas, V. I., Temenoff, J. S., and Mikos, A. G., "Biomaterials and bone mechanotransduction," *Biomaterials*, Vol. 22, 2001, pp. 2581–2593.

[13] Tabata, Y., "Tissue regeneration based on growth factor release," *Tissue Eng.*, Vol. 9, Suppl. 1, 2003, pp. S5–S15.

[14] Young, M. F., "Bone matrix proteins: Their function, regulation, and relationship to osteoporosis," *Osteoporos. Int.*, Vol. 14, Suppl. 3, 2003, pp. S35–S42.

[15] Krane, S. M., "Is collagenase necessary for bone and other connective tissue remodeling?" *Clin. Orthop. Relat. Res.*, Vol. 313, 1995, pp. 47–53.

[16] Sternlicht, M. D., and Werb, Z., "How matrix metalloproteinases regulate cell behavior," *Ann. Rev. Cell Dev. Biol.*, Vol. 17, 2001, pp. 463–516.

[17] Rosen, V., and Thies, R. S., "Adult skeletal repair," in *The Cellular and Molecular Basis of Bone Formation and Repair*, New York: Springer, 1995, pp. 97–142.

[18] Ueda, H., and Tabata, Y., "Polyhydroxyalkanonate derivatives in current clinical applications and trials," *Adv. Drug Deliv. Rev.*, Vol. 55, 2003, pp. 501–518.

[19] Rhee, S. H., "Bone-like apatite-forming ability and mechanical properties of poly(epsilon-caprolactone)/silica hybrid as a function of poly(epsilon-caprolactone) content," *Biomaterials*, Vol. 25, 2004, pp. 1167–1175.

[20] Suh, H., Song, M. J., Ohata, M., Kang, Y. B., and Tsutsumi, S., "Ex vivo mechanical evaluation of carbonate apatite-collagen-grafted porous poly-L-lactic acid membrane in rabbit calvarial bone," *Tissue Eng.*, Vol. 9, 2003, pp. 635–643.

[21] Karp, J. M., Shoichet, M. S., and Davies, J. E., "Bone formation on two-dimensional poly(DL-lactide-co-glycolide) (PLGA) films and three-dimensional PLGA tissue engineering scaffolds in vitro," *J. Biomed. Mater. Res.*, Vol. 64A, 2003, pp. 388–396.

[22] Kandziora, F., Schmidmaier, G., Schollmeier, G., Bail, H., Pflugmacher, R., Gorke, T., Wagner, M., Raschke, M., Mittlmeier, T., and Haas, N. P., "IGF-I and TGF-beta1 application by a poly-(D,L-lactide)-coated cage promotes intervertebral bone matrix formation in the sheep cervical spine," *Spine*, Vol. 27, 2002, pp. 1710–1723.

[23] Simon, C. G., Jr., Khatri, C. A., Wight, S. A., and Wang, F. W., "Preliminary report on the biocompatibility of a moldable, resorbable, composite bone graft consisting of calcium phosphate cement and poly(lactide-co-glycolide) microspheres," *J. Orthop. Res.*, Vol. 20, 2002, pp. 473–482.

[24] Matsuzak, K., Walboomers, F., de Ruijter, A., and Jansen, J. A., "Effect of microgrooved poly-l-lactic (PLA) surfaces on proliferation, cytoskeletal organization, and mineralized matrix formation of rat bone marrow cells," *Clin. Oral Implants Res.*, Vol. 11, 2000, pp. 325–333.

[25] Leinonen, S., Tiainen, J., Kellomaki, M., Tormala, P., Waris, T., Ninkovic, M., and Ashammakhi, N., "Holding power of bioabsorbable self-reinforced poly-L/DL-lactide 70/30 tacks and miniscrews in human cadaver bone," *J. Craniofac. Surg.*, Vol. 14, 2003, pp. 171–175.

[26] Nordstrom, P., Pohjonen, T., Tormala, P., and Rokkanen, P., "Shear-load carrying capacity of cancellous bone after implantation of self-reinforced polyglycolic acid and

poly-L-lactic acid pins: Experimental study on rats," *Biomaterials*, Vol. 22, 2001, pp. 2557–2561.

[27] Jabbari, E., and He, X., "Synthesis and material properties of functionalized lactide oligomers as in situ crosslinkable scaffolds for tissue regeneration," *Polym. Prepr.*, Vol. 47, No. 2, 2006, pp. 353–354.

[28] Chen, G., Sato, T., Ushida, T., Hirochika, R., and Tateishi, T., "Redifferentiation of dedifferentiated bovine chondrocytes when cultured in vitro in a PLGA-collagen hybrid mesh," *Febs. Lett.*, Vol. 542, Nos. 1–3, 2003, pp. 95–99.

[29] Poshusta, A. K., Burdick, J. A., Mortisen, D. J., Padera, R. F., Ruehlman, D., Yaszemski, M. J., and Anseth, K. S., "Histocompatibility of photocrosslinked polyanhydrides: A novel in situ forming orthopaedic biomaterial," *J. Biomed. Mater. Res.*, Vol. 64, No. 1, 2003, pp. 62–69.

[30] Anseth, K. S., Shastri, V. R., and Langer, R., "Photopolymerizable degradable polyanhydrides with osteocompatibility," *Nat. Biotech.*, Vol. 17, 1999, pp. 156–159.

[31] Fisher, J. P., Vehof, J. W., Dean, D., van der Waerden, J. P., Holland, T. A., Mikos, A. G., and Jansen, J. A., "Soft and hard tissue response to photocrosslinked poly(propylene fumarate) scaffolds in a rabbit model," *J. Biomed. Mater. Res.*, Vol. 59, 2002, pp. 547–556.

[32] Frazier, D. D., Lathi, V. K., Gerhart, T. N., and Hayes, W. C., "Ex vivo degradation of a poly(propylene glycol-fumarate) biodegradable particulate composite bone cement," *J. Biomed. Mater. Res.*, Vol. 35, 1997, pp. 383–389.

[33] Yaszemski, M. J., Payne, R. G., Hayes, W. C., Langer, R. S., and Mikos, A. G., "In vitro degradation of a poly(propylene fumarate)-based composite material," *Biomaterials*, Vol. 17, No. 22, 2006, pp. 2127–2130.

[34] Jabbari, E., Wang, S., Lu, L., Gruetzmacher, J. A., Ameenuddin, S., Hefferan, T. E., Currier, B. L., Windebank, A. J., and Yaszemski, M. J., "Synthesis, material properties and biocompatibility of a novel self-crosslinkable poly(caprolactone fumarate) as an injectable tissue engineering scaffold," *Biomacromolecules*, Vol. 6, 2005, pp. 2503–2511.

[35] Bennett, S., Connolly, K., Lee, D. R., Jiang, Y., Buck, D., Hollinger, J. O., and Gruskin, E. A., "Initial biocompatibility studies of a novel degradable polymeric bone substitute that hardens in situ," *Bone*, Vol. 19, 1996, 101S–107S.

[36] Elisseeff, J., Anseth, K. S., Sims, D., McIntosh, W., Randolph, M., and Langer, R., "Transdermal photopolymerization for minimally invasive implantation," *Proc. Natl. Acad. Sci.*, Vol. 96, 1999, pp. 3104–3107.

[37] Lee, K. Y., Alsberg, E., and Mooney, D. J., "Degradable and injectable poly(aldehyde guluronate) hydrogels for bone tissue engineering," *J. Biomed. Mater. Res.*, Vol. 56, 2001, pp. 228–233.

[38] Alsberg, E., Anderson, K. W., Albeiruti, A., Franceschi, R. T., and Mooney, D. J., "Cell-interactive alginate hydrogels for bone tissue engineering," *J. Dent. Res.*, Vol. 80, No. 11, 2001, pp. 2025–2029.

[39] He, X., and Jabbari, E., "Material properties and cytocompatibility of injectable MMP degradable poly(lactide ethylene oxide fumarate) hydrogel as a carrier for marrow stromal cells," *Biomacromolecules*, Vol. 8, 2007, pp. 780–792.

[40] Wintermantel, E., Mayer, J., Blum, J., Eckert, K.-L., Luscher, P., and Mathey, M., "Tissue engineering scaffolds using superstructures," *Biomaterials*, Vol. 17, 1996, pp. 83–91.

[41] Fujimura, K., Bessho, K., Okubo, Y., Segami, N., and Iizuka, T., "A bioactive bone cement containing Bis-GMA resin and A-W glass-ceramic as an augmentation graft material on mandibular bone," *Clin. Oral Implants Res.*, Vol. 14, 2003, pp. 659–667.

[42] van den Dolder, J., Wolke, J. G., and Jansen, J. A., "The cytocompatibility and early osteogenic characteristics of an injectable calcium phosphate cement," *Tissue Eng.*, Vol. 13, No. 3, 2007, pp. 493–500.

[43] TenHuisen, K. S., and Brown, P. W., "Formation of calcium-deficient hydroxyapatite from alpha-tricalcium phosphate," *Biomaterials*, Vol. 19, No. 23, 1998, pp. 2209–2217.

[44] Hutchens, S. A., Benson, R. S., Evans, B. R., O'Neill, H. M., and Rawn, C. J., "Biomimetic synthesis of calcium-deficient hydroxyapatite in a natural hydrogel," *Biomaterials*, Vol. 27, No. 26, 2006, pp. 4661–4670.

[45] Kasten, P., Vogel, J., Luginbuhl, R., Niemeyer, P., Tonak, M., Lorenz, H., Helbig, L., Weiss, S., Fellenberg, J., Leo, A., Simank, H. G., and Richter, W., "Ectopic bone formation associated with mesenchymal stem cells in a resorbable calcium deficient hydroxyapatite carrier," *Biomaterials*, Vol. 26, No. 29, 2005, pp. 5879–5889.

[46] Constantz, B. R., Ison, I. C., Fulmer, M. T., Poser, R. D., Smith, S. T., VanWagoner, M., Ross, J., Goldstein, S. A., Jupiter, J. B., and Rosenthal, D. I., "Skeletal repair by in situ formation of the mineral phase of bone," *Science*, Vol. 267, 1995, pp. 1796–1799.

[47] Liu, L., Zhang, L., Ren, B., Wang, F., and Zhang, Q., "Preparation and characterization of collagen-hydroxyapatite composite used for bone tissue engineering scaffold," *Artif. Cells Blood Substit. Immobil. Biotechnol.*, Vol. 31, No. 4, 2003, pp. 435–448.

[48] Rodrigues, C. V., Serricella, P., Linhares, A. B., Guerdes, R. M., Borojevic, R., Rossi, M. A., Duarte, M. E., and Farina, M., "Characterization of a bovine collagen-hydroxyapatite composite scaffold for bone tissue engineering," *Biomaterials*, Vol. 24, No. 27, 2003, pp. 4987–4997.

[49] Oliveira, J. M., Rodrigues, M. T., Silva, S. S., Malafaya, P. B., Gomes, M. E., Viegas, C. A., Dias, I. R., Azevedo, J. T., Mano, J. F., and Reis, R. L., "Novel hydroxyapatite/chitosan bilayered scaffold for osteochondral tissue-engineering applications: Scaffold design and its performance when seeded with goat bone marrow stromal cells," *Biomaterials*, Vol. 27, No. 36, 2006, pp. 6123–6137.

[50] Furuzono, T., Yasuda, S., Kimura, T., Kyotani, S., Tanaka, J., and Kishida, A., "Nano-scaled hydroxyapatite/polymer composite: IV. Fabrication and cell adhesion properties of a three-dimensional scaffold made of composite material with a silk fibroin substrate to develop a percutaneous device," *J. Artif. Org.*, Vol. 7, No. 3, 2004, pp. 137–144.

[51] Ajdukovic, Z., Najman, S., Dordevic, L. J., Savic, V., Mihailovic, D., Petrovic, D., Ignjatovic, N., and Uskokovic, D., "Repair of bone tissue affected by osteoporosis with hydroxyapatite-poly-L-lactide (HAp-PLLA) with and without blood plasma," *J. Biomater. Appl.*, Vol. 20, No. 2, 2005, pp. 179–190.

[52] Shikinami, Y., and Okuno, M., "Bioresorbable devices made of forged composites of hydroxyapatite (HA) particles and poly L-lactide (PLLA): II. Practical properties of miniscrews and miniplates," *Biomaterials*, Vol. 22, No. 23, 2001, pp. 3197–3211.

[53] Shikinami, Y., and Okuno, M., "Bioresorbable devices made of forged composites of hydroxyapatite (HA) particles and poly-L-lactide (PLLA): I. Basic characteristics," *Biomaterials*, Vol. 20, No. 9, 1999, pp. 859–877.

[54] Kim, S. S., Ahn, K. M., Park, M. S., Lee, J. H., Choi, C. Y., and Kim, B. S., "A poly(lactide-co-glycolide)/hydroxyapatite composite scaffold with enhanced osteoconductivity," *J. Biomed. Mater. Res.*, Vol. 80A, No. 1, 2007, pp. 206–215.

[55] Hakimimehr, D., Liu, D. M., and Troczynski, T., "In situ preparation of poly(propylene fumarate)/hydroxyapatite composite," *Biomaterials*, Vol. 26, No. 35, 2005, pp. 7297–7303.

[56] Wiria, F. E., Leong, K. F., Chua, C. K., and Liu, Y., "Poly(ε-caprolactone)/hydroxyapatite for tissue engineering scaffold fabrication via selective laser sintering," *Acta Biomater.*, Vol. 3, No. 1, 2007, pp. 1–12.

[57] Hu, Y., Zhang, C., Zhang, S., Xiong, Z., and Xu, J., "Development of a porous poly(L-lactic acid)/hydroxyapatite/collagen scaffold as a BMP delivery system and its use in healing canine segmental bone defect," *J. Biomed. Mater. Res.*, Vol. 67A, No. 2, 2003, pp. 591–598.

[58] Kruyt, M. C., Dhert, W. J., Oner, F. C., van Blitterswijk, C. A., Verbout, A. J., and de Bruijn, J. D., "Analysis of ectopic and orthotopic bone formation in cell-based tissue-engineered constructs in goats," *Biomaterials*, Vol. 28, No. 10, 2007, pp. 1798–1805.

[59] Hall, S. L., Lau, K. H., Chen, S. T., Felt, J. C., Gridley, D. S., Yee, J. K., and Baylink, D. J., "An improved mouse Sca-1+ cell-based bone marrow transplantation model for use in gene- and cell-based therapeutic studies," *Acta Haematol.*, Vol. 117, No. 1, 2007, pp. 24–33.

[60] Pelled, G., Tai, K., Sheyn, D., Zilberman, Y., Kumbar, S., Nair, L. S., Laurencin, C. T., Gazit, D., and Ortiz, C., "Structural and nanoindentation studies of stem cell-based tissue-engineered bone," *J. Biomech.*, Vol. 40, No. 2, 2007, pp. 399–411.

[61] Meyer, U., Wiesmann, H. P., Berr, K., Kubler, N. R., Handschel, J., "Cell-based bone reconstruction therapies—principles of clinical approaches," *Int. J. Oral Maxillofac. Implants*, Vol. 21, No. 6, 2006, pp. 899–906.

[62] Handschel, J., Wiesmann, H. P., Depprich, R., Kubler, N. R., and Meyer, U., "Cell-based bone reconstruction therapies: Cell sources," *Int. J. Oral Maxillofac. Implants*, Vol. 21, No. 6, 2006, pp. 890–898.

[63] Kruyt, M. C., Wilson, C. E., de Bruijn, J. D., van Blitterswijk, C. A., Oner, C. F., Verbout, A. J., and Dhert, W. J., "The effect of cell-based bone tissue engineering in a goat transverse process model," *Biomaterials*, Vol. 27, No. 29, 2006, pp. 5099–5106.

[64] Jabbari, E., Lu, L., and Yaszemski, M. J., "Synthesis and characterization of injectable and biodegradable composites for orthopedic applications," in S. K. Mallapragada and B. Narasimhan (eds.), *Handbook of Biodegradable Polymeric Materials and Their Applications*, Vol. 2, Stevenson Ranch, CA: American Scientific Publishers, 2004.

[65] Jabbari, E., Lu, L., Currier, B. L., Mikos, A. G., and Yaszemski, M. J., "Injectable polymers and hydrogels for orthopedic and dental applications," in L. J. Sandell and A. J.Grodzinsky (eds.), *Tissue engineering in musculoskeletal clinical practice*, Rosemont, IL: American Academy of Orthopedic Surgeons, 2004, Chapter 32.

[66] Buckwalter, J. A., M. J. Glimcher, R. R. Cooper, and R. Recker, "Bone biology I: Structure, blood supply, cells, matrix, and mineralization," *Instr. Course Lect.*, Vol. 45, pp. 371–386, 1996.

[67] Klenke, F. M., Y. Liu, H. Yuan, E. B. Hunziker, K. A. Siebenrock, and W. Hofstetter, "Impact of pore size on the vascularization and osseointegration of ceramic bone substitutes in vivo," *J. Biomed. Mater. Res. A*, Sep. 26, [Epub ahead of print], 2007.

[68] Laschke, M. W., K. Witt, T. Pohlemann, and M. D. Menger, "Injectable nanocrystalline hydroxyapatite paste for bone substitution: in vivo analysis of biocompatibility and vascularization. *J. Biomed. Mater. Res. B Appl. Biomater.*, Vol. 82, No. 2, pp. 494–505, 2007.

[69] Goldstein, A. S., T. M. Juarez, C. D. Helmke, M. C. Gustin, and A. G. Mikos. "Effect of convection on osteoblastic cell growth and function in biodegradable polymer foam scaffolds," *Biomaterials*, Vol. 22, No. 11, pp. 1279–1288, 2001.

[70] Kneser, U., L. Stangenberg, J. Ohnolz, O. Buettner, J. Stern-Straeter, D. Möbest, R. E. Horch, . B. Stark, and D. J. Schaefer, "Evaluation of processed bovine cancellous bone matrix seeded with syngenic osteoblasts in a critical size calvarial defect rat model," *J. Cell Mol. Med.*, Vol. 10, No. 3, pp. 695–707, 2006.

[71] Buckwalter, J. A., M. J. Glimcher, R. R. Cooper, and R. Recker, "Bone biology: Part I: Structure, blood supply, cells, matrix, and minerlization," *J. Bone Joint. Surg. A*, Vol. 77, No. 8, pp. 1256–1275, 1995.

[72] Meyer, U., U. Joos, and H. P. Wiesmann, "Biological and biophysical principles in extracorporal bone tissue engineering Part I," *Int. J. Oral Maxillofac. Surg.*, Vol. 33, pp. 325–332, 2004.

[73] Gronthos, S., and P. J. Simmons, "The biology and application of human bone marrow stromal cell precursors," *J. Hematother.*, Vol. 5, No. 1, pp. 15–23, 1996.

[74] Zhang, X., A. Naik, C. Xie, D. Reynolds, J. Palmer, A. Lin, H. Awad, R. Guldberg, E. Schwarz, and R. O'Keefe, "Periosteal stem cells are essential for bone revitalization and repair," *J. Musculoskelet. Neuronal Interact.*, Vol. 5, No. 4, pp. 360–362, 2005.

[75] Breitbart, A. S., D. A. Grande, R. Kessler, J. T. Ryaby, R. J. Fitzsimmons, and R. T. Grant, "Tissue engineered bone repair of calvarial defects using cultured periosteal cells," *Plast. Reconstr. Surg.*, Vol. 101, No. 3, pp. 567–576, 1998.

[76] Shen, F. H., J. M. Visger, G. Balian, S. R. Hurwitz, and D. R. Diduch, "Systemically administered mesenchymal stromal cells transduced with insulin-like growth factor-I localize to a fracture site and potentiate healing," *J. Orthop. Trauma*, Vol. 16, No. 9, pp. 651–659, 2002.

[77] Mödder, U. I., and S. Khosla, "Skeletal stem/osteoprogenitor cells: Current concepts, alternate hypotheses, and relationship to the bone remodeling compartment," *J. Cell Biochem.*, May 31, [Epub ahead of print], 2007.

[78] Kimelman, N., G. Pelled, G. A. Helm, J. Huard, E. M. Schwarz, and D. Gazit, "Review: gene- and stem cell-based therapeutics for bone regeneration and repair," *Tissue Eng.*, Vol. 13, No. 6, pp. 1135–1150, 2007.

[79] Mauney, J. R., V. Vladimir, and D. A. Kaplan, "Role of adult mesenchymal stem cells in bone tissue-engineering applications: current status and future prospects, *Tissue Eng.*, Vol. 11, No. 5-6, pp. 787-802, 2005.

[80] Wu, S., Y. Suzuki, Y. Ejiri, T. Noda, H. Bai, M. Kitada, K. Kataoka, M. Ohta, H. Chou, and C. Ide, "Bone marrow stromal cells enhance differentiation of cocultured neurosphere cells and promote regeneration of injured spinal cord," *J. Neurosci. Res.*, Vol. 72, pp. 343–351, 2003.

[81] Napoli, C., A. Balestrieri, and L. J. Ignarro, "Therapeutic approaches in vascular repair induced by adult bone marrow cells and circulating progenitor endothelial cells," *Curr. Pharm. Des.*, Vol. 13, No. 31, pp. 3245–3251, 2007.

[82] Fukiage, K., T. Aoyama, K. R. Shibata, S. Otsuka, M. Furu, Y. Kohno, K. Ito, Y. Jin, S. Fujita, S. Fujibayashi, M. Neo, T. Nakayama, T. Nakamura, and J. Toguchida, "Expression of vascular cell adhesion molecule-1 indicates the differentiation potential of human bone marrow stromal cells," *Biochem. Biophys. Res. Commun.*, Vol. 365, No. 3, pp. 406–412, 2008.

[83] Maniatopoulos, C., J. Sodek, and A. H. Melcher, "Bone formation in vitro by stromal cells obtained from bone marrow of young adult rats," *Cell tissue Res.*, Vol. 254, pp. 317–330, 1988.

[84] Temenoff, J. S., H. Park, E. Jabbari, T. L. Sheffield, R. G. LeBaron, C. G. Ambrose, and A. G. Mikos, "In-Vitro Osteogenic Differentiation of Marrow Stromal Cells Encapsulated in Biodegradable Hydrogels," *J. Biomed. Mater. Res.*, Vol. 70, pp. 235–244, 2004.

[85] Datta, N., Q. P. Pham, U. Sharma, V. I. Sikavitsas, J. A. Jansen, A. G. Mikos, "In vitro generated extracellular matrix and fluid shear stress synergistically enhance 3D osteoblastic differentiation," *Proc. Natl. Acad. Sci. USA*, Vol. 103, No. 8, 2488–2493, 2006.

[86] Jaiswal N., and S. P. Bruder, "The pleiotropic effects of dexamethasone on osteoblast differentiation depend on the developmental state of the responding cells," *J. Bone Miner. Res.*, Vol. 11, p. S259, 1996.

[87] Hefferan, T. E., E. Jabbari, A. Florschutz, R. M. Mardones, L. Lu, B. L. Currier, and M. J. Yaszemski, "rhBMP-2 enhances bone formation in a biodegradable scaffold," *J. Bone Mineral Res.*, Vol. 20, No. 9, p. S108, 2005.

[88] Donzelli, E., A. Salvade, P. Mimo, M. Vigano, M. Morrone, R. Papagna, F. Carini, A. Zaopo, M. Miloso, M. Baldoni, and G. Tredici, "Mesenchymal stem cells cultured on a collagen scaffold: In vitro osteogenic differentiation," *Arch. Oral Biol.*, Vol. 52, no. 1, pp. 64–73, 2007.

[89] Zaidi, N., A. J. Nixon, "Stem cell therapy in bone repair and regeneration," *Ann. N. Y. Acad. Sci.*, Vol. 1117, pp. 62–72, 2007.

[90] Gaballa, M. A., J. N. Sunkomat, H. Thai, E. Morkin, G. Ewy, and S. Goldman, "Grafting an acellular 3-dimensional collagen scaffold onto a non-transmural infarcted myocardium induces neo-angiogenesis and reduces cardiac remodeling," *J. Heart Lung Transplant.*, Vol. 25, No. 8, pp. 946–954, 2006.

[91] Farrell, E., F. G. O'Brien, P. Doyle, J. Fischer, I. Yannas, B. A. Harley, B. O'Connell, P. J. Prendergast, V. A. Campbell, "A collagen-glycosaminoglycan scaffold supports adult rat mesenchymal stem cell differentiation along osteogenic and chondrogenic routes," *Tissue Eng.*, Vol. 12, No. 3, pp. 459–468, 2006.

[92] He, X., and E. Jabbari, "Material properties and cytocompatibility of injectable mmp degradable poly(lactide ethylene oxide fumarate) hydrogel as a carrier for marrow stromal cells," *Biomacromolecules*, Vol. 8, pp. 780–792, 2007.

[93] Yost, M. J., C. F. Baicu, C. E. Stonerock, R. L. Goodwin, R. L. Price, J. M. Davis, H. Evans, P. D. Watson, C. M. Gore, J. Sweet, L. Creech, M. R. Zile, and L. Terracio, "A novel tubular scaffold for cardiovascular tissue engineering," *Tissue Eng.*, Vol. 10, No. 1-2, pp. 273–284, 2004.

[94] Valarmathi, M. T., J. D. Potts, M. J. Yost, R. L. Goodwin, and E. Jabbari, "A Novel Three-Dimensional Culture System to Study vasculogenesis and Osteogenic Differentiation of BMS Cells," *Trans. Soc. Biomaterials*, Vol. 1, p. 115, 2007.

[95] Ramakrishna, S., Mayer, J., Wintermantel, E., and Leong, K. W., "Biomedical applications of polymer-composite materials: A review," *Comp. Sci. Tech.*, Vol. 61, 2001, pp. 1189–1224.

[96] Mano, J. F., Sousa, R. A., Boesel, L. F., Neves, N. M., and Reis, R. L., "Bioinert, biodegradable and injectable polymeric matrix composites for hard tissue replacement: State of art and recent developments," *Comp. Sci. Tech.*, Vol. 64, 2004, pp. 789–817.

[97] Yaszemski, M. J., Payne, R. G., Hayes, W. C., Langer, R., and Mikos, A. G., "Evolution of bone transplantation: Molecular, cellular and tissue strategies to engineer human bone," *Biomaterials*, Vol. 17, 1996, pp. 175–185.

[98] Nazhat, S. N., Joseph, R., Wang, M., Smith, R., Tanner, K. E., and Bonfield, W., "Dynamic mechanical characterisation of hydroxyapatite reinforced polyethylene: Effect of particle size," *J. Mater. Sci. Mater. Med.*, Vol. 11, 2000, pp. 621–628.

[99] Deb, S., Wang, M., Tanner, K. E., and Bonfield, W., "Hydroxyapatite polyethylene composites: Effect of grafting and surface treatment of hydroxyapatite," *J. Mater. Sci. Mater. Med.*, Vol. 7, 1996, pp. 191–197.

[100] Wang, M., and Bonfield, W., "Chemically coupled hydroxyapatite polyethylene composites: Structure and properties," *Biomaterials*, Vol. 22, 2001, pp. 1311–1320.

[101] Wang, X., and Robertson, C. G., "Strain-induced nonlinearity of filled rubbers," *Phys. Rev.*, Vol. 72E, 2005, pp. 031406–031409.

[102] Sarvestani, A. S., He, X., and Jabbari, E., "The effect of osteonectin-derived peptide on the viscoelasticity of hydrogel/apatite nanocomposite scaffolds," *Biopolymers*, Vol. 85, No. 4, 2007, pp. 370–378.

[103] Sarvestani, A. S., and Jabbari, E., "Modeling the viscoelastic response of suspension of particles in polymer solution: The effect of polymer-particle interactions," *Macromol. Theory Simul.*, Vol. 16, 2007, pp. 378–385.

[104] Fantner, G. E., Birkedal, H., Kindt, J. H., Hassenkam, T., Weaver, J. C., Cutroni, J. A., Bosma, B. L., Bawazer, L., Finch, M. M., Cidade, G. A., Morse, D. E., Stucky, G. D., and Hansma, P. K., "Influence of the degradation of the organic matrix on the microscopic fracture behavior of trabecular bone," *Bone*, Vol. 35, 2004, pp. 1013–1022.

[105] Athanasiou, K. A., Zhu, C. F., Lanctot, D. R., Agrawal, C. M., and Wang, X., "Fundamentals of biomechanics in tissue engineering of bone," *Tissue Eng.*, Vol. 6, 2000, pp. 361–381.

[106] Boskey, A. L., "Mineral-matrix interactions in bone and cartilage: Clinical orthopedics and related research," *Clin. Orthop. Relat. Res.*, Vol. 281, 1992, pp. 244–274.

[107] Termine, J. D., Kleinman, H. K., Whitson, S. W., Conn, K. M., McGarvey, M. L., and Martin, G. R., "Osteonectin, a bone-specific protein linking mineral to collagen," *Cell*, Vol. 26, 1981, pp. 99–105.

[108] Doi, Y., Horiguchi, T., Kim, S. H., Moriwaki, Y., Wakamatsu, N., Adachi, M., Ibaraki, K., Moriyama, K., Sasaki, S., and Shimokawa, H., "Effects of non-collagenous proteins on

the formation of apatite in calcium beta-glycerophosphate solutions," *Arch. Oral Biol.*, Vol. 37, 1992, pp. 15–21.

[109] Nakase, T., Sugimoto, M., Sato, M., Kaneko, M., Tomita, T., Sugamoto, K., Nomura, S., Kitamura, Y., Yoshikawa, H., Yasui, N., Yonenobu, K., and Ochi, T., "Switch of osteonectin and osteopontin mRNA expression in the process of cartilage-to-bone transition during fracture repair," *Acta Histochem.*, Vol. 100, 1998, pp. 287–295.

[110] Sodek, J., Zhu, B., Huynh, M. H., Brown, T. J., and Ringuette, M., "Novel functions of the matricellular proteins osteopontin and osteonectin," *Connect. Tissue Res.*, Vol. 43, 2002, pp. 308–319.

[111] Xie, R.-L., and Long, G. L., "Elements within the first 17 amino acids of human osteonectin are responsible for binding to type V collagen," *J. Biol. Chem.*, Vol. 271, 1996, pp. 8121–8125.

[112] Boskey, A. L., Moore, D. J., Amling, M., Canalis, E., and Delany, A. M., "Infrared analysis of the mineral and matrix in bones of osteonectin-null mice and their wildtype controls," *J. Bone Miner. Res.*, Vol. 8, 2003, pp. 1005–1011.

[113] He, X., and Jabbari, E., "Solid-phase synthesis of reactive peptide crosslinker by selective deprotection," *Prot. Pept. Lett.*, Vol. 13, 2006, pp. 715–718.

[114] Sarvestani, A. S., He, X., and Jabbari, E., "Viscoelastic characterization and modeling of gelation kinetics of injectable in situ crosslinkable poly(lactide-ethylene oxide-fumarate) hydrogels," *Biomacromolecules*, Vol. 8, No. 2, 2007, pp. 406–415.

[115] Guth, E. J., "Theory of filler reinforcement," *J. Appl. Phys.*, Vol. 16, 1945, pp. 20–25.

CHAPTER 23

Technological Approaches to Renal Replacement Therapies

Vimalier Reyes-Ortiz, Mohammad R. Kaazempur Mofrad, Eli J. Weinberg, Joseph P. Vacanti, and Jeffrey T. Boresntein

> What is a man, when you come to think upon him, but a minutely set, ingenuous machine for turning, with infinite artfulness, the red wine of Shiraz into urine?
>
> —Isak Dinesen

23.1 Introduction

Kidneys are the organs in charge of the filtration-reabsorption process in the body. Their primary function is to filtrate the waste and excessive water from the body, keeping a necessary ion balance. This filtration process, in addition to other endocrine activities, is called renal functioning. A low percentage (<15 percent) of the total renal functioning is cataloged as an end-stage renal disease (ESRD). ESRD patients would not be able to survive without renal therapy, which is an alternative that compensates for the malfunctioning kidney. This could be either kidney transplant or dialysis. The latter is a blood-filtration process that can be done in two ways, using hemodialysis or peritoneal dialysis (PD) [1, 2].

Hemodialysis patients have to filter their blood through a dialysis machine [1–3], which contains a membrane and a secondary fluid called dialysate. Both are responsible for the filtration and carriage of solutes out of the blood. (Figure 23.1) [4–6]. Significant work has been done in order to find an adequate dialysis treatment [1, 2]. Mathematical approaches have become standard for dialysis quantification. These standards has been used as tools for the enhancement of the process, changing membrane aspects for a better separation of the solutes by size, charge, and concentration [4, 5]. Flow-membrane relations are considered factors as well.

Innovative approaches propose continuous hemodialysis and the achievement of a membraneless filtration by the application of nanotechnology and microfluid techniques [7, 8]. Kaazempur-Mofrad et al. [9] proposed a novel microelectromechanical system (MEMS)–based system for renal replacement. This less-invasive device offers significant improvement in ultrafiltration efficiency compared to other hemodialysis treatments [9]. Another promising approach seems to be the implementation of living membranes into a biohybrid hemodialysis treatment [1, 2]. Renal cells added to the filtration process supposedly cover for complex metabolic and endocrine activities that a membrane cannot perform by itself [1, 2].

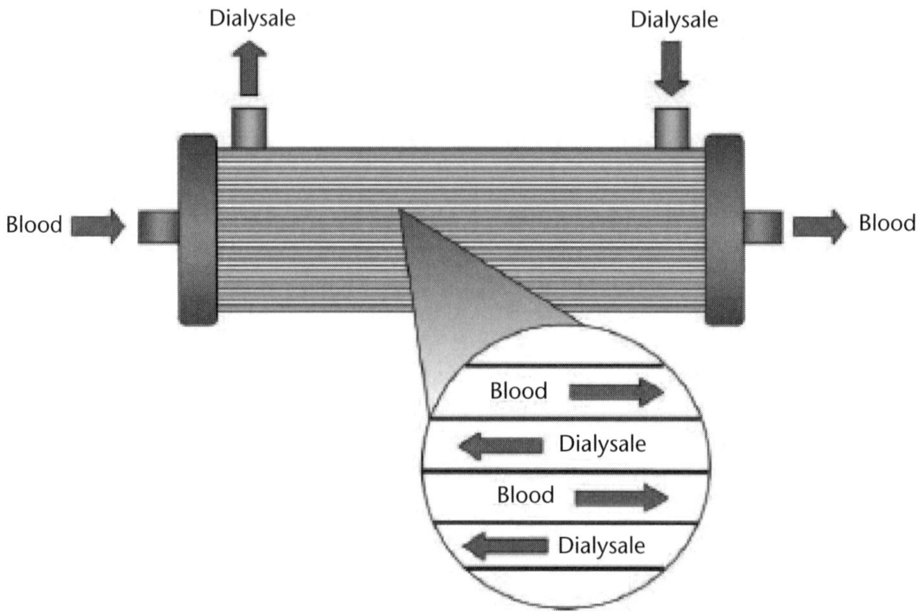

Figure 23.1 Basic drawing of an artificial kidney or dialyzer. Primary (blood) and secondary fluids (dialysate) have countercurrent flow [6].

The present chapter makes an effort to provide a historical perspective on the improvements made in the dialysis procedure since Kolff's rotational-drum dialyzer until recent nanotechnology-based and living-membrane hemodialysis systems. A brief description is provided of the vital function of the kidney, the ways to quantify its function, and the consequences of ESRD, which are the principal motivations for renal-technology research.

23.2 Kidney Functioning Overview

The separation functions of the kidney can be summarized into four key steps as follows: (1) eliminate the water-soluble nitrogenous end products of protein metabolism; (2) maintain electrolyte balance in body fluids and get rid of the excess electrolytes; (3) contribute to obligatory water loss and discharge excess water in the urine; and (4) maintain acid-base balance in body fluids and tissues [1]. However, it is better to understand a bit more about the functioning of the kidneys.

The kidneys are fist-sized organs whose main function is the filtration of wastes from the blood (Figure 23.2), while keeping a homeostatic ionic environment for the proper functioning of an organism. Some undesired molecules removed include urea, creatinine, uric acid, and medications [1, 10]. This involves a process called ultrafiltration that is accompanied by removal of excess water. Some of this water (together with other solutes) must be reabsorbed during the total filtration process in order to keep corporeal homeostasis.

Even when kidneys can be seen as blood filters, filtration is not their only function. Hormones such as rennin, erythropoietin, and calcitriol are also produced by

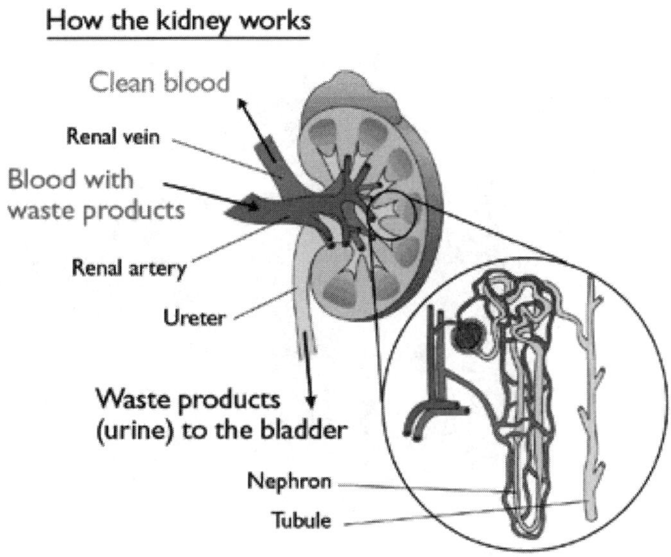

Figure 23.2 The kidney is an organ that filtrates wastes from the blood [11].

the kidney. The kidney is also related to the production of glucose from gluconeogenesis when required [1, 10]. Approximately 20 percent of the blood that comes from the heart is received in the kidneys via the renal artery (Figure 23.2). Filtered blood will later leave the kidney via the renal vein. During this filtration process, a plasma ultrafiltrate is formed by the glomerulus and modified into the tubules (Figure 23.2). Modification can include substrate/water reabsorption and/or secretion.

The glomerulus (responsible for this ultrafiltration) is a structure formed by many capillaries connected to a single blood supply (Figures 23.2 and 23.3) [1, 10]. Together with the tubules, the glomerulus forms part of the main filtration unit of the kidney, the nephron (Figure 23.2), which has also been described an arrangement of mass transfer units comprising a glomerulus, proximal tubule, and distal tubule [1]. There are approximately one million nephrons in an adult kidney.

The glomerular basement membrane (GBM) also contributes to the anchor, the architecture, and molecule restriction (filtration) of the complex. A slit diaphragm acts as the final glomerular filtration barrier. This slit structure has been taken into account for recent developments in the geometry of hemodialysis membranes [1, 2].Two principal pore sizes in the glomerular capillaries work out a great part of the size-filtration process. The most and least common solutes have approximate radii of 42Å and 70Å, respectively [1, 10]. Charge selectivity is provided by Heparan sulfate proteoglycans, synthesized by the glomerular endothelial and epithelial cells.

The composition of the urine will depend upon the glomerular filtration and its modification process through the kidney. Solutes and water excretion and reabsorption are strongly related to the glomerular filtration rate (GFR), which is directly related to the detection, progression, and treatment of kidney disease in patients (Table 23.1). The most commonly used molecule for GFR measurement is creatinine, which is produced from the metabolism of skeletal muscle creatine and

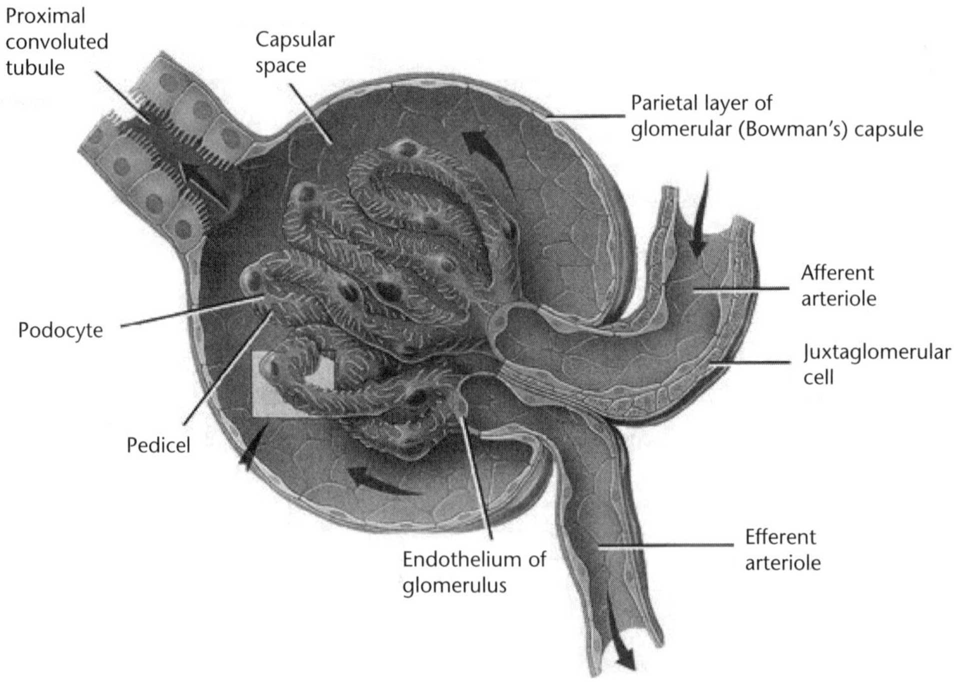

Figure 23.3 The glomerulus is structure formed by many capillaries connected to a single blood input [12].

Table 23.1 The Measurement of GFR Is Related to the Detection, Progression, and Treatment of Kidney Disease in Patients

GFR Results	*Stage*	*What It Means*
90–130, no protein inthe urine	Normal	Normal kidney function
90 or more, protein in the urine	1	Slight decrease in kidney function
60 to 89	2	Mild decrease in kidney function
30 to 59	3	Moderate decrease in kidney function
15 to 29	4	Severe decrease in kidney function
Less than 15	5	Kidney failure (dialysis or transplant will be needed soon)

Source: [13].

filtered by the glomerulus. However, the GFR measurement can be estimated by 10 to 20 percent because creatinine is also secreted in the proximal tubule. Therefore, the rate increase of tubular creatinine can be related to renal dysfunction.

23.3 Kidney Failure

One of the first question readers may ask themselves is, why do kidneys fail? Unfortunately, the exact causes of kidney failure are not completely understood. Kidney stones, polysyctic kidney disease, kidney infections, and kidney cysts are some of the common diseases that can lead to kidney failure. Furthermore, kidney failure rates are strongly related to diabetes and high blood pressure [1, 10]. This might indicate that the fluids affect kidneys more before entering them; therefore, an early-stage condition can be treated by electrolyte therapy, diet, and medication [1].

There are two types of renal failure: acute and chronic. Acute renal failure (ARF) refers to a sudden drop in kidney function, which can be caused by an injury where a significant amount of blood is lost or by a poisonous substance, among other causes. ARF might be reversible, and renal therapy replacement has been available for ARF since the 1940s. It was not until the 1960s that a treatment was developed for chronic renal disease [14, 15] Chronic kidney disease (CKD) is related to a slow and progressive decrease in renal function. CKD may cause patients complete and permanent renal failure. It is understood that the progressive nature of the disease is due not only to an initial disease but to a secondary process that perpetuates it [16]. Once an initial disease has irreversibly damaged the kidney, the organ will try to compensate for the loss of function. This short-term adaptation forms part of a cycle where every compensation will lead to further injury, progressing the condition to ESRD (Figure 23.4). ESRD patients cannot live without therapy replacement, either transplant or dialysis.

23.4 Treatments

23.4.1 Kidney Transplantation

Since the 1950s, kidney transplantation has been an option for ESRD patients. A donated kidney is placed inside the lower abdomen, connected to the renal artery and renal vein (Figure 23.5). Blood flows through the new kidney as it covers for the function of the failed kidneys. The old kidneys are left in place, unless they are prej-

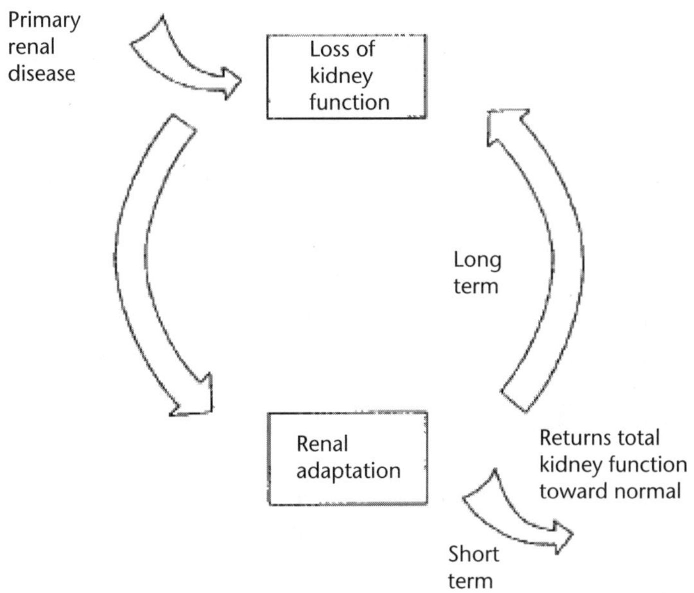

Figure 23.4 After the kidney suffers an injury, it might return to normal functioning. Otherwise, the organ can go through several cyclic adaptations previous to total failure [16].

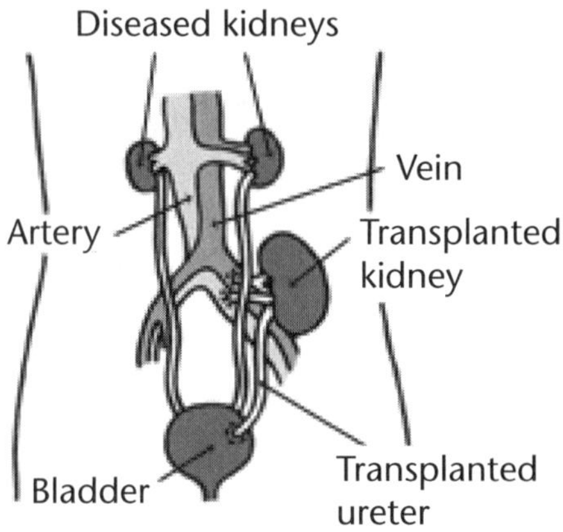

Figure 23.5 A transplanted kidney is placed inside the lower abdomen, connected to the renal artery and renal vein [17].

udicial to the patient's health. After transplantation, patients can regain a life similar to the one they had before their kidneys failed. Yet, this is not a total cure. Patients have to take medication for the rest of their lives to prevent rejection of the foreign kidney. The major risk is waiting for a donor, and it can take years before a donated kidney is found.

A kidney transplant can also fail. The body can still reject the kidney, making it necessary to go back on dialysis. Immunosuppressants can weaken the immune response, making the patient more susceptible to infections. Other secondary effects, like physical appearance, cataracts, diabetes, extra stomach acid, high blood pressure, bone disease, and liver damage, can take place in some patients. Before transplantation, most patients undergo dialysis, either hemodialysis or peritoneal dialysis.

23.4.2 Peritoneal Dialysis

Peritoneal dialysis is a more independent treatment that allows patients to give treatments to themselves almost anywhere [1, 6, 18]. A dextrose-based solution flows into the abdominal area through a catheter and is later drained after absorbing waste and extra fluids from the blood (Figure 23.6). The waste and extra fluids will flow from the blood to the solution though the peritoneum. The peritoneum is a two-layer serous membrane located in the abdomen that serves as a conduit for the blood. Because it is lined up with the abdominal area, a catheter can be placed into the abdomen and the peritoneal cavity for dialysis purposes. This catheter is a soft tube with polyester cuffs that merge with the tissue after it is placed. The end of the catheter inside the abdomen has holes for solution to flow through (Figure 23.6). The dextrose solution flows by gravity from a bag through the catheter and into the abdominal cavity, where wastes and extra fluids are absorbed. The solution will be drained, carrying out the absorbed materials. This total process and the time the solution remains in the cavity are called "exchange" and "dwell time," respectively. A regular PD patient goes through four to six exchanges a day with dwell times of four to six hours [1]. There are two ways this process can be executed, a continuous

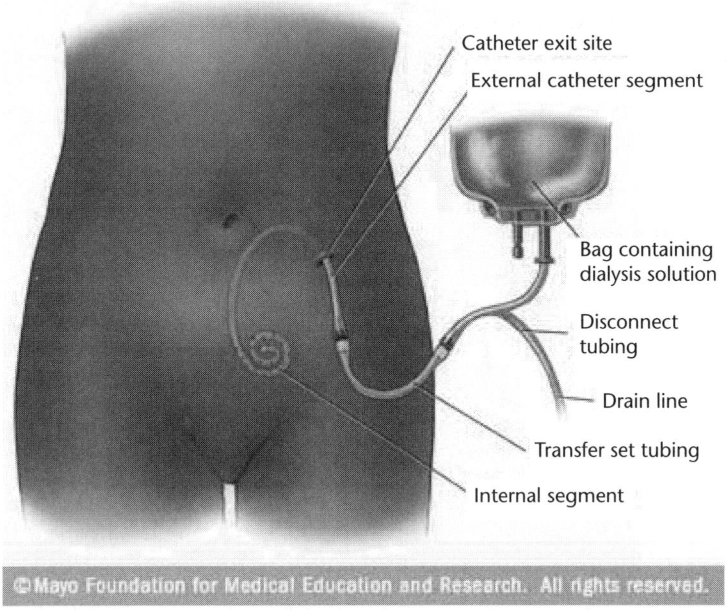

Figure 23.6 In peritoneal dialysis, the dextrose-based solution flows into the abdominal area through a catheter placed in the abdomen. Solution carrying blood wastes is later drained after the exchange and dwell time [19].

ambulatory peritoneal dialysis and a cycler-assisted peritoneal dialysis. The latter requires a machine (cycler) to fill and drain the solution, usually while the person is asleep.

23.4.3 Hemodialysis

Hemodialysis is the most common renal replacement that provides for blood-waste removal. Available for acute and chronic renal failure since the 1940s and 1960s, respectively, hemodialysis has become the most common treatment for dialysis patients [1, 6, 14, 15, 18, 20].

During hemodialysis, the patient's blood circulates from the patient through a filtering machine and, finally, to the patient again (Figure 23.7) [21]. A vascular access is used to take and return the blood to the patients. The blood is filtered in a dialyzer, which is a container with thousands of hollow fibers that the blood flows through. The dialysis solution flows in the opposite direction inside the hemodialyzer, outside of the fibers (Figure 23.1). Wastes and extra solutes diffuse from the blood through the membrane and into the dialysis solution.

Since 1971, special attention has been directed to dialysis efficiency and the impact on patients' health [18]. As result of extremely relevant efforts, two tests are used to monitor the concentration of waste in the blood. The major markers used are the urea prediction ratio (URR) and Kt/V, where K is a dimensionless number representative of the urea clearance, t is the time treatment, and V is the patient's total water volume. Adequate dialysis values have been standardized as 65 and 1.2 for URR and Kt/V, respectively [18, 22].

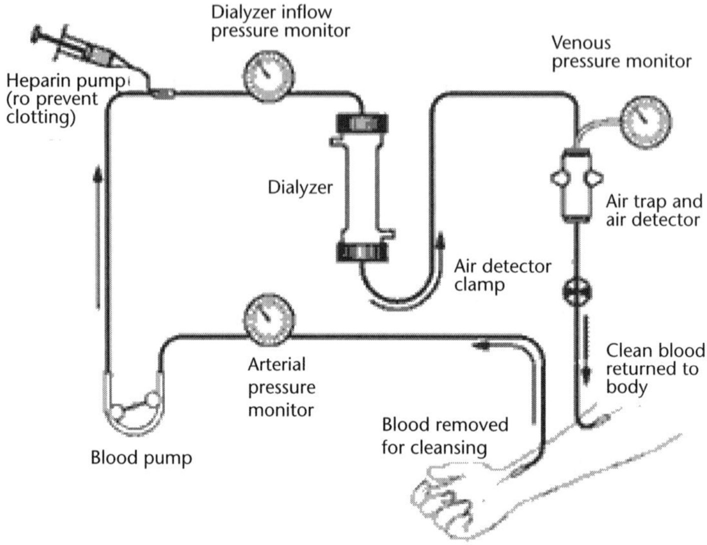

Figure 23.7 Since the 1960s, a hemodialysis therapy has allowed for repetitive blood circulation between the patient and a filtering machine through a U-shaped tube [21].

Advances in hemodialysis technology give patients the possibility of long life expectancy. However, only a kidney transplant can take over for the complete renal function. Current dialysis treatments can only mimic the principal filtration process (glomerular filtration) but cannot supplant the complex secretion and reabsorption activities. New approaches are trying to implement living membranes to current hemodialysis therapy. The term "living membranes" refers to the addition of renal cells, which are supposed to cover for the more complex secretion and metabolizing renal activities [5, 23]. This promising approach has been successfully tested in uremic animals and terminal human patients [5, 23].

23.5 History of Hemodialysis

A reliable long-term renal therapy for ESRD patients has been available since the 1960s. This dialysis-based process separates most of the waste molecules from the blood, preventing patients from suffering poisoning by their own metabolites. However, there are still high mortality and morbidity rates [17, 24], which are the main motivation for renal-technology enhancement.

No dialysis history would be complete without mentioning Thomas Graham, author of the first published description of the dialysis process (Figure 23.8) [25, 26]. Graham suspended a bell-shaped vessel filled with urine inside a container filled with distilled water. After removal of the bell-shaped vessel, the water in the container was boiled. After the "water" in the larger container was evaporated, the residue had a significant concentration of sodium chloride and urea. This was the product of urea diffusion from the urine, through the membrane, to the water.

Substantial artificial-membrane and dialysis [27, 28] research followed Graham's work, but the first-described dialysis procedures date from 1913, with the work done by Abel, Rowntree, and Turner [29]. In Abel's artificial kidney, the

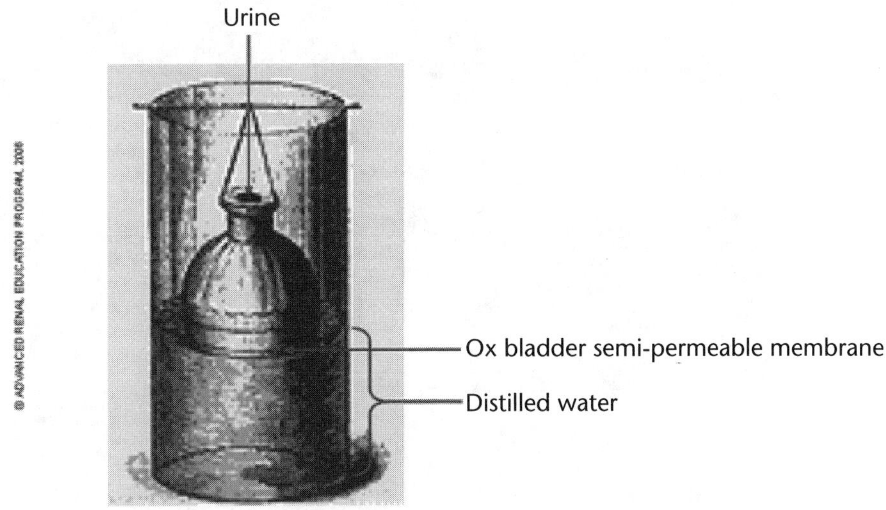

Based on: Graham T. *Philos Trans R Soc Lond* 144:117-128, 1854

Figure 23.8 Dialysis container: a bell-shaped vessel was filled with urine and put into the container [25, 26].

blood of dogs and rabbits had to run through porous colloid tubes, then through a salt solution, before returning to the animals. The anticoagulant applied back then was hirudin, which was obtained from leeches. In 1923, Haas attempted dialysis on humans with a similar mechanism, but the short duration of the treatment led to insignificant effects [30]. By 1937, Thalheimer had dialyzed uremic dogs with the first cellophane dialysis membrane [31]. This work served as a basis for Willem Kolff's first human dialyzer.

As the story goes, Kolff started his dialyzer work after observing a young man die in a uremic comma with no optional treatment [14]. Kolff developed the new dialyzer based on Abel's work. Substituting heparin for hirudin and replacing the colloid tube with a cellophane membrane, the first hemodialyzer (later known as the "rotational hemodialyzer") was made [20]. Blood was supposed to run from the patient through cellophane tubing ~130 feet long and wrapped around an aluminum drum. While running though that tube, the blood was exposed to a dialysis solution that was in contact with the rotating drum (Figure 23.9). Blood wastes were supposed to diffuse from the blood to the dialysis solution as happens in today's regular dialysis. Kolff treated sixteen patients before being successful with a sixty-seven-year-old woman who came out of uremic comma after eleven hours of hemodialysis [14]. The success led to a posterior stage of hemodialyzer manufacture and distribution, spreading it through the world.

Kolff took into account the mass of urea removed in the dialysate. However, this treatment was not enough for patients other that those suffering acute renal failure. After giving patients a number of therapies, physicians would no longer be able to connect them to the machine due to the damage that the veins suffered every time the treatment was given.

In 1960, Belding H. Scribner came up with a revolutionary idea that would let dialysis patients receive treatment without running out of locations to connect the

Figure 23.9 Kolff's rotational dialyzer, the first artificial kidney to treat ARF [6].

machine by establishing connection through a single place [14, 15, 18]. Scribner's plan consisted of inserting plastic tubes into an artery and a vein. The patient-machine connection would then take place through those tubes (Figure 23.7). When the patient was not connected to the machine, circulation would be maintained by an extracorporeal teflon U-shaped tube that allowed physicians to access the blood without making any incisions in the patient. The first patient under this new treatment was hemodialized for eleven years [14].

23.6 Dialyzer Improvements

After the long-term-treatment dialyzer was available, most hemodialysis research focused on improving Kolff's and Scribner's methods. An important contribution to the quantification of dialyzer performance was made in 1951, when Wolf et al. analyzed in vitro studies from Kolff's rotational drum dialyzer [32]. This analysis defined clinical terms such as "clearance" and "dialysance," measured the ratio of urea nitrogen clearance, and considered the Gibbs-Donnan phenomenon, Coulomb forces for ions diffusion, and the difficulties in ultrafiltration application with the studied apparatus [32]. Most of the terminologies and equations established by Wolf et al. have become standards for measuring dialyzer performance [32]. Wolf et al.'s work is one of the remarkable, fundamental contributions to the dialysis quantification area.

Observation of disparities in neuropathies in peritoneal dialysis and hemodialysis brought interest in looking at the possible relation between dialyzer performance and patient outcome. It seemed like the peritoneum was somehow more permeable to specifically middle-sized molecules than the hemodialysis membranes used then [18, 33]. In 1971, Babb et al. introduced the theory of the squared meter-hour [33].

Kinetics studies and modeling during the 1970s enabled the establishment of dialysis-efficiency standards, like *Kt*/*V* and the time-averaged concentration of urea (TAC_{urea}) [34]. These standards relate the theoretical dialysis doses to the mortality and morbidity in ESRD patients.

Simultaneous diffusion and ultrafiltration in the blood was later addressed in 1977 by Villarroel, Klein, and Holland [35]. Jaffrin, Ding, and Laurent have also provided for this correction when calculating for the mass transfer in a hemodialyzer, while considering changes on dialysate and blood flow due to ultrafiltration [36].

In the early 1980s through the 1990s, Malchesky et al. introduced what is considered the "golden standard" of dialysis-efficiency measurement, the direct dialysis quantification [37, 38]. This measurement actually requires collection of the entire used dialysate and calculates the clearance from pre- and postprocessed dialysate. This was followed by the urea reduction ratio, another dialysis standard introduced by Lowrie and Lew that does not account for the ultrafiltration [37].

Throughout the 1980s and 1990s, more mathematical approaches have been focused on the modification of these standards for the achievement of an "adequate dialysis." However, the technology available has allowed researchers to approach this goal from different perspectives. Modification of membrane structure and/or composition, application of microfluidics techniques, and even biohybridization of dialysis devices are some of the traditional and innovative approaches that move this research toward the achievement of adequate dialysis.

23.7 Innovative Hemodialysis Approaches

An engineering vision of the kidney as a flow-filtration apparatus allowed for this organ to be the first one whose extracorporeal substitution was successful. Kolff's simple prototype was based on the need to filter toxic solutes from the blood. Since then, research has focused on the improvement of this therapy, directing efforts toward the achievement of a glomerularlike filtration. Some of the parameters taken into account are membrane structure and composition, blood-membrane interaction, blood flow and dialyste flow, solute transport (e.g., convection and diffusion), and metabolic activity.

Modifications to a new cellulose membrane were motivated by concerns about the blood-membrane interaction, which brought up concepts like wetability and blood activation [39]. New materials and chemical modifications came along with the refinining of pore dimensions to enhance hydraulic permeability and middle molecule clearance. The hydroxyl groups were replaced or bonded with different chemicals, like acetyls and vitamin E, respectively [40]. These development allowed for the reduction of oxidant stresses [40]. An example of wetability has been the modification of the hydrophobixity and hydrophilicity of the inner lumen of fiber membranes with polymer alloys (Figure 23.10), which also benefits the diffusive properties of the membrane.

Other modifications have been focused on the dialysate flow rate (Q_B) rather than the blood flow. The target of these changes is a modification on the effective area of the membrane to facilitate Q_B during the dialysis process. Magnetic reso-

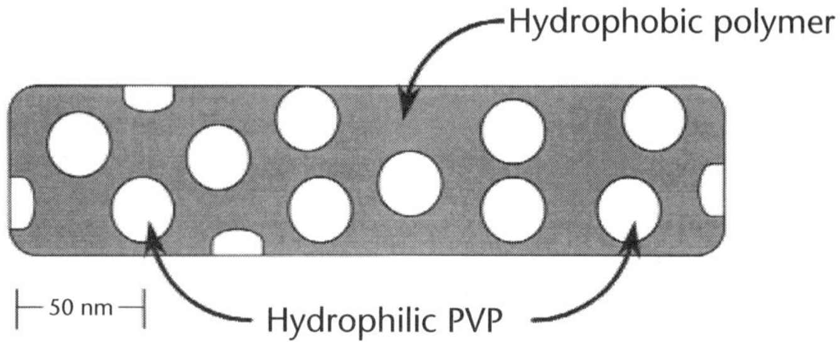

Figure 23.10 Modified membrane composition increasing wettability [39].

nance imaging studies indicate misdistribution of dialysate flow rates in its compartment (Figure 23.11) [39]. Misdistribution has been related to the packing density of the membrane in the dialyzer, which can prevent some hollow fibers from being effective. This prevents even dialysate perfusion through the tightly packed fibers. Yarns separators and moiré structures have been implemented in order to tackle this issue. The moiré structure configuration shows the most homogenous dialysate flow [39].

Flow is really important when it comes to one of the major difficulties of the filtration process, the removal of a wide range of toxin molecules, while retaining bigger molecules (e.g., albumin). There is no doubt that any improvement in this issue could enhance the treatment outcomes for ESRD patients. For this reason, the increase of pore size, "sharpening the molecular weight cutoff of high flux membranes to maximize removal of low molecular weight proteins," has been one of the main focuses of hemodialysis membrane developments [5]. From a different perspective, superflux or protein-leaking membranes have been designed for

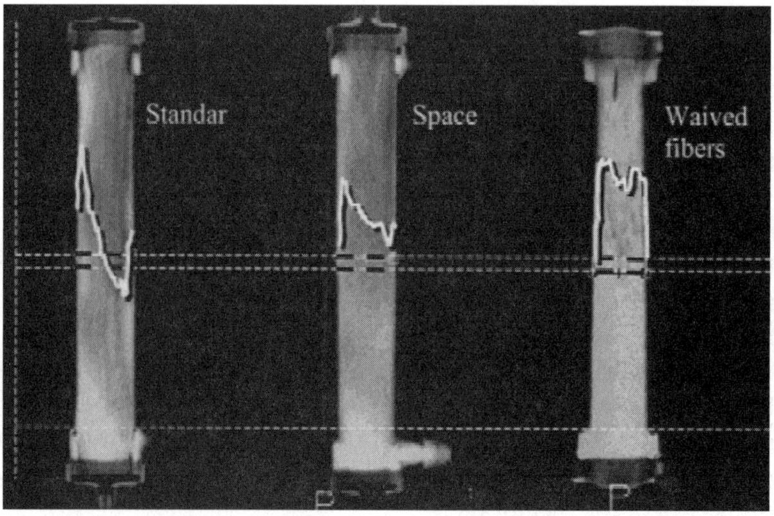

Figure 23.11 Injection of dye in the blood compartment of the dialyzer shows uneven distribution of flow in the apparatus. This distribution is due to the packing of the membranes in the dialyzer. Moiré (waived) structures have shown better results than yarns separators in generating an even flow distribution [39].

low-molecular-weight-protein clearance, but they also present higher albumin loss than the previous high-flux membranes [10].

Pore geometry is another concern for hemodialysis filtration. The round form of the pores strongly associates the hydraulic force with the pore radius as $Q / \Delta P = \pi r^4 / 8\mu L$, where Q is the volumetric flow, P is the hydrostatic pressure, μ is the viscosity, L is the length of the pore, and r is the radius of the pore. L is not necessarily the thickness of the membrane. It was mentioned that the glomerular base membrane has a slit-shaped diaphragm (Figure 23.3), which serves as a last barrier for glomerular filtration. This has inspired the development of a silicon membrane with slit-shaped pores [5] that provide for size and charge separation. The slit pores with dimensions of 10 to 100 nm by 45 nm are arranged on silicon chips (Figure 23.12). Different from the round pores, these slit-shaped pores relate the hydraulic permeability and steric hindrance to the solute passage with the long dimension of the pore (w) and the smallest dimension of the pore, respectively [5], as $Q / \Delta P = wh^3 / 12\mu L$, where the long dimension of the slit and thickness of the slit are represented by w and h, respectively. Artificial slit membranes are expected to present a better performance than normal round-pore membranes.

As previously mentioned, the electrostatic state of the membrane plays a key roll in solute separation [8, 41]. An electrical double layer (or Debye layer) can be formed by a "microfluidic substrate in contact with an aqueous solution" [5]. The thickness of this layer is of the same order [5] as the nanopores. Therefore, a pore can contribute to the charged-solute separation. A membrane containing slit-shaped pores is expected to regulate the electrostatic and steric hindrance as experienced with high hydraulic permeability [5]. Recent preliminary data from Humes and Fissell promise a feasible protein separation with this glomerularlike technology [5].

As discussed previously, the membrane seems to be the basis of dialysis treatment. Thus, a membraneless dialysis would be a really innovative approach [7, 8]. As the name suggests, the filtration process would be achieved without membrane

Figure 23.12 This slit-shaped silicon nanopore membrane mimics the GBM. The geometry also favors solute separation [5].

implementation. It is argued that even the complications of blood-membrane interaction can be avoided by the implementation of microfluidic techniques. This will eliminate the possibility of blood-activation reactions and membrane "fouling." Filtration relies on the diffusivity of molecules (in this case, the undesired solutes) between two miscible fluids (e.g., blood and dialysate). Both fluids may have the same overall thickness, direction, and flow velocity (Figure 23.13).

A continuously functioning artificial nephron claims to be the promise of nanotechnology in the dialysis area (Figure 23.14) [42]. The Human Nephron filter presents a two-stage filtration analogous to the processes carried out in the glomerulus and tubules. As in the glomerolus, the first process retains macromolecules and cells by a size-separation filtration. The second stage will be the reabsorption of solutes that will maintain the homeostasis of the body by the incorporation of synthetic ion channels and aquaporines. The reabsorption process remains unclear in the described process [5, 42].

In a recent work, Kaazempur-Mofrad, Krebs, and Borenstein proposed a novel MEMS-based system for renal replacement that incorporates fractal microvascular network designs [9]. This microfabricated device will offer a smaller, less invasive, and cheaper therapy for end-stage kidney failure. Their preliminary ultrafiltration studies showed that this micron-scaled hemofiltration system offers a significant improvement in ultrafiltration efficiency over other hemodialysis technologies. Additionally, the miniature aspect of this microfabricated system allows for it to be portable and potentially wearable. Most importantly, this new system can be used as a continuous renal replacement therapy. Kaazempur-Mofrad, Krebs, and Borenstein developed a computational algorithm to generate fractal microvascular networks that mimic proper anatomical and physiological conditions for flow and mass transport in the device (Figure 23.15) [9]. From these algorithms, consistent bilayer devices were fabricated using MEMS technologies and replica-molding

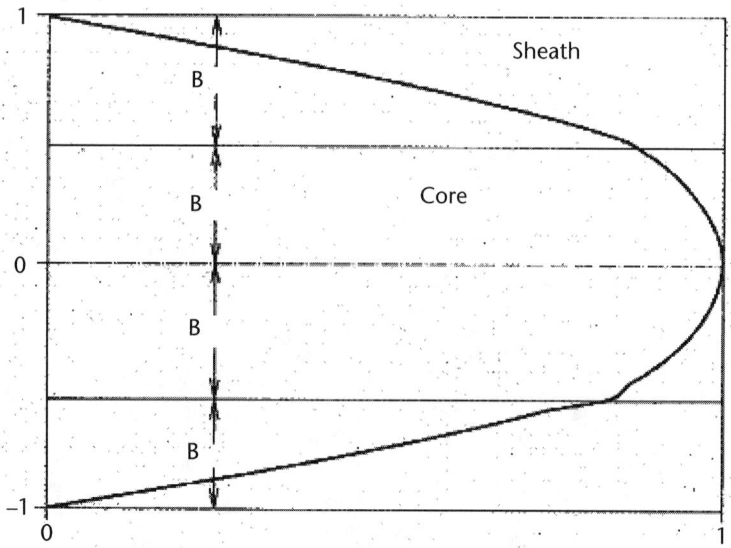

Figure 23.13 Bidimensional flow diagram of blood core sheathed by a secondary fluid above and below. The blood and the sheaths thickness should be 2*B* and *B*, respectively [7].

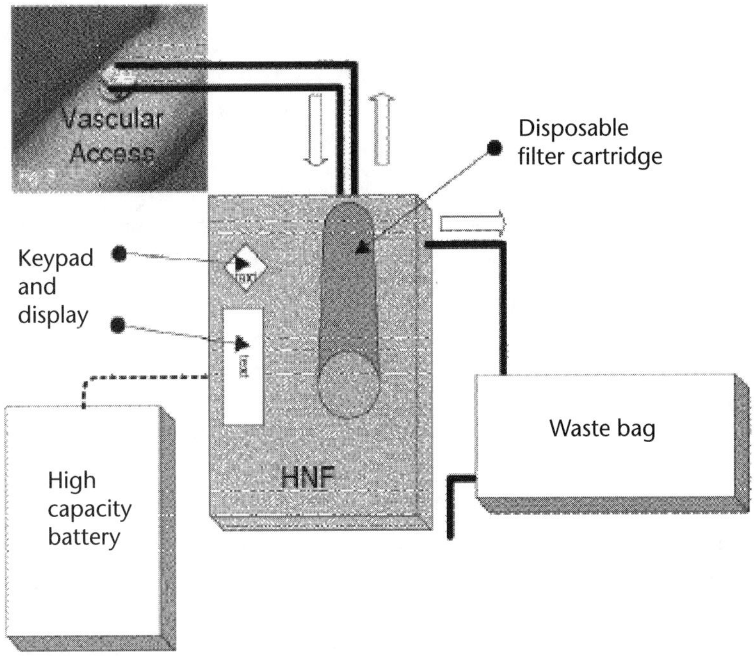

Figure 23.14 Wearble Human Neprhon device [42].

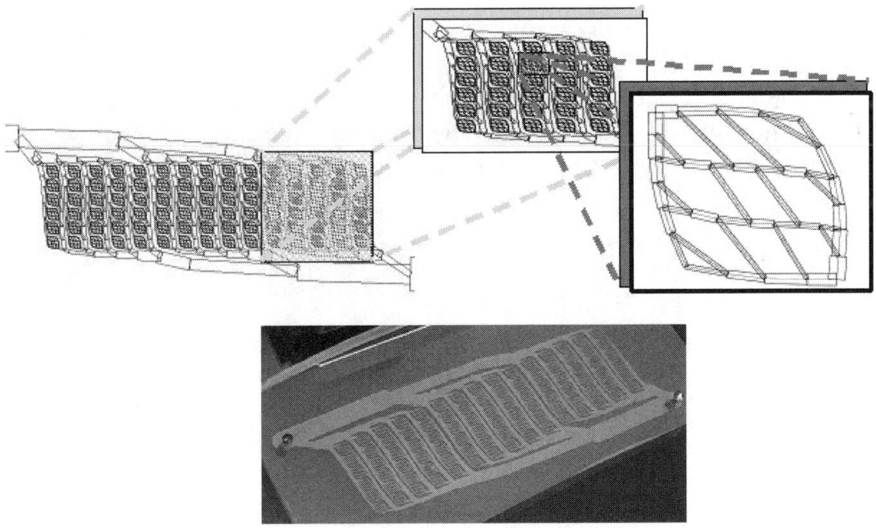

Figure 23.15 Schematics (top panels) and close-up (bottom panel) of the replica-molded microvascular channels [9].

processes (Figure 23.15). Each bilayer device comprised a microvascular and parenchymal network separated by a biocompatible polyethersulfone membrane (Figure 23.16). To study the ultrafiltration capabilities of this device, Kaazempur-Mofrad, Krebs, and Borenstein examined the clearances of urea and creatinine from the vascular/blood stream into the parenchymal/dialysate stream at

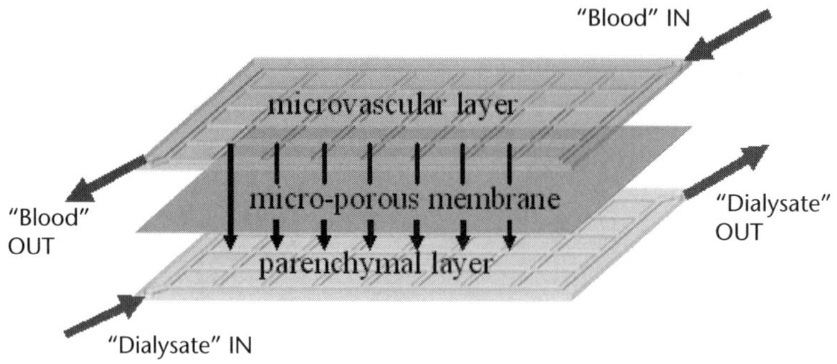

Figure 23.16 Experimental design set-up for the bilayer system [9].

varying flow rates under single-pass countercurrent conditions [9]. Fluid samples from each flow stream were taken at various time points up to five hours and analyzed for final concentration levels. These concentration values were used to calculate the individual solute clearances. The clearance values for urea and creatinine were normalized with respect to conventional hemodialysis parameters (surface area and flow rate) and showed superior performance in removing uremic solutes from an incoming vascular flow under optimized flow conditions. Improvements to this MEMS-based device will include incorporating renal proximal tubular cells to examine the readsorption of electrolytes, a critical hemostatic function not performed by standard hemodialysis treatment [9].

The last and most promising approach seems to be living membranes for dialysis [5, 23, 41, 43–46]. The term "living membranes" refers to an addition of kidney endothelial cells to the dialysis hollow fibers [23, 43, 44]. The main reason for this kind of approach is the fact that membranes cannot provide for the metabolic activities of the kidney cells. As mentioned before, the kidney is responsible not only for the filtration of metabolic products of the plasma but also for the metabolic and endocrine activities. Seeded cells on the fiber scaffold should cover for these activities, improving the current dialysis process and mimicking the renal function even more [5, 23, 41, 43–46].

In vitro testing done on a renal-tubule-cell assist device (RAD) demonstrated active transport, reabsorption, and metabolic and endocrine activities [23, 43, 44]. The Food and Drug Administration (FDA) approved phase I/II clinical trials on ten clinically ill patients with acute renal failure and multiple organ failure receiving continuous venous hemofiltration. The survival of ten patients past thirty days led to the phase II clinical trials [5, 46].

The membrane techniques discussed above have been focused on solute and fluid removal, which still provide incomplete substitution of renal functioning. The lacking metabolic, reabsorption, and endocrine renal functions could be provided by the implementation of renal tubule cells in current treatments. These tubule cells are expanded from adult kidney cells [42] and cultured along the inner surface of the fibers in a standard hemofiltration cartridge [41]. The high-flux membrane hemofiltration membrane is associated with an extracellular matrix coating that provides a scaffolding and immune barrier for the cells of the bioartificial tubule [5].

A potential hemocompatiblity and a hemofiltration enhancement are seen with the implantation of cell-seeded conduits along the filtration surfaces [5, 47]. Notable advances have been achieved in the fabrication of microcapillary networks with endothelial attachment [48]. Endothelial cells are expected to provide histocompatibility in these devices.

In vitro and ex vivo testing of the RAD have been conducted. The first showed differential properties in active transport, metabolic activities, and endocrine processes, which are all characteristic of renal tubule cells [41]. Porcine and human cells used for an ex vivo study met the results provided by the in vitro study [44]. Uremic dogs with bilateral nephrectomy [23, 44] were treated with RAD, resulting in improvements compared to animals treated with cell-free RAD controls. In the case of acute renal failure, plasma cytokine and the cardiovascular state of health were well modulated [43, 46, 49, 50].

This data led to an approval for the phase I/II clinical trial by the FDA. Ten patients with ARF and multiple organ failure (with predicted mortality rates averaging above 85 percent [5]) receiving continuous hemofiltration [47] were treated with RAD. The devices were seeded with human kidney cells isolated from organs.

It was demonstrated that the RAD treatment can be delivered and retain viability, durability, and functionality for twenty-four hours. Six of the ten patients survived past thirty days. This has moved the study to its current second stage of clinical trials [5].

23.8 Conclusion

The technology of renal replacement therapy has experienced steady progress since the time it was first applied on humans [5, 18]. Remarkable achievements have been made toward "adequate" dialysis. However, dialysis still is a crude procedure. ESRD patients still have to connect periodically to a filtering machine that does not yet fulfill their needs. Innovative approaches are being sought to develop a filtration process that provides better blood-membrane interaction, solute separation, continuity of the process, and, finally, renal endocrine activity. We hope these innovative steps in using nanotechnology for renal replacement will move this treatment toward what can be called an "ideal" dialysis therapy.

References

[1] Galletti, P. M., Colton, C. K., and Lysaght, M. J., "Artificial kidney," in E. J. D. B. B. R. C. P. LLC (ed.), *The Biomedical Engineering Handbook*, 2nd ed., Boca Raton, FL: CRC Press, 2000.

[2] Poh, C. K., j. L., Clark, W. R., and Gao, D., "Kidneys, artificial," in T. Francis (ed.), *Encyclopedia of Biomaterials and Biomedical Engineering*, New York: Marcek Dekker, 2006.

[3] NIH National Kidney and Urologic Diseases, *Information Clearinghouse, National Institute of Health*. 2006.

[4] Clark, W. R., and Gao, D., "Low-molecular weight proteins in end-stage renal disease: Potential toxicity and dialytic removal mechanisms," *J. Am. Soc. Nephrol.*, Vol. 13, No. 4, 2002, pp. S41–S47.

[5] Humes, H. D., and Fissell, W. H., T. K., "The future of hemodialysis membranes," *Kidney Int.*, Vol. 69, No, 2, 2006, pp. 1115–1119.

[6] Poh, C. K., Lu, J., Clark, W. R., and Gao, D., "Kidneys, artificial," in T. Francis (ed.), *Encyclopedia of Biomaterials and Biomedical Engineering*, New York: Marcel Dekker, 2006, pp. 1–22.

[7] Leonard, E. F., West, A. C., Shapley, N. C., and Larsen, M. U., "Dialysis without membranes: How and why?" *Blood Purif.*, Vol. 22, No. 5, 2004, pp. 92–100.

[8] Leonard, E. F., Cortell, S., and Vitale, N. G., "Membraneless dialysis—is it possible?" *Contrib. Nephrol.*, Vol. 149, No. 6, 200, pp. 343–353.

[9] Kaazempur-Mofrad, M. R., V. J., Krebs, N. J., and Borenstein, J. T., "A MEMS-based renal replacement system," *Proceedings of Solid-State Sensor, Actuator and Microsystems Workshop*," Vol. 10, No. 4, 2004, pp. 67–70.

[10] Ward, R. A., "Protein-leaking membranes for hemodialysis: A new class of membranes in search of an application?" *J. Am. Soc. Nephrol.*, Vol. 16, No. 2, 2005, pp. 2421–2430.

[11] Muskopf, S., "The urinary system and homeostasis," 2007.

[12] Greene, P. J., "The glomerulus (the kidney's what, where, and how)," 2007, Florida Golf Coast University.

[13] National Kidney Foundation, "Glomerular filtration rate," 2007.

[14] Albert Lasker Award, *J. Am. Soc. Nephrol.*, Vol. 13, 2002, pp. 3027–3030.

[15] Scribner, B. H., C. J., Buri, R., and Quinton, W., "The technique of continuous hemodialysis," *Trans. Am. Soc. Artif. Intern. Org.*, Vol. 6, 1960, pp. 88–103.

[16] Dworkin, L. D., "Why kidneys fail," *Mt. Sinai J. Med.* Vol. 59, No. 1, 1992, pp. 13–22.

[17] National Institute of Diabetes and Digestive and Kidney Diseases (NIDDKD), *Your kindneys and how they work*, NIH Publication No. 06-4241, 2005, pp. 1–12.

[18] Shinaberger, J. H., "Quantitation of dialysis: Historical perspective," *Semin. Dial.*, Vol. 14, No. 4, 2001, pp. 238–245.

[19] Mayo Foundation for Medical Education and Research, "Peritoneal dialysis: A flexible treatment option for kidney failure, in kidneys and urinary tract," 2007.

[20] Kolff, W., and Berk, H., "The artificial kidney: A dialyzer with a great area," *Acta Med. Scand.*, Vol. 117, 1944, pp. 121–134.

[21] National Institute of Diabetes and Digestive and Kidney Diseases (NIDDK), "Treatment methods for kidney failure," *Hemodialysis*, 2007.

[22] Gotch, F. A., and Sargent, J. A., "A mechanistic analysis of the National Cooperative Dialysis Study (NCDS)," *Kidney Int*, Vol. 28, 1985, pp. 526–534.

[23] Humes, H. D., Fissell, W. H., Weitzel, W. F., et al., "Metabolic replacement of renal function in uremic animals with a bioartificial kidney containing human cells," *Am. J. Kidney Dis.*, Vol. 39, No. 5, 2002, pp. 1078–1087.

[24] Unites States Renal Data System (USRDS), *Anual Data Report*. 2007.

[25] Graham, T., "Osmotic force," *Philos. Trans. R. Soc. Lond.*, Vol. 144, No. 1, 1854, pp. 177–228.

[26] Advanced Renal Education Program, "History of hemodialysis," 2006–2007, Fresenius Medical Care North America.

[27] Bigelow, G., "Collodion membranes," *J. Am. Chem. Soc.*, Vol. 29, No. 4, 1907, pp. 1576–1589.

[28] Richardson, B. W., "Practical studies in animal dialysis," *Asclepiad*, Vol. 6, No. 3, 1889, pp. 331–332.

[29] Abel, J. J., Rowntree, L. G., and Turner, B. B., "On the removal of diffusible substances from the circulating blood by means of dialysis," *Trans. Assoc. Am. Physic.*, Vol. 28, No. 2, 1913, pp. 14–19.

[30] Haas, G., "Dialysis of the flowing blood in the patient," *Klin Wochenschr*, Vol. 1888, No. 70, 1923.

[31] Thalheimer, W., "Experimental exchange transfusion for reducing azotemia: Use of the artificial kidney for this purpose," *Proc. Soc. Exp. Biol. Med.*, Vol. 37, No. 13, 1965, pp. 641–643.

[32] Wolf, A. V., Remp, D. G., Kieley, J. E., and Currie, G. D., "Artificial kidney function: Kinetics of hemodialysis," *J. Clin. Invert.*, Vol. 30, No. 1, 1951, pp. 1062–1070.

[33] Babb, A. L., Strand, M. J.,Uvelli, D. A., Milutinovic, J., and Scribner, B. H., "Quantitative description of dialysis treatment," *Dial. Index Kidney Int. Suppl.*, Vol. 5, No. 3, 1975, pp. 23–29.

[34] Sargent, J. A., and Gotch, F. A., "The analysis of concentration dependence of uremic lesions in clinical studies," *Kidney Int. Suppl.,* Vol. 5, No. 2, 1975, pp. 35–44.

[35] Villarroel, F., Klein, E., and Holland, F., "Solute flux in hemodialysis and hemofiltration membranes," *Trans. Am. Soc. Artif. Intern. Org.*, Vol. 23, No. 2, 1977, pp. 225–233.

[36] Jaffrin, M. Y., Ding, L. H., and Laurent, J. M., "Simultaneous convective and diffusive mass transfers in a hemodialyser," *J. Biomechan. Eng.*, Vol. 112, No. 2, 1990, pp. 212–219.

[37] Lowrie, E., and Lew, N., "The urea reduction ratio (URR): A simple method for evaluating hemodialysis treatment," *Contemp. Dial. Nephrol.*, Vol. 12, No. 3, 1991, pp. 11–20.

[38] Malcheski, P., Ellis, P., Nosse, C., Magnusson, C., Lankhorst, B., and Nakamoto, S., "Direct quantification of dialysis," *Dial. Transplant.*, Vol. 11, 1982, pp. 42–44.

[39] Ward, R. A., Leypoldt, J. K., Clark, W. R., et al., "What clinically important advances in understanding and improving dialyzer function have occurred recently?" *Semin. Dial.*, Vol. 14, No. 5, 2001, pp. 160–174.

[40] Libeta, C., Zucchi M., Gori E., Sepe V., Galli F., Maloni F., Milanesi F., and Dal Canton, A., "Vitamin E–loaded dialyzer resets PBMC-operated cytokine network in dialysis patients," *Kidney Int.*, Vol. 65, No. 4, 2004, pp. 1473–1481.

[41] Humes, H. D., MacKay, S. M., Funke, A. J., and Buffington, D. A., "Tissue engineering of a bioartificial renal tubule assist device: In vitro transport and metabolic characteristics," *Kidney Int.*, Vol. 55, No. 6, 1999, pp. 2502–2514.

[42] Nissenson, A. R., Ronco, C., Pergamit, G., et al., "Continuously functioning artificial nephron system: the promise of nanotechnology," *Hemodial. Int.*, Vol. 9, No. 3, 2005, pp. 210–217.

[43] Humes, H. D., Buffington, D. A., Lou. L., et al., "Cell therapy with a tissue-engineered kidney protects against the multi-organ consequences of septic shock," *Crit. Care Med.*, Vol. 31, No. 10, 2003, pp. 2421–2428.

[44] Humes, H. D., Buffington, D. A., MacKay, S. M., et al., "Replacement of renal function in uremic animals with a tissue-engineered kidney," *Nat. Biotechnol.*, Vol. 17, No. 5, 1999, pp. 451–455.

[45] Humes, H. D., Krauss, J. C., Cieslinski, D. A., and Funke, A. J., "Tubulogenesis from isolated single cells of adult mammalian kidney: Clonal analysis with a recombinant retrovirus," *Am. J. Physiol.: Renal Physiol.,* Vol. 271, No. 40, 1996, pp. F42–F49.

[46] Humes, H. D., Weitzel, W. F., Bartlett, R. H., et al., "Initial clinical results of the bioartificial kidney containing human cells in ICU patients with acute renal failure," *Kidney Int.*, Vol. 66, No. 2, 2004, pp. 1578–1588.

[47] Humes, H. D., "Bioartificial kidney for full renal replacement therapy," *Semin. Nephol.*, Vol. 20, No. 13, 2000, pp. 71–82.

[48] Shin, M., Matsuda, K., Ishii, O., et al., "Endothelialized networks with a vascular geometry in microfabricated poly(dimethyl siloxane)," *Biomed. Microdevices*, Vol. 6, No. 4, 2004, pp. 269–278.

[49] Fissell, W. H., Dyke, D. B., Buffington, D. A., et al., "Bioartificial kidney alters cytokine response and hemodynamics in endotoxin-challenged dogs," *Blood Purif.*, Vol. 20, No. 1, 2002, pp. 55–60.

[50] Fissell, W. H., Lou, L., Abrishami, S., et al., "Bioartificial kidney ameliorates Gram-negative bacteria-induced septic shock in uremic animals," *J. Am. Soc. Nephrol.*, Vol. 14, No. 2, 2003, pp. 454–461.

CHAPTER 24

Engineering Pulmonary Epithelia and Their Mechanical Microenvironments

Dongeun Huh, Yoko Kamotani, James B. Grotberg, and Shuichi Takayama

24.1 Introduction

The lung has an anatomically unique structure consisting of a branching network of conducting tubes that enable convective gas transport to and from alveolar compartments, where gas exchange occurs. Throughout development and adult life, the respiratory system experiences a variety of physical forces imposed by structural changes in surrounding tissue, continuous passage of fluids, and cyclic mechanical deformation of the basement membrane and extracellular matrix (ECM). As the most essential cellular constituent of the respiratory system, the epithelial cells comprising the luminal surface of airways and alveoli are known to sense and respond to this dynamic mechanical environment. During fetal lung development, mechanical forces generated by airway distension and intermittent fetal breathing movements have been shown to profoundly influence the proliferation, apoptosis, and differentiation of pulmonary epithelial cells [1, 2]. Mechanical perturbations in the mature lung also play a critical role in regulating the structure, function, and metabolism of the epithelial cells [3]. Unusual changes in the mechanical environment of the respiratory system often contribute to the progression and exacerbation of various pulmonary disorders by eliciting abnormal biological responses of epithelial tissue in airways and alveoli. For example, large deformations of the alveolar basement membrane due to overdistension of the lung during mechanical ventilation can induce injury of alveolar epithelial cells and upregulate inflammatory responses, which can lead to ventilator-induced lung injury [4]. In asthmatic airways, compressive mechanical forces resulting from smooth muscle constriction have been shown to cause airway epithelial cells to communicate with neighboring mesenchymal cells to initiate and amplify airway remodeling [5].

Over the past two decades, discovery and elucidation of these mechanosensitive events in various physiologic and pathologic situations have been of paramount importance and interest in respiratory biology and physiology. Experimentally investigating the effects of mechanical forces on pulmonary epithelial cells, however, is often challenging for several reasons: (1) Due to the complex architecture and mechanical properties of the lung and marked spatial heterogeneity of pulmonary epithelial cells, ex vivo and in vivo models based on whole-organ systems are difficult to use for studying the mechanical responses of airway and alveolar tissue at the cellular level. Instead, in vitro cell-culture systems can be used as an alternative approach. Unlike other traditional culture methods, however, in vitro culture of pulmonary epithelial cells often requires permeable supports for cell attachment and independent access to both basal and apical sides of a cellular monolayer to allow the cells to polarize. Moreover, after reaching confluence, the cells need to be grown at an air-liquid interface to differentiate and gain morphological and biochemical phenotypes that match those found in native tissue. (2) It is difficult to precisely reproduce in vivo–like dynamic mechanical microenvironments of pulmonary epithelial cells in vitro. This usually entails careful design, fabrication, and engineering of mechanical tools capable of generating different types of physical forces that can modulate the interactions of the epithelial cells with culture substrata or fluid flows to simulate various physiologic and pathologic conditions. Also, the need for the ability to systematically control the level of applied forces and to quantify actual mechanical stresses acting on the cells remains an important issue in such systems. (3) Mechanical forces in the respiratory system can induce not only immediately detectable changes in the morphology and viability of pulmonary epithelial cells but also very transient signaling events mediated by soluble molecules such as cytokines and growth factors. This can trigger complex intra- and intercellular signal transduction, leading to longer-lasting effects on the cells. Therefore, a detailed understanding of the cellular responses to mechanical stimulation requires accurate and quantitative measurements and analysis of the biochemical environments of pulmonary epithelial cells. Such tasks usually call for the development of biological assay systems with high sensitivity and specificity.

These challenges and requirements have limited our understanding of the behavior of lung cells in their mechanically active microenvironment and, in turn, driven the development of new methodologies to address the technical limitations arising from the lack of appropriate tools for cellular studies. In this chapter, we describe recent research efforts directed toward developing new experimental models that enable in vitro engineering of pulmonary epithelial cells, re-creation of their dynamic mechanical environment, and measurements of their response to various physiologic and pathologic mechanical forces. Specifically, we review the literature pertaining to cellular-level experimental studies based on in vitro systems that can: (1) induce and optimize the growth and differentiation of primary pulmonary epithelial cells in biomimetic culture environments; (2) recreate mechanical stresses originating from matrix deformation and tissue remodeling to investigate cell-surface or cell-cell interactions; (3) generate various pulmonary flows to simulate cell-fluid interactions; and (4) measure and characterize cytokine production induced by mechanically stimulating pulmonary epithelial cells.

Figure 24.1 shows the flow of research for microtechnology-based tissue engineering of the pulmonary epithelia and their microenvironments to analyze cellular-level responses to physiologic and pathologic mechanical stimuli. Hypotheses (dark grey block in the first row) concerning the effect of various mechanical microenvironments are formulated and classified into specific physical parameters (light grey blocks in the second row). These parameters are imposed upon the cells of interest (oval in the fourth row; for this chapter, the cells of interest are pulmonary epithelial cells) using microtools (white blocks in the third row). The responses of cells are also measured and analyzed by various microtools (white blocks in the fifth row) to characterize changes in biochemical and cellular parameters (light grey

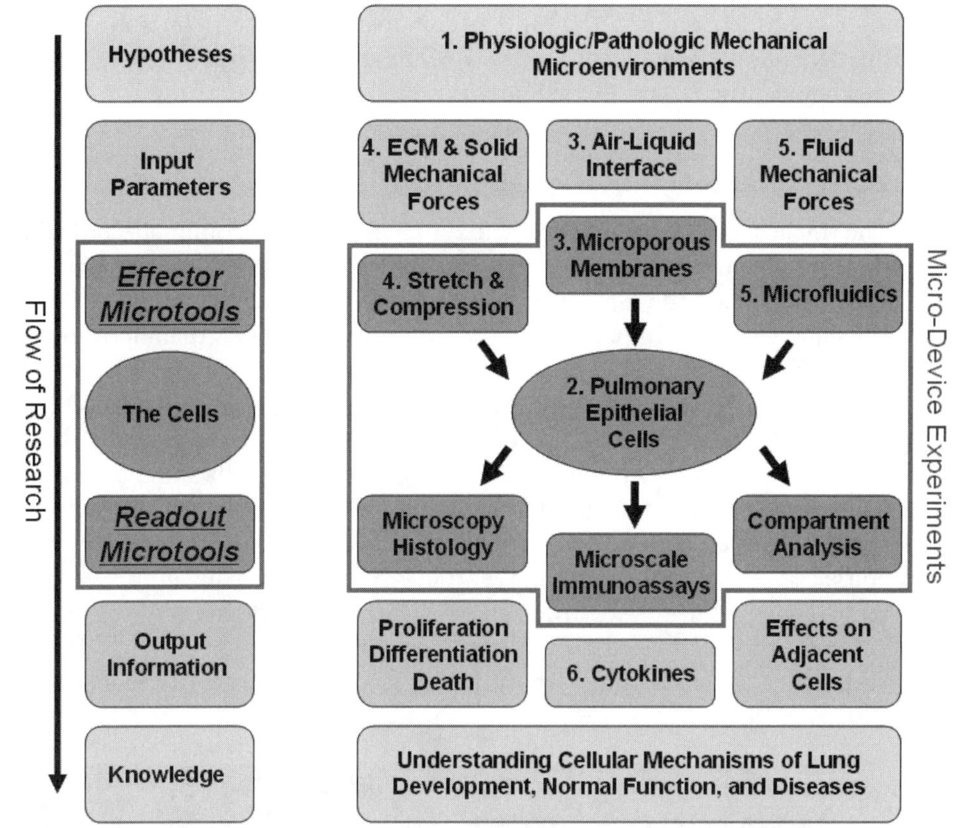

Figure 24.1 Microtechnology-based research approaches to engineering pulmonary epithelia and their mechanical microenvironments and to studying cellular responses to various mechanical stimuli. Numeric numbers represent section numbers.

blocks in the sixth row), ultimately leading to knowledge concerning the development, normal function, and diseases of the lung (dark grey block in the bottom row). The following sections of this chapter expand upon each of these topics represented in this scheme.

24.2 The Lung and Pulmonary Epithelial Cells

The complex hierarchy of branching airways is a major determinant of the internal structure of the lung that provides sufficient surface area for gas exchange to meet the need for oxygen uptake and carbon dioxide elimination. During inspiration, air passes through either the nose or mouth into the larynx, which opens into a long tube called the trachea. The trachea branches into two primary bronchi, one of which enters each lung. Within the lung, there are more than twenty generations of branching, resulting in the formation of intrapulmonary bronchi, bronchioles, alveolar ducts, and alveolar sacs. The lumens of these airways and alveoli consist of various types of epithelial cells that greatly differ in morphology and function, depending on their location within the airway tree. Also, there is considerable interspecies variation in the cellular composition of the pulmonary epithelium. In this section, we provide an overview of the different types of epithelial cells identified beyond the larynx in the human respiratory system.

24.2.1 Tracheobronchial Epithelial Cells

The trachea and bronchi have ciliated pseudostratified columnar epithelium with a thick basement membrane. This epithelium and its underlying layer of loose connective tissue called the lamina propria make up the tracheobronchial mucosa. Two principal cellular components of the tracheobronchial epithelium are ciliated cells and goblet cells [Figure 24.2(a)]. Ciliated cells represent approximately 80 percent of the epithelial cells residing on the luminal borders of large airways. They extend through the full thickness of the epithelium and are best characterized by beating cilia. Goblet cells are interdispersed among the ciliated cells and secrete mucus, a viscous fluid composed primarily of highly glycosylated proteins called mucins suspended in electrolyte solutions. Mucus acts as a fluid barrier to protect airway epithelium by trapping foreign particulates entering the lung, which are then removed from the airway toward the throat by a coordinated sweeping motion provided by the beating cilia of ciliated cells. This process is known as mucociliary clearance and serves as an important first line of defense in the respiratory system.

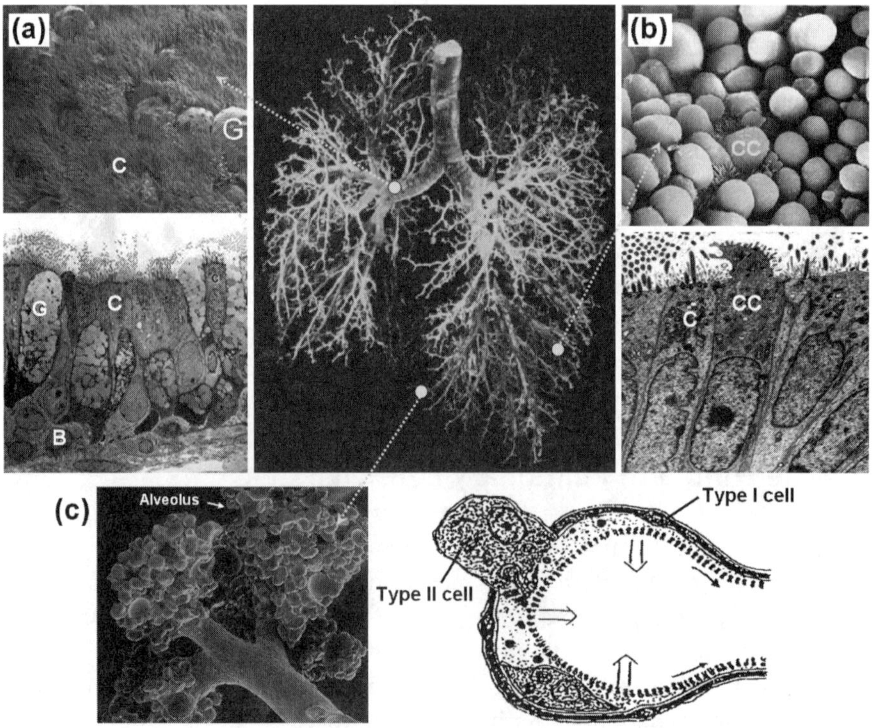

Figure 24.2 Pulmonary epithelial cells: (a) Scanning electron micrograph (top) of bronchial epithelium and its ultrastructure (bottom) viewed by transmission electron microscopy. G, C, and B represent goblet cells, ciliated cells, and basal cells, respectively. (b) Nonciliated and dome-shaped Clara cells (CC) are predominant in bronchiolar epithelium. Ciliated cells (C) are sporadically situated between Clara cells. (c) Alveolar epithelium consists of large cuboidal type II epithelial cells and thin squamous type I cells. (*Source:* [6–11], reproduced with permission.)

Other cell types present in tracheobronchial epithelium include basal cells and brush cells. Basal cells are distributed mainly along the basement lamina and are smaller in size compared to other cells. The main function of these cells is to attach the columnar epithelium to the basement membrane. They can also undergo mitosis and play an important role in the growth, maintenance, and replacement of the epithelial layer. Brush cells have a columnar structure and are characterized by the presence of microvilli on the apical surface. Their basal surface is in synaptic contact with afferent nerve endings, which suggests the possible role of these cells as chemoreceptors involved in general sensation.

24.2.2 Bronchiolar Epithelial Cells

Bronchioles are smaller branches of bronchi and distinguished by a prominent smooth muscle layer and the absence of cartilage around the outer walls. The branching of bronchioles eventually leads to the formation of terminal bronchioles, which then divide into respiratory bronchioles that have alveoli sporadically distributed on their walls. Bronchiolar epithelium consists mainly of ciliated cells and nonciliated secretory cells known as Clara cells [Figure 24.2(b)]. Ciliated cells have a simple columnar shape but become more cuboidal in distal bronchioles. With increasing airway generation, the number of basal cells, ciliated cells, and goblet cells decreases dramatically, and Clara cells become a predominant population in the bronchioles. Clara cells are dome-shaped, produce watery proteinaceous secretions called Clara cell secretory proteins, and usually protrude into the bronchiolar lumen. These secretory cells are unique to the bronchiolar region of the lung and differ from goblet cells in that their secretory granules are smaller in size and number and they contain abundant granular endoplasmic reticulum. Also, Clara cells are known as one of the most multifunctional and heterogeneous cell types in the mammalian lung, and their function has not yet been fully established [12]. Several lines of evidence suggest that they are involved in the production of surfactant proteins and may play an important role in the regulation of pulmonary inflammation. Also, it has been demonstrated that Clara cells are responsible for detoxifying harmful airborne substances inhaled into the lung using the isoenzymes called cytochrome p450 present in their smooth endoplasmic reticulum. Finally, they are mitotically active and have been shown to serve as the progenitor of ciliated cells as well as themselves, contributing to the regeneration of bronchiolar epithelium.

Although the presence of basal cells in the distal parts of the airway system is uncommon, they persist in small airways, but are absent in terminal bronchioles. Bronchiolar basal cells are flattened, triangular, or trapezoidal in shape and have different organelle features from those of bronchial basal cells. They act as progenitor cells for bronchiolar epithelium under certain conditions and can differentiate into Clara cells or ciliated cells.

24.2.3 Alveolar Epithelial Cells

The alveoli are the final branchings of the airway tree and serve as the gas exchange units of the lung. They have hollow sacs whose open ends are continuous with the lumens of the airways. The alveolar epithelium comprises two epithelial cell types with highly specialized functions: type I and type II alveolar epithelial cells [Figure 24.2(c)]. Type I cells are large (~40 μm in diameter) squamous cells with long and thin cytoplasmic extensions that spread out along the walls and comprise the thin alveolar epithelium; in regions of gas exchange in alveoli, these cells may measure as small as 0.1 μm in thickness. They constitute ~35 percent of all alveolar epithelial cells and cover ~93 percent of the alveolar epithelial surface area. Type I cells are responsible for gas exchange, and their small thickness greatly facilitates gas diffusion across the blood-air barrier. They are also known to regulate transport of physiologically important solutes and water between circulating blood and the alveolar space, contributing to maintaining alveolar fluid balance.

Type II cells are small cuboidal cells (~10 μm in diameter) that comprise ~7 percent of the alveolar epithelial surface area and ~65 percent of all alveolar epithelial cells. These cells are located in the alveolar corners and are identified by intracellular lamellar bodies that serve as storage granules for surfactant and by the presence of organelles and microvilli on their apical membrane. Their major function is to synthesize, secrete, and recycle surfactant, a surface-active lipoprotein that serves to reduce surface tension at the air-liquid interface in the alveolar epithelium. Other major functions include the synthesis of immune effector molecules and the regulation of fluid balance through transport of ions and water across transmembrane channels. Type II cells also maintain alveolar homeostasis by mediating repair to the injured alveolar epithelium; when type I cells are damaged and removed, type II cells can proliferate and differentiate into type I cells to replace the injured cells. Therefore, type II cells play a critical role in normal pulmonary function, as well as in the response of the lung to damage.

24.3 In Vitro Production and Engineering of Pulmonary Epithelium

In vitro culture of pulmonary epithelial cells has been an indispensable tool for advancing our understanding of the complex cellular and molecular processes underlying normal function and disease development in the respiratory system. The most important requirement for successful in vitro culture is the establishment of optimal conditions that enhance the proliferative capacity of cells and facilitate the expression of differentiated characteristics found in the native pulmonary epithelium. This is achieved mainly by defining the growth media, optimizing ECMs, and using an air-liquid interface for culture. Here, we discuss the use of these approaches to engineer in vitro culture of primary pulmonary epithelial cells and to reproduce the essential morphological and biochemical properties of in vivo airway or alveolar epithelium.

24.3.1 Engineering of Primary Airway Epithelial Cell Culture

A primary factor that affects the viability of freshly harvested airway epithelial cells is the culture media. The most widely used media formulations for primary airway epithelial cells are based on the Lechner's Laboratory of Human Carcinogenesis basal medium and the Dulbecco's modified Eagle's medium/Ham's F-12 [13]. For long-term culture, the exposure of the cells to serum is generally limited to short periods because blood-derived serum contains transforming growth factor (TGF)-β, which can inhibit the proliferation of primary airway epithelial cells [14, 15]. Small amounts of serum, however, have been shown to stimulate morphological differentiation of distal airway epithelial cells [16]. Other important supplements for modulating the proliferative capacity of airway epithelial cells include retinoic acid, epidermal growth factor (EGF), bombesin, and fibroblast growth factor [15, 17–21]. EGF also induces the morphological transformation of human bronchial epithelial cells in culture by causing them to generate filapodia extensions [22].

ECMs are also important in primary airway epithelial cell culture because they have a profound influence on cytoplasmic receptors and cellular architecture, which can change gene expression and eventually the presentation of specific phenotypes. Yamaya et al. studied the differentiated structure and function of primary human tracheal epithelial cells on different types of collagen culture supports, including human placental collagen, vitrogen gels, and ECM produced by bovine corneal endothelial cells [21]. Based on transepithelial electrical-resistance measurements, electron microscopic imaging, and immunohistochemistry, they showed that the cells cultured on vitrogen gel produced airway epithelium with cytological and electrical properties that closely mimicked native epithelium. Also, their method was effective for culturing primary cystic fibrosis epithelial cells and retaining their essential disease characteristics in vitro. Wu et al. investigated the differentiation of human tracheobronchial epithelial (HTBE) cells on collagen gels [23]. They demonstrated that the epithelial cells synthesized mucin and formed mucous granules in culture when they were maintained on collagen gel substrata but not on plastic tissue-culture plates or thin collagen-coated surfaces. In a follow-up study, Robinson and Wu also examine the effect of collagen gel (CG) thickness on the growth and differentiation of HTBE cells [24]. They found that the attachment and proliferation of the cells were independent of CG thickness, but the synthesis and secretion of mucin was elevated with an increasing CG thickness. A mixture of collagen, fibronectin, and bovine serum albumin has been used by Lechner et al. to induce clonal growth of human bronchial epithelial cells without using feeder cells [25]. Collagen gels can also be used to establish coculture of airway epithelial cells with lung fibroblasts. Infeld, Brennan, and Davis cocultured human lung fibroblasts between the layers of type I collagen gel with HTBE cells on the upper collagen lattice [26]. In this system, the fibroblasts were observed to migrate preferentially toward HTBE cells, suggesting that HTBE cells directed fibroblast migration. In another study using similar coculture systems, they also showed that HTBE cells penetrated collagen matrices when cultured with human fetal lung fibroblasts and that the invading clusters of the epithelial cells formed tubular structures undergoing dichotomous branching [27]. These studies provided important insights into airway morphogenesis during development, repair after lung injury, and the pathogenesis of bronchial neoplasms. Mio et al. also developed coculture of human bronchial epithelial cells (HBEC) and human lung fibroblasts in collagen gels to study connective-tissue contraction during wound healing and fibrosis [28]. Their results showed that HBEC might modulate fibroblast activities by causing them to generate higher levels of traction force and to induce the contraction of collagen gel.

The most critical step necessary to induce the differentiation of airway epithelial cells is to expose proliferated cells to an air-liquid interface (ALI). During this procedure, cells are grown on a semipermeable membrane that separates two independent chambers placed in a culture well, and medium is added to the basolateral side only. This configuration results in the formation of a layer of liquid with a thickness of ~15 μm and an increased level of aerobic respiration in the cells [29]. Dejong et al. demonstrated that ALI culture led to the formation and maturation of cilia in HBEC culture, regardless of the type of substratum used in the study (de-epidermized dermis and collagen membrane) [30]. Also, it was found that the cells maintained in submerged cultures showed no sign of ciliogenesis, evidencing the important role of ALI culture in morphological differentiation of airway epithelial cells. Using a canine bronchial epithelial cell-culture system, Johnson et al. showed that ALI culture conditions caused a significant increase in active transport of sodium ions compared to standard submerged culture methods [29]. It was also noted that cellular metabolism shifted from anaerobic to aerobic. These results indicate that increased oxygen availability to the epithelial cells in ALI culture produced a higher rate of transepithelial active sodium transport. Based on the use of ALI, Davidson et al. established a novel primary culture system of differentiated murine tracheal epithelial cells that can potentially facilitate the studies of tracheal epithelium from transgenic mouse models of human pulmonary disease such as cystic fibrosis [31]. Their technique allowed for the generation of confluent and polarized epithelial cultures with high transepithelial resistance (an index of tight junction formation), minimal contamination with nonepithelial cell types, differentiation into ciliated cells and goblet cells, and genetic expression/electrophysiological profile characteristic of murine tracheal epithelium. Similar examples can be found in the work by You et al., where they created primary culture of mouse tracheal epithelial cells using ALI that can undergo rapid proliferation and differentiation into ciliated cells, mucus-secreting cells, Clara cells, and progenitor-like cells [32]. Whitcutt, Adler, and Wu designed a simple and disposable culture chamber called a Whitcutt chamber where culture conditions could be switched between submerged culture and ALI culture by the movement of a transparent and permeable gelatin membrane [33]. Using this chamber, they demonstrated in vitro production of fully differentiated guinea pig tracheal epithelial cells exhibiting various mucociliary functions. Differentiated primary airway epithelial cells are known to rapidly dedifferentiate and lose their proliferative capacity and secretory/morphological properties during subculture [34–36]. To address this issue, Widdicombe et al. developed a simple technique for expanding primary culture of human tracheal epithelial cells and demonstrated that the epithelial cells could be passaged several times without losing the ALI-culture-induced differentiation characteristics [36].

24.3.2 Engineering of Primary Alveolar Epithelial Cell Culture

In vitro culture of primary type I alveolar epithelial cells is difficult to establish and maintain [37, 38]. For this reason, researchers have been dependent mainly upon primary culture of type II alveolar epithelial cells for cellular studies. It has been shown that primary type II cells acquire morphological and biochemical properties of type I cells over several days in culture [39, 40]; during differentiation, type II cells lose their cuboidal shape, form a monolayer with high transepithelial resistance, lose microvilli and lamellar bodies, and display thin cytoplasmic extensions away from a protruding nucleus. This process is also accompanied by the decreased synthesis of surfactant phospholipids/surfactant proteins and an increased reactivity to type I–specific membrane components. These type I–like cells have gained acceptance as an in vitro model of type I alveolar epithelium for investigating transport of solute, water, and drugs in alveoli [40]. Based on the extensive phenotypic changes of primary type II cells in culture, Elbert et al. established in vitro alveolar epithelium comprising type I–like cells [41]. The use of low-serum growth medium and polyester filter supports coated with fibronectin and collagen promoted the differentiation of type II cells into type I–like cells and enabled the production of monolayers with tight junctions and desmosomes. Fuchs et al. further characterized the differentiation process in the same culture system and showed the transition of a cell population consisting mainly of type II cells to an epithelial monolayer composed of both type I and type II cells [42]. They also found that the type I–like cells exhibited invaginations of plasma membrane resembling caveolae and the increased synthesis of the structural protein caveolin-1, both of which are believed to play an important role in macromolecule transport across the blood-air barrier form by type I cells in vivo [43]. The effect of ECMs on the differentiation of type II cells into type I–like cells was evaluated by Olsen et al. [44]. Using collagen, fibronectin, and laminin-5 (Ln5), they revealed that early in culture, type II cells lost their characteristic properties and developed intermediate type I–type II phenotypes, regardless of individual matrix components. Beyond this stage, the cells showed ECM-dependent behavior; type II cells on matrices consisting of collagen/fibronectin, collagen alone, or Ln5 alone continued a loss of their characteristics, whereas the cells on collagen/fibronectin/Ln5 or collagen/Ln5 regained their original phenotypes.

Under appropriate conditions, type II cells have been shown to maintain their phenotypic characteristics without differentiating into type I–like cells [39] and have been used as an in vitro model for examining the metabolism and function of type II alveolar epithelial cells. These techniques include the addition of growth factors, culture on various ECM substrata, and the use of an air-liquid interface. Borok et al. demonstrated that keratinocyte growth factor (KGF) could modulate and partially maintain the characteristic phenotype of type II cells by preventing and reversing the expression of aquaporin-5 (a water channel present in vivo on the apical surface of type I cells only) in primary culture of rat type II cells [45]. In another study, they also showed that the use of culture medium supplemented with KGF or rat serum inhibited or reversed the expression of T1α, a gene expressed exclusively by type I cells in the adult rat lung [46]. This verifies that the phenotype of type II cells can be retained in primary culture and that the differentiation between type I and type II cells is partially reversible. KGF was also shown to mediate epithelial-mesenchymal interactions in coculture of rat type II cells and lung fibroblasts, which then allowed the type II cells to maintain their ability to produce surfactant proteins [47]. Nonplastic culture substrata have proven effective for preserving the characteristics of type II cells. Geppert, Williams, and Mason used floating collagen membranes to prevent rat type II cells from spreading and undergoing subsequent morphological changes [48]. Laminin-rich matrix gel derived from Engelbroth-Holm-Swarm sarcoma was shown to prolong the retention of lamellar bodies [49] and to promote surfactant-protein synthesis [50] in type II cells. Organotypic culture systems based on gelatin sponge matrices enabled the formation of alveolar-like structures composed of surfactant-secreting rat type II cells with lamellar bodies, as well as long-term maintenance of their phenotype [51]. Although not as commonly practiced as in the case of primary airway epithelial cell culture, ALI culture contributes significantly to the maintenance of the differentiated characteristics of type II alveolar epithelial cells. Dobbs et al. corroborated this by showing that human type II cells cultured in ALI were cuboidal in shape, contained lamellar bodies, and had high content of mRNA for surfactant proteins, whereas the cells maintained in submerged culture differentiated into type I–like squamous cells [52]. Using ALI culture of human type II cells, Alcorn et al. demonstrated that increased oxygen availability to the cells due to the presence of ALI increased cell polarization and contributed to promotion and maintenance of type II cells without a loss of their essential properties, as well as the expression of surfactant proteins [53]. ALI culture combined with the use of defined serum-free medium and porous polycarbonate membranes coated with Matrigel was employed in coculture of rat type II cells with bovine microvascular endothelial cells to recreate the blood-air barrier in vitro [54].

24.4 Engineering of Cell-Matrix and Cell-Cell Interactions

Expansion and relaxation of the lung during respiration or mechanical ventilation generate mechanical deformation of the basement membrane of pulmonary epithelial cells, which in turn imposes mechanical stresses on the cells. Respiratory diseases with tissue remodeling often accompany chronic mechanical perturbations to the interactions between pulmonary epithelial cells themselves or between the cells and the underlying subepithelial layer. In this section, we categorize these mechanical forces and stimulations into different types depending on their essential features and discuss the resulting responses of pulmonary epithelial cells.

24.4.1 Cyclic Stretch in Two-Dimensional Culture

The Flexercell strain unit is a commercially available mechanical device that can impose cyclic stress (tension or compression) on cells by using vacuum to induce the deformation of an elastic membrane to which cells are attached [Figure 24.3(a)] [55]. Using fetal rat II alveolar epithelial cells cultured in this system, Sanchez-Esteban et al. demonstrated that cyclic stretch simulating breathing movements during lung development promoted the differentiation of fetal type II cells, as assessed by the increased production of surfactant phospholipids and the elevated expression of surfactant proteins B and C [56]. In primary culture of fetal rabbit type II alveolar cells, cyclic deformation also increased proliferation, surfactant-related phospholipid synthesis, and cAMP levels, suggesting that airway distension in the developing lung is involved in the maturation of type II cells [57]. In the edematous lung, the alveolar epithelium plays a crucial role in edema clearance by transporting Na^+ and liquid out of the air spaces. Waters et al. used the Flexercell system to examine the role of cyclic stretch in the regulation of Na^+-K^+-ATPase (mediator of active Na^+ transport) in murine alveolar epithelial cells and showed that short-term cyclic

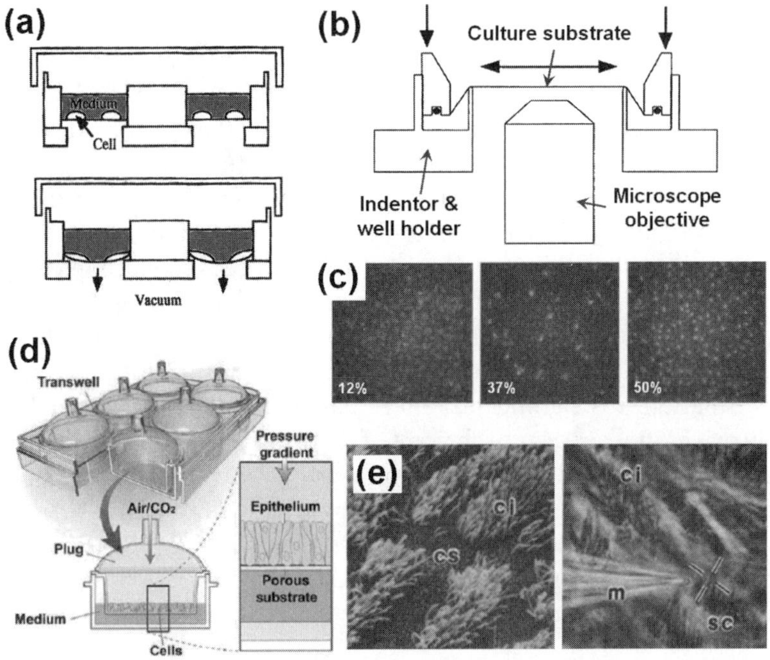

Figure 24.3 In vitro systems for generating mechanical forces that regulate cell-matrix and cell-cell interactions: (a) In the Flexercell strain unit, cells are cultured on an elastic membrane and stretched by negative pressure imposed by vacuum. (b) Cell-stretching system for exposing cells to uniform and equibiaxial mechanical strain. The device uses a silicone membrane as a cell-culture substratum deformed by the vertical motion of a circular indentor. (c) Alveolar epithelial cells exhibit injurious response to applied cyclic stretch in a dose-dependent manner (% represents the percent change in the membrane surface area). (d) Transwell-based cell-culture system used to apply an apical-to-basal transmembrane pressure gradient across bronchial epithelial cells grown at an air-liquid interface on a permeable membrane. (e) Scanning electron micrograph of cultured rabbit tracheal epithelial cells (left) and light micrograph showing mechanical stimulation of the apical surface of a single cell using a microprobe (ci: cilia, cs: cellsurface, m: microprobe, sc: stimulated cell). (*Source:* [5, 61–63], reproduced with permission.)

mechanical stress stimulated its activity by increasing intracellular Na^+ and recruiting Na^+-K^+-ATPase subunits to the basement membrane [58]. Short-term cyclic stretch simulated by the Flexercell also induced apoptosis and secretion of the major lipid component of pulmonary surfactant known as phosphatidylcholine in primary culture of rat alveolar type II cells [59]. Vlahakis et al. used A549 type II alveolar epithelial cells cultured on the Flexercell system and demonstrated that cyclic stretch enhanced the gene expression and release of interleukin (IL)–8, illustrating that epithelial deformation can trigger inflammatory signaling [38]. The Flexercell unit was also used to reveal that cyclic mechanical strain inhibited repair of epithelial wounds in human airway epithelial cell culture [60].

One drawback of the Flexercell system is that the cyclic inflation or deflation of its circular cell-culture substratum generates nonuniform loading conditions and causes the deformation of cells to vary significantly depending on its radial position [64, 65]. This limits quantitative analysis of cellular responses and prevents one from making detailed correlations between the amount of deformation and functional change. To overcome these limitations, Tschumperlin and Margulies developed a stretching device capable of subjecting cells to uniform equibiaxial strains [Figure 24.3(b)] [61]. Using this system, they demonstrated the injurious response of primary rat type II alveolar epithelial cells to applied cyclic stretch in a dose-dependent manner [Figure 24.3(c)]. Also, they studied the vulnerability of the cells at different stages of proliferation and revealed that the level of cellular injury decreased with increasing time elapsed after seeding. Their findings suggest that morphological and phenotypic alterations with time in culture fundamentally change the susceptibility of alveolar epithelial cells to mechanical deformation. In a separate study using the same type of cells and device, they showed that the extent of injury responses of the cells varied depending on the nature of applied membrane deformation, such as frequency, time, and magnitude [66]. It was noted that cyclic deformation was significantly more damaging than static stretch and that cellular injury was independent of the rate of deformation during single stretch and was elevated with increasing frequency and deformation amplitude. Waters et al. also constructed a pneumatically driven mechanical system to impose homogenous biaxial cyclic deformation to cells on elastomeric membranes [67]. In their study, they demonstrated that the inhibition of spreading and migration of airway epithelial cells during wound closure was dependent upon the magnitude of applied strain.

24.4.2 Static Stretch or Pressure in Two-Dimensional Culture

Wirtz and Dobbs devised a stretching system where hydrostatic pressure induced mechanical stretching of flexible silastic membranes [68]. Applying static stretch to primary rat type II alveolar epithelial cells grown in this system, they revealed the mechanism by which deep inflation of the lung stimulates surfactant secretion; mechanical stretch of type II cells caused a transient (<60 s) increase in cytosolic Ca^{2+}, and this was followed by a sustained (approximately fifteen- to thirty-minute) stimulation of surfactant secretion in a dose-dependent fashion.

The epithelium of asthmatic airways has a mechanically unique cellular microenvironment, where smooth muscle constriction leads to the buckling of the airway wall and the folding of airway epithelium into deep crevasses [5]. As a result, the epithelial cells along the folds are pushed up against each other and, thus, experience prolonged compressive mechanical stresses. Ressler et al. recreated this situation in vitro by applying static pressure selectively to the apical surface of primary rat tracheal epithelial cells cultured on a porous membrane in the Transwell system [Figure 24.3(d)] [69]. They found that the apical-to-basal transmembrane pressure gradient induced the expression of early growth factor–1 (transcription activator of airway remodeling genes), endothelin-1 (potent bronchoconstrictor), and TGF-β1 in a magnitude- and time-dependent manner. The same system was employed by Tschumperlin et al. to reveal that compressive mechanical stresses similar to those occurring in vivo during airway constriction gave rise to increased release of endothelin protein and TGF-β2 from HBECs [70]. These peptides were found to promote fibrotic protein synthesis by fibroblasts, which was consistent with subepithelial fibrosis characteristic of chronic asthma. In another study based on the same method, they also demonstrated that the pressure gradient applied across the airway epithelium at in vivo levels enhanced phosphorylation of extracellular signal-regulated kinase [71]. Swartz et al. created cocultures of HBECs and human lung fibroblasts in the Transwell system (epithelial cells in the upper chamber and fibroblasts in the lower chamber) and examined the communication of compressive forces between the two cell types [72]. Their study showed that the mechanical stresses applied to the airway epithelial cells were communicated to fibroblasts to induce integrated biological responses that mimicked the key features of airway remodeling found in asthma.

24.4.3 Mechanical Stretch in Three-Dimensional Culture

Skinner developed a solenoid-operated programmable stretch device where the movement of a metal bar induced by applying a magnetic field to the solenoid could generate the elongation of three-dimensional sponge matrices in a controllable manner [73]. Using this system, Liu et al. demonstrated that intermittent mechanical stretch simulating physical forces induced by fetal breathing movements enhanced DNA synthesis and division of fetal rat alveolar epithelial cells cultured in a gelatin sponge [74]. They also showed that the proliferative response of the fetal cells to mechanical stress did not take place in two-dimensional monolayer culture systems and was unique to three-dimensional culture [75]. In separate studies, they provided evidence that the strain-induced cellular proliferation was mediated by growth factors [76] and that mechanical stimulation upregulated the gene expression and protein synthesis of platelet-driven growth factor (PDGF)–B and PDGF β-receptor [77]. Intermittent mechanical strain inducing the proliferation of fetal lung cells was also shown to differentially regulate the expression of various ECM molecules and to selectively increase collagen and biglycan content [78]. Xu et al. used the same stretch device to examine the responses of fetal rat lung cells and fibroblasts cultured in a three-dimensional matrix, as well as the effect of the interactions between the two cell types on stretch-induced cell proliferation [79]. Their results showed that lung epithelial cells and fibroblasts responded to mechanical strain by increasing DNA synthesis when each cell type was subjected to strain separately, but DNA synthesis was suppressed when the epithelial cells were cocultured with fibroblasts having different gestational ages. The mechanical device used for stretching three-dimensional cell culture in these studies was developed into a commercially available computerized apparatus known as the Bio-Stretch System [80].

24.4.4 Direct Mechanical Stimulation of Single Cells

Precisely controlled motion of microstructures has been adopted as a unique strategy for mechanically stimulating pulmonary epithelial cells at the single-cell level to study their mechanical responses or cell-cell communication. Isakson et al. developed a method where a glass micropipette with a tip diameter of 1 μm was piezoelectrically deflected downward for a short period (~150 ms) to locally deform the apical membrane of a single cell [81]. Using this approach combined with coculture of rat type I and type II alveolar epithelial cells established on collagen/fibronectin/laminin-5 coated coverslips, they investigated intercellular signaling between the two cell types. From this study, it has been shown that mechanically induced Ca^{2+} signaling between type I and type II cells is coordinated by unique profiles of the gap-junction protein connexin. A similar approach to mechanical stimulation of single cells was reported by Sanderson and Dirksen [62]. Their system was also based on the piezoelectric vertical movement of a microprobe that caused the dimpling of a single cell [Figure 24.3(e)]. This mechanical actuation scheme was implemented to study the mechanosensitive response of ciliated airway epithelial cells. They demonstrated that mechanical stimulation resulted in a significant increase in ciliary beating frequency in a dose- and Ca^{2+}-dependent manner and that the locally applied mechanical stress could be transmitted to adjacent cells through cell-cell communication.

24.5 Engineering of Cell-Fluid Interactions

Pulmonary epithelial cells are constantly exposed to moving fluids in development, health, and disease. Recent evidence shows that during gestation, spontaneous peristaltic airway constrictions in the fetal lung propel liquid through the airway tree [2]. After birth, the microenvironment of pulmonary epithelial cells changes drastically due to the formation of an air-liquid interface over the pulmonary epithelium and convective transport of air generated by respiration. As a result of the formation of the air-liquid interface, surface tension becomes a dominant force in regulating the dynamics of pulmonary flows and the interaction of epithelial cells with fluids. In this section, we present in vitro systems that reproduce surface-tension-driven interfacial phenomena observed in the respiratory system to investigate the effect of the resulting fluid mechanical stresses on pulmonary epithelial cells. We also describe cell-culture devices that can recreate air-flow-induced shear stresses or apply fluid flows containing various gaseous compounds or soluble chemicals to pulmonary epithelial cells.

24.5.1 In Vitro Airway Reopening Models

Pulmonary airways are coated with a thin, viscous liquid film produced by secretory cells in the airway epithelium. In a wide range of respiratory diseases that can accompany surfactant deficiency or inactivation, the liquid layer lining the epithelial surface of small airways becomes more susceptible to surface-tension-driven instability that leads to the formation of an occluding liquid plug across the airway lumen and the obstruction of air flow in the peripheral respiratory units [82]. During respiration, the liquid plug propagates along the airway due to a pressure gradient until the plug volume decreases to the point at which it ruptures to reopen the blocked airway [83]. This reopening process, however, has been shown through computational simulations to generate abnormally high levels of mechanical stresses [84], which may exert deleterious mechanical forces on pulmonary epithelial cells.

To investigate the vulnerability of pulmonary epithelial cells to fluid mechanical-stress-induced injury experimentally, Bilek, Dee, and Gaver created an in vitro model of airway reopening where rat alveolar epithelial cells in a liquid-filled parallel-plate chamber were subjected to the steady progression of a semi-infinite air bubble [Figure 24.4(a)] [85]. Using this system, they showed the injurious response of the epithelial cells to mechanical stresses imposed by the movement of an air-liquid interface that simulated plug propagation during airway reopening. Also, they dem-

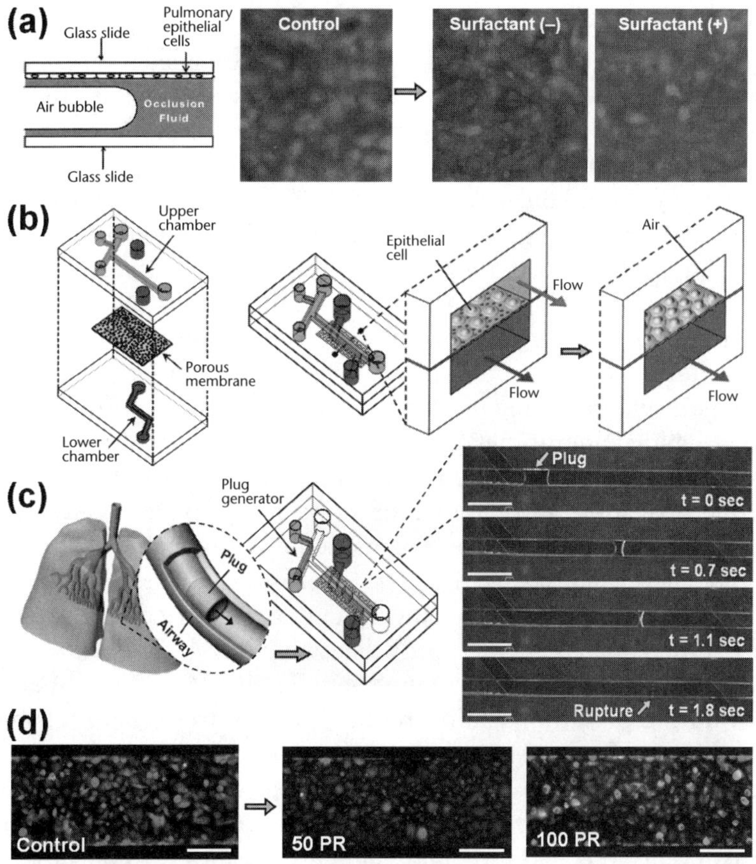

Figure 24.4 In vitro models of airway reopening: (a) Reopening of closed airway is modeled by the steady progression of a semi-infinite air finger through a liquid-filled parallel-plate chamber lined with pulmonary epithelial cells. The movement of an air-liquid interface at the bubble front results in cellular injury, which can be prevented by pulmonary surfactant. (*Source:* [85], reproduced with permission.) (b) Compartmentalized three-dimensional microfluidic small airway system created by soft-lithography-based microfabrication. The polymeric channel system enables proliferation and air-liquid-interface- induced differentiation of primary airway epithelial cells in a biomimetic culture environment. (c) More realistic in vitro re-creation of airway reopening is achieved by a computerized air-liquid two-phase microfluidic system integrated on-chip with microfluidic cell culture. This system produces propagation and rupture of a liquid plug with finite lengths by dynamically switching air-liquid two-phase flows in a plug generator. (d) Plug propagation and rupture cause injury of small airway epithelial cells in a dose-dependent fashion. PR represents propagation and rupture. Scale bars = 150 μm.

onstrated the role of pulmonary surfactant in mitigating the detrimental effect of fluid mechanical stresses on the cells. Through computational simulations, it was concluded that the steep pressure gradient near the front meniscus of the air bubble was the most likely cause of cellular injury. In their follow-up study based on the same system, it was found that human alveolar epithelial cells exhibited the same pattern of cellular injury in response to the propagation of semi-infinite air bubbles and that flow conditions mimicking repetitive airway reopening caused the denudation of the cellular monolayer [86]. More importantly, by varying the viscosity of liquid in the culture chamber, they showed that the magnitude of pressure gradient, not the duration of mechanical stimulation, determined the extent of cellular injury.

Although these studies provided important insights into the mechanism of airway injury during airway reopening, the experimental models looked only at the effect of the propagation of semi-infinite air fingers (this is relevant to the first breath of a new born) and failed to recreate the essential fluid dynamic characteristics of in vivo airway reopening, such as the propagation of liquid plugs with finite lengths and plug rupture. Also, they relied on submerged culture of transformed alveolar epithelial cells and thus had limited capabilities to reproduce in vivo–like airway epithelium. To address these issues, Huh et al. developed a compartmentalized three-dimensional microfluidic small airway system integrated with computerized air-liquid two-phase microfluidics [87]. This device produced fully differentiated airway epithelium through air-liquid-interface culture of primary human small airway epithelial cells in a biomimetic culture environment [Figure 24.4(b)] and recreated propagation and rupture of finite liquid plugs by dynamically switching air-liquid two-phase flows in polymeric microchannels [Figure 24.4(c)]. Using this system, they demonstrated that fluid mechanical stresses generated by plug propagation and rupture led to significant cellular damage in a dose-dependent manner, even with differentiated airway epithelia, which are structurally more robust than nondifferentiated epithelial monolayers [Figure 24.4(d)] and that the injury of the cells during plug propagation resulted mainly from the mechanical stresses produced by the front meniscus of a moving liquid plug. Furthermore, they revealed that plug rupture caused more severe damage than plug propagation alone and that transient pressure waves generated by plug rupture enabled acoustic detection of injury events as crackling sounds similar to respiratory crackles.

24.5.2 Application of Air-Flow-Induced Shear Stresses to Pulmonary Epithelial Cells

During tidal breathing, the flow of air through airway tubes exerts shear stresses on airway epithelial cells in vivo. Tarran et al. developed an in vitro airway epithelial cell-culture system where the physiological shear stresses were reproduced by the rotational motion of culture plates mounted on a motorized stage [88]. Using this device, they examined the effect of shear stress on the regulation of the volume of airway surface liquid (a critical component of the host defense mechanism mediated by mucociliary clearance) in normal and cystic fibrosis human nasal epithelial cell cultures. They found that the height of the surface liquid layer in normal epithelia increased significantly when the cells were subjected to shear stresses. This study also revealed that applied shear stresses induced the cellular release of ATP, which allowed cystic fibrosis airway epithelia to maintain airway surface liquid at physiological volumes and to sustain mucus transport over prolonged periods, whereas static culture without shear stresses resulted in the rapid depletion of the surface liquid layer.

Even-Tzur et al. created a modular device for subjecting airway epithelial cells cultured at an air-liquid interface to shear stresses produced by air flow [89]. In this system, airway epithelial cells are first cultured in custom-designed wells with permeable membranes that support proliferation and air-liquid-interface-induced differentiation. For mechanical stimulation, the differentiated cells are transferred to a flow chamber, where they are exposed to steady air flows that generate shear stresses uniformly distributed over the apical surface of the cellular monolayer. This system was used to demonstrate that wall shear stresses induced by physiological air flows affected mucus secretion and intracellular structures of differentiated human nasal epithelial cells.

24.5.3 Exposure of Pulmonary Epithelial Cells to Gaseous Compounds

Tarkington et al. created a flow system for generating and guiding ozone-containing gas streams to the apical surface of pulmonary epithelial cells cultured on a permeable membrane [90]. The operation of this device is based on serially connected flow vessels that can cooperatively control, monitor, and manipulate the pressure, ozone concentration, temperature, and humidity of gaseous samples delivered to the cells maintained in a culture incubator. Using this system, they demonstrated the deleterious effect of ozone exposure on human tracheaobronchial epithelial cells. They also revealed that the reduction in the viability of cells after ozone exposure was inversely proportional to the volume of liquid covering the apical side of a monolayer, illustrating the protective role of epithelial liquid film. Sun et al. used this flow system to demonstrate significantly decreased cellular viability, inhibition of cellular metabolism and replication, and compromised membrane integrity resulting from the exposure of HBECs to side-stream cigarette smoke [91].

Aufderheide and Mohr designed a new type of cell-culture system known as CULTEX that allowed for direct exposure of pulmonary epithelial cells maintained at an air-liquid interface to air flows containing various gaseous substances and complex mixtures [92]. This system consists of: (1) a modular culture unit based on the Transwell system where pulmonary epithelial cells are cultured on a porous membrane and intermittently exposed to an air-liquid interface at certain time intervals without a loss of viability; and (2) a specially designed tube system for transporting and uniformly distributing gaseous test samples to a monolayer of cells maintained at an air-liquid interface. This device is also equipped with computer-controlled sensing systems capable of regulating culture conditions such as temperature, CO_2 content, and humidity, eliminating the need for cell-culture incubators. This system was used to analyze cellular viability and metabolic activity after the exposure of HBECs to synthetic air, nitrogen dioxide, and ozone [93]. Knebel, Ritter, and Aufderheide investigated the effect of diesel exhaust fumes on HBECs and revealed dose-dependent cytotoxic responses of the cells using the CULTEX system directly connected to a diesel engine [94]. Wolz et al. extended the application of the CULTEX system to the examination of biological responses of HBECs to cigarette smoke [95]. They demonstrated that the exposure of the cells to side-stream cigarette smoke induced DNA damage in a dose-dependent manner. Based on the principle of sedimentation, Fukano, Yoshimura, and Yoshida modified this system to expose the cells to both vapor and particulate phases of mainstream smoke [96]. Using this system, they showed that cigarette-smoke exposure led to the increased expression of mRNA for the sensitive oxidative stress enzyme known as heme oxygenase–1, which has been implicated in the pathogenesis of various respiratory diseases.

24.5.4 Cell Culture Analog Bioreactors

Cell culture analog (CCA) bioreactors are in vitro culture systems that consist of interconnected compartments, each of which has different types of cells to represent different organs and to mimic their characteristic metabolism [97]. These systems are designed to allow multiple cell types to interact with each other through recirculating culture media. This makes CCA bioreactors more attractive for predicting integrated biological responses to chemicals or pharmaceuticals than conventional in vitro culture methods that are often limited in reproducing the communication between various cell types constituting different organs.

Researchers have used pulmonary epithelial cells as a cell type representative of the lung in several CCA devices and studied their response to metabolites delivered from other tissue compartments through fluid flows. Sweeney et al. designed a CCA bioreactor consisting of liver and lung chambers lined with rat hepatoma and alveolar epithelial cells, respectively [98]. Using this system, they showed that the circulation of reactive naphthalene metabolites from the liver to the lung compartments induced the cytotoxicity of the alveolar epithelial cells and the depletion of glutathione (GSH). Ghanem and Shuler improved this system to obtain physiologically more relevant operating conditions (e.g., liquid residence time, the ratio of lung to liver cells) by culturing cells on microbeads packed into tissue chambers [99].

Efforts to reduce the volume requirements of conventional CCA bioreactors have led to the development of microfabricated CCA devices. Compared to their macroscopic counterparts, microscopic CCA systems are more amenable to recapitulating physiological flow conditions and accurately modeling the physical parameters of physiological systems, such as the ratio of liquid to cells and liquid residence times in each organ [97]. Owing to miniaturization, they require much smaller amounts of reagents and cells, making them more advantageous for testing drugs or culturing cells that are expensive and limited in supply. Furthermore, the small size of microscale CCA makes it possible to realize high-throughput testing and analysis using multiple CCA devices operating in parallel. The development of a microfabricated CCA bioreactor with integrated oxygen sensors was reported by Sin et al. [100]. To enable real-time monitoring of cell metabolism, they constructed a microfluidic cell-culture system combined with fluorescence lifetime-based sensing of dissolved oxygen. The device comprises interconnected microscale chambers that are fabricated in silicon and sustain microfluidic culture of different types of cells such as lung epithelial cells and liver cells. Using this device, they demonstrated that the microscale CCA system provided a more physiologically realistic tissue-culture environment and might potentially enable more accurate and physiologically relevant toxicity testing. The microfabricated CCA was also shown to be capable of monitoring gas exchange of individual cell cultures in real time using integrated oxygen-sensing components, offering significant advantages over traditional cell-culture techniques that generally take endpoint measurements. Viravaidya et al. used this system to probe naphthalene toxicity and demonstrated that the exposure of alveolar epithelial cells to fluid flows containing naphthalene metabolites produced in the liver compartment led to the depletion of GSH in the epithelial cells [101]. In a separate study, the design of the microfabricated CCA system was modified to have four chambers to include culture of differentiated adipocytes that could mimic the key functions of fat tissue [102]. The new device was used to study the bioaccumulation, distribution, metabolism, and toxicity of compounds such as fluoranthene, naphthalene, and naphthoquinone. This investigation revealed that fluoranthene preferentially accumulated in the fat chamber and that the presence of adipocytes in the system significantly reduced GSH depletion of alveolar epithelial cells induced by naphthalene and naphthoquinone. Based on these findings, the microscale CCA system was suggested to hold great potential for assessing the absorption, distribution, metabolism, elimination, and toxicity characteristics of drug candidates prior to animal or human trials.

24.6 Measurements of Mechanically Induced Inflammatory Responses

Acute respiratory distress syndrome (ARDS) and acute lung injury are often fatal inflammatory lung conditions that occur when a traumatic event such as pneumonia, shock, sepsis, or massive aspiration leads to inflammation, increased pulmonary vascular permeability, and extravasation of fluid and inflammatory cells into the pulmonary interstitium and alveolar spaces [103]. Both conditions are characterized by a massive systemic inflammatory response involving a cascade of both pro- and anti-inflammatory molecules. In this section, we describe: (1) inflammatory mediators; and (2) the measurements of mechanically induced inflammatory responses.

24.6.1 Inflammatory Mediators

24.6.1.1 Cytokines

Cytokines are small, soluble, regulatory proteins (8 to 80 kDa) produced by cells during the inflammatory process and function as messengers in cell-to-cell communication. They are active at picomolar and femtomolar concentrations and principally act in an autocrine or paracrine fashion to initiate, amplify, and perpetuate local and systemic inflammatory responses. Cytokines are produced in cascades where initial signals are amplified by the target cells. Many cytokines are pleiotropic in that they have overlapping functions, with one cytokine having effects that are synergistic or antagonistic with other cytokines. Due to this interdependency, the composition of the "cytokine milieu" is more important than a single cytokine [104]. This sophisticated network involving complex feedback mechanisms is believed to play a central role in homeostasis of the immune system and the coordination of its responses [105].

24.6.1.2 Cytokines and Lung Injury

An imbalance in the production of cytokines can have profound effects on the body and is implicated in various diseases. Scientists have recognized that cytokine concentrations in biological fluids are often a useful indicator of the presence or severity of a disease. For example, in patients with ARDS, both pro- and anti-inflammatory cytokines are detected in the bronchoalveolar lavage fluid. Following lung injury, the early response cytokines [i.e., tumor necrosis factor (TNF)–α and IL-1β] are released by the alveolar macrophages [106]. These cytokines then act in a paracrine fashion to stimulate other cellular components of the alveolar-capillary wall to produce more effector molecules [107] that involve both lung injury (IL-8) and repair [TGF-β1, hepatocyte growth factor (HGF), IL-6, IL-10]. After the onset of ARDS, there is a major anti-inflammatory response that peaks and exceeds the proinflammatory response involving both cytokines (IL-10) and cytokine antagonists (IL-1 receptor antagonist, soluble IL-1 receptor II, soluble TNF receptors I and II) to downregulate and attenuate the proinflammatory response [108]. Often it is the balance of pro- and anti-inflammatory mediators that determines the outcome of ARDS.

The ability to measure these soluble mediators in immune regulation can help us to understand the role of cytokines in the immune system and disease pathogenesis, to identify markers to predict the outcome of diseases, and to develop preventive and curative therapies to treat lung injury [105]. The successful development of these therapeutic strategies to suppress harmful inflammatory responses depends on understanding the steps leading to the activation of the inflammatory response and an understanding of the regulatory mechanisms that are involved in the response.

24.6.2 Cytokine Release Induced by Mechanical Stimulation

During lung injury, the cells of the alveolar epithelium are stimulated to secrete IL-8, a potent chemoattractant for neutrophils. Mechanical signals, such as deformation, that distress the alveolar epithelium can also lead to an increase in IL-8 (Figure 24.5) [109. 110]. Neutrophils in the vascular component undergo extravasation (along the IL-8 gradient) into the tissue interstitium. Once there, the activated neutrophils release reactive oxygen metabolites, proteolytic enzymes, and additional cytokines in an acute inflammatory response [109]. This propagates into a cycle of inflammation leading to local tissue injury and biotrauma, where compartmentalization of the alveolar cytokines is lost, leading to their release into the pulmonary circulation and a subsequent systemic inflammatory response that can lead to multiple system organ failure and death [109].

It has been proposed that the structural disruption present is due to two factors: (1) lung overdistension in small airways and alveoli due to increased tidal volumes (volutrauma), and (2) shear forces generated from the repetitive opening and closing of atelectatic regions filled with fluid (edema), which causes severe wall stress and surfactant depletion (atelectrauma) [111]. Overdistension is mainly due to the end-inspiratory volume and is believed to degrade surfactant, disrupt the blood-gas barrier, and increase the levels of cytokines and inflammatory cells in the lung. The cyclic opening and reopening of injured alveoli can cause shear stress on the epithelial cell layers, and recent studies have suggested that the rapid rate of lung distension may be as important as the magnitude of distension in lung injury. Other factors to consider include the frequency of the stretch and the duration of stretch (inspiratory time) [112, 113].

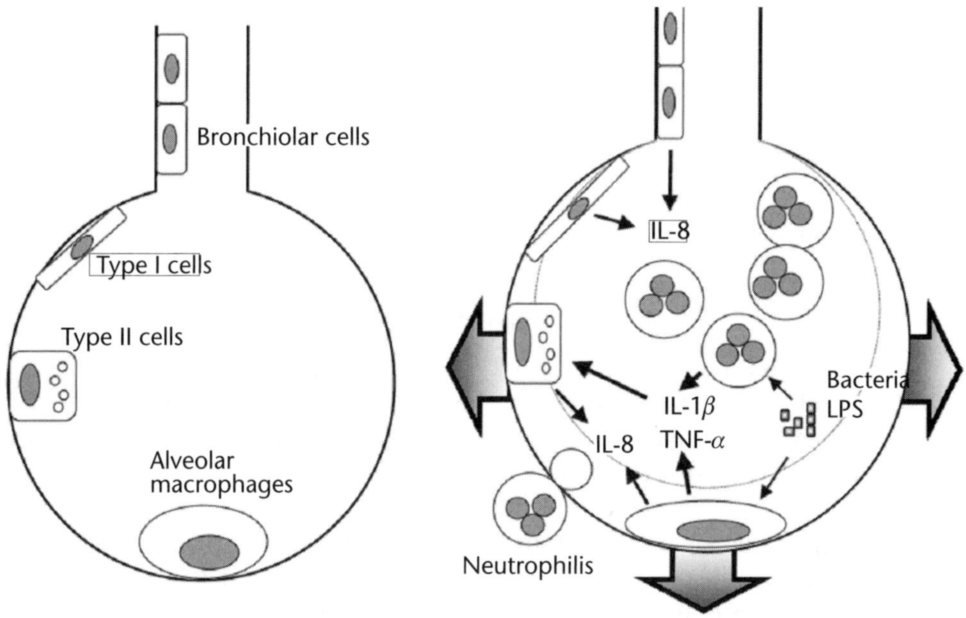

Figure 24.5 Mechanically induced inflammatory responses in lung injury. Mechanical stretch causes bronchiolar and alveolar epithelial cells to produce the neutrophil chemoattractant IL-8. Mechanical stimulation also activates the release of proinflammatory cytokines such as IL-1β and TNF-β from alveolar macrophages, which then act to stimulate alveolar epithelial cells to secrete more IL-8. Neutrophil recruitment into the alveolar space leads to the release of additional cytokines that propagate into an inflammatory cycle and eventually results in lung injury. (*Source:* [110], reprinted with permission.)

24.6.2.1 Cytokine Release by Alveolar Epithelial Cells

Many researchers have used A549 type II alveolar epithelial cells to represent the alveolar epithelium and its responses to mechanical stimulation. Vlahakis et al. used A549 cells cultured on the Flexercell system and concluded that cyclic stretch enhanced the gene expression and release of IL-8 in a dose-dependent manner when varying the amplitude of stretch [38]. Stimulating the cells with TNF-α, a proinflammatory signaling molecule known to cause gene expression and protein release by A549 cells [107, 114], further increased the release of IL-8 [38]. Yamamoto et al. also used the Flexercell system with A549 cells and demonstrated that cyclic stretch increased the expression of HGF, IL-8, and TGF-β1 at both the mRNA and protein levels [115, 116]. Quinn et al. used a cell-stretch device to apply a uniform biaxial strain to A549 cells grown on flexible fibronectin-coated silicone membranes [117]. Cycles of 5 percent and 15 percent strain were shown to induce IL-8 release, as well as transcription of IL-8 mRNA. Both were accompanied by an increase in the activity of the mitogen-activated protein kinases: stress-activated protein kinase and p38 [117].

Previous studies have also used primary type II alveolar cells subjected to mechanical stretch. Hammerschmidt et al. exposed type II cells to cyclic stretch using the Flexercell system and determined that cyclic mechanical stretch alters cell mediator release toward a proinflammatory pattern [118]. There was an increased production of eicosanoids, a decreased generation of the anti-inflammatory cytokine IL-10, and unchanged production of proinflammatory cytokines. Levine et al. isolated rat alveolar epithelial cells from healthy rats and from septic rats. The cells were exposed to high levels of cyclic equibiaxial stretch. The cells isolated from the septic rats were more vulnerable to the mechanical stretch and showed higher rates of cell death [119].

24.6.2.2 Cytokine Release by Alveolar Macrophages

Alveolar macrophages attach to the alveolar epithelium and can also play an important role in lung injury through the release of inflammatory mediators in response to mechanical stimulation (stretch) and chemical stimulation (inflammatory stimuli such as bacterial endotoxin). Pugin et al. developed an in vitro lung model where alveolar macrophages cultured on a silastic membrane are exposed to prolonged cyclic pressure-stretching strain. The macrophages responded by secreting IL-8 and matrix metalloproteinase–9 [120]. Lang, Barnetta, and Doylea cultured alveolar macrophages onto IgG-coated silastic membranes and applied a repetitive sinusoidal 30 percent mechanical strain [121]. Physical strain, in combination with lipopolysaccharide (LPS) enhanced macrophage secretion of TNF-α.

Because of the close proximity of alveolar macrophages to the pulmonary epithelium, it is believed that Type II alveolar epithelial cells respond to the alveolar macrophage-derived cytokines by releasing other pro- and anti-inflammatory cytokines such as IL-8 and IL-6. In fact, Standiford et al. have shown that LPS-stimulated alveolar macrophage media induce the increase of IL-8 in A549 epithelial cells [107]. Crestani et al. have found that A549 cells incubated with alveolar-macrophage-conditioned medium increased secretion of IL-6 [122].

24.7 Conclusion

Experimental studies outlined in this chapter highlight the remarkable breadth of mechanosensitive events mediated by pulmonary epithelial cells in the respiratory system. Undoubtedly, mechanical forces are essential for regulating the structure, function, and metabolism of pulmonary epithelial cells in a wide array of developmental, physiological, and pathological processes. As is evident from the studies described here, understanding various cellular responses to the dynamic mechanical environment of the respiratory system has been greatly facilitated by in vitro systems that can reproduce, manipulate, and measure the interaction of pulmonary epithelial cells with physiological and pathological mechanical forces.

Advancing our knowledge of the mechanically regulated behavior of lung cells will inevitably require careful design and engineering of new experimental tools that allow us to better mimic the geometrical and structural properties of the pulmonary epithelium and to precisely recreate the essential nature of tissue deformation and fluid flows occurring in vivo. Assay systems capable of detecting mechanotransduction signaling molecules with high sensitivity and specificity will also make a significant contribution to such efforts by providing a means to study the molecular mechanisms of observed mechanosensitive events. The development of these new tools will likely involve contributions from various scientific and engineering fields. Based on these predictions, we suggest the following as examples of important future research directions. (1) Existing in vitro models have limited capabilities to examine cellular responses to mechanical stresses originating from surface-tension-driven interfacial phenomena that are prevalent in small airways and alveoli. This is mainly due to the technical difficulties associated with recreating the dynamic and complex microscale flow configurations found in the distal parts of the lung. These limitations may be resolved by microfabricated systems that can exploit the advantages of small length scales and low Reynolds number to create in vivo–like microscale gas-liquid two-phase flows in a controllable and reproducible fashion. Also, such microsystems can be readily integrated with microfluidic cell culture to enable on-chip engineering and mechanical stimulation of pulmonary epithelial cells in a single device. (2) Most in vitro systems aimed to investigate the mechanosensitive behavior of lung cells often neglect to consider the effect of other important stimulating factors (e.g., inflammatory mediators, hormones, chemokines, and growth factors) that can act in concert with mechanical forces to strongly affect cellular responses. For example, it has been shown that the remodeling of asthmatic airways is increased by a synergistic combination of mechanical stresses with inflammatory mediators [123]. Therefore, future experimental models need to be designed to enable in vitro re-creation and study of the interplay between mechanical stimulation with other confounding factors in various combinations. Also, study of paracrine and autocrine effects may benefit from microscale culture systems that better mimic physiological cell-fluid ratios. (3) Recent progress in microtechnology offers new opportunities to develop microfluidics-based miniaturized assay systems with more advanced analytical and sample-handling capabilities. Assays performed in microfluidic systems are advantageous over standard assay methods for many reasons, including compatibility with flow-based systems, small reagent/sample consumption, simplified automation and procedures, parallel analysis of multiple samples, and low-cost mass production. These systems will greatly assist future studies by enabling rapid, efficient, and quantitative measurements of molecular profiles regulating the mechanosensitive response of pulmonary epithelial cells.

The ability to engineer and characterize pulmonary epithelia and their mechanically regulated microenvironments will enhance our fundamental understanding of the role of physical forces in normal lung function and the development of respiratory diseases. Because of various difficulties associated with studying this topic in vivo and using conventional in vitro systems, what is required are multidisciplinary approaches based on the engineering of integrated microtools capable of regulating multiple microenvironmental factors (integration of effector microtools in Figure 24.1) and performing real-time quantitative measurements of multiple cellular responses (integration of readout microtools in Figure 24.1). Although most of the components exist, integration is still a technological challenge as well as an area of opportunity. The effective design, construction, and application of such integrated systems will also require in-depth biological and clinical insights (top row in Figure 24.1), together with the bioengineering expertise to result in the highest impact and most useful knowledge (bottom row in Figure 24.1).

References

[1] Del Riccio, V., Van Tuyl, M., and Post, M., "Apoptosis in lung development and neonatal lung injury," *Pediatric Res.*, Vol. 55, No. 2, 2004, pp. 183–189.

[2] Schittny, J. C., Miserocchi, G., and Sparrow, M. P., "Spontaenous peristaltic airway contractions propel lung liquid through the bronchial tree of intact and fetal lung explants," *Am. J. Resp. Cell Mol. Biol.*, Vol. 23, No. 1, 2000, pp. 11–18.

[3] Riley, D. J., Rannels, D. E., Low, R. B., Jensen, L., and Jacobs. T. P., "Effect of physical forces on lung structure, function, and metabolism." *Am. Rev. Resp. Dis.*, Vol. 142, No. 4, 1990, pp. 910–914.

[4] Dos Santos, C. C., and Slutsky, A. S., "Mechanisms of ventilator-induced lung injury: A perspective," *J. Appl. Physiol.*, Vol. 89, No. 4, 2000, pp. 1645–1655.

[5] Tschumperlin, D. J., and Drazen, J. M., "Chronic effects of mechanical force on airways," *Ann. Rev. Physiol.*, Vol. 68, 2006, pp. 563–583.

[6] Kamm, R. D., "Airway wall mechanics," *Ann. Rev. Biomed. Eng.*, Vol. 1, 1999, pp. 47–72.

[7] Jeffery, P. K., and Li, D., "Airway mucosa: Secretory cells, mucus and mucin genes," *Eur. Resp. J.*, Vol. 10, No. 7, 1997, pp. 1655–1662.

[8] Voss, B., Kerenyi, T., Muller, K. M., and Wilhelm, "Scanning electron microscopical investigation of broncho-alveolar casts after intratracheal asbestos fibre instillation," *Int. J. Hyg. Environ. Health*, Vol. 203, No. 2, 2000, pp. 127–134.

[9] Scarpelli, E. M., "The alveolar surface network: A new anatomy and its physiological significance," *Anatom. Rec.*, Vol. 251, No. 4, 1998, pp. 491–527.

[10] Shahana, S., Bjornsson, E., Luviksdottir, D., Janson, C., Nettelbladt, O., Venge, P., and Roomans, G. M., "Ultrastructure of bronchial biopsies from patients with allergic and non-allergic asthma," *Resp. Med.*, Vol. 99, No. 4, 2005, pp. 429–443.

[11] Fortoul, T. I., Saldivar, L., Espejel-Mayab, G., Bazarro, P., Mussali-Galante, P., Avila-Casado, M. D., Colin-Barenque, L., and Avila-Costa, M. R., "Inhalation of cadmium, lead, or its mixture effects on the bronchial structure and its relation with metal tissue concentrations," *Environ. Toxicol. Pharmacol.*, Vol. 19, No. 2, 2005, pp. 329–334.

[12] Massaro, G. D., Singh, G., Mason, R., Plopper, C. G., Malkinson, A. M., and Gail, D. B., "Biology of the Clara cell," *Am. J. Physiol.*, Vol. 266, No. 1, 1994, pp. L101–L106.

[13] Gruenert, D. C., Finkbeiner, W. E., and Widdicombe, J. H., "Culture and transformation of human airway epithelial cells," *Am. J. Physiol.: Lung Cell. Mol. Physiol.*, Vol. 268, No. 3, 1995, pp. L347–L360.

[14] Gruenert, D. C., Basbaum, C. B., and Widdicombe, J. H., "Long-Term culture of normal and cystic fibrosis epithelial cells grown under serum-free conditions," *In Vitro Cell. Dev. Biol.*, Vol. 26, No. 4, 1990, pp. 411–418.

[15] Wu, R., Nolan, E., and Turner, C., "Expression of tracheal differentiated functions in serum-free, hormone-supplemented medium," *J. Cell. Physiol.*, Vol. 125, No. 2, 1985, pp. 167–181.

[16] Kitamura, H., Shibagaki, T., Inayama, Y., Ito, T., and Kanisawa, M., "Growth and differentiation of human distal airway epithelial cells in culture: Effects of small amounts of serum in defined medium," *Lab. Invest.*, Vol. 63, No. 3, 1990, pp. 420–428.

[17] Jetten, A. M., "Growth and differentiation factors in tracheobronchial epithelium," *Am. J. Physiol.*, Vol. 260, No. 6, 1991, pp. L361–L373.

[18] Lechner, J. F., Haugen, A., McClendon, I. A., and Shamsuddin, A. M., "Induction of squamous differentiation of normal human bronchial epithelial cells by small amounts of serum," *Differentiation*, Vol. 25, No. 3, 1984, pp. 229–237.

[19] Miller, L. A., Cheng, L. Z., and Wu, R., "Inhibition of epidermal growth factor–like growth factor secretion in tracheobronchial epithelial cells by vitamin A," *Cancer Res.*, Vol. 53, No. 11, 1993, pp. 2527–2533.

[20] Willey, J. C., Lechner, J. F., and Harris, C. C., "Bombesin and the C-terminal tetradecapeptide of gastrin releasing peptide are growth factors for normal human bronchial epithelial cells," *Exp. Cell Res.*, Vol. 153, No. 1, 1984, pp. 245–248.

[21] Yamaya, M., Finkbeiner, W. E., Chun, S. Y., and Widdicombe, J. H., "Differentiated structure and function of cultures from human tracheal epithelium," *Am. J. Physiol.*, Vol. 262, No. 6, 1992, pp. L713–L724.

[22] Beckmann, J. D., Stewart, A., Kai, M., and Keeton, T. P., "Controls of EGF-induced morphological transformation of human bronchial epithelial cells," *J. Cell. Physiol.*, Vol. 189, No. 2, 2001, pp. 171–178.

[23] Wu, R., Martin, W. R., Robinson, C. B., Stgeorge, J. A., Plopper, C. G., Kurland, G., Last, J. A., Cross, C. E., McDonald, R. J., and Boucher, R., "Expression of mucin synthesis and secretion in human tracheobronchial epithelial cells grown in culture," *Am. J. Resp. Cell Mol. Biol.*, Vol. 3, No. 5, 1990, pp. 467–478.

[24] Robinson, C. B., and Wu, R., "Mucin synthesis and secretion by cultured tracheal cells—effects of collagen gel substratum thickness," *In Vitro Cell. Dev. Biol.*, Vol. 29A, No. 6, 1993, pp. 469–477.

[25] Lechner, J. F., Haugen, A., McClendon, I. A., and Pettis, E. W., "Clonal growth of normal adult human bronchial epithelial cells in a serum-free medium, in vitro," *J. Tissue Cult. Assoc.*, Vol. 18, No. 7, 1982, pp. 633–642.

[26] Infeld, M. D., Brennan, J. A., and Davis, P. B., "Human tracheobronchial epithelial cells direct migration of lung fibroblasts in 3-dimensional collagen gels," *Am. J. Physiol.*, Vol. 262, No. 5, 1992, pp. L535–L541.

[27] Infeld, M. D., Brennan, J. A., and Davis, P. B., "Human fetal lung fibroblasts promote invasion of extracellular matrix by normal human tracheobronchial epithelial cells, in vitro—a model of early gland development," *Am. J. Resp. Cell Mol. Biol.*, Vol. 8, No. 1, 1993, pp. 69–76.

[28] Mio, T., Liu, X. D., Adachi, Y., Striz, I., Skold, C. M., Romberger, D. J., Spurzem, J. R., Illiq, M. G., Ertl, R., and Rennard, S. I., "Human bronchial epithelial cells modulate collagen gel contraction by fibroblasts," *Am. J. Physiol.: Lung Cell. Mol. Physiol.*, Vol. 274, No. 1, 1998, pp. L119–L126.

[29] Johnson, L. G., Dickman, K. G., Moore, K. L., Mandel, L. J., and Boucher, R. C., "Enhanced Na^+ transport in an air-liquid interface culture system," *Am. J. Physiol.*, Vol. 264, No. 6, 1993, pp. L560–L565.

[30] Dejong, P. M., Vansterkenburg, M. A. J. A., Hesseling, S. C., Kempenaar, J. A., Mulder, A. A., Mommaas, A. M., Dijkman, J. H., and Ponec, M., "Ciliogenesis in human bronchial epithelial cells cultured at the air-liquid interface," *Am. J. Resp. Cell Mol. Biol.*, Vol. 10, No. 3, 1994, pp. 271–277.

[31] Davidson, D. J., Kilanowski, F. M., Randell, S. H., Sheppard, D. N., and Dorin, J. R., "A primary culture model of differentiated murine tracheal epithelium," *Am. J. Physiol.: Lung Cell. Mol. Physiol.*, Vol. 279, No. 4, 2000, pp. L766–L778.

[32] You, Y., Richer, E. J., Huang, T., and Brody, S. L., "Growth and differentiation of mouse tracheal epithelial cells: Selection of a proliferative population," *Am. J. Physiol.: Lung Cell. Mol. Physiol.*, Vol. 283, No. 6, 2002, pp. L1315–L1321.

[33] Whitcutt, M. J., Adler, K. B., and Wu, R., "A biphasic chamber system for maintaining polarity of differentiation of cultured respiratory tract epithelial cells," *In Vitro Cell. Dev. Biol.*, Vol. 24, No. 5, 1988, pp. 420–428.

[34] Gray, T. E., Guzman, K., Davis, C. W., Abdullah, L. H., and Nettesheim, P., "Mucociliary differentiation of serially passaged normal human tracheobronchial epithelial cells," *Am. J. Resp. Cell Mol. Biol.*, Vol. 14, No. 1, 1996, pp. 104–112.

[35] Yoon, J. H., Kim, K. S., Kim, S. S., Lee, J. G., and Park, I. Y., "Secretory differentiation of serially passaged human nasal epithelial cells by retinoic acid: Expression of mucin and lysozyme," *Annals Otol. Rhinol. Laryngol.*, Vol. 109, No. 6, 2000, pp. 594–601.

[36] Widdicombe, J. H., Sachs, L. A., Morrow, J. L., and Finkbeiner, W. E., "Expansion of cultures of human tracheal epithelium with maintenance of differentiated structure and function," *Biotechniques*, Vol. 39, No. 2, 2005, pp. 249–255.

[37] Reid, L., Meyrick, B., Antony, V. B., Chang, L. Y., Crapo, J. D., and Reynolds, H. Y., "The mysterious pulmonary brush cell—a cell in search of a function," *Am. J. Resp. Crit. Care Med.*, Vol. 172, No. 1, 2005, pp. 136–139.

[38] Vlahakis, N. E., Schroeder, M. A., Limper, A. H., and Hubmayr, R. D., "Stretch induces cytokine release by alveolar epithelial cells in vitro," *Am. J. Physiol.: Lung Cell. Mol. Physiol.*, Vol. 277, No. 1, 1999, pp. L167–L173.

[39] Dobbs, L. G., "Isolation and culture of alveolar type II cells," *Am. J. Physiol.*, Vol. 258, No. 4, 1990, pp. L134–L147.

[40] Mathias, N. R., Yamashita, F., and Lee, V. H. L., "Respiratory epithelial cell culture models for evaluation of ion and drug transport," *Adv. Drug Rev.*, Vol. 22, Nos. 1–2, 1996, pp. 215–249.

[41] Elbert, K. J., Schafer, U. F., Schafers, H. J., Kim, K. J., Lee, V. H. L., and Lehr, C. M., "Monolayers of human alveolar epithelial cells in primary culture for pulmonary absorption and transport studies," *Pharma. Res.*, Vol. 15, No. 5, 1999, pp. 601–608.

[42] Fuchs, S., Hollins, A. J., Laue, M., Schaefer, U. F., Roemer, K., Gumbleton, M., and Lehr, C. M., "Differentiation of human alveolar epithelial cells in primary culture: Morphological characterization and synthesis of caveolin-1 and surfactant protein-C," *Cell Tissue Res.*, Vol. 311, No. 1, 2003, pp. 31–45.

[43] Gumbleton, M., Abulrob, A. N. G., and Campbell, L., "Caveolae: An alternative membrane transport compartment," *Pharma. Res.*, Vol. 17, No. 9, 2000, pp. 1035–1048.

[44] Olsen, C. E., Isakson, B. E., Seedorf, G. J., Lubman, R. L., and Boitano, S., "Extracellular matrix-driven alveolar epithelial cell differentiation in vitro," *Exp. Lung Res.*, Vol. 31, No. 5, 2005, pp. 461–482.

[45] Borok, Z., Lubman, R. L., Danto, S. I., Zhang, X. L., Zabski, S., M., King, L. S., Lee, D., M., Agre, P., and Crandall, E. D., "Keratinocyte growth factor modulates alveolar epithelial cell phenotype in vitro: Expression of aquaporin 5," *Am. J. Resp. Cell Mol. Biol.*, Vol. 18, No. 4, 1998, pp. 554–561.

[46] Borok, Z., Danto, S. I., Lubman, R. L., Cao, Y. X., Williams, M. C., and Crandall, E. D., "Modulation of T1 alpha expression with alveolar epithelial cell phenotype in vitro," *Am. J. Physiol.: Lung Cell. Mol. Physiol.*, Vol. 275, No. 1, 1998, pp. L155–L164.

[47] Mason, R. J., Lewis, M. C., Edeen, K. E., McCormick-Shannon, K., Nielsen, L. D., and Shannon, J. M., "Maintenance of surfactant protein A and D secretion by rat alveolar type II cells in vitro," *Am. J. Physiol.: Lung Cell. Mol. Physiol.*, Vol. 282, No. 2, 2002, pp. L249–L258.

[48] Geppert, E. F., Williams, M. C., and Mason, R. J., "Primary culture of rat alveolar type II cells on floating collagen membranes—morphological and biochemical observations," *Exp. Lung Res.*, Vol. 128, No. 2, 1980, pp. 363–374.

[49] Rannels, S. R., Yarnell, J. A., Fisher, C. S., Fabisiak, J. P., and Rannels, D. E., "Role of laminin in maintenance of type II pneumocyte morphology and function," *Am. J. Physiol.*, Vol. 253, No. 6, 1987, pp. C835–C845.

[50] Shannon, J. M., Emrie, P. A., Fisher, J. H., Kuroki, Y., Jennings, S. D., and Mason, R. J., "Effect of a reconstituted basement membrane on expression of surfactant apoproteins in cultures adult rat alveolar Type II cells," *Am. J. Resp. Cell Mol. Biol.*, Vol. 2, No. 2, 1990, pp. 183–192.

[51] Shami, S. G., Aghajanian, J. D., and Sanders, R. L., "Organotypic culture of fetal lung type II alveolar epithelial cells: Applications to pulmonary toxicology," *Environ. Health Persp.*, Vol. 56, 1984, pp. 87–94.

[52] Dobbs, L. G., Pian, M. S., Maglio, M., Dumars, S., and Allen, L., "Maintenance of the differentiated type II cell phenotype by culture with an apical air surface," *Am. J. Physiol.: Lung Cell. Mol. Physiol.*, Vol. 273, No. 2, 1997, pp. L347–L354.

[53] Alcorn, J. L., Smith, M. E., Smith, J. F., Margraf, L. R., and Mendelson, C. R., "Primary cell culture of human type II pneumocytes: Maintenance of a differentiated phenotype and transfection with recombinant adenoviruses," *Am. J. Resp. Cell Mol. Biol.*, Vol. 17, No. 6, 1997, pp. 672–682.

[54] Gueven, N., Glatthaar, B., Manke, H. G., and Haemmerle, H., "Co-cultivation of rat pneumocytes and bovine endothelial cells on a liquid-air interface," *Eur. Resp. J.*, Vol. 9, No. 5, 1996, pp. 968–975.

[55] Banes, A. J., Gilbert, J., Taylor, D., and Monbureau, O., "A new vacuum-operated stress-providing instrument that applies static or variable duration cyclic tension or compression cells in vitro," *J. Cell Sci.*, Vol. 75, 1985, pp. 35–42.

[56] Sanchez-Esteban, J., Cicchiello, L. A., Wang, Y. L., Tsai, S. W., Williams, L. K., Torday, J. S., and Rubin, L. P., "Mechanical stretch promotes alveolar epithelial type II cell differentiation," *J. Appl. Physiol.*, Vol. 91, No. 2, 2001, pp. 589–595.

[57] Scott, J. E., Yang, S. Y., Stanik, E., and Anderson, J. E., "Influence of strain on [H^3] thymidine incorporation, surfactant-related phospholipid synthesis, and cAMP levels in fetal type II alveolar cells," *Am. J. Resp. Cell Mol. Biol.*, Vol. 8, No. 3, 1993, pp. 258–265.

[58] Waters, C. M., Ridge, K. M., Sunio, G., Venetsanou, K., and Sznajder, J. I., "Mechanical stretching of alveolar epithelial cells increases Na^+-K^+-ATPase activity," *J. Appl. Physiol.*, Vol. 87, No. 2, 1999, pp. 715–721.

[59] Edwards, Y. S., Sutherland, L. M., Power, J. H. T., Nicholas, T. E., and Murray, A. W., "Cyclic stretch induces both apoptosis and secretion in rat alveolar type II cells," *Febs. Lett.*, Vol. 448, No. 1, 1999, pp. 127–130.

[60] Savla, U., and Waters, C. M., "Mechanical strain inhibits repair of airway epithelium in vitro," *Am. J. Physiol.: Lung Cell. Mol. Physiol.*, Vol. 274, No. 6, 1998, pp. L883–L892.

[61] Tschumperlin, D. J., and Margulies, S. S., "Equibiaxial deformation-induced injury of alveolar epithelial cells in vitro," *Am. J. Physiol.: Lung Cell. Mol. Physiol.*, Vol. 275, No. 6, 1998, pp. L1173–L1183.

[62] Sanderson, M. J., and Dirksen, E. R., "Mechanosensitivity of cultured ciliated cells from the mammalian respiratory tract: Implications for the regulation of mucociliary transport," *Proc. Natl. Acad. Sci. USA*, Vol. 83, No. 19, 1986, pp. 7302–7306.

[63] Yamaguchi, M., Shimizu, N., Shibata, Y., and Abiko, Y., "Effects of different magnitudes of tension force on alkaline phosphatase activity in periodontal ligament cells," *J. Dent. Res.*, Vol. 75, No. 3, 1996, pp. 889–894.

[64] Gilbert, J. A., Weinhold, P. S., Banes, A. J., Link, G. W., and Jones, G. L., "Strain profiles for circular cell culture plates containing flexible surfaces employed to mechanically deform cells in vitro," *J. Biomechan.*, Vol. 27, No. 9, 1994, pp. 1169–1177.

[65] Williams, J. L., Chen, J. H., and Belloli, D. M., "Strain fields on cell stressing devices employing clamped circular elastic diaphragms as substrates," *J. Biomechan. Eng.—Trans. ASME*, Vol. 114, No. 3, 1992, pp. 377–384.

[66] Tschumperlin, D. J., Oswari, J., and Margulies, S. S., "Deformation-induced injury of alveolar epithelial cells—effect of frequency, duration, and amplitude," *Am. J. Resp. Crit. Care Med.*, Vol. 162, No. 2, 2000, pp. 357–362.

[67] Waters, C. M., Glucksberg, M. R., Lautenschlager, E. P., Lee, C. W., Van Matre, R. M., Warp, R. J., Savla, U., Healy, K. E., Moran, B., Castner, D. G., and Bearinger, J. P., "A system to impose prescribed homogeneous strains on cultured cells," *J. Appl. Physiol.*, Vol. 91, No. 4, 2001, pp. 1600–1610.

[68] Wirtz, H., R. W., and Dobbs, L. G., "Calcium mobilization and exocytosis after one mechanical stretch of lung epithelial cells," *Science,* Vol. 250, No. 4985, 1990, pp. 1266–1269.

[69] Ressler, B., Lee, R. T., Randell, S. H., Drazen, J. M., and Kamm, R. D., "Molecular responses of rat tracehal epithelial cells to transmembrane pressure," *Am. J. Physiol.: Lung Cell. Mol. Physiol.*, Vol. 278, No. 6, 2000, pp. L1264–L1272.

[70] Tschumperlin, D. J., Shively, J. D., Kikuchi, T., and Drazen, J. M., "Mechanical stress triggers selective release of fibrotic mediators from bronchial epithelium," *Am. J. Resp. Cell Mol. Biol.*, Vol. 28, No. 2, 2003, pp. 142–149.

[71] Tschumperlin, D. J., Shively, J. D., Swartz, M. A., Silverman, E. S., Haley, K. J., Raab, G., and Drazen, J. M., "Mechanotransduction in the lung—bronchial epithelial compression regulates MAP kinase signaling and HB-EGF-like growth factor expression," *Am. J. Physiol.: Lung Cell. Mol. Physiol.*, Vol. 282, No. 5, 2002, pp. L904–L911.

[72] Swartz, M. A., Tschumperlin, D. J., Kamm, R. D., and Drazen, J. M., "Mechanical stress is communicated between different cell types to elicit matrix remodeling," *Proc. Natl. Acad. Sci. USA*, Vol. 98, No. 11, 2001, pp. 6180–6185.

[73] Skinner, S. J. M., "Fetal breathing movements: A mechanical stimulus for fetal lung cell growth and differentiation," in B. M. Johnson, and P. D. Gluckman (eds.), *Advances in fetal physiology*, Ithaca, NY: Perinatology, 1989, pp. 133–141.

[74] Liu, M., Skinner, S. J. M., Xu, J., Han, R. N. N., Tanswell, A. K., and Post, M., "Stimulation of fetal rat lung cell proliferation in vitro by mechanical stretch," *Am. J. Physiol.*, Vol. 263, No. 3, 1992, pp. L376–L383.

[75] Liu, M. Y., Xu, J., Souza, P., Tanswell, B., Tanswell, A. K., and Post, M., "The effect of mechanical strain on fetal rat lung cell proliferation: Comparison of two- and three-dimensional culture systems," *In Vitro Cell. Dev. Biol.: Anim.*, Vol. 31, No. 11, 1995, pp. 858–866.

[76] Liu, M. Y., Xu, J., Tanswell, A. K., and Post, M., "Stretch-induced growth-promoting activities stimulate fetal rat lung epithelial cell proliferation," *Exp. Lung Res.*, Vol. 19, No. 4, 1993, pp. 505–517.

[77] Liu, M. Y., Liu, J., Buch, S., Tanswell, A. K., and Post, M., "Antisense oligonucleotides for PDGF-B and its receptor inhibit mechanical strain-induced fetal lung cell growth," *Am. J. Physiol.: Lung Cell. Mol. Physiol.*, Vol. 269, No. 2, 1995, pp. L178–L184.

[78] Xu, J., Liu, M. Y., and Post, M., "Differential regulation of extracellular matrix molecules by mechanical strain of fetal lung cells," *Am. J. Physiol.: Lung Cell. Mol. Physiol.*, Vol. 276, No. 5, 1999, pp. L728–L735.

[79] Xu, J., Liu, M. Y., Tanswell, A. K., and Post, M., "Mesenchymal determination of mechanical strain-induced fetal lung cell proliferation," *Am. J. Physiol.: Lung Cell. Mol. Physiol.*, Vol. 275, No. 3, 1998, pp. L545–L550.

[80] Liu, M. Y., Montazeri, S., Jedlovsky, T., van Wert, R., Zhang, J., and Yan, J., "Bio-stretch, a compartmentalized cell strain apparatus for three-dimensional organotypic cultures," *In Vitro Cell. Dev. Biol.: Anim.*, Vol. 35, No. 2, 1999, pp. 87–93.

[81] Isakson, B. E., Seedorf, G. J., Lubman, R. L., Evans, W. H., and Boitano, S., "Cell-cell communication in heterocellular cultures of alveolar epithelial cells," *Am. J. Resp. Cell Mol. Biol.*, Vol. 29, No. 5, 2003, pp. 552–561.

[82] Cassidy, K. J., Halpern, D., Ressler, B. G., and Grotberg, J. B., "Surfactant effects in model airway closure experiments," *J. Appl. Physiol.*, Vol. 87, No. 1, 1999, pp. 415–427.

[83] Grotberg, J. B., "Respiratory fluid mechanics and transport processes," *Ann. Rev. Biomed. Eng.*, Vol. 3, 2001, pp. 421–457.

[84] Fujioka, H., and Grotberg, J. B., "Steady propagation of a liquid plug in a two-dimensional channel." *J. Biomechan. Eng.—Trans. ASME*, Vol. 126, No. 5, 2004, pp. 567–577.

[85] Bilek, A. M., Dee, K. C., and Gaver, D. P., "Mechanisms of surface-tension-induced epithelial cell damage in a model of pulmonary airway reopening," *J. Appl. Physiol.*, Vol. 94, No. 2, 2003, pp. 770–783.

[86] Kay, S. S., Bilek, A. M., Dee, K. C., and Gaver, D. P., "Pressure gradient, not exposure duration, determines the extent of epithelial cell damage in a model of pulmonary airway reopening." *J. Appl. Physiol.*, Vol. 97, No. 1, 2004, pp. 269–276.

[87] Huh, D., Fujioka, H., Tung, Y. C., Futai, N., Paine, R., III, Grotberg, J. B., and Takayama, S., "Acoustically detectable cellular-level lung injury induced by fluid mechanical stresses

in microfluidic airway systems," *Proc. Nat. Acad. Sci. U. S. A.*, Vol. 104, No. 48, 2007, pp. 18886–18891.

[88] Tarran, R., Button, B., Picher, M., Paradiso, A. M., Ribeiro, C. M., Lazarowski, E. R., Zhang, L., Collins, P. L., Pickles, R. J., Fredberg, J. J., and Boucher, R. C., "Normal and cystic fibrosis airway surface liquid homeostasis," *J. Biolog. Chem.*, Vol. 280, No. 42, 2005, pp. 35751–35759.

[89] Even-Tzur, N., Elad, D., Zaretsky, U., Randell, S. H., Haklai, R., and Wolf, M., "Custom-designed wells and flow chamber for exposing air-liquid interface cultures to wall shear stress," *Annals Biomed. Eng.*, Vol. 34, Vol. 12, 2006, pp. 1890–1895.

[90] Tarkington, B. K., Wu, R., Sun, W. M., Nikula, K. J., Wilson, D. W., and Last, J. A., "In vitro exposure of tracheobronchial epithelial cells and of tracheal explants to ozone," *Toxicology,* Vol. 88, Nos. 1–3, 1994, pp. 51–68.

[91] Sun, W. M., Wu, R., and Last, J. A., "Effects of exposure to environmental tabacco smoke on a human tracheobronchial epithelial cell line," *Toxicology,* Vol. 100, Nos. 1–3, 1995, pp. 163–174.

[92] Aufderheide, M., and Mohr, U., "CULTEX—an alternative technique for cultivation and exposure of cells of the respiratory tract to airborne pollutants at the air/liquid interface," *Exp. Toxicolog. Path.*, Vol. 52, No. 3, 2000, pp. 265–270.

[93] Ritter, D., Knebel, J. W., and Aufderheide, M., "In vitro exposure of isolated cells to native gaseous compounds: Development and validation of an optimized system for human lung cells," *Exp. Toxicolog. Path.*, Vol. 53, No. 5, 2001, pp. 373–386.

[94] Knebel, J. W., Ritter, D., and Aufderheide, M., "Exposure of human lung cells to native diesel motor exhaust—development of an optimized in vitro test strategy," *Toxicol. In Vitro,* Vol. 16, No. 2, 2002, pp. 185–192.

[95] Wolz, L., Krause, G., Scherer, G., Aufderheide, M., and Mohr, U., "In vitro genotoxicity assay of sidestream smoke using a human bronchial epithelial cell line," *Food Chem. Toxicol.*, Vol. 40, No. 6, 2002, pp. 845–850.

[96] Fukano, Y., Yoshimura, H., and Yoshida, T., "Heme oxygenase-1 gene expression in human alveolar epithelial cells (A549) following exposure to whole cigarette smoke on a direct in vitro exposure system," *Exp. Toxicolog. Path.*, Vol. 57, Nos. 5–6, 2006, pp. 411–418.

[97] Park, T. H., and Shuler, M. L., "Integration of cell culture and microfabrication technology," *Biotech. Prog.*, Vol. 19, No. 2, 2003, pp. 243–253.

[98] Sweeney, L. M., Shuler, M. L., Babish, J. G., and Ghanem, A., "A cell-culture analog of rodent physiology—application to naphthalene toxicology," *Toxicol. In Vitro,* Vol. 9, No. 3, 1995, pp. 307–316.

[99] Ghanem, A., and Shuler, M. L., "Characterization of a perfusion reactor utilizing mammalian cells on microcarrier beads," *Biotech. Prog.*, Vol. 16, No. 3, 2000, pp. 471–479.

[100] Sin, A., Chin, K. C., Jamil, M. F., Kostov, Y., Rao, G., and Shuler, M. L., "The design and fabrication of three-chamber microscale cell culture analog devices with integrated dissolved oxygen sensors," *Biotech. Prog.*, Vol. 20, No. 1, 2004, pp. 338–345.

[101] Viravaidya, K., Sin, A., and Shuler, M. L., "Development of a microscale cell culture analog to probe naphthalene toxicity," *Biotech. Prog.*, Vol. 20, No. 1, 2004, pp. 316–323.

[102] Viravaidya, K., and Shuler, M. L., "Incorporation of 3T3-L1 cells to mimic bioaccumulation in a microscale cell culture analog device for toxicity studies," *Biotech. Prog.*, Vol. 20, No. 2, 2004, pp. 590–597.

[103] Rinaldo, J. E., and Rogers, R. M., "Adult respiratory-distress syndrome—changing concepts of lung injury and repair," *New Eng. J. Med.*, Vol. 306, No. 15, 1982, pp. 900–909.

[104] Martin, T. R., "Lung cytokines and ARDS," *Chest,* Vol. 116, No. 1, 1999, pp. 2S–8S.

[105] Rodriguez-Caballero, A., Garcia-Montero, A. C., Bueno, C., Almeida, J., Varro, R., Chen, R., Pandiella, A., and Orfao, A., "A new simple whole blood flow cytometry-based method for simultaneous identification of activated cells and quantitative evaluation of cytokines released during activation," *Lab. Invest.*, Vol. 84, No. 10, 2004, pp. 1387–1398.

[106] Strieter, R. M., Kundel, S. L., Showell, H. J., Remick, D. G., Pha, S. H., Ward, P. A., and Marks, R. M., "Endothelial cells gene expression of a neutrophil chemotactic factor by TNF-α, LPS, and IL-1β." *Science,* Vol, 243, No. 4897, 1989, pp. 1467–1469.

[107] Standiford, T. J., Kunkel, S. L., Basha, M. A., Chensue, S. W., Lynch, J. P., III, Toews, G. B., Westwick, J., and Strieter, R. M., "Interleukin-8 gene expression by a pulmonary epithelial cell line," *J. Clin. Invest.*, Vol. 86, No. 6, 1990, pp. 1945–1953.

[108] Park, W. Y., Goodman, R. B., Steinberg, K. P., Ruzinski, J. T., Radella, F., II, Park, D., Pugin, J., Skerrett, S. H., Hudson, L. D., and Martin, T. R., "Cytokine balance in the lungs of patients with acute respiratory distress syndrome," *Am. J. Resp. Crit. Care Med.*, Vol. 164, No. 10, 2001, pp. 1896–1903.

[109] Tremblay, L., Valenza, F., Ribeiro, J., Li, J., and Slutsky, A. S., "Injurious ventilatory strategies increase cytokines and c-fos mRNA expression in an isolated rat lung model," *J. Clin. Invest.*, Vol. 99, No. 5, 1997, pp. 944–952.

[110] Goodman, R. B., Pugin, J., Lee, J. S., and Matthay, M. A., "Cytokine-mediated inflammation in acute lung injury," *Cytokine Growth Factor Rev.*, Vol. 14, No. 6, 2003, pp. 523–535.

[111] Bezzant, T. B., and Mortensen, J. D., "Risks and hazards of mechanical ventilation: A collective review of published literature," *DM Disease-a-Month*, Vol. 40, No. 11, 1994, pp. 581–640.

[112] Rich, P. B., Douillet, C. D., Hurd, H., and Boucher, R. C., "Effect of ventilatory rate on airway cytokine levels and lung injury," *J. Surg. Res.*, Vol. 113, No. 1, 2003, pp. 139–145.

[113] Ranieri, V. M., Zhang, H., Mascia, L., Aubin, M., Lin, C. Y., Mullen, J. B., Grasso, S., Binnie, M., Volgyesi, G. A., and Slutsky, A. S., "Pressure-time curve predicts minimally injurious ventilatory strategy in an insolated rat lung model," *Anesthesia,* Vol. 93, No. 5, 2000, pp. 1320–1328.

[114] Kwon, O. J., Au, B. T., Collins, P. D., Adcock, I. M., Mak, J. C., Robbins, R. R., Chung, K. F., and Barnes, P. J., "Tumor necrosis factor-induced interleukin-8 expression in cultured human airway epithelial cells," *Am. J. Physiol.: Lung Cell. Mol. Physiol.*, Vol. 267, No. 4, 1994, pp. L398–L405.

[115] Yamamoto, H., Teramoto, H., Uetani, K., Igawa, K., and Shimizu, E., "Stretch induces a growth factor in alveolar cells via protein kinase," *Resp. Physiol.*, Vol. 127, Nos. 2–3, 2001, pp. 105–111.

[116] Yamamoto, H., Teramoto, H., Uetani, K., Igawa, K., and Shimizu, E., "Cyclic stretch upregulates interleukin-8 and transforming growth factor–â1 production through a protein kinase C-dependent pathway in alveolar epithelial cells," *Respirology,* Vol. 7, No. 2, 2002, pp. 103–109.

[117] Quinn, D., Tager, A., Joseph, P. M., Bonventre, J. V., Force, T., and Hales, C. A., "Stretch-induced mitogen-activated protein kinase activation and interleukin-8 production in type II alveolar cells," *Chest,* Vol. 116, No. 1, 1999, pp. 89S–90S.

[118] Hammerschmidt, S., Kuhn, H., Sack, U., Schlenska, A., Gessner, C., Gillissen, A., and Wirtz, H., "Mechanical stretch alters alveolar type II cell mediator release toward a proinflammatory pattern," *Am. J. Resp. Cell Mol. Biol.*, Vol. 33, No. 2, 2005, pp. 203–210.

[119] Levine, G. K., Deutschman, C. S., Helfaer, M. A., and Margulies, S. S., "Sepsis-induced lung injury in rats increases alveolar epithelial vulnerability to stretch," *Crit. Care Med.*, Vol. 34, No. 6, 2006, pp. 1746–1751.

[120] Pugin, J., Dunn, I., Jolliet, P., Tassaux, D., Magnenat, J. L., Nicod, L. P., and Chevrolet, J. C., "Activation of human macrophages by mechanical ventilation in vitro," *Am. J. Physiol.: Lung Cell. Mol. Physiol.*, Vol. 275, No. 6, 1998, pp. L1040–L1050.

[121] Lang, C. J., Barnetta, E. K., and Doylea, I. R., "Stretch and CO2 modulate the inflammatory response of alveolar macrophages through independent changes in metabolic activity," *Cytokine,* Vol. 33, No. 6, 2006, pp. 346–351.

[122] Crestani, B., Cornillet, P., Dehoux, M., Rolland, C., Guenounou, M., and Aubier, M., "Alveolar type II epithelial cells produce interleukin-6 in vitro and in vivo—regulation by alveolar macrophage secretory products," *J. Clin. Invest.*, Vol. 94, No. 2, 1994, pp. 731–740.

[123] Choe, M. M., Sporn, P. H. S., and Swartz, M. A., "An in vitro airway wall model of remodeling," *Am. J. Physiol.: Lung Cell. Mol. Physiol.*, Vol. 285, No. 2, 2003, pp. L427–433.

CHAPTER 25

Microfabricated Systems for Analyzing Immune-Cell Functions

Darrell J. Irvine and Junsang Doh

25.1 Introduction

The immune system is a dynamic collection of cells distributed throughout the body during homeostasis between primary and secondary lymphoid organs (the dedicated tissue sites where adaptive immune responses are initiated), the bloodstream, and peripheral tissues [1, 2]. The immune system's roles (both beneficial and pathological) in response to infection, defense against tumors, autoimmunity, allergy, and tissue-/organ-transplant rejection demonstrate the broad importance of understanding immunity for public health. Given the critical role of the immune system in these diverse settings, strategies seeking to analyze and manipulate the immune system ex vivo and in vivo are of great interest for gaining a full understanding of immune-cell functions and, in some instances, providing new treatment modalities. The fine balance between protective immune responses and autoimmune destruction of healthy tissue requires exquisite regulation of the effector cells of the immune system; regulation of immune-cell function derives from an intricate web of extracellular (e.g., soluble cytokine) and cell-cell contact-derived signals. The complexity of these microenvironments in which the immune system functions makes the study and modulation of immune-cell functions a fascinating and daunting challenge, where many possibilities remain unexplored.

Here, we review several recently developed strategies to engineer the microenvironment of immune cells in vitro using micro- and nanotechnology and highlight some of the achievements and challenges ahead in the nascent field of immunity-focused bioengineering. In particular, we focus on three general areas of research: First, we describe strategies using micro- and nanopatterned surfaces as model substrates for the presentation of biological ligands to immune cells, where cell-surface contacts are used as models of naturally occurring cell-cell and cell-antigen contacts. Second, we describe the creation of single-cell arrays of immune cells using microfabricated substrates: live-cell "chips" for massive analysis and screening of lymphocyte functions at the single-cell level and ultrasensitive biosensing. Finally, we discuss new technologies based on microfluidics and controlled-release microspheres, which control the migration of immune cells in vitro and in vivo via tailored concentration gradients of soluble chemoattractants with defined spatial and temporal control in two and three dimensions. These systems have allowed chemoattractant-triggered immune-cell migration (chemotaxis, fugetaxis, and chemokinesis) to be dissected in ways not before possible and open up new possibilities for manipulating immune-cell migration in vivo for improved vaccines and immunotherapies. Due to space constraints, we have chosen to focus on these three particular areas of study, though numerous other approaches for manipulating immune cells are moving forward driven by micro- and nanotechnology, such as microfabricated "lab-on-a-chip" strategies for sorting, purifying, and analyzing lymphocytes [3–7]; antibody and peptide-MHC microarrays for analyzing lymphocyte antigen specificity [8–11]; peptide-MHC-decorated quantum dots for interrogating antigen-specific T cells [12], and the use of surfaces with defined micro- and nanoscale features to study the role of physical topography in immune-cell migration [13–16], to name a few. These diverse new applications of micro- and nanotechnology to "immunobioengineering" are greatly expanding the toolset available for in vitro dissection of immune-cell functions and in vivo modulation of immune responses.

25.2 Micro- and Nanopatterned Surfaces as Tools to Dissect Immune-Cell Functions

25.2.1 Engineered Surfaces as Ligand Presentation Surrogates

As outlined in the introduction, the fine balance required for the immune system to rapidly identify and destroy invading pathogens or transformed cells while avoiding autoimmunity requires exquisite regulation of the cellular actors in the immune response. One of the critical mechanisms of immune regulation is the controlled engagement of cell-surface receptors governing immune-cell activation. A detailed understanding of the molecular events controlling immune-cell activation is thus of major importance for understanding immunity and developing new immune system-targeting therapies [17, 18]. However, analysis of the dynamic molecular events governing immune-cell activation is a technically challenging task. For example, lymphocyte activation is regulated by cell-cell contacts, where one cell (call this the "stimulator" cell) triggers the activation of another (call this the "responder" cell) via the ligation of receptor-ligand pairs at the cell-cell junction [Figure 25.1(a)]. In this setting, straightforward questions such as the importance of the surface density of a given ligand on one cell for the response of its partner are not simple to address since the areal density of a given cell-surface protein is not trivial to modulate at will in resting primary cells, even in the present era of powerful genetic tools such as RNA interference. In other cases, events happening when receptors at the surface of immune cells bind to soluble ligands occur too quickly to be readily spatially and temporally resolved by traditional methods. As an alternative to such natural stimuli, several groups have pursued the development of model synthetic surfaces as surrogates to present physiological ligands to immune cells in well-defined, controlled contexts, allowing a cell-surface contact to mimic a cell-cell or cell-ligand interaction of interest [Figure 25.1(a)].

One of the first and most successful examples of using synthetic surfaces for ligand display that went beyond simply coating tissue-culture plates with recombinant proteins was the supported lipid membrane system developed in the early

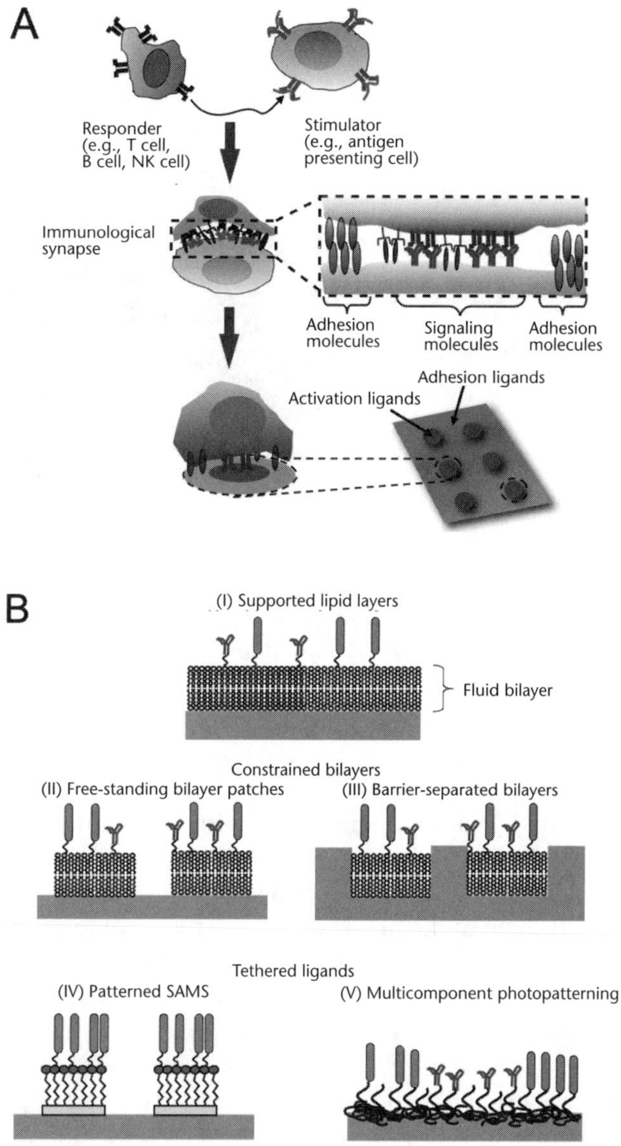

Figure 25.1 Using micro- and nanostructured surfaces to model cell-cell and cell-ligand interactions. (a) Example of using a cell-surface interaction to model a cell-cell interaction: A "responder" cell that is normally activated by interactions with a "stimulator" cell is seeded onto patterned synthetic substrates presenting ligands normally found on the surface of the stimulator cell. Ligand organization on the substrate can be used to induce receptor clustering on the responder cell in patterns mimicking those normally assembling in the cell-cell interface. (b) Strategies used to present engineer ligand presentation to immune cells from synthetic surfaces: (i) supported lipid bilayers with fully mobile ligands; (ii) and (iii) constrained bilayers; (iv) and (v) tethered ligands. (c) Control of ligand clustering on the nanometer scale by linking proteins to nanoparticles, which are subsequently patterned into micron-scale domains. Shown are fluorescence micrographs taken in two different fluorescence channels from the same region of a surface where 400 nm–diameter nanoparticles bearing fluorophores were patterned into segregated domains using strategy (v).

1980s by McConnell et al. [19]. In this approach, a monolayer or bilayer of phospholipids bearing membrane-anchored ligands (transmembrane or lipid-

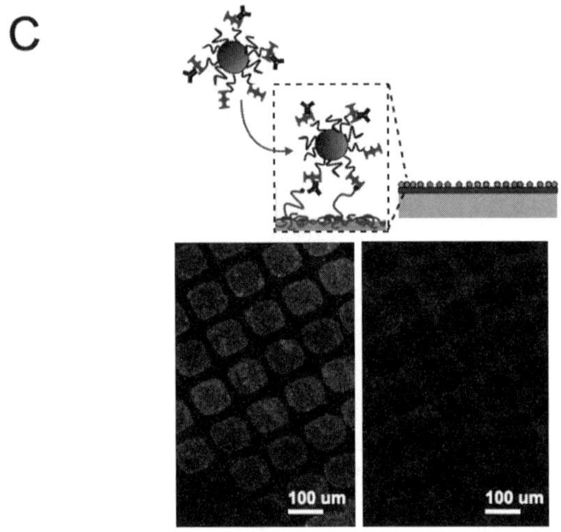

Figure 25.1 (continued)

anchored proteins) was deposited on a glass substrate. Cells making contact with these synthetic surfaces were activated in a ligand-specific manner determined by the composition of the supported membrane [20–22], and these ligand-presenting membranes were used to determine the minimal signals required for activation of innate immune cells such as macrophages [21], as well as $CD4^+$ and $CD8^+$ T cells [20, 23]. Building on these early examples, other groups developed refined systems based on supported lipid bilayers modified with lipid-anchored proteins, providing near-native mobility of ligands in the bilayer [Figure 25.1(b)(i)]. Dustin et al. have used such supported bilayer systems with great success in studies of T cell triggering and provided some of the first insights into the formation of the immunological synapse by T cells using this model system [24–30]. Because ligand is mobile in the two-dimensional plane of the supported membrane, cells seeded onto these bilayers bind and cluster ligand in a manner mimicking the clustering of ligand that occurs in the interface at immune-cell-cell contacts [27].

In the supported lipid bilayer model, the surface is not patterned per se and presents a uniform field of mobile ligands to cells seeded onto the bilayer. There are thus no constraints on how ligand clustering will occur (at micro or nano length scales), and clustering of receptors in the cell-surface interface occurs freely according to the physical chemistry of the ligand-receptor interactions and active transport of receptors by the cell [31]. Recently, the ability to engineer patterned surfaces with structural/chemical constraints on the display of ligands to immune cells has been exploited in several studies to gain control of interactions at the cell-surface interface. Two complementary approaches have been applied by several different groups [Figure 25.1(b)]. In the first approach, the classic fluid-supported lipid bilayer system is modified by patterning free-standing patches of lipid bilayer [Figure 25.1(b)(ii)] [32, 33] or confining bilayers by physical barriers [Figure 25.1(b)(iii)] [34], such that independent bilayer patches are confined to lithographically defined patterns on the surface. Patterning of supported bilayers into microscale domains was first demonstrated in the 1990s [35, 36] and has since been applied to characterizing immune-cell functions in a number of studies (described in more detail below). Ligands move freely within each bilayer microdomain but are unable to cross the edges of each bilayer patch, providing constraints on where ligands can move following engagement with receptors on the responder cell.

In the second strategy, ligands are bound to short polymer tethers in patterns on the surface, effectively fixing the ligands to the surface in defined patterns. One embodiment of this approach is via the binding of functionalized alkanethiols to patterned gold spots on silicon substrates, forming patterned self-assembled monolayers (SAMs) [Figure 25.1(b)(iv)]. SAMs are readily fabricated with a variety of terminal functional groups that can be used to link ligands to the alkyl chains before or after self-assembly. Using electron-beam lithography to pattern gold SAM binding sites on silicon, it has been shown that SAMs functionalized with ligands for mast cells (innate immune cells) can be created with individual ligand patches as small as 600 nm in diameter with high fidelity [37].

Patterning of two or more ligands into discrete physical sites can also be achieved. When cells are seeded on segregated patterns of two different ligands, the cell-surface receptors binding to these ligands are also physically isolated from one another in a manner determined by the surface pattern, allowing the segregation of cell-surface proteins observed during T cell [38], B cell [39], or NK cell activation [40] observed at immunological synapses to be mimicked [Figure 25.1(a)]. Photoresist materials have recently been developed that can be processed in all-aqueous conditions to maintain protein integrity, which also permits patterning of multiple ligands into defined domains on surfaces [41, 42]. Using these materials in a photolithographic "lift-off" process, ligands for immune cells patterned into defined, segregated microscale domains have been demonstrated as a model to interrogate the role of ligand clustering in T cell activation [43]. In addition, simultaneous control of ligand distributions on micron and submicron length scales is achievable by controlling the microscale segregation of two (or more) different ligand-presenting nanoparticles [Figure 25.1(c)].

These "constrained" model systems allow a number of key questions to be asked about receptor-ligand interactions occurring during immune-cell activation: First, as described above, constraints on the position of surface-bound proteins translate into constraints on the position of receptors that bind to these ligands. In this way, predetermined patterns of ligand on the substrate can enforce patterns of receptor clustering that deviate from what would occur in the "free" supported lipid bilayer model (or with live stimulator cells), allowing the importance of physical patterns of receptor clustering in the interface to be investigated. As discussed below, repatterning of "responder cell" surface receptors can in turn influence the clustering of intracellular signaling molecules that interact with these transmembrane proteins, leading to "templated" clustering of intracellular molecules driven by the pattern of ligands presented on the surface [33, 37, 43]. Questions such as the role of spatial position in the interface on cell signaling and the role of cluster size or microscale geometry on cell responses thus become accessible. Finally, artificial "compartmentalization" of ligands tethered to a surface can be used to isolate and stabilize receptor clustering and intracellular signaling events, allowing orchestrated biochemical steps of signaling occurring at the plasma membrane to be directly visualized.

25.2.2 Micron- and Submicron-Ligand Patterning to Model Cellular Activation by Immune Complexes

During an immune response, circulating antibodies bind to and crosslink soluble antigens, forming three-dimensional submicron- to micron-sized aggregates known as immune complexes. Antibodies also bind to Fc receptors (FcRs) expressed by innate immune cells: FcRs bind the Fc region (fragment crystallizable, the "back end") of antibodies. Mast cells, in particular, express high-affinity FcRs that allow these cells to be "sensitized" by binding to IgE antibody. If such IgE-decorated mast cells encounter antigen, binding of the cell surface-bound IgE to antigen leads to FcR crosslinking, triggering strong signaling in the mast cell and rapid internalization of the antigen-antibody complexes. Triggering of IgE-sensitized mast cells is of particular interest because the aberrant activation of these cells by harmless environmental antigen immune complexes is a characteristic feature of allergic reactions. Because FcR signaling normally leads to very rapid triggering and internalization of the bound antigen/receptor complex, it has been difficult to study the biochemical events involved in FcR signaling in mast cells. To overcome these difficulties, the groups of Baird, Craighead, and Ober collaborated to develop micro- and nanopatterned surfaces that presented noncell-internalizable surface-bound antigen. These surfaces allowed receptor clustering and local intracellular signaling responses to be analyzed in detail by dropping mast cells onto these model substrates and following the dynamics of cell-surface and intracellular protein clustering over well-defined surface-bound patches of ligand.

In initial studies, a strategy for patterning micron-scale patches of lipid bilayers bearing dinitrophenyl (DNP) ligands was developed using standard photolithographic procedures to create patterns of DNP bound to patterned bilayers as illustrated in Figure 25.1(b)(ii) [32]. DNP is a small molecule hapten for which antibodies are readily available. Mast cells sensitized to DNP by prebinding fluorescent anti-DNP antibody to their cell-surface FcRs reacted specifically to contact with the DNP surface arrays: the sensitized cells clustered FcRs over the bilayer patches, flattened their plasma membranes against the antibody-complexed substrates, and extended lamellipodia in response to the surface stimuli [32]. One of the proposed mechanisms for regulation of cell-surface signaling in many cells is the segregation of certain protein components at the cell surface into lipid "rafts," domains of the plasma membrane enriched in cholesterol and sphingomyelin that behave as dynamic gel-phase "islands" in the fluid bilayer [44]. During mast-cell activation, FcRs are thought to colocalize in lipid rafts enriched in certain key signaling molecules bound to the inner leaflet of the plasma membrane. Using patterned arrays of lipid-bilayer-bound DNP, Wu et al. directly tracked clustering of fluorescent signaling proteins at clusters of FcRs formed over DNP-bearing lipid patches and characterized the redistribution of signaling molecules bound to the inner leaflet of the plasma membrane, as well as lipid raft markers found in the outer leaflet of the plasma membrane [33]. These studies revealed actin-dependent differential clustering of inner- and outer-leaflet raft components at sites of FcR clustering over the DNP arrays, suggesting that the composition of lipid rafts is dynamically and separately controlled between the inner and outer face of the cell surface. Extending this model system, Senaratne et al. used electron-beam lithography to create submicron arrays of shallow gold-coated silicon posts 600 nm in diameter and 50 nm high with micron-scale pitches [37]. DNP-bearing self-assembled monolayers were assembled on these nanoscale gold platforms from solution to create submicron tethered clusters of DNP on substrates [Figure 25.1(b)(iv)]. By using such tiny ligand clusters, the clustering of fluorescent signaling molecules in response to a handful of engaged receptors at an extremely localized site (approaching the resolution limits of optical microscopy, ~0.1 μm) is possible, allowing the earliest signaling events proximal to receptor engagement to be interrogated. Results from this elegant model system are of broad interest because of the ubiquity of receptor clustering and intracellular signaling complex assembly for membrane-based signaling processes occurring in all cells.

25.2.3 Patterned Surfaces as Models to Probe Structure-Function Relationships of the T Cell Immunological Synapse

The mast-cell-patterned surface model systems just described are an example of using micro- and nanostructured surfaces to interrogate cell-surface signaling in response to macromolecules normally binding to the cell surface from soluble form in solution. A second set of exciting applications for engineered surfaces arises in the modeling of cell-cell junctions in immunology. Interestingly, though intercellular interactions are of ubiquitous importance in biology, the development of synthetic surface models for the study of cell-cell contacts has perhaps been most extensively explored in the context of immunobiology due to the paramount importance of cell-cell contacts in the regulation of immunity. Among the many cell-cell interactions regulating immune responses, the interaction between T cells and antigen-presenting cells (APCs—B cells, dendritic cells, or macrophages) is one of the most studied as it is one of the key initial events in adaptive immune responses. Motile T cells that encounter APCs bearing their cognate antigen halt their migration, repolarize the cell body toward the APC, and form long-lived contacts that are required for full T cell activation [45, 46]. It has long been known that signaling and adhesion proteins on both the T cell and APC polarize to the cell-cell contact zone during this conjugation [47], but in the late 1990s, reports of microscale patterning of signaling and adhesion proteins at the T-APC interface—the formation of what is now referred to as the immunological synapse (IS)—provided an entirely new perspective on the regulation of T cell activation [24, 38]. Strikingly, in these first studies, formation of a "mature" immunological synapse, defined by a characteristic "bull's eye" distribution of adhesion and signaling proteins in the interface [Figure 25.1(a)], correlated with full T cell activation. This finding raised the possibility that in addition to well-established parameters such as the density of foreign antigen displayed by the APC and the affinity of T cell receptors (TCRs) for these ligands, the physical organization of cell-surface receptors and intracellular signaling molecules in the interface could impact the outcome of T cell recognition. Subsequent investigations have revealed an entire zoology of patterns of receptor clustering in the T cell–APC interface in addition to the canonical "bull's eye" organization, and the immunological synapse structure is now known to vary depending on the maturation state of the T cell (thymocyte versus mature T cell), ligand quality/quantity, identity and activation status of the APC, and the temporal stage of T cell activation [48, 49]. Because of the diversity of experimental parameters and the difficulty of controlling the physical organization of cell-surface proteins interacting on live cells, it has been difficult to fully define the relationships between synapse structure/assembly-disassembly dynamics and T cell early/endpoint activation events.

In order to interrogate the role of synapse assembly on early T cell signaling events, both the micropatterned supported lipid bilayer [Figure 25.1(b)(iii)] [34, 50] and micropatterned tethered ligand approaches [Figure 25.1(b)(v)] [43] have been applied. In the early stages of synapse assembly, T cells actively redistribute TCRs engaged with peptide-MHC complexes on the surface of the APC to the center of the T-APC contact [24]. Mossman et al. used micropatterned bilayer surfaces presenting the adhesion protein intercellular cell-adhesion molecule-1 (ICAM-1) and peptide-Major Histocompatibility Complex (pMHC) molecules in micron-scale corrals to determine the effect of blocking this centralized clustering of engaged receptors [34]. ICAM-1 is a ligand for the key T cell integrin lymphocyte-function-associated antigen–1 (LFA-1), while pMHC is a native ligand for the T cell receptor. T cells seeded on these patterned surfaces formed microclusters of receptor-bound pMHC within each "corral" of the pattern, but the physical barrier at the edge of each corral prevented these clusters from coalescing. Using intracellular staining to identify the location of phosphorylated tyrosines in T cells interacting with these surfaces (as a surrogate marker of TCR triggering), active signaling after the first few minutes of T cell surface interactions was revealed to be concentrated in the microclusters of TCR/pMHC formed in the periphery of the cell-substrate interface rather than the center of the contact site, and signaling was found to be sustained in the peripheral clusters of these "repatterned synapses" relative to events occurring when T cells interacted with nonpatterned bilayers. In addition to exhibiting sustained tyrosine phosphorylation on the patterned surfaces, T cells were also found to exhibit elevated intracellular calcium signaling, a second downstream marker of T cell receptor signaling. These results suggest that early signaling events in T cell activation can be modulated by impeding the process of receptor clustering that normally occurs in the immunological synapse.

T cell responses to patterned surfaces have also been studied using a multicomponent "tethered-ligand" approach. Here, patterns of anti-CD3 (a surrogate ligand for T cells that stimulates the TCR/CD3 complex) and ICAM-1 were immobilized on surfaces using photolithography to create "immunological synapse arrays" composed of anti-CD3 activation sites and a surrounding ICAM-1 adhesion field [Figure 25.2(a)]. T cells seeded onto these surfaces adhered and crawled over the ICAM-1 adhesion field but stopped migration and formed synapses on contact with the anti-CD3 activation sites; the fraction of T cells receiving such a "stop signal" from the TCR ligand was dependent on the dose of ligand present on surfaces [Figure 25.2(b)]. Similar to the findings of Wu et al. studying mast-cell responses to patterned DNP ligands, T cell intracellular signaling molecules were templated in physical patterns reflecting the pattern of ligands displayed on the surface [Figure 25.2(c)]: on the cell surface, TCRs clustered over regions of immobilized anti-CD3, while LFA-1 was found to cluster in the adjacent ICAM-1-bearing regions. Inside the cell, protein kinase C-θ (PKC-θ), a molecule known to cluster at the center of the immunological synapse, colocalized with sites of TCR clustering, while talin, an adaptor molecule that interacts with the cytosolic tail of LFA-1, was found to cluster in regions where LFA-1 had accumulated. Changes in the surface ligand pattern

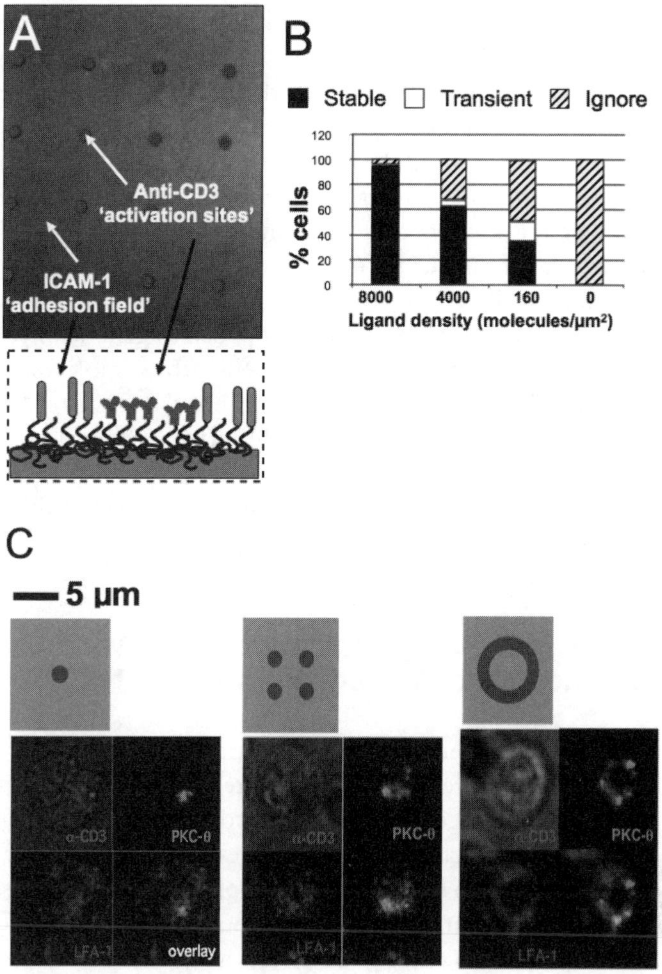

Figure 25.2 Using patterned surfaces to template immunological synapse formation by T cells: (a) Fluorescence overlay from a multicomponent "immunological synapse array" patterned surface bearing tethered anti-CD3 activation sites (6 µm in diameter) surrounded by an immobilized ICAM-1 adhesion field. (*Source:* [43] © 2006 National Academy of Sciences, U.S.A.) (b) T cells seeded on synapse arrays adhere and migrate; on encounter with activation sites, they will stably arrest in response to high areal densities of anti-CD3 in the activation sites. As ligand density is reduced, more cells either transiently halt migration (<20 min.) or completely ignore the activation sites. (c) T cells seeded onto synapse arrays and fixed/immunostained after twenty minutes of interaction with the surfaces demonstrated templating of cell-surface and intracellular signaling components in patterns mirroring the ligand pattern presented from surfaces. Shown are immunostaining results for cells interacting with synapse arrays of patterned anti-CD3 and ICAM-1 (geometries given by the red and green areas, respectively, of the pattern schematics): (top left) brightfield/anti-CD3 fluorescence overlay, (top right) PKC-θ fluorescence, (bottom left) LFA-1 fluorescence, (bottom right) PKC-θ/LFA-1 overlays. (*Source:* [43] © 2006 National Academy of Sciences, U.S.A.)

from "focal" activation sites (single spots of anti-CD3) to "multifocal" or annular patterns of anti-CD3 altered the clustering patterns of TCR, LFA-1, PKC-θ, and talin. For example, on annular anti-CD3 patterns, 25 percent of the T cells making full contact with activation sites exhibited an annular pattern of PKC-θ accumulation over the anti-CD3 region. Because the photolithographic process used allowed entire culture substrates to be patterned, the endpoint response of an entire population of T cells seeded on synapse-array surfaces could be measured. When the production of cytokines by T cells activated on synapse-array surfaces was measured, it was found that "focal" and annular anti-CD3 activation site patterns bearing the same total surface area (and thus, the same total quantity of TCR ligand) triggered T cells to secrete similar levels of the key autocrine growth factor interleukin-2, but production of the key effector cytokine interferon-γ was strongly suppressed on annular patterns of anti-CD3 [43].

The results of these first two studies using synthetic surfaces to physically alter immunological synapse formation, as opposed to genetically altering the T cell, suggest that the physical arrangement of cell-surface molecules on both micron and submicron length scales can have a significant impact on both the early and late events occurring in T cell activation. The extension of these approaches to address control over synapse structures at the nanoscale will no doubt lead to additional insights into the role of receptor clustering in governing this critical checkpoint in adaptive immunity.

25.3 Single-Cell Microarrays: Microwells and Microchambers for Assaying the Functions of Individual Lymphocytes

The surfaces described in the previous section were nominally chemically patterned, planar substrates, where immune cells were dropped into contact with the surface to analyze cellular or subcellular responses to the pattern of displayed ligands. A second class of microstructured surfaces that shows promise for the study of immune cells is single-cell microarrays, formed by physically confining single cells to defined two-dimensional or three-dimensional locations on a surface. Three approaches recently described for creating single-cell arrays of live lymphocytes are schematically illustrated in Figure 25.3(a). While micropatterning of anchorage-dependent cells is readily achieved by simply creating binary patterns of cell-adhesive and nonadhesive regions on surfaces [51], resting primary T- and B-lymphocytes exhibit little or no attachment to typical extracellular-matrix-adhesion proteins or serum-exposed culture surfaces. Thus, the creation of single-cell microarrays requires tethering lymphocytes to array sites via immobilized "capture" proteins [52], physical entrapment of single cells within microwells/microchambers [53, 54], or the combination of these two methods [55, 56]. Tethering via capture proteins can be achieved by immobilizing antibodies that bind to cell-surface proteins expressed by the target lymphocytes or by biotinylating live cells and tethering these modified cells to surfaces bearing immobilized streptavidin [52, 56]. Though both tethering and physical immobilization approaches appear to work in practice, physical immobilization in microwells or microchambers is particularly attractive as a simple and robust strategy for arraying live cells without triggering background stimulation of the cells.

Microwell-based approaches have recently been demonstrated for the creation of centimeter-sized microarrays of single T cells or B cells, where activation events occurring in hundreds of lymphocytes can be analyzed in parallel at the single-cell level both statically and in time lapse [53, 55, 56]. Yamamura et al. formed

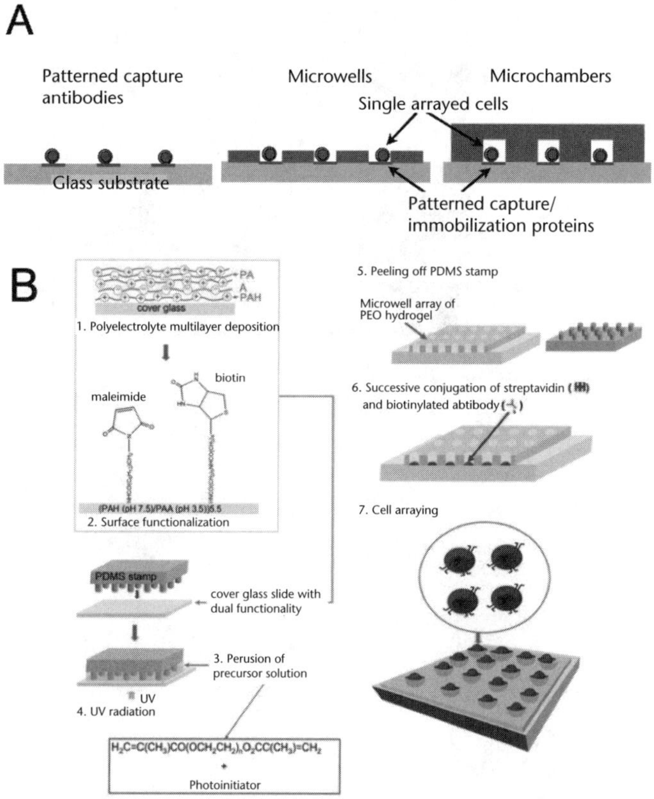

Figure 25.3 Single-cell lymphocyte microarrays: (a) Three approaches to assaying lymphocyte functions using multicellular or single-cell microarrays. (b) Soft lithographic strategy to fabricate protein-functionalized hydrogel microwells. (*Source:* [56] © 2006 Wiley-VCH.) (c) Example brightfield micrograph illustrating live B cells arrayed in a polyethylene glycol hydrogel microarray. (d) Schematic view of microarray T cell/B cell biosensor and quantitative time-lapse analysis of calcium signaling in arrayed T cells as a function of analyte peptide concentration added to the sensor. (*Source:* [56] © 2006 Wiley-VCH.)

microwells by electrodepositing nickel on a substrate bearing a photopatterned layer of poly(methyl methacrylate). Following nickel deposition, the nickel microstructure formed against the polymethylmethacrylate (PMMA) "master" was peeled away from the deposition substrate, revealing an array of nickel posts with the inverse microstructure of the desired microwell array. This metal microstructure was converted to a microwell array by molding of polystyrene over the nickel mold, providing a final polystyrene array "chip" comprising more than two hundred thousand wells, each 10 μm in diameter, 12 μm in depth, with an array pitch of 30 μm. Revzin et al. created poly(ethylene glycol) (PEG) hydrogel microwells on glass substrates by photopolymerizing a thin layer of PEG precursor solution on glass slides through a photomask [55, 57–59]. Following washing to remove unpolymerized precursor, the glass substrate "floors" of the microwells were modified by sequential adsorption of avidin and biotinylated antibodies to capture lymphocytes [55]. Kim et al. also prepared PEG microwells, using the soft-lithography-based approach outlined in Figure 25.3(b) to create microwell arrays where the base of each well was functionalized with antibodies chosen to aid in lymphocyte immobilization [56]. First, glass coverslips were functionalized with

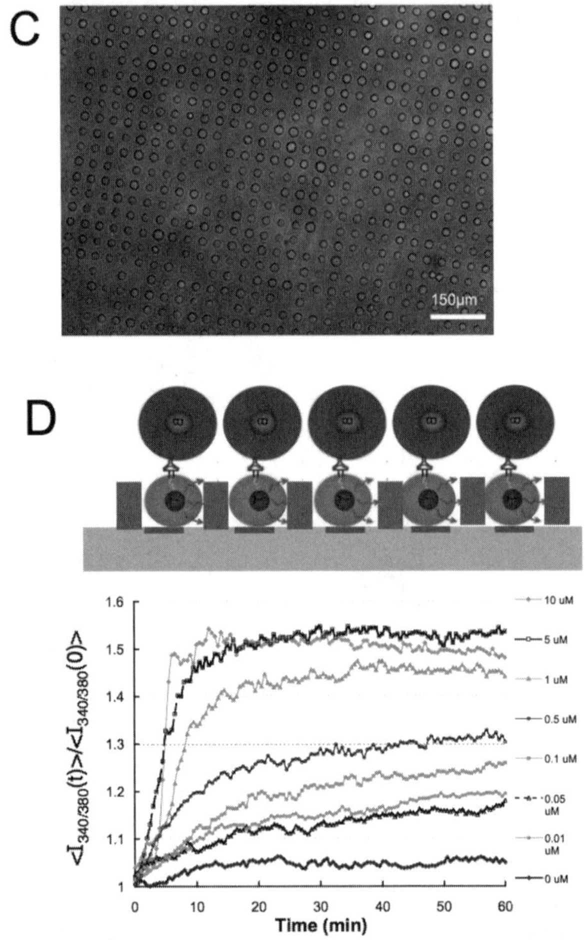

Figure 25.3 (continued)

maleimide and biotin groups. A poly(ethylene glycol)–based hydrogel was then photopolymerized on the coverslip using a polydimethylsiloxane (PDMS) stamp as a micromold; the stamp comprised a regular array of posts that defined the location and shape of the microwells formed. The stamp was then peeled away from the gel-coated substrate, revealing microwells bearing biotin groups at the base of each well. Biotinylated anti-CD44 antibodies were then immobilized to the base of each well via a streptavidin linker to allow anchoring of T or B lymphocytes to the wells via their cell-surface CD44 adhesion receptors. Each of the approaches described above allowed high-density single-cell arrays of lymphocytes to be created by simply sedimenting cell suspensions over the arrays followed by gentle washing. Notably, creation of microwells that closely matched the size of the cells to be arrayed and using tethering antibodies allowed for nearly perfect single-cell arrays to be created over regions of substrates encompassing several hundred array sites [Figure 25.3(c)] [56].

High-throughput optical microscopy analysis of microarrayed cells allows cell signaling events for hundreds of individual cells to be tracked over time. Yamamura et al. investigated B cell arrays exposed to activating stimuli to explore the possibility of using single-cell microarrays as a tool to screen peripheral blood lymphocytes for expression of antibodies specific for particular target antigens [53]. They demonstrated that individual B cells could be "picked" from cell arrays using a micromanipulator for further growth and analysis of individual B cell clones. In this way, antibody production by single B cells identified as responders in the microarray could be recovered for analysis of the DNA sequence of the antibodies they produced. Live lymphocyte arrays are also amenable to other methods of single-cell physical extraction, such as laser-capture microdissection [55]. To examine the potential of live lymphocytes for biosensing, Kim et al. used PEG-based microwells to array live T cells and overlaid T cell microarrays with a confluent layer of B cells [Figure 25.3(d)] [56]. Here, B cells served the role of capturing analytes from solution and presenting these peptides to the arrayed T cells. The T cells were loaded with a fluorescent calcium sensor dye, allowing optical recording of T cell triggering. When model peptide antigens were added in solution over the T-B microarrays, the peptides were captured by B cells on their surface MHC molecules and presented to T cells in the microarray, and T cell triggering was read out within seconds via fluorescence imaging of calcium signaling in the T cells. Quantitative analysis of the calcium signal of microarrayed T cells revealed that these T cell microarrays were highly sensitive detectors for their target peptides, able to reliably discriminate T cell calcium signaling in response to 10 nM solution concentrations of peptide from background calcium signals within ten minutes [Figure 25.3(d)]. Thus, microwell-arrayed lymphocytes may offer new strategies for diagnostic assays that can interrogate native lymphocyte antigen specificities or rapidly detect biomarkers associated with pathogens.

The "open" microwells just described are ideal for monitoring intracellular signaling events. However, what if we were instead interested in monitoring a secreted product from individual cells of the microarray? Love et al. recently reported an elegant strategy for capturing secreted antibodies from individual B cells in "sealed" microarrays fabricated by soft lithography [54]. In this approach [right-hand panel of Figure 25.3(a)], microwells were fabricated by molding PDMS against a negative microstructure silicon master. Individual B cells were seeded into these PDMS chambers, and the wells were closed by sealing the open face of the microwells against a glass coverslip. Using this system, it was shown that large collections of B cell clones (~20,000 cells) could be loaded and cultured individually in microchambers on a single microarray. Antibody secreted by single B cell clones was captured by secondary "capture" antibodies or target antigens coated on the coverslip floor of the microchamber array. On peeling of the PDMS microchambers away from the glass substrate, individual B cells were recovered, while the substrate bore a replica microarray of the secreted products from the cellular microchambers, allowing products secreted by each individual B cell to be characterized and the producing B cells of interest to be subsequently identified and recovered for further expansion/analysis [54]. This "microengraving" strategy, along with the complementary microwell studies of Yamamura et al. described above, may facilitate the rapid screening of large B cell clone libraries to identify and characterize antibody responses to antigens recovered from both mice and men. In addition, the capture substrate could detect a wide variety of secreted products, allowing other critical factors such as chemokines, cytokines, and immunomodulatory proteins produced by massive arrays of single cells to be characterized. Unlike similar single-cell analysis technologies like flow cytometry, the analysis of secreted products in the microchamber approach allows the secreting cells to be identified and recovered intact with minimal manipulation for further analysis or experimentation.

25.4 Control of Immune-Cell Migration in Model Microenvironments

The immune system is unique in that it's cellular actors are both dispersed throughout the body and highly mobile in order to perform their critical surveillance functions in protecting the host. Recruitment of cells to and from the blood and lymph into tissues, as well as their migration within the tissues, is thus a key step in many aspects of immune-cell function. Innate immune cells such as neutrophils and monocytes respond to acute infection by rapid extravasation from blood vessels at sites of inflammation and migration to the site of insult, while later in the adaptive immune response, a similar recruitment of effector T cells occurs [60]. Thus, regulated cell migration is central to the proper functioning of the immune system.

Chemotaxis of immune cells, directed migration of cells "up" a concentration gradient of a chemoattractant, has been studied for many years due to its importance in controlling the positioning of immune cells in the steady state and during immune responses [2, 60]. However, although many well-established assays for studying chemotaxis exist, such as the use of modified Boyden chambers [61], Dunn and Zigmond chambers [62], under-agarose migration [63], and collagen gel invasion assays [64], control over the concentration profiles of chemoattractants generated in these assays is highly limited [65, 66]. In most straightforward experimental systems, chemokine gradients are generated that are extremely short-lived and in which both the steepness of the gradient and the absolute concentration of the attractant vary significantly in time [65, 66]. Thus, many basic questions about how immune cells interpret the potentially complex (in space and time) chemoattractant gradients they may encounter in vivo remain unanswered. New microsystems for controlling the creation of soluble chemoattractant gradients, including the use of microfluidic chambers that allow controlled mixing of solutions on the microscale and microspheres that release chemoattractants, are beginning to shed new light on how immune cells chemotax and may provide new in vivo therapeutic avenues for vaccines and immunotherapy.

25.4.1 Creating Temporally Stable, Defined Chemoattractant Gradients Using Microfluidic Devices

As demonstrated by several of the examples discussed in the previous sections, soft-lithography-based microtechnologies are enabling many new kinds of analysis of immune-cell functions. Microfluidics for the study of immune-cell chemotaxis are another prime example of the impact of soft lithography on the engineering/immunobiology interface. Jeon et al. first demonstrated the use of microfluidics to create well-defined, temporally stable concentration gradients on chemoattractant molecules for the study of neutrophil chemotaxis [67]. This was achieved by connecting two or more solution inlets to a series of branching and recombining microfluidic channels, which allowed (for example) a solution of chemoattractant and buffer injected at two different inlet ports to be combined by controlled diffusive mixing into concentration profiles that had a well-defined behavior across the width of a downstream region of the microfluidic flow path. Linear, "hill-shaped," and "cliff" gradients of soluble factors were demonstrated with concentration profiles that could be maintained indefinitely in time via steady injection of the source solutions at the inlet ports. Using these stable concentration gradients (which are essentially impossible to achieve by any of the traditional assays described above) Jeon et al. showed that neutrophils migrate efficiently up concentration gradients of the inflammatory cytokine interleukin-8 (IL-8) but stop sharply on encountering an IL-8 "cliff," where the concentration of chemoattractant drops sharply. Interestingly, when neutrophils were instead challenged with a "hill" of IL-8 concentration, cells were capable of migrating over the peak of the hill distances of several cell diameters before altering their direction and turning to migrate back to the region of highest chemoattractant concentration. These responses suggest a sensitive discriminatory mechanism regulating chemotactic cell migration. Subsequent studies using more complex microfluidic systems that combine multiple gradient-generation channels with microfluidic valves allowed the investigation of neutrophil motility in response to rapid changes in chemoattractant gradients [68]. Using this system, Irimia et al. showed that neutrophils continue to chemotax up shallow IL-8 concentration gradients if the average concentration is rapidly increased while maintaining the concentration gradient [68]. In contrast, if an IL-8 gradient was rapidly "stepped down," maintaining a gradient but lowering the overall average concentration, neutrophils depolarized and exhibited an approximately five-minute delay before continuing to chemotax. Based on these findings, the authors propose a novel two-component temporal sensing model for neutrophil chemotaxis, where the cells detect both temporal changes in the average concentration of chemoattractant over the entire cell body and local concentration changes at the leading edge of the polarized cell [68]. Microfluidic concentration gradients have also been used to definitively demonstrate fugetaxis (migration of cells down a concentration gradient of a chemotactic factor) of neutrophils in response to steep IL-8 gradients [69]. Here, the ability of the gradient-generating microfluidic channel to create quantitatively precise linear concentration gradients revealed that neutrophils migrate up shallow concentration gradients of IL-8 but reverse direction and fugetax when faced with steep IL-8 gradients. Qualitatively opposite cellular responses to quantitative changes in chemoattractant concentration profiles may represent an important mechanism for regulation of recruitment of immune cells to inflammatory sites and may also be involved in the emigration of lymphocytes from the thymus [70–73]. Presuming that the ability of neutrophils to localize to well-defined regions and in some cases be strongly confined in space by the quantitative shape of chemoattractant concentration profiles is a general feature of immune-cell chemotaxis, these findings could have implications both for how immune cells navigate through inflammatory sites and how leukocytes are positioned and segregated within secondary lymphoid tissues.

25.4.2 Engineering Chemoattractant Gradients Using Controlled-Release Microspheres

The microfluidic systems just described for controlling the generation of soluble chemoattractant gradients are powerful approaches to the study of leukocyte chemotaxis in vitro. A second recently developed approach to controlling immune-cell migration, which is potentially applicable both in vitro and in vivo, is the use of controlled-release microspheres to generate defined chemoattractant gradients.

Microspheres fabricated using biodegradable polymers such as hydrolyzable polyesters have been studied for many years as potential vehicles for sustained and/or localized drug delivery: small molecule or macromolecular drugs are encapsulated within these microparticles and release by diffusion through aqueous pores formed in the degrading polymer matrix over time [74–76]. Parameters allowing the release rate of encapsulated materials to be tuned have been thoroughly studied, such as polymer matrix hydrophilicity, molecular weight, and crystallinity [77]. Typically, the objective of using controlled-release particles for drug delivery is to maintain a set level of a labile drug in the local tissue or systemic circulation without the need for repeated bolus injections. However, a tailored rate of release from microspheres can also be used to control the development of a chemoattractant profile around a slow-release microparticle.

This approach was used to control the chemotaxis of dendritic cells (DCs), the critical antigen-presenting cells involved in capturing antigens at sites of infection or vaccination for presentation to T and B cells in draining lymph nodes [78]. Two different types of chemoattractants, a low-molar-mass formyl peptide derived from bacteria and a host chemokine (CCL20), or macrophage inflammatory protein-3α, were encapsulated in poly(lactide-co-glycolide) (PLGA) microspheres prepared by common emulsion synthesis methods. As expected, the release rate for these factors could be tuned by control over fabrication parameters or the molecular weight of the matrix PLGA. Using a three-dimensional collagen gel migration assay where a "depot" of microspheres was suspended in a type I collagen matrix, surrounded by matrix containing immature (resting) DCs or monocytes (DC precursors), migration of cells in response to chemoattractants released from microspheres was tracked by videomicroscopy. Microsphere-generated chemoattractant gradients were capable of attracting cells over distances of up to 0.5 mm through the collagen gel. Interestingly, single isolated microspheres (~20 to 40 μm in diameter) releasing CCL20 were capable of acting as "point sources" and strongly attracted resting DCs from the surrounding collagen matrix for sustained periods of up to at least ten hours. Attracted cells were able to migrate directly to the point of contact with the attracting microspheres, without clear evidence of desensitization to the chemoattractant gradient. This simple approach provides both a useful tool for dissecting how leukocytes interpret three-dimensional soluble chemokine gradient in vitro and a potential platform for directing immune-cell migration in vivo. Chemoattractant-releasing microspheres may be of particular interest as a adjuvant for vaccines or DC-targeted immunotherapies in order to concentrate these target cells at the site of an injected antigen or immunomodulatory drug [79, 80].

25.5 Conclusions and Outlook

As illustrated by the examples covered in this chapter, micro- and nanotechnologies are rapidly being developed and applied to the study of innate and adaptive immune cells, allowing their complex individual functions, cell-cell interactions, and migration to be dissected in greater detail than ever before. This is made possible by the unprecedented control over the two-dimensional and three-dimensional microenvironments of individual cells offered by these novel approaches. In addition, these early "immunobioengineering" efforts suggest new biotechnological applications of immune cells in sensing, diagnostic assays, and the generation of immunological reagents such as monoclonal antibodies. The challenge facing this field is to gain temporal and spatial control over multiple microenvironment variables simultaneously in order to move from ultrareductionist in vitro models to systems that allow the complexity of real in vivo microenvironments to be captured and explored in a systematic fashion.

Acknowledgments

This work was supported by the National Science Foundation (award 0348259), the Arnold and Mabel Beckman Foundation, the DuPont-MIT Alliance, the Defense Advanced Research Projects Agency (award DAMD17-02-C-0130), and the National Institutes of Health (5R21AI062731-02).

References

[1] Luster, A. D., Alon, R., and von Andrian, U. H., "Immune cell migration in inflammation: Present and future therapeutic targets," *Nat. Immunol.*, Vol. 6, No. 12, 2005, pp. 1182–1190.

[2] Campbell, D. J., Kim, C. H., and Butcher, E. C., "Chemokines in the systemic organization of immunity," *Immunol. Rev.*, Vol. 195, 2003, pp. 58–71.

[3] Murthy, S. K., Sin, A., Tompkins, R. G., and Toner, M., "Effect of flow and surface conditions on human lymphocyte isolation using microfluidic chambers," *Langmuir*, Vol. 20, No. 26, 2004, pp. 11649–11655.

[4] Sekine, K., Revzin, A., Tompkins, R. G., and Toner, M., "Panning of multiple subsets of leukocytes on antibody-decorated poly(ethylene) glycol-coated glass slides," *J. Immunol. Methods*, Vol. 313, Nos. 1–2, 2006, pp. 96–109.

[5] Sethu, P., Moldawer, L. L., Mindrinos, M. N., Scumpia, P. O., Tannahill, C. L., Wilhelmy, J., Efron, P. A., Brownstein, B. H., Tompkins, R. G., and Toner, M., "Microfluidic isolation of leukocytes from whole blood for phenotype and gene expression analysis," *Anal. Chem.*, Vol. 78, No. 15, 2006, pp. 5453–5461.

[6] Sin, A., Murthy, S. K., Revzin, A., Tompkins, R. G., and Toner, M., "Enrichment using antibody-coated microfluidic chambers in shear flow: Model mixtures of human lymphocytes," *Biotechnol. Bioeng.*, Vol. 91, No. 7, 2005, pp. 816–826.

[7] Toner, M., and Irimia, D., "Blood-on-a-chip," *Ann. Rev. Biomed. Eng.*, Vol. 7, 2005, pp. 77–103.

[8] Belov, L., Huang, P., Barber, N., Mulligan, S. P., and Christopherson, R. I., "Identification of repertoires of surface antigens on leukemias using an antibody microarray," *Proteomics*, Vol. 3, No. 11, 2003, pp. 2147–2154.

[9] Ko, I. K., Kato, K., and Iwata, H., "Antibody microarray for correlating cell phenotype with surface marker," *Biomaterials,* Vol. 26, No. 6, 2005, pp. 687–696.

[10] Chen, D. S., Soen, Y., Stuge, T. B., Lee, P. P., Weber, J. S., Brown, P. O., and Davis, M. M., "Marked differences in human melanoma antigen-specific T cell responsiveness after vaccination using a functional microarray," *PLoS Med.*, Vol. 2, No. 10, 2005, p. e265.

11] Soen, Y., Chen, D. S., Kraft, D. L., Davis, M. M., and Brown, P. O., "Detection and characterization of cellular immune responses using peptide-MHC microarrays," *PLoS Biol.*, Vol. 1, No. 3, 2003, p. E65.

[12] Anikeeva, N., Lebedeva, T., Clapp, A. R., Goldman, E. R., Dustin, M. L., Mattoussi, H., and Sykulev, Y., "Quantum dot/peptide-MHC biosensors reveal strong CD8-dependent cooperation between self and viral antigens that augment the T cell response," *Proc. Natl. Acad. Sci. USA*, Vol. 103, No. 45, 2006, pp. 16846–16851.

[13] Tan, J., and Saltzman, W. M., "Topographical control of human neutrophil motility on micropatterned materials with various surface chemistry," *Biomaterials,* Vol. 23, No. 15, 2002, pp. 3215–3225.

[14] Tan, J., Shen, H., Carter, K. L., and Saltzman, W. M., "Controlling human polymorphonuclear leukocytes motility using microfabrication technology," *J. Biomed. Mater. Res.*, Vol. 51, No. 4, 2000, pp. 694–702.

[15] Tan, J., Shen, H., and Saltzman, W. M., "Micron-scale positioning of features influences the rate of polymorphonuclear leukocyte migration," *Biophys. J.*, Vol. 81, No. 5, 2001, pp. 2569–2579.

[16] Karlsson, M., Johansson, A., Tang, L., and Boman, M., "Nanoporous aluminum oxide affects neutrophil behaviour," *Microsc. Res. Tech.*, Vol. 63, No. 5, 2004, pp. 259–265.

[17] Huppa, J. B., and Davis, M. M., "T-cell-antigen recognition and the immunological synapse," *Nat. Rev. Immunol.*, Vol. 3, No. 12, 2003, pp. 973–983.

[18] Wulfing, C., Tskvitaria-Fuller, I., Burroughs, N., Sjaastad, M. D., Klem, J., and Schatzle, J. D., "Interface accumulation of receptor/ligand couples in lymphocyte activation: Methods, mechanisms, and significance," *Immunol. Rev.*, Vol. 189, 2002, pp. 64–83.

[19] McConnell, H. M., Watts, T. H., Weis, R. M., and Brian, A. A., "Supported planar membranes in studies of cell-cell recognition in the immune-system," *Biochimica et Biophysica Acta*, Vol. 864, No. 1, 1986, pp. 95–106.

[20] Brian, A. A., and McConnell, H. M., "Allogeneic stimulation of cytotoxic T cells by supported planar membranes," *Proc. Natl. Acad. Sci. USA*, Vol. 81, No. 19, 1984, pp. 6159–6163.

[21] Hafeman, D. G., von Tscharner, V., and McConnell, H. M., "Specific antibody-dependent interactions between macrophages and lipid haptens in planar lipid monolayers," *Proc. Natl. Acad. Sci. USA*, Vol. 78, No. 7, 1981, pp. 4552–4556.

[22] Weis, R. M., Balakrishnan, K., Smith, B. A., and McConnell, H. M., "Stimulation of fluorescence in a small contact region between rat basophil leukemia cells and planar lipid membrane targets by coherent evanescent radiation," *J. Biol. Chem.*, Vol. 257, No. 11, 1982, pp. 6440–6445.

[23] Watts, T. H., Brian, A. A., Kappler, J. W., Marrack, P., and McConnell, H. M., "Antigen presentation by supported planar membranes containing affinity-purified I-Ad," *Proc. Natl. Acad. Sci. USA*, Vol. 81, No. 23, 1984, pp. 7564–7568.

[24] Grakoui, A., Bromley, S. K., Sumen, C., Davis, M. M., Shaw, A. S., Allen, P. M., and Dustin, M. L., "The immunogical synapse: A molecular machine controlling T cell activation," *Science*, Vol. 285, 1999, pp. 221–227.

[25] Varma, R., Campi, G., Yokosuka, T., Saito, T., and Dustin, M. L., "T cell receptor-proximal signals are sustained in peripheral microclusters and terminated in the central supramolecular activation cluster," *Immunity*, Vol. 25, No. 1, 2006, pp. 117–127.

[26] Campi, G., Varma, R., and Dustin, M. L., "Actin and agonist MHC-peptide complex-dependent T cell receptor microclusters as scaffolds for signaling," *J. Exp. Med., Vol. 202*, No. 8, 2005, pp. 1031–1036.

[27] Groves, J. T., and Dustin, M. L., "Supported planar bilayers in studies on immune cell adhesion and communication," *J. Immunol. Methods,* Vol. 278, Nos. 1–2, 2003, pp. 19–32.

[28] Tseng, S. Y., Liu, M., and Dustin, M. L., "CD80 cytoplasmic domain controls localization of CD28, CTLA-4, and protein kinase Ctheta in the immunological synapse," *J. Immunol.,* Vol. 175, No. 12, 2005, pp. 7829–7836.

[29] Sumen, C., Dustin, M. L., and Davis, M. M., "T cell receptor antagonism interferes with MHC clustering and integrin patterning during immunological synapse formation," *J. Cell. Biol.,* Vol. 166, No. 4, 2004, pp. 579–590.

[30] Johnson, K. G., Bromley, S. K., Dustin, M. L., and Thomas, M. L., "A supramolecular basis for CD45 tyrosine phosphatase regulation in sustained T cell activation," *Proc. Natl. Acad. Sci. USA*, Vol. 97, No. 18, 2000, pp. 10138–10143.

[31] Qi, S. Y., Groves, J. T., and Chakraborty, A. K., "Synaptic pattern formation during cellular recognition," *Proc. Natl. Acad. Sci. USA*, Vol. 98, No. 12, 2001, pp. 6548–6553.

[32] Orth, R. N., Wu, M., Holowka, D. A., Craighead, H. G., and Baird, B. A., "Mast cell activation on patterned lipid bilayers of subcellular dimensions," *Langmuir,* Vol. 19, No. 5, 2003, pp. 1599–1605.

[33] Wu, M., Holowka, D., Craighead, H. G., and Baird, B., "Visualization of plasma membrane compartmentalization with patterned lipid bilayers," *Proc. Natl. Acad. Sci. USA*, Vol. 101, No. 38, 2004, pp. 13798–13803.

[34] Mossman, K. D., Campi, G., Groves, J. T., and Dustin, M. L., "Altered TCR signaling from geometrically repatterned immunological synapses," *Science*, Vol. 310, No. 5751, 2005, pp. 1191–1193.

[35] Groves, J. T., Ulman, N., and Boxer, S. G., "Micropatterning fluid lipid bilayers on solid supports," *Science*, Vol. 275, No. 5300, 1997, pp. 651–653.

[36] Groves, J. T., and Boxer, S. G., "Micropattern formation in supported lipid membranes," *Acct. Chem. Res.*, Vol. 35, No. 3, 2002, pp. 149–157.

[37] Senaratne, W., Sengupta, P., Jakubek, V., Holowka, D., Ober, C. K., and Baird, B., "Functionalized surface arrays for spatial targeting of immune cell signaling," *J. Am. Chem. Soc.*, Vol. 128, No. 17, 2006, pp. 5594–5595.

[38] Monks, C. R. F., Freiberg, B. A., Kupfer, H., Sciaky, N., and Kupfer, A., "Three-dimensional segregation of supramolecular activation clusters in T cells," *Nature*, Vol. 395, No. 6697, 1998, pp. 82–86.

[39] Batista, F. D., Iber, D., and Neuberger, M. S., "B cells acquire antigen from target cells after synapse formation," *Nature,* Vol. 411, No. 6836, 2001, pp. 489–494.

[40] Davis, D. M., Chiu, I., Fassett, M., Cohen, G. B., Mandelboim, O., and Strominger, J. L., "The human natural killer cell immune synapse," *Proc. Natl. Acad. Sci. USA*, Vol. 96, No. 26, 1999, pp. 15062–15067.

[41] Doh, J., and Irvine, D. J., "Photogenerated polyelectrolyte bilayers from an aqueous-processible photoresist for multicomponent protein patterning," *J. Am. Chem. Soc.,* Vol. 126, No. 30, 2004, pp. 9170–9171.

[42] Katz, J. S., Doh, J., and Irvine, D. J., "Composition-tunable properties of amphiphilic comb copolymers containing protected methacrylic acid groups for multicomponent protein patterning," *Langmuir*, Vol. 22, No. 1, 2006, pp. 353–359.

[43] Doh, J., and Irvine, D. J., "Immunological synapse arrays: Patterned protein surfaces that modulate immunological synapse structure formation in T cells," *Proc. Natl. Acad. Sci. USA*, Vol. 103, No. 15, 2006, pp. 5700–5705.

[44] Simons, K., and Ikonen, E., "Functional rafts in cell membranes," *Nature*, Vol. 387, No. 6633, 1997, pp. 569–572.

[45] Dustin, M. L., Bromley, S. K., Kan, Z. Y., Peterson, D. A., and Unanue, E. R., "Antigen receptor engagement delivers a stop signal to migrating T lymphocytes," *Proc. Natl. Acad. Sci. USA*, Vol. 94, No. 8, 1997, pp. 3909–3913.

[46] Negulescu, P. A., Krasieva, T. B., Khan, A., Kerschbaum, H. H., and Cahalan, M. D., "Polarity of T cell shape, motility, and sensitivity to antigen," *Immunity*, Vol. 4, No. 5, 1996, pp. 421–430.

[47] Kupfer, A., and Singer, S. J., "Cell biology of cyto-toxic and helper T-cell functions—immunofluorescence microscopic studies of single cells and cell couples," *Ann. Rev. Immunol.*, Vol. 7, 1989, pp. 309–337.

[48] Trautmann, A., and Valitutti, S., "The diversity of immunological synapses," *Curr. Opin. Immunol.*, Vol. 15, No. 3, 2003, pp. 249–254.

[49] Friedl, P., den Boer, A. T., and Gunzer, M., "Tuning immune responses: Diversity and adaptation of the immunological synapse," *Nat. Rev. Immunol.*, Vol. 5, No. 7, 2005, pp. 532–545.

[50] Groves, J. T., "Spatial mutation of the T cell immunological synapse," *Curr. Opin. Chem. Biol.*, Vol. 10, No. 6, 2006, pp. 544–550.

[51] Chen, C. S., Mrksich, M., Huang, S., Whitesides, G. M., and Ingber, D. E., "Geometric control of cell life and death," *Science*, Vol. 276, No. 5317, 1997, pp. 1425–1428.

[52] Kim, H., Doh, J., Irvine, D. J., Cohen, R. E., and Hammond, P. T., "Large area two-dimensional B cell arrays for sensing and cell-sorting applications," *Biomacromolecules*, Vol. 5, No. 3, 2004, pp. 822–827.

[53] Yamamura, S., Kishi, H., Tokimitsu, Y., Kondo, S., Honda, R., Rao, S. R., Omori, M., Tamiya, E., and Muraguchi, A., "Single-cell microarray for analyzing cellular response," *Anal. Chem.*, Vol. 77, No. 24, 2005, pp. 8050–8056.

[54] Love, J. C., Ronan, J. L., Grotenbreg, G. M., van der Veen, A. G., and Ploegh, H. L., "A microengraving method for rapid selection of single cells producing antigen-specific antibodies," *Nat. Biotechnol.*, Vol. 24, No. 6, 2006, pp. 703–707.

[55] Revzin, A., Sekine, K., Sin, A., Tompkins, R. G., and Toner, M., "Development of a microfabricated cytometry platform for characterization and sorting of individual leukocytes," *Lab Chip*, Vol. 5, No. 1, 2005, pp. 30–37.

[56] Kim, H., Cohen, R. E., Hammond, P. T., and Irvine, D. J., "Live lymphocyte arrays for biosensing," *Adv. Funct. Mater.*, Vol. 16, No. 10, 2006, pp. 1313–1323.

[57] Revzin, A., Rajagopalan, P., Tilles, A. W., Berthiaume, F. O., Yarmush, M. L., and Toner, M., "Designing a hepatocellular microenvironment with protein microarraying and poly(ethylene glycol) photolithography," *Langmuir*, Vol. 20, No. 8, 2004, pp. 2999–3005.

[58] Revzin, A., Russell, R. J., Yadavalli, V. K., Koh, W. G., Deister, C., Hile, D. D., Mellott, M. B., and Pishko, M. V., "Fabrication of poly(ethylene glycol) hydrogel microstructures using photolithography," *Langmuir*, Vol. 17, No. 18, 2001, pp. 5440–5447.

[59] Revzin, A., Tompkins, R. G., and Toner, M., "Surface engineering with poly(ethylene glycol) photolithography to create high-density cell arrays on glass," *Langmuir*, Vol. 19, No. 23, 2003, pp. 9855–9862.

[60] Luster, A. D., "Chemokines—chemotactic cytokines that mediate inflammation," *New Eng. J. Med.*, Vol. 338, No. 7, 1998, pp. 436–445.

[61] Dunzendorfer, S., Kaser, A., Meierhofer, C., Tilg, H., and Wiedermann, C. J., "Dendritic cell migration in different micropore filter assays," *Immunol. Lett.*, Vol. 71, No. 1, 2000, pp. 5–11.

[62] Zigmond, S. H., "Ability of polymorphonuclear leukocytes to orient in gradients of chemotactic factors," *J. Cell. Biol.*, Vol. 75, No. 2, Pt. 1, 1977, pp. 606–616.

[63] Heit, B., and Kubes, P., "Measuring chemotaxis and chemokinesis: The under-agarose cell migration assay," *Sci. STKE*, Vol. 2003, No. 170, 2003, p. PL5.

[64] Wilkinson, P. C., "Cell locomotion and chemotaxis: Basic concepts and methodologies," *Methods*, Vol. 10, 1996, pp. 74–81.

[65] Frow, E. K., Reckless, J., and Grainger, D. J., "Tools for anti-inflammatory drug design: In vitro models of leukocyte migration," *Med. Res. Rev.*, Vol. 24, No. 3, 2004, pp. 276–298.

[66] Lauffenburger, D. A., Tranquillo, R. T., and Zigmond, S. H., "Concentration gradients of chemotactic factors in chemotaxis assays," *Methods Enzymol.*, Vol. 162, 1988, pp. 85–101.

[67] Jeon, N. L., Baskaran, H., Dertinger, S. K., Whitesides, G. M., Van de Water, L., and Toner, M., "Neutrophil chemotaxis in linear and complex gradients of interleukin-8 formed in a microfabricated device," *Nat. Biotechnol.*, Vol. 20, No. 8, 2002, pp. 826–830.

[68] Irimia, D., Liu, S. Y., Tharp, W. G., Samadani, A., Toner, M., and Poznansky, M. C., "Microfluidic system for measuring neutrophil migratory responses to fast switches of chemical gradients," *Lab Chip*, Vol. 6, No. 2, 2006, pp. 191–198.

[69] Tharp, W. G., Yadav, R., Irimia, D., Upadhyaya, A., Samadani, A., Hurtado, O., Liu, S. Y., Munisamy, S., Brainard, D. M., Mahon, M. J., Nourshargh, S., van Oudenaarden, A., Toner, M. G., and Poznansky, M. C., "Neutrophil chemorepulsion in defined interleukin-8 gradients in vitro and in vivo," *J. Leukoc.* Biol., Vol. 79, No. 3, 2006, pp. 539–554.

[70] Poznansky, M. C., Olszak, I. T., Foxall, R., Evans, R. H., Luster, A. D., and Scadden, D. T., "Active movement of T cells away from a chemokine," *Nat. Med.*, Vol. 6, No. 5, 2000, pp. 543–548.

[71] Cyster, J. G., "Chemorepulsion and thymocyte emigration," *J. Clin. Invest.*, Vol. 109, No. 8, 2002, pp. 1011–1012.

[72] Zlatopolskiy, A., and Laurence, J., "'Reverse gear' cellular movement mediated by chemokines," *Immunol. Cell. Biol.*, Vol. 79, No. 4, 2001, pp. 340–344.

[73] Poznansky, M. C., Olszak, I. T., Evans, R. H., Wang, Z., Foxall, R. B., Olson, D. P., Weibrecht, K., Luster, A. D., and Scadden, D. T., "Thymocyte emigration is mediated by active movement away from stroma-derived factors," *J. Clin. Invest.*, Vol. 109, No. 8, 2002, pp. 1101–1110.

[74] Sinha, V. R., and Trehan, A., "Biodegradable microspheres for protein delivery," *J. Control. Release*, Vol. 90, No. 3, 2003, pp. 261–280.

[75] Tamber, H., Johansen, P., Merkle, H. P., and Gander, B., "Formulation aspects of biodegradable polymeric microspheres for antigen delivery," *Adv. Drug Deliv. Rev.*, Vol. 57, No. 3, 2005, pp. 357–376.

[76] Gombotz, W. R., and Pettit, D. K., "Biodegradable polymers for protein and peptide drug delivery," *Bioconjug. Chem.*, Vol. 6, No. 4, 1995, pp. 332–351.

[77] Okada, H., Yamamoto, M., Heya, T., Inoue, Y., Kamei, S., Ogawa, Y., and Toguchi, H., "Drug-delivery using biodegradable microspheres," *J. Control.* Release, Vol. 28, Nos. 1–3, 1994, pp. 121–129.

[78] Zhao, X., Jain, S., Benjamin Larman, H., Gonzalez, S., and Irvine, D. J., "Directed cell migration via chemoattractants released from degradable microspheres," *Biomaterials*, Vol. 26, No. 24, 2005, pp. 5048–5063.

[79] Furumoto, K., Soares, L., Engleman, E. G., and Merad, M., "Induction of potent antitumor immunity by in situ targeting of intratumoral DCs," *J. Clin. Invest.*, Vol. 113, No. 5, 2004, pp. 774–783.

[80] Fushimi, T., Kojima, A., Moore, M. A., and Crystal, R. G., "Macrophage inflammatory protein 3alpha transgene attracts dendritic cells to established murine tumors and suppresses tumor growth," *J. Clin. Invest.*, Vol. 105, No. 10, 2000, pp. 1383–1393.

CHAPTER 26

Microscale Hepatic Tissue Engineering

Rohit Jindal, Martin L. Yarmush, Mehmet Toner, and Arno W. Tilles

26.1 Introduction

The liver is the largest solid organ in the body, normally weighing between 1.2 and 1.8 kg [1]. The basic structural unit of the liver is the hepatic parenchymal cell; these cells, hepatocytes, account for approximately two-thirds of the total liver cell mass [2]. Structurally, hepatocytes are arranged in a plate-like configuration with the space between the plates occupied by sinusoids. The major nonparenchymal cells that occupy the sinusoid include endothelial cells, Kupffer cells, and stellate cells [3]. The arrangement of hepatocytes and nonparenchymal cells in the sinusoid is shown in Figure 26.1. The endothelial cells in the sinusoid are fenestrated and form the sinusoidal lining within the space of Disse, which separates them from the hepatocytes. These cells are responsible for regulating the exchange of macromolecules, solutes, and fluid between the blood and the surrounding tissues [4]. Stellate cells, the fat-storing cells in the liver, reside in the space of Disse between the hepatocytes and endothelial cells. In the normal liver, stellate cells play a role in the storage and transport of retinoid, along with maintenance of the extracellular

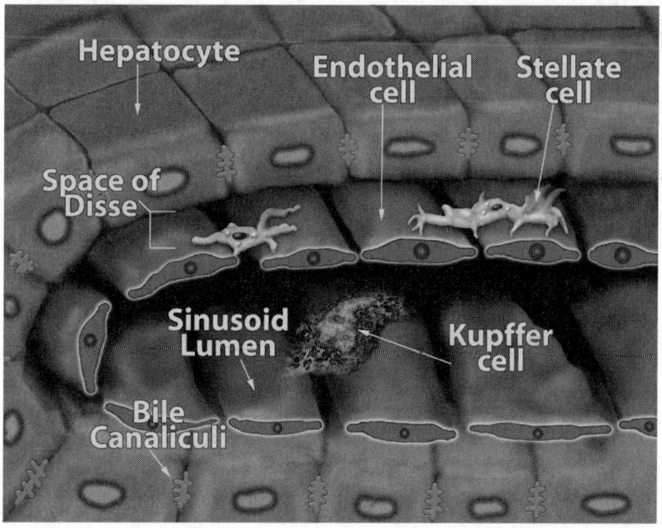

Figure 26.1 Schematic of the arrangement of parenchymal and nonparenchymal cells in the sinusoid.

matrix (ECM) [5]. Kupffer cells are the resident macrophages of the liver, playing a significant role in the endocytosis of particulates, cells, and foreign matter from the blood [6]. Although nonparenchymal cells carry out important tasks involved in the normal functioning of the liver, hepatocytes perform most of the vital metabolic liver functions.

The liver is the principal organ in the body involved in the metabolism of amino acids, carbohydrates, lipids, and vitamins. It converts dietary sources of energy to substrates that can be oxidized by other tissues for their energy needs [7]. The liver is also responsible for the synthesis of plasma proteins such as albumin and other carrier proteins. In addition, hepatocytes express large numbers of cytochrome P450 and other enzymes that can convert xenobiotics, including drugs and toxins, to inactive, less lipophilic metabolites that can subsequently be excreted into the bile and thereby eliminated from the body [8]. Consequently, a sudden and significant loss in the metabolic capacity of the liver for detoxification and bile secretion has profound, and potentially life threatening, consequences for the patient.

In 2003, more than twenty-seven thousand people died from chronic liver disease in the United States alone [9], and in spite of the advances made in the management of liver disease, the only viable option for patients suffering from end-stage liver failure is whole-organ transplantation. Unfortunately, the severe shortage of liver donors presents a significant problem for those awaiting transplant. Currently, more than seventeen thousand patients in the United States await a new liver [10]. In order to cope with the shortage of liver donors, efforts must be focused on alternative strategies to address this problem, such as the development of a bioartificial liver (BAL) device and cell transplantation [11].

This chapter describes the current microfabrication, microfluidic, and tissue engineering technologies being used to develop new liver-cell microarrays. The first section describes methods for the maintenance of stable hepatocyte cultures in vitro. This includes the culturing of cells in a three-dimensional layered structure and micropatterning to accommodate nonparenchymal cells. The next section describes the various designs for BAL devices and the use of polymer scaffolds to develop hepatic constructs for transplantation. Finally, the focus shifts toward the development of liver-cell microarrays for high-throughput screening applications. A summary and concluding remarks follows.

26.2 Strategies for Developing Stable Hepatocyte Culture Models

The ability to culture hepatocytes outside the body is central to the development of BAL devices. The in vitro models of hepatocyte cultures are useful for the screening of drugs for toxicological and pharmacological applications. They are also suitable for gaining mechanistic insight into the role of various intracellular and extracellular factors in causing liver injury since conditions in cell culture can be carefully controlled in comparison to animal studies. After isolation, hepatocytes rapidly lose their differentiated function [12, 13] unless they are cultured under specific conditions, which might include coculture with nonparenchymal cells [14–16], the use of specific ECM configurations such as a collagen sandwich [17–19], or culturing them as three-dimensional spheroids [20, 21]. Table 26.1 lists devices that have utilized

Table 26.1 Various Device Configurations for Diagnostic and Bioartificial Liver-Support Applications

Device	*Cell Culture Configuration*	*Cell Type*	*Potential Application*	*References*
Multilayered PDMS device	Monolayer	Hep G2	BAL support	[66]
SU-8 microgrooves on glass substrate	Monolayer	Primary rat hepatocytes	BAL support	[71]
Glass substrate with micropatterned collagen	Monolayer	Primary rat hepatocytes	Diagnostic	[89, 90]
Circular cavities in polystyrene substrate	3-D spheroid	Primary rat hepatocytes	Diagnostic	[91]
Glass substrate with micropatterned PEG	3-D encapsulation in PEG-based polymer	Primary rat hepatocytes	Diagnostic	[92]
Two-layered PDMS device with channels for perfusing medium and oxygen	Monolayer	Primary rat hepatocytes	Diagnostic	[95]
Silicon scaffold with filter underneath for holding the cells	3-D spheroid	Primary rat hepatocytes	Diagnostic and BAL support	[96]
PMMA scaffold with laser drilled pores for perfusing medium	3-D spheroid	Primary rat hepatocytes	Diagnostic and BAL support	[97]
Two-layered PDMS device integrating microvalves for seeding multiple cell types and evaluating multiple input	Monolayer	H-35	Diagnostic	[98]

various cell-culture configurations for therapeutic and diagnostic applications. Despite the progress made in maintaining hepatocytes in cell culture, it remains an active area of research to develop better in vitro models to satisfy the biological requirements of a BAL device. In this section, we describe the various approaches that have been taken toward the stabilization of hepatocyte function in vitro, with special emphasis on the use of novel materials and various microfabrication techniques for achieving this goal.

26.2.1 Effect of Three-dimensional Layering

Recent studies have demonstrated the importance of culturing cells in a three-dimensional tissue-like microenvironment [22–25]. It has been shown that culturing cells in three dimensions alters the expression of receptors [26], changes the migration and morphology of the cells [22], and modifies gene-expression patterns [27] when compared to two-dimensional cultures. The beneficial effect of culturing hepatocytes in three dimensions was demonstrated by Dunn et al. [18]. Isolated hepatocytes cultured on collagen gel (two-dimensional culture) progressively stop expressing liver-specific markers and lose viability. In contrast, overlaying a second layer of collagen gel on top of hepatocytes cultured on collagen gel results in a gradual increase in their expression of liver-specific functions (as measured by urea- and albumin-synthesis rates), which are stable for several weeks [28]. Furthermore, this three-dimensional layering allows hepatocytes to adopt a morphology and intracellular organization closely mimicking that found in vivo [29].

Three-dimensional, layered cell-culture models are especially relevant for studying hepatocytes and endothelial-cell interactions. In the liver, endothelial cells and hepatocytes are arranged in layers with the space of Disse separating the two cell types. Development of cell-culture systems that mimic this organization provides a more physiologically relevant model. Various methods for layering endothelial cells on top of hepatocytes include the use of thermoresponsive polymers [30, 31], magnetic liposome technology [32], and polyelectrolyte multilayers [33]. Okano et al. introduced poly(N-isopropylacrylamide) (PIPAAm) as a thermoresponsive polymer for layering endothelial cells on top of hepatocytes [30]. The basic concept is illustrated in Figure 26.2(a). Endothelial cells are cultured on top of PIPAAm at 37°C, which is above the lower critical solution temperature (LCST). Above the LCST,

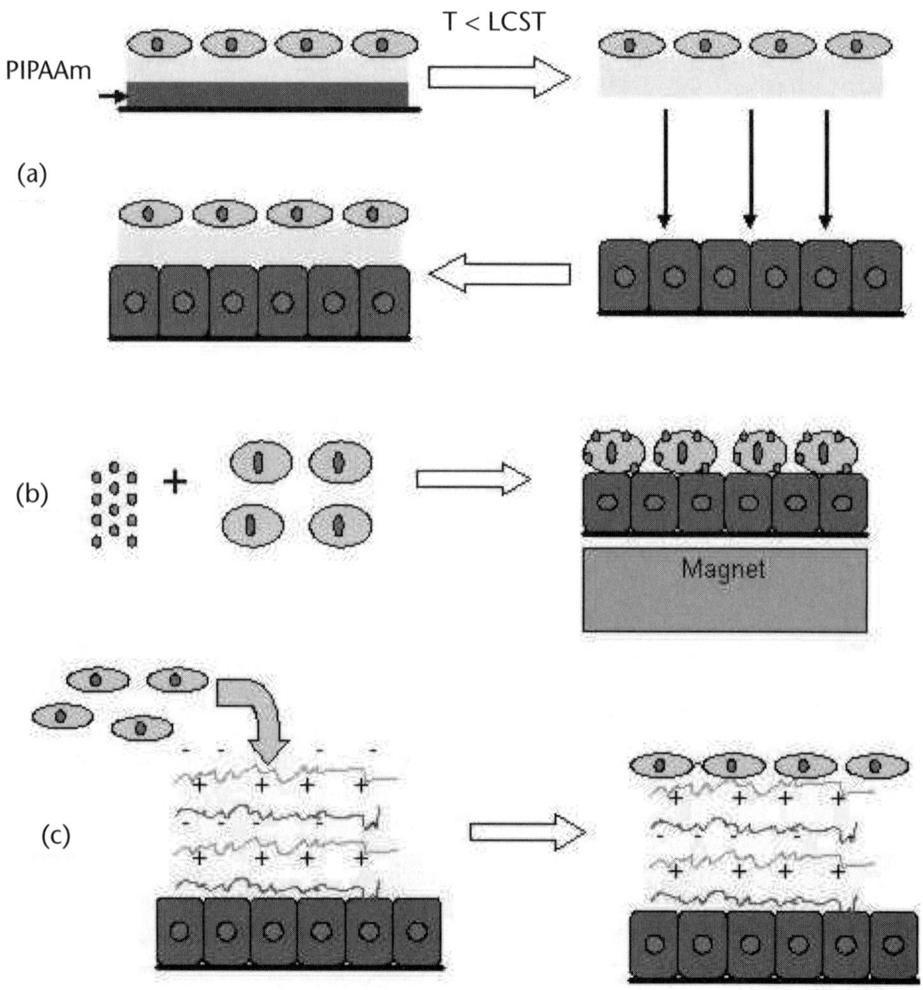

Figure 26.2 Schematic of different approaches used for three-dimensional layering of endothelial cells on top of hepatocytes: (a) Endothelial cells are grown to confluence on thermoresponsive polymer (PIPAAm). Sheet of endothelial cells supported by ECM is recovered by reducing the temperature below the LCST of the polymer. Layering is achieved by placing the sheet of endothelial cells on top of hepatocytes. (b) Endothelial cells are labeled with magnetic particles, and layering is achieved by positioning a magnet under the culture dish. (c) Endothelial cells are overlaid on top of the polyelectrolyte multilayers assembled on top of hepatocytes.

PIPAAm is hydrophobic, and the cells attach to the polymer. Once the endothelial cells have grown to confluence and the cells have secreted enough ECM to support the cell sheet, the temperature is reduced below the LCST. This results in the detachment of the cell sheet as PIPAAm becomes hydrophilic at the lower temperature. The detached endothelial cell sheet, supported by ECM, is introduced on top of the hepatocytes, resulting in the layered coculture. In this layered system, hepatocytes maintained albumin expression for more than forty-one days of culture, while hepatocytes cultured without the endothelial cell sheet lost function after ten days of culture [30].

Magnetic liposome technology for layering endothelial cells on top of hepatocytes was employed by Ito et al. [32]. This method relied on the labeling of endothelial cells with magnetite particles, which were added to the hepatocyte monolayer. A magnet was used for holding the endothelial cells in place on top of the hepatocytes, as shown in Figure 26.2(b). In the layered coculture, albumin secretion was much higher in comparison to the control. Rajagopalan et al. successfully layered hepatocytes and endothelial cells using alternating layers of cationic and anionic polyelectrolytes as shown in Figure 26.2(c) [33]. Specifically, alternating layers of chitosan (cationic) and DNA (anionic) were loaded on top of the monolayer of hepatocytes cultured on collagen gel. Next, endothelial cells were cultured on top of the polyelectrolyte multilayers. The same approach was also used for layering of fibroblasts and hepatocytes on top of hepatocytes. In all cases, enhanced liver-specific function was observed for the multilayered structure in comparison to the control [33].

26.2.2 Micropatterning of Hepatocytes and Nonparenchymal Cells

A number of studies have established that culturing hepatocytes with other cell types increases the stability of primary hepatocytes in culture. Although the exact mechanism by which this stabilization is achieved is not known, soluble factors [34] and insoluble ECM molecules [35–37] are likely mediators of increasing hepatocyte stability in cocultures. In traditional cocultures of hepatocytes and nonparenchymal cells, the two cell types are arranged randomly with no control over the heterotypic contact. Combining cell-culture techniques with micropatterning provides tools for developing in vitro coculture models in which cell-cell interaction can be manipulated in a controlled fashion. Such models can also provide better insight into the role played by cell-cell communication in enhancing hepatocyte stability.

Bhatia, Yarmush, and Toner developed a method for the micropatterning of hepatocytes and fibroblasts on a glass substrate [38]. The basic approach involved the patterning of collagen on the glass substrate using photolithography, as shown in Figure 26.3 [39]. After patterning the collagen, the hepatocytes were seeded, and it was observed that they selectively attached to the collagen patterns [40]. Next, fibroblasts were seeded, and they attached to the unmodified region of the glass substrate. Using this method, experiments were designed that could control the variation of the heterotypic contact area, the homotypic contact area, and the cell number independently. In one set of experiments [41], the heterotypic contact area between hepatocytes and fibroblasts was varied independently of the number of cells in each subpopulation. This was achieved by varying both the size and cen-

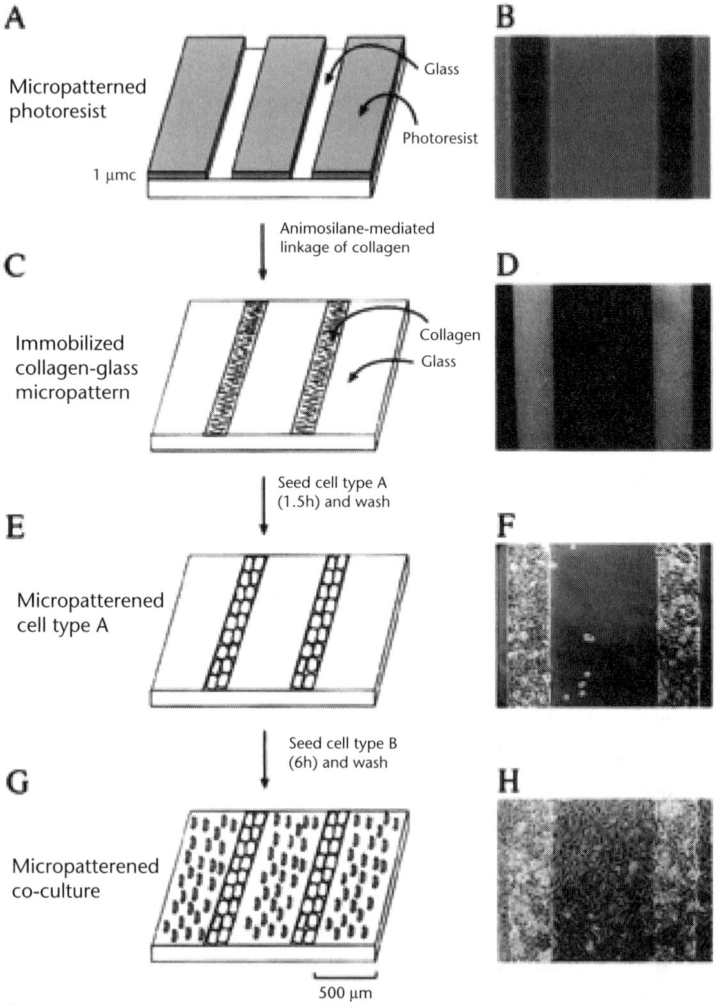

Figure 26.3 Schematic of method for micropatterning of hepatocytes and fibroblasts [39]: (a) Photoresist was spin-coated on glass substrate, and patterning was achieved using photolithography. (b) Epifluorescence image showing pattern of photoresist on glass. (c) Collagen was immobilized, followed by removal of photoresist. (d) Immunofluorescence image showing pattern of collagen on the glass substrate. (e) Hepatocytes were introduced on patterned glass substrate where they attached selectively to collagen-modified glass substrate. (f) Phase-contrast image demonstrating attachment of hepatocytes to collagen pattern. (g) Micropatterned hepatocytes were exposed to fibroblasts, which attached to the exposed region of the glass. (h) Phase contrast image showing micropatterns of hepatocytes and fibroblast on the glass substrate.

ter-to-center spacing between individual collagen islands (patterns) on the glass surface while keeping the total surface area of the modified collagen and unmodified region constant. The experimental results showed that the higher heterotypic contact area led to higher function of hepatocytes. Furthermore, hepatocytes near the heterotypic surface showed higher albumin staining, underscoring the importance of the local microenvironment of cells in modulating cell function.

The same micropatterning technique was used to investigate the effect of fibroblast density on hepatocyte function without changing the heterotypic contact area. This was achieved by keeping the size of individual islands of collagen the same, while changing the overall surface area of the unmodified region of glass [42]. It was observed that the increase in fibroblast number resulted in higher hepatocyte function. Since the homotypic contact area of hepatocytes and heterotypic contact area in the coculture were kept constant, the increase in function was attributed to higher homotypic interactions of fibroblasts. Recently Zinchenko et al. carried out similar experiments for investigating the effect of Kupffer cells on hepatocytes in micropatterned coculture [43]. In contrast to fibroblast coculture, a higher density of Kupffer cells was found to be detrimental to hepatocyte function. The best results were obtained when the seeding ratio of hepatocytes to Kupffer cells was 5:1 for urea and 10:1 for albumin function.

26.3 Bioartificial Liver Devices

Bioartificial liver devices have the potential to provide temporary hepatic support to patients suffering from liver failure who are awaiting liver transplantation. In some cases, given the liver's tremendous capacity to regenerate [44], the hepatic support provided by the BAL device may allow spontaneous recovery, thereby averting a liver transplant. The beneficial effects of BAL devices have been demonstrated in various animal-injury models [45–47] and clinical trials [48, 49]. Early devices were based on nonbiological systems incorporating hemodialysis or hemofiltration units for providing temporary liver support [50–53]. Since then, various systems that incorporate primary hepatocytes or hepatoma cell lines have been developed. These devices process blood or plasma of the patient by providing metabolic support and detoxification functions. Details of these systems describing the source of cells, the system configuration, and the preclinical and clinical trial results can be found in several reviews [11, 54–57]. Here, we focus our attention on how the microfabrication and microfluidic technology has been used to address some of the specific demands presented by the BAL devices.

A BAL device should satisfy a number of design criteria before it can be implemented in a clinical setting. In order to provide meaningful liver support, the device should be able to provide a functional capacity equivalent to 10 percent of the liver cell mass [58]. Thus, the device should have the capacity to culture a large number of hepatocytes and to maintain their viability and function. Hepatocytes are involved in a number of metabolic activities and thus have high oxygen-uptake rates. The BAL design should allow for a sufficient supply of oxygen and nutrients to cells. By culturing cells in devices that incorporate microfluidic channels, the demand for oxygen and nutrients can be met by perfusing through these channels.

Since the inception of technology for the development of microfluidic devices, a number of substrates have been used for fabrication, including glass [59, 60], poly(dimethylsiloxane) (PDMS) [61, 62], and silicon [63]. PDMS in particular has several desirable properties, such as high oxygen permeability [64] and amenability to bonding with different types of surfaces [65]. Leclerc, Sakai, and Fujii exploited both of these properties to develop a multilayered PDMS device for culturing Hep G2 cells, as shown in Figure 26.4(a) [66]. In their design, the potential for scale up was demonstrated by stacking ten layers of PDMS to form four cell-culture cham-

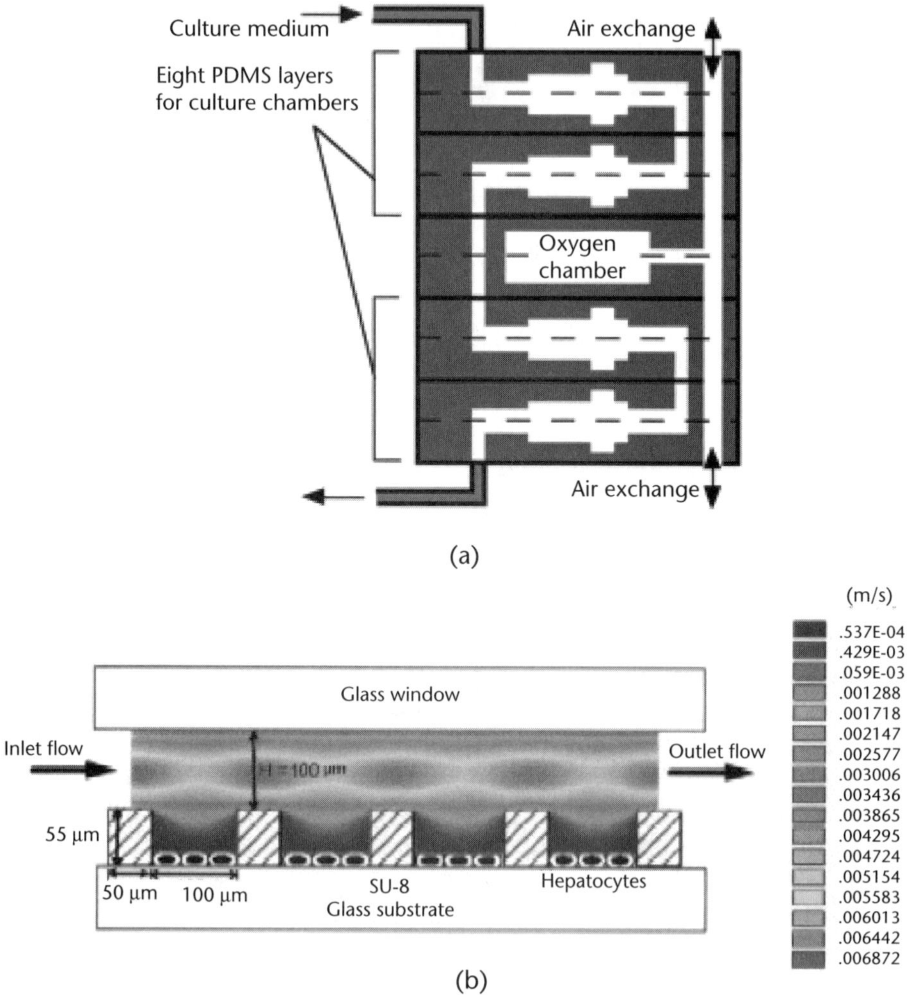

Figure 26.4 Various designs of BAL devices: (a) structure of a multilayer bioreactor with multiple cell-culture chambers and an oxygen chamber [66], and (b) schematic of hepatocytes in a microgrooved bioreactor and contour plot of velocity vector field in channel [67].

bers with one oxygen chamber in the middle. Each cell-culture chamber consisted of a network of microchannels for uniform distribution of medium and microstructures for attachment of cells. The design also included a thin PDMS wall that separated the oxygen supply chamber from the microfluidic channels. The authors showed that the oxygenation chamber was able to support approximately ten times more cells compared to a bioreactor without the oxygenation chamber.

Although a number of bioreactor designs [68–70] have included internal oxygenators to alleviate oxygen limitations, integration of an oxygenator complicates the design of the device and can potentially limit the ability to scale up the device. Integration of medium flow where oxygen is supplied by convective transport can support hepatocytes without an additional means of oxygenation. Since hepatocytes are sensitive to shear stress, the medium flow rate through the device must be carefully controlled to allow sufficient oxygenation without compromising function. Exposing hepatocytes to shear stresses greater than 5 dynes/cm^2 continuously for three days has been shown to detrimentally affect their albumin- and urea-synthesis rates [71]. In order to protect hepatocytes from shear stress at high medium-flow rates, Park et al. fabricated microgrooves on glass substrates using photolithography [67]. Figure 26.4(b) shows the schematic of hepatocytes seeded in the microgrooves and the simulated contour plot of the velocity vectors in the channel. Using finite element analysis, the authors found that the velocity gradient change near the cell surface (i.e., bottom of the microgrooves) was smaller then that near the top surface of the flow channel, indicating that the microgrooves would provide protection to the attached hepatocytes from the mechanical effects of the flowing medium. In addition, it was demonstrated that the albumin- and urea-synthesis rates of hepatocytes remained stable over five days of perfusion in the microgrooved-substrate bioreactor, whereas in the flat-substrate bioreactor, they decreased over the same period.

26.4 Hepatic Constructs for Transplantation

The success of direct hepatocyte transplantation will provide a simple and less-expensive therapeutic alternative to whole-organ transplantation. It will enable the use of a single donor organ for multiple recipients. Initial studies of hepatocyte transplantation in the animal models of acute [72] and chronic liver failure [73] and liver-based metabolic disease [74] had shown promising results with prolonged animal survival and improved function. Encouraged by the success in animal models, hepatocyte transplantation has been studied in the clinical setting for both liver failure [75, 76] and liver-based metabolic disorders [77]. However, with the exception of one case [78], definite evidence is lacking for long-term benefits of hepatocyte transplantation in humans [79].

Various approaches for increasing the viability and function of transplanted hepatocytes have been developed. Initial studies in which hepatocytes were attached to collagen-coated dextran microcarrier beads resulted in significant improvements in survival and function of transplanted hepatocytes in two mutant rat strains [80], suggesting the importance of a support matrix. Transplantation of constructs in which hepatocytes are integrated within a support matrix presents several advantages. (1) Encapsulation in the matrix has been used for immunoisolation of transplanted hepatocytes. The basic approach behind immunoisolation involves designing the matrix in such a way that it permits exchange of nutrients and oxygen across the matrix while higher-molecular-weight substances such as antibodies and leukocytes that elicit the immune response are excluded from the cells by the matrix [11]. Immunoisolation enables the use of cells from a xenogeneic source and thus expands the cell source without the added complexity of using immunosuppressive drugs. Various matrix materials that have been used for encapsulation of hepatocytes include microporous polypropylene membranes [81], alginate-polylysine membranes [82], hollow fibers of cellulose, polysulfone, polyvinylidine [83], and hydrogel hollow fibers made of an acrylonitrile-sodium methallylsulfonate copolymer [84]. (2) A matrix allows for greater flexibility in optimization of those properties that can enhance the function of the transplanted hepatocytes.

(3) A biodegradable matrix provides the mechanical support for structure development while allowing native tissue to gradually replace the degrading matrix. Thus, the use of a matrix promotes enhanced engraftment of transplanted cells. Synthetic, biodegradable matrix materials that have been used for seeding hepatocytes include poly(lactic acid) and poly(lactic-co-glycolic acid) [85]. However, such constructs are limited to the use of allogeneic cell sources as immunoisolation of cells is lost. (4) The use of matrix allows for the immobilization of growth factors [86, 87], which further improves the function and survival of transplanted hepatocytes.

One limitation of incorporating cells in the matrix is the problem of delivering nutrients and oxygen to the cells. Limited access to oxygen and nutrients increases the vulnerability of cells in the center of the construct. Enhanced mass transport into the construct has been achieved by designing highly porous scaffold matrix materials [83] or by the inclusion of growth factors such as endothelial growth factor [86] and basic fibroblast growth factor [87], both which promote vascularization. However, there is a trade-off between the design of a highly porous scaffold and the use of a xenogeneic cell source. The design should not expose the xenogeneic cells to the immunomodulators but at the same time must allow for adequate oxygen and nutrient transport.

26.5 Liver-Cell Microarrays

The development of liver-cell microarrays that facilitate high-throughput toxicology screening will greatly benefit drug development. The liver's unique ability to metabolize compounds leaves it particularly vulnerable to the toxic effects of drugs routinely ingested. Many drugs have been withdrawn from the market due to their toxic effects on the liver [88]. Pharmaceutical companies invest billions of dollars to bring a drug to the market. Liver-cell microarrays can greatly facilitate the identification of harmful compounds and thereby reduce the overall costs associated with animal testing and clinical trials. Microarray technology can also be used to optimize cell-culture conditions in a cost-effective manner through screening of the various medium formulations in combination with the ECM support.

26.5.1 Hepatic Microarrays for Screening Under Static Conditions

Traditionally, micropatterning and surface-engineering techniques have been employed for creating microarrays. Typically, adhesive regions are patterned on a single substrate, and microarrays of desired molecules are formed by selective attachment of molecules to the adhesive region. In the case of cells, patterning of ECM proteins such as collagen, fibronectin, or laminin facilitates arraying of cells as they attach selectively to the regions modified by the ECM proteins. Microarrays of hepatocytes have been created by patterning collagen using various methods, including photolithography [39, 89] and robotic protein printing [90]. Photolithographic patterning of collagen has been achieved using both traditional photoresist- and polyethylene glycol (PEG)–based polymer [89]. The photoresist-based method for patterning of collagen has already been discussed in Section 26.2.2. The patterning of collagen using PEG is similar to the conventional photoresist-based method except that after patterning of collagen, PEG is not removed and serves an additional purpose of providing a cell-repellant surface. Thus, better confinement of cells within the pattern is achieved using this method. Figure 26.5(a) shows the high-density array of hepatocytes formed using PEG-based patterning of collagen. Even though photolithography-based methods have been successfully used for the patterning of collagen, they are complex, multistep processes requiring the use of a clean room facility. In contrast, robotic printing is a simple method that relies on directly dispensing the protein solution on the surface in the desired pattern. It has been successfully used for patterning of collagen and fabrication of hepatocyte microarrays, as shown in Figure 26.5(b) [90].

Although the protein patterning methods discussed so far allow for preparing microarrays of hepatocytes, typically they require the use of nonparenchymal cells for maintaining hepatocytes in culture. Another method that permits microarraying of hepatocytes without using nonparenchymal cells is culturing hepatocytes as spherical multicellular aggregates (spheroids). Fukuda and Nakazawa prepared an array of hepatocyte spheroids on a polystyrene substrate in which microscale circular cavities were created using micromilling techniques [91]. The diameter of the cavity was used for controlling the size of the spheroid and the application of rotary conditions facilitated the formation of the spheroid geometry. The long-term maintenance of hepatocyte function was evaluated using cytochrome P-450 activity, and it was observed that hepatocytes cultured for ten days in the spheroid microarray chips retained higher activity when compared to the monolayer culture [91].

Stable microarrays of hepatocytes have also been created by encapsulating cells in three-dimensional PEG-based hydrogels [92]. This basic approach involves introducing cells suspended in a photosensitive prepolymer solution to be patterned by photocrosslinking the gel under a mask, thereby entrapping the cells in arrays within the gel. This method was used for preparing patterns with a resolution of approximately 100 μm [93]. Further improvement in resolution was achieved by combining dielectrophoretic (DEP) electropatterning with photopatterning [92]. DEP electropatterning manipulates an electric field, which is controlled by

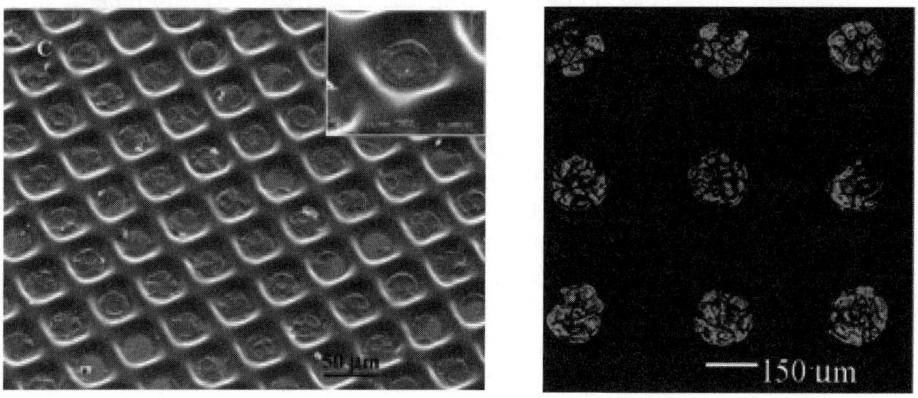

(a) (b)

Figure 26.5 Microarrays of hepatocytes prepared using different methods: (a) optical image of hepatocyte microarray prepared using PEG-based photolithography method [89], and (b) optical image of hepatic microarray prepared using robotic protein printing [90].

micropatterned electrodes, to place the cells in particular locations. DEP electropatterning permits micrometer-scale control over the localization of cells. In this method, the position of the cells is controlled by the local electric field, while the polymer crosslinks on top of entrapping the cells in the desired pattern. The long-term maintenance of patterns of hydrogels encapsulating cells has been reported demonstrating the potential of this method for fabricating hepatocyte microarrays [92].

26.5.2 Microfluidics Incorporated Hepatic Microarrays

Integration of microfluidics in liver-cell microarrays will greatly enhance the utility of these arrays for high-throughput screening applications. Microfluidics facilitates greater flexibility in terms of control over the input to the cultured cells. Various experimental conditions, including different combinations, concentrations, and temporal control of molecules, can be simultaneously evaluated with relative ease [94]. Although microdevices that incorporate individual interrogation of each array element and also permit flexibility over the input control have not been realized, some progress has been made toward achieving this goal.

Kane et al. developed a sixty-four element microarray device for culturing primary hepatocytes [Figure 26.6(a)] [95]. The device incorporated microfluidic channels for independently supplying medium and oxygen to cells in the array as shown in Figure 26.6(b). Soft lithography techniques and photolithographic patterning were combined for fabricating the two layered device and then seeding it with micropatterns of hepatocytes and fibroblasts. The two-layered device was fabricated in PDMS with one layer consisting of wells for seeding cells and microchannels for supplying perfusion medium, while the other layer was used for supplying oxygen. The first layer of PDMS was bonded to glass, and a coculture of hepatocytes and fibroblasts was micropatterned in each well using the same method as described in Section 26.2.2 (Figure 26.3). Figure 26.6(c) shows the micropattern of hepatocytes within a single array element prior to seeding of fibroblasts. Next, a second layer of PDMS was clamped to the first layer with a thin PDMS layer separating the medium-perfusion channel from the oxygen-supply channel. The feasibility of this approach for creating a microarray of primary hepatocytes in a microfluidic device was evaluated by measuring albumin and urea function. The results were encouraging with stable function observed over a thirty-day period.

In another design, microfabrication was used to create a three-dimensional scaffold for culturing hepatocytes in microarrays under flow [96]. Specifically, the scaffold consisted of holes in silicon with filters for holding the cells in the holes. The design allowed a secondary cross-flow across the cells in addition to a primary flow on top of the wells as shown in Figure 26.7(a). This system supported primary hepatocytes, which were grown as spheroids prior to seeding in the microarrays, without any external oxygenator. The oxygen demand was met by perfusing medium while keeping the shear stress below the physiological range. A similar design that employed a microstructured scaffold for holding the cells in the array format was developed by Eschbach et al. [97]. The scaffold consisted of microcontainers [Figure 26.7(b, c)] that were fabricated by injection molding of poly(methyl methacrylate). The bottom of the scaffold consisted of pores that

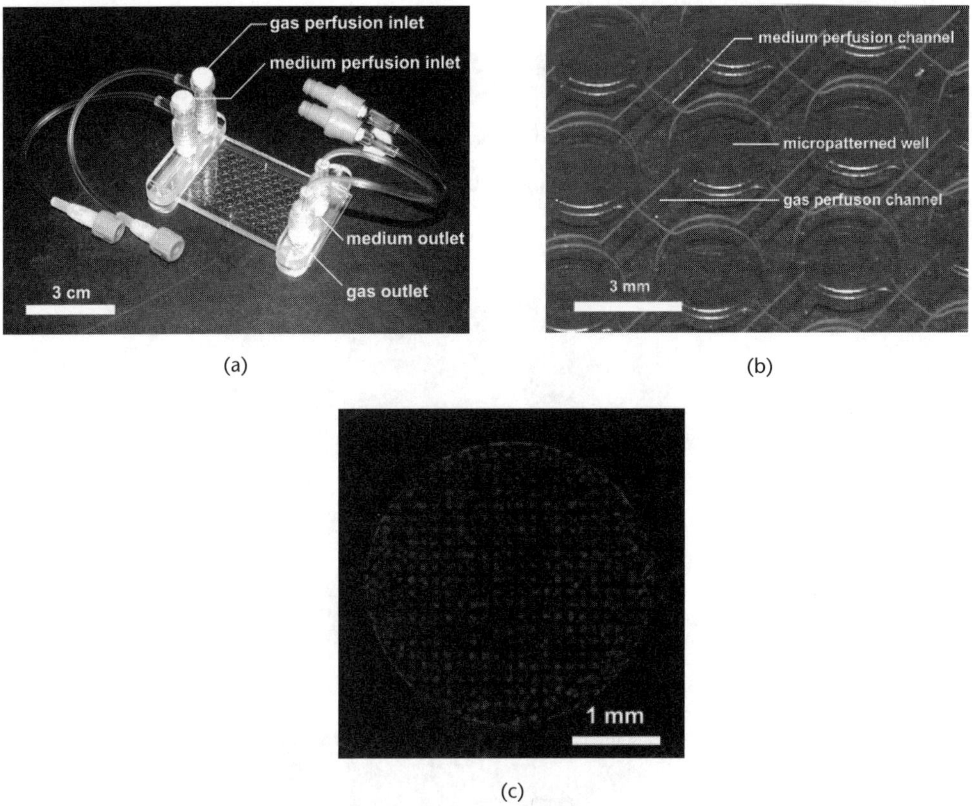

Figure 26.6 An 8 × 8 array device for fabricating hepatocyte microarray [95]: (a) Photograph of the device, (b) photograph of the gas- and medium-perfusion channels within the array, and (c) fluorescence image of micropatterned hepatocytes, labeled with a cell-tracker dye, within a single array element.

allowed medium to pass through while retaining the cells in the microcontainers. Thus, the design allowed for perfusion across the cells in addition to flow on both sides of the cells. Hepatocytes were cultured as spheroids whose size was controlled by the microcontainer dimensions. This system supported stable albumin and cytochrome P450 function for two weeks in culture.

King et al. further refined the fabrication of the microarray of hepatocytes to allow independent evaluation of array elements [98]. The device consisted of an array of 256 elements that were connected with microfluidic channels as shown in Figure 26.8(a). Integration of microvalves in the device enabled seeding of different cells in each row and a different stimulant in each column. Thus, the design not only enabled simultaneous evaluation of multiple inputs but also provided for seeding of different cell types in a predefined array format. The basic operation of the microvalve, the state of the valves during seeding of cells, and the introduction of stimulant are shown in Figure 26.8. The device was fabricated using multilevel soft lithography and incorporated a two-layer PDMS structure bonded to glass. The first layer of PDMS consisted of the cell-seeding chambers and the microchannels connecting the chambers. The second layer consisted of the actuation channels that were used for applying positive and negative pressure in order to close or open the

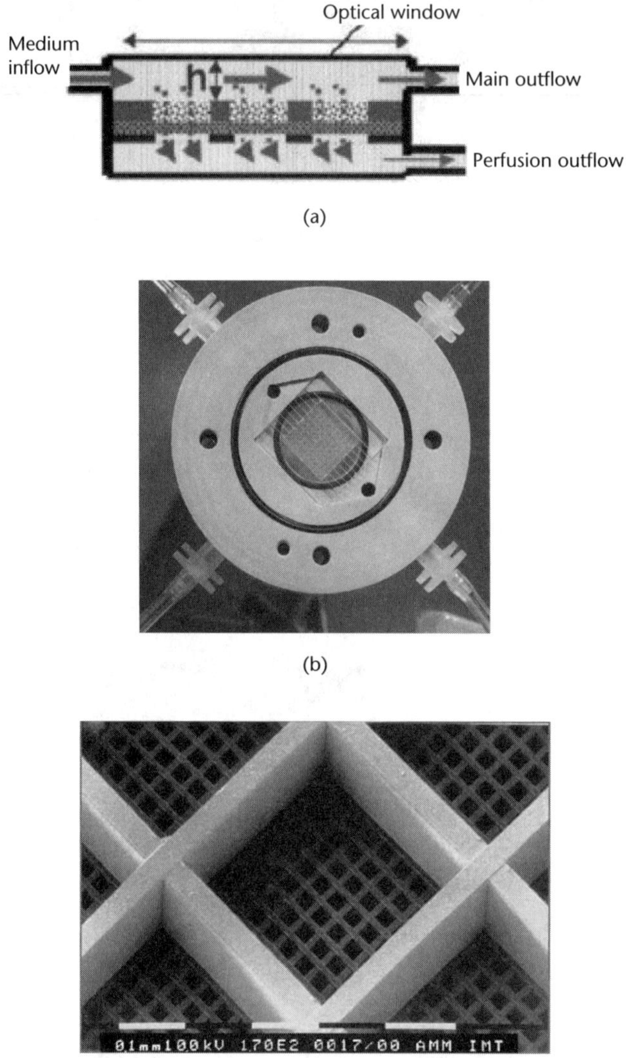

(a)

(b)

(c)

Figure 26.7 Various designs of scaffolds for culturing hepatocytes in array format: (a) schematic of scaffold that allows perfusion across the cells in addition to flow on top of the cells [96], (b) photograph of the polymethylmethacrylate (PMMA) scaffold in the bioreactor assembly that enables perfusion across the cells [97], and (c) scanning electron micrograph image of the PMMA scaffold showing 8 × 8 array of pores in each microcontainer [97].

valves. As opposed to the approach of Kane et al., cells were seeded after the whole device was assembled.

The usefulness of this device for conducting high-throughput experiments was further enhanced by the utilization of reporter cell lines. These cell lines were created by transfecting rat hepatoma cells with a plasmid encoding for enhanced green fluorescence protein (eGFP) under the control of a transcription factor response element. The intensity of eGFP reflected the degree of binding of the transcription factor to its response element [99]. Thus, the use of cell lines permitted quantification of the

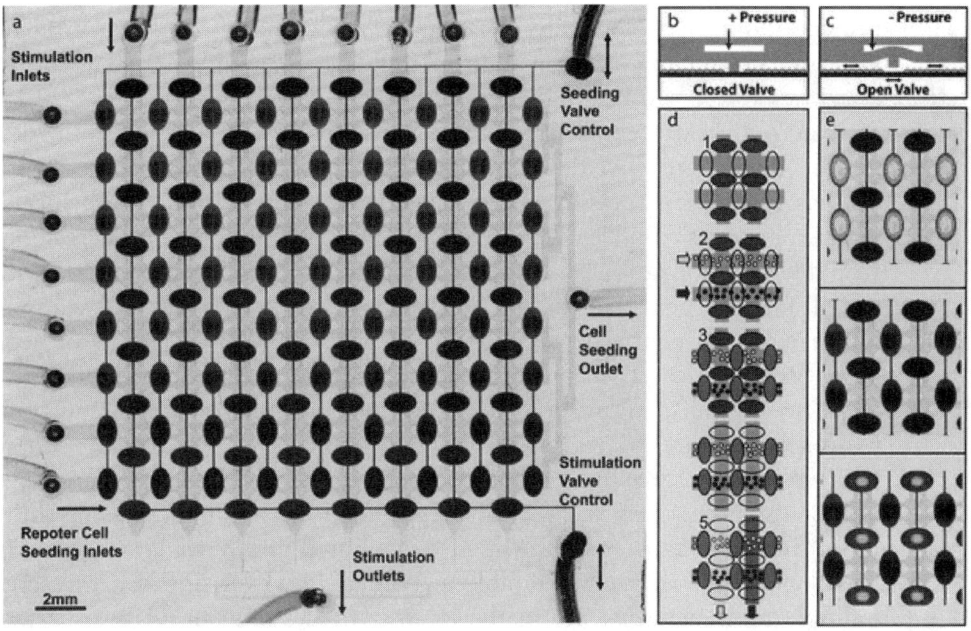

Figure 26.8 Hepatic living cell array device [98]: (a) Image of the two-layered PDMS device with tubing connections for seeding cells, stimulant introduction, and valve operation. The first layer of the device (shown in yellow) consisted of 256 array elements connected by microchannels. The second layer incorporated the microvalves and channels connecting them to apply positive and negative pressure. (b, c) Schematic showing closed and open state of the valve upon application of positive and negative pressure, respectively. (d) Schematic showing state of valves during an experiment: (1) opening of cell-seeding valves, (2) seeding of cells, (3) closing cell-seeding valves, (4) opening of stimulation control valves, and (5) introduction of stimulus. (e) Image of dye-filled device during different steps of an experiment: top image, steps 1 and 2; middle image, step 3; bottom image, steps 4 and 5.

dynamics of transcription factor binding in a single experiment by monitoring the intensity of eGFP over time. The utility of this device was evaluated by seeding eight cell lines that were stimulated with eight different soluble factor inputs for studying the dynamics of transcription factors in the context of liver inflammation. The high-throughput capability of the device was demonstrated by carrying out approximately five thousand time point measurements in a single thirty-six hour experiment.

26.6 Summary

A considerable amount of progress has been made in our ability to culture and manipulate hepatocytes outside the body. Use of cell-patterning techniques has enabled studies that can separate the effect of parenchymal-to-nonparenchymal cell ratio from heterotypic interface in modulating hepatocyte function. Incorporation of new materials in cell-culture experiments has facilitated the development of new in vitro models that not only stabilize hepatocyte function but also closely mimic the in vivo–like morphology of hepatic tissue. With advances in our ability to engineer hepatic tissue in terms of the cellular microenvironment, in vitro tissue models will become more physiologically relevant for investigating the basic mechanisms of liver injury, studying liver metabolism, and the screening of drugs for toxicity and pharmacology.

Microfabrication and microfluidic technologies have been used for developing improved designs of BAL devices that address the issues of minimizing substrate limitations to the hepatocytes and device scale-up. Integration of polymer scaffolds with hepatocytes could lead to the development of immunoisolation techniques that would permit utilization of xenogeneic hepatocytes for transplantation without the use of immunosuppressor drugs. Furthermore, the use of scaffolds can potentially improve engraftment efficiency of hepatocytes. Stable microarrays of hepatocytes have been fabricated using various techniques, such as patterning with non-parenchymal cells, culturing as three-dimensional spheroids, and encapsulating in hydrogel structures. Integration of microfluidics has significantly enhanced the high-throughput screening capability of the hepatic microarrays. Microfluidics technology also offers the possibility of integrating downstream processing with cell culture. Thus, one can envision a highly integrated and automated device in which hepatocytes are arranged in a microarray that allows for simultaneous screening of multiple molecules (inputs) in the context of desired output measurements (for example albumin- and urea-synthesis rates) using a single array.

References

[1] Millward-Sadler, G. H., Wright, R., and Arthur, M. J. P. (eds.), *Wright's Liver and Biliary Disease: Pathophysiology, Diagnosis and Management*, Vol. 1, London: Saunders, 1992.

[2] Junqueira, W. B., L. C., and Carneiro, J., *Basic Histology: Text and Atlas*, 11th ed., McGraw-Hill, 2005.

[3] McCuskey, R. S., "Anatomy of the liver," in T. D. Boyer et al. (eds.), *Hepatology: A Textbook of Liver Disease*, Philadelphia: Saunders, 2006, pp. 3–21.

[4] Wisse, E., De Zanger, R. B., Charels, K., Van Der Smissen, P., and McCuskey, R. S., "The liver sieve: Considerations concerning the structure and function of endothelial fenestrae, the sinusoidal wall and the space of Disse," *Hepatology*, Vol. 5, No. 4, 1985, pp. 683–692.

[5] Li, D., and Friedman, S. L., "Hepatic stellate cells: Morphology, function, and regulation," in I. M. Arias et al. (eds.), *The Liver, Biology and Pathobiology*, Philadelphia: Lippincott Williams and Wilkins, 2001, pp. 455–468.

[6] Wisse, E., Braet, F., Luo, D., De Zanger, R., Jans, D., Crabbe, E., and Vermoesen, A., "Structure and function of sinusoidal lining cells in the liver," *Toxicol. Pathol.*, Vol. 24, No. 1, 1996, pp. 100–111.

[7] van den Berghe, G., "The role of the liver in metabolic homeostasis: Implications for inborn errors of metabolism," *J. Inherited Metab. Dis.*, Vol. 14, No. 4, 1991, pp. 407–420.

[8] Jakoby, W. B., *Enzymatic Basis of Detoxication*, Vol. 1, New York: Academic Press, 1980.

[9] www.cdc.gov/nchs/data/hestat/finaldeaths03_tables.pdf.

[10] www.optn.org/latestData/rptData.asp.

[11] Chan, C., Berthiaume, F., Nath, B. D., Tilles, A. W., Toner, M., and Yarmush, M. L., "Hepatic tissue engineering for adjunct and temporary liver support: Critical technologies," *Liver Transplant.*, Vol. 10, No. 11, 2004, pp. 1331–1342.

[12] Enat, R., Jefferson, D. M., Ruiz-Opazo, N., Gatmaitan, Z., Leinwand, L. A., and Reid, L. M., "Hepatocyte proliferation in vitro: Its dependence on the use of serum-free hormonally defined medium and substrata of extracellular matrix," *Proc. Natl. Acad. Sci. USA*, Vol. 81, No. 5, 1984, pp. 1411–1415.

[13] Guguen-Guillouzo, C., and Guillouzo, A., "Modulation of functional activities in cultured rat hepatocytes," *Mol. Cell. Biochem.*, Vol. 53–54, Nos. 1–2, 1983, pp. 35–56.

[14] Guguen-Guillouzo, C., Clement, B., Baffet, G., Beaumont, C., Morel-Chany, E., Glaise, D., and Guillouzo, A., "Maintenance and reversibility of active albumin secretion by adult rat hepatocytes co-cultured with another liver epithelial cell type," *Exp. Cell Res.*, Vol. 143, No. 1, 1983, pp. 47–54.

[15] Morin, O., and Normand, C., "Long-term maintenance of hepatocyte functional activity in co-culture: Requirements for sinusoidal endothelial cells and dexamethasone," *J. Cell. Physiol.*, Vol. 129, No. 1, 1986, pp. 103–110.

[16] Rojkind, M., Novikoff, P. M., Greenwel, P., Rubin, J., Rojas-Valencia, L., de Carvalho, A. C., Stockert, R., Spray, D., Hertzberg, E. L., and Wolkoff, A. W., "Characterization and functional studies on rat liver fat-storing cell line and freshly isolated hepatocyte coculture system," *Am. J. Path.*, Vol. 146, No. 6, 1995, pp. 1508–1520.

[17] Dunn, J. C., Tompkins, R. G., and Yarmush, M. L., "Long-term in vitro function of adult hepatocytes in a collagen sandwich configuration," *Biotech. Prog.*, Vol. 7, No. 3, 1991, pp. 237–245.

[18] Dunn, J. C., Yarmush, M. L., Koebe, H. G., and Tompkins, R. G., "Hepatocyte function and extracellular matrix geometry: Long-term culture in a sandwich configuration," *Faseb J.*, Vol. 3, No. 2, 1989, pp. 174–177.

[19] Dunn, J. C., Tompkins, R. G., and Yarmush, M. L., "Hepatocytes in collagen sandwich: Evidence for transcriptional and translational regulation," *J. Cell. Biol.*, Vol. 116, No. 4, 1992, pp. 1043–1053.

[20] Koide, N., Sakaguchi, K., Koide, Y., Asano, K., Kawaguchi, M., Matsushima, H., Takenami, T., Shinji, T., Mori, M., and Tsuji, T., "Formation of multicellular spheroids composed of adult rat hepatocytes in dishes with positively charged surfaces and under other nonadherent environments," *Exp. Cell Res.*, Vol. 186, No. 2, 1990, pp. 227–235.

[21] Ma, M., Xu, J., and Purcell, W. M., "Biochemical and functional changes of rat liver spheroids during spheroid formation and maintenance in culture: I. Morphological maturation and kinetic changes of energy metabolism, albumin synthesis, and activities of some enzymes," *J. Cell. Biochem.*, Vol. 90, No. 6, 2003, pp. 1166–1175.

[22] Cukierman, E., Pankov, R., Stevens, D. R., and Yamada, K. M., "Taking cell-matrix adhesions to the third dimension," *Science,* Vol. 294, No. 5547, 2001, pp. 1708–1712.

[23] Ishii, I., Tomizawa, A., Kawachi, H., Suzuki, T., Kotani, A., Koshushi, I., Itoh, H., Morisaki, N., Bujo, H., Saito, Y., Ohmori, S., and Kitada, M., "Histological and functional analysis of vascular smooth muscle cells in a novel culture system with honeycomb-like structure," *Atherosclerosis*, Vol. 158, No. 2, 2001, pp. 377–384.

[24] Weaver, V. M., Petersen, O. W., Wang, F., Larabell, C. A., Briand, P., Damsky, C., and Bissell, M. J., "Reversion of the malignant phenotype of human breast cells in three-dimensional culture and in vivo by integrin blocking antibodies," *J. Cell. Biol.*, Vol. 137, No. 1, 1997, pp. 231–245.

[25] Wolf, K., Mazo, I., Leung, H., Engelke, K., von Andrian, U. H., Deryugina, E. I., Strongin, A. Y., Brocker, E. B., and Friedl, P., "Compensation mechanism in tumor cell migration: Mesenchymal-amoeboid transition after blocking of pericellular proteolysis," *J. Cell. Biol.*, Vol. 160, No. 2, 2003, pp. 267–277.

[26] Anders, M., Hansen, R., Ding, R. X., Rauen, K. A., Bissell, M. J., and Korn, W. M., "Disruption of 3-D tissue integrity facilitates adenovirus infection by deregulating the coxsackievirus and adenovirus receptor," *Proc. Natl. Acad. Sci. USA*, Vol. 100, No. 4, 2003, pp. 1943–1948.

[27] Abbott, A., "Cell culture: Biology's new dimension," *Nature*, Vol. 424, No. 6951, 2003, pp. 870–872.

[28] Berthiaume, F., Moghe, P. V., Toner, M., and Yarmush, M. L., "Effect of extracellular matrix topology on cell structure, function, and physiological responsiveness: Hepatocytes cultured in a sandwich configuration," *Faseb J.*, Vol. 10, No. 13, 1996, pp. 1471–1484.

[29] Moghe, P. V., Berthiaume, F., Ezzell, R. M., Toner, M., Tompkins, R. G., and Yarmush, M. L., "Culture matrix configuration and composition in the maintenance of hepatocyte polarity and function," *Biomaterials*, Vol. 17, No. 3, 1996, pp. 373–385.

[30] Harimoto, M., Yamato, M., Hirose, M., Takahashi, C., Isoi, Y., Kikuchi, A., and Okano, T., "Novel approach for achieving double-layered cell sheets co-culture: Overlaying endothelial cell sheets onto monolayer hepatocytes utilizing temperature-responsive culture dishes," *J. Biomed. Mater. Res.*, Vol. 62, No. 3, 2002, pp. 464–470.

[31] Hirose, M., Yamato, M., Kwon, O. H., Harimoto, M., Kushida, A., Shimizu, T., Kikuchi, A., and Okano, T., "Temperature-responsive surface for novel co-culture systems of hepatocytes with endothelial cells: 2-D patterned and double layered co-cultures," *Yonsei Med. J.*, Vol. 41, No. 6, 2000, pp. 803–813.

[32] Ito, A., Takizawa, Y., Honda, H., Hata, K., Kagami, H., Ueda, M., and Kobayashi, T., "Tissue engineering using magnetite nanoparticles and magnetic force: Heterotypic layers of cocultured hepatocytes and endothelial cells," *Tissue Eng.*, Vol. 10, Nos. 5–6, 2004, pp. 833–840.

[33] Rajagopalan, P., Shen, C. J., Berthiaume, F., Tilles, A. W., Toner, M., and Yarmush, M. L., "Polyelectrolyte nano-scaffolds for the design of layered cellular architectures," *Tissue Eng.*, Vol. 12, No. 6, 2006, pp. 1553–1563.

[34] Schrode, W., Mecke, D., and Gebhardt, R., "Induction of glutamine synthetase in periportal hepatocytes by cocultivation with a liver epithelial cell line," *Eur. J. Cell. Biol.*, Vol. 53, No. 1, 1991, pp. 35–41.

[35] Baffet, G., Loyer, P., Glaise, D., Corlu, A., Etienne, P. L., and Guguen-Guillouzo, C., "Distinct effects of cell-cell communication and corticosteroids on the synthesis and distribution of cytokeratins in cultured rat hepatocytes," *J. Cell Sci.*, Vol. 99, Pt. 3, 1991, pp. 609–615.

[36] Goulet, F., Normand, C., and Morin, O., "Cellular interactions promote tissue-specific function, biomatrix deposition and junctional communication of primary cultured hepatocytes," *Hepatology*, Vol. 8, No. 5, 1988, pp. 1010–1018.

[37] Loreal, O., Levavasseur, F., Fromaget, C., Gros, D., Guillouzo, A., and Clement, B., "Cooperation of Ito cells and hepatocytes in the deposition of an extracellular matrix in vitro," *Am. J. Path.*, Vol. 143, No. 2, 1993, pp. 538–544.

[38] Bhatia, S. N., Yarmush, M. L., and Toner, M., "Controlling cell interactions by micropatterning in co-cultures: Hepatocytes and 3T3 fibroblasts," *J. Biomed. Mater. Res.*, Vol. 34, No. 2, 1997, pp. 189–199.

[39] Bhatia, S. N., Balis, U. J., Yarmush, M. L., and Toner, M., "Effect of cell-cell interactions in preservation of cellular phenotype: Cocultivation of hepatocytes and nonparenchymal cells," *Faseb. J.*, Vol. 13, No. 14, 1999, pp. 1883–1900.

[40] Bhatia, S. N., Toner, M., Tompkins, R. G., and Yarmush, M. L., "Selective adhesion of hepatocytes on patterned surfaces," *Annals NY Acad. Sci.*, Vol. 745, 1994, pp. 187–209.

[41] Bhatia, S. N., Balis, U. J., Yarmush, M. L., and Toner, M., "Probing heterotypic cell interactions: Hepatocyte function in microfabricated co-cultures," *J. Biomat. Sci. Polym. Ed.*, Vol. 9, No. 11, 1998, pp. 1137–1160.

[42] Bhatia, S. N., Balis, U. J., Yarmush, M. L., and Toner, M., "Microfabrication of hepatocyte/fibroblast co-cultures: Role of homotypic cell interactions," *Biotech. Prog.*, Vol. 14, No. 3, 1998, pp. 378–387.

[43] Zinchenko, Y. S., Schrum, L. W., Clemens, M., and Coger, R. N., "Hepatocyte and kupffer cells co-cultured on micropatterned surfaces to optimize hepatocyte function," *Tissue Eng.*, Vol. 12, No. 4, 2006, pp. 751–761.

[44] Michalopoulos, G. K., and DeFrances, M. C., "Liver regeneration," *Science*, Vol. 276, No. 5309, 1997, pp. 60–66.

[45] Shito, M., Tilles, A. W., Tompkins, R. G., Yarmush, M. L., and Toner, M., "Efficacy of an extracorporeal flat-plate bioartificial liver in treating fulminant hepatic failure," *J. Surg. Res.*, Vol. 111, No. 1, 2003, pp. 53–62.

[46] Shinoda, M., Tilles, A. W., Kobayashi, N., Wakabayashi, G., Takayanagi, A., Totsugawa, T., Harada, H., Obara, H., Suganuma, K., Berthiaume, F., Shimazu, M., Shimizu, N., Tanaka, N., Kitajima, M., Tompkins, R. G., Toner, M., and Yarmush, M. L., "A bioartificial liver device secreting interleukin-1 receptor antagonist for the treatment of hepatic failure in rats," *J. Surg. Res.*, Vol. 137, No. 1, 2007, pp. 130–140.

[47] Shinoda, M., Tilles, A. W., Wakabayashi, G., Takayanagi, A., Harada, H., Obara, H., Suganuma, K., Berthiaume, F., Shimazu, M., Shimizu, N., Kitajima, M., Tompkins, R. G., Toner, M., and Yarmush, M. L., "Treatment of fulminant hepatic failure in rats using a bioartificial liver device containing porcine hepatocytes producing interleukin-1 receptor antagonist," *Tissue Eng.,* Vol. 12, No. 5, 2006, pp. 1313–1323.

[48] Sussman, N. L., Gislason, G. T., and Kelly, J. H., "Extracorporeal liver support: Application to fulminant hepatic failure," *J. Clin. Gastroent.*, Vol. 18, No. 4, 1994, pp. 320–324.

[49] Watanabe, F. D., Mullon, C. J., Hewitt, W. R., Arkadopoulos, N., Kahaku, E., Eguchi, S., Khalili, T., Arnaout, W., Shackleton, C. R., Rozga, J., Solomon, B., and Demetriou, A. A., "Clinical experience with a bioartificial liver in the treatment of severe liver failure: A phase I clinical trial," *Annals Surg.*, Vol. 225, No. 5, 1997, pp. 484–491.

[50] Stange, J., Hassanein, T. I., Mehta, R., Mitzner, S. R., and Bartlett, R. H., "The molecular adsorbents recycling system as a liver support system based on albumin dialysis: A summary of preclinical investigations, prospective, randomized, controlled clinical trial, and clinical experience from 19 centers," *Artific. Org.*, Vol. 26, No. 2, 2002, pp. 103–110.

[51] Covic, A., Goldsmith, D. J., Gusbeth-Tatomir, P., Volovat, C., Dimitriu, A. G., Cristogel, F., and Bizo, A., "Successful use of Molecular Absorbent Regenerating System (MARS) dialysis for the treatment of fulminant hepatic failure in children accidentally poisoned by toxic mushroom ingestion," *Liver Int.*, Vol. 23, Suppl. 3, 2003, pp. 21–27.

[52] Ash, S. R., Caldwell, C. A., Singer, G. G., Lowell, J. A., Howard, T. K., and Rustgi, V. K., "Treatment of acetaminophen-induced hepatitis and fulminant hepatic failure with extracorporeal sorbent-based devices," *Advances in Renal Replac. Ther.*, Vol. 9, No. 1, 2002, pp. 42–53.

[53] Laleman, W., Wilmer, A., Evenepoel, P., Verslype, C., Fevery, J., and Nevens, F., "Review article: Non-biological liver support in liver failure," *Aliment. Pharm. Therapeut.*, Vol. 23, No. 3, 2006, pp. 351–363.

[54] Allen, J. W., Hassanein, T., and Bhatia, S. N., "Advances in bioartificial liver devices," *Hepatology*, Vol. 34, No. 3, 2001, pp. 447–455.

[55] Strain, A. J., and Neuberger, J. M., "A bioartificial liver—state of the art," *Science,* Vol. 295, No. 5557, 2002, pp. 1005–1009.

[56] van de Kerkhove, M. P., Hoekstra, R., Chamuleau, R. A., and van Gulik, T. M., "Clinical application of bioartificial liver support systems," *Annals Surg.*, Vol. 240, No. 2, 2004, pp. 216–230.

[57] Court, F. G., Wemyss-Holden, S. A., Dennison, A. R., and Maddern, G. J., "Bioartificial liver support devices: Historical perspectives," *ANZ J. Surg.,* Vol. 73, No. 9, 2003, pp. 739–748.

[58] Yarmush, M. L., Dunn, J. C., and Tompkins, R. G., "Assessment of artificial liver support technology," *Cell Transplant.*, Vol. 1, No. 5, 1992, pp. 323–341.

[59] Harrison, D. J., Fluri, K., Seiler, K., Fan, Z. H., Effenhauser, C. S., and Manz, A., "Micromachining a miniaturized capillary electrophoresis-based chemical-analysis system on a chip," *Science,* Vol. 261, 1993, pp. 895–897.

[60] Culbertson, C. T., Jacobson, S. C., and Ramsey, J. M., "Microchip devices for high-efficiency separations," *Anal. Chem.*, Vol. 72, No. 23, 2000, pp. 5814–5819.

[61] McDonald, J. C., Duffy, D. C., Anderson, J. R., Chiu, D. T., Wu, H., Schueller, O. J., and Whitesides, G. M., "Fabrication of microfluidic systems in poly(dimethylsiloxane)," *Electrophoresis*, Vol. 21, No. 1, 2000, pp. 27–40.

[62] Duffy, D. C., McDonald, J. C., Schueller, O. J. A., and Whitesides, G. M., "Rapid prototyping of microfluidic systems in poly(dimethylsiloxane)," *Anal. Chem.*, Vol. 70, 1998, pp. 4974–4984.

[63] Manz, A., Graber, N., and Widmer, H. M., "Miniaturized total chemical-analysis systems—a novel concept for chemical sensing," *Sens. Actuators B Chem.*, Vol. 1, 1990, pp. 244–248.

[64] Charati, S. G., and Stern, S. A., "Diffusion of gases in silicone polymers: Molecular dynamics simulations," *Macromolecules*, Vol. 31, No. 16, 1998, pp. 5529–5535.

[65] McDonald, J. C., and Whitesides, G. M., "Poly(dimethylsiloxane) as a material for fabricating microfluidic devices," *Acct. Chem. Res.*, Vol. 35, No. 7, 2002, pp. 491–499.

[66] Leclerc, E., Sakai, Y., and Fujii, T., "Microfluidic PDMS (polydimethylsiloxane) bioreactor for large-scale culture of hepatocytes," *Biotech. Prog.*, Vol. 20, No. 3, 2004, pp. 750–755.

[67] Park, J., Berthiaume, F., Toner, M., Yarmush, M. L., and Tilles, A. W., "Microfabricated grooved substrates as platforms for bioartificial liver reactors," *Biotechnol. Bioeng.*, Vol. 90, No. 5, 2005, pp. 632–644.

[68] De Bartolo, L., Jarosch-Von Schweder, G., Haverich, A., and Bader, A., "A novel full-scale flat membrane bioreactor utilizing porcine hepatocytes: Cell viability and tissue-specific functions," *Biotech. Prog.*, Vol. 16, 2000, pp. 102–108.

[69] Flendrig, L. M., la Soe, J. W., Jorning, G. G., Steenbeek, A., Karlsen, O. T., Bovee, W. M., Ladiges, N. C., te Velde, A. A., and Chamuleau, R. A., "In vitro evaluation of a novel bioreactor based on an integral oxygenator and a spirally wound nonwoven polyester matrix for hepatocyte culture as small aggregates," *J. Hepatol.*, Vol. 26, No. 6, 1997, pp. 1379–1392.

[70] Tilles, A. W., Balis, U. J., Baskaran, H., Yarmush, M. L., and Toner, M., "Internal membrane oxygenation removes substrate oxygen limitations in a small-scale flat-plate hepatocyte bioreactor," in Y. Ikada and N. Ohshima (eds.), *International symposium on tissue engineering for therapeutic use*, Vol. 5, Elsevier: Amsterdam, 2001, pp. 59–71.

[71] Tilles, A. W., Baskaran, H., Roy, P., Yarmush, M. L., and Toner, M., "Effects of oxygenation and flow on the viability and function of rat hepatocytes cocultured in a microchannel flat-plate bioreactor," *Biotechnol. Bioeng.*, Vol. 73, No. 5, 2001, pp. 379–389.

[72] Gupta, S., and Chowdhary, J. R., "Hepatocyte transplantation: Back to the future," *Hepatology*, Vol. 15, No. 1, 1992, pp. 156–162.

[73] Ribeiro, J., Nordlinger, B., Ballet, F., Cynober, L., Coudray-Lucas, C., Baudrimont, M., Legendre, C., Delelo, R., and Panis, Y., "Intrasplenic hepatocellular transplantation corrects hepatic encephalopathy in portacaval-shunted rats," *Hepatology*, Vol. 15, No. 1, 1992, pp. 12–18.

[74] Yoshida, Y., Tokusashi, Y., Lee, G. H., and Ogawa, K., "Intrahepatic transplantation of normal hepatocytes prevents Wilson's disease in Long-Evans cinnamon rats," *Gastroenterology*, Vol. 111, No. 6, 1996, pp. 1654–1660.

[75] Bilir, B. M., Guinette, D., Karrer, F., Kumpe, D. A., Krysl, J., Stephens, J., McGavran, L., Ostrowska, A., and Durham, J., "Hepatocyte transplantation in acute liver failure," *Liver Transplant.*, Vol. 6, No. 1, 2000, pp. 32–40.

[76] Mito, M., and Kusano, M., "Hepatocyte transplantation in man," *Cell Transplant.*, Vol. 2, No. 1, 1993, pp. 65–74.

[77] Horslen, S. P., McCowan, T. C., Goertzen, T. C., Warkentin, P. I., Cai, H. B., Strom, S. C., and Fox, I. J., "Isolated hepatocyte transplantation in an infant with a severe urea cycle disorder," *Pediatrics*, Vol. 111, No. 6, Pt. 1, 2003, pp. 1262–1267.

[78] Fox, I. J., Chowdhury, J. R., Kaufman, S. S., Goertzen, T. C., Chowdhury, N. R., Warkentin, P. I., Dorko, K., Sauter, B. V., and Strom, S. C., "Treatment of the Crigler-Najjar syndrome type I with hepatocyte transplantation," *New Eng. J. Med.*, Vol. 338, No. 20, 1998, pp. 1422–1426.

[79] Fox, I. J., and Chowdhury, J. R., "Hepatocyte transplantation," *Am. J. Transplant.*, Vol. 4, Suppl. 6, 2004, pp. 7–13.

[80] Demetriou, A. A., Whiting, J. F., Feldman, D., Levenson, S. M., Chowdhury, N. R., Moscioni, A. D., Kram, M., and Chowdhury, J. R., "Replacement of liver function in rats by transplantation of microcarrier-attached hepatocytes," *Science*, Vol. 233, No. 4769, 1986, pp. 1190–1192.

[81] Takebe, K., Shimura, T., Munkhbat, B., Hagihara, M., Nakanishi, H., and Tsuji, K., "Xenogeneic (pig to rat) fetal liver fragment transplantation using macrocapsules for immunoisolation," *Cell Transplant.*, Vol. 5, No. 5, Suppl. 1, 1996, pp. S31–S33.

[82] Bruni, S., and Chang, T. M., "Encapsulated hepatocytes for controlling hyperbilirubinemia in Gunn rats," *Int. J. Artific. Org.*, Vol. 14, No. 4, 1991, pp. 239–241.

[83] Yang, M. B., Vacanti, J. P., and Ingber, D. E., "Hollow fibers for hepatocyte encapsulation and transplantation: Studies of survival and function in rats," *Cell Transplant.*, Vol. 3, No. 5, 1994, pp. 373–385.

[84] Gomez, N., Balladur, P., Calmus, Y., Baudrimont, M., Honiger, J., Delelo, R., Myara, A., Crema, E., Trivin, F., Capeau, J., and Nordlinger, B., "Evidence for survival and metabolic activity of encapsulated xenogeneic hepatocytes transplanted without immunosuppression in Gunn rats," *Transplantation*, Vol. 63, No. 12, 1997, pp. 1718–1723.

[85] Mooney, D. J., Sano, K., Kaufmann, P. M., Majahod, K., Schloo, B., Vacanti, J. P., and Langer, R., "Long-term engraftment of hepatocytes transplanted on biodegradable polymer sponges," *J. Biomed. Mater. Res.*, Vol. 37, No. 3, 1997, pp. 413–420.

[86] Elcin, Y. M., Dixit, V., Lewin, K., and Gitnick, G., "Xenotransplantation of fetal porcine hepatocytes in rats using a tissue engineering approach," *Artific. Org.*, Vol. 23, No. 2, 1999, pp. 146–152.

[87] Lee, H., Cusick, R. A., Browne, F., Ho Kim, T., Ma, P. X., Utsunomiya, H., Langer, R., and Vacanti, J. P., "Local delivery of basic fibroblast growth factor increases both angiogenesis and engraftment of hepatocytes in tissue-engineered polymer devices," *Transplantation*, Vol. 73, No. 10, 2002, pp. 1589–1593.

[88] Gershell, L. J., and Atkins, J. H., "A brief history of novel drug discovery technologies," *Nat. Rev. Drug Discov.*, Vol. 2, No. 4, 2003, pp. 321–327.

[89] Revzin, A., Tompkins, R. G., and Toner, M., "Surface engineering with poly(ethylene glycol) photolithography to create high-density cell arrays on glass," *Langmuir*, Vol. 19, No. 23, 2003, pp. 9855–9862.

[90] Revzin, A., Rajagopalan, P., Tilles, A. W., Berthiaume, F., Yarmush, M. L., and Toner, M., "Designing a hepatocellular microenvironment with protein microarraying and poly(ethylene glycol) photolithography," *Langmuir*, Vol. 20, No. 8, 2004, pp. 2999–3005.

[91] Fukuda, J., and Nakazawa, K., "Orderly arrangement of hepatocyte spheroids on a microfabricated chip," *Tissue Eng.*, Vol. 11, Nos. 7–8, 2005, pp. 1254–1262.

[92] Albrecht, D. R., Tsang, V. L., Sah, R. L., and Bhatia, S. N., "Photo- and electropatterning of hydrogel-encapsulated living cell arrays," *Lab Chip*, Vol. 5, No. 1, 2005, pp. 111–118.

[93] Liu, V. A., and Bhatia, S. N., "Three-dimensional photopatterning of hydrogels containing living cells," *Biomed. Microdevices*, Vol. 4, No. 4, 2002, pp. 257–266.

[94] Thompson, D. M., King, K. R., Wieder, K. J., Toner, M., Yarmush, M. L., and Jayaraman, A., "Dynamic gene expression profiling using a microfabricated living cell array," *Anal. Chem.*, Vol. 76, No. 14, 2004, pp. 4098–4103.

[95] Kane, B. J., Zinner, M. J., Yarmush, M. L., and Toner, M., "Liver-specific functional studies in a microfluidic array of primary mammalian hepatocytes," *Anal. Chem.*, Vol. 78, No. 13, 2006, pp. 4291–4298.

[96] Powers, M. J., Domansky, K., Kaazempur-Mofrad, M. R., Kalezi, A., Capitano, A., Upadhyaya, A., Kurzawski, P., Wack, K. E., Stolz, D. B., Kamm, R., and Griffith, L. G., "A microfabricated array bioreactor for perfused 3-D liver culture," *Biotechnol. Bioeng.*, Vol. 78, No. 3, 2002, pp. 257–269.

[97] Eschbach, E., Chatterjee, S. S., Noldner, M., Gottwald, E., Dertinger, H., Weibezahn, K. F., and Knedlitschek, G., "Microstructured scaffolds for liver tissue cultures of high cell density: Morphological and biochemical characterization of tissue aggregates," *J. Cell. Biochem.*, Vol. 95, No. 2, 2005, pp. 243–255.

[98] King, K. R., Wang, S., Irimia, D., Jayaraman, A., Toner, M., and Yarmush, M. L., "A high-throughput microfluidic real-time gene expression living cell array," *Lab Chip*, Vol. 7, No. 1, 2007, pp. 77–85.

[99] Wieder, K. J., King, K. R., Thompson, D. M., Zia, C., Yarmush, M. L., and Jayaraman, A., "Optimization of reporter cells for expression profiling in a microfluidic device," *Biomed. Microdevices*, Vol. 7, No. 3, 2005, pp. 213–222.

CHAPTER 27

Nano- and Microtechnologies for the Development of Engineered Skin Substitutes

Katie A. Bush and George D. Pins

27.1 Overview

The use of engineered skin substitutes for the treatment of damaged skin has been successful in some clinical applications. However, incomplete or prolonged healing times for regenerated skin and mechanically induced graft failure remain persistent problems. At present, there are no skin substitutes that completely restore the anatomy, physiology, biological stability, or aesthetic nature of uninjured skin after treatment and healing of the wound. To address the limitations of engineered skin substitutes requires an understanding of the mechanisms by which cellular nano- and microenvironments, including the three-dimensional architecture and biochemical composition of tissue scaffolds, modulate cellular adhesion, migration, proliferation, and differentiation, as well as regeneration of epidermal and dermal tissues. In this chapter, we present an overview of nano- and microfabrication technologies that have been used to design (1) epidermal, (2) dermal, and (3) composite skin substitutes with precisely tailored cellular environments. For each design strategy, we discuss the use of various fabrication technologies, addressing the merits, challenges, and future considerations associated with directing cell functions to enhance regeneration of injured skin. Additionally, we discuss the use of these fabrication technologies in creating nano- and microenvironments to study the underlying cell biology that governs skin regeneration.

27.1.1 Clinical Need for Engineered Skin Substitutes

Every year in the United States 2.4 million burn injuries are reported and approximately twelve thousand people are burned severely enough to require skin grafting [1]. Diabetic and venous ulcers, as well as pressure sores, affect an additional three to four million people in the United States annually [2, 3]. It is anticipated that these figures will continue to increase with the rise of the average age of the population and the incidences of diabetes. Prompt restoration of skin integrity for burns, ulcers, and pressure sores is critical for preventing infection and water loss, scarring, amputation, and death. While autografts can be harvested to treat localized skin injuries, a lack of available donor sites, as well as donor site morbidity, routinely hinders the recovery of patients with large burns or skin traumas [4–6]. As such, the successful development of engineered skin substitutes that provide permanent coverage to the wound site and regenerate the anatomy and physiology of uninjured skin will have significant impact on the treatment of patients with serious skin injuries [5–7].

27.1.2 Current Strategies for Engineered Skin Substitutes

Engineered skin substitutes offer the prospect of treating severe burn injuries and skin traumas with an off-the-shelf therapy that will promote the repair and restoration of native skin properties after placement into the wound bed [7]. Conceptually, researchers have adopted three strategies for designing engineered skin substitutes that recapitulate a subset of the functional properties of native skin, fabricating tissue analogs from: (1) cultured cells, (2) biomaterials, or (3) composites of biomaterials and cells to promote rapid tissue regeneration (for reviews, see [4–6, 8, 9]). For example, Rheinwald and Green established a method of in vitro cultivation and serial subculture of epidermal cells to produce viable keratinocyte sheets to facilitate epidermal regeneration [10]. Clinical trials demonstrated that these autologous sheets of cells were a major surgical innovation, providing permanent coverage to large surface-area burns as well as to acute and chronic wounds [11, 12]. While these epithelial sheets provide acceptable cosmetic results and eventually regenerate dermal tissue, they are thin and fragile. Furthermore, blistering, scarring, and wound contraction can occur after grafting onto excised burn wound beds [13–15]. To promote dermal regeneration, Yannas et al. pioneered the use of a collagen-glycosaminoglycan (GAG) sponge coupled with a silicone membrane barrier that acts as a temporary wound covering [16, 17]. The collagen-GAG sponge promotes vascularized neodermal formation, and the silicone layer is eventually replaced with a thin epidermal autograft [18]. Although this dermal analog has achieved some clinical success, a second surgical procedure is necessary to reepithelialize the surface, and the cosmetic appearance of the regenerated skin is often flat and featureless. Boyce et al. designed a composite skin substitute by seeding fibroblasts into a porous collagen-GAG sponge and culturing a flat epidermal layer on the laminated surface of the sponge [19, 20]. In clinical studies, these composite skin substitutes showed some success for the treatment of full-thickness burns, although engraftment rates were suboptimal [21, 22]. Another skin equivalent has been produced by culturing an epidermal layer on the surface of a fibroblast-populated collagen gel [23–25]; these have been approved by the Food and Drug Adminstration for the treatment of venous and diabetic leg ulcers [26]. Recently, experimental skin equivalents with capillary-like networks have also been produced [27, 28]. While each of these design strategies has produced engineered tissue analogs that achieved some clinical success in restoring damaged skin, prolonged healing times, scarring, and limited tissue functionality remain persistent problems. As such, future tissue analogs must adopt smart materials design strategies, emulating native tissue structures and biochemical signaling cues at the nano- and microscale to better direct cell-mediated tissue regeneration.

27.2 Nano- and Microscale Approaches to Producing Engineered Skin Substitutes

27.2.1 Engineered Skin Substitutes: Clinical and Design Considerations

To improve the ability of engineered skin substitutes to restore the properties of native tissue, they must satisfy several clinical and engineering-design criteria. These tissue analogs should be readily available off-the-shelf and inexpensive, possess physical and mechanical properties that facilitate easy handling, suturing, and adhesion to the wound bed, prevent water loss, and act as a microbial barrier. To minimize scarring and to actively promote the regeneration of native skin functionality, engineered skin substitutes should also integrate with native host tissue, delivering growth factors, cytokines, and extracellular matrix (ECM) components that direct host cells to facilitate angiogenesis, neodermal, and neoepidermal regeneration. Specifically, the design of future skin substitutes must direct cellular migration, proliferation, and differentiation, as well as cellular processes that mediate the morphogenesis of cells into analogs of functional skin that promote rapid barrier formation, reduce healing times, and restore the native tissue physiology.

Significant effort has been devoted to producing biocompatible scaffolds with defined pore sizes that help ensure proper cell-cell contacts, cell-matrix interactions and preserve cellular function [29–31]. For example, in collagen sponges, the average pore diameter (70 to 200 μm) and orientation (random) are critical for optimal scaffold activity [16, 17, 32]. Additionally, cell-attachment, migration, and proliferation phenomena have been examined on substrates derived from ECM components and synthetic biomaterials [2, 20, 33–36]. To improve the design and performance of engineered skin substitutes, future scaffolds must promote physiologic tissue regeneration by mimicking three-dimensional tissue meso-, micro- and nanostructures with organ-scale cellular resolution (10–1,000 μm), cellular microenvironments (1–10 μm), and ECM cues (1–100 nm) that mediate cellular function to regenerate dermal and epidermal tissues (Figure 27.1) [37–42].

27.2.2 Engineered Epidermal Substitutes

27.2.2.1 Cultured Epithelial Autografts

Restoring an epidermal layer that provides a barrier against fluid loss and infection is critical to wound healing and skin regeneration. In the mid-1970s, Rheinwald et al. developed in vitro techniques to subculture keratinocytes into large confluent sheets of epithelial tissues [10, 43]. These cultured epithelial autografts (CEAs) are enzymatically removed from the culture substrate and placed onto the wound of the patient. This technique provides clinicians with large areas of epithelial wound dressing from small pieces of biopsy tissue. However, there are several limitations associated with these grafts. CEAs require biopsies from the patients and a two- to three-week interval to produce sufficient quantities of graft material. Additionally, CEAs are thin, fragile, expensive, and they require custom preparation. They also fail to provide adequate epithelial coverage, blistering often occurs, and engraftment rates are suboptimal [5, 12–14]. One of the primary reasons for suboptimal graft take is the enzymatic treatment required to remove the cultured CEA from the culture plate. This treatment disrupts keratinocyte interactions with

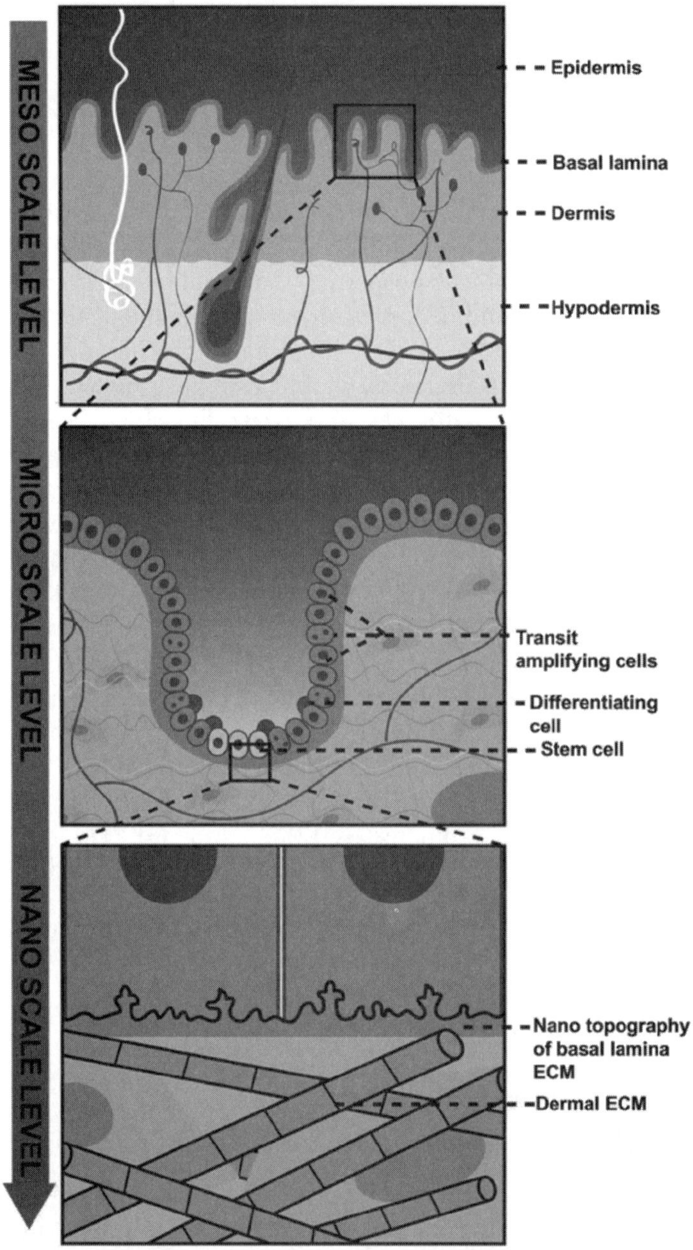

Figure 27.1 Cartoon illustrating the nano- and microenvironments of skin. The top panel shows mesoscale features of skin, highlighting the principal structural layers. The center panel shows microscale features of skin. This illustration focuses on the microtopography of the basal lamina and highlights the basal layer of keratinocytes with transiently amplified, differentiating, and stem cells labeled. The bottom panel shows the nanoscale features of skin, illustrating the nanotopography of the ECM in the basal lamina (collagenous, and noncollagenous proteins) and the dermis (mainly composed of collagen type I fibrils).

the underlying ECM proteins that are deposited on the substrate and cause contraction of the CEA. Furthermore, this process decreases mechanical stability of the grafts and hinders their attachment to wound beds [4, 13].

27.2.2.2 Microfabricated Carrier Devices for Keratinocyte Delivery

To improve the durability and the engraftment rate of CEAs, several investigators have attempted to create carrier devices that support keratinocyte culture that can be transferred to the wound bed, eliminating the need for enzymatic treatment prior to implantation. Microfabrication techniques have been used to create support scaffolds with microenvironments that emulate native tissue properties and promote keratinocyte attachment, as well as the deposition of cell-mediated ECM proteins. This ECM provides the biochemical signaling cues that direct keratinocyte proliferation and differentiation, as well as the development of a mature epidermis and a basement membrane that anchors the cellular layer to the underlying substrate.

27.2.2.2.1 Laserskin

Fidia Advanced Biopolymers, s.r.L., developed a 100 percent benzyl esterified hyaluronic acid derivative (HYAFF-11) that can be manufactured into thin transparent membranes with thicknesses of 20 or 200 μm (Laserskin). These membranes are a promising scaffold for tissue regeneration because (1) they are composed of hyaluronic acid, a glycoprotein known to increase wound healing, and (2) they contain laser-perforated microholes (40 to 60 μm in diameter) that allow for keratinocyte migration from the membrane to the wound environment. Additionally, they include a series of 0.5 mm–thick holes between arrays to facilitate the drainage of fluids after placement onto the wound site [44].

To examine the utility of Laserskin as a carrier device for CEAs, keratinocytes were plated onto the Laserskin membrane in the presence of a feeder layer of nonproliferating 3T3 fibroblasts. Keratinocytes were observed in the holes on the membrane after forty-eight to seventy-two hours of culture [45]. After allowing the cultures to become confluent or near confluent, the Laserskin membranes were placed on dermal supports. In vitro studies where CEAs were placed on human de-epidermalized dermis and nonwoven meshes of HYAFF-11 showed that a well-differentiated epithelial layer forms [45, 46]. In the dermal equivalent containing HYAFF-11, the ECM proteins fibronectin, laminin-1, and types I, III, and IV collagens were located at the dermal-epidermal junction (DEJ). These proteins support keratinocyte attachment and provide signaling cues responsible for directing reepithelialization [46]. In vivo studies also investigated using CEAs and Laserskin in combination with the collagen-based dermal substitute, Integra (Johnson and Johnson) [47]. Integra was grafted onto full-thickness wounds in rats, and after fourteen to twenty days, the silicone membrane was removed, and keratinocytes cultured on Laserskin for four to six days (Laserskin + CEA) were placed onto the wound site. The rate of graft take of Laserskin + CEA on Integra was 70 percent, a marked improvement over previous results investigating graft take of Laserskin + CEA on freshly excised, full-thickness wounds (40 percent) [44]. Research has also been performed to investigate the affect of culturing keratinocytes on the surface of Laserskin membranes for two days and then inversely applying the cultured grafts onto full-thickness wounds in athymic mice [48]. Results from this investigation indicate that the keratinocytes formed an epithelialized layer after a three-week period. These findings demonstrate that epithelial layers cultured on Laserskin can be transferred to wound beds in a shorter time period, substantially reducing the waiting time required for cultured autologous epithelial cells.

27.2.2.2.2 Microtextured Hydrogels for Use as a Biological Carrier Device

Researchers have also developed microfabricated hydrogel sheets as carrier devices to graft keratinocytes onto recipient substrata without the use of enzymatic treatment. Using photolithography, 100×100, 250×250, and 500×500 μm^2 square-shaped pits and columns (depths of 200, 10, or 5 μm) were patterned onto the surface of polyvinyl alcohol/polyvinyl acetate copolymer hydrogel sheets. The microtopographic features on the surfaces of the hydrogel sheets were designed to (1) serve as conduits for cell migration from the vehicles to the host surfaces, and (2) prevent adherent cells from being mechanically sheared off the surfaces of the scaffolds during application [49]. Keratinocytes and fibroblasts were cultured on the surfaces of the micropatterned hydrogel sheets and inverted onto either acellular dermis or tissue-culture plastic. The efficiency of cell transfer onto acellular dermis was greater than that onto flat tissue-culture plastic. This result is attributed to the topographic features on the surfaces of the dermal scaffolds that merge with the surfaces of the hydrogels as well as the ECM proteins that bind the keratinocytes to the surfaces. Cell transfer from flat hydrogel surfaces was also compared to surfaces containing microtopography. This analysis indicated a higher efficiency of grafting using the pit and column surfaces, irrespective of the feature sizes. An additional analysis, characterizing the effect of shear stress on cell viability, showed that all microtopographic hydrogels provided more protection from mechanically induced damage than flat surfaces. However, hydrogels with 250 and 500 μm features performed most effectively, maintaining 80 percent of the initial cell population [49].

27.2.2.3 Future of Epithelial Substitutes

The results of studies using CEAs to transplant epithelial layers indicate that presence of a dermal substitute pregrafted in the wound site enhances the function and the quality of the regenerating tissue. When full-thickness wounds are treated with CEAs alone or with CEAs on carrier devices, engraftment rates are less effective than when CEAs are applied to a dermal wound bed that is prepared with an allograft, autologous dermal tissue, or an engineered dermal substitute [5, 44, 47, 50–54]. A functioning dermal layer allows for increased vascularization and improves initial engraftment rates, as well as long-term durability. Additionally, dermal components that contain fibroblasts provide a regulatory environment that contains both biochemical and structural cues to direct the growth and differentiation of keratinocytes in CEAs [55, 56]. These findings suggest that future research efforts should focus on the development of cell-seeded dermal substitutes to augment the regenerative properties of epidermal scaffolds.

27.2.3 Engineered Dermal Substitutes

27.2.3.1 Structures and Biochemistries of Dermis at Nano- and Microscale Levels

The design of engineered dermal analogs has centered on developing structural materials that emulate the collagen-rich ECM of native tissue and also promote neodermal regeneration. In native tissue, the dermal layer is composed of an ECM arranged as a basket-woven meshwork of collagen, elastic (elasticity), and reticular (support) fibers interspersed with noncollagenous glycoproteins and proteoglycans

(Figure 27.1, bottom panel). The major fiber-forming collagens in the dermis are type I and type III, with type I collagen accounting for 80 to 90 percent of the total collagen content [57–59]. The type I collagen fibrils are uniform in size and are approximately 50 nm in diameter, whereas type III collagen fibrils are less homogenous in size and are approximately 30 to 130 nm in diameter [60]. This ECM provides physical strength, elasticity, and scaffolding that supports the tissue's microvascular network, nerve endings, and adnexal structures, such as sweat glands and hair follicles. The nanoscale network of structural fibrils also plays a key role in directing ligand-mediated cell adhesions that guide cell migration, proliferation, differentiation, and apoptosis, as well as tissue organization and matrix remodeling [61]. Additionally, this network of fibrils is responsible for facilitating nutrient transport from the microvascular network to the host cells and binding cytokines and growth factors that regulate tissue regeneration [41].

Considerable effort has focused on developing porous three-dimensional scaffolds that serve as analogs of the dermal matrix. To facilitate in vitro studies of cell-matrix interactions and to induce tissue regeneration upon implantation, these bioactive dermal substitutes must possess (1) a degradation rate that allows the scaffold to remain intact for a critical period, (2) a chemical composition containing ligands appropriate for cell binding, and (3) a mean pore size that allows cells to migrate into the scaffold [62]. A previous study of dermal regeneration showed that collagen-GAG scaffolds were inactive when the mean pore size was either lower than 20 μm or higher than 120 μm [32]. These findings suggest that the pores in degradable polymeric scaffolds must be large enough to permit cell migration into the structure but small enough to provide a high surface area that facilitates sufficient ligand binding densities to adhere a critical number of cells to direct tissue regeneration [62, 63]. Recently, several research groups have implemented nano- and microfabrication technologies to maximize the functional surface areas for tissue scaffolds in an effort to enhance the performance of dermal substitutes.

27.2.3.2 Electrospinning Nanofibrous Dermal Substitutes

Electrospinning is a technique that facilitates the fabrication of polymers into a highly porous three-dimensional network containing fibers ranging from 3 nm to greater than 5 μm in diameter [64]. Although electrospinning was first described over seventy years ago, it has gained popularity in the field of tissue engineering because it is an extremely efficient, rapid, and inexpensive method to produce nano- and microscale scaffolds. Depending on conditions during the electrostatic spinning process, key scaffold design parameters, such as the shape of the scaffolding, fiber diameter size, fiber orientation, mechanical properties, degradation rate, and pore size, can be controlled with a high degree of precision [65–67].

Studies investigating synthetic biocompatible polymers and copolymers for use as dermal substitutes found that when these materials are electrospun, they form highly porous structures. These membranes contain randomly oriented fibers of varying dimensions that create a very large surface area to volume ratio. Studies investigating poly(ε-caprolactone) (PCL) and poly(D,L-lactide-co-glycolide) (PLGA) report fiber diameters ranging from 250 to 800 nm and porosity values of 90 percent or greater [68, 69]. When dermal fibroblasts were seeded on PCL,

PLGA, or PLGA/dextran scaffolds, the cells integrated with the surrounding fibers to form three-dimensional cellular networks. Analyses of fibroblasts on PLGA/dextran scaffolds showed that cytoskeletal architecture was unorganized [70]. Additionally, cells cultured on electrospun PCL scaffolds exhibited decreased cell proliferation when compared to fibroblasts cultured on polystyrene tissue-culture plastic [67, 71]. These cellular interactions on synthetic biomaterials are primarily directed by chemical bonds such as covalent, hydrophobic, and hydrogen bonds. To properly direct tissue regeneration, cellular processes must be governed by ligand-integrin binding interactions in a manner that emulates native tissue processes [72]. Thus, although electrospun biocompatible polymers have the ability to mimic the structure of native ECM and support cell attachment, proliferation, and migration, it is unlikely that these scaffolds will completely replace the signaling cues of native ECM without further biochemical modification.

Natural polymers are of great interest for producing engineered dermal substitutes because they provide enhanced biocompatibility as well as ECM cues that are critical in directing cellular functions. Considerable effort has been devoted to adapting the electrospinning process to produce scaffolds composed of collagen nanofibers as well as composites containing synthetic polymers and collagen in order to mimic both the structural and biological properties of native skin. Collagen is the most-studied natural polymer because it is the principal structural element of the ECM. Matthews et al., first described a novel processing strategy of electrospinning that allows for the fabrication of collagen filaments [66]. This study indicated that the structural properties of electrospun collagen, such as filament diameter and shape of scaffolding, vary with the tissue of origin, the isotype, and the concentration of collagen used in the electrospinning process. Additionally, it was shown that by controlling fiber orientation, it is possible to tailor mechanical properties of the nanofibrous scaffolds [66]. The ultimate tensile strength of hydrated type I collagen electrospun scaffolds (7.4 MPa) is similar to native skin (5 to 30 MPa) [73, 74]. Electrospun collagen-r-PCL composite scaffolds, produced through the development of a coaxial electrospinning technique, have been compared to electrospun collagen scaffolds, electrospun PCL scaffolds, and electrospun PCL scaffolds coated with collagen after electrospinning [71]. These composite scaffolds facilitate cell-proliferation rates that are comparable to electrospun collagen scaffolds and are significantly greater than collagen-coated PCL electrospun scaffolds [71]. Cellular behavior has also been evaluated on electrospun collagen scaffolds as well as other native polymers, such as chitin and silk, that have been coated with the ECM proteins laminin, fibronectin, and type I collagen [74–76]. These studies showed that type I collagen and laminin-coated scaffolds promote adhesion and spreading of both fibroblasts and keratinocytes.

Electrospun scaffolds also serve as a valuable platform technique for developing model systems to study the complex interactions of multiple cells types within engineered dermal substitutes. Tissue-culture-grade polystyrene was electrospun into fibers of approximately 10 μm with interfiber spaces of 50 to 150 μm, and keratinocytes, fibroblasts, and endothelial cells were seeded on the culture surfaces. Immunohistological analyses of these scaffolds showed that the cells supported each other's survival, infiltrated the matrix, and exhibited a degree of native spatial three-dimensional organization. These findings suggest electrospun dermal scaffolds have the potential to serve as in vitro model systems that may be used to investigate cell-cell interactions in dermal tissue or as engineered skin substitutes to conduct dermatotoxicity screening assays [77].

27.2.3.3 Fused Deposition Modeling of Dermal Substitutes

Several research groups have investigated the rapid prototyping technique known as fused deposition modeling (FDM) to investigate the role of scaffold morphology (size and shape of fibers) in dermal tissue formation. This technology allows for the fabrication of three-dimensional scaffolds with highly controllable and reproducible scaffold morphology, as well as an interconnected pore network that is free of residual solvents [78]. FDM creates the scaffold by layering a 0° to 90° pattern of liquefied polymer from a ~280 nm nozzle using a computer controlled x-y-z table. Fibers that are produced can be laid down with defined spacing and pore size at the microscale. Analyses of cellular responses to poly(ethyleneglycol terephthalate)-poly(butylene terephthalate) copolymer scaffolds showed that the distance between fibers greatly affected the interconnectivity of dermal fibroblasts. Homogenous and dense tissue formation was reported on scaffolds with fiber spacing less than 273 +/– 55 μm seeded with fibroblasts. When the spacing reached 400 μm, fibroblasts were still able to bridge the spaces between fibers; however, there was a lack of ECM organization. As the fiber space increased, the fibroblasts could not bridge the space, and tissue formation was only noted around the fibers [79].

27.2.3.4 Future of Dermal Substitutes

The results of these studies suggest that electrospinning and FDM technologies offer promising new approaches to fabricating nano- and microstructured scaffolds for dermal regeneration. Many investigators have used electrospinning to produce tissue analogs with mechanical properties and microarchitectures that are comparable to native skin. Furthermore, electrospun nanofibrous scaffolds offer the potential of creating structures with large surface-area-to-volume ratios. These surfaces serve as templates that can facilitate high-density binding of ligands to direct cellular functions. When these scaffolds present ECM cues such as type I collagen and laminin, cellular responses are similar to those observed on native tissue scaffolds. Additionally, highly porous electrospun scaffolds may facilitate high gas permeability with controlled evaporative water loss and an enhanced fluid drainage ability [67, 68]. FDM technology provides a new method to create engineered dermal substitutes with precisely tailored scaffold shapes and interconnected pore structures that may control cellular responses and tissue permeability. To fully assess the potential of these approaches for designing tissue analogs, future studies on electrospun and FDM fabricated scaffolds must focus on assessing their ability to integrate with host tissue and to direct microvascular and dermal regeneration. The design of future nano- and microfabricated dermal substitutes must also consider strategies for incorporating epidermal layers. This design iteration is essential to overcome the need for additional surgical procedures to reepithelialize the surfaces of dermal scaffolds.

27.2.4 Composite Skin Substitutes

Several composite skin substitutes have been investigated to treat patients suffering from severe burns and chronic ulcers. These tissue analogs are produced by culturing epidermal layers on the surfaces of either fibroblast-seeded collagen-GAG sponges [80, 81] or fibroblast-populated collagen lattices [23–25, 82]. In clinical studies, these skin substitutes demonstrated some success for the treatment of full-thickness burns [83], as well as venous and diabetic ulcers [26], but suboptimal engraftment rates, prolonged healing times, and mechanically induced graft failure remain persistent problems. A design feature common to each of these skin substitutes is a flat interface between the epidermal and dermal components. In native skin, an integral component of the tissue microarchitecture is the basal lamina, a thin membranous layer of connective tissue found at the interdigitated interface between the dermis and the epidermis.

27.2.4.1 Structures and Biochemistries of Basal Lamina at Nano- and Microscale Levels

In skin, the basal lamina maintains physiological functions by providing an interface between cells and the ECM, modulating cellular differentiation and preserving tissue architecture and organization (Figure 27.1, bottom panel). The basal lamina in skin is approximately 50 to 90 nm thick and contains a variety of collagenous and noncollagenous ECM molecules, including type IV collagen, laminin, fibronectin, and heparin sulfate proteoglycans (reviewed in [84]). It acts as a barrier that prevents fibroblasts in the dermis from directly contacting epidermal keratinocytes, but it does not prevent diffusion of macromolecules into and out of the epidermis; nor does it prevent the innervation of the epidermis. The destruction of the basal lamina may result in scar formation and loss of tissue function during wound healing. The basal lamina acts as a guide and template that helps control cell polarity, migration, and differentiation [85]. Skin regeneration studies on athymic mice showed that the rate of keratinocyte differentiation and epidermal maturation was enhanced when transplanted composite skin grafts contained basal lamina components [86]. In vitro studies comparing epidermal regeneration of different dermal analogs indicated that the epidermal layer is mechanically unstable in the absence of basal lamina components [87]. Nolte et al., proposed that the deposition of basal lamina components during tissue regeneration promotes the formation of a highly differentiated epidermal layer, prevents blistering, and improves graft stability and persistence [88]. Additionally, a preliminary clinical study of cultured skin grafts suggested that early formation of the basal lamina at the DEJ prevents skin blistering and ulceration [89].

The basal lamina in skin is not a simple flat plane of connective tissue; rather, it conforms to a series of ridges and invaginations known as rete ridges and papillary projections (Figure 27.1, top and center panels). The complex topography of the DEJ provides structural stability to the epidermis. In areas of the skin exposed to excessive friction (plantar and palmer surfaces), the dermal papillae and epidermal ridges are longer and more numerous, suggesting that increased interfacial area between the epidermis and dermis helps provide additional mechanical stability [90]. Molecular studies indicate that keratinocyte attachment to the basal lamina is mediated by integrin binding to ECM proteins associated with the DEJ [91–93]. Together, these findings suggest that the design of microfabricated basal lamina analogs that mimic the complex topographic features and ECM composition of native tissue will improve the mechanical and structural properties of engineered skin substitutes.

27.2.4.2 Composite Skin Substitutes with Microtextured Basal Lamina Analogs

To address some of the limitations associated with current composite skin substitutes, our laboratory developed a series of microfabrication strategies to produce basal lamina analogs with precisely defined features comparable to those found in native tissue structures. Laser micromachining and photolithography have been used to create master patterns, negative replicates, and collagen-GAG membranes containing ridges and channels (e.g., grooves 50 to 400 μm in width and depth) of length scales similar to the invaginations found in native tissue at the DEJ. These membranes were incorporated into dermal analogs and seeded with keratinocytes to produce skin substitutes with precisely microtextured cell-matrix interfaces (Figure 27.2) [94, 95].

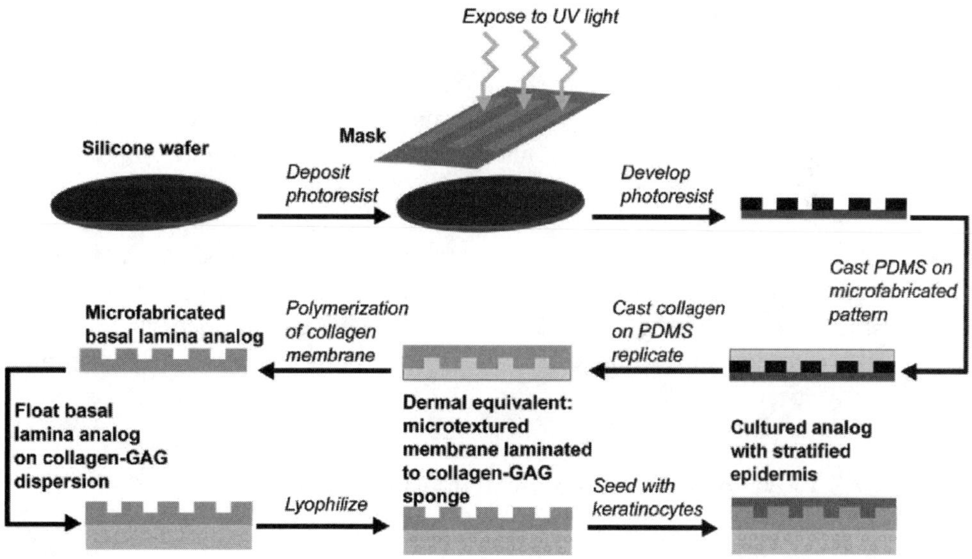

Figure 27.2 Schematic of process to produce microtextured skin equivalents. A high-resolution transparency mask was created by printing two-dimensional drawings onto acetate film using a high-resolution laser photoplotter. To create master patterns with varied depths, the transparency masks were aligned on the surface of silicone wafers, which were precoated with SU-8 photoresist and exposed to ultraviolet (UV) light. The unexposed regions were dissolved, and the UV-crosslinked regions yielded the patterned features on the surfaces of the silicone wafers. Polydimethylsiloxane silicone elastomer (PDMS) was allowed to polymerize on the master patterns and then removed, creating negative replicates that contained surfaces with inverse topographical features of the master patterns. GAG coprecipitate was cast onto the PDMS-negative replicates, allowed to polymerize, and then separated from the surfaces, creating thin microfabricated basal lamina analogs. These basal lamina analogs were laminated to the surfaces of collagen-GAG sponges, and the surfaces were seeded with keratinocytes producing microtextured skin substitutes with stratified epidermal layers.

To analyze the relationship between keratinocyte function and the characteristic topography at the DEJ of native skin, keratinocytes were seeded on the surfaces of basal lamina analogs, and histological analyses were performed after seven days of tissue culture at the air-liquid interface. The keratinocytes formed a differentiated and stratified epidermis that conformed to the features of the microtextured membranes. Morphometric analyses of immunostained skin equivalents indicated that the rates of keratinocyte stratification and differentiation increased as channel depth increased and channel width decreased. This trend was most pronounced in channels with the highest depth-to-width aspect ratios (i.e., 200 μm deep, 50 μm wide) [94, 95]. These findings suggest that incorporating microtextured basal lamina analogs into the design of skin substitutes will enhance the performance of the tissue analogs, promoting the rapid regeneration of robust epidermal layers on the surfaces of the precisely engineered dermal-epidermal interfaces. Future studies should focus on leveraging the utility of these membranes to systematically analyze distinct three-dimensional microarchitectural and biochemical cues that direct the cellular functions and promote the regeneration of highly functional dermal and epidermal tissue on the surfaces of microtextured skin substitutes.

27.3 Nano- and Microscale Approaches for Controlling Cellular Microenvironments

Improving the design and performance of future skin substitutes requires a fundamental understanding of the mechanisms by which the three-dimensional structural features and the biochemical cues on tissue scaffolds modulate the cellular functions that direct tissue regeneration. Nano- and microscale technologies afford opportunities to design high-precision analytical tools to elucidate specific cell-cell and cell-substrate interactions that control cellular phenotypes, cellular organization, and subcellular processes on the surfaces of tissue scaffolds. The results of these studies will provide a wealth of data that can be used to identify design parameters to create skin substitutes with structural and mechanical properties comparable to native tissue.

27.3.1 Nano- and Microfabricated Surfaces to Characterize Cell Function

At the nano- and microscale, the basal lamina provides an environment that influences keratinocyte spatial organization, as well as epidermal structural and mechanical stability, through interactions with ECM proteins (Figure 27.1, center and bottom panels). These interactions allow for bidirectional transmission of signals between the ECM protein and the cytoplasm of the cell, leading to a series of intracellular signaling pathways that regulate cell binding and cellular functions [96]. The ECM proteins in the basal lamina also form a complex nanotopography of pores, fibers, and ridges [97–99]. Recent work has shown that the topography of the ECM found in basal lamina modulates cellular functions [100]. Together, these findings highlight the importance of understanding the cell-signaling roles of ECM features at the cellular level (1 to 100 μm) and subcellular level (100 to 500 nm).

27.3.1.1 Cellular Responses to Biochemical and Nano- and Microtopographical Features

The epidermis undergoes constant regeneration, with a turn over of approximately four weeks. Highly proliferative keratinocytes continue to replicate when they are in direct contact with the basal lamina, then undergo terminal differentiation as they move upward through the layers of the epidermis. Microfabrication techniques such as photolithography enable investigations correlating relationships between cell geometry and keratinocyte proliferation and terminal differentiation. Surfaces with precisely defined adhesive rectangular islands ($300 \times 750\ \mu m$) or circular islands (ranging from 400 to 5000 μm^2) were distributed at increments that defined ranges of cell-substratum contacts but prevented cell-cell contacts [101, 102]. Changes in keratinocyte-substratum contact area were shown to effect DNA synthesis and involucrin expression. Smaller keratinocyte-substratum contact areas (rounded morphology) decreased the percentage of cells undergoing DNA synthesis and increased the percentage of cells expressing involucrin. Restricted contact area has also shown changes in the patterns of expression of keratins and components of the ECM.

In an effort to characterize cellular responses to nano- and microtopographic features, several investigators have studied fibroblast and epithelial cell functions on lithographically modified surfaces containing synthetic topographic features with dimensions that mimic those found on native membrane and dermis structures [38, 103–108]. For example, Curtis and Wilkinson cultured fibroblasts on micro-grooved silicone substrates and found that for ridge widths of $\leq 4.0\ \mu m$, cells were highly orientated and elongated along the surface grooves [104]. Furthermore, results showed that ridge widths affected cell function more than groove depth. In contrast, BHK cells cultured on fused silica substrates exhibited alignment that was most pronounced with respect to groove depth [109]. Recently, Nealey et al. examined corneal epithelial cell functions on the surfaces of synthetic substrates and showed that cells elongated and aligned along patterns of grooves and ridges with feature dimensions as small as 70 nm in width, 150 nm in depth, and 400 nm in pitch spacing [110, 111]. This alignment was most pronounced in grooves that were at least 800 nm in width and 600 nm in depth. To explain these findings, one hypothesis suggests that cellular alignment is directed by cell–topographic feature interactions that orient focal-adhesion contacts [109, 112–115]. However, other studies suggest that cellular alignment occurs in the absence of focal-adhesion formation [110]. These apparently disparate reports on the most important topographic features highlight the complexity of the problem in understanding cell-topography interactions.

While these studies have provided insight into the topographic factors that modulate cellular adhesion, proliferation, and differentiation, few investigators have created nano- and microstructured three-dimensional scaffolds directly from crosslinked collagens and other ECM proteins to examine how biochemical and structural cues direct cell functionality. To evaluate these features, Campagnola et al. developed a nano-/microfabrication method based on multiphoton-excited (MPE) photochemical crosslinking of proteins and polymers [116, 117]. These three-dimensional protein matrices created from collagen types I, II, and IV, bovine serum albumin (BSA), fibronectin (FN), and fibrinogen (FG) retain biological activity following MPE crosslinking [118–120]. Dermal fibroblast adhesion and morphology were evaluated on MPE crosslinked proteins patterned at 10 or 40 μm increments having individual widths of 600 to 700 nm and heights of 3 μm [120, 121]. BSA lines were used to examine the role of topography in the absence of biochemistry, and FN and FG were used to examine the biochemistry since it is known that fibroblasts bind directly to both proteins at arginine-glycine-aspartic acid (RGD) sites through known integrins [122, 123]. Morphological analyses showed statistical differences between cells on BSA patterns and those on FN and FG patterns. Fibroblasts cultured on BSA patterns at 10 μm were elongated and conformed to the lines, whereas at 10 μm for FN and FG, fibroblasts spread across multiple patterns. Fibroblasts on 40 μm lines showed similar area, elongation, and orientation; however, when evaluating focal adhesions, it was found that the protein vinculin corresponded to attachment on FN and FG lines but on BSA lines. Together, these findings suggest that crosslinked ECM protein structures direct cell adhesion and spreading, but topographic and ECM cues mediate different forms of guidance [121].

27.3.2 Microtextured Surfaces to Characterize Keratinocyte Function in Stem Cell Niches

Epidermal stem cells are of great importance because they are responsible for the assembly and maintenance of the epidermis, as well as the rapid regeneration of damaged tissue [124, 125]. These multipotent stem cells exhibit a high degree of spatial organization and cell clustering, as well as a high capacity for proliferation and a strong adhesion to the underlying basal lamina (Figure 27.1, center panel) [124, 126–129]. In human palmar epidermis, the highest percentage of proliferating basal and suprabasal cells are located in the deeper rete ridges [92, 126–128]. It has also been shown that expression of $\alpha 2\beta 1$ integrin, a marker of epidermal stem cells, varies with topography of the basal lamina. These observations suggest that microfabricated basal lamina analogs with three-dimensional features and ECM cues that mimic the cellular microenvironments of the DEJ may be a valuable tool to systematically examine the relationships between the dimensions of topographic features and the proliferation and differentiation of basal keratinocytes in the epidermis [94]. These membranes may be able to elucidate cellular microenvironments that recapitulate the epidermal stem cells niches observed in native tissue. The results of these studies will identify design parameters to create the next generation of engineered skin substitutes with microtextured DEJs that promote stem cell clustering and promote the rapid regeneration of a robust epidermal layer. Finally, these precisely tailored basal lamina analogs promise to serve as valuable in vitro model systems to analyze the three-dimensional and biochemical cues that regulate keratinocyte stem cell proliferation and differentiation in native tissue.

27.4 Future Considerations

The design of future engineered skin substitutes must focus on developing scaffolds with three-dimensional structures, ECM cues, and cytokines expressed during wound healing to better direct cell-mediated regeneration of dermal and epidermal tissues. Widespread implementation of nano- and microscale technologies such as electrospinning and photolithography have led to the development of innovative new approaches for creating engineered biomaterials with pore sizes, fibrillar microarchitectures, mechanical properties, and three-dimensional cellular microenvironments similar to native epithelial tissues. Microscale technologies have also provided opportunities to devise novel analytical tools to characterize fundamental principles of cell biology, such as the cell-cell, cell-substrate, and cell-soluble factor interactions that guide tissue regeneration. Together, these advances offer the promise of creating tissue scaffolds with precisely engineered cellular nano- and microenvironments to guide the regeneration of highly functional skin.

Acknowledgments

This work was supported in part by a Whittaker Foundation Grant (G.D.P.) and NIH grant R15 EB-005645 (G.D.P.)

References

[1] *Burn incidence and treatment in the United States: 1999 fact sheet*, Philadelphia: The Burn Foundation, 1999.

[2] Clark, R. A. F., and Singer, A. J., "Wound repair: Basic biology to tissue engineering," in R. P. Lanza, R. Langer, and J. Vacanti (eds.), *Principles of tissue engineering*, San Diego: Academic Press, 2000, pp. 857–878.

[3] Morgan, J. R., Sheridan, R. L., Tompkins, R. G., Yarmush, M. L., and Burke, J. F., "Burn dressings and skin substitutes," in B. D. Ratner et al. (eds.), *Biomaterials Science: An Introduction to Materials in Medicine*, San Diego: Academic Press, 2004, pp. 360–370.

[4] Jones, I., Currie, L., and Martin, R., "A guide to biological skin substitutes," *Br. J. Plast. Surg.*, Vol. 55, No. 3, 2002, pp. 185–193.

[5] Sheridan, R. L., and Tompkins, R. G., "Skin substitutes in burns," *Burns*, Vol. 25, No. 2, 1999, pp. 97–103.

[6] Supp, D. M., and Boyce, S. T., "Engineered skin substitutes: Practices and potentials," *Clin. Dermatol.*, Vol. 23, No. 4, 2005, pp. 403–412.

[7] Boyce, S. T., "Design principles for composition and performance of cultured skin substitutes," *Burns*, Vol. 27, No. 5, 2001, pp. 523–533.

[8] Phillips, T., "New skin for old: Developments in biological skin substitutes," *Arch. Dermatol.*, Vol. 134, 1998, pp. 344–349.

[9] Shakespeare, P. G., "The role of skin substitutes in the treatment of burn injuries," *Clin. Dermatol.*, Vol. 23, No. 4, 2005, pp. 413–418.

[10] Rheinwald, J. G., and Green, H., "Formation of a keratinizing epithelium in culture by a cloned cell line derived from a teratoma," *Cell*, Vol. 6, No. 3, 1975, pp. 317–330.

[11] Gallico, G. G., III, O'Connor, N. E., Compton, C. C., Remensnyder, J. P., Kehinde, O., and Green, H., "Cultured epithelial autografts for giant congenital nevi," *Plast. Reconstruct. Surg.*, Vol. 84, 1989, pp. 1–9.

[12] Gallico, G. G., III, O'Connor, N. E., Compton, C. C., Kehinde, O., and Green, H., "Permanent coverage of large burn wounds with autologous cultured human epithelium," *N. Engl. J. Med.*, Vol. 311, 1984, pp. 448–451.

[13] Compton, C. C., Gill, J. M., Bradford, D. A., Regauer, S., Gallico, G. G., and O'Connor, N. E., "Skin regenerated from cultured epithelial autografts on full-thickness burn wounds from 6 days to 5 years after grafting," *Lab. Invest.*, Vol. 60, 1989, pp. 600–612.

[14] Sheridan, R. L., and Tompkins, R. G., "Cultured autologous epithelium in patients with burns of ninety percent or more of the body surface," *J. Trauma*, Vol. 38, 1995, pp. 48–50.

[15] Woodley, D. T., Peterson, H. D., Herzog, S. R., Stricklin, G. P., Burgeson, R. E., Briggaman, R. A., Cronce, D. J., and O'Keefe, E. J., "Burn wounds resurfaced by cultured epidermal autografts show abnormal reconstitution of anchoring fibrils," *J. Amer. Med. Assoc.*, Vol. 258, 1988, p. 2566.

[16] Yannas, I. V., Burke, J. F., Orgill, D. P., and Skrabut, E. M., "Wound tissue can utilize a polymeric template to synthesize a functional extension of skin," *Science,* Vol. 215, No. 4529, 1982, pp. 174–176.

[17] Yannas, I. V., Burke, J. F., Gordon, P. L., Huang, C., and Rubenstein, R. H., "Design of an artificial skin: II. Control of chemical composition," *J. Biomed. Mater. Res.*, Vol. 14, No. 2, 1980, pp. 107–132.

[18] Heimbach, D., Luterman, A., Burke, J., Cram, A., Herndon, D., Hunt, J., Jordan, M., McManus, W., Solem, L., Warden, G., et al., "Artificial dermis for major burns: A multi-center randomized clinical trial," *Ann. Surg.*, Vol. 208, No. 3, 1988, pp. 313–320.

[19] Boyce, S. T., Christianson, D. J., and Hansbrough, J. F., "Structure of a collagen-GAG dermal skin substitute optimized for cultured human epidermal keratinocytes," *J. Biomed. Mater. Res.*, Vol. 22, 1988, pp. 939–957.

[20] Boyce, S. T., and Hansbrough, J. F., "Biologic attachment, growth and differentiation of cultured human epidermal keratinocytes on a graftable collagen and chondroitin-6-sulfate substrate," *Surgery*, Vol. 103, 1988, pp. 421–430.

[21] Hansbrough, J. F., Boyce, S. T., Cooper, M. L., and Foreman, T. J., "Burn wound closure with cultured autologous keratinocytes and fibroblasts attached to a collagen-glycosaminoglycan substrate," *JAMA*, Vol. 262, No. 15, 1989, pp. 2125–2130.

[22] Boyce, S. T., Goretsky, M. J., Greenhalgh, D. G., Kagan, R. J., Rieman, M. T., and Warden, G. D., "Comparative assessment of cultured skin substitutes and native skin autograft for treatment of full-thickness burns," *Ann. Surg.*, Vol. 222, 1995, pp. 743–752.

[23] Bell, E., Sher, S., Hull, B., Merrill, C., Rosen, S., Chamson, A., Asselineau, D., Dubertret, L., Coulomb, B., Lapiere, C., Nusgens, B., and Neveux, Y., "The reconstitution of living skin," *J. Invest. Dermatol.*, Vol. 81, 1983, pp. 2S–10S.

[24] Bell, E., Ehrlich, H. P., Buttle, D. J., and Nakatsuji, T., "Living tissue formed in vitro and accepted as skin-equivalent tissue of full thickness," *Science,* Vol. 211, No. 4486, 1981, pp. 1052–1054.

[25] Wilkins, L. M., Watson, S. R., Prosky, S. J., Meunier, S. F., and Parenteau, N. L., "Development of a bilayered living skin construct for clinical applications," *Biotech. Bioeng.*, Vol. 43, 1994, pp. 747–756.

[26] Falanga, V., Margolis, D., Alvarez, O., Auletta, M., Maggiacomo, F., Altman, M., Jensen, J., Sabolinski, M., and Jardin-Young, J., "Rapid healing of venous ulcers and lack of clinical rejection with an allogeneic cultured human skin equivalent," *Arch. Dermatol.*, Vol. 134, 1998, pp. 293–300.

[27] Black, A. F., Berthod, F., L'Heureux, N., Germain, L., and Auger, F. A., "In vitro reconstruction of a human capillary-like network in a tissue-engineered skin equivalent," *Faseb J.*, Vol. 12, 1998, pp. 1331–1340.

[28] Hudon, V., Berthod, F., Black, A. F., Damour, O., Germain, L., and Auger, F. A., "A tissue-engineered endothelialized dermis to study the modulation of angiogenic and angiostatic molecules on capillary-like tube formation in vitro," *Br. J. Dermatol.*, Vol. 148, No. 6, 2003, pp. 1094–1104.

[29] Yannas, I. V., Burke, J. F., Gordon, P. L., Huang, C., and Rubenstein, R. H., "Design of an artificial skin: II. Control of chemical composition," *J. Biomed. Mater. Res.*, Vol. 14, 1980, pp. 107–131.

[30] Yannas, I. V., Burke, J. F., Orgill, D. P., and Skrabut, E. M., "Wound tissue can utilize a polymeric template to synthesize a functional extension of skin," *Science,* Vol. 215, 1982, pp. 174–176.

[31] Yannas, I. V., Lee, E., Orgill, D. P., Skrabut, E. M., and Murphy, G. F., "Synthesis and characterization of a model extracellular matrix that induces partial regeneration of adult mammalian skin," *Proc. Natl. Acad. Sci. USA*, Vol. 86, 1989, pp. 933–937.

[32] Yannas, I. V., Lee, E., Orgill, D. P., Skrabut, E. M., and Murphy, G. F., "Synthesis and characterization of a model extracellular matrix that induces partial regeneration of adult mammalian skin," *Proc. Natl. Acad. Sci. USA*, Vol. 86, No. 3, 1989, pp. 933–937.

[33] Guo, M., Toda, K.-I., and Grinnell, F., "Activation of human keratinocyte migration on type I collagen and fibronectin," *J. Cell. Sci.*, Vol. 96, 1990, pp. 197–205.

[34] Tinois, E., Tiollier, J., Gaucherand, M., Dumas, H., Tardy, M., and Thivolet, J., "In vitro and post-transplantation differentiation of human keratinocytes grown on the human type IV collagen film of a bilayered dermal substitute," *Exp. Cell Res.*, Vol. 193, 1991, pp. 310–319.

[35] Grzesiak, J. J., Pierschbacher, M. D., Amodeo, M. F., Malaney, T. I., and Glass, J. R., "Enhancement of cell interactions with collagen/glycosaminoglycan matrices by RGD derivatization," *Biomaterials*, Vol. 18, 1997, pp. 1625–1632.

[36] Tjia, J. S., Aneskievich, B. J., and Moghe, P. V., "Substrate-absorbed collagen and cell secreted fibronectin concertedly induce cell migration on poly(lactide-glycolide)substrates," *Biomaterials*, Vol. 20, 1999, pp. 2223–2233.

[37] Bhatia, S. N., and Chen, C. S., "Tissue engineering at the micro-scale," *Biomed. Microdevices*, Vol. 2, 1999, pp. 131–144.

[38] Flemming, R. G., Murphy, C. J., Abrams, G. A., Goodman, S. L., and Nealey, P. F., "Effects of synthetic micro- and nano-structured surfaces on cell behavior," *Biomaterials*, Vol. 20, 1999, pp. 573–588.

[39] Desai, T. A., "Micro- and nanoscale structures for tissue engineering constructs," *Med. Eng. Phys.*, Vol. 22, 2000, pp. 595–606.

[40] Khademhosseini, A., Langer, R., Borenstein, J., and Vacanti, J., "Microscale technologies for tissue engineering and biology," *Proc. Natl. Acad. Sci. USA*, Vol. 103, 2006, pp. 2480–2487.

[41] Griffith, L. G., and Swartz, M. A., "Capturing complex 3-D tissue physiology in vitro," *Nat. Rev.: Mol. Cell Biol.*, Vol. 7, 2006, pp. 211–224.

[42] Sniadecki, N., Desai, R., Ruiz, S., and Chen, C. S., "Nanotechnology for cell-substrate interactions," *Annals Biomed. Eng.*, Vol. 34, 2006, pp. 59–74.

[43] Green, H., Kehinde, O., and Thomas, J., "Growth of cultured human epidermal cells into multiple epithelia suitable for grafting," *Proc. Natl. Acad. Sci. USA*, Vol. 76, No. 11, 1979, pp. 5665–5668.

[44] Lam, P. K., Chan, E. S. Y., Liew, C. T., Lau, C. H., Yen, S. C., and King, W. W. K., "Development and evaluation of a new composite laserskin graft," *J. Trauma*, Vol. 47, 1999, pp. 918–922.

[45] Pianigiani, E., Andreassi, A., Taddeucci, P., Alessandrini, C., Fimiani, M., and Andreassi, L., "A new model for studying differentiation and growth of epidermal cultures on hyaluronan-based carrier," *Biomaterials*, Vol. 20, No. 18, 1999, pp. 1689–1694.

[46] Brun, P., Cortivo, R., Zavan, B., Vecchiato, N., and Abatangelo, G., "In vitro reconstructed tissues on hyaluronan-based temporary scaffolding," *J. Mater. Sci. Mater. Med.,* Vol. 10, Nos. 10–11, 1999, pp. 683–688.

[47] Lam, P. K., Chan, E. S. Y., Liew, C. T., Lau, C. H., Yen, S. C., and King, W. W. K., "Combination of a new composite biocompatible skin graft on the neodermis of artificial skin in an animal model," *ANZ J. Surg.*, Vol. 72, 2002, pp. 360–363.

[48] Rennekampff, H. O., Kiessig, V., and Hansbrough, J. F., "Current concepts in the development of cultured skin replacements," *J. Surg. Res.,* Vol. 62, No. 2, 1996, pp. 288–295.

[49] Smith, A. G., Din, A., Denyer, M., Crowther, N. J., Eagland, D., Vowden, K., Vowden, P., and Britland, S. T., "Microengineered surface topography facilitates cell grafting from a prototype hydrogel wound dressing with antibacterial capability," *Biotechnol. Prog.*, Vol. 22, No. 5, 2006, pp. 1407–1415.

[50] Cuono, C., Langdon, R., and McGuire, J., "Use of cultured epidermal autografts and dermal allografts as skin replacement after burn injury," *Lancet*, Vol. 1, No. 8490, 1986, pp. 1123–1124.

[51] Compton, C. C., Hickerson, W., Nadire, K., and Press, W., "Acceleration of skin regeneration from cultured epithelial autografts by transplantation to homograft dermis," *J. Burn Care Rehabil.*, Vol. 14, No. 6, 1993, pp. 653–662.

[52] Orgill, D. P., Butler, C., Regan, J. F., Barlow, M. S., Yannas, I. V., and Compton, C. C., "Vascularized collagen-glycosaminoglycan matrix provides a dermal substrate and improves take of cultured epithelial autografts," *Plast. Reconstr. Surg.*, Vol. 102, No. 2, 1998, pp. 423–429.

[53] Kangesu, T., Navsaria, H. A., Manek, S., Shurey, C. B., Jones, C. R., Fryer, P. R., Leigh, I. M., and Green, C. J., "A porcine model using skin graft chambers for studies on cultured keratinocytes," *Br. J. Plast. Surg.*, Vol. 46, No. 5, 1993, pp. 393–400.

[54] Navsaria, H. A., Kangesu, T., Manek, S., Green, C. J., and Leigh, I. M., "An animal model to study the significance of dermis for grafting cultured keratinocytes on full thickness wounds," *Burns*, Vol. 20, Suppl. 1, 1994, pp. S57–S60.

[55] Coulomb, B., Lebreton, C., and Dubertret, L., "Influence of human dermal fibroblasts on epidermalization," *J. Invest. Dermatol.*, Vol. 92, No. 1, 1989, pp. 122–125.

[56] Coulomb, B., Friteau, L., Baruch, J., Guilbaud, J., Chretien-Marquet, B., Glicenstein, J., Lebreton-Decoster, C., Bell, E., and Dubertret, L., "Advantage of the presence of living dermal fibroblasts within in vitro reconstructed skin for grafting in humans," *Plast. Reconstr. Surg.*, Vol. 101, No. 7, 1998, pp. 1891–1903.

[57] Uitto, J., Perejda, A. J., Abergel, R. P., Chu, M. L., and Ramirez, F., "Altered steady-state ratio of type I/III procollagen mRNAs correlates with selectively increased type I procollagen biosynthesis in cultured keloid fibroblasts," *Proc. Natl. Acad. Sci. USA*, Vol. 82, No. 17, 1985, pp. 5935–5939.

[58] Bauer, E. A., and Uitto, J., "Collagen in cutaneous diseases," *Int. J. Dermatol.*, Vol. 18, No. 4, 1979, pp. 251–270.

[59] Weber, L., Kirsch, E., Muller, P., and Krieg, T., "Collagen type distribution and macromolecular organization of connective tissue in different layers of human skin," *J. Invest. Dermatol.*, Vol. 82, No. 2, 1984, pp. 156–160.

[60] Keene, D. R., Sakai, L. Y., Bachinger, H. P., and Burgeson, R. E., "Type III collagen can be present on banded collagen fibrils regardless of fibril diameter," *J. Cell. Biol.*, Vol. 105, No. 5, 1987, pp. 2393–2402.

[61] Bornstein, P., and Sage, E., "Matricellular proteins: Extracellular modulators of cell function," *Curr. Opin. Cell. Biol.*, Vol. 14, 2002, pp. 608–616.

[62] Yannas, I. V., *Tissue and Organ Regeneration in Adults*, New York: Springer, 2001.

[63] O'Brien, F., Harley, B., Yannas, I. V., and Gibson, L., "The effect of pore size on cell adhesion in collagen-GAG scaffolds," *Biomaterials*, Vol. 26, 2005, pp. 433–441.

[64] Subbiah, T., Bhat, G. S., Tock, R. W., Pararneswaran, S., and Ramkumar, S. S., "Electrospinning of nanofibers," *J. Appl. Polym. Sci.*, Vol. 96, 2005, p. 557.

[65] Pham, Q. P., Sharma, U., and Mikos, A. G., "Electrospinning of polymeric nanofibers for tissue engineering applications: A review," *Tissue Eng.*, Vol. 12, No. 5, 2006, pp. 1197–1211.

[66] Matthews, J. A., Wnek, G. E., Simpson, D. G., and Bowlin, G. L., "Electrospinning of collagen nanofibers," *Biomacromolecules*, Vol. 3, No. 2, 2002, pp. 232–238.

[67] Venugopal, J. R., Zhang, Y., and Ramakrishna, S., "In vitro culture of human dermal fibroblasts on electrospun polycaprolactone collagen nanofibrous membrane," *Artif. Org.*, Vol. 30, No. 6, 2006, pp. 440–446.

[68] Venugopal, J., and Ramakrishna, S., "Biocompatible nanofiber matrices for the engineering of a dermal substitute for skin regeneration," *Tissue Eng.,* Vol. 11, Nos. 5–6, 2005, pp. 847–854.

[69] Li, W. J., Laurencin, C. T., Caterson, E. J., Tuan, R. S., and Ko, F. K., "Electrospun nanofibrous structure: A novel scaffold for tissue engineering," *J. Biomed. Mater. Res.*, Vol. 60, No. 4, 2002, pp. 613–621.

[70] Pan, H., Jiang, H., and Chen, W., "Interaction of dermal fibroblasts with electrospun composite polymer scaffolds prepared from dextran and poly lactide-co-glycolide," *Biomaterials*, Vol. 27, No. 17, 2006, pp. 3209–3220.

[71] Zhang, Y. Z., Venugopal, J., Huang, Z. M., Lim, C. T., and Ramakrishna, S., "Characterization of the surface biocompatibility of the electrospun PCL-collagen nanofibers using fibroblasts," *Biomacromolecules*, Vol. 6, No. 5, 2005, pp. 2583–2589.

[72] Ratner, B. D., Hoffman, A. S., Schoen, F. J., and Lemons, J. E., *Biomaterials Science: An Introduction to Materials in Medicine*, 2nd ed., Amsterdam: Elsevier, 2004.

[73] Edwards, C., and Marks, R., "Evaluation of biomechanical properties of human skin," *Clin. Dermatol.*, Vol. 13, No. 4, 1995, pp. 375–380.

[74] Rho, K. S., Jeong, L., Lee, G., Seo, B. M., Park, Y. J., Hong, S. D., Roh, S., Cho, J. J., Park, W. H., and Min, B. M., "Electrospinning of collagen nanofibers: Effects on the behavior of normal human keratinocytes and early-stage wound healing," *Biomaterials*, Vol. 27, No. 8, 2006, pp. 1452–1461.

[75] Noh, H. K., Lee, S. W., Kim, J. M., Oh, J. E., Kim, K. H., Chung, C. P., Choi, S. C., Park, W. H., and Min, B. M., "Electrospinning of chitin nanofibers: Degradation behavior and cellular response to normal human keratinocytes and fibroblasts," *Biomaterials*, Vol. 27, No. 21, 2006, pp. 3934–3944.

[76] Min, B. M., Lee, G., Kim, S. H., Nam, Y. S., Lee, T. S., and Park, W. H., "Electrospinning of silk fibroin nanofibers and its effect on the adhesion and spreading of normal human keratinocytes and fibroblasts in vitro," *Biomaterials*, Vol. 25, Nos. 7–8, 2004, pp. 1289–1297.

[77] Sun, T., Mai, S., Norton, D., Haycock, J. W., Ryan, A. J., and MacNeil, S., "Self-organization of skin cells in three-dimensional electrospun polystyrene scaffolds," *Tissue Eng.*, Vol. 11, Nos. 7–8, 2005, pp. 1023–1033.

[78] Hutmacher, D. W., "Polymeric scaffolds in tissue engineering bone and cartilage," *Biomaterials*, Vol. 21, 2000, pp. 2529–2543.

[79] Wang, H., Pieper, J., Peters, F., van Blitterswijk, C. A., and Lamme, E. N., "Synthetic scaffold morphology controls human dermal connective tissue formation," *J. Biomed. Mater. Res. A*, Vol. 74, No. 4, 2005, pp. 523–532.

[80] Boyce, S. T., and Hansbrough, J. F., "Biologic attachment, growth, and differentiation of cultured human epidermal keratinocytes on a graftable collagen and chondroitin-6-sulfate substrate," *Surgery*, Vol. 103, No. 4, 1988, pp. 421–431.

[81] Boyce, S. T., Christianson, D. J., and Hansbrough, J. F., "Structure of a collagen-GAG dermal skin substitute optimized for cultured human epidermal keratinocytes," *J. Biomed. Mater. Res.*, Vol. 22, No. 10, 1988, pp. 939–957.

[82] Parenteau, N., Sabolinski, M., Prosky, S., Nolte, C., Oleson, M., Kriwet, K., and Bilbo, P., "Biological and physical factors influencing the successful engraftment of a cultured skin substitute," *Biotech. Bioeng.*, Vol. 52, 1996, pp. 3–14.

[83] Boyce, S. T., Goretsky, M. J., Greenhalgh, D. G., Kagan, R. J., Rieman, M. T., and Warden, G. D., "Comparative assessment of cultured skin substitutes and native skin autograft for treatment of full-thickness burns," *Ann Surg.,* Vol. 222, No. 6, 1995, pp. 743–752.

[84] Burgeson, R. E., "Basement membranes," in T. B. Fitzpatrick et al. (eds.), *Dermatology in General Medicine*, New York: McGraw-Hill, 1987, pp. 288–303.

[85] Vracko, R., "Basal lamina scaffold-anatomy and significance for maintenance of orderly tissue structure," *Am. J. Pathol*, Vol. 77, 1974, pp. 313–346.

[86] Medalie, D. A., Eming, S. A., Collins, M. E., Tompkins, R. G., Yarmush, M. L., and Morgan, J. R., "Differences in dermal analogs influence subsequent pigmentation, epidermal differentiation, basement membrane, and rete ridge formation of transplanted composite skin grafts," *Transplantation*, Vol. 64, No. 3, 1997, pp. 454–465.

[87] Guo, M., and Grinnell, F., "Basement membrane and human epidermal differentiation in vitro," *J. Invest. Dermatol.*, Vol. 93, 1989, pp. 372–378.

[88] Nolte, C. J. M., Oleson, M. A., Hansbrough, J. F., Morgan, J., Greenleaf, G., and Wilkins, L., "Ultrastructural features of composite skin cultures grafted onto athymic mice," *J. Anat.*, Vol. 185, 1994, pp. 325–333.

[89] Harriger, M. D., Warden, G. D., Greenhalgh, D. G., Kagan, R. J., and Boyce, S. T., "Pigmentation and microanatomy of skin regenerated from composite grafts of cultured cells and bioplymers applied to full-thickness burn wounds," *Transplantation*, Vol. 59, 1995, pp. 702–707.

[90] Odland, G. F., "The morphology of the attachment between the dermis and the epidermis," *Anat. Rec.*, Vol. 108, 1950, pp. 399–413.

[91] Carter, W. G., Kaur, P., Gil, S. G., Gahr, P. J., and Wayner, E. A., "Distinct functions for integrins $\alpha3\beta1$ in focal adhesions and $\alpha6\beta4$/bullous perphigoid antigen in a new stable anchoring contact (SAC) of keratinocytes: Relation to hemidesmosomes," *J. Cell. Biol.*, Vol. 111, 1990, pp. 3141–3154.

[92] Carter, W. G., Symington, B. E., and Kaur, P., "Cell adhesion and the basement membrane in early epidermal morphogenesis," in T. P. Fleming (ed.), *Epithelial Organization and Development*, London: Chapman and Hall, 1992, pp. 299–327.

[93] Burgeson, R. E., "Dermal-epidermal adhesion in skin," in D. H. Rohrbach and R. Timpl (eds.), *Molecular and Cellular Aspects of Basement Membranes*, New York: Academic Press, 1993, pp. 49–66.

[94] Pins, G. D., Toner, M., and Morgan, J. R., "Microfabrication of an analog of the basal lamina: Biocompatible membranes with complex topographies," *Faseb J.*, Vol. 14, No. 3, 2000, pp. 593–602.

[95] Downing, B. R., Cornwell, K., Toner, M., and Pins, G. D., "The influence of microtextured basal lamina analog topography on keratinocyte function and epidermal organization," *J. Biomed. Mater. Res. A*, Vol. 72, No. 1, 2005, pp. 47–56.

[96] Abrams, G. A., Teixeira, A. I., Nealey, P. F., and Murphy, C. J., "Effects of substratum topography on cell behavior," in A. K. Dillow and A. M. Lowman (eds.), *Biomimetic Materials and Design*, New York: Marcel Dekker, 2002.

[97] Abrams, G. A., Goodman, S. L., Nealey, P. F., Franco, M., and Murphy, C. J., "Nanoscale topography of the basement membrane underlying the corneal epithelium of the rhesus macaque," *Cell Tissue Res.*, Vol. 299, No. 1, 2000, pp. 39–46.

[98] Abrams, G. A., Goodman, S., and Murphy, C. J., "Nanoscale topography of the extracellular matrix underlying the corneal epithelium of the primate," *Invest. Ophthalmol. Vis. Sci.*, Vol. 38, 1997, p. S505.

[99] Abrams, G. A., Schaus, S. S., Goodman, S. L., Nealey, P. F., and Murphy, C. J., "Nanoscale topography of the corneal epithelial basement membrane and Descemet's membrane of the human," *Cornea*, Vol. 19, No. 1, 2000, pp. 57–64.

[100] Karuri, N. W., Liliensiek, S., Teixeira, A. I., Abrams, G., Campbell, S., Nealey, P. F., and Murphy, C. J., "Biological length scale topography enhances cell-substratum adhesion of human corneal epithelial cells," *J. Cell. Sci.*, Vol. 117, Pt. 15, 2004, pp. 3153–3164.

[101] O'Neill, C. H., Jordan, P., and Ireland, G., "Evidence for two distinct mechanisms of anchorage stimulation in freshly explanted and 3T3 Swiss mouse fibroblasts," *Cell*, Vol. 44, 1986, pp. 489–496.

[102] Watt, F. M., Jordan, P. W., and O'Neill, C. H., "Cell shape controls terminal differentiation of human epidermal keratinocytes," *Proc. Natl. Acad. Sci. USA*, Vol. 85, No. 15, 1988, pp. 5576–5580.

[103] Goodman, S. L., Sims, P. A., and Albrecht, R. M., "Three-dimensional extracellular matrix textured biomaterials," *Biomaterials*, Vol. 17, No. 21, 1996, pp. 2087–2095.

[104] Curtis, A., and Wilkinson, C., "Topographical control of cells," *Biomaterials*, Vol. 18, 1997, pp. 1573–1583.

[105] Dalby, M. J., Riehle, M. O., Sutherland, D. S., Agheli, H., and Curtis, A. S., "Use of nanotopography to study mechanotransduction in fibroblasts—methods and perspectives," *Eur. J. Cell. Biol.*, Vol. 83, No. 4, 2004, pp. 159–169.

[106] Uttayarat, P., Toworfe, G. K., Dietrich, F., Lelkes, P. I., and Composto, R. J., "Topographic guidance of endothelial cells on silicone surfaces with micro- to nanogrooves: Orientation of actin filaments and focal adhesions," *J. Biomed. Mater. Res. A*, Vol. 75, No. 3, 2005, pp. 668–680.

[107] Fleischmajer, R., Utani, A., MacDonald, E. D., II, Perlish, J. S., Pan, T.-C., Chu, M.-L., Nomizu, M., Ninomiya, Y., and Yamada, Y., "Initiation of skin basement membrane formation at the epidermo-dermal interface involves assembly of laminins through binding to cell membrane receptors," *J. Cell. Sci.*, Vol. 111, 1998, pp. 1929–1940.

[108] den Braber, E. T., Ruijter, J. E., Smits, J., H. T., Ginsel, L. A., A. F., and Jansen, v. R., J. A., "Effect of parallel surface microgrooves and surface energy on cell growth," *J. Biomed. Mater. Res.*, Vol. 29, 1995, pp. 511–518.

[109] Britland, S., Morgan, H., Wojiak-Stodart, B., Riehle, M., Curtis, A., and Wilkinson, C., "Synergistic and hierarchical adhesive and topographic guidance of BHK cells," *Exp. Cell Res.*, Vol. 228, No. 2, 1996, pp. 313–325.

[110] Teixeira, A. I., Abrams, G. A., Bertics, P. J., Murphy, C. J., and Nealey, P. F., "Epithelial contact guidance on well-defined micro- and nanostructured substrates," *J. Cell. Sci.*, Vol. 116, 2003, pp. 1881–1892.

[111] Teixeira, A. I., Nealey, P. F., and Murphy, C. J., "Responses of human keratocytes to micro- and nanostructured substrates," *J. Biomed. Mater. Res.*, Vol. 71A, No. 3, 2004, pp. 369–376.

[112] Ohara, P. T., and Buck, R. C., "Contact guidance in vitro: A light, transmission, and scanning electron microscopic study," *Exp. Cell Res.*, Vol. 121, No. 2, 1979, pp. 235–249.

[113] den Braber, E. T., de Ruijter, J. E., Ginsel, L. A., von Recum, A. F., and Jansen, J. A., "Orientation of ECM protein deposition, fibroblast cytoskeleton, and attachment complex components on silicone microgrooved surfaces," *J. Biomed. Mater. Res.*, Vol. 40, 1998, pp. 291–300.

[114] Kooten, T. G. v., and Recum, A. F. v., "Cell adhesion to textured silicone surfaces: The influence of time of adhesion and texture on focal contact and fibronectin fibril formation," *Tissue Eng.*, Vol. 5, 1999, pp. 223–240.

[115] den Braber, E. T., de Ruijter, J. E., Ginsel, L. A., von Recum, A. F., and Jansen, J. A., "Quantitative analysis of fibroblast morphology on microgrooved surfaces with various grove and ridge dimensions," *Biomaterials*, Vol. 17, 1996, pp. 2037–2044.

[116] Pitts, J. D., Campagnola, P. J., Epling, G. A., and Goodman, S. L., "Reaction efficiencies for sub-micron multi-photon freeform fabrications of proteins and polymers with applications in sustained release," *Macromolecules*, Vol. 33, 2000, pp. 1514–1523.

[117] Campagnola, P. J., Howell, A. R., Delguidas, D., Epling, G. A., Pitts,, J. D., and Goodman, S. L., "3-dimensional sub-micron polymerization of acrylamide by multi-photon excitation of xanthene dyes," *Macromolecules*, Vol. 33, 2000, pp. 1511–1513.

[118] Basu, S., and Campagnola, P. J., "Properties of crosslinked protein matrices for tissue engineering applications synthesized by multiphoton excitation," *J. Biomed. Mater. Res.*, Vol. 71A, No. 2, 2004, pp. 359–368.

[119] Basu, S., and Campagnola, P. J., "Enzymatic activity of alkaline phosphatase inside protein and polymer structures fabricated via multiphoton excitation," *Biomacromolecules*, Vol. 5, No. 2, 2004, pp. 572–579.

[120] Basu, S., Cunningham, L. P., Bush, K. A., Pins, G. D., Howell, A. R., and Campagnola, P. J., "Multi-photon excited fabrication of collagen matrices crosslinked by a modified

benzophenone dimer: Bioactivity and enzymatic degradation," *Biomacromolecules*, Vol. 6, No. 3, 2005, pp. 1465–1467..

[121] Pins, G. D., Bush, K. A., Cunningham, L. P., and Campagnola, P. J., "Multiphoton excited fabricated nano and micro patterned extracellular matrix proteins direct cellular morphology," *J. Biomed. Mater. Res. A*, Vol. 78, No. 1, 2006, pp. 194–204.

[122] Gailit, J., Clarke, C., Newman, D., Tonnesen, M. G., Mosesson, M. W., and Clark, R. A., "Human fibroblasts bind directly to fibrinogen at RGD sites through integrin alpha(v)beta3," *Exp. Cell Res.,* Vol. 232, No. 1, 1997, pp. 118–126.

[123] Cutler, S. M., and Garcia, A. J., "Engineering cell adhesive surfaces that direct integrin alpha5beta1 binding using a recombinant fragment of fibronectin," *Biomaterials*, Vol. 24, No. 10, 2003, pp. 1759–1770.

[124] Watt, F. M., "Stem cell fate and patterning in mammalian epidermis," *Curr. Opin. Genet. Dev.*, Vol. 11, No. 4, 2001, pp. 410–417.

[125] Slack, J. M., "Stem cells in epithelial tissues," *Science,* Vol. 287, No. 5457, 2000, pp. 1431–1433.

[126] Jones, P. H., Harper, S., and Watt, F. M., "Stem cell patterning and fate in human epidermis," *Cell*, Vol. 80, No. 1, 1995, pp. 83–93.

[127] Jensen, U. B., Lowell, S., and,Watt, F. M., "The spatial relationship between stem cells and their progeny in the basal layer of human epidermis: A new view based on whole-mount labelling and lineage analysis," *Development*, Vol. 126, No. 11, 1999, pp. 2409–2418.

[128] Watt, F. M., "Epidermal stem cells: Markers, patterning and the control of stem cell fate," *Philos. Trans. R. Soc. Lond. B Biol. Sci.*, Vol. 353, No. 1370, 1998, pp. 831–837.

[129] Blanpain, C., Lowry, W. E., Geoghegan, A., Polak, L., and Fuchs, E., "Self-renewal, multipotency, and the existence of two cell populations within an epithelial stem cell niche," *Cell*, Vol. 118, 2004, pp. 635–648.

About the Editors

Ali Khademhosseini is an assistant professor of medicine and health sciences and technology at Harvard-MIT's Division of Health Sciences and Technology and the Harvard Medical School. His research is based on developing micro- and nanoscale technologies to control cellular behavior with particular emphasis in developing microscale biomaterials and engineering systems for tissue engineering and drug discovery. He has published more than 70 peer-reviewed papers, 80 abstracts, and 14 issued or pending patents. He has received multiple awards, including the TR35, awarded to the top Young Innovators by the Technology Review Magazine (2007), the Coulter Foundation Early Career (2006), BMW Group Scientific Award finalist (2007), outstanding research mentor at MIT (2004), outstanding researcher in polymer science by OMNOVA/MIT (2005), and outstanding research by Biomedical Engineering Society (2005). He received his Ph.D. in bioengineering from MIT (2005), and an M.A.Sc. (2001) and a B.A.Sc. (1999) from University of Toronto both in chemical engineering.

Jeffrey Borenstein is currently director of the Biomedical Engineering Center and is a Distinguished Member of the Technical Staff at the Charles Stark Draper Laboratory in Cambridge, Massachusetts. There he serves as the Technical Director for Draper's new biomedical engineering program, with projects in drug delivery, clinical diagnostics, regenerative medicine, and biodefense. He is also program leader for biomaterials and tissue engineering and the Draper Laboratory Liaison for the Center for the Integration of Medicine and Innovative Technology (CIMIT). Dr. Borenstein is a principal investigator for several programs in tissue engineering and drug delivery with support from the NIH and other sources. Dr. Borenstein has a Ph.D. in physics from the University at Albany and 25 years of experience in micro-electronics, microfabrication technology, and biomedical devices. Dr. Borenstein's current work is focused on the application of microsystems technologies towards organ assist and organ replacement for transplant medicine and drug discovery, diagnostics and sensing applications, and drug delivery. He has ten issued patents, as well as twenty published patent applications and over eighty peer-reviewed journal articles and conference proceedings.

Mehmet Toner, PhD, is a world-renowned cryobiologist and a professor of surgery at the Harvard Medical School and a professor of biomedical engineering at the Harvard-MIT Division of Health Sciences and Technology (HST). Toner was born in Istanbul, Turkey in July, 1958. He obtained his undergraduate degree in Mechanical Engineering at the Istanbul Technical University in 1983, and his master's degree and doctorate in Mechanical Engineering and Medical Engineering at MIT in 1989. Prof. Toner has made valuable contributions to the specific fields of cryobiology and biopreservation and to the wider field of biomedical engineering in the form of thirteen patented inventions, numerous books chapters, over 200 journal publications, and over 350 scientific presentations. Professor Toner is currently located at the Massachusetts General Hospital and Shriners Hospital for Children, the Center for Engineering in Medicine(CEM)is regarded as one of the best biomedical research groups in the world. Dr. Toner currently serves as its associate director, as well as the director of the CEM-affiliated BioMEMS Resource Center.

Shuichi Takayama was born in Tokyo Japan, in 1968. He received his PhD in chemistry from The Scripps Research Institute at La Jolla, California, in 1998, and performed further postdoctoral studies as a Leukemia and Lymphoma Society Fellow at Harvard University from 1998 to 2000. At the end of 2000 he joined the Biomedical Engineering Department at the University of Michigan as an assistant professor. Currently he is an associate professor of biomedical engineering in the Macromolecular Science and Engineering Program. His current research interests are: microfluidics, nanofluidics, biomaterials and surface engineering, Microscale Tissue Engineering, And Cellular Micro-environomics, more specifically Braille display-based microfluidic systems, air–liquid two-phase microfluidic systems, tunable nanofluidics, fracture fabrication and materials patterning, in vitro fertilization on a chip, microfluidic tissue engineering of small airway injuries, subcellular signaling, and microscale stem cell niche engineering.

List of Contributors

G. Balasundarum
184 Hope Street
Division of Engineering
Brown University
Providence, RI 02917

David J. Beebe
1550 Engineering Drive, Rm. 2142
Department of Biomedical Engineering
University of Wisconsin-Madison
Madison, WI 53706
E-mail: djbeebe@wisc.edu

Jeffrey T. Borenstein
Biomedical Engineering Center
The Charles Draper Laboratory
555 Technology Square
Cambridge, MA, 02139-3563
E-mail: jborenstein@draper.com

Mark-Anthony Bray
Department of Biomedical Engineering
Harvard University
Pierce Hall, Room 322A
29 Oxford St.
Cambridge, MA 02138

Katie A. Bush
Graduate School of Biomedical Sciences
University of Massachusetts Medical School
Department of Biomedical Engineering
Worcester Polytechnic Institute
100 Institute Road
Worcester, MA 01609
E-mail: kbush@wpi.edu

Richard L. Carpenedo
313 Ferst Drive
Atlanta, GA 30332-0535

Christopher S. Chen
Department of Bioengineering
University of Pennsylvania
510 Skirkanich Hall
210 South 33rd Street
Philadelphia, PA 19104
E-mail: chrischen@seas.upenn.edu

Jan-Hung Chen
Department of Mechanical and Industrial Engineering
Institute of Biomaterials and Biomedical Engineering
University of Toronto
5 King's College Road
Toronto, ON M5S 3G8
Canada

Bong Geun Chung
Natural Science 2 Room 3109
Department of Biomedical Engineering
University of California, Irvine
Irvine, CA 92697

Utkan Demairci
Harvard-Massachusetts Institute of Technology Division of Health Sciences and Technology
Harvard Medical School/Brigham and Women's Hospital
65 Landsdowne Street, Rm. 267
Cambridge, MA 02139
E-mail: udemirci@rics.bwh.harvard.edu

Tejal Desai
Director, Laboratory of Therapeutic Micro and Nanotechnology
Department of Physiology
Division of Bioengineering
University of California, San Francisco MC 2520
Byers Hall, Rm. 203C
1700 4th Street
San Francisco, CA 94158-2330
E-mail: Tejal.desai@ucsf.edu

Junsang Doh
Department of Mechanical Engineering/I-Bio Program
Pohang University of Science and Technology
Hyoja-Dong, Nam-Gu, Pohang, Gyeongbuk, Korea 790-784

N. Gozde Durmus
Harvard-Massachusetts Institute of Technology Division of Health Sciences and Technology
Harvard Medical School/Brigham and Women's Hospital
65 Landsdowne Street, Rm. 267
Cambridge, MA 02139

Jennifer H. Elisseeff
Whitaker Biomedical Engineering Institute
Department of Biomedical Engineering
Johns Hopkins University
3400 N. Charles Street/Clark Hall 106
Baltimore, MD 21218
E-mail: jhe@jhu.edu

Adam Feinberg
Department of Biomedical Engineering
Harvard University
Pierce Hall, Room 322A
29 Oxford St.
Cambridge, MA 02138

Albert Folch
Department of Bioengineering
William H. Foege Building, Rm. N430-N
1705 NE Pacific St.
Campus Box 355061
University of Washington
Seattle, WA 98195
E-mail: afolch@u.washington.edu

Junji Fukuda
Institute of Materials Science
University of Tsukuba
Tsukuba, Ibaraki, Japan, 305-8573

Nicholas Geisse
Department of Biomedical Engineering
Harvard University
Pierce Hall, Room 322A
29 Oxford St.
Cambridge, MA 02138

James B. Grotberg
Department of Biomedical Engineering
Department of Surgery
University of Michigan, Ann Arbor
2125 Gerstacker Bldg.
2200 Bonisteel Blvd.
Ann Arbor, MI 48109-2099
E-mail: grotberg@umich.edu

Edward Haeggstrom
Harvard-Massachusetts Institute of Technology Division of Health Sciences and Technology
Harvard Medical School/Brigham and Women's Hospital
65 Landsdowne Street, Rm. 267
Cambridge, MA 02139

Seungpyo Hong
Department of Chemical Engineering
Massachusetts Institute of Technology
77 Massachusetts Ave. E25-342
Cambridge, MA 02139

Hossein Hosseinkhani
International Center for Young Scientist (ICYS)
National Institute for Materials Science (NIMS)
Tsukuba, Ibaraki 305-0044, Japan

Mohsen Hosseinkhani
Harvard-Massachusetts Institute of Technology Division of Health Sciences and Technology
Harvard Medical School/Brigham and Women's Hospital
65 Landsdowne Street, Rm. 265
Cambridge, MA 02139

Dongeun Huh
Departments of Surgery and Pathology
Harvard Medical School & Children's Hospital
Karp Family Research Laboratories, #11.005C
300 Longwood Ave.
Boston, MA 02115
E-mail: dan.huh@childrens.harvard.edu

Darrell J. Irvine
MIT
Room 8-425
77 Massachusetts Avenue
Cambridge, MA 02139

Rohin K. Iyer
Institute of Biomaterials and Biomedical Engineering
University of Toronto
164 College St., Rm. 407
Toronto, Ontario
M5S 3G9
Canada

Esmaiel Jabbari
Department of Chemical Engineering
Swearingen Engineering Center, Rm. 2C11
University of South Carolina
301 South Main Street
Columbia, SC 29208
E-mail: jabbari@engr.sc.edu

Roshan James
Department of Biomedical Engineering
PO Box 800374
University of Virginia
Charlottesville, VA 22908
E-mail: rj8b@virginia.edu

Noo Li Jeon
Natural Science 2 Room 3109
Department of Biomedical Engineering
University of California, Irvine
Irvine, CA 92697
E-mail: njeon@uci.edu

Xingyu Jiang
National Center for NanoScience and Technology
2 ZhongGuanCun North 1st Street
Beijing, 100080 China

Rohit Jindal
BioMEMS Resource Center
Massachusetts General Hospital
Harvard Medical School
114 16th Street, Room 1239
Charlestown, MA 02129-4404

Yoko Kamotani
Department of Biomedical Engineering
University of Michigan, Ann Arbor
3725 Green Brier Blvd., Apt. 203C
Ann Arbor, MI 48105
E-mail: ykamotani@umich.edu

Jeffrey M. Karp
Harvard-MIT Division of Health Sciences and Technology
Brigham and Women's Hospital
Harvard Medical School Room 313, PRB
65 Landsdowne Street
Cambridge, MA 02139
E-mail: jkarp@rics.bwh.harvard.edu

Emre Kayaalp
Assistant Professor of Medicine
Faculty of Medicine
Yeditepe University
26 Aðustos Yerleþimi, 34755 Kayýþdaðý
Istanbul, Turkey

Ali Khademhosseini
Harvard-Massachusetts Institute of Technology Division of Health Sciences and Technology
Harvard Medical School/Brigham and Women's Hospital
65 Landsdowne Street, Rm. 265
Cambridge, MA 02139
E-mail: alik@rics.bwh.harvard.edu

Deok-Ho Kim
Department of Biomedical Engineering
Johns Hopkins University
Baltimore, MD 21218
E-mail: dhkim@jhu.edu

Sun Min Kim
Department of Mechanical Engineering
Inha University
253 Young Hyun-dong, Nam-gu
Incheon, Republic of Korea, 402-751

Sangamesh Kumbar
Department of Orthopaedic Surgery
PO Box 400741
University of Virginia
Charlottesville, VA 22903
E-mail: sgk4t@virginia.edu

Po-ling Kuo
Department of Biomedical Engineering
Harvard University
Pierce Hall, Room 322A
29 Oxford St.
Cambridge, MA 02138

Kristen Elizabeth LaFlamme
MedErgy Marketing, Inc.
790 Township Line Road
Suite 200
Yardley, PA 19067-4200
klaflamme@medergymarketing.com

Professor Robert Langer
Department of Chemical Engineering
Massachusetts Institute of Technology
77 Massachusetts Avenue
Room E25-342
Cambridge, MA 02139-4307
E-mail: rlanger@mit.edu

Cato T. Laurencin
Lillian T. Pratt Distinguished Professor
Department of Orthopaedic Surgery
Department of Biomedical Engineering and Chemical Engineering
400 Ray C. Hunt Drive, Suite 330
Charlottesville, VA 22903
E-mail: laurencin@virginia.edu

Grace Lee
Harvard-Massachusetts Institute of Technology Division of Health Sciences and Technology
Harvard Medical School/Brigham and Women's Hospital
65 Landsdowne Street, Rm. 267
Cambridge, MA 02139

Andre Levchenko
Department of Biomedical Engineering
Johns Hopkins University
Baltimore, MD 21218
E-mail: alev@jhu.edu

Nianzhen Li
Department of Bioengineering
William H. Foege Building, Rm. N430-N
1705 NE Pacific St.
Campus Box 355061
University of Washington
Seattle, WA 98195

Francis Lin
School of Medicine
Stanford University
Stanford, CA 95305
E-mail: flin2@stanford.edu

Richard Lin
Harvard-Massachusetts Institute of Technology Division of Health Sciences and Technology
Harvard Medical School/Brigham and Women's Hospital
65 Landsdowne Street, Rm. 267
Cambridge, MA 02139

Carlos Lopez
25 Griggs Place
Allston, MA 02134

Todd C. McDevitt
The Wallace H. Coulter Department of Biomedical Engineering
Georgia Institute of Technology and Emory University
U. A. Whitaker Building
313 Ferst Drive, Suite 2102
Atlanta, GA 30332-0535
E-mail:todd.mcdevitt@bme.gatech.edu

Jordan Miller
Department of Bioengineering
Rice University
6100 Main Street, MS-142
Houston, TX 77005

SangJun Moon
Harvard-Massachusetts Institute of Technology Division of Health Sciences and Technology
Harvard Medical School/Brigham and Women's Hospital
65 Landsdowne Street, Rm. 267
Cambridge, MA 02139

Shashi K. Murthy
Department of Chemical Engineering
Northeastern University
360 Huntington Ave., 342 SN
Boston, MA 02115

Lakshmi S. Nair
Department of Orthopaedic Surgery
PO Box: 800759
University of Virginia
Charlottesville, VA 22908
E-mail: nair@virginia.edu

James J. Norman
Stanford University
Department of Cardiology, Division of Pediatrics, and
Department of Mechanical Engineering
496 Lomita Mall
MC 4040
Stanford, CA 95305-404
E-mail: jjnorman@stanford.edu

Kevin Kit Parker
Department of Biomedical Engineering
Harvard University
Pierce Hall, Room 322A
29 Oxford St.
Cambridge, MA 02138
E-mail: kkparker@seas.harvard.edu

Gokhan Percin
Harvard-Massachusetts Institute of Technology Division of Health Sciences and Technology
Harvard Medical School/Brigham and Women's Hospital
65 Landsdowne Street, Rm. 267
Cambridge, MA 02139

George D. Pins
Associate Professor of Biomedical Engineering
Department of Biomedical Engineering
Worcester Polytechnic Institute
100 Institute Road
Worcester, MA 01609
E-mail: gpins@wpi.edu

Brian Plouffe
Department of Chemical Engineering
Northeastern University
360 Huntington Ave., 342 SN
Boston, MA 02115

Ketul C. Popat
Department of Mechanical Engineering
School of Biomedical Engineering
Colorado State University
1374 Campus Delivery
Fort Collins, CO 80523-1374
E-mail: Ketul.popat@colostate.edu

Milica Radisic
Institute of Biomaterials and Biomedical Engineering
Department of Chemical Engineering and Applied Chemistry
University of Toronto
164 College St., Rm. 407
Toronto, Ontario
M5S 3G9
milica@chem-eng.utoronto.ca

Vimalier Reyez
Biomedical Engineering Center
The Charles Draper Laboratory
555 Technology Square
Cambridge, MA 02139-3563

Mohammad Reza Kaazempur Mofrad
Department of Bioengineering
208A Stanley Hall #1762
University of California
Berkeley, CA 94720-1762
E-mail: mofrad@berkeley.edu

Sami Alom Ruiz
Department of Bioengineeering
University of Pennsylvania
510 Skirkanich Hall
210 South 33rd Street
Philadelphia, PA 19104
E-mail: salom@seas.upenn.edu

Wajeeh Saadi
Natural Science 2 Room 3109,
Department of Biomedical Engineering
University of California, Irvine
Irvine, CA 92697

Alireza S. Sarvestani
Department of Chemical Engineering
Swearingen Engineering Center
University of South Carolina
301 South Main Street
Columbia, SC 29208
E-mail: sarvesta@engr.sc.edu

Sean Sheehy
Department of Biomedical Engineering
Harvard University
Pierce Hall, Room 322A
29 Oxford St.
Cambridge, MA 02138
E-mail: ssheehy@seas.harvard.edu

Craig A. Simmons
Department of Mechanical and Industrial Engineering, Faculty of Dentistry, and Institute of Biomaterials and Biomedical Engineering
University of Toronto
5 King's College Road
Toronto, ON M5S 3G8
Canada
E-mail: c.simmons@utoronto.edu

S. Sivirisoot
184 Hope Street
Division of Engineering
Brown University
Providence, RI 02917

Gary D. Smith
Obstetrics and Gynecology
University of Michigan
6410 Medical Science 1
1301 E Catherine St
Ann Arbor, MI 48109-0617
E-mail: smithgd@umich.edu

Karthikeyan Subramani
Institute for Nanoscale Science and Technology (INSAT)
University of Newcastle upon Tyne
Newcastle upon Tyne, NE1 7ER, United Kingdom

Kahp-Yang Suh
School of Mechanical and Aerospace Engineering
Seoul National University, Kwanak-gu, Shinlim-dong
San 56-1, Bldg. 301, Rm. 1219, 151-744, Seoul, Republic of Korea.
E-mail: sky4u@snu.ac.kr

Shuichi Takayama
Department of Biomedical Engineering
Department of Macromolecular Science and Engineering
University of Michigan, Ann Arbor
2115 Gerstacker Bldg.
2200 Bonisteel Blvd.
Ann Arbor, MI 48109-2099
E-mail: takayama@umich.edu

Sarah Tao
The Charles Stark Draper Laboratory, Inc.
555 Technology Square, MS 37
Cambridge, MA 02139
E-mail: stao@draper.com

Rahul Thakar
Dept. of Physiology
Division of Bioengineering
University of California, San Francisco MC 2520
Byers Hall
1700 4th Street
San Francisco, CA 94158-2330
E-mail: thakar@gmail.com

Arno W. Tilles
BioMEMS Resource Center
Massachusetts General Hospital
Harvard Medical School
114 16th Street, Room 1239
Charlestown, MA 02129-4404

Mehmet Toner
Director, BioMEMS Resource Center
Massachusetts General Hospital
Harvard Medical School
114 16th Street, Room 1239
Charlestown, MA 02129-4404
E-mail: mtoner@hms.harvard.edu

Anna Tourovskaia
Department of Bioengineering
William H. Foege Building, Rm. N430-N
1705 NE Pacific St.
Campus Box 355061
University of Washington
Seattle, WA 98195

Joseph P. Vacanti
John Homans Professor of Surgery, Harvard Medical School
Chief of Pediatric Surgery
Director, Laboratory for Tissue Engineering and Organ Fabrication
Massachusetts General Hospital
Harvard Medical School
Warren Building 1157
Boston, MA
E-mail: jvacanti@partners.org

Thomas J. Webster
Divisions of Engineering and Orthopaedics
182 Hope Street
Brown University
Providence, RI 02917
E-mail: Thomas_Webster@brown.edu

Eli J. Weinberg
Biomedical Engineering Center
The Charles Draper Laboratory
555 Technology Square
Cambridge, MA, 02139-3563

Jennifer West
W100E GRB Hall
Department of Bioengineering
Rice University
6100 Main Street
Houston, TX, 77251
E-mail; jwest@rice.edu

Martin L. Yarmush
Director, The Center for Engineering in Medicine
Massachusetts General Hospital
Harvard Medical School
114 16th Street, Room 1402
Charlestown, MA 02129-4404

Cindy Y.Y. Yip
Institute of Biomaterials and Biomedical Engineering
University of Toronto
164 College Street
Toronto, ON, M5S 3G5
Canada

L. Zhang
184 Hope Street
Division of Engineering
Brown University
Providence, RI 02917

Shuguang Zhang
Center for Biomedical Engineering
Room 56-341
Massachusetts Institute of Technology
Cambridge, MA 02139-4307
E-mail: Shuguang@MIT.EDU

Index

F

H

M

N

O

R

T

Related Titles from Artech House

Biological Database Modeling, Jake Chen and Amandeep S. Sidhu, editors

Biomolecular Computation for Bionanotechnology, Jian-Qin Liu and Katsunori Shimohara

Electrotherapeutic Devices: Principles, Design, and Applications, George D. O'Clock

Fundamentals and Applications of Microfluidics, Second Edition, Nam-Trung and Steven T. Wereley

Genome Sequencing Technology and Algorithms, Sun Kim, Haixu Tang, and Elaine R. Mardis, editors

Life Science Automation Fundamentals and Applications, Mingjun Zhang, Bradley Nelson, and Robin Felder, editors

Matching Pursuit and Unification in EEG Analysis, Piotr Durka

Micro and Nano Manipulations for Biomedical Applications, Tachung C. Yih and Ilie Talpasanu, editors

Microfluidics for Biotechnology, Jean Berthier and Pascal Silberzan

Systems Bioinformatics: An Engineering Case-Based Approach, Gil Alterovitz and Marco F. Ramoni, editors

Text Mining for Biology and Biomedicine, Sophia Ananiadou and John McNaught, editors